REFERENCE
DO NOT REMOVE
FROM LIBRARY

**WITHDRAWN FROM
TSC LIBRARY**

Mechanical Engineers' Handbook

Mechanical Engineers' Handbook
Third Edition

Manufacturing and Management

Edited by
Myer Kutz

JOHN WILEY & SONS, INC.

This book is printed on acid-free paper. ∞

Copyright © 2006 by John Wiley & Sons, Inc. All rights reserved.

Published by John Wiley & Sons, Inc., Hoboken, New Jersey.
Published simultaneously in Canada.

No part of this publication may be reproduced, stored in a retrieval system, or transmitted in any form or by any means, electronic, mechanical, photocopying, recording, scanning, or otherwise, except as permitted under Section 107 or 108 of the 1976 United States Copyright Act, without either the prior written permission of the Publisher, or authorization through payment of the appropriate per-copy fee to the Copyright Clearance Center, Inc., 222 Rosewood Drive, Danvers, MA 01923, (978) 750-8400, fax (978) 750-4470, or on the web at www.copyright.com. Requests to the Publisher for permission should be addressed to the Permissions Department, John Wiley & Sons, Inc., 111 River Street, Hoboken, NJ 07030, (201) 748-6011, fax (201) 748-6008, or online at http://www.wiley.com/go/permission.

Limit of Liability/Disclaimer of Warranty: While the publisher and author have used their best efforts in preparing this book, they make no representations or warranties with respect to the accuracy or completeness of the contents of this book and specifically disclaim any implied warranties of merchantability or fitness for a particular purpose. No warranty may be created or extended by sales representatives or written sales materials. The advice and strategies contained herein may not be suitable for your situation. The publisher is not engaged in rendering professional services, and you should consult a professional where appropriate. Neither the publisher nor author shall be liable for any loss of profit or any other commercial damages, including but not limited to special, incidental, consequential, or other damages.

For general information on our other products and services, please contact our Customer Care Department within the United States at (800) 762-2974, outside the United States at (317) 572-3993 or fax (317) 572-4002.

Wiley also publishes its books in a variety of electronic formats. Some content that appears in print may not be available in electronic books. For more information about Wiley products, visit our web site at www.wiley.com.

Library of Congress Cataloging-in-Publication Data:
Mechanical engineers' handbook/edited by Myer Kutz.—3rd ed.
      p. cm.
  Includes bibliographical references and index.
  ISBN-13 978-0-471-44990-4
  ISBN-10 0-471-44990-3 (cloth)
 1. Mechanical engineering—Handbooks, manuals, etc. I. Kutz, Myer.
  TJ151.M395 2005
  621—dc22

2005008603

Printed in the United States of America.

10 9 8 7 6 5 4 3 2

To Alan and Nancy, now and forever

Contents

Preface   ix
Vision Statement   xi
Contributors   xiii

PART 1   MANUFACTURING   1

1. Product Design for Manufacturing and Assembly (DFM&A)   3
   Gordon Lewis

2. Achieving Enterprise Goals with New Process Technology   22
   Steve W. Tuszynski

3. Classification Systems   68
   Dell K. Allen

4. Production Planning   110
   Bhaba R. Sarker, Dennis B. Webster, and Thomas G. Ray

5. Production Processes and Equipment   173
   Magd E. Zohdi, William E. Biles, and Dennis B. Webster

6. Metal Forming, Shaping, and Casting   245
   Magd E. Zohdi and William E. Biles

7. Mechanical Fasteners   286
   Murray J. Roblin, updated by Anthony Luscher

8. Statistical Quality Control   315
   Magd E. Zohdi

9. Computer-Integrated Manufacturing   328
   William E. Biles and Magd E. Zohdi

10. Material Handling   349
    William E. Biles, John S. Usher, and Magd E. Zohdi

11. Coatings and Surface Engineering: Physical Vapor Deposition   396
    Allan Matthews and Suzanne L. Rohde

12. Product Design and Manufacturing Processes for Sustainability   414
    I. S. Jawahir, P. C. Wanigarathne, and X. Wang

PART 2   MANAGEMENT, FINANCE, QUALITY, LAW, AND RESEARCH   445

13. Managing Projects in Engineering Organizations Using Interorganizational Teams   447
    Karen L. Higgins and Joseph A. Maciarello

14. Managing People   484
    Hans J. Thamain

15. Finance and the Engineering Function   505
    William Brett
16. Detailed Cost Estimating   531
    Rodney D. Stewart
17. Investment Analysis   564
    Byron W. Jones
18. Total Quality Management, Six Sigma, and Continuous Improvement   583
    Jack B. ReVelle and Robert Alan Kemerling
19. Registrations, Certifications, and Awards   616
    Jack B. ReVelle and Cynthia M. Sabelhaus
20. Safety Engineering   639
    Jack B. ReVelle
21. What the Law Requires of the Engineer   701
    Alvin S. Weinstein, and Martin S. Chizek
22. Patents   725
    David A. Burge and Benjamin D. Burge
23. Electronic Information Resources: Your Online Survival Guide   758
    Robert N. Schwarzwalder, Jr.
24. Sources of Mechanical Engineering Information   777
    Fritz Dusold and Myer Kutz

Index   785

Preface

The third volume of the Third Edition of the *Mechanical Engineers' Handbook* comprises two parts: Manufacturing and Management. Each part contains 12 chapters. Contributors include business owners, consultants, lawyers, librarians, and academics from all around the United States.

Part 1 opens with a chapter from the second edition on Product Design for Manufacturing and Assembly (DFM&A). The centerpiece of Part 1 includes the chapters that in earlier editions of the handbook have been called "the handbook within the handbook." Developed by a team at Louisiana State University and the University of Louisville, these six chapters, which have been updated, span manufacturing topics from production planning, production processes and equipment, metal forming, shaping, and casting, statistical quality control, computer-integrated manufacturing, to material handling. The chapter on classification systems remains unchanged from earlier editions; the chapter on mechanical fasteners has been revised extensively. Part 1 has three chapters entirely new to the handbook: a chapter on physical vapor deposition, one on environmentally conscious manufacturing, and one on a new approach to dealing with process technology in the context of design, tooling, manufacturing, and quality engineering. The latter chapter is indicative of how much contributors can give of themselves. Its content is the lifeblood of its author's consulting practice.

Part 2 covers a broad array of topics. The 12 chapters can be broken down into four groups. The first two chapters cover project and people management. The first of these chapters, on project management, deals with a subject that has appeared in previous editions, but the chapter is entirely new, to reflect advances in this field. The people management chapter has been revised. The following three chapters deal with fundamentals of financial management and are unchanged. The next three chapters, contributed by a team led by Jack ReVelle, treat a set of management issues, including Total Quality Management; registrations, certifications, and awards; and safety engineering. Two chapters cover legal issues of interest to engineers, including patents. The final two chapters cover online and print information sources useful to mechanical engineers in their daily work. The chapter on online sources is a new version of the chapter that appeared originally in 1998.

Vision for the Third Edition

Basic engineering disciplines are not static, no matter how old and well established they are. The field of mechanical engineering is no exception. Movement within this broadly based discipline is multidimensional. Even the classic subjects on which the discipline was founded, such as mechanics of materials and heat transfer, continue to evolve. Mechanical engineers continue to be heavily involved with disciplines allied to mechanical engineering, such as industrial and manufacturing engineering, which are also constantly evolving. Advances in other major disciplines, such as electrical and electronics engineering, have significant impact on the work of mechanical engineers. New subject areas, such as neural networks, suddenly become all the rage.

In response to this exciting, dynamic atmosphere, the *Mechanical Engineers' Handbook* is expanding dramatically, from one volume to four volumes. The third edition not only is incorporating updates and revisions to chapters in the second edition, which was published in 1998, but also is adding 24 chapters on entirely new subjects as well, incorporating updates and revisions to chapters in the *Handbook of Materials Selection,* which was published in 2002, as well as to chapters in *Instrumentation and Control,* edited by Chester Nachtigal and published in 1990.

The four volumes of the third edition are arranged as follows:

Volume I: *Materials and Mechanical Design*—36 chapters
    Part 1. Materials—14 chapters
    Part 2. Mechanical Design—22 chapters

Volume II: *Instrumentation, Systems, Controls, and MEMS*—21 chapters
    Part 1. Instrumentation—8 chapters
    Part 2. Systems, Controls, and MEMS—13 chapters

Volume III: *Manufacturing and Management*—24 chapters
    Part 1. Manufacturing—12 chapters
    Part 2. Management, Finance, Quality, Law, and Research—12 chapters

Volume IV: *Energy and Power*—31 chapters
    Part 1: Energy—15 chapters
    Part 2: Power—16 chapters

The mechanical engineering literature is extensive and has been so for a considerable period of time. Many textbooks, reference works, and manuals as well as a substantial number of journals exist. Numerous commercial publishers and professional societies, particularly in the United States and Europe, distribute these materials. The literature grows continuously, as applied mechanical engineering research finds new ways of designing, controlling, measuring, making and maintaining things, and monitoring and evaluating technologies, infrastructures, and systems.

Most professional-level mechanical engineering publications tend to be specialized, directed to the specific needs of particular groups of practitioners. Overall, however, the mechanical engineering audience is broad and multidisciplinary. Practitioners work in a variety of organizations, including institutions of higher learning, design, manufacturing, and consulting firms as well as federal, state, and local government agencies. A rationale for an expanded general mechanical engineering handbook is that every practitioner, researcher, and bureaucrat cannot be an expert on every topic, especially in so broad and multidisciplinary a field, and may need an authoritative professional summary of a subject with which he or she is not intimately familiar.

Starting with the first edition, which was published in 1986, our intention has always been that the *Mechanical Engineers' Handbook* stand at the intersection of textbooks, research papers, and design manuals. For example, we want the handbook to help young engineers move from the college classroom to the professional office and laboratory where they may have to deal with issues and problems in areas they have not studied extensively in school.

With this expanded third edition, we have produced a practical reference for the mechanical engineer who is seeking to answer a question, solve a problem, reduce a cost, or improve a system or facility. The handbook is not a research monograph. The chapters offer design techniques, illustrate successful applications, or provide guidelines to improving the performance, the life expectancy, the effectiveness, or the usefulness of parts, assemblies, and systems. The purpose is to show readers what options are available in a particular situation and which option they might choose to solve problems at hand.

The aim of this expanded handbook is to serve as a source of practical advice to readers. We hope that the handbook will be the first information resource a practicing engineer consults when faced with a new problem or opportunity—even before turning to other print sources, even officially sanctioned ones, or to sites on the Internet. (The second edition has been available online on knovel.com.) In each chapter, the reader should feel that he or she is in the hands of an experienced consultant who is providing sensible advice that can lead to beneficial action and results.

Can a single handbook, even spread out over four volumes, cover this broad, interdisciplinary field? We have designed the third edition of the *Mechanical Engineers' Handbook* as if it were serving as a core for an Internet-based information source. Many chapters in the handbook point readers to information sources on the Web dealing with the subjects addressed. Furthermore, where appropriate, enough analytical techniques and data are provided to allow the reader to employ a preliminary approach to solving problems.

The contributors have written, to the extent their backgrounds and capabilities make possible, in a style that reflects practical discussion informed by real-world experience. We would like readers to feel that they are in the presence of experienced teachers and consultants who know about the multiplicity of technical issues that impinge on any topic within mechanical engineering. At the same time, the level is such that students and recent graduates can find the handbook as accessible as experienced engineers.

Contributors

Dell K. Allen
Brigham Young University
Provo, Utah

William E. Biles
University of Louisville
Louisville, Kentucky

William Brett
New York, New York

Benjamin D. Burge
Intel Americas, Inc.
Chantilly, Virginia

David A. Burge
David A. Burge Co., L.P.A.
Cleveland, Ohio

Martin S. Chizek
Weinstein Associates International
Delray Beach, Florida

Fritz Dusold
New York, New York

Karen L. Higgins
NAVAIR Weapons Division
China Lake, California

I. S. Jawahir
University of Kentucky
Lexington, Kentucky

Byron W. Jones
Kansas State University
Manhattan, Kansas

Robert Alan Kemerling
Ethicon Endo-Surgery, Inc.
Cincinnati, Ohio

Myer Kutz
Myer Kutz Associates, Inc.
Delmar, New York

Gordon Lewis
Digital Equipment Corporation
Maynard, Massachusetts

Anthony Luscher
The Ohio State University
Columbus, Ohio

Joseph A. Maciariello
Claremont Graduate University
Claremont, California

Allan Matthews
Sheffield University
Sheffield, United Kingdom

Thomas G. Ray
Louisiana State University
Baton Rouge, Louisiana

Jack B. Revelle
ReVelle Solutions, LLC
Santa Ana, California

Murray J. Roblin
California State Polytechnic University
Pomona, California

Suzanne L. Rohde
The University of Nebraska
Lincoln, Nebraska

Cynthia M. Sabelhaus
Raytheon Missile Systems
Tucson, Arizona

Bhaba R. Sarker
Louisiana State University
Baton Rouge, Louisiana

Robert N. Schwarzwalder, Jr.
University of Hawaii at Mānoa
Honolulu, Hawaii

Rodney D. Stewart (deceased)
Mobile Data Services
Huntsville, Alabama

Hans J. Thamain
Bentley College
Waltham, Massachusetts

Steve W. Tuszynski
Algoryx, Inc.
Los Angeles, California

John S. Usher
University of Louisville
Louisville, Kentucky

X. Wang
University of Kentucky
Lexington, Kentucky

P. C. Wanigarathne
University of Kentucky
Lexington, Kentucky

Dennis B. Webster
Louisiana State University
Baton Rouge, Louisiana

Alvin S. Weinstein
Weinstein Associates International
Delray Beach, Florida

Magd E. Zohdi
Louisiana State University
Baton Rouge, Louisiana

Mechanical Engineers' Handbook

PART 1
MANUFACTURING

CHAPTER 1
PRODUCT DESIGN FOR MANUFACTURING AND ASSEMBLY (DFM&A)

Gordon Lewis
Digital Equipment Corporation
Maynard, Massachusetts

1	INTRODUCTION	3	2.3 The DFM&A Road Map	14
2	DESIGN FOR MANUFACTURING AND ASSEMBLY	4	3 WHY IS DFM&A IMPORTANT?	21
	2.1 What Is DFM&A?	5	REFERENCES	21
	2.2 Getting the DFM&A Process Started	12		

1 INTRODUCTION

Major changes in product design practices are occurring in all phases of the new product development process. These changes will have a significant impact on how all products are designed and the development of the related manufacturing processes over the next decade. The high rate of technology changes has created a dynamic situation that has been difficult to control for most organizations. There are some experts who openly say that if we have no new technology for the next five years, corporate America might just start to catch up. The key to achieving benchmark time to market, cost, and quality is in up-front technology, engineering, and design practices that encourage and support a wide latitude of new product development processes. These processes must capture modern manufacturing technologies, piece parts that are designed for ease of assembly, and parts that can be fabricated using low-cost manufacturing processes. Optimal new product design occurs when the designs of machines and of the manufacturing processes that produce those machines are congruent.

The obvious goal of any new product development process is to turn a profit by converting raw material into finished products. This sounds simple, but it has to be done efficiently and economically. Many companies do not know how much it costs to manufacture a new product until well after the production introduction. Rule #1: The product development team must be given a cost target at the start of the project. We will call this cost the *unit manufacturing cost* (UMC) target. Rule #3: The product development team must be held accountable for this target cost. What happened to rule #2? We'll discuss that shortly. In the meantime, we should understand what UMC is.

$$UMC = BL + MC + TA$$

where BL = burdened assembly labor rate per hour; this is the direct labor cost of labor, benefits, and all appropriate overhead cost
MC = material cost; this is the cost of all materials used in the product

TA = tooling amortization; this is the cost of fabrication tools, molds and assembly tooling, divided by the forecast volume build of the product

UMC is the direct burdened assembly labor (direct wages, benefits, and overhead) plus the material cost. Material cost must include the cost of the transformed material plus piece part packaging plus duty, insurance, and freight (DIF). Tooling amortization should be included in the UMC target cost calculation, based on the forecast product life volume.

Example UMC Calculation **BL + MC + TA**

Burdened assembly labor cost calculation (BL)

$$BL = (\underbrace{\$18.75 + 138\%}_{\text{Wages+Benefits overhead}})^{\text{Labor}} = \$44.06/\text{hr}$$

Burdened assembly labor is made up of the direct wages and benefits paid to the hourly workers, plus a percentage added for direct overhead and indirect overhead. The overhead added percentage will change from month to month based on plant expenses.

Material cost calculation (MC)

$$\begin{aligned} &\quad\quad\text{(Part cost + Packaging) + DIF + Mat. Acq. cost =} \\ MC &= (\$2.45 + \$.16) + 12\% + 6\% = \\ MC &= \$2.61 \quad\quad\quad\quad + \$.31 + \$.15 \quad\quad = \$3.07 \\ &\quad\quad\quad\quad\quad\quad\quad\quad\quad\quad\quad\quad\quad\quad\quad\text{Material FOB assm. plant} \end{aligned}$$

Material cost should include the cost of the parts and all necessary packaging. This calculation should also include a percent adder for duty, insurance, and freight (DIF) and an adder for the acquisition of the materials (Mat. Acq.). DIF typically is between 4 and 12% and Mat. Acq. typically is in the range of 6 to 16%. It is important to understand the MC because material is the largest expense in the UMC target.

Tooling amortization cost calculations (TA)

$$\begin{aligned} TA &= \overbrace{TC}^{\text{(Tool cost)}} / \overbrace{PL}^{\text{\# of parts}} \\ TA &= \$56,000/10,000 = \$5.60 \text{ per assembly} \end{aligned}$$

TC is the cost of tooling and PL is the estimated number of parts expected to be produced on this tooling. Tooling cost is the total cost of dies and mold used to fabricate the component parts of the new product. This also should include the cost of plant assembly fixtures and test and quality inspection fixtures.

The question is, "How can the product development team quickly and accurately measure UMC during the many phases of the project?" What is needed is a tool that provides insight into the product structure and at the same time exposes high-cost areas of the design.

2 DESIGN FOR MANUFACTURING AND ASSEMBLY

Designing for Manufacturing and Assembly (DFM&A) is a technique for reducing the cost of a product by breaking the product down into its simplest components. All members of the design team can understand the product's assembly sequence and material flow early in the design process.

DFM&A tools lead the development team in reducing the number of individual parts that make up the product and ensure that any additional or remaining parts are easy to handle and insert during the assembly process. DFM&A encourages the integration of parts and processes, which helps reduce the amount of assembly labor and cost. DFM&A efforts include programs to minimize the time it takes for the total product development cycle, manufacturing cycle, and product life-cycle costs. Additionally, DFM&A design programs promote team cooperation and supplier strategy and business considerations at an early stage in the product development process.

The DFM&A process is composed of two major components: *design for assembly* (DFA) and *design for manufacturing* (DFM). DFA is the labor side of the product cost. This is the labor needed to transform the new design into a customer-ready product. DFM is the material and tooling side of the new product. DFM breaks the parts fabrication process down into its simplest steps, such as the type of equipment used to produce the part and fabrication cycle time to produce the part, and calculates a cost for each functional step in the process. The program team should use the DFM tools to establish the material target cost before the new product design effort starts.

Manufacturing costs are born in the early design phase of the project. Many different studies have found that as much as 80% of a new product's cost is set in concrete at the first drawing release phase of the product. Many organizations find it difficult to implement changes to their new product development process. The old saying applies: "only wet babies want to change, and they do it screaming and crying." Figure 1 is a memo that was actually circulated in a company trying to implement a DFM&A process. Only the names have been changed.

It is clear from this memo that neither the engineering program manager nor the manufacturing program manager understood what DFM&A was or how it should be implemented in the new product development process. It seems that their definition of concurrent engineering is, "Engineering creates the design and manufacturing is forced to concur with it with little or no input." This is not what DFM&A is.

2.1 What Is DFM&A?

DFM&A is not a magic pill. It is a tool that, when used properly, will have a profound effect on the design philosophy of any product. The main goal of DFM&A is to lower product cost by examining the product design and structure at the early concept stages of a new product. DFM&A also leads to improvements in serviceability, reliability, and quality of the end product. It minimizes the total product cost by targeting assembly time, part cost, and the assembly process in the early stages of the product development cycle.

The life of a product begins with defining a set of product needs, which are then translated into a set of product concepts. Design engineering takes these product concepts and refines them into a detailed product design. Considering that from this point the product will most likely be in production for a number of years, it makes sense to take time out during the design phase to ask, "How should this design be put together?" Doing so will make the rest of the product life, when the design is complete and handed off to production and service, much smoother. To be truly successful, the DFM&A process should start at the early concept development phase of the project. True, it will take time during the hectic design phase to apply DFM&A, but the benefits easily justify additional time.

DFM&A is used as a tool by the development team to drive specific assembly benefits and identify drawbacks of various design alternatives, as measured by characteristics such as total number of parts, handling and insertion difficulty, and assembly time. DFM&A converts time into money, which should be the common metric used to compare alternative designs, or redesigns of an existing concept. The early DFM&A analysis provides the product

6 Product Design for Manufacturing and Assembly

Memorandum: *Ajax Bowl Corporation*

DATE: January 26, 1997
TO: Manufacturing Program Manager, Auto Valve Project
FROM: Engineering Program Manager, Auto Valve Project
RE: Design for Manufacturing & Assembly support for Auto Valve Project
CC: Director, Flush Valve Division

Due to the intricate design constraints placed on the Auto Valve project engineering feels they will not have the resources to apply the Design for Manufacturing and Assembly process. Additionally, this program is strongly schedule driven. The budget for the project is already approved as are other aspects of the program that require it to be on-time in order to achieve the financial goals of upper management.

In the meeting on Tuesday, engineering set down the guidelines for manufacturing involvement on the Auto Valve project. This was agreed to by several parties (not manufacturing) at this meeting.

The manufacturing folks wish to be tied early into the Auto Valve design effort:

1. This will allow manufacturing to be familiar with what is coming.
2. Add any ideas or changes that would reduce overall cost or help schedule.
3. Work vendor interface early, manufacturing owns the vendor issues when the product comes to the plant, anyways.

Engineering folks like the concept of new ideas, but fear:

1. Inputs that get pushed without understanding of all properly weighted constraints.
2. Drag on schedule due to too many people asking to change things.
3. Spending time defending and arguing the design.

PROPOSAL—Turns out this is the way we will do it.

Engineering shall on a few planned occasions address manufacturing inputs through one manufacturing person. Most correspondence will be written and meeting time will be minimal. It is understood that this program is strongly driven by schedule, and many cost reduction efforts are already built into the design so that the published budget can be met.

The plan for Engineering:

- When drawings are ready, Engineering Program Manager (EPM) will submit them to Manufacturing Program Manager (MPM).
- MPM gathers inputs from manufacturing people and submits them back in writting to EPM. MPM works questions through EPM to minimize any attention units that Engineering would have to spend.
- EPM submits suggestions to Engineering, for one quick hour of discussion/acceptance/veto.
- EPM submits written response back to MPM and works any Design continues under ENG direction.
- When a prototype parts arrives, the EPM will allow the MPM to use it in manufacturing discussions.
- MPM will submit written document back to EPM to describe issues and recommendations.
- Engineering will incorporate any changes that they can handle within the schedule that they see fit.

Figure 1

development team with a baseline to which comparisons can be made. This early analysis will help the designer to understand the specific parts or concepts in the product that require further improvement, by keeping an itemized tally of each part's effect on the whole assembly. Once a user becomes proficient with a DFM&A tool and the concepts become second nature, the tool is still an excellent means of solidifying what is by now second nature to DFA veterans, and helps them present their ideas to the rest of the team in a common language: cost.

DFM&A is an interactive learning process. It evolves from applying a specific method to a change in attitude. Analysis is tedious at first, but as the ideas become more familiar and eventually ingrained, the tool becomes easier to use and leads to questions: questions about the assembly process and about established methods that have been accepted or existing design solutions that have been adopted. In the team's quest for optimal design solutions, the DFM&A process will lead to uncharted ways of doing things. Naturally, then, the environment in which DFA is implemented must be ripe for challenging pat solutions and making suggestions for new approaches. This environment must evolve from the top down, from upper management to the engineer. Unfortunately, this is where the process too often fails.

Figure 2 illustrates the ideal process for applying DFM&A. The development of any new product must go through four major phases before it reaches the marketplace: concept, design, development, and production. In the concept phase, product specifications are created and the design team creates a design layout of the new product. At this point, the first design for assembly analysis should be completed. This analysis will provide the design team with a theoretical minimum parts count and pinpoint high-assembly areas in the design.

At this point, the design team needs to review the DFA results and adjust the design layout to reflect the feedback of this preliminary analysis. The next step is to complete a design for manufacturing analysis on each unique part in the product. This will consist of developing a part cost and tooling cost for each part. It should also include doing a producibility study of each part. Based on the DFM analysis, the design team needs to make some

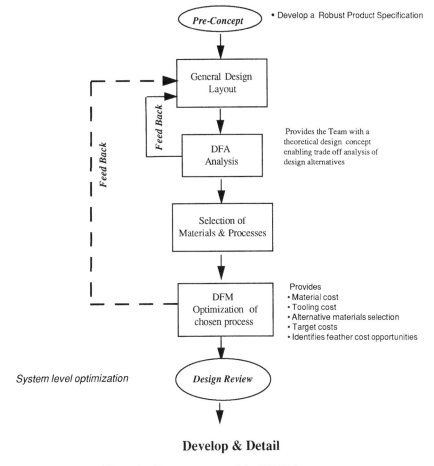

Figure 2 Key components of the DFM&A process.

additional adjustments in the design layout. At this point, the design team is now ready to start the design phase of the project. The DFM&A input at this point has developed a preliminary bill of material (BOM) and established a target cost for all the unique new parts in the design. It has also influenced the product architecture to improve the sequence of assembly as it flows through the manufacturing process.

The following case study illustrates the key elements in applying DFM&A. Figure 3 shows a product called the *motor drive assembly*. This design consists of 17 parts and assemblies. Outwardly it looks as if it can be assembled with little difficulty. The product is made up of two sheet metal parts and one aluminum machined part. It also has a motor assembly and a sensor, both bought from an outside supplier. In addition, the *motor drive assembly* has nine hardware items that provide other functions—or do they?

At this point, the design looks simple enough. It should take minimal engineering effort to design and detail the unique parts and develop an assembly drawing. Has a UMC been developed yet? Has a DFM&A analysis been performed? The DFA analysis will look at each process step, part, and subassembly used to build the product. It will analyze the time it takes to "get" and "handle" each part and the time it takes to insert each part in the assembly (see Table 1). It will point out areas where there are difficulties handling, aligning, and securing each and every part and subassembly. The DFM analysis will establish a cost for each part and estimate the cost of fabrication tooling. The analysis will also point out high-cost areas in the fabrication process so that changes can be made.

At this point, the DFA analysis suggested that this design could be built with fewer parts. A review of Table 2, column 5, shows that the design team feels it can eliminate the bushings, stand-offs, end-plate screws, grommet, cover, and cover screws. Also by replacing the end plate with a new snap-on plastic cover, they can eliminate the need to turn the

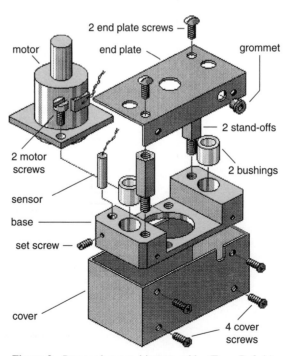

Figure 3 Proposed motor drive assembly. (From Ref. 1.)

Table 1

Motor Drive Assembly	
Number of parts and assemblies	19
Number of reorientation or adjustment	1
Number of special operations	2
Total assembly time in seconds	213.4
Total cost of fabrication and assembly tooling	$3590
Tool amortization at 10K assemblies	$ 0.36
Total cost of labor at $74.50/hr	$ 4.42
Total cost of materials	$ 42.44
Total cost of labor and materials	$ 46.86
Total UMC	$ 47.22

(reorientation) assembly over to install the end plate and two screws. Taking the time to eliminate parts and operations is the most powerful part of performing a DFA analysis. This is rule #2, which was left out above: DFM&A is a team sport. Bringing all members of the new product development team together and understanding the sequence of assembly, handling, and insertion time for each part will allow each team member to better understand the function of every part.

DFM Analysis

The DFM analysis provided the input for the fabricated part cost. As an example, the base is machined from a piece of solid aluminum bar stock. As designed, the base has 11 different holes drilled in it and 8 of them require taping. The DFM analysis (see Table 3) shows that it takes 17.84 minutes to machine this part from the solid bar stock. The finished machined base costs $10.89 in lots of 1000 parts. The ideal process for completing a DFM analysis might be as follows.

In the case of the base, the design engineer created the solid geometry in Matra Data's Euliked CAD system (see Fig. 4). The design engineer then sent the solid database as an STL file to the manufacturing engineer, who then brought the STL file into a viewing tool called *Solid View* (see Fig. 5). SolidView allowed the ME to get all the dimensioning and geometry inputs needed to complete the Boothroyd Dewhurst design for manufacturing machining analysis of the base part. SolidView also allowed the ME to take cut sections of the part and then step through it to insure that no producibility rules had been violated.

Today all of the major CAD supplies provide the STL file output format. There are many new CAD viewing tools like SolidView available, costing about $500 to $1000. These viewing tools will take STL or IGS files. The goal is to link all of the early product development data together so each member can have fast, accurate inputs to influence the design in its earliest stage.

In this example, it took the ME a total of 20 minutes to pull the STL files into SolidView and perform the DFM analysis. Engineering in the past has complained that DFM&A takes too much time and slows the design team down. The ME then analyzes the base as a die casting part, following the producibility rule. By designing the base as a die casting, it is possible to mold many of the part features into the part. This net shape die cast design will reduce much of the machining that was required in the original design. The die cast part will still require some machining. The DFM die casting analysis revealed that the base

Table 2 Motor Drive Assembly: Design for Assembly Analysis

1	2	3	4	5	6	7	8	9	10	11	12	13	14	15	16	17
Name	Sub No. Entry No.	Type	Repeat Count	Minimum Items	Tool Fetching Time, sec	Handling Time, sec	Insertion or Op'n Time, sec	Total Time, sec	Labor Cost, $	Ass'y Tool or Fixture Cost, $	Item Cost, $	Total Item Cost, $	Manuf. Tool Cost, $	Target Cost, $	Part Number	Description
Base	1.1	Part	1	1	0	1.95	1.5	3.45	0.07	500	10.89	10.89	950	7.00	1P033-01	Add base to fixture
Bushing	1.2	Part	2	0	0	1.13	6.5	15.26	0.32	0	1.53	3.06	0	0.23	16P024-01	Add & press fit
Motor	1.3	Sub	1	1	0	7	6	13	0.27	0	18.56	18.56	0	12.00	121S021-02	Add & hold down
Motor screw	1.4	Part	2	2	2.9	1.5	9.6	25.1	0.52	0	0.08	0.16	0	0.08	112W0223-06	Add & thread
Sensor	1.5	Sub	1	1	0	5.6	6	11.6	0.24	0	2.79	2.79	0	2.79	124S223-01	Add & hold down
Set screw	1.6	Part	1	1	2.9	3	9.2	15.1	0.31	0	0.05	0.05	0	0.05	111W0256-02	Add & thread
Stand-off	1.7	Part	2	0	2.9	1.5	9.6	25.1	0.52	0	0.28	0.56	0	0.18	110W0334-07	Add & thread
End plate	1.8	Part	1	1	0	1.95	5.2	7.15	0.15	0	2.26	2.26	560	0.56	15P067-01	Add & hold down
End plate screw	1.9	Part	2	0	2.9	1.8	5.7	17.9	0.37	0	0.03	0.06	0	0.03	110W0777-04	Add & thread
Grommet	1.1	Part	1	0	0	1.95	11	12.95	0.27	0	0.12	0.03	0	0.12	116W022-08	Add & push fit
Dress wires—grommet	1.11	Oper	2	2	—	—	18.79	18.79	0.39	0	0.00	0.00	0	0.00		Library operation
Reorientation	1.12	Oper	1	0	—	—	4.5	4.5	0.09	350	0.00	0.00	0	0.00		Reorient & adjust
Cover	1.13	Part	1	0	0	2.3	8.3	10.6	0.22	0	3.73	3.73	1230	1.20	2P033-01	Add
Cover screw	1.14	Part	4	0	2.9	1.8	5.7	32.9	0.68	0	0.05	0.18	0	0.05	112W128-03	Add & thread
Totals =			22	9				213.4	4.42	850		42.33	2740	24.28		
								UMC =	47.22							

Production life volume = 10,000
Annual build volume = 3000
Assm. labor rate $/hr = $74.50

Note: The information presented in this table was developed from the Boothroyd Dewhurst DFA software program, version 8.0.[2]

Table 3 Machining Analysis Summary Report

Setups	Time Minutes	Cost $
Machine Tool Setups		
Setup	0.22	0.10
Nonproductive	10.63	4.87
Machining	6.77	3.10
Tool wear	—	0.31
Additional cost/part	—	0.00
Special tool or fixture	—	0.00
Library Operation Setups		
Setup	0.03	0.02
Process	0.20	0.13
Additional cost/part	—	0.03
Special tool or fixture	—	0.00
Material	—	2.34
Totals	17.84	10.89
Material	Gen aluminum alloy	
Part number	5678	
Initial hardness	55	
Form of workpiece	Rectangular bar	
Material cost, $/lb	2.75	
Cut length, in.	4.000	
Section height, in.	1.000	
Section width, in.	2.200	
Product life volume	10,000	
Number of machine tool setups	3	
Number of library operation setups	1	
Workpiece weight, lb	0.85	
Workpiece volume, cu in.	8.80	
Material density, lb/cu in.	0.097	

casting would cost $1.41 and the mold would cost $9050. Table 4 compares the two different fabrication methods.

This early DFM&A analysis provides the product development team with accurate labor and material estimates at the start of the project. It removes much of the complexity of the assembly and allows each member of the design team to visualize every component's function. By applying the basic principles of DFA, such as

- Combining or eliminating parts
- Eliminating assembly adjustments
- Designing part with self-locating features
- Designing parts with self-fastening features
- Facilitating handling of each part
- Eliminating reorientation of the parts during assembly
- Specifying standard parts

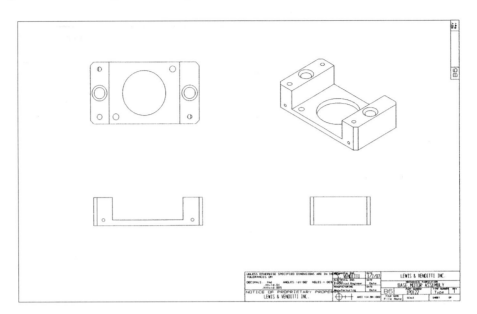

Figure 4

the design team is able to rationalize the motor drive assembly with fewer parts and assembly steps. Figure 6 shows a possible redesign of the original motor drive assembly. The DFM&A analysis (Table 5) provided the means for the design team to question the need and function of every part. As a result, the design team now has a new focus and an incentive to change the original design.

Table 6 shows the before-and-after DFM&A results.

If the motor drive product meets its expected production life volume of 10,000 units, the company will save $170,100. By applying principles of DFM&A to both the labor and material on the motor drive, the design team is able to achieve about a 35% cost avoidance on this program.

2.2 Getting the DFM&A Process Started

Management from All of the Major Disciplines Must Be on Your Side

In order for the DFM&A process to succeed, upper management must understand, accept, and encourage the DFM&A way of thinking. They must *want* it. It is difficult, if not impossible, for an individual or group of individuals to perform this task without management support, since the process requires the cooperation of so many groups working together. The biggest challenge of implementing DFM&A is the cooperation of so many individuals towards a common goal. This does not come naturally, especially if it is not perceived by the leaders as an integral part of the business's success. In many companies, management does not understand what DFM&A is. They believe it is a manufacturing process. It is not; it is a new product development process, which *must* include all disciplines (engineering, service, program managers, and manufacturing) to yield significant results. The simplest method to achieve cooperation between different organizations is to have the team members work in a common location (co-located team). The new product development team needs some nur-

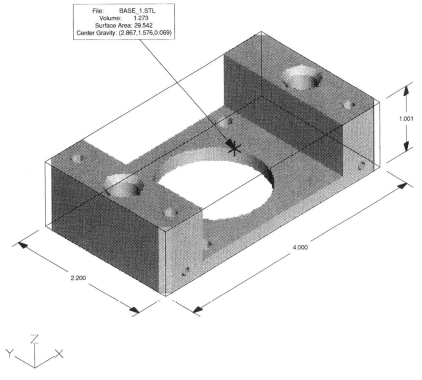

Figure 5

turing and stimulation to become empowered. This is an area where most companies just don't understand the human dynamics of building a high-performance team. Table 7 should aid in determining whether you are working in a team environment or a work group environment.

Many managers will say that their people are working in a team environment, but they still want to have complete control over work assignments and time spent supporting the team. In their mind, the team's mission is secondary to the individual department manager's

Table 4

	Die Cast and Machined	Machined from Bar Stock
Stock cost		$2.34
Die casting	$1.41	
$9050 die casting tooling/10,000	$0.91	
Machining time, min	3.6	17.84
Machining cost	$3.09	$8.55
Total cost	5.41	$10.89

14 Product Design for Manufacturing and Assembly

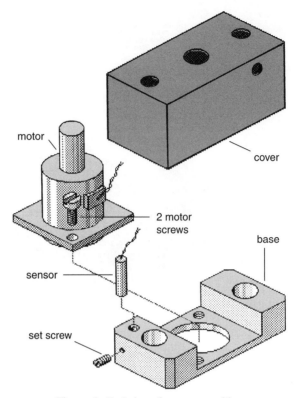

Figure 6 Redesign of motor assembly.

goals. This is not a team; it is a work group. The essential elements of a high-performance team are

- A clear understanding of the team's goals (a defined set of goals and tasks assigned to each individual team member)
- A feeling of openness, trust, and communication
- Shared decision-making (consensus)
- A well-understood problem-solving process
- A leader who legitimizes the team-building process

Management must recognize that to implement DFM&A in their organization, they must be prepared to change the way they do things. Management's reluctance to accept the need for change is one reason DFM&A has been so slow to succeed in many companies. Training is one way of bringing DFM&A knowledge to an organization, but training alone cannot be expected to effect the change.

2.3 The DFM&A Road Map

The DFM&A Methodology (A Product Development Philosophy)

- Form a multifunctional team
- Establish the product goals through competitive benchmarking

Table 5 Redesign of Motor Drive Assembly: Design for Assembly Analysis

1	2	3	4	5	6	7	8	9	10	11	12	13	14	15	16	17
Name	Sub No. Entry No.	Type	Repeat Count	Minimum Items	Tool Fetching Time, sec	Handling Time, sec	Insertion or Op'n Time, sec	Total Time, sec	Labor Cost, $	Ass'y Tool or Fixture Cost, $	Item Cost, $	Total Item Cost, $	Manuf. Tool Cost, $	Target Cost, $	Part Number	Description
Base, casting	1.1	Part	1	1	0	1.95	1.5	3.45	0.07	850	3.09	3.09	9,050	3.09	1P033-02	Add base to fixture
Motor	1.2	Sub	1	1	0	3	6	9	0.19		18.56	18.56		15.00	121S021-02	Add & hold down
Motor screw	1.3	Part	2	2	2.9	1.5	9.6	25.1	0.52		0.08	0.16		0.08	112W0223-06	Add & thread
Sensor	1.4	Sub	1	1	0	5.6	6	11.6	0.24		2.79	2.79		2.79	124S223-01	Add & hold down
Set screw	1.5	Part	1	1	2.9	2.55	9.2	14.65	0.30		0.05	0.05		0.05	111W0256-02	Add & thread
Push/pull wire—easy	1.6	Oper	2	2	—	—	—	18.79	0.39		0.00	0.00		0.00		Library operation
Cover	1.7	Part	1	1	0	1.95	1.8	3.75	0.08		1.98	1.98	7,988	1.70	2P033-02	Add & snap fit
Totals =			9	9				86.34	1.79	850		26.63	17,038	22.71		
							UMC =	30.21								

Production life volume = 10,000
Annual build volume = 3000
Assm. labor rate = $74.50

Note. The information presented in this table was developed from the Boothroyd Dewhurst DFA software program, version 8.0.[2]

Table 6 Comparison of DFM&A Results

	Motor Drive Assembly	Redesign of Motor Drive Assembly
Number of parts and assemblies	19	7
Number of reorientation or adjustment	1	0
Number of special operations	2	2
Total assembly time in seconds	213.4	86.34
Total cost of labor at $74.50/hr	$4.42	$1.79
Total cost of materials	$42.44	$26.63
Total cost of labor and material	$46.86	$28.42
Total cost of fabrication tooling	$3590	$17,888
Tool amortization at 10K assemblies	$0.36	$1.79
Total UMC	$47.22	$30.21
Savings =	$17.01	

- Perform a design for assembly analysis
- Segment the product into manageable subassemblies or levels of assembly
- As a team, apply the design for assembly principles
- Use creativity techniques to enhance the emerging design
- As a team, evaluate and select the best ideas
- Ensure economical production of every piece part
- Establish a target cost for every part in the new design
- Start the detailed design of the emerging product
- Apply design for producibility guidelines
- Reapply the process at the next logical point in the design
- Provide the team with a time for reflection and sharing results

Table 7 Human Factors Test (check the box where your team fits)

Yes	Team Environment	Yes	Work Group Environment
	Are the team members committed to the group's common goals?	✓	Are members loyal to outside groups with conflicting interests (functional managers)?
	Is there open communication with all members of the team?	✓	Is information unshared?
	Is there flexible, creative leadership?	✓	Is there a dominating leadership?
	Is the team rewarded as a group?	✓	Is there individual recognition?
	Is there a high degree of confidence and trust between members?	✓	Are you unsure of the group's authority?

This DFM&A methodology incorporates all of the critical steps needed to insure a successful implementation.

Develop a multifunctional team of all key players before the new product architecture is defined. This team must foster a creative climate that will encourage ownership of the new product's design. The first member of this team should be the project leader, the person who has the authority for the project. This individual must control the resources of the organization, should hand-pick the people who will work on the team, and should have the authority to resolve problems within the team.

The team leader should encourage and develop a creative climate. It is of utmost importance to assemble a product development team that has the talent to make the right decisions, the ability to carry them out, and the persistence and dedication to bring the product to a successful finish. Although these qualities are invaluable, it is of equal importance that these individuals be allowed as much freedom as possible to germinate creative solutions to the design problem as early as possible in the product design cycle.

The product development team owns the product design and the development process. The DFM&A process is most successful when implemented by a multifunctional team, where each person brings to the product design process his or her specific area of expertise. The team should embrace group dynamics and the group decision-making process for DFM&A to be most effective.

Emphasis has traditionally been placed on the design team as the people who drive and own the product. Designers need to be receptive to team input and share the burden of the design process with other team members.

The team structure depends on the nature and complexity of the product. Disciplines that might be part of a product team include

- Engineering
- Manufacturing
- Field service and support
- Quality
- Human factors or ergonomics
- Purchasing
- Industrial design and packaging
- Distribution
- Sales
- Marketing

Although it is not necessary for all of these disciplines to be present all of the time, they should have an idea of how things are progressing during the design process.

Clearly, there can be drawbacks to multidisciplinary teams, such as managing too many opinions, difficulty in making decisions, and factors in general that could lengthen the product development cycle. However, once a team has worked together and has an understanding of individual responsibilities, there is much to gain from adopting the team approach. Groups working together can pool their individual talents, skills, and insight so that more resources are brought to bear on a problem. Group discussion leads to a more thorough understanding of problems, ideas, and potential solutions from a variety of standpoints. Group decision-making results in a greater commitment to decisions, since people are more motivated to support and carry out a decision that they helped make. Groups allow individuals to improve existing skills and learn new ones.

Having the team located together in one facility makes the process work even better. This co-location improves the team's morale and also makes communication easier. Remembering to call someone with a question, or adding it to a meeting agenda, is more difficult than mentioning it when passing in the hallway. Seeing someone reminds one of an issue that may have otherwise been forgotten. These benefits may seem trivial, but the difference that co-location makes is significant.

As a team, establish product goals through a competitive benchmarking process: concept development. Competitive benchmarking is the continuous process of measuring your own products, services, and practices against the toughest competition, or the toughest competition in a particular area. The benchmarking process will help the team learn who the "best" are and what they do. It gives the team a means to understand how this new product measures up to other products in the marketplace. It identifies areas of opportunities that need changing in the current process. It allows the team to set targets and provides an incentive for change. Using a DFM&A analysis process for the competitive evaluation provides a means for relative comparison between those of your products and those of your competitors. You determine exactly where the competition is better.

Before performing a competitive teardown, decide on the characteristics that are most important to review, what the group wants to learn from the teardown, and the metrics that will be noted. Also keep the teardown group small. It's great to have many people walk through and view the results, but a small group can better manage the initial task of disassembly and analysis. Ideally, set aside a conference room for several days so the product can be left out unassembled, with a data sheet and metrics available.

Perform a design for assembly analysis of the proposed product that identifies possible candidate parts for elimination or redesign and pinpoints high-cost assembly operations. Early in this chapter, the motor drive assembly DFM&A analysis was developed. This example illustrates the importance of using a DFA tool to identify, size, and track the cost-savings opportunities. This leads to an important question: Do you need a formal DFA analysis software tool? Some DFM&A consultants will tell you that it is not necessary to use a formal DFA analysis tool. It is my supposition that these consultants want to sell you consulting services rather than teach the process. It just makes no sense *not* to use a formal DFA tool for evaluating and tracking the progress of the new product design through its evolution. The use of DFA software provides the team with a focus that is easily updated as design improvements are captured. The use of DFA software does not exclude the need for a good consultant to get the new team off to a good start. The selection of a DFA tool is a very important decision. The cost of buying a quality DFA software tool is easily justified by the savings from applying the DFA process on just one project.

At this point, the selection of the manufacturing site and type of assembly process should be completed. Every product must be designed with a thorough understanding of the capabilities of the manufacturing site. It is thus of paramount importance to choose the manufacturing site at the start of product design. This is a subtle point that is frequently overlooked at the start of a program, but to build a partnership with the manufacturing site, the site needs to have been chosen! Also, manufacturing facilities have vastly different processes, capabilities, strengths, and weaknesses, that affect, if not dictate, design decisions. When selecting a manufacturing site, the process by which the product will be built is also being decided.

As a team, apply the design for assembly principles to every part and operation to generate a list of possible cost opportunities. The generic list of DFA principles includes the following:

- Designing parts with self-locating features
- Designing parts with self-fastening features
- Increasing the use of multifunctional parts
- Eliminating assembly adjustments
- Driving standardization of fasteners, components, materials, finishes, and processes

It is important for the team to develop its own set of DFA principles that relate to the specific product it is working on. Ideally, the design team decides on the product characteristics it needs to meet based on input from product management and marketing. The product definition process involves gathering information from competitive benchmarking and teardowns, customer surveys, and market research. Competitive benchmarking illustrates which product characteristics are necessary.

Principles should be set forth early in the process as a contract that the team draws up together. It is up to the team to adopt many principles or only a few, and how lenient to be in granting waivers.

Use brainstorming or other creativity techniques to enhance the emerging design and identify further design improvements. The team must avoid the temptation to start engineering the product before developing the DFM&A analysis and strategy. As a team, evaluate and select the best ideas from the brainstorming, thus narrowing and focusing the product goals.

With the aid of DFM software, cost models, and competitive benchmarking, establish a target cost for every part in the new design. Make material and manufacturing process selections. Start the early supplier involvement process to ensure economical production of every piece part. Start the detailed design of the emerging product. Model, test, and evaluate the new design for fit, form, and function. Apply design for producibility guidelines to the emerging parts design to ensure that cost and performance targets are met.

Provide the team with a time for reflection and sharing results. Each team member needs to understand that there will be a final review of the program, at which time members will be able to make constructive criticism. This time helps the team determine what worked and what needs to be changed in the process.

Use DFM&A Metrics

The development of some DFM&A metrics is important. The team needs a method to measure the before-and-after results of applying the DFM&A process, thus justifying the time spent on the project. Table 8 shows the typical DFM&A metrics that should be used to compare your old product design against a competitive product and a proposed new redesign.

The total number of parts in an assembly is an excellent and widely used metric. If the reader remembers only one thing from this chapter, let it be to strive to reduce the quantity of parts in every product designed. The reason limiting parts count is so rewarding is that when parts are reduced, considerable overhead costs and activities that burden that part also disappear. When parts are reduced, quality of the end product is increased, since each part that is added to an assembly is an opportunity to introduce a defect into the product. Total assembly time will almost always be lowered by reducing the quantity of parts.

Table 8 DFM&A Metrics

	Old Design	Competitive	New Design
Number of Parts & Assemblies			
Number of Separate Assm. Operations			
Total Assembly Time			
Total Material Cost			
Totals			

Table 9 DFM&A New Products Checklist

Design for Manufacturing and Assembly Consideration	Yes	No
Design for assembly analysis completed:	☐	☐
Has this design been analyzed for minimal part count?	☐	☐
Have all adjustments been eliminated?	☐	☐
Are more than 85% common parts and assemblies used in this design?	☐	☐
Has assembly sequence been provided?	☐	☐
Have assembly and part reorientations been minimized?	☐	☐
Have more than 96% preferred screws been used in this design?	☐	☐
Have all parts been analyzed for ease of insertion during assembly?	☐	☐
Have all assembly interferences been eliminated?	☐	☐
Have location features been provided?	☐	☐
Have all parts been analyzed for ease of handling?	☐	☐
Have part weight problems been identified?	☐	☐
Have special packaging requirements been addressed for problem parts?	☐	☐
Are special tools needed for any assembly steps?	☐	☐

Ergonomics Considerations	Yes	No
Does design capitalize on self-alignment features of mating parts?	☐	☐
Have limited physical and visual access conditions been avoided?	☐	☐
Does design allow for access of hands and tools to perform necessary assembly steps?	☐	☐
Has adequate access been provided for all threaded fasteners and drive tooling?	☐	☐
Have all operator hazards been eliminated (sharp edges)?	☐	☐

Wire Management	Yes	No
Has adequate panel pass-through been provided to allow for easy harness/cable routing?	☐	☐
Have harness/cable supports been provided?	☐	☐
Have keyed connectors been provided at all electrical interconnections?	☐	☐
Are all harnesses/cables long enough for ease of routing, tie down, plug in, and to eliminate strain relief on interconnects?	☐	☐
Does design allow for access of hands and tools to perform necessary wiring operations?	☐	☐
Does position of cable/harness impede air flow?	☐	☐

Design for Manufacturing and Considerations	Yes	No
Have all unique design parts been analyzed for producibility?	☐	☐
Have all unique design parts been analyzed for cost?	☐	☐
Have all unique design parts been analyzed for their impact of tooling/mold cost?	☐	☐

Assembly Process Consideration	Yes	No
Has assembly tryout been performed prior to scheduled prototype build?	☐	☐
Have assembly views and pictorial been provided to support assembly documentation?	☐	☐
Has opportunity defects analysis been performed on process build?	☐	☐
Has products cosmetics been considered (paint match, scratches)?	☐	☐

A simple method to test for potentially unnecessary parts is to ask the following three questions for each part in the assembly:

1. During the products operation, does the part move relative to all other parts already assembled?   (*answer yes or no*)
2. Does the part need to be made from a different material or be isolated from all other parts already assembled?   (*answer yes or no*)
3. Must the part be separate from all other parts already assembled because of necessary assembly or disassembly of other parts?   (*answer yes or no*)

You must answer the questions above for each part in the assembly. If your answer is "no" for all three questions, then that part is a candidate for elimination.

The total time it takes to assemble a product is an important DFM&A metric. Time is money, and the less time needed to assemble the product, the better. Since some of the most time-consuming assembly operations are fastening operations, discrete fasteners are always candidates for elimination from a product. By examining the assembly time of each and every part in the assembly, the designer can target specific areas for improvement. Total material cost is self-explanatory.

The new product DFM&A checklist (Table 9) is a good review of how well your team did with applying the DFM&A methodology. Use this check sheet during all phases of the product development process; it is a good reminder. At the end of the project you should have checked most of the *yes* boxes.

3 WHY IS DFM&A IMPORTANT?

DFM&A is a powerful tool in the design team's repertoire. If used effectively, it can yield tremendous results, the least of which is that the product will be easy to assemble! The most beneficial outcome of DFM&A is to reduce part count in the assembly, which in turn will simplify the assembly process, lower manufacturing overhead, reduce assembly time, and increase quality by lessening the opportunities for introducing a defect. Labor content is also reduced because with fewer parts, there are fewer and simpler assembly operations. Another benefit to reducing parts count is a shortened product development cycle because there are fewer parts to design. The philosophy encourages simplifying the design and using standard, off-the-shelf parts whenever possible. In using DFM&A, renewed emphasis is placed on designing each part so it can be economically produced by the selected manufacturing process.

REFERENCES

1. G. Boothroyd, P. Dewhurst, and W. Knight, *Product Design for Manufacturing and Assembly,* Marcel Dekker, New York, 1994.
2. Boothroyd Dewhurst Inc., *Design for Assembly Software,* Version 8.0, Wakefield, RI, 1996.

CHAPTER 2

ACHIEVING ENTERPRISE GOALS WITH NEW PROCESS TECHNOLOGY

Steve W. Tuszynski
Algoryx, Inc.
Los Angeles, California

1 INTRODUCTION	**22**	
1.1 A Historical Perspective on Technological Development	22	
1.2 The Traditional Approach	23	
1.3 Problems with the Traditional Approach	26	
1.4 Inefficiencies with the Traditional Approach	31	
1.5 A Summary of the Problems	31	
2 THE NEW TECHNOLOGY	**32**	
2.1 History	32	
2.2 What the New Technology Is Not	32	
2.3 Tuszynski's Relational Algorithm (TRA)	33	
3 CONCLUSION	**58**	
4 IMPLEMENTATION	**60**	
APPENDIX A: TUSZYNSKI'S PROCESS LAWS	**60**	
APPENDIX B: DEFINITIONS	**63**	
APPENDIX C: NONTECHNICAL STATISTICAL GLOSSARY	**64**	

1 INTRODUCTION

1.1 A Historical Perspective on Technological Development

The main thrust of technological development has been to explore the relationships between causes and effects. We do this for many reasons, two of which are prime. The first reason is so that we can understand the natural processes that surround us and comprise our environment. The second reason is the basic premise that if we understand what the relationships are between causes and effects, we will be able to produce the effects we want by activating the causes that produce the results we want and minimizing the causes that detract from the results we want. This has been true from the earliest glimmerings of technology developed by mankind.

For example, what engineer has not learned Newton's law that $F = ma$? If we know the acceleration (a) that we want and the mass (m) of the item to be accelerated, then we can compute the force (F) required. If the mass of the item is fixed, then there will be only two variables and the acceleration will be proportional to the force applied. In this instance, if the force is doubled the acceleration is doubled. If the force is quadrupled, so is the acceleration. What we learn from Newton is that acceleration is related to force. In this instance, the two variables are "co-related." We also say, with the same meaning, that they are "correlated."

When did we start thinking this way? Perhaps it was when our early ancestors first threw rocks at animals to defend themselves or get food, or perhaps even earlier when early humans tried to figure out what kind of behavior it took to survive or reproduce. This approach is ingrained in the human mentality and has been the foundation on which we have built our technology. This approach has been necessary, useful, and productive. It is hard to imagine our existence without our understanding the linkages between causes and effects.

Converting raw material to useful products in a manufacturing process involves converting inputs into outputs. More generally, the result of any natural process is change. The purpose of man-made processes is to produce change toward a desired goal or objective.

1.2 The Traditional Approach

Manufacturing Process Flow Diagram

The balance of this chapter discusses process technology in the context of design, tooling, manufacturing, and quality engineering.* Figure 1 is a manufacturing process flow diagram. It shows inputs into and output from the manufacturing process. Control and noise variables influence the output for any given input.

Inputs

Inputs are items input into the process and can be physical or nonphysical. Inputs are usually physical items when the process produces physical output, but they can be nonphysical items where the process is an algorithmic or computational process. Nonphysical items can be of many types. In a manufacturing or simulation process, inputs are usually either variable or attribute data. Inputs can also be combinations of physical and nonphysical items.

The manufacturing process is a single step or a series of sequential steps that modifies the inputs to the process. Each stage in a manufacturing process will have an input and an output.† The input of any one stage will be the output of the preceding stage and the output of any one stage will be the input to the subsequent stage.

Control Variables

Control variables are the process parameters‡ controlled by an operator or process engineer. In essence, these are the knobs and dials, whether manually or automatically controlled, on

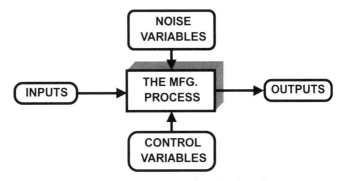

Figure 1 The traditional manufacturing flow diagram.

---

*Hopefully, those individuals schooled in sciences and technologies other than engineering and manufacturing will see applications from this chapter to their respective areas of expertise.
†For ease of reading, multiple inputs and/or outputs are referred to here as an input or an output.
‡Pressures, temperatures, speeds, times, chemical concentrations, orientations, power settings, frequencies, intensities, agitation levels, etc. are typical process control settings.

the manufacturing equipment that are adjusted to produce conforming parts.* Figure 2 shows typical process control variables such as pressures, temperatures, times, and speeds. These are the variables that are controlled by the process operator.

Noise Variables

Noise variables are those variables that influence the output of the process. Figure 2 shows various noise variables for a particular process. Noise variables are not controlled because either they cannot be controlled or we choose to not control them. Noise variables may be left uncontrolled for numerous reasons, including circumstances where

- They are unknown.
- They are too expensive to control.
- They are too time-consuming to control.
- It is not possible to control them.
- Controlling them would not make an appreciable difference in the quality or producibility of the manufactured part.

Output Variables

The output of the process consists of the item to be manufactured or produced. The output of the process will have different characteristics. In the instance where parts are being man-

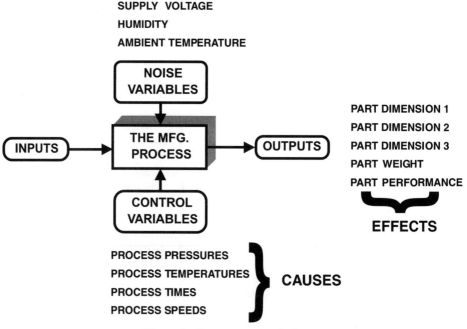

Figure 2 Process causes and effects.

---

*Conforming parts meet engineering specification or drawing values. Conforming parts are defined as "good parts" and nonconforming parts are defined as "bad parts."

ufactured, the characteristics will be referred as part characteristics. Figure 2 shows various typical part characteristics such as dimensions, weight, or performance as outputs of the manufacturing process.

For purposes of this chapter, part characteristics are divided, arbitrarily and for convenience, into four categories:

1. Variable characteristics
2. Attribute characteristics
3. Material characteristics
4. Performance characteristics

Variable characteristics are most typically dimensions. Dimensions are usually subcategorized into critical and noncritical dimensions.*

Attribute characteristics are data describing a part characteristic not measurable on a number scale. Typical attribute characteristics are on or off, the presence or absence of undesirable characteristics, supplier A, B, or C, material type m, n, or o, etc. Some attributes, such as color, can be converted to variable data (for example, a combination of red, green, and blue) if it is worth the cost and effort. In the context of this chapter, visual attribute characteristics can be thought of as the presence or absence of some desirable or undesirable part characteristic as determined through visual inspection.

Material characteristics are the physical properties of the manufactured part. Typical material characteristics could be tensile strength, surface hardness, density, or reflectivity. Material characteristics are usually variable data. The categorization of part characteristics into these first three categories is not crucial, but is a matter of convenience.

Performance characteristics refer to those characteristics that are measures of how well the part performs relative to its functional requirements. Performance characteristics are usually variable characteristics but can also be attribute characteristics.

Causes and Effects
Figure 2 identifies the control variables as causes and the output characteristics as effects. The foundation of the traditional approach is to relate causes and effects.

The Traditional Approach to Cause and Effect
Figure 3 shows how the essence of the traditional approach is to determine the linkage between process causes and effects. This approach is premised on the logical belief that if we adequately understand the relationships between causes and the effects, then we should be able to set the control variables to values that produce the desired part characteristics. In essence, the traditional approach looks for the correlations that model the relationships between causes and effects.

The Traditional Approach to Determining Correlations
The traditional approach to determining the correlations between causes and effects is to evaluate the part characteristics when parts are manufactured under different control settings (causal conditions). In some instances, the control settings will be deliberately changed to induce variation in the manufactured parts. In other instances, there may be enough natural

---

*Critical dimensions are those considered by the design engineer to be critical to form, fit, or function (performance).

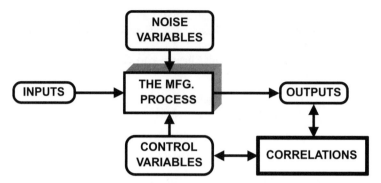

Figure 3 The traditional approach to cause and effect.

variation in the control settings that, over time, enough data with enough variation will be generated so that the correlations can be determined.

Prior to the invention and application of efficient statistical methods, variation in part characteristics was usually induced by changing one control variable at a time and then determining how each part characteristic changed. Sequentially changing each control setting one at a time is very time and cost inefficient and can lead to erroneous conclusions.

Design of Experiments
Design of experiments (DOE) is a statistical methodology that has greatly improved the efficiency of determining the relationships between causes and effects. DOE is a large step forward in improving the time and cost efficiencies and in reducing erroneous conclusions. However, as discussed below, DOE does not universally explain all of the relationships between causes and effects.

1.3 Problems with the Traditional Approach

As useful as the traditional approach is, there are many situations in which it is difficult, impossible, or uneconomical to determine the relationships between causes and effects. These situations can occur irrespective of whether DOE is used. This section discusses several situations that make determining the relationships between causes and effects impractical.

More Than a Few Control Variables—Many Relationships
Some processes have relatively few control variables, while others can have many. As shown in Fig. 4, plastic injection-molding processes, for example, can have over 20 control variables. When many control variables are involved, there are many cause-and-effect relationships to be evaluated and understood.

More Than a Few Control Variables—Many Control Setting Combinations
Further, as the number of control variables increases, the possible number of combinations of control variables increases geometrically. For most process variables, there are an infinite (analog) or large (digital) number of settings for each control variable. However, the situation can get complex even when there are only two or three levels chosen for each control variable. For example, if there are 20 control variables and each control variable is examined at only two settings—a high and a low value—then there are over one million possible combinations. If each of the 20 control variables is examined at three settings—a high, a nominal, and a low value—then there are over three billion possible combinations.

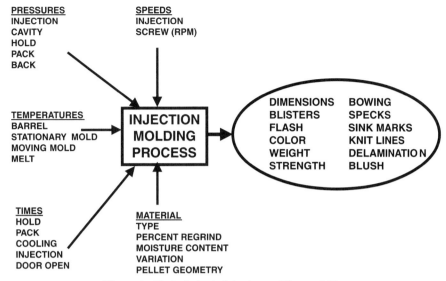

Figure 4 Typical plastic injection molding variables.

More Than a Few Part Characteristics
For products with multiple part characteristics, process engineers and operators have the difficult task of attempting to adjust process settings to produce parts with all dimensions simultaneously at target values. Some parts have a large number of critical characteristics. For example, the single plastic injection-molded part shown in Fig. 5 has 42 critical dimensions. In this instance, one must determine the relationship between each control variable and each of the 42 critical dimensions.

Different Responses to Control Variable Changes
Different part characteristics can have different responses to changes in control settings. Figure 6 shows how the length of a part increases when the process temperature setting is increased. However, the diameter of the part decreases as the process temperature setting is increased. One cannot increase both the length and diameter of the part by changing the process control setting.

Multiunit Processes
Some processes produce multiple parts for each process cycle. Injection-molded parts, for example, are frequently made with multicavity molds. Figure 7 illustrates this point for a part that has 5 critical dimensions and is manufactured in an 8-cavity mold. It is not unusual for molded plastic parts to be made with 8-, 16-, or 32-cavity molds. Molded rubber parts

Figure 5 Typical plastic injection molding variables: a single part with 42 critical dimensions.

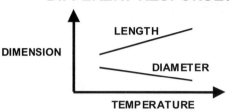

Figure 6 Responses can be different to changes in control settings.

can be made with molds having hundreds of cavities. Semiconductor wafer fabrication processes can result in thousands of circuits on each wafer.

Multiunit processes can get quite complex when each part has many part characteristics. For example, when a part that has 42 critical dimensions is produced in a 32-cavity mold, each machine cycle produces 1344 separate critical dimensions.

Simple Interactions

Figure 8 illustrates a simple interaction between two control variables. In this example, if the pressure control variable is at setting level 1, the length of the part increases as the temperature control variable increases. However, if the pressure control variable is at setting level 2, the length of the part decreases as the temperature increases. Put more simply, the response of the length to changes in temperature depends on the value of the pressure. Simple interactions are common in many manufacturing processes.

Complex Interactions

Figure 9 illustrates a complex interaction between three control variables. In this instance, if the pressure control variable is at setting level 1, the length of the part

- Decreases as the temperature increases when the speed is at level 1.
- Remains unchanged as the temperature increases when the speed is at level 2.
- Increases as the temperature increases when the speed is at level 3.

A different set of three response curves will also exist for the pressure control variable setting at level 2. Put more simply, the response of the length to changes in temperature depends not only on the value of the pressure but also on the value of the speed. Although

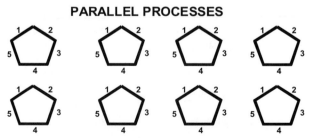

Figure 7 Multiunit processes greatly increase the number of part characteristics.

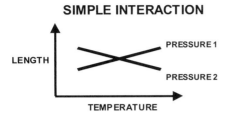

Figure 8 Simple interactions make it difficult to determine responses.

complex interactions are not as common as simple interactions, they do occur occasionally in manufacturing processes.

Nonlinear Responses

Figure 10 shows a nonlinear response between a part characteristic and a process control variable. The length increases, reaches a maximum, and then decreases as the temperature control variable increases. If one is sampling only two temperature levels to determine the response of the length to changes in temperature, one could conclude, depending on the two points chosen, that

- Length increases as temperature increases.
- Length is insensitive to changes in temperature. Or
- Length decreases as temperature increases.

Two types of errors can occur. The first is the linear approximation of a nonlinear response. The second is that a reversal, as noted immediately above, can occur, which would invalidate the conclusion on how much and in which direction to change the temperature.

DOE May Not Be Helpful

Design of experiments has proven to be a useful tool for circumstances where (1) one part or performance characteristic needs to be optimized and (2) there are few process complexities. When the first criterion is not met, DOE can and generally does give conflicting results. When the second criterion is not met, it is difficult to model the process and get useful results.

Small or Nonexistent Producibility Windows

The producibility window may be nonexistent, i.e., it can be impossible to produce good parts. Operators and process engineers can waste significant time learning this. Even if a set

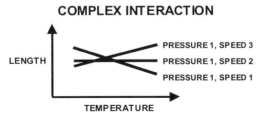

Figure 9 Complex interactions make it even more difficult to determine responses.

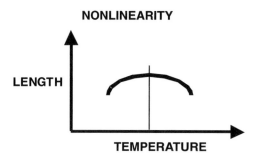

Figure 10 Nonlinear responses can lead to errors.

of process settings can be found that produces good parts, the producibility window may be so small that even minor variations in process settings result in bad parts (ones that do not meet specifications).

Operator Technique
Small or nonexistent producibility windows frequently occur prior to achieving a good first article. This can make it difficult or impossible to produce good parts. The path of least resistance is for the operator to try to produce good parts by adjusting process settings.* Also, the techniques that worked for an operator in the past may not work for the current part.

Tolerance Relaxation Dependency on Operator Process Settings
When the producibility window is small or nonexistent, process engineers frequently ask the design engineer for tolerance relaxation on the problematic part characteristics (dimensions, for example). There is a potential problem with the method currently used to determine which design tolerances need to be relaxed and by how much. The problem occurs because the part characteristics clearly are dependent on the process settings selected by the operator. A subsequent change to the process settings can invalidate the need for the original tolerance relaxations and create the need for new tolerance relaxations on different part characteristics. Based on the parts produced by operator A, the process engineer could ask, for example, for an increase in the upper (+) tolerance on dimension X. Based on the parts produced by operator B (or operator A at a different point in time with different process settings), the process engineer could ask for a decrease in the lower (−) tolerance on dimension Y. No prior art technology has solved this problem or given the design engineer a ranking of the order in which to relax design tolerances or the size of the required tolerance relaxations independent of process settings.

Tooling Modification Dependency on Operator Process Settings
A similar but usually more grievous situation can occur with tooling, molds, and fixtures. A preproduction mold, for example, might go through 5–8 different modifications before it is

---

*Changing process settings is usually the first technique used to try to produce a good first article. If a good first article cannot be produced, the next step, usually after extensive "fiddling" by the process operator, is often to ask the design engineer for tolerance relaxation. If the design engineer does not relax the tolerances, the next option is usually to modify the tooling. Tooling modification is usually the last of these three options because of time, cost, and risk considerations. Finally, process variables can be monitored and controlled through various equipment.

qualified to produce good parts. Tool and fixture modifications can be risky and time-consuming. For all processes, dimensional results depend on the values selected for the process settings. This creates a problem for the tooling engineer when deciding how to modify the tooling or fixture. Use of different process settings can invalidate previously made tool, mold, or fixture changes. Tooling and fixture engineers refer to this as the "tyranny of the operator." No prior art technology has solved this problem or given the tooling engineer, in a single step and independent of process settings, the tooling modifications required to produce parts at design targets.

Trial and Error, Iteration, and Guesswork
The absence of a scientific method for eliminating the preceding complexities has resulted in trial-and-error, iteration, and guesswork attempts to produce good first articles and to produce the highest quality parts during production.

Changing Process Technology
When tooling or fixtures cannot be adequately modified and/or tolerances can not be adequately relaxed for a given material, and/or inadequate process control exists, then a more capable process must be used to produce the part. No prior art technology has enabled the design, tooling/fixture, process, and quality engineers to easily determine when this is the case.

1.4 Inefficiencies with the Traditional Approach

Current Statistical Process Control Studies Are Inefficient
During part development (first article, qualification, and certification), statistical process control (SPC) studies, when they are done, are usually done on all dimensions. During production, SPC studies, when they are done, are usually done on either all or a subset of dimensions. As will be shown later, this is inefficient and incurs unnecessary cost.

Current Process Capability Studies Are Inefficient
In a similar fashion, process capability studies are done on all dimensions. This is also inefficient and incurs unnecessary cost.

Shipping and Receiving Inspections Are Inefficient
In a similar fashion, shipping and receiving inspections are usually performed by sampling a subset of parts and measuring all critical dimensions. This incurs unnecessary costs for both the customer and supplier.

1.5 A Summary of the Problems

Historical gains in the quality of manufactured parts over the last two decades have been significantly eroded by increases in measurement and recording costs and by statistical process control (SPC) and process capability (Cpk, Ppk) analysis costs. The efforts of engineers to modify preproduction tooling and fixtures to production tooling and fixtures are frustrated by operator changes to process settings. Process engineers have difficulty determining the values of process settings and design engineers find it difficult to design for producibility when there are multiple part characteristics. The use of standard design tolerances increases manufacturing costs. Determining tolerance relaxations is a trial-and-error process compounded by operator changes to process settings. Replacing obsolete materials can be problematic. The inability to produce parts at design target values decreases product performance.

Design, process, tooling, and quality parameters are interrelated because they co-jointly influence part characteristics and consequently how the part performs and decision-making on whether or not the part is a conforming part. Prior art has not provided engineers and decision makers with an integrated system of technology that incorporates these interrelationships.

2 THE NEW TECHNOLOGY

2.1 History

In 2000, a large, international original equipment manufacturer (OEM) was having difficulty producing good first articles for a new product line of injection-molded plastic parts. These problems stimulated the development of innovative technology that solved the problems. The new technology has been proven in several widely diverse industries* with numerous case studies.† No changes are required to the manufacturing process. Huge reductions in measurement and analysis costs have been achieved, as well as increased quality, increased productivity, and reduced time to market. The bottom line has been significantly increased profits and return of investment (ROI). The new technology is called Tuszynski's relational algorithm (TRA).

2.2 What the New Technology Is Not

It is sometimes easier to introduce new technology by stating it is not. Most importantly, as shown in Fig. 11, TRA does not attempt to determine the relationships (correlations) between causes (process settings) and effects (part characteristics), so TRA bypasses the process complexities and inefficiencies mentioned above.

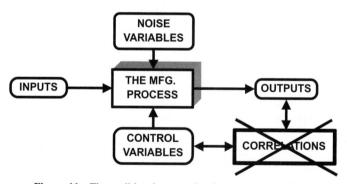

Figure 11 The traditional approach relates cause and effect.

---

*This new technology has been proven for plastic injection molding, sheet metal punching, sheet metal forming, CNC laser cutting, and semiconductor wafer fabrication. Case studies are underway or planned for plastic extrusion, rubber molding, hot and cold metal heading, plating, etching, and wire forming.
†The author would welcome feedback on applications of the algorithms presented in this chapter to new processes and industries.

The following are lists of what TRA is not and what it does not do. TRA is not any of the following types of computer programs:

- A statistical process control (SPC) program
- A process capability (Cpk) analysis program
- A design of experiments (DOE) program
- A finite element analysis (FEA) program
- A plastic flow simulation program

TRA is not used for

- Designing tooling, molds, and fixtures
- Designing mold runner and gate systems

TRA does not use iterative procedures.

2.3 Tuszynski's Relational Algorithm (TRA) (Fig. 12)

The New Algorithms

TRA is a pioneering system of interrelated algorithms that has led to breakthrough insights into manufacturing processes and to powerful new computational software. This state-of-the-art technology reduces the cost of enterprise quality management (EQM) operations and facilitates the implementation of lean strategies (LS) by achieving huge reductions in inspection, statistical process control, and process capability analysis (Cpk) costs, by providing optimized tooling, and by eliminating multiple redundancies in six sigma programs and quality operating systems (QOS).

Foundation of the New Algorithms

TRA is based on the fact that although the relationships between causes (process settings*) and effects (part characteristics†) may be difficult or impossible to determine, the relationships between effects for many processes are consistent and predictable irrespective of changes in the process settings. One of the part characteristics is selected as the predictor characteristic.‡ The predictor characteristic is the characteristic that is the statistically best predictor of all other part characteristics.

When Does TRA Work?

TRA works when there is correlation between part characteristics. This is generally true when the process adds or subtracts material or changes the shape or form of the material. If there is no correlation between part characteristics, TRA will not work.

A Graphical Illustration of TRA

Figure 13 illustrates a condition where there are four interrelated dimensions on a part—A, B, C, and P—where P has been selected as the predictor dimension and A, B, and C are the

---

*Pressures, temperatures, times, speeds, etc.
†Part characteristics include dimensions, weights, material characteristics, and performance characteristics.
‡The predictor characteristic is be referred to as the predictor dimension. Most applications of TRA to date have been with dimensional part characteristics.

Tuszynski's Relational Algorithm (TRA)

Fastest Time to Market – Best Design, Best Tooling, Best Process, Least Analysis

Use TRA™ to Achieve Lean and Six Sigma Goals

- Reduce Time-to-Market
- Improve Profits and ROI
- Maximize Quality
- Reduce Scrap, Rejects and Rework
- Eliminate Repeated Tooling and Mold Modifications
- Locate the Process Sweet Spot
- Simultaneously Optimize All Cpks
- Optimize Design Targets for Producibility
- Optimize Design Tolerances for Producibility
- Simulate Tooling Changes without Time, Cost or Risk
- Improve CNC Programming
- Slash Inspection Costs
- Eliminate Destructive Inspection
- Slash SPC Analysis Costs
- Slash Cpk Analysis Costs
- Integrate Cross-Functional Decision-Making
- Improve Customer-Supplier Communication

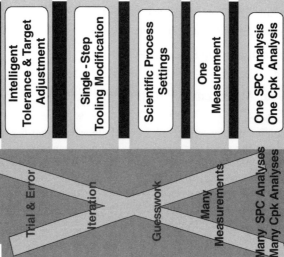

Preliminary Design
Pre-production Tooling
Initial Process Settings

Historical Process
- Trial & Error
- Iteration
- Guesswork
- Many Measurements
- Many SPC Analyses
- Many Cpk Analyses

TRA
- Intelligent Tolerance & Target Adjustment
- Single-Step Tooling Modification
- Scientific Process Settings
- One Measurement
- One SPC Analysis
- One Cpk Analysis

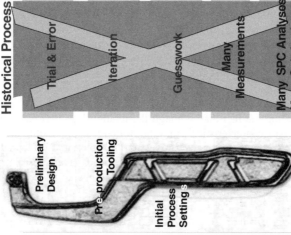

Final Design
Production Tooling
Best Process Settings
Reduced Measurement
Reduced SPC/Cpk Analyses

© 2004-2005

- Plastic Injection Molding
- Plastic Extrusion
- Rubber Molding

- Semiconductor Wafer Fab
- Thermoforming
- Cold and Hot Heading

- CNC Laser Cutting
- Sheet Metal Forming
- Sheet Metal Punching

ALGORYX, INC.™

750 S. Bundy Drive, Ste. 304
Los Angeles, CA 90049
310-820-0987
www.algoryx.com
steve@algoryx.com

Figure 12 An illustration of the application of TRA across engineering functionalities.

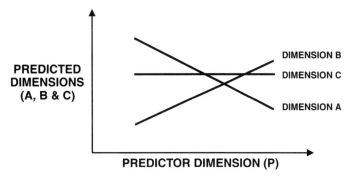

Figure 13 A part with four dimensions.

predicted dimensions. The relationships between the dimensions are defined by regression lines fitted through data generated from manufactured parts.*

TRA Process Conclusions

Figure 13 is simple, yet sophisticated and complex at the same time. Figure 13 leads us to the following rather startling conclusion:

> Even though the relationships between causes (process settings) and effects (part characteristics) may be difficult or impossible to determine, the relationships between effects are consistent and predictable irrespective of changes in the process settings!!!

This conclusion is startling from at least three perspectives. First, it does not matter what the complexities were or how many complexities were present when the data were generated. The relationships between part characteristics can be determined irrespective of the complexities. In essence, process complexities are eliminated. Second, the relationships between part characteristics are simple, understandable, and fixed.† The relationships can be easily visualized when they are presented graphically. Third, it does not matter which particular combination of process settings was used to determine any single point on any regression line.‡ The relationships between part characteristics are fixed in spite of the multiplicity of combinations of process settings that can result in a specific dimensional outcome.

The relationships shown in Fig. 13 represent the entire universe of possible part relationships!!!

All 3 billion (20 process settings at 3 levels) possible process setting combinations are encompassed by Fig. 13.

A Single Degree-of-Freedom System

The system of regression lines (and part characteristics) shown in Figs. 13 and 14 is also a single-degree-of-freedom system. Figure 14 shows that when the value of the predictor dimension (P) is known, the values for all predicted dimensions (A, B, and C) are also known.

---

*The regression lines shown in Fig. 13 assume perfect correlation. The assumption of perfect correlation is used here to simplify the graphs and the discussion. It will be removed later in the chapter.

†The relationships are fixed as long as tooling is fixed and materials are unchanged. If the tooling dimensions change, the changes in relationships are predictable.

‡For certain processes, it is possible to produce a part with a specified part characteristic (or with a specified set of part characteristics) by selecting different combinations of process settings.

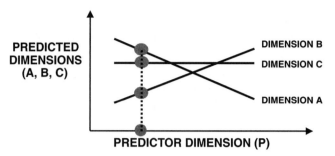

Figure 14 The regression lines constitute a single-degree-of-freedom system.

When the value of the predictor characteristic is known, the value of all predicted characteristics is known.

In essence, the entire system of part dimensions is collapsed into a single predictor dimension. Instead of measuring all dimensions, only the predictor dimension needs to be measured. Instead of performing SPC analysis on all dimensions, SPC analysis needs to be performed only on the predictor dimension.* Instead of performing process capability (Cpk) analysis on all dimensions, process capability analysis needs to be performed only on the predictor dimension. If destructive measurement is being done, there is potential for eliminating the destructive measurement by measuring only the predictor dimension. Any point on Figs. 13 and 14 can usually be obtained by alternative combinations of process settings, meaning that any given point in Figs. 13 and 14 is not restricted to one unique combination of process settings.

Operating Point Adjustments
The operating point is defined as a point on any of the regression lines. The operating point is adjusted by changing one or more process settings. This is analogous to sliding a bead along a wire.

Generating Input Data
Input data for TRA processing is obtained from

1. Existing manufacturing data if there is adequate variation in the data
2. Existing DOE data
3. New data generated by inducing variation in the manufacturing process

In some cases, there may be sufficient existing manufacturing data to generate the correlation charts. In many companies, DOE studies exist on parts and manufacturing processes. Existing DOE data are a treasure-trove of untapped data. It can be further analyzed with the TRA computational software to obtain the many benefits outlined in this paper. In this context, the new technology is "one large step beyond DOE."

Figure 15 shows a table of five setups (runs) with five parts manufactured for each setup. This creates a sample of 25 parts. The relevant dimensions are measured for each part. The

---

*Important restrictions may apply to multiunit processes. Refer to later discussions on in-control versus out-of-control conditions. Even with these restrictions, huge savings are still obtainable.

| Sample | Process Variable | | |
Part No.	Pressure	Speed	Time
1 a,b,c,d,e	Low	Low	High
2 a,b,c,d,e	High	Low	Low
3 a,b,c,d,e	Low	High	Low
4 a,b,c,d,e	High	High	High
5 a,b,c,d,e	Nominal	Nominal	Nominal

Figure 15 Inducing variation to make 25 sample parts.

author recommends that a minimum of 25–30 data points (sample parts) be used for the correlation study.

If the part is a new part or there is no existing DOE, then data must be generated for the correlation study. In this instance, variation is induced into the manufacturing process, and parts are made and measured. A noncritical dimension may be the best statistical predictor of the critical dimensions. Consequently, noncritical dimensions can be included in the correlation study. The usual practice to induce variation has been to select the three process parameters that have the greatest impact on part dimensionality and/or performance. These three parameters are then varied in accordance with an experiment designed to meet the user's requirements.

The five setups in Fig. 15 are a DOE L_4 with a nominal (center) point. Once the entire system of dimensions is reduced to a single predictor dimension, this facilitates, if one so chooses, learning the relationships between the predictor and the process settings. This enables getting useful results from a DOE program. Alternatively, a 3-factor, 2-level, full factorial with a centerpoint and 3 repetitions or replications per run can used.* This design generates 27 data points (parts) and gives all main effects, all interactions, and an indication of linearity if a follow-on DOE study is done. Additional alternative data gathering designs can be used depending on the circumstances. It is not necessary, at all, to do a DOE follow-on study† to learn the relationships shown in Fig. 13. The results from a follow-on DOE study are a by-product, albeit a very useful by-product, from the correlation study.‡

Figure 16 shows the part characteristic data when formatted into a TRA input table. The input table comprises the vehicle for inputting the data into the TRA computational software.

Selecting the Predictor

Figure 17 shows the rankings table generated by the TRA computational software. The software ranks all part characteristics from statistically best to statistically worst "Predictor Characteristic" or "Predictor Dimension" using a proprietary algorithm. A substantial number of computations, depending on the number of part characteristics—typically 2,000 to 50,000—are involved. The predictor dimension can be a critical dimension on the engi-

---

*Some users start (run 0) and end (run 9) with a centerpoint run. This gives an indication of process drift during the course of the study.
†The author wishes to stress that we do not have to use DOE to get useful results. In this context, any information we learn about main effects or interactions is "free" or "bonus" information that is a *by-product* of the correlation study.
‡The potential for having a follow-on DOE study is why a TRA correlation study is occasionally mistaken for a DOE study.

Run	Cav1.150	Cav1.358 gate	Cav1.358 90°	Cav1.455 gate	Cav1.455 90°	Cav1.318	Cav1.540	Cav1.478	Cav2.150	Cav2.358 gate	Cav2.358 90°	Cav2.455 gate	Cav2.455 90°	Cav2.318	Cav2.540	Cav2.478
1a	0.1490	0.3575	0.3540	0.4565	0.4520	0.3165	0.5415	0.4795	0.1490	0.3575	0.3540	0.4565	0.4525	0.3150	0.5400	0.4800
1b	0.1490	0.3575	0.3540	0.4565	0.4520	0.3165	0.5415	0.4795	0.1490	0.3575	0.3535	0.4565	0.4525	0.3150	0.5400	0.4800
1c	0.1490	0.3580	0.3540	0.4565	0.4520	0.3165	0.5410	0.4795	0.1490	0.3575	0.3535	0.4560	0.4525	0.3150	0.5405	0.4800
1d	0.1490	0.3575	0.3540	0.4565	0.4520	0.3165	0.5415	0.4790	0.1490	0.3575	0.3540	0.4565	0.4525	0.3150	0.5400	0.4800
1e	0.1490	0.3575	0.3540	0.4560	0.4520	0.3165	0.5415	0.4790	0.1490	0.3570	0.3540	0.4565	0.4525	0.3150	0.5400	0.4800
2a	0.1485	0.3575	0.3540	0.4560	0.4520	0.3160	0.5405	0.4795	0.1485	0.3575	0.3535	0.4560	0.4520	0.3150	0.5400	0.4795
2b	0.1485	0.3575	0.3540	0.4565	0.4520	0.3165	0.5405	0.4790	0.1485	0.3575	0.3535	0.4565	0.4520	0.3150	0.5400	0.4795
2c	0.1485	0.3575	0.3540	0.4565	0.4520	0.3160	0.5405	0.4795	0.1485	0.3575	0.3535	0.4565	0.4520	0.3150	0.5400	0.4795
2d	0.1485	0.3575	0.3540	0.4560	0.4520	0.3160	0.5410	0.4790	0.1485	0.3575	0.3535	0.4565	0.4520	0.3150	0.5395	0.4795
2e	0.1485	0.3575	0.3540	0.4560	0.4520	0.3165	0.5410	0.4795	0.1485	0.3575	0.3535	0.4565	0.4520	0.3150	0.5395	0.4795
3a	0.1490	0.3575	0.3530	0.4555	0.4515	0.3160	0.5405	0.4785	0.1485	0.3570	0.3530	0.4560	0.4525	0.3150	0.5390	0.4790
3b	0.1490	0.3575	0.3530	0.4555	0.4515	0.3160	0.5405	0.4785	0.1485	0.3570	0.3535	0.4560	0.4520	0.3150	0.5390	0.4790
3c	0.1485	0.3570	0.3530	0.4555	0.4515	0.3160	0.5405	0.4785	0.1485	0.3570	0.3535	0.4560	0.4520	0.3150	0.5390	0.4790
3d	0.1485	0.3570	0.3530	0.4560	0.4515	0.3160	0.5405	0.4785	0.1485	0.3570	0.3535	0.4555	0.4520	0.3145	0.5390	0.4790
3e	0.1490	0.3570	0.3530	0.4560	0.4515	0.3160	0.5405	0.4785	0.1485	0.3570	0.3530	0.4555	0.4520	0.3145	0.5390	0.4785
4a	0.1490	0.3575	0.3530	0.4565	0.4520	0.3160	0.5400	0.4785	0.1485	0.3570	0.3530	0.4565	0.4520	0.3145	0.5390	0.4785
4b	0.1485	0.3575	0.3530	0.4560	0.4520	0.3160	0.5400	0.4785	0.1485	0.3570	0.3530	0.4565	0.4520	0.3145	0.5390	0.4785
4c	0.1490	0.3575	0.3535	0.4560	0.4515	0.3160	0.5405	0.4785	0.1485	0.3570	0.3530	0.4560	0.4520	0.3145	0.5390	0.4790
4d	0.1485	0.3570	0.3530	0.4560	0.4515	0.3160	0.5405	0.4785	0.1485	0.3570	0.3530	0.4560	0.4520	0.3145	0.5390	0.4790
4e	0.1490	0.3570	0.3530	0.4560	0.4515	0.3160	0.5405	0.4785	0.1485	0.3570	0.3535	0.4560	0.4520	0.3145	0.5390	0.4790
5a	0.1485	0.3570	0.3530	0.4560	0.4515	0.3155	0.5400	0.4780	0.1485	0.3565	0.3530	0.4555	0.4520	0.3140	0.5390	0.4780
5b	0.1485	0.3570	0.3530	0.4560	0.4515	0.3155	0.5400	0.4780	0.1485	0.3570	0.3530	0.4560	0.4520	0.3140	0.5390	0.4780
5c	0.1485	0.3575	0.3530	0.4560	0.4515	0.3155	0.5405	0.4780	0.1485	0.3570	0.3530	0.4560	0.4520	0.3140	0.5390	0.4780
5d	0.1485	0.3575	0.3530	0.4560	0.4515	0.3155	0.5405	0.4780	0.1485	0.3570	0.3530	0.4565	0.4520	0.3140	0.5390	0.4780
5e	0.1485	0.3570	0.3530	0.4555	0.4515	0.3155	0.5405	0.4780	0.1485	0.3565	0.3530	0.4555	0.4520	0.3140	0.5390	0.4780

Figure 16 Measure and record the dimensions.

Rankings Table

Col. No.	Unranked Variable	Metric	Ranked Variable	Metric	Best Predictor Data Column No.	User Predictor Data Column No.
1	Var1	98.0	Var1	98.0	1	5
2	Var2	97.8	Var5	97.9		
3	Var3	97.2	Var2	97.8	Best Predictor Variable	User Predictor Variable
4	Var4	97.7	Var7	97.8	Var1	Var5
5	Var5	97.9	Var4	97.7		
6	Var6	97.5	Var6	97.5		
7	Var7	97.8	Var3	97.2		
8	Var8	96.5	Var8	96.5		
9	Var9	95.7	Var10	96.4		
10	Var10	96.4	Var9	95.7		

Figure 17 The rankings table specifies the predictor dimension.

neering drawing or not. In practice, more than one predictor dimension may be selected. In this case, one predictor is selected for each data subset.

Predictive statistical capabilities are based on a comparison of correlation coefficients between all possible combinations of part characteristics. Figures 18 and 19 illustrate possible part characteristic combinations for a single-unit and a multiunit process, respectively. The user has the option of overriding the software to select the second or third best statistical predictor. This alternative is provided in the event the statistically best predictor is difficult, unreliable, or noneconomical to measure. In the example shown in Fig. 17, variable 1 in data column 1 was the statistically best predictor. However, the user elected to use variable 5 as the predictor because variable 1 required cutting the part open to access it.

Generating the Correlation Charts

Figure 20 shows conceptually how a correlation chart is created. Each data point represents one part. One correlation chart is generated for each predicted dimension. For a linear data

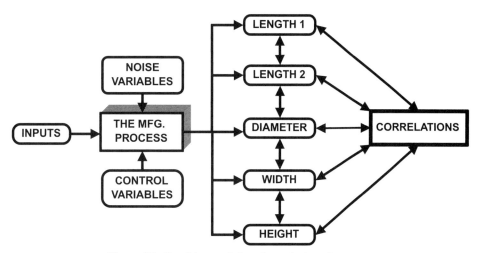

Figure 18 Possible correlations for a single-unit process.

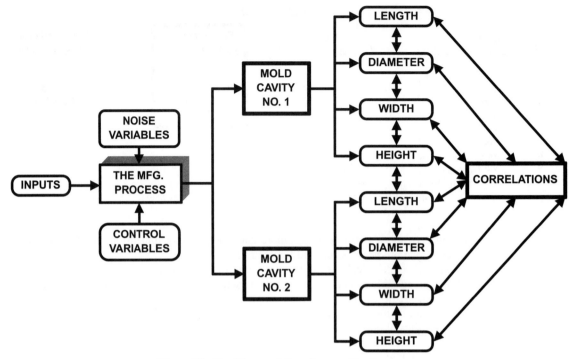

Figure 19 Possible correlations for a multiunit process.

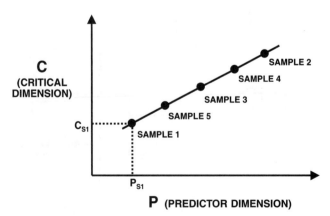

Figure 20 Generating a correlation chart.

set, a linear regression line is fitted through the data set using a least-squares curve-fitting technique. Nonlinear data sets require nonlinear regression lines.*

Adding Design Information to the Correlation Charts
Figure 21 shows how design information is added to the correlation charts. There are only two dimensions for each correlation chart. One is the predictor dimension; the other is the critical (predicted) dimension. The critical dimension has a target value (C-TARGET) and the predictor dimension has a target value (P-TARGET). The intersection of these two values is defined as the target intersection.

Figure 21 adds variable names and structure to the correlation chart. The horizontal dotted lines are the upper (USLc) and lower (LSLc) specification limits for the critical dimension. The vertical dotted lines are the upper (USLp) and lower (LSLp) specification limits for the predictor dimension.

Region of Conformance
The region of conformance is defined as the area bounded by and within the four specification limits in Fig. 22. Good parts lie within the region of conformance. Bad parts lie outside the region of conformance. In practice, the boundaries of the region of conformance are, as a matter of convenience, referred to as the "specification box" or "spec. box." The regression line can have four basic orientations relative to the region of conformance. Each of the four basic orientations is defined as a condition.

Five Relationship Conditions
Condition 1—Robust Critical. Figure 23 illustrates condition 1, which is defined as "robust."† The regression line is relatively flat and the correlation coefficient is close to zero. The critical dimension will be within specification limits regardless of whether or not the

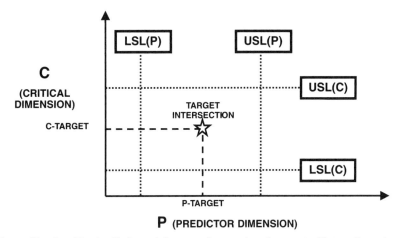

Figure 21 Specification limits and the target intersection: (one part with two dimensions).

---

*Surprisingly, at least on first examination, all data sets examined to date exhibit linear relationships.
†The robust relationship is a subset of a nonconstraining relationship (condition 2).

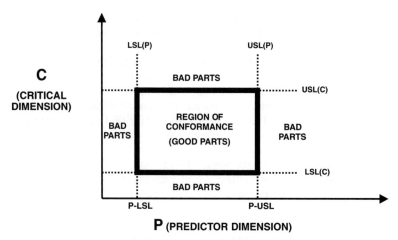

Figure 22 The region of conformance.

predictor dimension is within specification limits.* When the predictor dimension is within specification limits, then the paired dimension values fall within the region of conformance, the part will be a good part. The robust critical dimension never needs to be measured.

Condition 2—Nonconstraining Critical. Figure 24 illustrates condition 2, which is defined as "nonconstraining." When the predictor dimension is within specification limits, then both the critical and predicted dimensions will be within specification limits. Therefore, when the

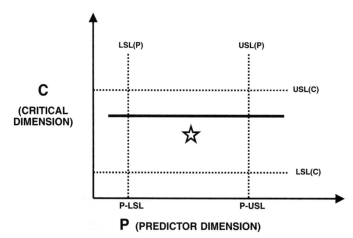

Figure 23 A robust critical dimension (condition 1).

---

*The correlation coefficient is 0.0 for a zero slope regression line. In this case, the regression relationship is usually considered to be of no practical use, However, TRA uses the zero slope regression line by recognizing that the predicted dimension is insensitive to changes in process settings.

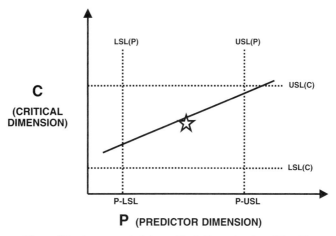

Figure 24 A nonconstraining critical dimension (condition 2).

predictor dimension is within its specification limits, the operating point will be in the region of conformance and the part will be a good part. The nonconstraining critical dimension never has to be measured.

Condition 3—Constraining Critical. Figure 25 illustrates condition 3, which is defined as "constraining." When the predictor dimension is between Pmin and Pmax, then the constraining critical dimension will be within its specification limits. When this is the case, the constraining critical dimension never has to be measured.

For a regression line with a positive slope, Pmin is located at the intersection of the regression line and the lower specification limit of the critical dimension. For a regression line with a negative slope, Pmin is located at the intersection of the regression line and the upper specification limit of the critical dimension.

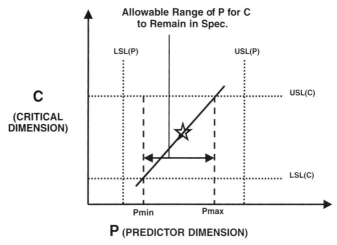

Figure 25 A constraining critical dimension (condition 3).

For a regression line with a positive slope, Pmax is located at the intersection of the regression line and the upper specification limit of the critical dimension. For a regression line with a negative slope, Pmax is located at the intersection of the regression line and the lower specification limit of the critical dimension.

The preceding discussion can be summarized as follows:

> When the predictor is between the greater of LSLp and Pmin and the lesser of USLp and Pmax, the constraining critical dimension never needs to be measured.

IMPERFECT CORRELATION. In the real manufacturing world, perfect correlation seldom occurs. Figure 26 illustrates imperfect correlation conceptually. Measurement error is one of the main contributors to imperfect correlation. The condition of imperfect correlation is resolved by bounding the regression lines in Figs. 23–25 with upper and lower (three-sigma) prediction intervals shown in Fig. 26.

Figures 27 and 28 are examples of real world data showing imperfect correlation. The critical dimension is robust when

- The slope of the regression line is relatively flat
- The upper and lower prediction intervals are within the region of conformance when the predictor is in between it's upper and lower specification limits.

The robust critical dimension still never needs to be measured even though there is imperfect correlation. The critical dimension is nonconstraining when

- The upper and lower prediction intervals are within the region of conformance when the predictor is in between it's upper and lower specification limits.

The nonconstraining critical dimension still never needs to be measured even though there is imperfect correlation.

Figure 29 illustrates a constraining condition with imperfect correlation. When the predictor dimension is between Pmin and Pmax, then the constraining critical dimension will be within its specification limits. When this is the case, the constraining critical dimension never has to be measured.

For a regression line with a positive slope

- Pmin is located at the intersection of the lower prediction interval and the lower specification limit of the critical dimension

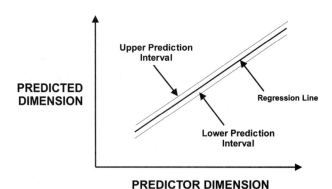

Figure 26 Dispersion of data points (primarily measurement variability).

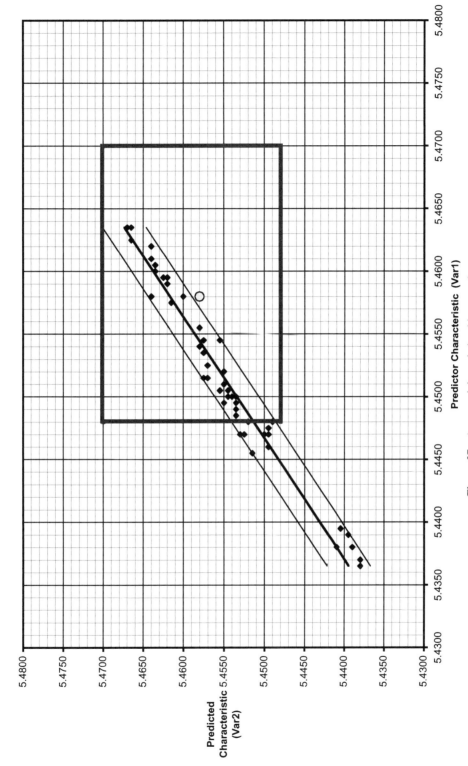

Figure 27 A constraining relationship example.

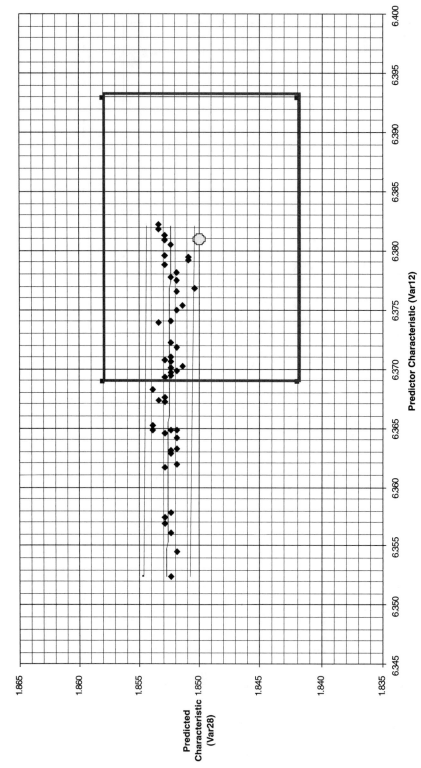

Figure 28 A robust dimension example.

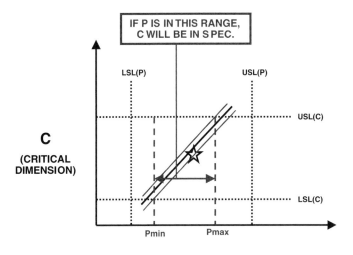

Figure 29 Upper and lower constraints on the predictor.

- Pmax is located at the intersection of the upper prediction interval and the upper specification of the critical dimension.

For a regression line with a negative slope

- Pmin is located at the intersection of the lower prediction interval and the upper specification limit of the critical dimension.
- Pmax is located at the intersection of the upper prediction interval and the lower specification limit of the critical dimension.

OPERATING LIMITS. Figure 30 illustrates a part that has three dimensions—two critical dimensions (C1 and C2) and the predictor dimension (P). With Fig. 25 as a reference, both C1 and C2 in Fig. 30 are constraining critical dimensions. For the part to be a good part, the predictor must be in between the most constraining (largest) of the Pmin's, which is defined as Pmin,* and the most constraining (smallest) of the Pmax's, which is defined as Pmax*.

OPERATING RANGE. The operating range is the distance between Pmin* and Pmax*. The lower operating limit is the greater of LSLp or Pmin*. The upper operating limit is the lesser of USLp or Pmax*. Figure 31 shows more clearly that the operating range consists of the range of values for the predictor where both C1 and C2 are within specification limits.

OPERATING TARGET. Figures 30 and 31 also identify the location of the operating target. For symmetrical process output, the operating target is located at the center of the operating range. For nonsymmetrical process output, the operating target is best selected as the point at which there is equal area in each of the tails of the process output distribution outside of the operating range. Alternative schemes can be used to determine the operating target.

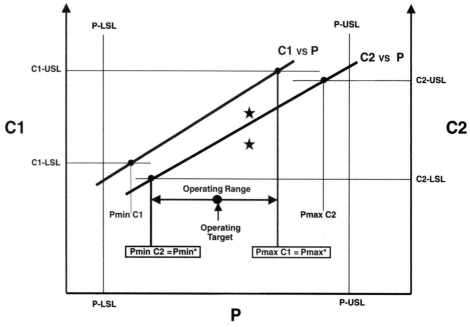

Figure 30 Generalizing to three or more dimensions.

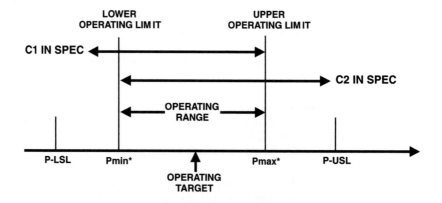

Figure 31 The operating range and operating target.

THE CONSTRAINT TABLE. The constraint table shown in Fig. 32 is a continuation of the previous example shown in Fig. 17. The constraint table is generated by the TRA computational software. Here are the key elements from the constraint table:

- Variable 5 is the predictor dimension.
- The lower operating limit is 5.4617 in. and is created by variable 9
- The upper operating limit is 5.4647 in. and is created by variable 2
- The operating range is 0.0030 in.
- The operating target is 5.4632 in.

Condition 4—Defects—Fixes Needed. Figure 33 illustrates condition 4, which is defined, as "defects." Regardless of the value of the predictor dimension in Fig. 33, the critical dimension will be outside of specification limits and defects will be produced. Similarly, regardless of the value of the predictor dimension in Fig. 34, a portion of the critical dimensions are likely to be out of specification and defects are likely to be produced. Figures 33 and 34 can be summarized as follows:

> When at least one prediction interval lies outside the conformance area, defective parts are likely to be produced and the defect critical dimension never needs to be measured. Instead, other action must be taken.

When the predicted dimension is

1. Robust,
2. Nonconstraining, or
3. Constraining and the predictor dimension is measured

then the predicted dimension does not have to be measured and no further action is required. If none of the above three conditions exist, then action must be taken to produce good parts:

Condition 5—Cannot Produce Parts at Design Target. Figure 35 illustrates a situation where it is not possible to produce parts at the target intersection, meaning that it is not possible to produce a part where both the critical and predictor dimensions are at their design target

Constraint Table

Variable Name	Data Column	Pmin	Pmax	Lower Op. Limit	Upper Op. Limit
Var1	1	5.4538	5.4660	5.4617	5.4647
Var2	2	5.4512	5.4647	9	2
Var3	3	5.4531	5.4660		
Var4	4	5.4497	5.4650	**Operating Range**	
Var5	5	Predictor	Predictor	0.0030	
Var6	6	5.4511	5.4660		
Var7	7	5.4512	5.4660	**Operating Target**	
Var8	8	5.4565	5.4658	5.4632	
Var9	9	5.4617	5.4660		
Var10	10	5.4584	5.4660	**Data is Constrained**	

Figure 32 The constraint table.

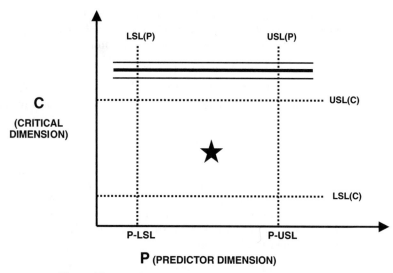

Figure 33 A defect condition requires change (condition 4).

values. This is because the target intersection does not lie within the prediction intervals that bound the data scatter around the regression line.

Fixes

Fix 1—Eliminate Defects by Changing a Design Tolerance. Figure 36 illustrates fix 1. The critical dimension design tolerances are increased to the point where the specification limits encompass the upper and lower prediction intervals. In this instance, the critical dimension's upper (+) design tolerance is increased. The critical dimension design target is unchanged. Referring back to Fig. 30, we see that the operating range will increase as Pmin* moves to the left. Pmin* is moved to the left by moving Pmin-C2 to the left. Pmin-C2 is moved to

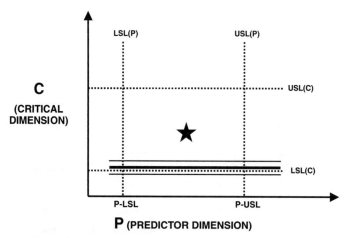

Figure 34 A defect condition requires change (condition 4A).

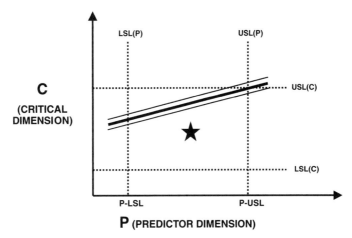

Figure 35 Cannot produce parts at target (condition 5).

the left by reducing (lowering) the lower specification limit for C2 (C2-LSL). The lower specification limit for C2 is reduced by increasing the value of the lower (−) tolerance on C2. At some point as Pmin-C2 moves to the left, Pmin-C1 will also become a constraining critical dimension. To continue to increase the operating range beyond this point, the lower tolerances must be relaxed on both C1 and C2.

The same logic applies when there are more than the two predicted dimensions, as illustrated in Fig. 30. Whenever a new constraining dimension is encountered as the operating range is increased, the tolerance on that new constraining dimension will be also have to be relaxed. For this reason, a successively larger number of tolerances will have to be relaxed to achieve successively larger increases in the operating range (producibility window). The Pmin tolerance relaxation table is a triangular table for this reason. Figure 37 illustrates a sample Pmin tolerance relaxation table. There is a similar table for Pmax. The Pmin and Pmax tolerance relaxation tables provide the design engineer with

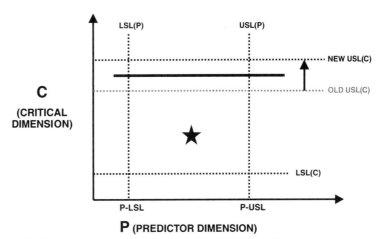

Figure 36 Fix 1: expand the specification box (change a specification limit, same target).

Pmin Tolerance Relaxation Table

Limiting Variable	Old Pmin	New Pmin	Individ. Gain	Cumul. Gain	New Tolerances Var9	Var10	Var8	Var1	Var3	Var2	Var7	Var6	Var4	Var5
Var9	5.4617	5.4584	0.0033	0.0033	6.3655									
Var10	5.4584	5.4565	0.0019	0.0052	6.3635	6.3668								
Var8	5.4565	5.4538	0.0027	0.0079	6.3607	6.3637	5.4451							
Var1	5.4538	5.4531	0.0007	0.0087	6.3599	6.3628	5.4444	5.4472						
Var3	5.4531	5.4512	0.0018	0.0105	6.3580	6.3607	5.4424	5.4453	5.4461					
Var2	5.4512	5.4512	0.0000	0.0105	6.3580	6.3606	5.4424	5.4453	5.4460	5.4479				
Var7	5.4512	5.4511	0.0001	0.0107	6.3578	6.3605	5.4423	5.4451	5.4459	5.4478	5.4479			
Var6	5.4511	5.4497	0.0014	0.0121	6.3564	6.3589	5.4408	5.4437	5.4444	5.4463	5.4464	5.4467		
Var4	5.4497	5.4480	0.0017	0.0137	6.3546	6.3570	5.4390	5.4419	5.4427	5.4444	5.4447	5.4452	5.4463	
Var5	5.4480	5.4480	0.0000	0.0137	N/A	N/A	N/A	N/A	N/A	N/A	N/A	N/A	N/A	N/A

THE FIRST Pmin TOLERANCE RELAXATION DOUBLES THE OPERATING RANGE IN THIS CASE

Var9 original lower spec limit = 6.369"
Var8 original lower spec limit = 5.448"
There is a similar table for the Pmax's.

Figure 37   Pmin tolerance relaxation table.

- A prioritized ranking of the order in which design tolerances must be relaxed
- The required increase in design tolerances required to achieve any specified increase in the producibility window

The Pmin and Pmax tolerance relaxation tables eliminate design engineer dependency on operator process settings when determining tolerance relaxations.

Fix 2—Eliminate Defects by Changing the Design Target. Figure 38 illustrates fix 2. The critical dimension target value is changed. This moves the entire specification box in the desired direction. In this instance, the design target is increased, which shifts the specification box upward to the point where the specification box encompasses the upper and lower prediction intervals. The design tolerances are unchanged. The size of the specification box is unchanged.

Design engineers find it practical to change the design tolerance when

- There is no impact on fit or function, or
- The target value of a mating dimension can be changed.

In some cases, tooling engineers find it easier to change the tooling on a mating part. This also is a situation where the design target can be changed.

Fix 3—Eliminate Defects by Shifting the Regression Line. Figure 39 illustrates a shift in a robust regression line. The regression line is shifted by modifying tooling or other preprocess dimensions so that the regression line passes through the target intersection. The offset is defined in Fig. 39 as the vertical (or other) distance between the regression line and the target intersection. Figure 40 illustrates a shift for a nonzero slope regression line.

Figure 41 shows the offset table. The offset table is generated by the TRA computational software. It specifies, for each predicted dimension, the vertical shift required for the regression line to pass through the target intersection. The offset table eliminates

- Tooling engineer dependency on operator process settings
- Multiple cycles of tooling, mold, and fixture changes when going from preproduction to production tooling

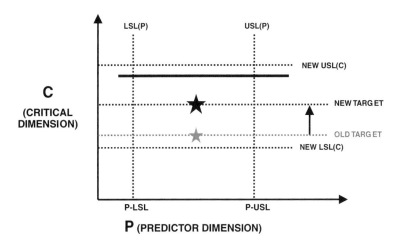

Figure 38 Fix 2: shift the specification box (change the target, same specification limits).

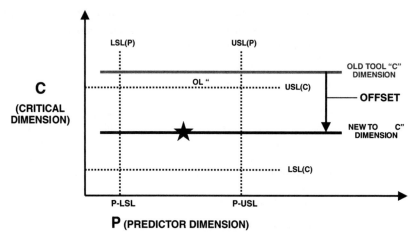

Figure 39 Fix 3: shift the regression line (change the tooling, same target, same specification limits).

Fix 4—Eliminate Defects by Improving Measurement Accuracy. Data scatter about the regression line can be reduced by increasing measurement accuracy. The reduced scatter reduces the size of the prediction intervals. This has the potential of moving both prediction intervals inside of the specification box.

Fix 5—Produce Parts at Target. As shown in Fig. 35, condition 5 exists when it is possible to produce good parts but it is not possible to produce parts where both the critical and predictor dimensions are at the target value. This is because the target intersection lies outside of the prediction intervals. The target intersection can be brought within the prediction intervals by changing the design target(s) or shifting the regression line.

Material Selection

Figure 42 shows the producibility window for three different materials. The larger the producibility window, the easier it is to produce the part. Replacements for obsolete material

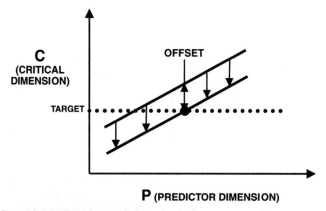

Figure 40 Determine tooling changes independently from operator changes to process settings.

Offset Table

Variable Name	Vertical Offset
Var1	-0.0026
Var2	0.0010
Var3	-0.0011
Var4	0.0017
Var5	Predictor
Var6	-0.0012
Var7	-0.0001
Var8	-0.0023
Var9	-0.0096
Var10	-0.0072

Figure 41 The offset table specifies optimum changes to tooling.

can be evaluated. Design and manufacturing engineers can evaluate the trade-off between producibility and cost. In best case, the material with the greatest producibility window has the least cost.

Conflict Reduction
TRA reduces conflict and increases collaboration and communication between

- Design engineers
- Tooling engineers
- Process engineers
- Quality engineers
- Customers and suppliers

Improved Decision-making
TRA greatly improves decision-making by

- Eliminating the effect of process complexities
- Visual presentation of results

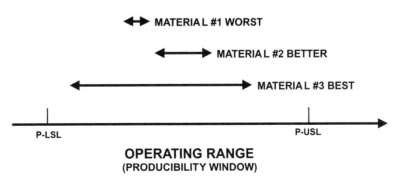

Figure 42 Different materials have different producibility windows.

- Using a scientific method instead of trial and error, iteration, and guesswork
- A systems approach that integrates all of the engineering functions

TRA Process Flowchart

Prior art techniques use trial and error, iteration, and guesswork to transition from a preproduction condition where good parts can be difficult or impossible to produce to a production condition where good parts can be produced. Figure 43 is a process flowchart of TRA. Tooling, design tolerances, design targets, and material selection are optimized in a single step with TRA. Figure 43 also shows the relationship of the TRA computational software to existing measurement, SPC, process capability, and DOE software. TRA does not replace this software. Instead, the TRA computational software is complementary to the two major functions of existing software as follows:

- The TRA computation software greatly reduces the amount of measurement, SPC, and Cpk analysis that must be performed by the existing software.
- The TRA computational software enables DOE software to be used in situations where there are complexities that normally prevent the use of DOE to understand the relationship between causes (process settings) and effects (part characteristics).

When Should TRA Be Used?

Figure 44 shows that the earlier that TRA is used, the quicker the time savings, the financial savings, and the quality improvement will be realized. TRA yields the greatest benefits if it is used when parts are first produced by the process.

Must the Process Be in Control during Production?

The process does *not* have to be in control during production when

- The process is a single-unit process.
- The process is a dependent multiunit process.
- The out-of-control situation is runs, periodicity, trends, hugging, outside of the three sigma limits, etc.

The process *does* have to be in control during production if the process is an independent multiunit process, which means that the out-of-control situation could affect only one or several of the multiunit processes. Figure 45 shows that there is a substantial savings in measurement and analysis costs in either event.

Simulation Benefits

The computation software also simulates the operation of the manufacturing system used to produce the parts and the measurement system used to measure them. In the simulation mode, the software makes it possible to assess the impact of contemplated changes to design targets, design tolerances, and tooling on manufactured part dimensions without incurring the cost, time, and risk of changing tooling or design parameters and then producing, measuring, and analyzing the new parts.

The left-hand chart in Fig. 46 shows data points generated by a CNC laser cutter.* As can be seen, some of the data points are right on the edge of being bad parts. In this instance,

---

*These data were generated under manufacturing conditions, meaning that variation was not induced. The data scatter represents the normal process variability.

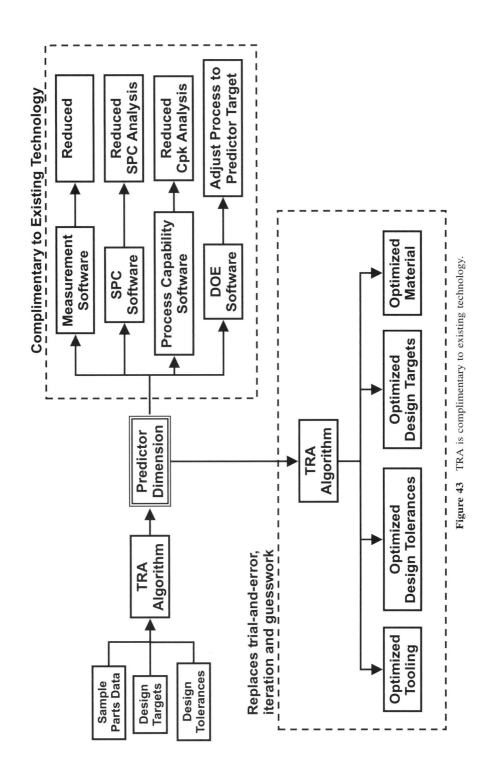

Figure 43 TRA is complimentary to existing technology.

57

58 Achieving Enterprise Goals with New Process Technology

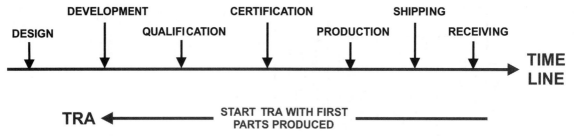

Figure 44 TRA gives cost savings at all stages of development and production.

the CNC programming was set to produce parts at the target intersection. The TRA computational software identified biases in the CNC laser cutter. These biases were fed back into the TRA software. The input data was adjusted to simulate the results of the changes that would be made to eliminate the machine/programming biases. The results are shown in the right-hand chart in Fig. 46. The data points are now clustered around the target intersection. The TRA computational software has simulated changes to the CNC programming. Changing CNC programming is relatively low-cost and risk-free. That may not the case for changing molds, fixtures, and other hard tooling.

3 CONCLUSION

Figures 47–49 show the historical evolution of

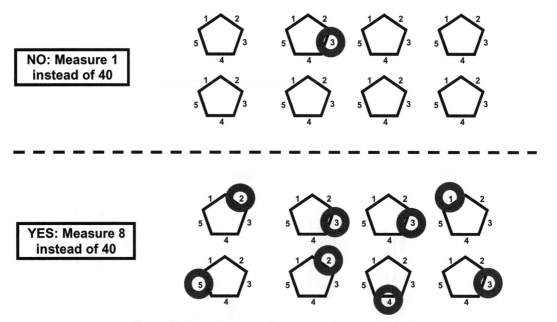

Figure 45 Must the process be in control during production?

3 Conclusion 59

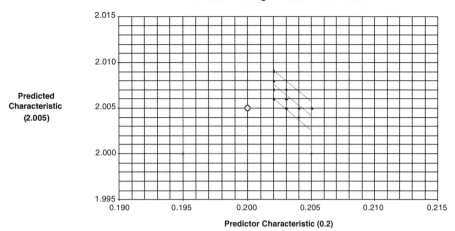

BEFORE

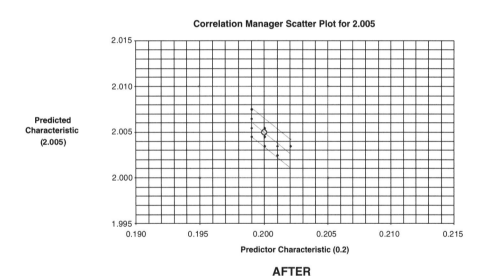

AFTER

Figure 46 TRA computational software can be used to simulate changes.

- Quality
- Manufacturability
- Reduction in time-to-market

relative to the contributions made by TRA. The algorithms discussed in this paper are the new model for the manufacture of products in many industries. They provide an immediate increase in profitability and market competitiveness by reducing cost and increasing quality during production and reducing time, cost and risk during development.

60 Achieving Enterprise Goals with New Process Technology

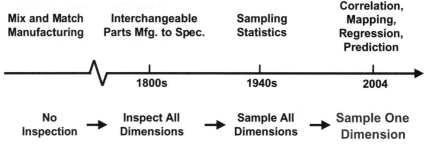

Figure 47 The evolution of quality.

4 IMPLEMENTATION

The TRA algorithms and computational software have received U.S. patent no. 6,687,558. Other patent applications are pending. A demonstration version of the software is available at www.algoryx.com. TRA provides useful insights into manufacturing processes. However, once more than a few part characteristics are involved, the computations and graphical analysis become tedious and labor intensive. The computation software is available from Algoryx, Inc. at the following address: Algoryx, Inc., 750 S. Bundy Drive, Ste. 304, Los Angeles, CA 90049, 310-820-0987, steve@algoryx.com.

APPENDIX A: TUSZYNSKI'S PROCESS LAW

1. Every process has causes and effects.
2. Causes can be categorized into (i) slider variables, (ii) shifter variables, and (iii) noise variables.
3. Although the relationships between slider variables and effects may be difficult, economically infeasible, or impossible to predict, the relationships between effects are consistent and predictable.
4. Slider variables can be used to reposition the operating points within relationships.
5. Shifter variables can be used to change the relationships between effects.
6. The system of relationships has one degree of freedom; one effect can be used to predict one or more other effects.

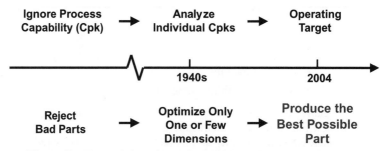

Figure 48 The evolution of manufacturability and product performance.

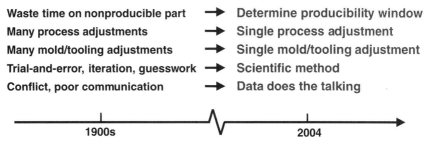

Figure 49 The evolution in reducing time-to-market.

7. The effect best able to predict all other effects is the predictor effect.
8. The relationship between any two effects, including the predictor, is unique.
9. When effects have tolerances that bound their region of conformance, each effect is, relative to the predictor, either
    a. Robust
    b. Nonconstraining
    c. Constraining, or
    d. Nonconforming
10. Effects never have to be measured when they are
    a. Robust
    b. Nonconstraining and the predictor is conformal
    c. Constraining and the predictor is within the operating range
    d. Nonconforming
11. Process performance is optimal when the predictor is at the operating target.
12. Effects can be produced at the target intersection only when the target intersection is within the regression area.
13. Nonconforming effects can be made conforming by any combination of

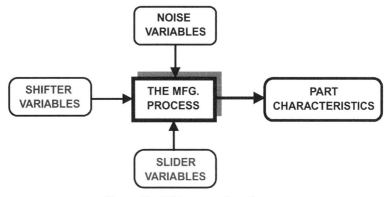

Figure 50 TRA process flow diagram.

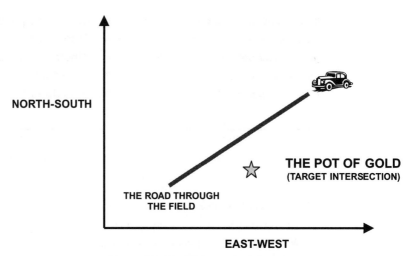

Figure 51 You can't get there from here.

      a. Relaxing tolerances
      b. Changing design targets
      c. Changing shifter variables
      d. Constraining slider variables, or
      e. Reducing measurement error

14. If shifter variables change predictably with time, the number of process cycles or any other variable, the resultant changes in relationships are predictable.

15. Tuszynski's process laws apply to
      a. Single-unit processes whether the process is in control or not
      b. All units jointly of multiunit dependent processes whether the process is in control or not

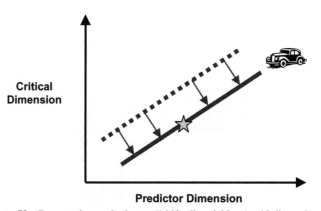

Figure 52 Reroute the road: change "shifter" variables (mold dimensions).

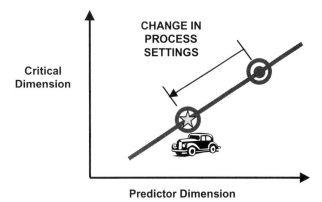

Figure 53 Drive along the road and grab the gold: change "slider" variables (process settings).

      c. All units jointly of multiunit processes when the process is in control

      d. Individual units severably of a multiunit process when the processing units are independent and the process is out of control.

APPENDIX B: DEFINITIONS

Cause: A factor that brings about an effect—typically a process setting.

Constraining effects: Constraining effects place upper and lower constraints on the predictor effect. These constraints are constant and are inside the predictor effect tolerances.

Degrees of freedom: The number of parameters (variables) that can be independently varied.

Effect: A result—more specifically, a feature or performance characteristic.

Noise variables: Variables that are uncontrolled because either we do not have the ability to control them or we chose to not control them.

Nonconforming effects: Nonconforming effects are nonconforming when the predictor effect is conformal.

Nonconstraining effects: Nonconstraining effects are conformal when the predictor effect is conformal.

Operating points: Operating points are the points on the regression-prediction models (lines).

Operating range: The operating range is the region between the maximum lower constraint and the minimum upper constraint on the predictor effect.

Operating target: The operating target is the point in the operating range that maximizes the percentage of process output in the operating range. For symmetrical processes, the operating target is located at the center of the operating range.

Prediction intervals: Upper and lower boundaries around the regression-prediction line that bound the data set—typically at plus/minus three sigma values.

Regression area: The area around the regression model bounded by the prediction intervals.

Relationships: The relationships (correlations) between variables—typically a regression-prediction model, which can be linear or nonlinear.

Robust effects: Robust effects are conformal irrespective of the value of the predictor effect.
Shifter variables: Shifter variables change the relationships.
Slider variables: Variables that move the operating points along the relationships. Also called process variables.
Target intersection: The intersection of design target values for effects.

APPENDIX C: NONTECHNICAL STATISTICAL GLOSSARY

Attribute data: Usually refers to a data describing a part characteristic not measurable on a number scale, such as on/off, present/absent, supplier A, B, or C, material type m, n, or o, etc. Some attributes, such as color, can be converted to variable data (a combination of red-blue-green) if it is worth the cost and effort. Also, see *variable data* in this glossary.

Causation: The act or agency that produces an effect. In some manufacturing processes, there is a direct relationship between a process control variable and the process output. In other manufacturing processes, the relationship between process control variables and the process output can be extremely complex. Also, see *relationship vs. causation* in this glossary.

C-LSL: The lower specification limit of a critical dimension.

C-USL: The upper specification limit of a critical dimension.

Coefficient of determination: The square of the correlation coefficient. It represents the percentage of variation in Y that is accounted for by variation in X. If the coefficient of determination is 75% (0.75), then 25% of the variation in Y is caused by a factor or factors other than X. Also, see *correlation coefficient* in this glossary.

Complex interaction: The effect of a change in one process control variable on a part characteristic depends on the level at which two or more different process control variables are set. Also, see *simple interaction* in this glossary.

Constraint table: A table showing, for each critical characteristic, Pmax and Pmin or no constraints (P-USL and/or P-LSL).

Control limits: Limits drawn on a control chart that are a function of the sample size and the variability within the data.

Correlation coefficient: A number ranging between $+1.0$ and -1.0 that represents the degree of relationship between two variables from perfect positive to zero to perfect negative correlation. Plus means a positive slope and minus means a negative slope. Zero or very low correlation can be caused by either a regression line with zero slope or by two variables that exhibit no relationship.

Critical dimension: A dimension that the designer feels is critical to the form, fit, or function of the part. Also called a "critical" or "criticals."

C-target: The nominal or target value of the critical dimension set by the design engineer that presumably optimizes form, fit or function. A point.

Design of experiments (DOE): A set of techniques to extract useful information from a relatively small number of experimental runs instead of testing all possible combinations of variables. Used to (1) optimize the process that makes a product, (2) make the manufacturing process robust, (3) optimize product performance, and (4) make product performance robust.

Go/no-go gage: Gives an accept/reject-good part/bad part decision without needing to measure the actual numerical value of the predictor dimension (or any other dimension). This saves time and money.

Graphical (visual) linear regression: Drawing a straight line (also called the regression line) through the center of the data points on a scatter plot. The line is visually positioned so that the data points are equally centered above and below the line.

Graphical prediction intervals: Typically drawn as two straight lines equidistant from and parallel to the regression line through the outermost data points.

Hot measurement: Allows the measurement of a part shortly after it is produced instead of waiting hours or days. Allows production to be stopped or the process to be modified if bad parts are being produced.

In-control process: A process where all (99.74%) of the data points are within the control limits and there are no non-normal patterns such as trends or shifts. See also *out-of-control process* in this glossary.

Joint operating position: A point on the regression line (or within the bounds of the prediction intervals).

Mathematical linear regression: Drawing a straight line through the center of the data points on a scatter plot. Typically done by computing the slope of the line and one of the intercepts (Y or X).

Mathematical prediction intervals: Bounds on the scatter of the data points around the regression line calculated using statistical formulas.

Natural process limits: Typically the plus and minus three sigma points on the distribution representing the actual process output.

Nominal process settings: Process settings that result in all critical dimensions being within specification limits. Ignores the difference between the average process output(s) and the design target(s) and does not attempt to minimize variation between parts or between cavities. See also *optimum process settings* in this glossary.

Noncritical dimension: A dimension that the designer feels is not critical to the form, fit, or function of the part.

Nonlinear regression lines: A linear (first-order) regression line is of the form $Y = a + bX$. A second-order regression is of the form $Y = a + bX + cX^2$. Regression lines are seldom fitted higher than third order.

Nonlinearities: Increasing a process control variable can cause a dimension to initially increase and then decrease.

Offset table: A table showing how many units the regression line is above or below the target intersection.

Optimum process settings: Process settings that minimize the difference between the average process output(s) and the design target(s) and minimize variation between parts or between cavities. See also *nominal process settings* in this glossary.

Out-of-control process: A process where data points are outside of the control limits and/or there are non-normal patterns such as trends or shifts. See also *in-control process* in this glossary.

Part characteristic: Something physical about a part, such as a dimension, weight, color, or hardness. Can be variable or attribute data. Also referred to as a dimension for ease of reading.

P-LSL: The lower specification limit of the predictor dimension.

Pmax.: The point on the X axis—the predictor dimension—determined by the intersection of the upper prediction interval and the upper specification limit of the critical dimension. A point.

Pmax*: The smallest (lowest) value Pmax from the constraint table. The most constraining of the Pmax's. A point.

Pmin*: The point on the X axis—the predictor dimension—determined by the intersection of the lower prediction interval and the lower specification limit of the critical dimension. A point.

Pmin*: The largest (highest) value Pmin from the constraint table. The most constraining of the Pmin's. A point.

Population: The complete set of items. For example, (1) all the parts of the same type produced by a process during a given production run, or (2) all the parts of the same type produced by a process during the life of that process.

P-range*: The difference between Pmin* and Pmax*. A distance.

Prediction intervals: Two lines that are equidistant above and below the regression line. The prediction intervals bound a specified percentage of the data points on a scatter diagram. The upper prediction interval is above the regression line. The lower prediction interval is below the regression line.

Prediction using regression lines: Given X, Y can be predicted. Given Y, X can be predicted. The prediction intervals give the uncertainty range of the predicted variable.

Predictor characteristic: A part characteristic selected to predict other part characteristics. Also called a "predictor."

Predictor dimension: A predictor characteristic that happens to be a dimension. Although called a "predictor dimension" in some of my material, the more accurate description would be predictor characteristic. For example, the predictor dimension could be a weight (which is a characteristic, not a dimension). The use of the word "dimension" is used to improve readability.

Press setting: A control setting on an injection-molding press that controls some element of the press operation. These are typically pressures, temperatures, times, and speeds.

Process characteristic: Something physical about the process that makes the part. Typically process characteristics are pressure, temperature, time, speed, chemical concentration, part orientation, etc. Can be variable or attribute data. Some process characteristics are controlled. Others are not ("noise"). Also called process variables or process control variables or process conditions. Press settings are examples of process variables.

Process capability study: A comparison of the distance between the process upper and lower natural limits (plus and minus three sigma) to the distance between the specification limits (TOL). The offset of the average process output from the design target can either be considered (Cpk) or ignored (Cp).

Process control variable: A process variable that is controlled or set by the operator. A press setting for an injection-molding press. Changing a process control variable typically shifts the joint operating position along the regression line.

Process input: Anything that influences the output of a process other than process control or process noise variables. The mold cavity dimensions would be examples of process inputs for an injection-molding process. The preplated dimensions of a part would be examples of process inputs for a plating process. Changing a process input typically shifts the regression line.

P-target: The nominal or target value of the predictor dimension set by the design engineer that presumably optimizes form, fit or function. Equivalent to C-target. A point.

P-target*: The value at which it is desired to have the average output of the process. This is typically the center of P-range* for a process with a symmetrical output distribution. A point.

P-USL: The upper specification limit of the predictor dimension.

Range: The difference between the highest and lowest values in a sample.

Relationship versus causation: Scatter plots and correlation coefficients are used to illustrate and determine the degree of relationship between two variables. Scatter plots and correlation coefficients are not used to say that a change in X caused a change in Y. Causation may or may not be present but must be proved by other means.

Robust: To make something robust is to make it resistant to outside influences.

Sample: A subset of a population, hopefully taken using valid sampling methods (randomized, unbiased, etc.). When sampling from a production line, samples are gathered in subgroups. The samples within a subgroup should be taken as close together in time (or in order of manufacturing sequence) as possible. The time between subgroup samples is determined by the total number of samples to be taken for a given period of time.

Scatter plot: Also known as scatter or correlation or X-Y plots. Can be called plots or diagrams or charts. Used to graphically (visually) illustrate the degree of relationship between two variables (X and Y).

Sigma: Shorthand for one standard deviation unit. One sigma unit is a distance. Also known as a "Z" unit for a nondimensional normal curve.

Simple interaction: The effect of a change in one process control variable on a part characteristic depends on the level at which a different process control variable is set. For example, increasing pressure when temperature is low will increase a dimension, while increasing pressure when temperature is high will decrease that dimension.

Specification limits: The upper and lower limits within which a part characteristic must be manufactured in order for the part to adequately meet the requirements of form, fit, and function. These are specified by the design engineer. They are points.

Standard deviation: A measure of the "spread" or "width" of a distribution. A distance. Other measures of the variability of a distribution are the range, average deviation, variance, and semi-interquartile range.

Target: The nominal value of a part characteristic established by the design engineer that optimizes form, fit, or function. A point. Manufacturing first tries to produce good parts (within specification) and then to produce parts that are on target with minimum variation.

Target intersection: The point on a scatter chart determined by the intersection of the Y-axis (critical) target and the X-axis (predictor) target.

Tolerance (TOL): The difference between the upper and lower specification limits. A distance.

Variability (VAR): The difference between the upper and lower natural process limits. Typically taken as the distance between the plus and minus three sigma values of the process output. A distance.

Variable data: Refers to data describing a part characteristic measurable on a continuous number scale. Examples are dimensions, weights, hardness, tensile strength, etc.

CHAPTER 3
CLASSIFICATION SYSTEMS

Dell K. Allen
Manufacturing Engineering Department
Brigham Young University
Provo, Utah

1	**PART FAMILY CLASSIFICATION AND CODING**	**68**		3.2 Process Divisions	91
				3.3 Process Taxonomy	92
	1.1 Introduction	68		3.4 Process Code	93
	1.2 Application	69		3.5 Process Capabilities	94
	1.3 Classification Theory	71			
	1.4 Part Family Code	73	4	**FABRICATION EQUIPMENT CLASSIFICATION**	**95**
	1.5 Tailoring the System	77		4.1 Introduction	95
2	**ENGINEERING MATERIALS TAXONOMY**	**77**		4.2 Standard and Special Equipment	97
				4.3 Equipment Classification	97
	2.1 Introduction	77		4.4 Equipment Code	99
	2.2 Material Classification	81		4.5 Equipment Specification Sheets	100
	2.3 Material Code	83			
	2.4 Material Properties	84	5	**FABRICATION TOOL CLASSIFICATION AND CODING**	**103**
	2.5 Material Availability	88		5.1 Introduction	103
	2.6 Material Processability	88		5.2 Standard and Special Tooling	104
3	**FABRICATION PROCESS TAXONOMY**	**89**		5.3 Tooling Taxonomy	105
				5.4 Tool Coding	106
	3.1 Introduction	89		5.5 Tool Specification Sheets	107

1 PART FAMILY CLASSIFICATION AND CODING

1.1 Introduction

History

Classification and coding practices are as old as the human race. They were used by Adam, as recorded in the Bible, to classify and name plants and animals, by Aristotle to identify basic elements of the earth, and in more modern times to classify concepts, books, and documents. But the classification and coding of manufactured pieceparts is relatively new. Early pioneers associated with workpiece classification are Mitrafanov of the former USSR, Gombinski and Brisch, both of the United Kingdom, and Opitz of Germany. In addition, there are many who have espoused the principles developed by these men, adapted them and enlarged upon them, and created comprehensive workpiece classification systems. It is reported that over 100 such classification systems have been created specifically for machined parts, others for castings or forgings, and still others for sheet metal parts, and so on. In the United States there have been several workpiece classification systems commercially developed and used, and a large number of proprietary systems created for specific companies.

Why are there so many different part-classification systems? In attempting to answer this question, it should be pointed out that different workpiece classification systems were initially developed for different purposes. For example, Mitrafanov apparently developed his system to aid in formulating group production cells and in facilitating the design of standard tooling packages; Opitz developed his system for ascertaining the workpiece shape/size distribution to aid in designing suitable production equipment. The Brisch system was developed to assist in design retrieval. More recent systems are production-oriented.

Thus, the intended application perceived by those who have developed workpiece classification systems has been a major factor in their proliferation. Another significant factor has been personal preferences in identification of attributes and relationships. Few system developers totally agree as to what should or should not be the basis of classification. For example: Is it better to classify a workpiece by function as "standard" or "special" or by geometry as "rotational" or "nonrotational"? Either of these choices makes a significant impact on how a classification system will be developed.

Most classification systems are hierarchal, proceeding from the general to the specific. The hierarchical classification has been referred to by the Brisch developers as a monocode system. In an attempt to derive a workpiece code that addressed the question of how to include several related, but nonhierarchical, workpiece features, the feature code or polycode concept was developed. Some classification systems now include both polycode and monocode concepts.

A few classification systems are quite simple and yield a short code of five or six digits. Other part-classification systems are very comprehensive and yield codes of up to 32 digits. Some part codes are numeric and some are alphanumeric. The combination of such factors as application, identified attributes and relationships, hierarchal versus feature coding, comprehensiveness, and code format and length have resulted in a proliferation of classification systems.

1.2 Application

Identification of intended applications for a workpiece classification system are critical to the selection, development, or tailoring of a system.

It is not likely that any given system can readily satisfy both known present applications and unknown future applications. Nevertheless, a classification system can be developed in such a way as to minimize problems of adaptation. To do this, present and anticipated applications must be identified. It should be pointed out that development of a classification system for a narrow, specific application is relatively straightforward. Creation of a classification system for multiple applications, on the other hand, can become complex and costly.

Figure 1 is a matrix illustrating this principle. As the applications increase, the number of required attributes also generally increases. Consequently, system complexity also increases, but often at a geometric or exponential rate, owing to the increased number of combinations possible. Therefore, it is important to establish reasonable application requirements first while avoiding unnecessary requirements and, at the same time, to make provision for adaptation to future needs.

In general, a classification system can be used to aid (1) design, (2) process planning, (3) materials control, and (4) management planning. A brief description of selected applications follows.

Design Retrieval

Before new workpieces are introduced into the production system, it is important to retrieve similar designs to see if a suitable one already exists or if an existing design may be slightly altered to accommodate new requirements. Potential savings from avoiding redundant designs range in the thousands of dollars.

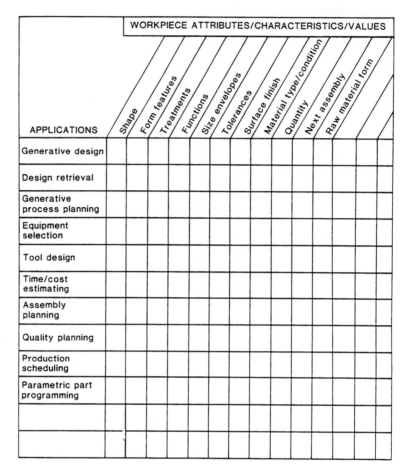

Figure 1 Attribute selection matrix.

Design retrieval also provides an excellent starting point for standardization and modularization. It has been stated that "only 10–20% of the geometry of most workpieces relates to the product function." The other 80–90% of the geometric features are often a matter of individual designer taste or preference. It is usually in this area that standardization could greatly reduce production costs, improve product reliability, increase ease of maintenance, and provide a host of other benefits.

One potential benefit of classification is in meeting the product liability challenge. If standard analytic tools are developed for each part family, and if product performance records are kept for those families, then the chances of negligent or inaccurate design are greatly reduced.

The most significant production savings in manufacturing enterprise begin with the design function. The function must be carefully integrated with the other functions of the company, including materials requisition, production, marketing, and quality assurance. Otherwise, suboptimization will likely occur, with its attendant frequent redesign, rework, scrap, excess inventory, employee frustration, low productivity, and high costs.

Generative Process Planning

One of the most challenging and yet potentially beneficial applications of workpiece classification is that of process planning. The workpiece class code can provide the information required for logical, consistent process selection and operation planning.

The various segments of the part family code may be used as keywords on a comprehensive process-classification taxonomy. Candidate processes are those that satisfy the conditions of the given basic shape *and* the special features *and* the size *and* the precision *and* the material type *and* the form *and* the quality/time requirements.

After outputting the suitable processes, economic or other considerations may govern final process selection. When the suitable process has been selected, the codes for form features, heat treatments, coatings, surface finish, and part tolerance govern computerized selection of fabrication and inspection operations. The result is a generated process plan.

Production Estimating

Estimating of production time and cost is usually an involved and laborious task. Often the results are questionable because of unknown conditions, unwarranted assumptions, or shop deviations from the operation plan. The part family code can provide an index to actual production times and costs for each part family. A simple regression analysis can then be used to provide an accurate predictor of costs for new parts falling in a given part family. Feedback of these data to the design group could provide valuable information for evaluating alternative designs prior to their release to production.

Parametric and Generative Design

Once the product mix of a particular manufacturing enterprise has been established, high-cost, low-profit items can be singled out. During this sorting and characterization process, it is also possible to establish tabular or parametric designs for each basic family. Inputting of dimensional values and other data to a computer graphics system can result in the automatic production of a drawing for a given part. Taking this concept back one more step, it is conceivable that merely inputting a product name, specifications, functional requirements, and some dimensional data would result in the generation of a finished design drawing. Workpiece classification offers many exciting opportunities for productivity improvement in the design arena.

Parametric Part Programming

A logical extension of parametric design is that of parametric part programming. Although parametric part programming or family of parts programming has been employed for some time in advanced numerical control (NC) work, it has not been tied effectively to the design database. It is believed that workpiece classification and coding can greatly assist with this integration. Parametric part programming provides substantial productivity increases by permitting the use of common program modules and reduction of tryout time.

Tool Design Standardization

The potential savings in tooling costs are astronomical when part families are created and when form features are standardized. The basis for this work is the ability to adequately characterize component pieceparts through workpiece classification and coding.

1.3 Classification Theory

This section outlines the basic premises and conventions underlying the development of a Part Family Classification and Coding System.

Basic Premises

The first premise underlying the development of such a system is that a workpiece may be best characterized by its most apparent and permanent attribute, which is its basic shape. The second premise is that each basic shape may have many special features (e.g., holes, slots, threads, coatings) superimposed upon it while retaining membership in its original part family. The third premise is that a workpiece may be completely characterized by (1) basic shape, (2) special features, (3) size, (4) precision, and (5) material type, form, and condition. The fourth premise is that code segments can be linked to provide a humanly recognizable code, and that these code segments can provide pointers to more detailed information. A fifth premise is that a short code is to be adequate for human monitoring, and linking to other classification trees but that a bitstring (0's, 1's) that is computer-recognizable best provides the comprehensive and detailed information required for retrieval and planning purposes. Each bit in the bitstring represents the presence or absence of a given feature and provides a very compact, computer-processable representation of a workpiece without an excessively long code. The sixth premise is that mutually exclusive workpiece characteristics can provide unique basic shape families for the classification, and that common elements (e.g., special features, size, precision, and materials) should be included only once but accessed by all families.

E-tree Concept

Hierarchical classification trees with mutually exclusive data (E-trees) provide the foundation for establishing the basic part shape (Fig. 2). Although a binary-type hierarchical tree is preferred because it is easy to use, it is not uncommon to find three or more branches.

It should be pointed out, however, that because the user must select only one branch, more than two branches require a greater degree of discrimination. With two branches, the user may say, "Is it this or that?" With five branches, the user must consider, "Is it this or this or this or this or this?" The reading time and error rate likely increase with the number of branches at each node. The E-tree is very useful for dividing a large collection of items into mainly exclusive families or sets.

N-tree Concept

The N-tree concept is based on a hierarchal tree with nonmutually exclusive paths (i.e., all paths may be selected concurrently). This type of tree (Fig. 3) is particularly useful for representing the common attributes mentioned earlier (e.g., form features, heat treatments, surface finish, size, precision, and material type, form, and condition).

In the example shown in Fig. 3, the keyword is Part Number (P/N) 101. The attributes selected are shown by means of an asterisk (*). In this example the workpiece is characterized as having a "bevel," a "notch," and a "tab."

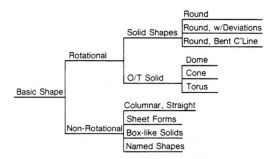

Figure 2 E-tree concept applied to basic shape classification.

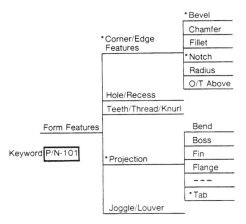

Figure 3 N-tree concept applied to form features.

Bitstring Representation
During the traversal of either an E-tree or an N-tree, a series of 1's and 0's are generated, depending on the presence or absence of particular characteristics or attributes. The keyword (part number) and its associated bitstring might look something like this:

$$P/N\text{-}101 = 100101 \cdots 010$$

The significance of the bitstring is twofold. First, one 16-bit computer word can contain as many as 16 different workpiece attributes. This represents a significant reduction in computer storage space compared with conventional representation. Second, the bitstring is in the proper format for rapid computer processing and information retrieval. The conventional approach is to use lists and pointers. This requires relatively large amounts of computation and a fast computer is necessary to achieve a reasonable response time.

Keywords
A keyword is an alphanumeric label with its associated bitstring. The label may be descriptive of a concept (e.g., stress, speed, feed, chip-thickness ratio), or it may be descriptive of an entity (e.g., cutting tool, vertical mill, 4340 steel, P/N-101). In conjunction with the Part Family Classification and Coding System, a number of standard keywords are provided. To conserve space and facilitate data entry, some of these keywords consist of one- to three-character alphanumeric codes. For example, the keyword code for a workpiece that is rotational and concentric, with two outside diameters and one bore diameter, is "B11." The keyword code for a family of low-alloy, low-carbon steels is A1. These codes are easy to use and greatly facilitate concise communication. They may be used as output keys or input keys to provide the very powerful capability of linking to other types of hierarchal information trees, such as those used for process selection, equipment selection, or automated time standard setting.

1.4 Part Family Code

Purpose
Part classification and coding is considered by many to be a prerequisite to the introduction of group technology, computer-aided process planning, design retrieval, and many other manufacturing activities. Part classification and coding is aimed at improving productivity,

reducing unnecessary variety, improving product quality, and reducing direct and indirect cost.

Code Format and Length

The part family code shown in Fig. 4 is composed of a five-section alphanumeric code. The first section of the code gives the basic shape. Other sections provide for form features, size, precision, and material. Each section of the code may be used as a pointer to more detailed information or as an output key for subsequent linking with related decision trees. The code length is eight digits. Each digit place has been carefully analyzed so that a compact code would result that is suitable for human communication and yet sufficiently comprehensive for generative process planning. The three-digit basic shape code provides for 240 standard families, 1160 custom families, and 1000 functional or named families. In addition, the combination of 50 form features, 9 size ranges, 5 precision classes, and 79 material types makes possible 2.5×10^{71} unique combinations! This capability should satisfy even the most sophisticated user.

Basic Shape

The basic shapes may be defined as those created from primitive solids and their derivatives (Fig. 5) by means of a basic founding process (cast, mold, machine). Primitives have been divided into rotational and nonrotational shapes. Rotational primitives include the cylinder, sphere, cone, ellipsoid, hyperboloid, and toroid. The nonrotational primitives include the cube (parallelepiped), polyhedron, warped (contoured) surfaces, free forms, and named shapes. The basic shape families are subdivided on the basis of predominant geometric characteristics, including external and internal characteristics.

The derivative concentric cylinder shown in Fig. 5 may have several permutations. Each permutation is created by merely changing dimensional ratios as illustrated or by adding form features. The rotational cylindrical shape shown may be thought of as being created from the intersection of a negative cylinder with a positive cylinder.

Figure 5a, with a length/diameter (L/D) ratio of 1:1, could be a spacer; Fig. 5b, with an L/D ratio of 0.1:1, would be a washer; and Fig. 5c, with an L/D ratio of 5:1, could be a thin-walled tube. If these could be made using similar processes, equipment, and tooling, they could be said to constitute a family of parts.

Name or Function Code

Some geometric shapes are so specialized that they may serve only one function. For example, a crankshaft has the major function of transmitting reciprocating motion to rotary motion. It is difficult to use a crankshaft for other purposes. For design retrieval and process planning purposes, it would probably be well to classify all crankshafts under the code name "crankshaft." Of course, it may still have a geometric code such as "P75," but the descriptive code will aid in classification and retrieval. A controlled glossary of function codes with

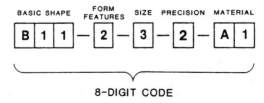

Figure 4 Part family code.

1 Part Family Classification and Coding 75

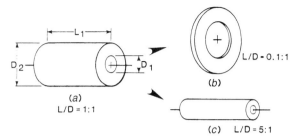

Figure 5 Permutations of concentric cylinders.

cross references, synonyms, and preferred labels would aid in using name and function codes and avoid unnecessary proliferation.

Special Features
To satisfy product design requirements, the designer creates the basic shape of a workpiece and selects the engineering material of which it is to be made. The designer may also require special processing treatments to enhance properties of a given material. In other words, the designer adds special features. Special features of a workpiece include form features heat treatments, and special surface finishes.

Form features may include holes, notches, splines, threads, and so on. The addition of a form feature does not change the basic part shape (family), but does enable it to satisfy desired functional requirements. Form features are normally imparted to the workpiece subsequent to the basic founding process.

Heat treatments are often given to improve strength, hardness, and wear resistance of a material. Heat treatments, such as stress relieving or normalizing, may also be given to aid in processing the workpiece.

Surface finishing treatments, such as plating, painting, and anodizing, are given to enhance corrosion resistance, improve appearance, or meet some other design requirement.

The special features are contained in an N-tree format with an associated complexity-evaluation and classification feature. This permits the user to select many special features while still maintaining a relatively simple code. Basically, nine values (1–9) have been established as the special feature complexity codes. As the user classifies the workpiece and identifies the special features required, the number of features is tallied and an appropriate complexity code is stored. Figure 6 shows the number count for special features and the associated feature code.

The special feature complexity code is useful in conveying to the user some idea of the complexity of the workpiece. The associated bitstring contains detailed computer-interpretable information on all features. (Output keys may be generated for each individual feature.) This information is valuable for generative process planning and for estimating purposes.

Size Code
The size code is contained in the third section of the part family code. This code consists of one numeric digit. Values range from 1 to 9, with 9 representing very large parts (Fig. 7). The main purpose of the size code is to give the code user a feeling for the overall size envelope for the coded part. The size code is also useful in selecting production equipment of the appropriate size.

FEATURE COMPLEXITY CODE	NO. SPECIAL FEATURES
1	1
2	2
3	3
4	5
5	8
6	13
7	21
8	34
9	GT 34

Figure 6 Complexity code for special features.

Precision Class Code

The precision class code is contained in the fourth segment of the part family code. It consists of a single numeric digit with values ranging from 1 to 5. Precision in this instance represents a composite of tolerance and surface finish. Class 1 precision represents very close tolerances and a precision-ground or lapped-surface finish. Class 5, on the other hand, represents a rough cast or flame-cut surface with a tolerance of greater than 1/32 in. High precision is accompanied by multiple processing operations and careful inspection operations. Production costs increase rapidly as closer tolerances and finer surface finishes are specified. Care is needed by the designer to ensure that high precision is warranted. The precision class code is shown in Fig. 8.

Material Code

The final two digits of the part family code represent the material type. The material form and condition codes are captured in the associated bitstring.

SIZE CODE	MAXIMUM DIMENSION		DESCRIPTION	EXAMPLES
	ENGLISH (in.)	METRIC (mm)		
1	.5	10	Sub-miniature	Capsules
2	2	50	Miniature	Paper clip box
3	4	100	Small	Large match box
4	10	250	Medium-small	Shoe box
5	20	500	Medium	Bread box
6	40	1000	Medium-large	Washing machine
7	100	2500	Large	Pickup truck
8	400	10000	Extra-large	Moving van
9	1000	25000	Giant	Railroad box-car

Figure 7 Part family size classification.

CLASS CODE	TOLERANCE	SURFACE FINISH
1	LE .0005"	LE 4 RMS
2	.0005"–.002"	4–32 RMS
3	.002"–.010"	32–125 RMS
4	.010"–.030"	125–500 RMS
5	GT .030"	GT 500 RMS

Figure 8 Precision class code.

Seventy-nine distinct material families have been coded (Fig. 9). Each material family or type is identified by a two-digit code consisting of a single alphabetic character and a single numeric digit.

The stainless-steel family, for example, is coded "A6." The tool steel family is "A7." This code provides a pointer to specification sheets containing comprehensive data on material properties, availability, and processability.

The material code provides a set of standard interface codes to which may be appended a given industry class code when appropriate. For example, the stainless-steel code may have appended to it a specific material code to uniquely identify it as follows: "A6-430" represents a chromium-type, ferritic, nonhardenable stainless steel.

1.5 Tailoring the System

Nearly all classification systems must be customized to meet the needs of each individual company or user. This effort can be greatly minimized by starting with a general system and then tailoring it to satisfy unique user needs. The Part Family Classification and Coding System permits this customizing. It is easy to add new geometric configurations to the existing family of basic shapes. It is likewise simple to add additional special features or to modify the size or precision class ranges. New material codes may be readily added if necessary.

The ability to modify easily an existing classification system without extensively reworking the system is one test of its design.

2 ENGINEERING MATERIALS TAXONOMY

2.1 Introduction

Serious and far-reaching problems exist with traditional methods of engineering materials selection. The basis for selecting a material is often tenuous and unsupported by defensible selection criteria and methods. A taxonomy of engineering materials accompanied by associated property files can greatly assist the designer in choosing materials to satisfy a design's functional requirements as well as procurement and processing requirements.

Material Varieties
The number of engineering materials from which a product designer may choose is staggering. It is estimated that over 40,000 metals and alloys are available, plus 250,000 plastics,

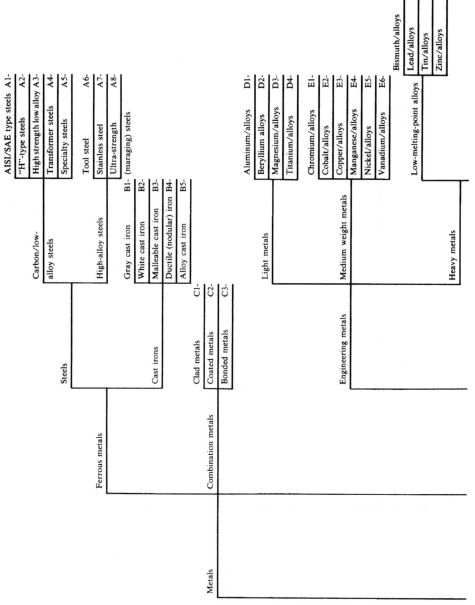

Figure 9 Engineering materials.

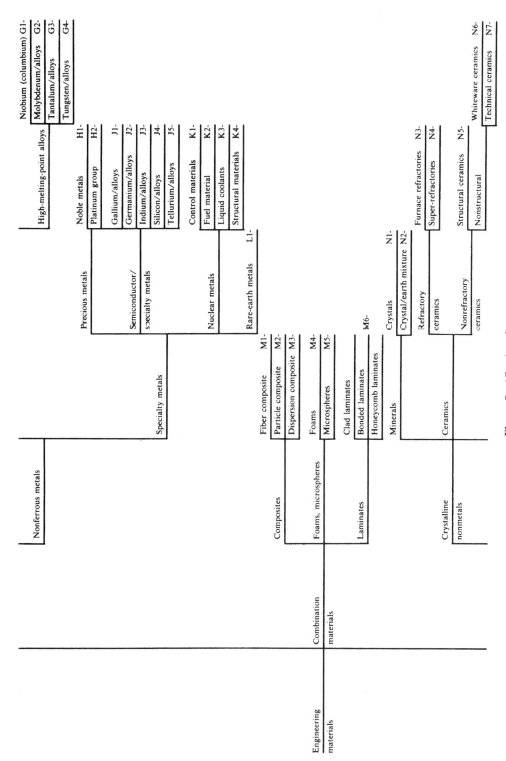

Figure 9 (*Continued*)

```
Nonmetals and          Crystalline glass    N8-
compounds
               Fibrous materials
                                            Natural woods       P1-
                                                Treated wood    P2-
                       Wood/products                                  Layered/jointed wood  P3-
                                            Processed wood            Fibrous-felted (ASTM) P4-
                                                                                                Particle board  P5-
                                                                      Particle products
                                                                                                Molded wood     P6-
                                            Cork                P7-
                                            Cellulose fiber paper    Q1-
                       Paper/products       Inorganic fiber paper    Q2-
                                            Special papers/          Q3-
                                            products
                       Textile fiber        Natural fibers      R1-
                       products             Manmade fibers      R2-
                       Glasses              Commercial glass    S1-
                                            Technical glass     S2-
               Amorphous                    Thermoplastics      T1-
               materials    Plastics        Thermoset plastics  T2-
                                            Natural rubber      U1-
                            Rubber/elastomers  Synthetic rubber  U2-
                                            Elastomers          U3-
```

Figure 9 (*Continued*)

uncounted composites, ceramics, rubbers, wood products, and so on. From this list, the designer must select the one for use with the new product. Each of these materials can exhibit a wide range of properties, depending on its form and condition. The challenge faced by the designer in selecting optimum materials can be reduced by a classification system to aid in identifying suitable material families.

Material Shortages
Dependency on foreign nations for certain key alloying elements, such as chromium, cobalt, tungsten and tin, points up the critical need for conserving valuable engineering materials and for selecting less strategic materials wherever possible. The recyclability of engineering materials has become another selection criterion.

Energy Requirements
The energy required to produce raw materials, process them, and then recycle them varies greatly from material to material. For example, recycled steel requires 75% less energy than steel made from iron ore, and recycled aluminum requires only about 10% of the energy of primary aluminum. Energy on a per-volume basis for producing ABS plastic is 2×10^6 Btu/in.^{3}, whereas magnesium requires 8×10^6 Btu/in.^{3}

2.2 Material Classification

Although there are many specialized material classification systems available for ferrous and nonferrous metals, there are no known commercial systems that also include composites and nonmetallics such as ceramic, wood, plastic, or glass. To remedy this situation, a comprehensive classification of all engineering materials was undertaken by the author. The resulting hierarchal classification or taxonomy provides 79 material families. Each of these families may be further subdivided by specific types as desired.

Objectives
Three objectives were established for developing an engineering materials classification system: (1) minimizing search time, (2) facilitating materials selection, and (3) enhancing communication.

Minimize Search Time. Classifying and grouping materials into recognized, small subgroups having similar characteristic properties (broadly speaking) minimizes the time required to identify and locate other materials having similar properties. The classification tree provides the structure and codes to which important procedures, standards, and critical information may be attached or referenced. The information explosion has brought a superabundance of printed materials. Significant documents and information may be identified and referenced to the classification tree to aid in bringing new or old reference information to the attention of users.

Facilitate Materials Selection. One of the significant problems confronting the design engineer is that of selecting materials. The material chosen should ideally meet several selection criteria, including satisfying the design functional requirements, producibility, availability, and the more recent constraints for life-cycle costing, including energy and ecological considerations.

Materials selection is greatly enhanced by providing materials property tables in a format that can be used manually or that can be readily converted to computer usage. A secondary goal is to reduce material proliferation and provide for standard materials within an organization, thus reducing unnecessary materials inventory.

Enhance Communication. The classification scheme is intended to provide the logical grouping of materials for coding purposes. The material code associated with family of materials provides a pointer to the specific material desired and to its condition, form, and properties.

Basis of Classification

Although it is possible to use a fairly consistent basis of classification within small subgroups (e.g., stainless steels), it is difficult to maintain the same basis with divergent groups of materials (e.g., nonmetals). Recognizing this difficulty, several bases for classification were identified, and the one that seemed most logical (or that was used industrially) was chosen. This subgroup base was then cross-examined relative to its usefulness in meeting objectives cited in the preceding subsection.

The various bases for classification considered for the materials taxonomy are shown in Fig. 10. The particular basis selected for a given subgroup depends on the viewpoint chosen. The overriding viewpoint for each selection was (1) Will it facilitate material selection for design purposes? and (2) Does it provide a logical division that will minimize search time in locating materials with a predominant characteristic or property?

Taxonomy of Engineering Materials

An intensive effort to produce a taxonomy of engineering materials has resulted in the classification shown in Fig. 11. The first two levels of this taxonomy classify all engineering materials into the broad categories of metals, nonmetals and compounds, and combination

Base		Example
A.	State	Solid–liquid–gas
B.	Structure	Fibrous–crystalline–amorphous
C.	Origin	Natural–synthetic
D.	Application	Adhesive–paint–fuel–lubricant
E.	Composition	Organic–inorganic
F.	Structure	Metal–nonmetal
G.	Structure	Ferrous–nonferrous
H.	Processing	Cast–wrought
I.	Processing response	Water-hardening–oil-hardening–air-hardening, etc.
J.	Composition	Low alloy–high alloy
K.	Application	Nuclear–semiconducting–precious
L.	Property	Light weight–heavy
M.	Property	Low melting point–high melting point
N.	Operating environment	Low-temperature–high-temperature
O.	Operating environment	Corrosive–noncorrosive

Figure 10 Basis for classifying engineering materials.

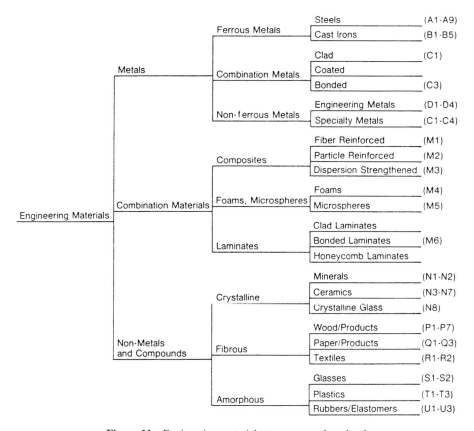

Figure 11 Engineering materials taxonomy—three levels.

materials. Metals are further subdivided into ferrous, "nonferrous," and combination metals. Nonmetals are classified as crystalline, fibrous, and amorphous.

Combination materials are categorized as composites, foams, microspheres, and laminates. Each of these groups is further subdivided until a relatively homogeneous materials family is identified. At this final level a family code is assigned.

Customizing

The Engineering Materials Taxonomy may be easily modified to fit a unique user's needs. For example, if it were desirable to further subdivide "fiber-reinforced composites," it could easily be done on the basis of type of filament used (e.g., boron, graphite, glass) and further by matrix employed (polymer, ceramic, metal). The code "M1," representing fiber-reinforced composites, could have appended to it a dash number uniquely identifying the specific material desired. Many additional material families may also be added if desired.

2.3 Material Code

As was mentioned earlier, there are many material classification systems, each of which covers only a limited segment of the spectrum of engineering materials available. The purpose of the Engineering Materials Taxonomy is to overcome this limitation. Furthermore, each of the various materials systems has its own codes. This creates additional problems. To solve this coding compatibility problem, a two-character alphanumeric code is provided as a standard interface code to which any industry or user code may be appended. This provides a very compact standard code so that any user will recognize the basic material family even though perhaps not recognizing a given industry code.

Material Code Format

The format used for the material code is shown in Fig. 12. The code consists of four basic fields of information. The first field contains a two-character interface code signifying the material family. The second field is to contain the specific material type based on composition or property. This code may be any five-character alphanumeric code. The third field contains a two-digit code containing the material condition (e.g., hot-worked, as-cast, ¾-hard). The fourth and final field of the code contains a one-digit alphabetic code signifying the material form (e.g., bar, sheet, structural shape).

Material Families

Of the 79 material families identified, 13 are ferrous metals, 30 are nonferrous metals, 6 are combination materials (composites, foams, laminates), and 26 are nonmetals and compounds.

The five-digit code space reserved for material type is sufficient to accommodate the UNS (Unified Numbering System) recently developed by ASTM, SAE, and others for metals and alloys. It will also accommodate industry or user-developed codes for nonmetals or combination materials. An example of the code (Fig. 10) for an open-hearth, low-carbon steel would be "A1-C1020," with the first two digits representing the steel family and the last five digits the specific steel alloy.

Material Condition

The material condition code consists of a two-digit code derived for each material family. The intent of this code is to reflect processes to which the material has been subjected and its resultant structure. Because of the wide variety of conditions that do exist for each family of materials, the creation of a D-tree for each of the 79 families seems to be the best approach. The D-tree can contain processing treatments along with resulting grain size, microstructure, or surface condition if desired. Typical material condition codes for steel family "A1" are given in Fig. 13.

Material form code consists of a single alphabetic character to represent this raw material form (e.g., rod, bar, tubing, sheet, structural shape). Typical forms are shown in Fig. 14.

2.4 Material Properties

Material properties have been divided into three broad classes: (1) mechanical properties, (2) physical properties, and (3) chemical properties. Each of these will be discussed briefly.

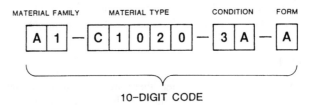

Figure 12 Format for engineering materials code.

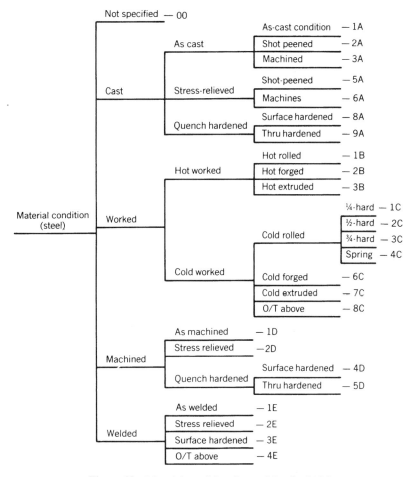

Figure 13 Material condition for steel family "A1."

Mechanical Properties
The mechanical properties of an engineering material describe its behavior or quality when subjected to externally applied forces. Mechanical properties include strength, hardness, fatigue, elasticity, and plasticity. Figure 15 shows representative mechanical properties. Note that each property has been identified with a unique code number to reduce confusion in communicating precisely which property is intended. Confusion often arises because of the multiplicity of testing procedures that have been devised to assess the value of a desired property. For example, there are at least 15 different penetration hardness tests in common usage, each of which yields different numerical results from the others. The code uniquely identifies the property and the testing method used to ascertain it.

Each property of a material is intimately related to its composition, surface condition, internal condition, and material form. These factors are all included in the material code. A modification of any of these factors, either by itself or in combination, can result in quite different mechanical properties.

Thus, each material code combination is treated as a unique material. As an example of this, consider the tensile strength of a heat-treated 6061 aluminum alloy: in the wrought

O—Unspecified

Rotational Solids

    A—Rod/wire
    B—Tubing/pipe

Flat Solids

    C—Bar, flats
    D—Hexagon/octagon
    E—Sheet/plate

Structural Shapes

    F—Angle
    G—T section
    H—Channel
    I—H, I sections
    J—Z sections
    K—Special sections (extruded, rolled, etc.)

Fabricated Solid Shapes

    L—Forging
    M—Casting/ingot
    N—Weldment
    P—Powder metal
    Q—Laminate
    R—Honeycomb
    S—Foam

Special Forms

    T—Resin, liquid, granules
    U—Fabric, roving, filament
    V—Putty, clay
    W—Other
    Y—Reserved
    Z—Reserved

Figure 14 Raw material forms.

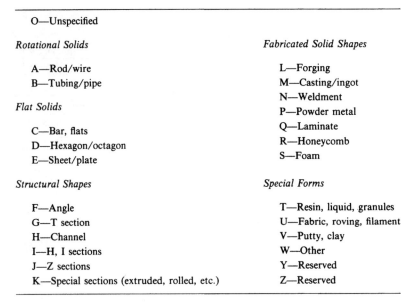

Mechanical Properties: D 1 - 0 6 0 6 1 - 1 B - C

Material Family/Type: Aluminum 6061-T6

Code	Description	Value	Units
11.02	Brinell hardness number	95	HB
12.06	Yield strength, 0.2% offset	40,000	psi
12.11	Ultimate tensile strength	45,000	psi
12.20	Ultimate shear (bearing) strength	30,000	psi
12.30	Impact energy (Charpy V-notch)		ft-lb
12.60	Fatigue (endurance limit)	14,000	psi
12.70	Creep strength		psi
13.01	Modulus of elasticity (tensile)	10.0×10^6	psi
13.02	Modulus of elasticity (compressive)	10.2×10^6	psi
13.20	Poisson's ratio	—	—
14.02	Elongation	15	%
14.10	Reduction of area	—	%
14.30	Strain hardening coefficient	—	%
14.40	Springback	—	%

Figure 15 Representative mechanical properties.

condition, the ultimate tensile strength is 19,000 psi; with the T4-temper, the ultimate tensile strength is 35,000 psi; and in the T913 condition, the ultimate tensile strength is 68,000 psi.

Physical Properties

The physical properties of an engineering material have to do with the intrinsic or structure-insensitive properties. These include melting point, expansion characteristics, dielectric strength, and density. Figure 16 shows representative physical properties.

Again, each property has been coded to aid in communication. Magnetic properties and electrical properties are included in this section for the sake of simplicity.

Chemical Properties

The chemical properties of an engineering material deal with its reactance to other materials or substances, including its operating environment. These properties include chemical reactivity, corrosion characteristics, and chemical compatibility. Atomic structure factors, chemical valence, and related factors useful in predicting chemical properties may also be included in the broad category of chemical properties. Figure 17 shows representative chemical properties.

Physical Properties	D 1 - 0 6 0 6 1		
Material Family/Type: Aluminum 6061-T6			
Prepared by:	Date:	Approved by:	Date:
Revision No./Date:			

Code	Description	Value	Units
21.01	Coefficient of linear expansion	13×10^{-6}	in./in./°F
21.05	Thermal conductivity	1070	Btu/in./ft²/°F/hr
21.40	Minimum service temperature	−320	°F
21.50	Maximum service temperature	700	°F
21.66	Melting range	1080–1200	°F
21.80	Recrystallization temperature	650	°F
21.90	Annealing temperature	775	°F, 2–3 hr
21.92	Stress-relieving temperature	450	°F, 1 hr
21.95	Solution heat treatment	970	°F
21.96	Precipitation heat treatment	350	°F, 6–10 hr
22.01	Electrical conductivity (weight)	40	%
22.02	Electrical conductivity (volume)	135	%
22.10	Electrical resistivity (volume)	26	ohms mil, ft
26.01	Specific weight	0.098	lb/in.³
26.03	Specific gravity	270	g/cm³
26.35	Crystal (lattice) system	f.c.c.	—
26.70	Damping index	0.03	Very low
26.71	Strength-to-weight ratio		
26.72	Basic refining energy	100,000	Btu/lb
26.73	Recycling energy	10,000	Btu/lb

Figure 16 Representative physical properties.

88 Classification Systems

Chemical Properties	D	1	-	0	6	0	6	1

Material Family/Type: Aluminum 6061-T6

Prepared by:	Date:	Approved by:	Date:

Revision No./Date:

Code	Description	Value[a]	Units
32.01	Resistance to high-temperature corrosion	C	
32.02	Resistance to stress corrosion cracking	C	
32.03	Resistance to corrosion pitting	B	
32.04	Resistance to intergranular corrosion	B	
32.10	Resistance to fresh water	A	
32.11	Resistance to salt water	A	
32.15	Resistance to acids	A	
32.20	Resistance to alkalies	C	
32.25	Resistance to petrochemicals	A	
32.30	Resistance to organic solvents	A	
32.35	Resistance to detergents	B	
33.01	Resistance to weathering	A	

[a] *Key:* A = fully resistant; B = slightly attacked; C = unsatisfactory.

Figure 17 Representative chemical properties.

2.5 Material Availability

The availability of an engineering material is a prime concern in materials selection and use. Material availability includes such factors as stock shapes, sizes, and tolerances; material condition and finish; delivery; and price.

Other factors of increasing significance are energy requirements for winning the material from nature and recyclability. Figure 18 shows representative factors for assessing material availability.

2.6 Material Processability

Relative processability ratings for engineering materials in conjunction with material properties and availability can greatly assist the engineering designer in selecting materials that will meet essential design criteria. All too often, the processability of a selected engineering material is unknown to the designer. As likely as not, the materials may warp during welding or heat treatment and be difficult to machine, which may result in undesirable surface stresses because of tearing or cracking during drawing operations. Many of these problems could be easily avoided if processability ratings of various materials were ascertained, recorded, and used by the designer during the material selection process. Figure 19 shows relative processability ratings. These ratings include machinability, weldability, castability, moldability, formability, and heat-treatability. Relative ratings are established through experience for each family. Ratings must not be compared between families. For example, the machinability rating of two steels may be compared, but they should not be evaluated against brass or aluminum.

| Availability | D | 1 | - | 0 | 6 | 0 | 6 | 1 |

| Material Family/Type: Aluminum 6061-T6 |

| Prepared by: | Date: | Approved by: | Date: |

Revision No./Date:

Surface Condition
    Cold worked
    Hot worked
    Cast
    Clad
    Peened
    Chromate
    Anodized
    Machined

Internal Condition
    Annealed
    Solution treated—naturally aged
    Solution treated—artificially aged
    Stress relieved
    Cold worked

Forms Available
    Sheet
    Plate
    Bar
    Tubing
    Wire
    Rod
    Extrusions
    Ingot

Figure 18 Factors relating to material availability.

3 FABRICATION PROCESS TAXONOMY

3.1 Introduction

Purpose

The purpose of classifying manufacturing processes is to form families of related processes to aid in process selection, documentation of process capabilities, and information retrieval. A taxonomy or classification of manufacturing processes can aid in process selection by providing a display of potential manufacturing options available to the process planner.

| Processability | D | 1 | - | 0 | 6 | 0 | 6 | 1 | - | 1 | B | - | C |

| Material Family/Type: | Aluminum 6061-T6 |

| Prepared by: | Date: | Approved by: | Date: |

Revision No./Date:

Processability Type	Rating			
	Poor 1	Fair 2	Good 3	Excellent 4
Machinability				X
Grindability (silicon carbide adhesive)			X	
Shear behavior				X
EDM rating	X			
Chemical etch factor		X		
Forgeability			X	
Extrudability				
Formability			X	
Weldability				X
Heat-treatability				X

Figure 19 Relative processability ratings.

Documentation of process capabilities can be improved by providing files containing the critical attributes and parameters for each classified process. Information retrieval and communication relative to various processes can be enhanced by providing a unique code number for each process. Process information can be indexed, stored, and retrieved by this code.

Classification and coding is an art and, as such, it is difficult to describe the steps involved, and even more difficult to maintain consistency in the results. The anticipated benefits to users of a well-planned process classification outweigh the anticipated difficulties, and thus the following plan is being formulated to aid in uniform and consistent classification and coding of manufacturing processes.

Primary Objectives
There are three primary objectives for classifying and coding manufacturing processes: (1) facilitating process planning, (2) improving process capability assessment, and (3) aiding in information retrieval.

Facilitate Process Selection. One of the significant problems confronting the new process planner is process selection. The planner must choose, from many alternatives, the basic process, equipment, and tooling required to produce a given product of the desired quality and quantity in the specified time.

Although there are many alternative processes and subprocesses from which to choose, the process planner may be well acquainted with only a small number of them. The planner may thus continue to select these few rather than become acquainted with many of the newer and more competitive processes. The proposed classification will aid in bringing to the attention of the process planner all the processes suitable for modifying the shape of a material or for modifying its properties.

Improve Process Capability Assessment. One of the serious problems facing manufacturing managers is that they can rarely describe their process capabilities. As a consequence, there is commonly a mismatch between process capability and process needs. This may result in precision parts being produced on unsuitable equipment, with consequent high scrap rates, or parts with no critical tolerances being produced on highly accurate and expensive machines, resulting in high manufacturing costs.

Process capability files may be prepared for each family of processes to aid in balancing capacity with need.

Aid Information Retrieval. The classification and grouping of manufacturing processes into subgroups having similar attributes will minimize the time required to identify and retrieve similar processes. The classification tree will provide a structure and branches to which important information may be attached or referenced regarding process attributes, methods, equipment, and tooling.

The classification tree provides a logical arrangement for coding existing processes as well as a place for new processes to be added.

3.2 Process Divisions

Manufacturing processes can be broadly grouped into two categories: (1) shaping processes and (2) nonshaping processes. Shaping processes are concerned primarily with modifying the shape of the plan material into the desired geometry of the finished part. Nonshaping processes are primarily concerned with modifying material properties.

Shaping Processes

Processes available for shaping the raw material to produce a desired geometry may be classified into three subdivisions: (1) mass-reducing processes, (2) mass-conserving processes, and (3) mass-increasing or joining processes. These processes may then be further subdivided into mechanical, thermal, and chemical processes.

Mass-reducing processes include cutting, shearing, melting or vaporizing, and dissolving or ionizing processes. Mass-conserving processes include casting, molding, compacting, deposition, and laminating processes. Mass-increasing or, more commonly, joining, processes include pressure and thermal welding, brazing, soldering, and bonding. The joining processes are those that produce a megalithic structure not normally disassembled.

Nonshaping Processes

Nonshaping processes that are available for modifying material properties or appearance may be classified into two broad subdivisions: (1) heat-treating processes and (2) surface-finishing processes.

Heat-treating processes are designed primarily to modify mechanical properties, or the processability ratings, of engineering materials. Heat-treating processes may be subdivided into (1) annealing (softening) processes, (2) hardening processes, and (3) other processes. The "other" category includes sintering, firing/glazing, curing/bonding, and cold treatments. Annealing processes are designed to soften the work material, relieve internal stresses, or change the grain size. Hardening treatments, on the other hand, are often designed to increase strength and resistance to surface wear or penetration. Hardening treatments may be applied to the surface of a material or the treatments may be designed to change material properties throughout the section.

Surface-finishing processes are those used to prepare the workpiece surface for subsequent operations, to coat the surface, or to modify the surface. Surface-preparation processes include descaling, deburring, and degreasing. Surface coatings include organic and inorganic; metallic coatings applied by spraying, electrostatic methods, vacuum deposition, and electroplating; and coatings applied through chemical-conversion methods.

Surface-modification processes include burnishing, brushing, peening, and texturing. These processes are most often used for esthetic purposes, although some peening processes are used to create warped surfaces or to modify surface stresses.

3.3 Process Taxonomy

There are many methods for classifying production processes. Each may serve unique purposes. The Fabrication Process Taxonomy is the first known comprehensive classification of all processes used for the fabrication of discrete parts for the durable goods manufacturing industries.

Basis of Classification

The basis for process classification may be the source of *energy* (i.e., mechanical, electrical, or chemical); the *temperature* at which the processing is carried out (i.e., hot-working, cold-working); the type of material to be processed (i.e., plastic, steel, wood, zinc, or powdered metal); or another basis of classification.

The main purpose of the hierarchy is to provide functional groupings without drastically upsetting recognized and accepted families of processes within a given industry. For several reasons, it is difficult to select only one basis for classification and apply it to all processes and achieve usable results. Thus, it will be noted that the fabrication process hierarchy has several bases for classification, each depending on the level of classification and on the particular family of processes under consideration.

Classification Rules and Procedures

Rule 1. Processes are classified as either shaping or nonshaping, with appropriate mutually exclusive subdivisions.

Rule 2. Processes are classified as independent of materials and temperature as possible.

Rule 3. Critical attributes of various processes are identified early to aid in forming process families.

Rule 4. Processes are subdivided at each level to show the next options available.

Rule 5. Each process definition is in terms of relevant critical attributes.

Rule 6. Shaping process attributes include

   6.1 Geometric shapes produced

   6.2 Form features or treatments imparted to the workpiece

   6.3 Size, weight, volume, or perimeter of parts

   6.4 Part precision class

   6.5 Production rates

   6.6 Setup time

6.7 Tooling costs

6.8 Relative labor costs

6.9 Scrap and waste material costs

6.10 Unit costs versus quantities of 10, 100, 1K, 10K, 100K

Rule 7. All processes are characterized by

7.1 Prerequisite processes

7.2 Materials that can be processed, including initial form

7.3 Basic energy source: mechanical, thermal, or chemical

7.4 Influence of process on mechanical properties such as strength, hardness, or toughness

7.5 Influence of process on physical properties such as conductivity, resistance, change in density, or color

7.6 Influence of process on chemical properties such as corrosion resistance

Rule 8. At the operational level, the process may be fully described by the operation description and sequence, equipment, tooling, processing parameters, operating instructions, and standard time.

The procedure followed in creating the taxonomy was first to identify all the processes that were used in fabrication processes. These processes were then grouped on the basis of relevant attributes. Next, most prominent attributes were selected as the parent node label. Through a process of selection, grouping, and classification, the taxonomy was developed. The taxonomy was evaluated and found to readily accommodate new processes that subsequently were identified. This aided in verifying the generic design of the system. The process taxonomy was further cross-checked with the equipment and tooling taxonomies to see if related categories existed. In several instances, small modifications were required to ensure that the various categories were compatible. Following this method for cross-checking among processes, equipment, and tooling, the final taxonomy was prepared. As the taxonomy was subsequently typed and checked, and large charts were developed for printing, small remaining discrepancies were noted and corrected. Thus, the process taxonomy is presented as the best that is currently available. Occasionally, a process is identified that could be classified in one or more categories. In this case, the practice is to classify it with the preferred group and cross-reference it in the second group.

3.4 Process Code

The process taxonomy is used as a generic framework for creating a unique and unambiguous numeric code to aid in communication and information retrieval.

The process code consists of a three-digit numeric code. The first digit indicates the basic process division and the next two digits indicate the specific process group. The basic process divisions are as follows:

000	Material identification and handling
100	Material removal processes
200	Consolidation processes
300	Deformation processes
400	Joining processes

94 Classification Systems

500	Heat-treating processes
600	Surface-finishing processes
700	Inspection
800	Assembly
900	Testing

The basic process code may be extended with the addition of an optional decimal digit similar to the Dewey Decimal System. The process code is organized as shown in Fig. 20.

The numeric process code provides a unique, easy-to-use shorthand communication symbol that may be used for manual or computer-assisted information retrieval. Furthermore, the numeric code can be used on routing sheets, in computer databases, for labeling of printed reports for filing and retrieval purposes, and for accessing instructional materials, process algorithms, appropriate mathematical and graphical models, and the like.

3.5 Process Capabilities

Fundamental to process planning is an understanding of the capabilities of various fabrication processes. This understanding is normally achieved through study, observation, and industrial experience. Because each planner has different experiences and observes processes through different eyes, there is considerable variability in derived process plans.

Fabrication processes have been grouped into families having certain common attributes. A study of these common attributes will enable the prospective planner to learn quickly the significant characteristics of the process without becoming confused by the large amount of factual data that may be available about the given process.

Also, knowledge about other processes in a given family will help the prospective planner learn about a specific process by inference. For example, if the planner understands that "turning" and "boring" are part of the family of single-point cutting operations and has learned about cutting-speed calculations for turning processes, the planner may correctly infer that cutting speeds for boring operations would be calculated in a similar manner, taking into account the rigidity of each setup. It is important at this point to let the prospective planner know the boundaries or exceptions for such generalizations.

A study of the common attributes and processing clues associated with each of these various processes will aid the planner. For example, an understanding of the attributes of a given process and recognition of process clues such as "feed marks," "ejector-pin marks," or "parting lines" can help the prospective planner to identify quickly how a given part was produced.

Figures 21 and 22 show a process capability sheet that has been designed for capturing information relative to each production process.

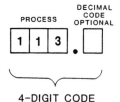

Figure 20 Basic process code.

Process: Turning/Facing	Code: 101

Prepared by:	Date:	Approved by:	Date:

Revision No. & Date:

Schematic:

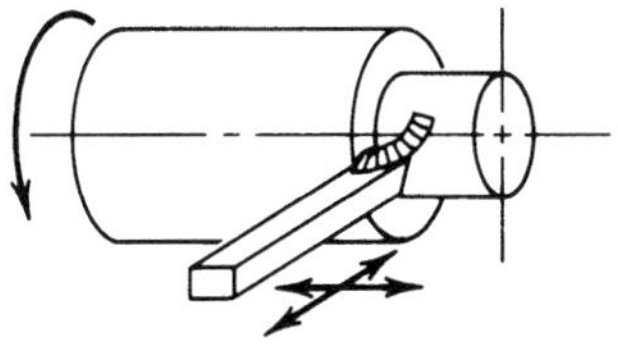

Attributes:
- Single point cutting tool
- Chips removed from external surface
- Helical or annular (tree-ring) feed marks are present.

Basic Shapes Produced:
Surfaces of revolution (cylindrical, tapered, spherical) or flat shoulders or ends. May have discontinuities in surfaces (interrupted cut).

Form Features or Treatments:
Bead, boss, chf'r, groove, lip, radius, thread

Size Range:
1–6

Precision Class:
1–4

Raw Material Type: Steel, cast iron, light metals, non-ferrous engineering metals, low-m.p. metals, refractory metals, nuclear metals, composites, refractories, wood, polymers, rubbers and elastometers

Figure 21 Process capability sheet.

4 FABRICATION EQUIPMENT CLASSIFICATION

4.1 Introduction

Utilization of Capital Resources
One of the primary purposes for equipment classification systems is to better utilize capital resources. The amount of capital equipment and tooling per manufacturing employee has been reported to range from $30,000 to $50,000. An equipment classification system can be

Process: Turning/Facing				Code: 101			
Raw Material Condition: Hot-rolled, cold-rolled, forged, cast							
Raw Material Form: rod, tubing, forgings, castings							
Production Rate		1 A	10 B	100 C	1,000 D	10,000 E	100,000 F
Tooling Costs	High-3						
	Med-2						
	Low-1						
Set-up Time	High-3						
	Med-2						
	Low-1						
Labor Costs	High-3						
	Med-2						
	Low-1						
Scrap & Waste Material Costs	High-3						
	Med-2						
	Low-1						
Unit Costs	High-3						
	Med-2						
	Low-1						
Prerequisite Processes: Hot-rolling, cold rolling, forging, casting, p/m compacting							
Influence on Mechanical Properties: Creates very thin layer of stressed work material. Grains may be slightly deformed, and built-up edge may be present on work surface.							
Influence on Physical Properties: N/A							
Influence on Chemical Properties: Highly stressed work surface may promote corrosion.							

Figure 22 Process capability sheet.

a valuable aid in capacity planning, equipment selection, equipment maintenance scheduling, equipment replacement, elimination of unnecessary equipment, tax depreciation, and amortization.

Equipment Selection

A key factor in equipment selection is a knowledge of the various types of equipment and their capabilities. This knowledge may be readily transmitted through the use of an equipment classification tree showing the various types of equipment and through equipment specification sheets that capture significant information regarding production capabilities.

Equipment selection may be regarded as matching—the matching of production needs with equipment capabilities. Properly defined needs based on current and anticipated requirements, when coupled with an equipment classification system, provide a logical, consistent strategy for equipment selection.

Manufacturing Engineering Services
Some of the manufacturing engineering services that can be greatly benefitted by the availability of an equipment classification system include process planning, tool design, manufacturing development, industrial engineering, and plant maintenance. The equipment classification code can provide an index pointer to performance records, tooling information, equipment specification sheets (mentioned previously), and other types of needed records.

Quality Assurance Activities
Acceptance testing, machine tool capability assessment, and quality control are three important functions that can be enhanced by means of an equipment classification and coding system. As before, the derived code can provide a pointer to testing and acceptance procedures appropriate for the given family of machines.

4.2 Standard and Special Equipment

The classification system described below can readily accommodate both standard and special equipment. *Standard fabrication equipment* includes catalog items such as lathes, milling machines, drills, grinders, presses, furnaces, and welders. Furthermore, they can be used for making a variety of products. Although these machines often have many options and accessories, they are still classified as standard machines. *Special fabrication machines,* on the other hand, are custom designed for a special installation or application. These machines are usually justified for high-volume production or special products that are difficult or costly to produce on standard equipment. Examples of special machines include transfer machines, special multistation machines, and the like.

4.3 Equipment Classification

The relationship among the fabrication process, equipment, and tooling is shown graphically in Fig. 23 The term *process* is basically a concept and requires equipment and tooling for its physical embodiment. For example, the grinding process cannot take place without equipment and tooling. In some instances, the process can be implemented without equipment, as in "hand-deburring." The hierarchical relationships shown in Fig. 23 between the process, equipment, and tooling provide a natural linkage for generative process planning. Once the required processes have been identified for reproducing a given geometric shape and its associated form features and special treatments, the selection of equipment and tooling is quite straightforward.

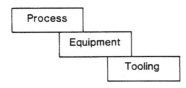

Figure 23 Relationships between process, equipment, and tooling.

Rationale

Two major functions of an equipment classification system are for process planning and tool design. These functions are performed each time a new product or piecepart is manufactured. Consequently, the relationships between processes, equipment, and tooling have been selected as primary in development of the equipment classification system.

The equipment taxonomy parallels the process taxonomy as far as possible. Primary levels of classification include those processes whose intent is to change the form of the material—for example, shaping processes and those processes whose intent is to modify or enhance the material properties. These nonshaping processes include heat treatments and coating processes, along with attendant cleaning and deburring.

As each branch of a process tree is traversed, it soon becomes apparent that there is a point at which an equipment branch must be grafted in. It is at this juncture that the basis for equipment classification must be carefully considered. There are a number of possible bases for classification of equipment, including

1. Form change (shaping, nonshaping)
2. Mass change (reduction, consolidation, joining)
3. Basic process (machine, cast, forge)
4. Basic subprocess (deep hole drill, precision drill)
5. Machine type (gang drill, radial drill)
6. Energy source (chemical, electrical, mechanical)
7. Energy transfer mechanism (mechanical, hydraulic, pneumatic)
8. Raw material form (sheet metal, forging, casting)
9. Shape produced (gear shaper, crankshaft lathe)
10. Speed of operation (high speed, low speed)
11. Machine orientation (vertical, horizontal)
12. Machine structure (open-side, two-column)
13. General purpose/special purpose (universal mill, spar mill)
14. Kinematics/motions (moving head, moving bed)
15. Control type (automatic, manual NC)
16. Feature machined (surface, internal)
17. Operating temperature (cold rolling, hot rolling)
18. Material composition (plastic molding, aluminum die casting)
19. Machine size (8-in. chucker, 12-in. chucker)
20. Machine power (600 ton, 100 ton)
21. Manufacturer (Landis, Le Blond, Gisholt)

In reviewing these bases of classification, it is apparent that some describe fundamental characteristics for dividing the equipment population into families, and others are simply attributes of a given family. For example, the features of "shaping," "consolidation," and "die casting" are useful for subdivision of the population into families (E-tree), whereas attributes such as "automatic," "cold-chamber," "aluminum," "100-ton," and "Reed-Prentice" are useful for characterizing equipment within a given family (N-tree). "Automatic" is an attribute of many machines; likewise, "100-ton" could apply to general-purpose presses, forging presses, powder-metal compacting presses, and so on. Similarly, the label

"Reed-Prentice" could be applied equally well to lathes, die-casting machines, or injection molders. In other words, these terms are not very useful for development of a taxonomy but are useful for characterizing a family.

Equipment Taxonomy
The first major division, paralleling the processes, is shaping or nonshaping. The second level for shaping is (1) mass reducing, (2) mass conserving, and (3) mass increasing. The intent of this subdivision, as was mentioned earlier, is to classify equipment whose intent is to change the form or shape of the workpiece. The second level for nonshaping equipment includes (1) heat treating and (2) surface finishing. The intent of this subdivision is to classify equipment designed to modify or enhance material properties or appearance. The existing taxonomy identifies 257 unique families of fabrication equipment.

Customizing
As with other taxonomies described herein, the equipment taxonomy is designed to readily accommodate new classes of machines. This may be accomplished by traversing the tree until a node point is reached where the new class or equipment must appropriately fit. The new equipment with its various subclasses may be grafted in at this point and an appropriate code number assigned. It should be noted that code numbers have been intentionally reserved for this purpose.

4.4 Equipment Code

The code number for fabrication equipment consists of a nine-character code. The first three digits identify the basic process, leaving the remaining six characters to identify uniquely any given piece of equipment. As can be seen in Fig. 24, the code consists of four fields. Each of these fields will be briefly described in the following paragraphs.

Process Code
The process code is a three-digit code that refers to one of the 222 fabrication processes currently classified. For example, code 111 = drilling and 121 = grinding. Appended to this code is a code for the specific type of equipment required to implement the given process.

Equipment Family Code
The code for equipment type consists of a one-character alphabetic code. For example, this provides for up to 26 types of turning machines, 101-A through 101-Z, 26 types of drilling machines, and so on. Immediately following the code for equipment type is a code that uniquely identifies the manufacturer.

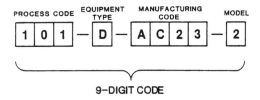

Figure 24 Fabrication equipment code.

Manufacturer Code

The manufacturer code consists of a four-digit alphanumeric code. The first character in the code is an alphabetic character (A–Z) representing the first letter in the name of the manufacturing company. The next three digits are used to identify uniquely a given manufacturing company. In-house-developed equipment would receive a code for your own company.

Model Number

The final character in the code is used to identify a particular manufacturer model number. Thus, the nine-digit code is designed to provide a shorthand designation as to the basic process, type of equipment, manufacturer, and model number. The code can serve as a pointer to more detailed information, as might be contained on specification sheets, installation instructions, maintenance procedures, and so on.

4.5 Equipment Specification Sheets

In the preceding subsections the rationale for equipment classification was discussed, along with the type of information useful in characterizing a given piece of equipment. It has been found that this characterization information is best captured by means of a series of equipment specification sheets.

The philosophy has been that each family must have a tailored list of features or specifications to characterize it adequately. For example, the terms *swing, center-distance, rpm,* and *feed per revolution* are appropriate for a lathe family but not for forming presses, welding machines, electrical discharge machines, or vibratory deburring machines. Both common attributes and selected ones are described in the following paragraphs and shown in Figs. 25 and 26.

Equipment Family/Code

The equipment family label consists of the generic family name, subgroup name (e.g., "drill," "radial"), and the nine-digit equipment code.

Equipment Identification

Equipment name, make, model, serial number, and location provide a unique identification description for a given piece of equipment. The equipment name may be the familiar name given to the piece of equipment and may differ from the generic equipment family name.

Acquisition Data

The acquisition data include capital cost, date acquired, estimated life, and year of manufacture. This information can be used for amortization and depreciation purposes.

Facilities/Utilities

Facilities and utilities required for installing and operating the equipment include power (voltage, current, phases, frequency), and other connections, floor space, height, and weight.

Specifications

The specifications for functionality and capabilities for each family of machines must be carefully defined to be useful in selecting machines to meet needs of intended applications.

Operation Codes

A special feature of the equipment specification sheet is the section reserved for operation codes. The operation codes provide an important link between workpiece requirements and

Equipment Family: Mill, NC, vertical	Code: 113-K
Identification:	
Name: Cintimatic, Single Spindle Make/Model: Cincinnati Milling Machine Co.	Serial No.: 364 Location: 115 SNLB
Acquisition:	
Capital Cost: $25,000 Date Acquired: 15 Aug. 1963	Estimated Life: 15 yr Year of Manufacture: 1963
Maintenance:	
Condition: U2 Date: 15 Aug. 1963	Reevaluation Date: 15 January 1981
Facilities:	
Voltage: 230 volts, 3 ph, 60 Hz Current: 3 hp Other Connections: Air, 40 psi	Floor Space: 63 in. × 74 in. Height: 101 in. Weight:
Specifications:	
Working surface .... 22 in. × 36 in. Throat .... 16¼ in. Table top to spindle .... 14–24 in. (8 in. travel) Weight capacity, max .... 1000 lb Spindle: Axes .... One axis Range .... 85–3800 rpm Rate .... 1–40 ipm Table: Axes .... Two axes Range .... 15 in. × 25 in. Rate .... Feed—0–40 ipm, rapid travel 200 ipm Motor hp: Drive motor .... 3.0 hp Feed .... Hydraulic servo motors Coolant .... Air mist spray Spindle taper .... #40 NMTB T-slots .... 3 in X axis, $^{11}/_{16}$ in. wide Accuracy .... ±0.001 in. in 24 in. Control type .... Accramatic Series 200 control	

Figure 25 Equipment specifications.

equipment capability. For example, if the workpiece is a rotational, machined part with threads, grooves, and a milled slot, and if a lathe is capable of operations for threading and grooving but not a milling slot, then it follows that either two machines are required to produce the part or a milling attachment must be installed on the lathe. A significant benefit of the operation code is that it can aid process planners in selecting the minimum number of machines required to produce a given workpiece. This fact must, of course, be balanced with production requirements and production rates. The main objective is to reduce transportation and waiting time and minimize cost. (See Fig. 26.)

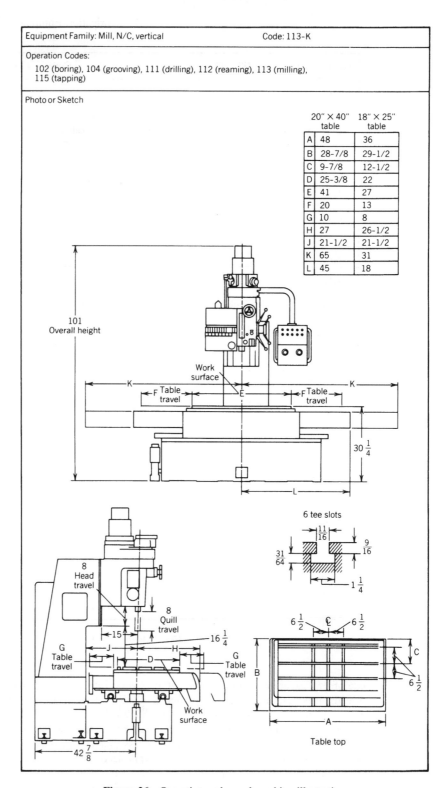

Figure 26 Operation codes and machine illustration.

Photograph or Sketch

A photograph or line drawing provides considerable data to aid in plant layout, tool design, process planning, and other production planning functions. Line drawings often provide information regarding T-slot size, spindle arbor size, limits of machine motion, and other information useful in interfacing the machine with tooling, fixtures, and the workpiece.

5 FABRICATION TOOL CLASSIFICATION AND CODING

5.1 Introduction

Because standard and special tooling represent a sizable investment, it is prudent to minimize redundant tooling, to evaluate performance of perishable tooling, and to provide good storage and retrieval practices to avoid excessive tool-float and loss of valuable tooling.

The use of a standard tool-classification system could provide many benefits for both the supplier and the user. The problem is to derive a comprehensive tool-classification system that is suitable for the extremely wide variety of tools available to industry and that is agreeable to all suppliers and users. Although no general system exists, most companies have devised their own proprietary tool-classification schemes. This has resulted in much duplicate effort. Because of the difficulty of developing general systems that are expandable to accommodate new tooling categories, many of these existing schemes for tool classification are found to be inadequate.

This section describes a new classification and coding system for fabrication tooling. Assembly, inspection, and testing tools are not included. This new system for classifying fabrication tools is a derivative of the work on classifying fabrication processes and fabrication equipment. Furthermore, special tooling categories are directly related through a unique coding system to the basic shape of the workpieces they are used to fabricate.

Investment

The investment a manufacturing company must make for standard and special tooling is usually substantial. Various manufacturing companies may carry in stock from 5000 to 10,000 different tools and may purchase several thousand special tools, as required. As a rule of thumb, the investment in standard tooling for a new machine tool is often 20–30% of the basic cost of the machine. Special tooling costs may approach or even exceed the cost of certain machines. For instance, complex die-casting molds costing from $50,000 to $250,000 are quite commonplace.

The use of a tool classification system can aid a manufacturing enterprise by helping to get actual cost data for various tooling categories and thus begin to monitor and control tooling expenditures. The availability of good tooling is essential for economical and productive manufacturing. The intent of monitoring tooling costs should be to ensure that funds are available for such needed tooling and that these funds are wisely used. The intent should not be the miserly allocation of tooling money.

Tool Control

Tool control is a serious challenge in almost all manufacturing enterprises. Six important aspects that must be addressed in any good tool control system include*

---

*"Small Tools Planning and Control," in *Tool Engineers Handbook,* McGraw-Hill, New York, 1959, Section 3.

104 Classification Systems

Tool procurement	Tool dispensing
Tool storage	Tool performance measurement
Tool identification and marking	Tool maintenance

The availability of a standard, comprehensive tool classification and coding system can greatly aid each of these elements of tool control. For example, tool-procurement data may be easily cross-indexed with a standard tool number, thus reducing problems in communication between the user and the purchasing department.

Tool storage and retrieval may be enhanced by means of standard meaningful codes to identify tools placed in a given bin or at a given location. This problem is especially acute with molds, patterns, fixtures, and other special tooling.

Meaningful tool identification markings aid in preventing loss or misplacement of tools. Misplaced tools can quickly be identified and returned to their proper storage locations. Standard tool codes may also be incorporated into bar codes or other machine-readable coding systems if desirable.

The development of an illustrated tooling manual can be a great asset to both the user and to the tool crib personnel in identifying and dispensing tools. The use of a cross-referenced standard tool code can provide an ideal index to such a manual.

Tool-performance measures require the use of some sort of coding system for each type of tool to be evaluated. Comparison of tools within a given tool family may be facilitated by means of expanded codes describing the specific application and the various types of failures. Such extended codes may be easily tied to the standard tool family code.

Tool-maintenance and repair costs can be best summarized when they are referenced to a standard tool family code. Maintenance and acquisition costs could be easily reduced to obtain realistic life-cycle costs for tools of a given family or type.

In summary, tool control in general may be enhanced with a comprehensive, meaningful tool classification and coding system.

5.2 Standard and Special Tooling

Although tooling may be classified in many ways, such as "durable or perishable tooling," "fabrication or assembly tooling," and "company-owned or customer-owned tooling," the fabrication tooling system described below basically classifies tooling as "standard tooling" or "special tooling."

Definition of Terms
Standard tooling: Standard tooling is defined as that which is basically off the shelf and may be used by different users or a variety of products. Standard tooling is usually produced in quantity, and the cost is relatively low.

Special tooling: Special tooling is that which is designed and built for a specific application, such as a specific product or family of products. Delivery on such tooling may be several weeks, and tooling costs are relatively high.

Examples
Examples of standard tooling are shown in Fig. 27. Standard tooling usually includes cutting tools, die components, nozzles, certain types of electrodes, rollers, brushes, tool holders, laps, chucks, mandrels, collets, centers, adapters, arbors, vises, step-blocks, parallels, angle-plates, and the like.

Examples of special tooling are shown in Fig. 28. This usually includes dies, molds, patterns, jigs, fixtures, cams, templates, N/C programs, and the like. Some standard tooling

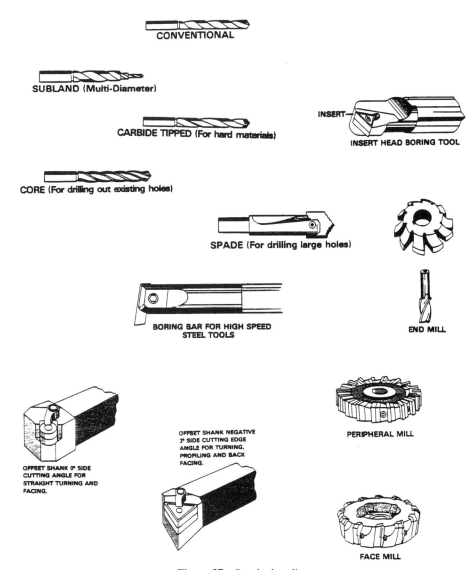

Figure 27 Standard tooling.

may be modified to perform a special function. When this modification is performed in accordance with a specified design, then the tooling is classified as special tooling.

5.3 Tooling Taxonomy

The tooling taxonomy is based on the same general classification system used for fabrication processes and for fabrication equipment. The first-level divisions are "shaping" and "nonshaping." The second-level divisions for shaping are "mass reduction," "mass conserving," and "mass increasing" (joining, laminating, etc.). Second-level divisions for nonshaping are

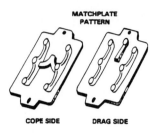

Figure 28 Special tooling.

"heat treatment" and "surface finishing." Third-level subdivisions are more variable but include "mechanical," "chemical," and "thermal," among other criteria for subdivisions.

Rationale
The basic philosophy has been to create a tooling classification system that is related to fabrication processes, to fabrication equipment, and to fabrication products insofar as possible. The statement "Without the process, there is no product" has aided in clarifying the importance of a process classification to all phases of manufacturing. It was recognized early that "process" is really a concept and that only through the application of "equipment and tooling" could a process ever be implemented. Thus, this process taxonomy was used as the basis of both equipment classification and this tooling classification. Most tooling is used in conjunction with given families of equipment and in that way is related to the equipment taxonomy. Special tooling is also related to the workpiece geometry through a special coding system that will be explained later. Standard tooling may be applied to a number of product families and may be used on a variety of different machine tools.

5.4 Tool Coding

The tool code is a shorthand notation used for identification and communication purposes. It has been designed to provide the maximum amount of information in a short, flexible code. Complete tool information may be held in a computer database or charts and tables. The code provides a pointer to this information.

Tool Code Format
The format used for the tool code is shown in Fig. 29. The code consists of three basic fields of information. The first field contains a three-digit process code that identifies the process for which the tooling is to be used. The second field consists of a one-digit code that indicates the tool type. Tool types are explained in the next subsection. The last field

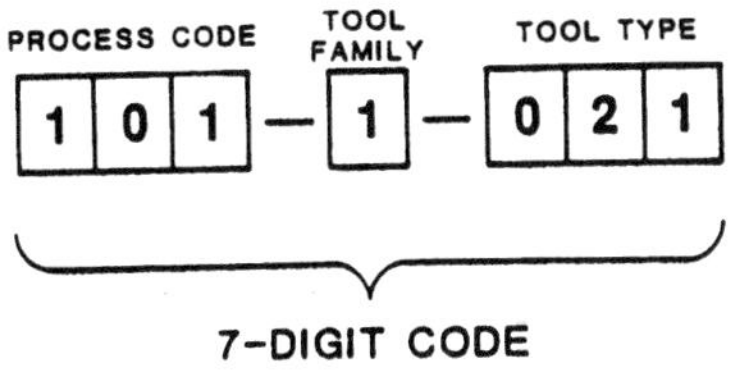

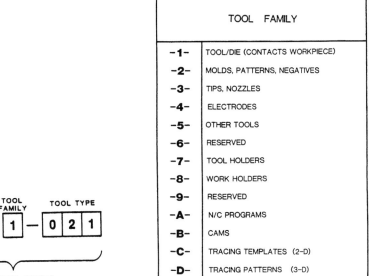

Figure 29 Fabrication tool code format.

TOOL FAMILY	
-1-	TOOL/DIE (CONTACTS WORKPIECE)
-2-	MOLDS, PATTERNS, NEGATIVES
-3-	TIPS, NOZZLES
-4-	ELECTRODES
-5-	OTHER TOOLS
-6-	RESERVED
-7-	TOOL HOLDERS
-8-	WORK HOLDERS
-9-	RESERVED
-A-	N/C PROGRAMS
-B-	CAMS
-C-	TRACING TEMPLATES (2-D)
-D-	TRACING PATTERNS (3-D)

Figure 30 Tool types.

consists of either a three-digit numeral code for standard tooling or a three-digit alphanumeric code for special tooling. Standard tool codes have been designed to accommodate further subdivision of tool families if so desired. For example, the tool code for single-point turning inserts is 101-1-020. The last three digits could be amplified for given insert geometry (e.g., triangular, −021; square, −022; round, −023). A dash number may further be appended to these codes to uniquely identify a given tool, as shown on the tool specification sheets.

Special tool codes are identified in the charts by a box containing three small squares. It is intended that the first three digits of this part family code will be inserted in this box, thus indicating the basic shape family for which the tool is designed. This way it will be possible to identify tool families and benefit from the application of group technology principles.

Tool Types

A single-digit alphanumeric code is used to represent the tool type. A code type of −1, for example, indicates that the tool actually contacts the workpiece, while a −2 indicates that the tool is used indirectly in shaping the workpiece. A foundry patten is an example of this: the pattern creates this mold cavity into which molten metal is introduced. The various tool type codes are shown in Fig. 30.

5.5 Tool Specification Sheets

The tool classification system is used to identify the family to which a tool belongs. The tool specification sheet is used to describe the attributes of a tool within the family. Figure 31 shows a sample tool specification sheet for a standard tool. Special tooling is best described with tool drawings and will not be discussed further.

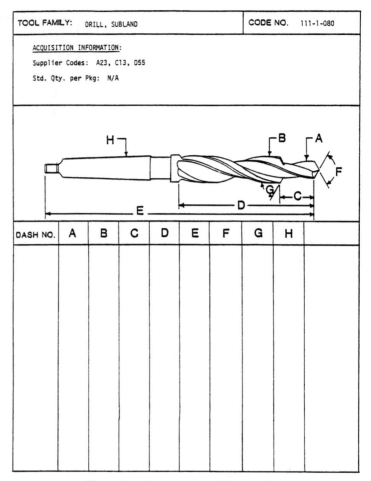

Figure 31 Sample tool specification sheet.

Tool Identification
Tool identification consists of the tool name and the tool code number. The general tool family name is written first, followed by a specific qualifying label if applicable (e.g., "drill, subland, straight-shank"). The tool code consists of the seven-digit code described previously.

Acquisition Information
Information contained in this acquisition section may contain identifying codes for approved suppliers. This section may also contain information relating to the standard quantity per package, if applicable, special finish requirements, or other pertinent information.

Tool Sketches
Tool sketches are a valuable feature of the tool specification sheet. Prominent geometric relationships and parameters are shown in the sketch. Information relative to interfacing this tool to other devices or adapters should also be shown. This may include type and size or shape, key slot size, and tool capacity.

Tool Parameters

Tool parameters are included on the tool specification sheet to aid selection of the most appropriate tools. Because it is expensive to stock all possible tools, the usual practice is to identify preferred tools and to store these. Preferred tools may be so indicated by a special symbol such as an asterisk (*) in the dash-number column.

Tool parameters must be selected that are appropriate for each tool family and that match product requirements. Tool parameters must also be identified that will aid in interfacing tools with fabrication equipment, as was explained earlier. Typical tool parameters for a subland drill are shown in Fig. 31.

CHAPTER 4
PRODUCTION PLANNING

Bhaba R. Sarker
Dennis B. Webster
Thomas G. Ray
Department of Industrial Engineering
Louisiana State University
Baton Rouge, Louisiana

1	**INTRODUCTION**	110
2	**FORECASTING**	111
	2.1 General Concepts	111
	2.2 Qualitative Forecasting	112
	2.3 Causal Methods	115
	2.4 Methods of Analysis of Time Series	117
	2.5 Forecasting Error Analysis	120
	2.6 Conclusions on Forecasting	120
3	**INVENTORY MODELS**	120
	3.1 General Discussion	120
	3.2 Types of Inventory Models	122
	3.3 The Modeling Approach	123
4	**AGGREGATE PLANNING— MASTER SCHEDULING**	131
	4.1 Alternative Strategies to Meet Demand Fluctuations	131
	4.2 Aggregate Planning Costs	131
	4.3 Approaches to Aggregate Planning	132
	4.4 Levels of Aggregation and Disaggregation	133
	4.5 Aggregate Planning Dilemma	134
5	**MATERIALS REQUIREMENTS PLANNING**	135
	5.1 Procedures and Required Inputs	135
	5.2 Lot Sizing Techniques	139
	5.3 Beyond MRP	144
6	**JOB SEQUENCING AND SCHEDULING**	144
	6.1 Structure of the General Sequencing Problem	145
	6.2 Single Machine Problem	146
	6.3 Flow Shops	148
	6.4 Job Shops	150
	6.5 Heuristics/Priority Dispatching Rules	152
	6.6 Assembly Line Balancing	156
7	**JAPANESE MANUFACTURING PHILOSOPHY**	162
	7.1 Just-in-Time Philosophy/Kanban Mechanism	162
	7.2 Time-Based Competition	165
8	**SUPPLY CHAIN MANAGEMENT**	166
	8.1 Distribution Logistics	167
	8.2 Applications of Kanban Mechanism to Supply Chain	168
	8.3 General Remarks	169
	REFERENCES	169
	BIBLIOGRAPHY	171

1 INTRODUCTION

The more manufacturing changes, the more it stays the same. Certainly the rapid introduction of newer technology and better approaches to management have led to unprecedented increases in productivity, but at its heart, the objective of manufacturing is still to provide the right product in the right quantity at the right time with the right quality at the right price to its customers.

The topics discussed in this chapter are related to how manufacturing organizations strive to meet this objective. Forecasting provides the manufacturer with a basis for anticipating customer demand so as to have adequate product on hand when it is demanded. Of course, the preferred approach would be to wait for an order and then produce and ship immediately when the order arrives. This approach is, for practical purposes, impossible for products with any significant lead time in manufacturing, raw materials, or component supply. Consequently, most manufacturing facilities develop raw materials, in-process, and finished goods inventories, which have to be established and managed.

Aggregate planning approaches establish overall production requirements, and materials requirements planning techniques provide a methodology for ensuring that adequate inventory is available to complete the work required on products needed to meet forecasted customer demands. Job-sequencing methodologies are used to develop shop schedules for production processes to reduce the time for manufacturing products or meet other performance objectives. Although the basic functions in manufacturing are the same, the manner in which they are implemented drastically affects the effectiveness of the outcome. Improvements in technology have greatly increased the productivity that is achieved. Likewise, significant increases in reduction of costs and improved customer service are being achieved by changing management philosophies. Flowing from the *just-in-time* (JIT) concepts developed by the Japanese are practices designated as lean and/or agile manufacturing, and the use of enterprise resource planning. More recently, the term supply chain management has been used to describe the inherent linkages among all of the functions of a manufacturing enterprise. The following materials are presented to introduce the reader to what these terms mean and how these approaches are related.

2 FORECASTING

2.1 General Concepts

The function of production planning and control is based on establishing a plan, revising the plan as required, and adhering to the plan to accomplish desired objectives. Plans are based on a forecast of future demand for the related products or services. Good forecasts are a requirement for a plan to be valid and functionally useful. When managers are faced with forecasts, they need to plan what actions must be taken to meet the requirements of the forecast. The actions taken thusly prepare the organization to cope with the anticipated future state of nature that is predicated upon the forecast.[1-3]

Forecasting methods are traditionally grouped into one of three categories: qualitative techniques, time-series analysis, or causal methods. The qualitative techniques are normally based on opinions or surveys. The basis for time-series analysis is historical data and the study of trends, cycles, and seasons. Causal methods are those that try to find relationships between independent and dependent variables, determining which variables are predictive of the dependent variable of concern. The method selected for forecasting must relate to the type of information available for analysis.

Definitions
Deseasonalization: The removal of seasonal effects from the data for the purpose of further study of the residual data.
Error analysis: The evaluation of errors in the historical forecasts done as a part of forecasting model evaluation.
Exponential smoothing: An iterative procedure for the fitting of polynomials to data for use in forecasting.
Forecast: Estimation of a future outcome.
Horizon: A future time period or periods for which a forecast is required.

Index number: A statistical measure used to compare an outcome, which is measured by a cardinal number with the same outcome in another period of time, geographic area, profession, etc.

Moving average: A forecasting method in which the forecast is an average of the data for the most recent n periods.

Qualitative forecast: A forecast made without using a quantitative model.

Quantitative forecast: A forecast prepared by the use of a mathematical model.

Regression analysis: A method of fitting a mathematical model to data by minimizing the sums of the squares of the data from a theoretical line.

Seasonal data: Data that cycle over a known seasonal period such as a year.

Smoothing: A process for eliminating unwanted fluctuations in data, which is normally accomplished by calculating a moving average or a weighted moving average.

Time-series analysis: A procedure for determining a mathematical model for data that are correlated with time.

Time-series forecast: A forecast prepared with a mathematical model from data that are correlated with time.

Trend: Underlying patterns of movement of historic data that becomes the basis for prediction of future forecasts.

2.2 Qualitative Forecasting

These forecasts are normally used for purposes other than production planning. Their validity is more in the area of policy making or in dealing with generalities to be made from qualitative data. Among these techniques are the Delphi method, market research, consensus methods, and other techniques based on opinion or historical relationships other than quantitative data. The Delphi method is one of a number of nominal group techniques. It involves prediction with feedback to the group that gives the predictor's reasoning. Upon each prediction the group is again polled to see if a consensus has been reached. If no common ground for agreement has occurred the process continues moving from member to member until agreement is reached. Surveys may be conducted of relevant groups and their results analyzed to develop the basis for a forecast. One group appropriate for analysis is customers. If a company has relatively few customers this select number can be an effective basis for forecasting. Customers are surveyed and their responses combined to form a forecast. Many other techniques are available for nonquantitative forecasting. An appropriate area to search if these methods seem relevant to a subjective problem at hand is the area of *nominal group techniques*. Quantitative forecasting involves working with numerical data to prepare a forecast. This area is further divided into two subgroups of techniques according to the data type involved. If historical data are available and it is believed that the dependent variable to be forecast relates only to time, an approach called time-series analysis is used. If the data available suggest relationships of the dependent variable to be forecast to one or more independent variables then the techniques used fall into the category of causal analysis. The most commonly used method in this group is regression analysis.

Moving Average

A moving average can normally be used to remove the seasonal or cyclical components of variation. This removal is dependent on the choice of a moving average that contains sufficient data points to bridge the season or cycle. For example, a five-period centered moving average should be sufficient to remove seasonal variation from monthly data. A disadvantage to the use of moving averages is the loss of data points due to the inclusion of multiple points into the calculation of a single point.

Example 1 Computation of Moving Average. A five-year *simple* moving average in column 4 of Table 1 represents the forecast based on data of five recent past periods. Note that the five-year *centered* moving average lost four data points—two on each end of the data series (last two columns in Table 1). Observation of the moving average indicated a steady downward trend in the data. The raw data had fluctuations that might tend to confuse an observer initially due to the apparent positive changes from time to time.

Weighted Moving Average
A major disadvantage of the moving average method—the effect of extreme data points—can be overcome by using a weighted moving average for N periods (Table 2). In this average the affect of the extreme data points may be decreased by weighing them less that the data points at the center of the group. There are many ways for this to be done.

One method would be to weigh the center point of the N-period (in this case a five-period) average as 50% of the total with the remaining points weighted for the remaining 50%. For $N = 5$, the total is $60.0 + 56.6 + 53.0 + 54.6 + 51.2 = 275.3$ and the five-period total less the centered value is $275.3 - 53.0 = 222.2$. Hence, weighted average is $0.5(222.3/4) + 0.5(53.0) = 54.3$. Similar to the calculation for Example 1, this would yield the results as shown in Table 2.

Example 2 Weighted Moving Average. Table 3 displays the two forecasts. The results are comparable to the weighted average forecast, distinguishing a slight upswing from period 5 to period 6, which was ignored by the moving average method.

Exponential Smoothing
This method determines the forecast (F) for the next period as the weighted average of the last forecast and the current demand (D). The current demand is weighted by a constant, α, and the last forecast is weighted by the quantity $1 - \alpha$ ($0 \leq \alpha \leq 1$). New forecast = α (demand for current period) + $(1 - \alpha)$ forecast for current period. This can be expressed symbolically as

Table 1 Moving Average Computation

Year	Data	5-Year Moving Total	5-Year Moving Average	5-Year Centered Moving Total	5-Year Centered Moving Average
1	60.0	—	—	—	—
2	56.5	—	—	—	—
3	53.0	—	—	275.3	55.06
4	54.6	—	—	269.2	53.24
5	51.2	—	—	261.1	52.20
6	53.9	275.3	55.06	257.2	51.44
7	48.4	269.2	59.24	250.9	50.18
8	49.1	261.1	52.20	242.1	48.42
9	48.3	257.2	51.44	232.8	46.56
10	42.4	250.9	50.18	—	—
11	44.6	242.1	48.42	—	—

Table 2 Weighted Moving Average

Year	Data	5-Year Moving Total	5-Year Total Less the Center Value	Weighted Average 0.5(Col 4)/4 + 0.5(Col 2)
1	60.0	—	—	—
2	56.5	—	—	—
3	53.0	275.3	222.3	54.3
4	54.6	269.2	214.6	54.1
5	51.2	261.1	209.9	51.8
6	53.9	257.2	203.3	52.4
7	48.4	250.9	202.5	49.5
8	49.1	242.1	193.0	48.7
9	48.3	232.8	184.5	47.2
10	42.4	—	—	—
11	44.6	—	—	—

$$F_t = \alpha D_{t-1} + (1 - \alpha) F_{t-1} \tag{1}$$

The forecast F_t is the one-step ahead forecast for the period t made in period $t - 1$. Using the similarity as in Eq. (1), we can write

$$F_{t-1} = \alpha D_{t-2} + (1 - \alpha) F_{t-2} \tag{2}$$

Substituting Eq. (2) in Eq. (1), we have

$$F_t = \alpha D_{t-1} + \alpha(1 - \alpha) D_{t-2} + (1 - \alpha)^2 F_{t-2} \tag{3}$$

$$F_t = \alpha D_{t-1} + \alpha(1 - \alpha) D_{t-2} + \alpha(1 - \alpha)^2 D_{t-3} + (1 - \alpha)^3 F_{t-3} \tag{4}$$

and, in general,

$$F_t = \sum_{i=0}^{\infty} \alpha(1 - \alpha)^i D_{t-i-1} = \sum_{i=0}^{\infty} a^i D_{t-i-1} \tag{5}$$

Obviously, the exponential weights are $\alpha(1 - \alpha)^i = a_0 > a_1 > a_2 > \cdots > a_{i-1} > a_i$, where $\sum_{i=1}^{\infty} a_i = \sum_{i=1}^{\infty} \alpha(1 - \alpha)^i = 1$, indicating that exponential smoothing technique applies a declining set of weights to all past data. For further treatment of the exponential smoothing techniques, readers may refer to Nahmias[4] or Bedworth and Bailey.[1]

Table 3 Forecasts by Simple Moving Average and Weighted Moving Average Methods

Period	Moving Average Forecast	Weighted Average Forecast
3	55.1	54.3
4	53.8	54.1
5	52.2	51.8
6	51.4	52.4
7	50.2	49.5
8	48.4	48.7
9	46.6	47.2

Normally the forecast for the first period is taken to be the actual demand for that period (i.e., forecast and demand are the same for the initial data point). The smoothing constant is chosen as a result of analysis of error by a method such as mean absolute deviation coupled with the judgment of the analyst. A high value of α makes the forecast very responsive to the occurrence in the last period. Similarly, a small value would lead to a lack of significant response to the current demand. Evaluations must be made in light of the cost effects of the errors to determine what value of α is best for a given situation. Example 3 shows the relationship between actual data and forecasts for various values of α.

Example 3 Exponential Smoothing. Table 4 provides the forecasts by exponential smoothing method using different values of α.

2.3 Causal Methods

This category of methods falls within the second group of quantitative forecasting methods mentioned earlier. These methods assume that there are certain factors that have a cause–effect relationship with the outcome of the quantity to be forecast and that knowledge of these factors will allow a more accurate prediction of the dependent quantity. The statistical models of regression analysis fall within this category of forecasting.

Basic Regression Analysis
The simplest model for regression analysis is the linear model. The basic approach involves the determination of a theoretical line that passes through a group of data points that appear to follow a linear relationship. The desire of the modeler is to determine the equation for the line that would minimize the sums of the squares of the deviations of the actual points from the corresponding theoretical points. The values for the theoretical points are obtained by substituting the values of the independent variable, Y_i, into the functional relationship

$$\hat{Y}_i = a + bx_i \qquad (6)$$

The difference between the data and the forecasted value of point i is

$$e_i = Y_i - \hat{Y}_i \qquad (7)$$

Squaring this value and summing the relationship over the N related points yields the total error, E, as

Table 4 Forecasts F_t for Various α Values by Exponential Method

Period	Demand	$\alpha = 0.1$	$\alpha = 0.2$	$\alpha = 0.3$
1	85	85.0	85.0	85.0
2	102	85.0	85.0	85.0
3	110	86.7	88.4	90.1
4	90	89.0	92.7	96.1
5	105	89.1	92.2	94.3
6	95	90.7	94.8	97.5
7	115	91.1	94.8	96.8
8	120	93.5	98.8	102.3
9	80	96.2	103.0	107.6
10	95	94.6	98.4	99.3

$$E = \sum_{i=1}^{N} (Y_i - \hat{Y}_i)^2 \tag{8}$$

Substituting the functional relationship for the forecasted value of Y gives

$$E = \sum_{i=1}^{N} (Y_i - \hat{a} - \hat{b}x_i)^2 \tag{9}$$

By using this relationship, taking the partial derivatives of E with respect to a and b, and solving the resulting equations simultaneously, we obtain the normal equations for least squares for the linear regression case:

$$\Sigma Y = aN + b\Sigma X$$

$$\Sigma XY = a\Sigma X + b\Sigma X^2$$

Solving these equations yields values for a and b:

$$b = \frac{N\Sigma XY - \Sigma X \Sigma Y}{N\Sigma X^2 - (\Sigma X)^2} \tag{10}$$

and

$$a = \overline{Y} - b\overline{X} \tag{11}$$

The regression equation is then $Y_i = a + bx_i$, and the correlation coefficient, r, which gives the relative importance of the relationship between x and y, is

$$r = \frac{N\Sigma XY - \Sigma X \Sigma Y}{\sqrt{[N\Sigma X^2 - (\Sigma X)^2]}\sqrt{[N\Sigma Y^2 - (\Sigma Y)^2]}} \tag{12}$$

This value of r can range from $+1$ to -1. The plus sign would indicate a positive correlation (i.e., large values of x are associated with large values of y; a negative correlation implies that large values of x are associated with small values of y) and the negative sign would imply a negative correlation.

Example 4 Simple Linear Regression. From the X and Y data set in Table 5, and using the Eqs. (10)–(12), the following computational results yield

$$b = \frac{3(47) - 9(15)}{3(29) - 81} = 1 \quad \text{and} \quad a = 5 - 1(3) = 2$$

and the linear forecast model is $Y = 2 + X$. The correlation coefficient is

Table 5 Linear Forecasting

	Y	X	XY	X²	Y²
	4	2	8	4	16
	5	3	15	9	25
	6	4	24	16	36
Σ	15	9	47	29	77
Mean	5	3	—	—	—

$$r = \frac{3(47) - 9(15)}{\sqrt{[3(29) - 81]}\sqrt{[3(77) - 225]}} = 1 \tag{13}$$

which indicates that the *X-Y* data are 100% positively correlated.

Quadratic Regression
This regression model is used when the data appear to follow a simple curvilinear trend and the fit of a linear model is not adequate. The procedure for deriving the normal equations for quadratic regression is similar to that for linear regression. The quadratic model has three parameters that must be estimated, however: the constant term, a; the coefficient of the linear term, b; and the coefficient of the square term, c. The model is

$$Y_i = a + bx_i + cx_i^2 \tag{14}$$

Its normal equations are

$$\Sigma Y = Na + b\Sigma X + c\Sigma X^2$$
$$\Sigma XY = a\Sigma X + b\Sigma X^2 + c\Sigma X^3$$
$$\Sigma X^2 Y = a\Sigma X^2 + b\Sigma X^3 + c\Sigma X^4$$

The normal equations for least squares for a cubic curve, quartic curve, etc. can be generalized from the expressions for the linear and quadratic models.

2.4 Methods of Analysis of Time Series

We now discuss in general several methods for analysis of time series. These methods provide ways of removing the various components of the series, isolating them, and providing information for their consideration should it be desired to reconstruct the time series from its components.

The movements of a time series are classified into four types: long-term or *trend* movements, *cyclical* movements, *seasonal* movements, and *irregular* movements. Each of these components can be isolated or analyzed separately. Various methods exist for the analysis of the time series. These methods decompose the time series into its components by assuming that the components are either multiplicative or additive. If the components are assumed to be multiplicative, the following relationship holds:

$$Y = T \times C \times S \times I \tag{15}$$

where Y is the outcome of the time series, T is the trend value of the time series, and C, S, and I are indices, respectively, for cyclical, seasonal, and irregular variations.

To process data for this type of analysis it is best to first plot the raw data to observe its form. If the data are yearly, they need no deseasonalization. If they are monthly or quarterly data, they can be converted into yearly data by summing the data points that would add to a year before plotting. (Seasonal index numbers can be calculated to seasonalize the data later if required.) By plotting yearly data the period of apparent data cycles can be determined or approximated. A centered moving average of appropriate order can be used to remove the cyclical effect in the data. Further, cyclical indices can be calculated when the order of the cycle has been determined. At this point the data contain only the trend and irregular components of variation. Regression analysis can be used to estimate the trend component of the data, leaving only the irregular, which is essentially a forecasting error.

118 Production Planning

Index numbers are calculated by grouping data of the same season together, calculating the average over the season for which the index is to be prepared, and then calculating the overall average of the data over each of the seasons. Once the seasonal and overall averages are obtained, the seasonal index is determined by dividing the seasonal average by the overall average.

Example 5 Average Forecast. A business has been operational for 24 months. The sales data in thousands of dollars for each of the monthly periods are given in Table 6. The overall total is $282 + 302 = 584$ and the average is $584/24 = 24.333$. The index for January would be

$$I_{Jan} = 0.5\ (20 + 24)/24.333 = 0.904$$

For the month of March the index would be

$$I_{Mar} = 0.5\ (28 + 30)/24.333 = 1.191$$

To use the index, a trend value for the year's sales would be calculated, the average monthly sales would be obtained, and then this figure would be multiplied by the index for the appropriate month to give the month's forecast.

Note that a season can be defined as any period for which data are available for appropriate analysis. If there are seasons within a month, i.e., four weeks in which the sales vary considerably according to a pattern, a forecast could be indexed within the monthly pattern also. This would be a second indexing within the overall forecast. Further, seasons could be chosen as quarters rather than months or weeks. This choice of the period for the analysis is dependent on the requirements for the forecast.

Data given on a seasonal basis can be deseasonalized by dividing them by the appropriate seasonal index. Once this has been done they are labeled *deseasonalized data*. These data still contain the trend, cyclical, and irregular components after this adjustment.

Seasonal Forecasts

Given a set of quarterly data for three years as given in Table 7, four more tables (Tables 7a–d)] are generated for seasonal and yearly average, seasonal indices, deseasonalized forecast, and seasonalized forecasts.

Table 6 Sales Data for Two Years

Month	Year 1	Year 2	Total	Average	Index
January	20	24	44	22.0	0.904
February	23	27	50	25.0	1.026
March	28	30	58	29.0	1.191
April	32	35	67	33.5	1.375
May	35	36	71	35.5	1.456
June	26	28	54	27.0	1.117
July	25	27	52	26.0	1.066
August	23	23	46	23.0	0.944
September	19	17	36	18.0	0.737
October	21	22	43	21.5	0.882
November	18	19	37	18.5	0.750
December	12	14	26	13.5	0.552
TOTAL	282	302	584	292[a]	12.014

[a] Monthly average = 24.333.

2 Forecasting

Table 7 Seasonal Sales Data

Year	Q_1	Q_2	Q_3	Q_4
1	520	730	820	530
2	590	810	900	600
3	650	900	1000	650

Table 7a Seasonal Data

Year	Q_1	Q_2	Q_3	Q_4	Total	Average
1	520	730	820	530	2600	650
2	590	810	900	600	2900	725
3	650	900	1000	650	3200	800
Total	1760	2440	2720	1780	—	—
Average	586.0	813.6	906.7	593.3	—	—

Table 7b Seasonal Index

Year	Q_1	Q_2	Q_3	Q_4	Total	Average
1	0.800[1]	1.123	1.261	0.815	4.0	—
2	0.813[2]	1.117	1.241	0.828	4.0	—
3	0.812	1.125	1.250	0.812	4.0	—
Total	2.425	3.365	3.752	2.455	—	—
Average	0.808	1.122	1.251	0.818	4.0	—

1: 520/650 = 0.800, 2: 590/725 = 0.813

Table 7c Deseasonalized Data (Unadjusted) and Forecast

Year, t	Q_1	Q_2	Q_3	Q_4	Total	Average
1	643[1]	650[3]	655	648	2596	649
2	730[2]	720	719	733	2902	725
3	804	802	799	794	3199	800
4						876.67
5		$F_t = 572.67 + 76t$				952.57

1: 520/0.808 = 643, 2: 590/0.808 = 730, 3: 730/1.125 = 650

Table 7d Reseasonalized Forecast (Adjusted) from Deseasonalized Data

Year, t (Index)	Q_1 (0.808)	Q_2 (1.122)	Q_3 (1.251)	Q_4 (0.818)	Total	Average
4	707[1]	982[3]	1095	716	—	876.67
5	769[2]	1068	1191	778	—	952.57

1: 876.67 × 0.808 = 707, 2: 952.67 × 0.808 = 769, 3: 876.67 × 1.122 = 982

2.5 Forecasting Error Analysis

One common method of evaluating of forecast accuracy is termed mean absolute deviation (MAD) from the procedure used in its calculation. For each available data point a comparison of the forecasted value is made to the actual value. The absolute value of the differences is calculated. This absolute difference is then summed over all values and its average calculated to give the evaluation:

$$\text{MAD} = \frac{\text{Sum of the absolute deviations}}{\text{Number of deviations}} = \frac{1}{N} \sum |Y_i - \hat{Y}_i| \quad (16)$$

Alternative forecasts can be analyzed to determine the value of MAD and a comparison made using this quantity as an evaluation criteria. Other criteria can also be calculated. Among these are the mean square of error (MSE) and the standard error of the forecast (S_{xy}). These evaluation criteria are calculated as

$$\text{MSE} = \frac{1}{N} \sum_{i=1}^{N} (Y_i - \hat{Y}_i)^2 \quad (17)$$

and

$$S_{xy} = \frac{\sqrt{\sum Y^2 - a\sum Y - b\sum XY}}{N - 2} \quad (18)$$

In general, these techniques are used to evaluate the forecast and then the results of the various evaluations together with the data and forecasts are studied. Conclusions may then be drawn as to which method is preferred or the results of the various methods compared to determine what they in effect distinguish.

2.6 Conclusions on Forecasting

A number of factors should be considered in choosing a method of forecasting. One of the most important factors is *cost*. The problem of valuing an accurate forecast is presented. If the question "how will the forecast help and in what manner it will save money?" can be answered, a decision can be made regarding the allocation of a percentage of the savings to the cost of the forecasting process. Further, concern must be directed to the required *accuracy* of a forecast to achieve desired cost reductions. Analysis of past data and the testing of the proposed model using this historical data provide a possible scenario for hypothetical testing of the effects of cost of variations of actual occurrences from the plan value (forecast).

In many cases an inadequate database will prohibit significant analysis. In others the database may not be sufficient for the desired projection into the future. The answers to each of these questions are affected by the type of product or service for which the forecast is to be made as well as the value of the forecast to the planning process.

3 INVENTORY MODELS

3.1 General Discussion

Normally, items waiting to be purchased or sold are considered to be in inventory. One of the most pressing problems in the manufacturing and sale of goods is the control of this inventory. Many companies experience financial difficulties each year due to a lack of adequate control in this area. Whether it is raw material used to manufacture a product or products waiting to be sold, problems arise when too many or too few items are available. The greatest number of problems arises when too many items are held in inventory.

The primary factor in the reduction of inventory costs is deciding when to order, how much to order, and if back-ordering is permissible. Inventory control involves decisions by management as to the source from which the inventory is to be procured and as to the quantity to be procured at the time. This source could be from another division of the company handled as an intrafirm transfer, outside purchase from any of a number of possible vendors, or manufacture of the product in house.

The basic decisions to be made once a source has been determined are how much to order and when to order. Inherent in this analysis is the concept of demand. Demand can be known or unknown, probabilistic or deterministic, constant or lumpy. Each of these characteristics affects the method of approaching the inventory problem.

For the *unknown demand* case a decision must be made as to how much the firm is willing to risk. Normally, the decision would be to produce some k units for sale and then determine after some period of time to produce more or to discontinue production due to insufficient demand. This amounts to the reduction of the unknown demand situation to one of a lumpy demand case after the decision has been made to produce the batch of a finite size. Similarly, if a decision is made to begin production at a rate of n per day until further notice, the unknown demand situation has been changed to a constant known demand case.

Lumpy demand, or demand that occurs periodically with quantities varying, is frequently encountered in manufacturing and distribution operations. It is distinguished from the *known demand* case. This second case is that of a product which has historic data from which forecasts of demand can be prepared. A factor of concern in these situations is the lead time and the unit requirement on a periodic basis. The following are the major factors to be considered in the modeling of the inventory situation.

Demand is the primary stimulus on the procurement and inventory system and it is the justification for its existence. Specifically the system may exist to meet the demand of customers, the spare parts demand of an operational weapons system, the demand of the next step in a manufacturing process, etc. The characteristic of demand, although independent of the source chosen to replenish inventories, will depend on the nature of the environment giving rise to the demand.

The simplest demand pattern may be classified as deterministic. In this special case, the future demand for an item may be predicted with certainty. Demand considered in this restricted sense is only an approximation of reality. In the general case, demand may be described as a random variable that takes on values in accordance with a specific probability distribution.

Procurement quantity is the order quantity, which in effect determines the frequency of ordering and is related directly to the maximum inventory level.

Maximum shortage is also related to the inventory level.

Item cost is the basic purchase cost of a unit delivered to the location of use. In some cases delivery cost will not be included if that cost is insignificant in relation to the unit cost. (In these cases the delivery cost will be added to overhead and not treated as a part of direct material costs.)

Holding costs are incurred as a function of the quantity on hand and the time duration involved. Included in these costs are the real out-of-pocket costs, such as insurance, taxes, obsolescence, and warehouse rental and other space charges, and operating costs, such as light, heat, maintenance, and security. In addition, capital investment in inventories is unavailable for investment elsewhere. The rate of return forgone represents a cost of carrying inventory.

The inventory holding cost per unit of time may be thought of as the sum of several cost components. Some of these may depend on the maximum inventory level incurred. Others may depend on the average inventory level. Still others, like the cost of capital invested, will depend on the value of the inventory during the time period. The determination of holding cost per unit for a specified time period depends on a detailed analysis of each cost component.

Ordering cost is the cost incurred when an order is placed. It is composed of the cost of time, materials, and any expense of communication in placing an order. In the case of a manufacturing model it is replaced by *setup cost*. Setup cost is the cost incurred when a machine's tooling or jigs and fixtures must be changed to accommodate the production of a different part or product.

Shortage cost is the penalty incurred for being unable to meet a demand when it occurs. This cost does not depend on the source chosen to replenish the stock but is a function of the number of units short and the time duration involved.

The specific dollar penalty incurred when a shortage exists depends on the nature of the demand. For instance, if the demand is that of customers of a retail establishment, the shortage cost will include the loss of good will. In this case the shortage cost will be small relative to the cost of the item. If, however, the demand is that of the next step of a manufacturing process, the cost of the shortage may be high relative to the cost of the item. Being unable to meet the requirements for a raw material or a component part may result in lost production or even closing of the plant. Therefore, in establishing shortage cost, the seriousness of the shortage condition and its time duration must be considered.[5]

3.2 Types of Inventory Models

Deterministic models assume that quantities used in the determination of relationships for the model are all known. These quantities are such things as demand per unit of time, lead time for product arrival, and costs associated with such occurrences as a product shortage, the cost of holding the product in inventory, and that cost associated with placing an order for product.

Constant demand is one case that can be analyzed within the category of deterministic models. It represents very effectively the case for some components or parts in an inventory which are used in multiple parents, these multiple parent components having a composite demand which is fairly constant over time.

Lumpy demand is varying demand that occurs at irregular points in time. This type of demand is normally a dependent demand that is driven by an irregular production schedule affected by customer requirements. Although the same assumptions are made regarding the knowledge of related quantities, as in the constant demand case, this type of situation is analyzed separately under the topic of materials requirements planning (MRP). This separation of methodology is due to the different inputs to the modeling process in that the knowledge about demand is approached by different methods in the two cases.

Probabilistic models consider the same quantities as do the deterministic models but treat the quantities that are not cost related as random variables. Hence, demand and lead time have their associated probability distributions. The added complexity of the probabilistic values requires that these models be analyzed by radically different methods.

Definitions of Terms
The following terms are defined to clarify their usage in the material related to inventory that follows. Where appropriate, a literal symbol is assigned to represent the term.

Inventory (*I*): Stock held for the purpose of meeting a demand either internal or external to the organization.

Lead time (*L*): The time required to replenish an item of inventory by either purchasing from a vendor or manufacturing the item in-house.

Demand (*D*): The number of units of an inventory item required per unit of time.

Reorder point (*r*): The point at which an order must be placed for the procured quantity to arrive at the proper time or, for the manufacturing case, the finished product to begin flowing into inventory at the proper time.

Reorder quantity (*Q*): The quantity for which an order is placed when the reorder point is reached.

Demand during lead time (D_L): This quantity is the product of lead time and demand. It represents the number of units that will be required to fulfill demand during the time that it takes to receive an order that has been placed with a vendor.

Replenishment rate (*P*): This quantity is the rate at which replenishment occurs when an order has been placed. For a purchase situation it is infinite (when an order arrives, in an instant the stock level rises from 0 to *Q*). For the manufacturing situation it is finite.

Shortage: The units of unsatisfied demand that occur when there is an out-of-stock situation.

Back-order: One method of treating demand in a shortage situation when it is acceptable to the customer. (A notice is sent to the customer saying that the item is out of stock and will be shipped as soon as it becomes available.)

Lumpy demand: Demand that occurs in an aperiodic manner for quantities whose volume may or may not be known in advance. Constant demand models should normally never be used in a lumpy demand situation. The exception would be a component that is used for products that experience lumpy demand but itself experiences constant demand. The area of MRP (materials requirements planning) was developed to deal with the lumpy demand situations.

3.3 The Modeling Approach

Modeling in operations research involves the representation of reality by the construction of a model in one of several alternative ways. These models may be iconic, symbolic, or mathematical. For inventory models the latter is normally the selection of choice. The model is developed to represent a concept whose relationships are to be studied. As much detail can be included in a particular model as is required to effectively represent the situation. The detail omitted must be of little significance to its effect on the model. The model's fidelity is the extent to which it accurately represents the situation for which it is constructed.

Inventory modeling involves building mathematical models to represent the interactions of the variables of the inventory situation to give results adequate for the application at hand. In this section treatment is limited to deterministic models for inventory control. Probabilistic or stochastic models may be required for some analysis. References 2, 5, and 6 may be consulted if more sophisticated models are required.

General

Using the terminology defined above a basic logic model of the general case inventory situation will be developed. The objective of inventory management will normally be to determine an operating policy that will provide a means to reduce inventory costs. To reduce costs a determination must first be made as to what costs are present. The general model is as follows:

$$\text{Total cost} = \begin{pmatrix}\text{Cost of}\\ \text{items}\end{pmatrix} + \begin{pmatrix}\text{Cost of}\\ \text{ordering}\end{pmatrix} + \begin{pmatrix}\text{Cost of holding}\\ \text{items in stock}\end{pmatrix} + \begin{pmatrix}\text{Cost of}\\ \text{shortage}\end{pmatrix}$$

This cost is stated without a base period specified. Normally it will be stated as a per-period cost with the period being the same period as the demand rate (D) period.

Models of Inventory Situations

Purchase Model with Shortage Prohibited. This model is also known as a infinite replenishment rate model with infinite storage cost. This latter name results from the slope of the replenishment rate line (it is vertical) when the order arrives. The quantity on hand instantaneously changes from zero to Q. The shortage condition is preempted by the assignment of an infinite value to storage cost (see Fig. 1).

For this case, for unit purchase price of C_i dollars/unit, the item cost per period is symbolically $C_i D$. If the ordering cost is C_p dollars/order, the lot ordering cost is $(C_p D)/Q$. The shortage cost is zero since shortage is prohibited and the inventory holding cost is $(C_h Q)/2$. The equation for total cost is then

$$\text{TC}(Q) = C_i D + \left(\frac{D}{Q}\right)C_p + \left(\frac{Q}{2}\right)C_h \tag{19}$$

Analysis of this model reveals that the first component of cost, the cost of items, does not vary with Q. (Here we are assuming a constant unit cost; purchase discounts models are covered later.) The second component of cost, the cost of ordering, will vary on a per-period basis with the size of the order (Q). For larger values of Q, the cost will be smaller since fewer orders will be required to receive the fixed demand for the period. The third component of cost, cost of holding items in stock, will increase with increasing order size Q and conversely decrease with smaller order sizes. The fourth component of cost, cost of shortage, is affected by the reorder point. It is not affected by the order size and for this case shortage is not permitted.

This equation is essentially obtained by determining the cost of each of the component costs on a per cycle basis and then dividing that expression by the number of periods per cycle (Q/D). To obtain the extreme point(s) of the function it is necessary to take the

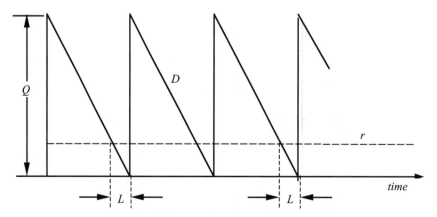

Figure 1 Basic inventory model with instantaneous replenishment.

derivative of TC(Q) with respect to Q, equate this quantity to zero, and solve for the corresponding value(s) of Q. This yields

$$0 = -C_p D/Q^2 + C_h/2.$$

or

$$\hat{Q} = \sqrt{\frac{2C_p D}{C_h}} \qquad (20)$$

and

$$L = DT \qquad (21)$$

Inspection of the sign of the second derivative of this function reveals that the extreme point is a minimum. This fits the objective of the model formulation. The quantity to be ordered at any point in time is then $\hat{Q}$ and the time to place the order will be when the inventory level drops to r (the units consumed during the lead time for receiving the order).

Purchase Model with Shortage Permitted. This model is also known as an infinite replenishment rate model with finite shortage costs (see Fig. 2). For this model the product cost and the ordering cost are the same as for the previous model:

$$C_i D + (C_p D)/Q$$

The holding cost is different, however. It is given by

$$C_h [Q - (DL - r)]^2 / 2Q$$

This represents the unit periods of holding per cycle times the holding cost per unit period. The unit periods of holding is obtained from the area of the triangle whose altitude is $Q - (DL - r)$ and whose base is the same quantity divided by the slope of the hypotenuse. In the same manner the unit periods of shortage is calculated. For that case the altitude is $(DL - r)$ and the base is $(DL - r)$ divided by D. The shortage cost component is then

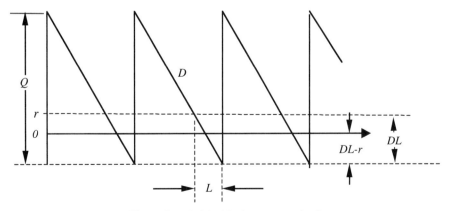

Figure 2 Model with shortage permitted.

$$C_s(DL - r)^2/2Q$$

The total cost per period is given by

$$TC(Q, DL - r) = C_iD + \left(\frac{D}{Q}\right)C_p + \left\{\frac{[Q - (DL - r)]^2}{2Q}\right\}C_h$$
$$+ \left\{\frac{[Q - (DL - r)]^2}{2Q}\right\}C_s \quad (22)$$

Note that the quantity $DL - r$ is used as a variable. This is done for the purposes of amplifying the equations that result when the partial derivatives are taken for the function. Taking these derivatives and solving the resulting equations simultaneously for the values of Q and $DL - r$ yields the following relationships:

$$\hat{Q} = \sqrt{\frac{(2C_pD)}{C_h} + \frac{C_pD}{C_s}} \quad (23)$$

$$\hat{L} = DL - \sqrt{\frac{2C_hC_pD}{C_s(C_h + C_s)}} \quad (24)$$

Manufacturing Model with Shortage Prohibited. This model is also known as a finite replenishment rate model with infinite storage costs. Figure 3 illustrates the situation, expressed as

$$\hat{Q} = \sqrt{\frac{2C_pD}{C_h(1 - D/P)}} \quad (25)$$

$$r = DL \quad (26)$$

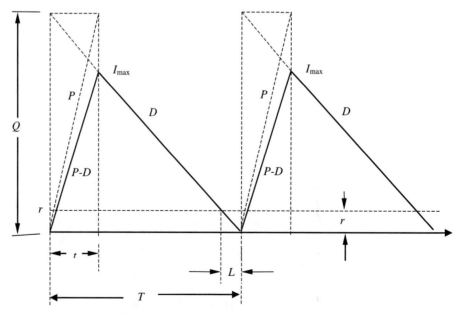

Figure 3 Manufacturing model with no shortage.

Manufacturing Model with Shortage Permitted. This model is also known as a finite replenishment rate model with finite shortage costs. It is the most complex of the models treated here as it is the general case model. All of the other models can be obtained from it by properly defining the replenishment rate and shortage cost. For example, the purchase model with shortage prohibited is obtained by defining the manufacturing rate and the storage cost as infinite. Upon doing this, the equations reduce to those appropriate for the stated situation (see Fig. 4). For this model the expressions for Q and r are

$$\hat{Q} = \sqrt{\frac{1}{1 - D/P}} \sqrt{\frac{2C_p D}{C_h} + \frac{2C_p D}{C_s}} \qquad (27)$$

$$\hat{r} = DL \sqrt{\frac{2C_p D(1 - D/P)}{C_s(1 + C_s/C_h)}} \qquad (28)$$

Models for Purchase Discounts

MODELS FOR PURCHASE DISCOUNTS WITH FIXED HOLDING COST. In this situation, the holding cost (C_w) is assumed to be fixed, not a function of unit costs. A supplier offers a discount for ordering a larger quantity. The normal situation is as shown in Table 8.

The decision maker must apply the appropriate economic order quantity (EOQ) purchase model, either finite or zero shortage (infinite storage) cost. Upon choice of the appropriate model the following procedure will apply:

1. Evaluate Q and calculate $TC(\hat{Q})$.
2. Evaluate $TC(q_{k+1})$, where $q_{k=1}$ is the smallest quantity in the price break interval above that interval where q lies.
3. If $TC(\hat{q}) < TC(q_{k+1})$ the ordering quantity will be $\hat{q}$. If not, go to step 4.

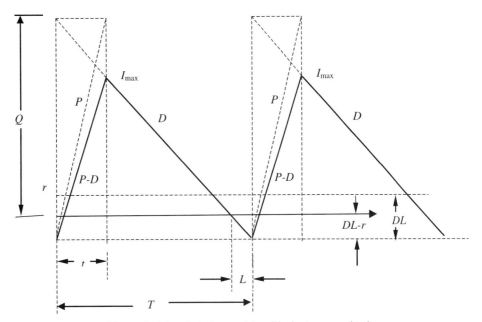

Figure 4 Manufacturing models with shortage permitted.

Table 8 Purchased Quantity for Different Price

Range of Quantity Purchased	Price, C_p
$1 - q_1$	C_1
$q_1 = 1 - q_2$	C_2
$q_2 = 1 - q_3$	C_3
—	—
$q_{m-1} = 1 - q_m$	C_m

4. Since the total cost of the minimum quantity in the next interval above that interval containing $\hat{q}$ is a basic amount, a evaluation must be made successively of total costs of the minimum quantities in the succeeding procurement intervals until one reflects on increase in cost or the last choice is found to be the minimum.

Example 6   Discounted Inventory Model. In a situation where shortage is not permitted the ordering cost is $50, the holding cost is $1 per unit year, and the demand is 10,000 units/year:

$$\hat{Q} = \sqrt{\frac{2C_p D}{C_W}} = \sqrt{\frac{2(\$50)10,000}{1}} = 1000 \text{ units}$$

$$TC(Q) = C_i D + \frac{C_n Q}{2} + \frac{D}{Q} C_p$$

$$TC(\hat{Q}) = \$20(10,000) + \$1\left(\frac{1000}{2}\right) + \$50\left(\frac{\$810,000}{1,000}\right) = \$201,000/\text{year}$$

The question is whether the smaller quantity in the next discount interval (1200–1799) gives a lower total cost (see Table 9):

$$TC(1200) = \$18(10,000) + \$1\left(\frac{1200}{2}\right) + \$50\left(\frac{10,000}{1800}\right) = \$181,033$$

Since this is a lower cost, an evaluation must be made of the smallest quantity in the next interval, 1800:

$$TC(1800) = 16.50(10,000) + \$1\left(\frac{1800}{2}\right) + \$50\left(\frac{10,000}{1800}\right) = \$166,175$$

Since there are no further intervals for analysis this is the lowest total cost and its associated q, 1800, should be chosen as the optimal $\hat{Q}$. The total cost function for this model is shown in Fig. 5.

Table 9 Discounted Price Range

Q	C_i
0–500	22.00
501–1199	20.00
1200–1799	18.00
1800–∞	16.50

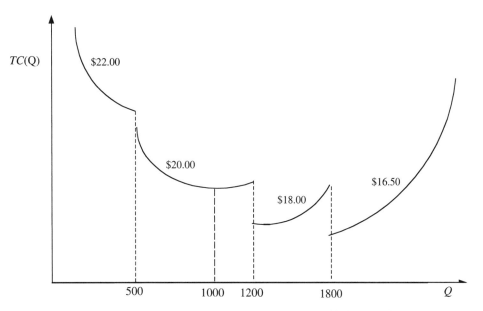

Figure 5 Total cost function for a quantity discount model with fixed holding cost.

QUANTITY DISCOUNT MODEL WITH VARIABLE HOLDING COST. In this case, the holding cost is variable with unit cost, i.e., $C_w = KC_i$. Again, the appropriate model must be chosen for shortage conditions. For the zero shortage case

$$\hat{Q} = \sqrt{\frac{2C_p D}{KC_i}} \qquad (29)$$

To obtain the optimal value of Q in this situation the following procedure must be followed:

1. Evaluate $\hat{Q}$ using the expression above and the item cost for the first interval.
    a. If the value of $\hat{Q}$ falls within the interval for C_i use this $TC(\hat{Q})$ for the smallest cost in the interval.
    b. If $\hat{Q}$ is greater than the maximum quantity in the interval, use Q_{max}, where Q_{max} is the greatest quantity in the interval, and evaluate $TC(Q_{max})$ as the lowest cost point in the interval.
    c. If $\hat{Q}$ is less than the smallest Q in the interval uses Q_{min}, where Q_{min} is the smallest quantity in the interval, as the best quantity and evaluate $TC(Q_{min})$.
2. For each cost interval follow the steps of part 1 of the procedure to evaluate best values in the interval.
3. Choose the minimum total cost from the applications of steps 1 and 2.

Example 7 Quantity Discounted Model with Variable Holding Cost. Using the same data as the Example 6, assume $C_h = 0.05C_1$:

$$\hat{Q} = \sqrt{\frac{2(50)10,000}{(0.05)(22)}} \cong 990 \text{ units/order}$$

Since $990 > 500$, the smallest cost in the interval will be at

$$TC(1000) = 22.00(10{,}000) + \$1\left(\frac{500}{2}\right) + \$50\left(\frac{10{,}000}{500}\right)$$

$$= 220{,}000 + 250 + 1000 = \$221{,}250$$

Using the second interval

$$\hat{Q} = \sqrt{\frac{2(50)10{,}000}{0.05(20)}} = 1000 \text{ units/order}$$

Since this value falls within the interval, TC(1000) is calculated

$$TC(1000) = 20(10{,}000) + \$1\left(\frac{1000}{2}\right) + 50\left(\frac{10{,}000}{1000}\right) = \$201{,}000$$

For the next interval (1201–1799), $C_3 = 18.00$ and

$$\hat{Q} = \sqrt{\frac{2(50)10000}{0.05(18.00)}} \approx 1050 \text{ units/order}$$

(which falls outside the interval to the left). Hence, the smallest quantity in the interval will be used. This is lower than either interval previously evaluated. For the next interval

$$\hat{Q} = \sqrt{\frac{2(50)10{,}000}{0.05(16.50)}} \cong 1101 \text{ units.}$$

Hence, the smallest quantity in the interval must be used:

$$TQ(1800) = 16.50(10{,}000) + \$1\left(\frac{1800}{2}\right) + \$50\left(\frac{10{,}000}{1800}\right)$$

$$= 165{,}000 + 900 + 272 = \$166{,}172$$

This is the last interval for evaluation and yields the lowest total cost; hence, it is chosen. Where the previous total cost function for fixed holding cost was a segmented curve with offsets, this is a combination of different curves, each valid over a specific range. In the case of the fixed holding cost, it had only one minimum point, yet the offsets in the total cost due to changes in applicable unit cost in an interval could change the overall minimum. In this situation, there are different values of $\hat{Q}$ for each unit price. The question becomes whether the value of $\hat{Q}$ falls within the price domain. If it is does, the total cost function is evaluated at that point; if not, a determination must be made as to whether the value of $\hat{Q}$ lies to the left or right of the range. If it is to the left the smallest value in the range is used to determine the minimum cost in the range. If it is to the right, the maximum value in the range is used.

Conclusions Regarding Inventory Models

The discussion here has covered only a small percentage of the class of deterministic inventory models, although these models represent a large percentage of applications. Should the models discussed here not adequately represent the situation under study, further research should be directed at finding a model with improved fidelity for the situation. Other models are covered in Refs. 2 and 7.

4 AGGREGATE PLANNING—MASTER SCHEDULING

Aggregate planning is the process of determining overall production, inventory, and work-force levels that are required to meet forecasted demand over a finite time horizon while trying to minimize the associated production, inventory, and work-force costs. Inputs to the aggregate planning process are forecasted demand for the products (either aggregated or individual); outputs from aggregate planning after desegregation into the individual products are the scheduled end products for the master production schedule. The time horizon for aggregate planning normally ranges from six to eighteen months, with a twelve-month average.

The difficulties associated with aggregate planning are numerous. Product demand forecasts vary widely in their accuracy; the process of developing a suitable aggregate measure to use for measuring the value or quantity of production in a multiple product environment is not always possible; actual production does not always meet scheduled production; unexpected events, including material shortages, equipment breakdowns, and employee illness, occur. Nevertheless, some form of aggregate planning is often required because seldom is there a match between the timing and quantity for product demand versus product manufacture. How the organization should staff and produce to meet this imbalance between production and fluctuating demand is what aggregate planning is about.[8]

4.1 Alternative Strategies to Meet Demand Fluctuations

Manufacturing managers use numerous approaches to meet changes in demand patterns for both short and intermediate time horizons. Among the more common are the following:

1. Produce at a constant production rate with a constant work force, allowing inventories to build during periods of low demand, and supplying demand from inventories during periods of high demand. This approach is used by firms with tight labor markets, but customer service may be adversely affected and levels of inventories may widely fluctuate between being excessively high to being out of stock.

2. Maintain a constant work force, but vary production within defined limits by using overtime, scheduled idle time, and potential subcontracting of production requirements. This strategy allows for rapid reaction to small or modest changes in production when faced with similar demand changes. It is the approach generally favored by many firms, if overall costs can be kept within reasonable limits.

3. Produce to demand letting the work force fluctuate by hiring and firing, while trying to minimize inventory carrying costs. This approach is used by firms that typically use low-skilled labor where the availability of labor is not an issue. Employee morale and loyalty, however, will always be degraded if this strategy is followed.

4.2 Aggregate Planning Costs

Aggregate planning costs can be grouped into one or more of the following categories:

1. *Production costs.* These costs include all of those items that are directly related to or necessary in the production of the product, such as labor and material costs. Supplies, equipment, tooling, utilities, and other indirect costs are also included, generally through the addition of an overhead term. Production costs are usually

divided into fixed and variable costs, depending on whether the cost is directly related to production volume.

2. *Inventory costs.* These costs include the same ordering, carrying, and shortage costs discussed in inventory models.

3. *Costs associated with work force and production rate changes.* These costs are in addition to the regular production costs and include the additional costs incurred when new employees are hired and exiting employees are fired or paid overtime premiums. They may also include costs when employees are temporarily laid off or given alternative work that underutilizes their skills, or production is subcontracted to an outside vendor.

4.3 Approaches to Aggregate Planning

Researchers and practitioners alike have been intrigued by aggregate planning problems, and numerous approaches have been developed over the decades. Although difficult to categorize, most approaches can be grouped as in Table 10.

Optimal formulations take many forms. Linear programming models are popular formulations and range from the very basic, which assume deterministic demand, a fixed work force, and no shortages, to complicated models that use piecewise linear approximations to quadratic cost functions, variable demand, and shortages.[9–13] The linear decision rule (LDR) technique was developed in an extensive project and is one of the few instances where the approach was implemented.[7,14–16] Nevertheless, due to the very extensive data collection, updating, and processing requirements to develop and maintain the rules, no other implementation has been reported. Lot size models usually are either of the capacitated (fixed capacity) or uncapacitated (variable capacity) variety.[17,18] Although a number of lot size models have been developed and refined, including some limited implementation, computational complexity constrains consideration to relatively small problems.[19] Goal programming models are attempts at developing more realistic formulations by including multiple goals and objectives. Essentially these models possess the same advantages and disadvantages of LP models, with the additional benefit of allowing trade-offs among multiple objectives.[20,21] Other optional approaches have modeled the aggregate planning problem using queueing,[22] dynamic programming,[23–25] and Lagrangian techniques.[26,27]

Nonoptional approaches have included the use of search techniques (ST), simulation models (SM), production switching heuristics (PSH), and management coefficient models (MCM). STs involve first the development of a simulation model that describes the system under study to develop the system's response under various operating conditions. A standard search technique is then used to find the parameter settings that maximize or minimize the desired response.[28,29] SMs also develop a model describing the firm, which is usually run

Table 10 Classification of Aggregate Planning Approaches

Original	Nonoptimal
Linear programming	Search techniques
Linear decision rule	Simulation models
Lot size models	Production switching heuristics
Goal programming	Management coefficient models
Other analytical aspects	—

Source. Modified from Ref. 7.

using a restricted set of schedules to see which performs best. SMs allow the development of very complex systems, but computationally may be so large as to disallow exhaustive testing.[30]

PSHs were developed to avoid frequent rescheduling of work force sizes and production rates. For example, the production rate P in period t is determined by Hwang and Cha[31]:

$$P_t = \begin{cases} L & \text{if } F_t - I_{t-1} < N - C \\ H & \text{if } F_t - I_{t-1} > N - A \\ N & \text{if otherwise} \end{cases} \quad (30)$$

where F_t = demand forecast for period t
I_{t-1} = net inventory level (inventory on hand minus backorder) at the beginning of period t
L = low level production rate
N = normal level production rate
H = high level production rate
A = minimum acceptable target inventory level
C = maximum acceptable target inventory level

Although this example shows three levels of production, fewer or more levels could be specified. The fewer the levels, the less rescheduling and vice versa. However, with more levels, the technique should perform better, because of its ability to better track fluctuations in demand and inventory levels. MCMs were developed by attempting to model and duplicate management's past behavior.[11] However, consistency in past performance is required before valid models can be developed, and it has been argued that if consistency is present, the model is not required.[32]

4.4 Levels of Aggregation and Disaggregation

It should be obvious from the previous discussion that different levels of aggregation and disaggregation can be derived from use of the various models. For example, many of the linear programming formulations assume aggregate measures for multiple production and demand units such as production hours, and provide output in terms of the number of production hours that must be generated per planning period. For the multiple-product situation, therefore, a scheduler at the plant level would have to disaggregate this output into the various products by planning periods to generate the master production schedule. However, if data were available to support it, a similar, albeit more complex, model could be developed that considered the individual products in the original formulation, doing away with the necessity of desegregation. This is not often done because of the increased complexity of the resultant model, the increased data requirements, and the increased time and difficulty in solving the formulation. Also, aggregate forecasts that are used as input to the planning process are generally more accurate than forecasts for individual products.

A major task facing the planner, therefore, is determining the level of aggregation and desegregation required. Normally this is determined by the following:

1. The decision requirements and the level of detail required. Aggregate planning at a corporate level is usually more gross than that done at a division level.
2. The amount, form, and quality of data available to support the aggregate planning process. The better the data, the better the likelihood that more complex models can be supported. Complex aggregate models may also require less desegregation.

3. The timing frequency and resources available to the planner. Generally, the more repetitive the planning, the simpler the approach becomes. Data and analysis requirements as well as analyst's capabilities significantly increase as the complexity of the approach increases.

4.5 Aggregate Planning Dilemma

Although aggregate planning (AP) models have been available since 1955 and many variations have been developed in the ensuing decades, few implementations of these models have been reported. Aggregate planning is still an important production planning process, but many managers are unimpressed by the modeling approach. Why is that? One answer is that aggregate planning occurs throughout the organizational structure, but is done by different individuals at different levels in the organization for different purposes. For example, a major aggregate planning decision is that of plant capacity, which is a constraint on all lower-level aggregate planning decisions. Determinations of if and when new plant facilities are to be added are generally corporate decisions and are made at that level. However, input for the decision comes from both division and plant levels. Division-level decision makers may then choose between competing plant facilities in their aggregate planning process in determining which plants will produce which quantity of which products within certain time frames, with input from the individual plant facilities. Plant-level managers may aggregate plan their production facilities for capacity decisions, but then must disaggregate these into a master production schedule for their facility that is constrained by corporate and division decisions.

Most models developed to date do not explicitly recognize that aggregate planning is a hierarchical decision-making process performed on different levels by different people. Therefore, AP is not performed by one individual in the organization, as implicitly assumed by many modeling approaches, but by many people with different objectives in mind.

Other reasons that have been given for the lack of general adoption of AP models include

1. The AP modeling approach is viewed as a top-down process, whereas many organizations operate AP as a bottom-up process.
2. The assumptions to use many of the models, such as linear cost structures, the aggregation of all production into a common measure, or all workers are equal, are too simplistic or unrealistic.
3. Data requirements are too extensive or costly to obtain and maintain.
4. Decision makers are intimidated by or unwilling to deal with the complexity of the models' formulations and required analyses.

Given this, therefore, it is not surprising that few modeling approaches have been adopted in industrial settings. Although research continues on AP, there is little to indicate any significant modeling breakthrough in the new future that will dramatically change this situation. One direction, however, is to recognize the hierarchical decision-making structure of AP, and to design modeling approaches that utilize it. These systems may be different for different organizations and will be difficult to design, but currently appear to be one approach for dealing with the complexity necessary in the aggregate planning process if a modeling approach is to be followed. For a comprehensive discussion of hierarchical planning systems, see Ref. 33.

5 MATERIALS REQUIREMENTS PLANNING

Materials requirement planning is a procedure for converting the output of the aggregate planning process, the master production schedule, into a meaningful schedule for releasing orders for component inventory items to vendors or to the production department as required to meet the delivery requirements of the master production schedule. It is used in situations where the demand for a product is irregular and highly varying as to the quantity required at a given time. In these situations the normal inventory models for quantities manufactured or purchased do not apply. Recall that those models assume a constant demand and are inappropriate for the situation where demand is unknown and highly variable. The basic difference between the independent and dependent demand systems is the manner in which the product demand is assumed to occur. For the constant demand case it is assumed that the daily demand is the same. For dependent demand a forecast of required units over a planning horizon is used. Treating the dependent demand situation differently allows the business to maintain a much lower inventory level in general than would be required for the same situation under an assumed constant demand. This is so because the average inventory level will be much less in the case where MRP is applied. With MRP, the business will procure inventory to meet high demand just in advance of the requirement and at other times maintain a much lower level of average inventory.

Basic Definitions of Terms

Available units: Units of stock that are in inventory and are not in the category of buffer or safety stock and are not otherwise committed.

Gross requirements: The quantity of material required at a particular time that does not consider any available units.

Inventory unit: A unit of any product that is maintained in inventory.

Lead time: The time requirement for the conversion of inventory units into required sub assemblies or the time required to order and receive an inventory unit.

MRP (materials requirements planning): A method for converting the end item schedule for a finished product into schedules for the components that make up the final product.

MRP-II (manufacturing resources planning): A procedural approach to the planning of all resource requirements for the manufacturing firm.

Net requirements: The units of a requirement that must be satisfied by either purchasing or manufacturing.

Product structure tree: A diagram representing the hierarchical structure of the product. The trunk of the tree would represent the final product as assembled from the subassemblies and inventory units that are represented by level one, which come from sub-subassemblies and inventory units that come from the second level and so on ad infinitum.

Scheduled receipts: Material that is scheduled to be delivered in a given time bucket of the planning horizon.

Time bucket: The smallest distinguishable time period of the planning horizon for which activities are coordinated.

5.1 Procedures and Required Inputs

The *master production schedule* is a schedule devised to meet the production requirements for a product during a given planning horizon. It is normally prepared from fixed orders in the short run and product requirements forecasts for the time past that for which firm product orders are available. This master production schedule together with information regarding inventory status and the product structure tree and/or the bill-of-materials are used to produce a planned order schedule. An example of a master production schedule is shown in Table 11.

An *MRP schedule* is the basic document used to plan the scheduling of requirements for meeting the NTS. An example is shown in Table 12. Each horizontal section of this schedule is related to a single product, part, or subassembly from the product structure tree. The first section of the first form would be used for the parent product. The following sections of the form and required additional forms would be used for the children of this parent. This process is repeated until all parts and assemblies are listed.

To use the MRP schedule it is necessary to complete a schedule first for the parent part. Upon completion of this level-zero schedule the *bottom line* becomes the input into the schedule for each child of the parent. This procedure is followed until such time each component, assembly, or purchased part has been scheduled for ordering or production in accordance with the time requirements and other limitations that are imposed by the problem parameters. Note that if a part is used at more than one place in the assembly or manufacture of the final product, it has only one MRP schedule, which is the sum of the requirements at the various levels. The headings of the MRP schedule are as follows:

Item code. The company-assigned designation of the part or subassembly as shown on the product structure tree or the bill-of-materials.

Level code. The level of the product structure tree at which the item is introduced into the process. The completed product is designated level 0, subassemblies or parts that go together to make up the completed product are level 1, sub-subassemblies and parts that make up level 1 subassemblies are level 2, etc.

Lot size. The size of the lot that is purchased when an order is placed. This quantity may be an economic order quantity or a lot-for-lot purchase. (This later expression is used for a purchase quantity equal to the number required and no more.)

Lead time. The time required to receive an order from the time the order is placed. This order may be placed internally for manufacturing or externally for purchase.

On hand. The total of all units of stock in inventory.

Safety stock. Stock on hand that is set aside to meet emergency requirements.

Allocated stock. Stock on hand that has been previously allocated for use such as for repair parts for customer parts orders.

Table 11 A Master Production Schedule for a Given Product

Part Number	Quantity Needed	Due Date
A000	25	3
A000	30	5
A000	30	8
A000	30	10
A000	40	12
A000	40	15

Table 12 Example MRP Schedule Format

Item Code	Level Code	Lot Size	Lead Time (weeks)	On Hand	Saftey Stock	Allocated		1	2	3	4	5	6	7	8	9	10	11	12
							Gross requirements												
							Scheduled receipts												
							Available												
							Net requirements												
							Planned order receipts												
							Planned order releases												
							Gross requirements												
							Scheduled receipts												
							Available												
							Net requirements												
							Planned order receipts												
							Planned order releases												
							Gross requirements												
							Scheduled receipts												
							Available												
							Net requirements												
							Planned order receipts												
							Planned order releases												
							Gross requirements												
							Scheduled receipts												
							Available												
							Net requirements												
							Planned order receipts												
							Planned order releases												

The rows related to a specific item code are designated as follows:

Gross requirements. The unit requirements for the specific item code in the specific time bucket, which is obtained from the master production schedule for the level code 0 items. For item codes at levels other than level code 0, the gross requirements are obtained from the planned order releases for the parent item. Where an item is used at more than one level in the product, its gross requirements would be the summation of the planned order releases of the items containing the required part.

Scheduled receipts. This quantity is defined at the beginning of the planning process for products that are on order at that time. Subsequently, it is not used.

Available. Those units of a given item code that are not safety stock and are not dedicated for other uses.

Net requirements. For a given item code this is the difference between gross requirements and the quantity available.

Planned order receipts. An order quantity sufficient to meet the net requirements, which are determined by comparing the net requirements to the lot size (ordering quantity) for the specific item code. If the net requirements are less than the ordering quantity, an order of the size shown as the lot size will be placed; if the lot size is LFL (lot-for-lot), a quantity equal to the net requirements will be placed.

Planned order releases. This row provides for the release of the order discussed in planned order receipts to be released in the proper time bucket such that it will arrive appropriately to meet the need of its associated planned order receipt. Note also that this planned order release provides the input information for the requirements of those item codes that are the children of this unit in subsequent generations if such generations exist in the product structure.

Example 8 Material Requirement Planning (see Ref. 1). To offer realistic problems, consider the following simple product. If you were a cub scout, you may remember building and racing a little wooden race car. Such cars come 10 in a box. Each box has 10 preformed wood blocks, 40 wheels, 40 nails for axles, and a sheet of 10 vehicle number stickers. The problem is the manufacture and boxing of these race car kits. An assembly explosion and manufacturing tree are given in Figs. 6 and 7.

Studying the tree indicates four operations. The first is to cut 50 rough car bodies from a piece of lumber. The second is to plane and slot each car body. The third is to bag 40 nails and wheels. The fourth is to box materials for 10 race cars.

The information from the production structure tree for the model car together with available information regarding lot sizes, lead time, and stock on hand is posted in the MRP schedule format to provide information for analysis of the problem. In the problem, no safety stock was prescribed and no stock was allocated for other use. This information allowed the input into the MRP format of all information shown below for the eight item codes of the product. The single input into the right side of the problem format is the MPS for the parent product, A000. With this information each of the values of the MRP schedule can be calculated. Note that the output (planned order releases) of the level 0 product multiplied by the requirements per parent unit (as shown in parentheses at the top right corner of the *child* component, in the product structure tree) becomes the *gross requirements* for the (or each) *child* of the parent part.

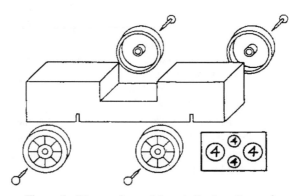

Figure 6 Diagram for model car indicating all parts.[1]

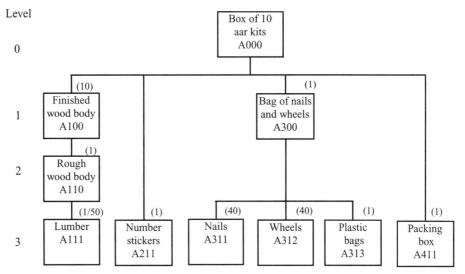

Figure 7 Product structure tree.

Calculations. As previously stated, the *gross requirements* come from either the MPS (for the parent part) or the calculation of the *planned order releases* for the parent part *times* the per unit requirement of the current *child,* per parent part. The *scheduled receipts* are receipts scheduled from a previous MRP plan. The *available* units are those on hand from a previous period plus the *scheduled* receipts from previous MRP. The *net requirements* are *gross requirements* less the *available* units. If this quantity is negative, indicating that there is more than enough, it is set to *zero*. If it is positive, it is necessary to include an order in a previous period of quantity equal to or greater than the lot size, sufficient to meet the current need. This is accomplished by backing up a number of periods equal to the lead time for the component and placing an order in the *planned order releases* now that is equal to or greater than the lot size for the given component.

Scheduled receipts and *planned order receipts* are essentially the arrival of product. The distinction between the two is that scheduled receipts are orders that were made on a previous MRP plan. The *planned order receipts* are those that are scheduled on the current plan. Further, to keep the system operating smoothly, the MRP plan must be reworked as soon as new information becomes available regarding demand for the product for which the MPS is prepared. This essentially, provides an ability to respond and to keep materials in the *pipeline* for delivery. Without updating, the system becomes cumbersome and unresponsive. For example, most of the component parts are exhausted at the end of the 15-week period; hence, to respond in the sixteenth week would require considerable delay if the schedule were not updated. The results of this process are shown in Tables 13, 14, and 15.

The planned order release schedule (Table 16) is the result of the Materials Requirements Planning procedure. It is essentially the summation of the bottom lines for the individual components from the MRP schedules. It displays an overall requirement for meeting the original master production schedule.

5.2 Lot Sizing Techniques

Several techniques are applicable to the determination of the lot size for the order. If there are many products and some components are used in several products, it may be that demand

Table 13

Item Code	Level Code	Lot Size	Lead Time (weeks)	On Hand	Safety Stock	Allocated		1	2	3	4	5	6	7	8	9	10	11	12	13	14	15
A000	0	50	1	20	0	0	Gross requirements	20	20	25	—	30	15	15	30	35	30	—	40	15	15	40
							Schedule receipts			—		—			—		35	5	—			15
							Available			20	45	45			15		—		5			15
							Net requirements			5		—			15		—		35			25
							Planned order receipts			50		—			50		—		50			50
							Planned order releases		50					50				50			50	
A100	1	50	1	100	0	0	Gross requirements	100	500					500				500			500	
							Scheduled receipts															
							Available		100	0	0	0	0	0	0	0	0	0	0	0	0	0
							Net requirements	400	400				500	500			500	500		500	500	
							Planned order receipts	400	400				500	500			500	500		500	500	
							Planned order releases	400				500				500			500			
A300	1	50	1	150	0	0	Gross requirements	150	50				100	50			50	50			50	
							Scheduled receipts		150	100	100	100	100	100	50	50	50	50				
							Available		0	0	0	0	0	0	0	0	0	0	0	50	50	0
							Net requirements		—					0				0			50	
							Planned order receipts		—					—								
							Planned order releases					500			500			500				
A110	2	100	1	200	0	0	Gross requirements	400	0	0	0	0	500	0	0	0	500	0	0	500	0	
							Scheduled receipts	200*														
							Available	400	0	0	0	0	0	0	0	0	500	0	0	500	0	
							Net requirements	0					500				500			500		
							Planned order receipts	0					500				500			500		
							Planned order releases	0				500				500			500			

*Ordered on a previous schedule.

Table 14

Item Code	Level Code	Lot Size	Lead Time (weeks)	On Hand	Safety Stock	Allocated		1	2	3	4	5	6	7	8	9	10	11	12	13	14	15
A111	3	10	3	5	0	0	Gross requirements					10				10			10			
							Scheduled receipts	5	5	5	5	5	5	5	5	5	5	5	5	5	5	5
							Available					5	5			5			5			
							Net requirements					5				5			5			
							Planned order receipts		10			10				10			10			
							Planned order releases					—	10			10						
A211	3	500	10	500	0	0	Gross requirements		50					50				50			50	
							Scheduled receipts	500	500	450	450	450	450	450	400	400	400	400	350	350	350	300
							Available															
							Net requirements															
							Planned order receipts															
							Planned order releases															
A311	3	500	2	300	0	0	Gross requirements															
							Scheduled receipts	300	300	300	300	300	300	300	300	300	300	300	300	300	300	300
							Available													2000		
							Net requirements													300	300	300
							Planned order receipts											2000		1700		
							Planned order releases													2000		
A312	3	500	2	200	0	0	Gross requirements															
							Scheduled receipts	200	200	200	200	200	200	200	200	200	200	200	200	200	200	200
							Available													2000		
							Net requirements													200	200	200
							Planned order receipts											2000		1800		
							Planned order releases													2000		

Table 15

Item Code	Level Code	Lot Size	Lead Time (weeks)	On Hand	Safety Stock	Allocated		1	2	3	4	5	6	7	8	9	10	11	12	13	14	15
A313	3	500	3	30	0	0	Gross requirements														50	
							Scheduled receipts															
							Available	30	30	30	30	30	30	30	30	30	30	30	30	30	30	
							Net requirements														20	
							Planned order receipts														500	
							Planned order releases															480
A411	3	500	5	40	0	0	Gross requirements		50					50				50			50	
							Scheduled receipts															
							Available	40	40	490	490	490	490	490	440	440	440	440	390	390	390	340
							Net requirements		10													
							Planned order receipts		500*													
							Planned order releases															

*Ordered on a previous schedule.

Table 16 Planned Ordered Release Schedule

	Week														
	1	2	3	4	5	6	7	8	9	10	11	12	13	14	15
A000		50					50				50			50	
A100	400					500				500			500		
A300														50	
A110					500				500				500		
A111		10					10			10					
A211															
A311											2000				
A312											2000				
A313											500				
A411															

Note: An advance order of 200 units of item 110 would have to have been made on a previous MRP schedule.

for that common component is relatively constant. If that is the case, EOQ models such as those used in the topic on inventory can be applied.

The POQ (periodic order quantity) is a variant of the EOQ where a nonconstant demand over a planning horizon is averaged. This average is then assumed to be the constant demand. Using this value of demand the EOQ is calculated. The EOQ is divided into the total demand if demand is greater than EOQ. This resultant figure gives the number of inventory cycles for the planning horizon. The actual forecast is then related to the number of inventory cycles and the order sizes are determined.

Example 9 Periodic Order Quantity. Given the data in Table 17 for a product that is purchased, assume that holding cost is $10 per unit year. Order cost is $25. Calculate the POQ. No shortage is permitted.

Using the basic EOQ formula,

$$\hat{Q} = \sqrt{\frac{2C_p D}{C_n}} = \sqrt{\frac{2(\$25)(29 \times 52)}{\$10}} = 86.9 \approx 87 \text{ units/order}$$

$$N = \frac{\text{Demand (Units per 12 weeks)}}{\text{Units per order}} = \frac{348}{87} = 4 \text{ orders per 12 weeks}$$

Lot for lot (LFL) is the approach to the variable demand situation, which merely requires that an order size equal to the required number of products be placed. The first order would be $25 + 29 + 34 = 88$ units. The second would be $26 + 24 + 32 = 82$ units. The third and fourth orders would be 81 and 97, respectively.

It is coincidental that the number of orders turned out to be an integer. Had a non-integer occurred it could be rounded to the nearest integer. An economic evaluation can be made if costs are significant of which rounding (up or down) would yield the lower cost option. Other methods exist in the area of lot sizing.

Table 17

Week	1	2	3	4	5	6	7	8	9	10	11	12	$\bar{D}$
Demand	25	29	34	26	24	32	28	25	28	35	32	30	29

5.3 Beyond MRP

The desired output of MRP systems is the creation of a master schedule showing the time-phased production of all parts and components as well as the timing of ordering of all raw materials and components necessary to meet the forecasted product demand. Unfortunately, MRP systems do not have capacity planning capabilities to ensure that generated schedules are feasible. What evolved to remedy this deficiency was the development of manufacturing resource planning systems (referred to as MRP-II, or closed-loop systems), which make it possible to consider production capacity and supply constraints in generating integrated, feasible production schedules. Further enhancements often include shop floor control systems, which can track and monitor the execution of the production schedules and develop more effective and cost efficient production systems. More recently (see Rondeau and Littral[34]), these additions have included manufacturing execution systems (MES), which are interfaces between shop floor control systems and the MRP-II system to provide not only control but some optimization capabilities with respect to the use of equipment and resultant schedules.

The growth in the features provided by these MRP-II systems with or without enhancements, however, does come at a price, in both actual dollars and added complexity. There is also the problem that while these systems are very powerful, the flexibility and capability to react to changes in the production environment because of changes in the product, improvements in materials, or stability of vendors and customers is less than that required for many modern, agile production environments. Global competition likewise has demanded that manufacturing firms be flexible and capable of quickly modifying product specifications and processes. Thus, the same or similar products may be produced in multiple facilities and used by customers from different cultures with different expectations.

To be successful in this environment, the most commonly found management outlook today appears to be customer oriented and directed toward what is known as supply chain management (SCM). SCM is used to describe the inherent linkages that exist among all of the activities of a manufacturing organization and systems and procedures that can be used to align all units of the organization so that it can be managed for overall common objectives.

A tool used in managing supply chains is enterprise resource planning (ERP). This software system is designed to integrate and optimize a variety of business processes across the entire manufacturing firm, which may include multiple sites, products, and facilities. This includes activities such as managing human resources, financial and accounting functions, sales and distribution, and inventory and manufacturing.[35] ERP evolved from MRP-II and generally contains three key features that go beyond MRP-II. First, ERP includes a variety of business functions, such as purchasing, sales, manufacturing, distribution, and accounting. Second, the functions are integrated so that information or data used by one activity is linked to any other activities that use that same data. More importantly, when data are entered or changed in one of the functions, it also is changed in any other function that may need or use that data. Third, each of the functions is modular in nature and can be implemented in any combination with any other business function.[35] This modularity and integration provides a stable information technology platform, which allows for relatively easy expansion, contraction, changes, and updates in information requirements as the business itself changes. None of this, of course, is simple, inexpensive, or easy to implement. Commitment to implement an ERP system is a strategic business decision and is done only after extensive study with decisions being made at the highest level in the organization.[36]

6 JOB SEQUENCING AND SCHEDULING

Sequencing and scheduling problems are among the most common situations found in service and manufacturing facilities. Determining the order and deciding when activities or tasks should be done are part of the normal functions and responsibilities of management and, increasingly, of the employees themselves. These terms are often used interchangeably, but it is important to note the difference. *Sequencing* is determining the order of a set of activities to be performed, whereas *scheduling* also includes determining the specific times when each activity will be done. Thus, scheduling includes sequencing, i.e., to be able to develop a schedule for a set of activities you must also know the sequence in which those activities are to be completed.[37]

6.1 Structure of the General Sequencing Problem

The job sequencing problem is usually stated as follows: Given n jobs to be processed on n machines, each job having a *setup time, processing time,* and *due date* for the completion of the job, and requiring processing on one or more of the machines, determine the sequence for processing the jobs on the machines to optimize the *performance criterion.* The factors used to describe a sequencing problem are

1. The number of machines in the shop, m
2. The number of jobs, n
3. The type of shop or facility, i.e., job shop or flow shop
4. The manner in which jobs arrive at the shop, i.e., static or dynamic
5. The performance criterion used to measure the performance of the shop

Usual assumptions for the sequencing problem include

1. Setup times for the jobs on each machine are independent of sequence and can be included in the processing times.
2. All jobs are available at time zero to begin processing.
3. All setup times, processing times, and due dates are known and are deterministic.
4. Once a job begins processing on a machine, it will not be preempted by another job on that machine.
5. Machines are continuously available for processing, i.e., no breakdowns occur.

Commonly used performance criteria include the following:

1. Mean flow time (F)—the average time a set of jobs spends in the shop, which includes processing and waiting times
2. Mean idle time of machines—the average idle time for the set of machines in the shop
3. Mean lateness of jobs (L)—the difference between the actual completion time (C_j) for a job and its due date (d_j), i.e., $L_j = C_j - d_j$. A negative value means that the job is completed early. Therefore,

$$\bar{L} = \sum_{j=1}^{n} (C_j - d_j)/n \qquad (31)$$

4. Mean tardiness of jobs (T)—the maximum of 0 or its value of lateness, i.e., $T_j = \max\{0, L_j\}$. Therefore,

$$\bar{T} = \sum_{j=1}^{n} \max\{0, L_j\}/n \qquad (32)$$

5. Mean number of jobs late
6. Percentage of jobs late
7. Mean number of jobs in the system
8. Variance of lateness (S_L^2)—for a set of jobs and a given sequence, the variance calculated for the corresponding L_j's; i.e.,

$$\sum_{j=1}^{n} (L_j - \bar{L})^2/(n-1) \qquad (33)$$

The following material covers the broad range of sequencing problems from the simple to the complex. The discussion begins with the single machine problem and progresses through multiple machines. It includes quantitative and heuristic results for both flow shop and job shop environments.

6.2 Single Machine Problem

In many instances the single machine sequencing problem is still a viable problem. For example, if one were trying to maximize production through a bottleneck operation, consideration of the bottleneck as a single machine might be a reasonable assumption. For the single machine problem, i.e., n jobs/one machine, results include the following.

Mean Flow Time

To minimize the mean flow time, jobs should be sequenced so that they are in increasing shortest processing time (SPT) order, that is,

$$t_{[1]} \leq t_{[2]} \leq \cdots \leq t_{[n]} \qquad (34)$$

Example 10 Shortest Processing Time (SPT). Given are the following jobs and processing times (t_j's) for the jobs (see Table 18).

If the jobs are processed in the shortest processing time order, i.e., (4, 2, 1, 3), then the completion times are given in Table 19. Therefore, $\bar{F} = 60/4 = 15$ days. Any other sequence will only increase $\bar{F}$. Proof of this is available in references 1, 37, and 38.

Mean Lateness

As a result of the definition of lateness, SPT sequencing will minimize mean lateness ($\bar{L}$) in the single machine shop.

Weighted Mean Flow Time

The above results assumed that all jobs were of equal importance. What if, however, jobs should be weighted according to some measure of importance. Some jobs may be more

Table 18

Job j	t_j (days)
1	7
2	6
3	8
4	5

Table 19

Job j	T_j (days)	C_j (days)
4	5	5
2	6	11
1	7	18
3	8	26
$\Sigma C_j =$		60

important because of customer priority or profitability. If this importance can be measured by a weight assigned to each job, a weighted mean flow time measure, F_w, can be defined as

$$\overline{F_w} = \sum_{j=1}^{n} w_j F_j / \sum_{j=1}^{n} w_j \tag{35}$$

To minimize weighted mean flow time $(-F_w)$, jobs should be sequenced in increasing order of weighted shortest processing time, i.e.,

$$\frac{t_{[1]}}{w_{[1]}} \leq \frac{t_{[2]}}{w_{[2]}} \leq \cdots \leq \frac{t_{[n]}}{w_{[n]}} \tag{36}$$

where the brackets indicate the first, second, etc., jobs in sequence.

Example 11 Weighted Shortest Processing. Consider the data in Table 20. If jobs 2 and 6 are considered 3 times as important as the rest of the job, what sequence should be selected?

Solution

Job	1	2	3	4	5	6
w_j	1	3	1	1	1	3
T_j/w_j	20	9	16	6	15	8

Therefore, the job processing sequence should be 4, 6, 2, 5, 3, and 1.

Maximum Lateness/Maximum Tardiness

Other elementary results given without proof or example include the following. To minimize the maximum job lateness (L_{max}) or the maximum job tardiness (T_{max}) for a set of jobs, the jobs should be sequenced in order of nondecreasing due dates, i.e.,

$$d_{[1]} \leq d_{[2]} \leq \ldots \leq d_{[n]} \tag{37}$$

Table 20

Job	1	2	3	4	5	6
t_j (days)	20	27	16	6	15	24

Minimize the Number of Tardy Jobs

If the sequence above, known as the earliest due date sequence, results in zero or one tardy job, then it is also an optional sequence for the number of tardy jobs, N_T. In general, however, to find an optional sequence minimizing N_T, an algorithm attributed to Moore and Hodgson[39] can be used. The algorithm divides all jobs into two sets:

1. Set E, where all the jobs are either early or on time
2. Set T, where all the jobs are tardy

The optional sequence then consists of set E jobs followed by set T jobs. The algorithm is as follows:

Step 1. Begin by placing all jobs in set E in nondecreasing due date order, i.e., earliest due date order. Note that set T is empty.

Step 2. If no jobs in set E are tardy, stop; the sequence in set E is optional. Otherwise, identify the first tardy job in set E, labeling this job k.

Step 3. Find the job with the longest processing time among the first k jobs in sequence in set E. Remove this job from set E and place it in set T. Revise the job completion times of the jobs remaining in set E and go back to step 2 above.

Example 12 Tardy Jobs. Consider the data in Table 21.

Solution

Step 1. $E = \{3, 1, 4, 2\}; T = \{\Phi\}$.

Step 2. Job 4 is first late job.

Step 3. Job 1 is removed from E:

$$E = \{3, 4, 2\}; \quad T = \{1\}$$

Step 2. Job 2 is first late job.

Step 3. Job 2 is removed from

$$E = \{3, 4\}; \quad T = \{1, 2\}$$

Step 2. No jobs in E are now late.

Therefore, optional sequences are (3, 4, 1, 2) and (3, 4, 2, 1)

6.3 Flow Shops

General flow shops can be depicted as in Fig. 8. All products being produced through these systems flow in the same direction without backtracking. For example, in a four-machine

Table 21

Job	t_j (days)	d_j (days)
1	10	14
2	18	27
3	2	4
4	6	16

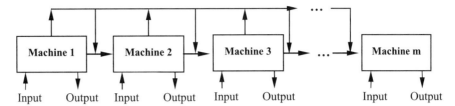

Figure 8 Product flow in a general flow shop.

general flow shop, product 1 may require processing on machines 1, 2, 3, and 4; product 2 may require machines 1, 3, and 4; while product 3 may require machines 1 and 2 only. Thus, a flow shop processes jobs much like a production line but because it often processes jobs in batches, it may look more like a job shop.

Two Machines and n Jobs

The most famous result in sequencing literature is concerned with two machine flow shops and is known as *Johnson's sequencing algorithm*.[40] This algorithm will develop an optional sequence using makespan as the performance criterion. *Makespan* is defined as the time required to complete the set of jobs through all machines.

Steps for the algorithm are as follows:

1. List all processing times for the job set for machines 1 and 2.
2. Find the minimum processing time for all jobs.
3. If the minimum processing time is on machine 1, place the job first or as early as possible in the sequence. If it is on machine 2, place that job last or as late as possible in the sequence. Remove that job for further consideration.
4. Continue, by going back to step 2, until all jobs have been sequenced.

As an example, consider the five-job problem in Table 22. Applying the algorithm will give an optional sequence of (2, 4, 5, 3, 1) through the two machines, with a makespan of 26 time units.

Three Machines and n Jobs

Johnson's sequencing algorithm can be extended to a three-machine flow shop and may generate an optional solution with makespan as the criterion.

The extension consists of creating a two-machine flow shop from the three machines by summing the processing times for all jobs for the first two machines for artificial machine 1, and, likewise, summing the processing times for all jobs for the last two machines for

Table 22

Job	1	2	3	4	5
Machine 1 (t_{ij})	5	1	8	2	7
Machine 2 (t_{2j})	3	4	5	6	6

artificial machine 2. Johnson's sequencing algorithm is then used on the two artificial machine flow shop problem.

Example 13 Flow Shop Sequencing (Johnson's Algorithm). Consider the three-machine flow shop problem in Table 23. Forming the five jobs, two artificial machine problem gives the results in Table 24.

Therefore, the sequence using Johnson's sequencing algorithm is (3, 1, 4, 5, 2). It has been shown that the sequence obtained using this extension is optimal with respect to makespan if any of the following conditions hold:

1. min t_{1j} > max t_{2j}
2. min t_{3j} > max t_{2j}
3. If the sequence using $\{t_{1j}, t_{2j}\}$ only, i.e., the first two machines, is the same sequence as that using only $\{t_{2j}, t_{3j}\}$, i.e., only the last two machines, as two, two-machine flow shops.

The reader should check to see that the sequence obtained above is optimal using these conditions.

More Than Three Machines

Once the number of machines exceeds three, there are few ways to find optimal sequences in a flow shop environment. Enumeration procedures, like branch and bound, are generally the only practical approach that has been successfully used, and then only in problems with five or less machines. The more usual approach is to develop heuristic procedures or to use assignment rules such as priority dispatching rules. See the section on heuristics/priority dispatching rules for more details.

6.4 Job Shops

General job shops can be represented as in Fig. 9. Products being produced in these systems may begin with any machine or process, followed by a succession of processing operations on any other sequence of machines. It is the most flexible form of production, but experience has shown that it is also the most difficult to control and to operate efficiently.

Two Machines and n Jobs

Johnson's sequencing algorithm can also be extended to a two-machine job shop to generate optimal schedules when makespan is the criterion.

Table 23

Job	Processing Times		
	t_{1j}	t_{2j}	t_{3j}
1	1	3	8
2	4	1	3
3	1	2	3
4	7	2	7
5	6	1	5

Table 24

Job	t^1_{aj}	T^1_{bj}
1	4	11
2	5	4
3	3	5
4	9	9
5	7	6

The steps to do this are as follows:

Step 1. Divide the job set into four sets:

Set {*A*}—jobs that require only one processing operation and that on machine 1.

Set {*B*}—jobs that require only one processing operation and that on machine 2.

Set {*AB*}—jobs that require two processing operations, the first on machine 1, the second on machine 2.

Set {*BA*}—jobs that require two processing operations, the first on machine 2, the sccond on machine 1.

Step 2. Sequence jobs in set {AB} and set {BA} using Johnson's sequencing algorithm (note that in set {BA}, machine 2 is the first machine in the process).

Step 3. The optional sequence with respect to makespan is on machine 1. Process the {AB} jobs first, then the {A} jobs, then the {BA} jobs (note that the {A} jobs can be sequenced in any order within the set). On machine 2, process the {BA} jobs first, then the {B} jobs, then the {AB} jobs (note that the {B} jobs can be sequenced in any order within the set).

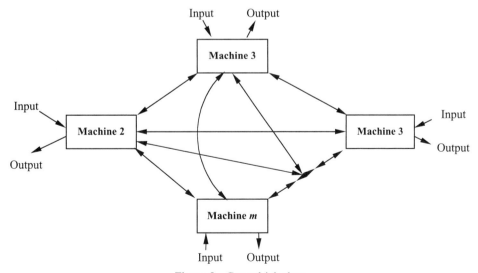

Figure 9 General job shop.

For example, from Table 25 the optical sequences are

Machine 1: 6, 5, 4, 1, 8, 9, 7, 10
Machine 2: 8, 9, 7, 10, 2, 3, 6, 5, 4

giving a makespan of 42 time units.

Machines/n Jobs

Once the problem size exceeds two machines in a job shop, optimal sequences are difficult to develop even with makespan as the criterion. If optimal sequences are desired, the only options are usually enumeration techniques like branch and bound, which attempt to take account of the special structure that may exist in the particular problem. However, because of the complexity involved in these larger problems, sequencing attention generally turns away from seeking the development of optimal schedules to the development of feasible schedules through the use of heuristic decision rules called priority dispatching rules.

6.5 Heuristics/Priority Dispatching Rules

A large number of these rules have evolved, each with their proponents, given certain shop conditions and desired performance criteria. Some of the more commonly found ones are

FCFS—select the job on a first come-first served basis

SPT—select the job with the shortest processing time

EDD—select the job with the earliest due date

STOP—select the job with the smallest ratio of remaining slack time to the number of remaining operations

LWKR—select the job with the least amount of work remaining to be done

Rules such as these are often referred to as either local or global rules. A local rule is one that is applied from the perspective of each machine or processing operation, whereas a global view is applied from the perspective of the overall shop. For example, SPT is a local rule, since deciding which of the available jobs to process next is determined by each machine or process operator. On the other hand, LWKR is global rule, since it considers all

Table 25

Job Set	Job	Processing Times	
		t_{1j}	t_{2j}
{A}	1	3	—
{B}	2	—	2
	3	—	4
{AB}	4	4	2
	5	6	5
	6	3	7
{BA}	7	3	8
	8	4	1
	9	7	9
	10	2	4

remaining processing that must be done on the job. Therefore, LWKR can be considered the global equivalent of SPT. Some rules, like FCFS can be used in either local or global applications. The choice of local or global use is often a matter of whether shop scheduling is done in a centralized or decentralized manner and whether the information system will support centralized scheduling. Implicit within these concepts is the fact that centralized scheduling requires more information to be distributed to individual workstations, and is inherently a more complex scheduling environment requiring more supervisory oversight. Global scheduling intuitively should produce better system schedules, but empirical evidence seems to indicate that local rules are generally more effective.

Whichever rule may be selected, priority assignments are used to resolve conflicts. As an example of this consider the following three-machine, four-job sequencing problem in Table 26. If we assume that all jobs are available at time zero, the initial job loading is shown in Fig. 10. As shown, there is no conflict on machines 1 and 2, so the first operation for jobs 1 and 2 would be assigned to these machines. However, there is a conflict on machine 3. If SPT were being used, the first operation for job 3 would be assigned on machine 3, and the earliest that the first operation of job 4 could be assigned to machine 3 is at time equal to 2 days.

Continuing this example following these assignments, the situation shown in Fig. 11 would then exist. If we continue to use SPT, we would assign the second operation of job 2 to machine 1 to resolve the conflict. (*Note:* If FCFS were being used, the second operation of job 3 would have been assigned.) Even though there is no conflict at this stage on machine 2, no job would normally be assigned at this time because it would introduce idle time unnecessarily within the schedule. The first operation for job 4, however, would be assigned to machine 3.

With these assignments, the schedule now appears in Fig. 12. The assignments made at this stage would include

1. Second operation, job 4 to machine 2
2. Third operation, job 2 to machine 3, since no other job could be processed on machine, 3 during the idle time from 3 to 6 days

Note that the second operation for job 3 may or may not be scheduled at this time because the third operation for job 4 would also be available to begin processing on machine 1 at the 6th day. Because of this, there would be a conflict at the beginning of the 6th day, and if SPT is being used, the third operation for job 4 would be selected over the second operation for job 3.

Several observations from this partial example can be made:

1. Depending on the order in which conflicts are resolved, two people using the same priority dispatching rule may develop different schedules.

Table 26

Job	Processing Times (Days)			Operation Sequence		
	t_{1j}	t_{2j}	t_{2j}	M/C-1	M/C-2	M/C-3
1	4	6	8	1	2	3
2	2	3	4	2	1	3
3	4	2	1	2	3	1
4	3	3	2	3	2	1

154 Production Planning

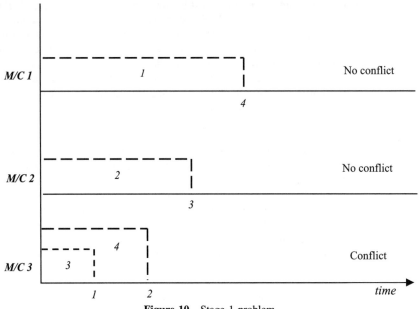

Figure 10 Stage 1 problem.

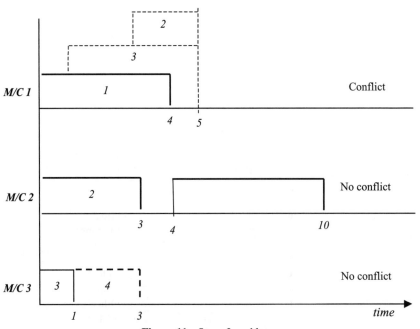

Figure 11 Stage 2 problem.

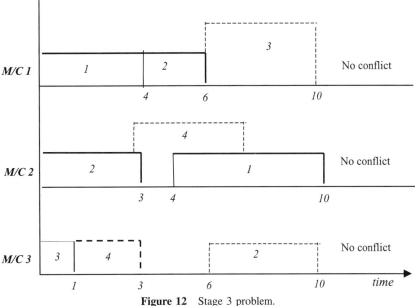

Figure 12 Stage 3 problem.

2. Developing detailed schedules is a complex process. Almost all large-scale scheduling environments would benefit from the use of computer aids.
3. Determining the effectiveness of a dispatching rule is difficult in the schedule generating process, because of the precedence relationships that must be maintained in processing.

Testing the effectiveness of dispatching rules is most often done by means of simulation studies. Such studies are used to establish the conditions found in the shop of interest for testing various sequencing strategies that management may believe to be worth investigating. For a good historical development of these as well as general conclusions that have attempted to be drawn (see Ref. 41).

Two broad classifications of priority dispatching rules seemed to have emerged:

1. Those trying to reduce the flow time in which a job spends in the system, i.e., by increasing the speed going through the shop or reducing the waiting time
2. Those due date-based rules, which may also manifest themselves as trying to reduce the variation associated with the selected performance measure

Although simulation has proven to be effective in evaluating the effectiveness of dispatching rules in a particular environment, few general conclusions have been drawn. When maximum throughput or speed is the primary criterion, SPT is often a good rule to use, even in situations when the quality of information in poor concerning due dates and processing times. When due date rules are of interest, selection is much more difficult. Results have been developed showing that when shop loads are heavy, SPT still may do well; when shop loads are moderate, STOP was preferable. Other research has shown that the manners in which due dates are set as well the tightness of the due dates can greatly affect the performance of the rule. Overall, the following conclusions which can be drawn:

1. It is generally more difficult to select an effective due date-based rule than a flow time-based rule.
2. If time and resources are available, the best course of action is to develop a valid model of the particular shop of interest, and experiment with the various candidate rules to determine which are most effective, given that situation.

6.6 Assembly Line Balancing

Assembly lines are viewed as one of the purest forms of production lines. A usual form is visualized as shown in Figs. 13 and 14, where work moves continuously by means of a powered conveyor through a series of workstations where the assigned work is performed.

Definitions

Cycle time (C): The time available for a workstation to perform its assigned work, assumed to be the same for each workstation. The cycle time must be greater than or equal to the longest work element for the product. Note that it is also the time between successive completions of units of product on the line.

Balance delay of a workstation: The difference between the cycle time (C) and the station time (S_j) for a workstation, i.e., the idle time for the station ($C - S_j$).

Station time (S_j): The total amount of work assigned to station j, which consists of one or more of the work elements necessary for completion of the product. Note that each S_j must be less than or equal to C.

Work element (i): An amount of work necessary in the completion of a unit of product ($i = 1, 2, \ldots, I$). It is usually considered indivisible. I is the total number of work elements necessary to complete one unit of product.

Work element time (t_i): The amount of time required to complete work element i. Therefore, the sum of all of the work elements, i.e., the total work content,

$$T = \sum_{i=1}^{I} t_i$$

is the time necessary to complete one unit of product.

Workstation (j): A location on the line where assigned work elements on the product are performed ($1 \leq j \leq J$).

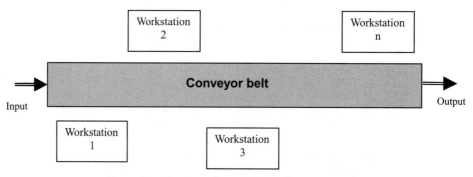

Figure 13 Structure of assembly line (conveyor type).

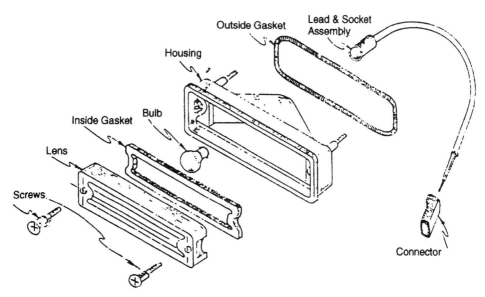

Figure 14 Automobile backup-light assembly.[42]

Structure of the Assembly Line Balancing Problem

The objective of assembly line balancing is to assign work elements to the workstations so as to minimize the total balance delay (total idle time) on the line. The problem is normally presented by means of a listing of the work elements and a precedence diagram, which show the relationships that must be maintained in the assembling of the product. See Figs. 14 and 15 for work content data and precedence relationship. In designing the assembly line, therefore, the work elements must be assigned to the workstations while adhering to these precedence relationships.

Note that if the balance delay is summed over the entire production line, the total balance delay (total idle time) is equal to

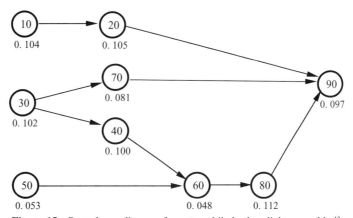

Figure 15 Precedence diagram for automobile backup-light assembly.[42]

$$\sum_{j=1}^{J} (C - S_j)$$

So, to minimize the total balance delay is the same as

$$\text{Min} \sum_{j=1}^{J} (C - S_j) = JC - \sum_{j=1}^{J} S_j$$

$$= JC - (\text{Total work content for 1 unit of product}) \qquad (38)$$

$$= JC - \text{a constant}$$

Therefore, minimizing the total balance delay is equivalent to

1. Keeping the number of workstations constant and minimizing the cycle time, or
2. Keeping the cycle time constant and minimizing the number of work stations, or
3. Jointly trying to minimize the product of cycle time and number of work stations

Which approach might be followed could depend on the circumstances. For example, if production space was constrained, approach 1 above might be used to estimate the volume the line would be capable of producing. Approach 2 might be used if the primary concern was ensuring a certain volume of product could be produced in a certain quantity of time. Approach 3 could be used in developing alternative assignments by trading off faster line speed (shorter cycle times, more workstations, and greater production) for slower line speeds (fewer workstations, longer cycle times, and less production).

Designing the Assembly Line

Given the above structure and definitions, the following must hold:

1. $(\max t_j \leq C \leq T)$
2. Minimum number of work stations = $[T/C]$, where the brackets indicate the value is rounded to the next largest integer
3. C_{max} = Product time available/Product volume required

(C_{max} is the maximum value the cycle time can be if the line is to generate the specified quantity in the specified time). As an example, consider the data provided in Table 27 and Figs. 14 and 15.

Designing a line to produce 2000 units in a 7½-hour shift would give

From condition 3: $\quad C = \dfrac{(7.5 \text{ hr/shift} (60 \text{ min/hr})}{2000 \text{ units(shift)}} = 0.225 \text{ min/unit}$

From condition 2: $\quad$ Minimum number of work stations = $\dfrac{0.802}{0.225} = [3.56] = 4$

Also note that condition 1 is satisfied, i.e., $0.112 \leq 0.225 \leq 0.802$.

Line Balancing Techniques

Efforts have been made to optimally model variations of these problems, but currently no procedures exist that guarantee optimal solutions to these types of problems. Practitioners and researchers, therefore, have developed a variety of heuristic procedures.[43] A general approach in making the assignment of work elements to workstations is to select a cycle time and to start assigning work elements where precedence restrictions are satisfied to the

Table 27

Workstation	Work Elements Assigned	Station Time	Balance Delay Time
1	10 20	0.104 0.105 0.209	0.016
2	30 70	0.102 0.081 0.183	0.042
3	40 50 60	0.100 0.053 0.048 0.201	0.024
4	80 90	0.112 0.097 0.209	0.016

first workstation. Combinations of work elements may be explored to reduce the idle time present to the lowest level possible, before going to the next workstation and repeating the procedure. This process is continued until all work elements have been assigned.

Example 14 Line Balancing. Applying this procedure to the data in Table 27 and Figure 14 would give the solution for a cycle time of 0.225 minutes. Also note that the cycle time could be reduced to 0.209 minutes with this assignment and would theoretically reduce the total balance delay by 0.016 minutes $\times$ 4 workstations = 0.064 minutes, resulting in a production increase to a total of

$$\frac{(7.5 \text{ hr/shift})(60 \text{ min/shift})}{0.209 \text{ min}} = 2153 \text{ units/shift}$$

Mixed-Model Assembly Lines

The above discussion is predicated on the premise that only one product is being manufactured on the line. Many production lines are designed to produce a variety of products. Good examples of these are assembly lines that may produce several models of the same automobile with a wide variety of options. These are often referred to as mixed-model assembly lines. Similar examples are applicable in filing cabinet manufacturing, as shown Figure 16.[44,45] These assembly lines are significantly more complex to design than the single model line because of two problem areas:

1. The assignment of work elements to the workstations
2. The ordering or sequencing of models on the line

One usual approach taken in designing a mixed-model line uses the same general objective of minimizing the total balance delay (or idle time) in the assignment of work elements to workstations. However, in the mixed-model case a production period has to be defined and the assignments are made so as to minimize the total amount of idle time for all stations and models for the production period, rather than for the cycle time as in the single-model case. To use this approach the designer must define all of the work elements

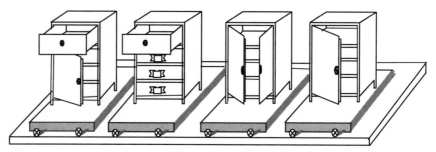

Figure 16 A filing cabinet manufacturing line (mixed model).[44,45]

for all of the models and determine the quantity of each of the models being assembled within the specified production period. Once the total work content and the time allowed for production are known, work elements are assigned to workstations usually based on similarity of the work elements, tooling or equipment required, and time to perform the tasks. If the stations on the mixed-model line are not tightly linked and small in-process inventory buffers are allowed to exist between workstations, this approach seems satisfactory. However, if the stations are tightly linked where no in-process inventory is allowed between stations or if the line is operating as a just-in-time (JIT) system, this approach may not produce satisfactory assignments without analysts being especially diligent in determining the sequence of models being produced.

Determining the order of models to produce on the line is generally more difficult than the problem of assigning work elements to workstations, because it has to be done in a constantly changing environment. This difficulty stems from two interrelated subproblems:

1. Trying to fully utilize the resources of the line, so that no station is idled due to a bottleneck or lack of product to work on
2. Trying to even out the flow of component parts to the line from upstream manufacturing or subassembly operations feeding the line, so that these resources are also fully utilized or have a relatively constant amount of work

The first of these, i.e., fully utilizing the resources of the line, is the easier of the two, especially if some flexibility is present in the system, such as producing in a make-to-stock environment or allowing small buffer, in-process inventories between workstations or variable-time workstations on the line. Examples of how some of this flexibility can be built in can be found in Ref. 46.

Smoothing out the flow of components into the assembly line is a very difficult problem, especially if the facility is operating in a JIT environment. One of the earliest approaches, which discussed how this problem was handled in the Toyota company, is known as a *goal chasing method*.[47,48] The procedure has since evolved into a newer version, and since it has the capability of handling multiple goals, it is called a *goals-coordinating method*.[48] This procedure has two main components: (1) appearance ratio control, and (2) continuation and interval controls.

Appearance ratio control is a heuristic that determines the sequence of models on the line by attempting to minimize the variances of the components used for those products, i.e., minimize the actual variation in component usage around a calculated average usage. A production schedule of end products is built by starting with the first end product to be scheduled, then working toward the last end product. For each step in determining the sequence, the following is calculated for each product, with the minimum D determining the next product to be produced:

$$D_{Ki} = \sqrt{\sum_{j=1}^{\beta} \left(\frac{KN_j}{Q} - X_{j,K-1} - b_{ij} \right)^2} \qquad (39)$$

where D_{Ki} = the distance to be minimized for sequence number K and for end product i
β = the number of different components required
K = the sequence number of the current end product in the schedule
N_j = the total number of components j required for all products in the final schedule
Q = the total production quantity of all end products in the final schedule
$X_{i,K}$ = the cumulative number of components actually used through assembly sequence K
b_{ij} = the number of components required to make one unit of end product i

However, while this approach results in a smoothed production for the majority of the schedule, it will potentially cause uneven use of components during the final phases of the day's schedule. To prevent this, continuation and interval controls are applied as constraints, which may override the appearance ratio control and introduce other type models on the line. Continuation controls ensure that no more than a designated number of consecutive end products that use a particular component are scheduled (a maximum sequencing number condition), whereas interval controls ensure that at least a designated minimum number of certain end products are scheduled between other end products that require a particular component (a minimum sequencing condition).

The overall sequencing selection process then works as follows.

Step 1. Appearance ration control is used to determine the first (or next) end product in the sequence.

Step 2. If the selected end product also satisfies the continuation and interval controls, the end product is assigned that position in the sequence. Unless all end products have been scheduled, go to step 1. Otherwise, stop, the schedule is complete.

Step 3. If the selected end product does not satisfy both the continuation and interval controls, the appearance ratio control is applied to the remaining end products, while ignoring the component that violated the continuation and/or interval controls. Out of the end products that do not require the component in question, the end product that minimizes the amount of total deviation in the following formula would be selected as the next (Kth) in sequence (j = component number).

$$\sum_{j=1}^{n} \left(\frac{\text{Total number of end product of the specified component } i}{\text{Total number of end product}} \right) \times K - \left(\begin{array}{c} \text{Accumulated number} \\ \text{of component } j \\ \text{up to } (K-1)\text{th} \end{array} \right)$$
$$+ \left(\begin{array}{c} \text{Number of component} \\ j \text{ of } K\text{th additional} \\ \text{end product} \end{array} \right)$$

Unless all end products have been scheduled, go to step 1. Otherwise, stop, the schedule is complete.

As the number of models and components increases, the difficulties of developing satisfactory solutions for leveling production for mixed-model lines also increase. As this occurs the response is often to shorten the scheduled time period from, say, a day to every hour, to reduce the number of alternatives being investigated. On the one hand, this may seem desirable, particularly if the facility is operating in a JIT environment, but there is a danger that the resulting schedules will become so inefficient that they will degrade the overall performance of the line. The leveling of production on mixed-model lines remains an active research topic, with much of the research focusing on developing better or more efficient heuristic scheduling procedures.[49,50]

Parallel Line Balancing

When the task time of one or more elements in an assembly line becomes more than the specified cycle time, the concept of parallel line balance becomes pronounced. Helgeson and Birnie[51] developed the ranked positional weight (RPW) technique and Moodie and Young[52] developed a two-phase procedure for a serial line balancing by using a largest candidate rules and trade and transfer of elements between uneven stations. Many researchers studied the parallel line balancing problem from various scopes of the balance and techniques, but the most recent studies by Sarker and Shanthikumar[53] provided a more general heuristic to balance such a line applicable for both serial and parallel lines.

If the work element times are t_i, $i = 1, 2, \ldots, n$, and C is the cycle time for the line, then the line configuration is given by the following relationship:

$$\text{Configuration} = \begin{cases} \text{Parallel line} & \text{if } \max\{t_i, i = 1, 2, \ldots, n\} > C \\ \text{Serial line} & \text{if } \max\{t_i, i = 1, 2, \ldots, n\} \leq C \end{cases} \quad (40)$$

Figure 17*a* provides a 7-stage balanced parallel line configuration for a specified cycle time of $C = 8$ minutes, but for an operational cycle time of $6C = 48$ minutes.[51] The station time of the balanced line is provided at the top of each block in Figure 17*a* and the number at the bottom indicates the output/48 minutes at the end of each line, the number of parallel lines, and the total output at the end of each parallel configuration as shown in the figure. Sarker and Shanthikumar[51] provided a complete exercise of the heuristic. An equivalent serial line configuration of this parallel line configuration is depicted in Fig. 17*b*.

7 JAPANESE MANUFACTURING PHILOSOPHY

Within the arena of manufacturing, a number of new approaches have revolutionized thinking toward designing and controlling manufacturing organizations. Foremost among this thinking have been the Japanese, who have developed and perfected a whole new philosophy. Some of these more important concepts related to production planning and control are presented below.

7.1 Just-in-Time Philosophy/Kanban Mechanism

The central concepts are to design manufacturing systems that are as simple as possible and then to design simple control procedures to control them. This does not mean that Japanese manufacturing systems are simple, but it does mean that the design is well engineered to perform the required functions and the system is neither overdesigned nor underdesigned.

Central to this philosophy is the *just-in-time* (*JIT*) concept. JIT is a group of beliefs and management practices that attempt to eliminate all forms of waste in a manufacturing enterprise, where waste is defined as anything not necessary in the manufacturing organization. Waste in practice may include inventories, waiting times, equipment breakdowns, scrap, defective products, and excess equipment changeover times. The elimination of waste and the resulting simplification of the manufacturing organization are the results of implementing the following related concepts usually considered as defining or making up JIT.

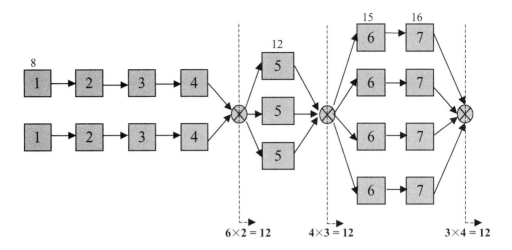

(a) Parallel lines

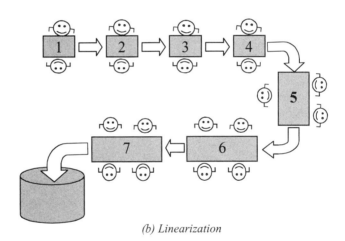

(b) Linearization

Figure 17 Parallel assembly line.

1. *Kanban* (the word means card) is used to control the movement and quantity of inventory through the shop, since a kanban card must be attached to each container of parts. The amount of production and in-process inventory, therefore, is controlled by the number of cards that are issued to the plant floor. An additional, major benefit of using kanban is the very significant reduction in the information system that has to be used to control production.

Various forms of kanban exist, but the most frequently encountered are variations of the single-card or two-card system. One example of a two-card kanban system is that presented in Fig. 18. This example consists of two workstations, A and B. For simplicity, it is assumed that the production from workstation A is used at workstation B. The containers that move

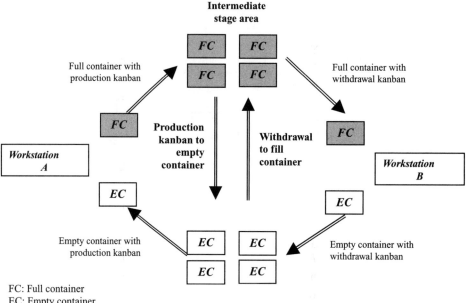

Figure 18 Two-stage kanban system.

between these workstations have been sized to hold only a certain quantity of product. The two different types of kanbans used are a *withdrawal* and a *production* kanban. To control the amount of production for a given period of time, say one day, workstation B is issued a predetermined number of withdrawal kanbans. The system operates as follows:

    a. When workstation B needs parts, the operator takes an empty container, places a withdrawal kanban on it, and takes it to the storage area.

    b. The full containers in the storage area each have a production kanban on them. The worker removes the production kanban from a full container and places it on the empty container, and removes the withdrawal kanban and places it on the full container.

    c. The worker then transports the full container (now with the withdrawal kanban) back to workstation B.

    d. Workstation A checks the production kanbans (on the empty containers) when checking for work to do. If a production kanban is present, this is the signal to begin production. If no production kanbans are present, workstation A does not continue to produce parts.

The materials in a JIT system are kept minimal at the raw materials site, the work-in-process inventory, and also the finished goods warehouse. Raw materials are procured in shipments and consumed over time more or less at a constant rate, whereas in both work-in-process and the finished goods warehouse, the products are demanded right on time instantaneously, resulting in smooth buildup followed by lumpy demand as reflected in Fig. 19. In the figure such an operation is shown with the flow of kanbans, either full or empty or partially empty.

For this system to work, certain rules have to be adhered to:

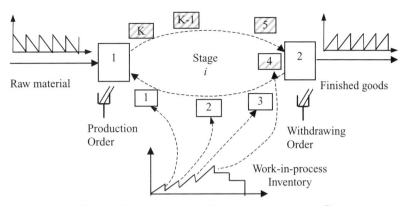

Figure 19 A single-stage JIT operational mechanics.[54]

    i. Each workstation works as long as there are *parts to work on* and a *container in which to put them*. If one or the other is missing, production stops.

    ii. There must be the same number of kanban cards as there are containers.

    iii. Containers are conveyed either full with only their standard quantities or empty.

2. *Lot size reduction* is used to reduce the amount of in-process inventory in concert with kanban, by selecting the proper size containers to use, and to increase the flexibility of the shop to change from one product to another. Overall benefits from using reduced lot sizes include shorter throughput times for product, and thus smaller lead times are required in satisfying customer orders.

3. *Scheduling* is used to schedule small lot production to increase the flexibility of the shop in reacting to changes in demand, and to produce the quantity of goods just in time to when they are needed.

4. *Setup time reduction* is used to reduce the times required for machines to change from one product to another so as to allow lot size reduction and JIT scheduling. Reducing changeover times between products is critical to operating the production facility more like a flow shop and less like a job shop.

5. *Total quality management and maintenance* is used to reduce the disturbances to the manufacturing system by attempting to eliminate the making of defective products and breakdown of equipment. Central to the Japanese manufacturing philosophy is an obsession with maintenance and quality issues. For such a tightly controlled system to work it is imperative that equipment function when it is supposed to, and components and products be produced that meet or exceed customer requirements. Unexpected breakdowns or the production of bad parts is considered waste, and causes of such happenings are always high on the list for elimination in the quest for continuous improvement of the manufacturing processes.

6. *Employee cross-training* is used to provide flexibility in the work force to allow the organization to be able to react to changes in product demand and its resultant effect on the type and quantity of employee skills required. Multiskilled workers are necessary prerequisites in any form of JIT implementation.

7.2 Time-Based Competition

Following on the heels of JIT and the Japanese manufacturing philosophy is a business strategy called time-based competition. The successes of these earlier approaches were primarily grounded in providing the customer with better, more consistent quality products, which might also be less expensive in certain cases. Quality and cost were the major attributes of competitiveness for the organizations that successfully employed these techniques. Although being competitive in quality and cost will always be important, some industries are finding that this alone is not enough to maintain an edge over their competitors, since many of their competitors have also gained benefits by implementing JIT and related concepts. A third element is being introduced—that of time. Time-based competition (TBC) seeks a competitive advantage by the reduction of lead times associated with getting product to customers. TBC attempts to achieve reductions in the times required to design, manufacture, sell, and deliver products for its customers by analyzing and redesigning the processes that perform these functions.

TBC is seen as a natural evolution of JIT in that the implementation of JIT was most often found in production. Because time spent on the shop floor represents less than one-half of the time it takes to get a product to the customer for most industries, TBC is a form of extension of JIT to the rest of the manufacturing organization, including such areas as design, sales, and distribution. Wherever in the organization lead times exist that lengthen the time it takes to get the desired product to the customer, the TBC approach seeks to reduce them.

Two forms of TBC exist: first to market for new products (FM) and first to customer for existing products (FC). Companies that seek to gain a competitive advantage through FM tend to be in dynamic industries, such as those that manufacture automobiles and consumer products. For these industries, new innovations, developments, and improvements are important for their product's image, and are necessary to maintain and increase product sales. Companies employing FC as a competitive advantage tend to be in more stable industries, where innovations and new product developments are less frequent and dramatic. Thus, the products that competitors sell are very similar and competitive in terms of features, price, and quality. Here the emphasis is on speed—reducing the time it takes to get the product in the customer's hands from the time at which it is ordered. There is nothing, of course, that prevents a company from employing both FM and FC approaches, and in the continuous improvement context, both approaches will be necessary if the full benefits of TBC are to be realized.

8 SUPPLY CHAIN MANAGEMENT

A concept that has gained increasing acceptance in the past ten years is the realization that the traditional areas of procurement, production, and logistics functions that every manufacturing organization has are interrelated and should be managed as one system. The result has been the development of the term supply chain management (SCM) with further refinements and expansions so that the term now includes virtually any activity in which a manufacturing organization is involved. Common activities that go under the name of SCM include the following.

1. *Purchasing.* This may be as simple as deciding which vendor to purchase goods and services from or as complex as looking at the procurement function as a strategic business decision. Commonly embedded in purchasing when approached as an SCM component are the ideas of reducing the number of suppliers by setting up selection and certification processes, developing long-term contractual relationships that are mutually beneficial, and setting up well-defined lines of communication for supplier involvement in the organization's decisions that affect them.

2. *Production.* Activities considered SCM in internal production commonly include establishing procedures for supplier and customer involvement. Suppliers may assist in such areas as reducing quality problems associated with materials and components, developing more compatible lot sizes for production equipment, or reducing inventory requirements. Customers may assist in improving product design and quality, or suggest better packaging and shipping alternatives. Development of long-term relationships with vendors and customers may also include setting up formal committees or teams of personnel from the interested parties to work on mutually beneficial solutions to problems affecting the working relationship.

3. *Logistics.* Movement of product to customers has always been an area of great importance, but now the term is also being used to describe the process of getting materials and components to manufacturing as well as intra- and interfacilities movement. Transportation costs are one of the largest costs in making the product available to customers. Again, vendors and customers are often involved. Just-in-time scheduling of deliveries to reduce inventories for both manufacturer and customer, multiple receiving and shipping docks for receipt and shipment of items closest to point of use or convenience, receipt of items at first production process, JIT scheduling for downstream production operations, and vendor-managed inventories are some of the approaches that may be used to reduce costs and improve service.

8.1 Distribution Logistics

Fixed-interval deliveries of the manufactured goods require a reliable manufacturing system with regular supply of raw materials to ensure the production and delivery of finished goods to the customers. A scenario encountered in a production–inventory system such as a supply chain logistics system in electronics industries is depicted in Fig. 20. A silicon wafer vendor supplies wafers to company M, which manufactures Power PC chips, which, in turn, are delivered to several outside customers, such as companies A, I, and M itself. To keep the buyers' demands satisfied at different time intervals, the manufacturing company (company M in this case) has to maintain its production at regular pace by procuring silicon wafers at

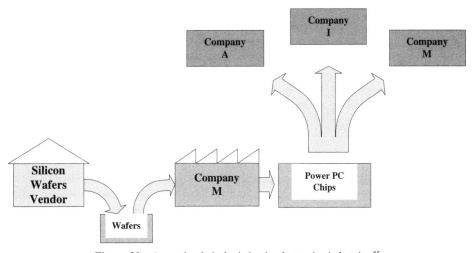

Figure 20 A supply chain logistics in electronics industries.[55]

168 Production Planning

regular intervals of time. Therefore, both the manufacturing company and the finished goods customers need to operate in a harmonic logistics, and to keep the wafer production–inventory system operative at minimal cost, the supply chain logistics of raw materials (silicon wafers) and finished products (Power PC chips) should be efficient (see Fig. 20).

8.2 Applications of Kanban Mechanism to Supply Chain

The concept of kanban system can be applied to interplant transportation for a supply system as shown for two consecutive workstations in Fig. 21. A kanban may be conceived as a tag with a container, which is basically a truck or any other conveyance. This concept can also be extended to multistage transportation in a supply chain system as shown in Fig. 22 for a 3-stage supply system. The production and withdrawl kanban posts are marked with *P* and *W* following by plant number in the parentheses.

Some of the factors for decision making in this system are itemized below:

- Transportation occurs between plants and distribution centers within the plants.
- Transporters serve as in-transit inventories that affect the MRP and inventory valuation.
- The material flow from stage to stage could be converging (assembly types) and diverging (distributive or fabrication type).
- Safety stock, inventory, and demand fluctuations play an important role in the reliability of the supply system.
- The integrated control mechanism should be synchronized for all components and parts between plants and workstations for smooth supply between any two or more points.
- On-line control through modern techniques is essential for better service and customer relation. Also, an exchange of information promotes centralized vs. decentralized storage and mode of transportation.
- Transporters and their schedules.
- Transportation types and modes require decision-making.
- Control of interplant supplies.
- Bulk deliveries by trucks or other intermodal mechanisms.
- Congestion problems in supply mode and warehouses.

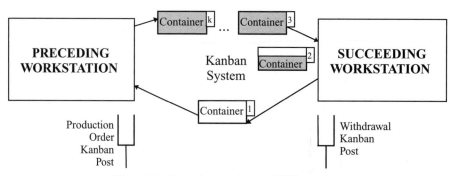

Figure 21 A single-stage kanban (SSK) system.

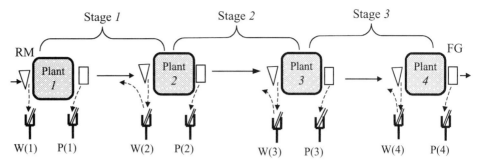

Figure 22 A 3-stage supply transportation system.

- Delivery problems due to unscheduled shipments or dispatching.
- Lack of logistics capability analysis and optimal strategy contribute to poor performance of the system.

8.3 General Remarks

One of the underlying requirements for SCM to work is trust and commitment among the parties involved, i.e., vendors, manufacturer, and customers. As a prerequisite for successful implementation, many SCM improvements may require that significant time and resources be spent in developing these interorganizational relationships, but it is equally important that the intraorganizational relationships be developed. Getting manufacturing to work with marketing or with the financial people may represent some of the greatest challenges and obstacles. It is also imperative that effective working relationships, once established, be nourished to ensure continuing open lines of communication.

As with most management approaches applied to manufacturing, the objectives of SCM are to reduce costs and improve customer service. What is relatively new in the view of SCM is that by linking all of the business, manufacturing, and logistics functions, a competitive advantage in the marketplace may be achieved. The difficulty, of course, is that while this is conceptually feasible, in reality, the overall system is large and complex, with many subsystems whose goals are not necessarily compatible with each other. Thus, while the global goal may be total systems integration, current SCM efforts are less comprehensive and focus on opportunities where potential for improvements are the greatest, such as those mentioned above.[56] Although total systems integration is unlikely to occur in the foreseeable future, continuing to improve the linking of functions and aligning of goals of the units of the manufacturing organization can do much to sustain the continuing drive to reduce costs and improve customer service.

REFERENCES

1. D. D. Bedworth and J. E. Bailey, *Integrated Production Control Systems,* Wiley, New York, 1982.
2. E. A. Silver and R. Peterson, *Decision Systems for Inventory Management and Production Planning,* Wiley, New York, 1985.
3. Johnson, L. A., and D. C. Montgomery, *Operations Research in Production Planning: Scheduling and Inventory Control,* Wiley, New York, 1974.

4. S. Nahmias, *Production and Operations Analysis,* 5th ed., Irwin/McGraw-Hill, New York, 2005.
5. A. H. Taha, *Operations Research,* 4th ed., MacMillian, New York, 1998.
6. D. P. Gover and G. L. Thompson, *Programming and Probability Models for Operations Research,* Wadsworth, 1973.
7. S. Nam and R. Logendran, "Aggregate Production Planning—A Survey of Models and Methodologies," *European Journal of Operations Research,* **61,** 255–272 (1992).
8. Hax, A. C., "Aggregate Production Planning," in *Production Handbook,* 4th ed., J. White (ed.), Wiley, New York, 1987, pp. 3.116–3.127.
9. A. Charnes, W. W. Cooper, and Mellon, B., "A Model for Optimizing Production by Reference to Cost Surrogates," *Econometrics,* **23,** 307–323 (1955).
10. E. H. Bowman, "Production Scheduling by the Transportation Method of Linear Programming," *Operations Research,* **4,** 100–103 (1956).
11. E. H. Bowman, "Consistency and Optimality in Managerial Decision Making," *Management Science,* **9,** 310–321 (1963).
12. M. E. Posner and W. Szwarc, "A Transportation Type Aggregate Production Model with Backordering," *Management Science,* **29,** 188–199 (1983).
13. K. Singhal and V. Adlakha, "Cost and Shortage Trade-offs in Aggregate Production Planning," *Decision Science,* **20,** 158–164 (1989).
14. C. C. Holt, F. Modigliani, and H. A. Simon, "A Linear Decision Rule for Production and Employment Scheduling," *Management Science,* **2,** 1–30 (1955).
15. C. C. Holt, F. Modigliani, and J. F. Muth, "Derivation of a Linear Decision Rule for Production and Employment," *Management Science,* **2,** 159–177 (1956).
16. C. C. Holt, F. Modigliani, J. F. Muth, and H. A. Simon, *Planning Production Inventories and Work Force,* Prentice-Hall, Englewood Cliffs, NJ, 1960.
17. A. S. Manne, "Programming of Economic Lot Sizes," *Management Science,* **4,** 115–135 (1958).
18. H. M. Wagner and T. M. Whitin, "Dynamic Version of the Economic Lot Size Model," *Management Science,* **5,** 89–96 (1958).
19. S. Gorenstein, "Planning Tire Production," *Management Science,* **17,** B72–B82 (1970).
20. S. M. Lee and L. J. Moore, "A Practical Approach to Production Scheduling," *Journal of Production and Inventory Management,* **15,** 79–92 (1974).
21. R. F. Deckro and J. E. Hebert, "Goal Programming Approaches to Solving Linear Decision Rule Based Aggregate Production Planning Models," *IIE Transactions,* **16,** 308–315 (1984).
22. D. P. Gaver, "Operating Characteristics of a Simple Production, Inventory-control Model," *Operations Research,* **9,** 635–649 (1961).
23. W. I. Zangwill, "A Deterministic Multiproduct, Multifacility Production and Inventory Model," *Operations Research,* **14,** 486–507 (1966).
24. W. I. Zangwill, "A Deterministic Multiperiod Production Scheduling Model with Backlogging," *Management Science,* **13,** 105–119 (1966).
25. W. I. Zangwill, "Production Smoothing of Economic Lot Sizes with Non-decreasing Requirements," *Management Science,* **13,** 191–209 (1966).
26. G. D. Eppen and F. J. Gould, "A Lagrangian Application to Production Models," *Operations Research,* **16,** 819–829 (1968).
27. D. R. Lee and D. Orr, "Further Results on Planning Horizons in the Production Smoothing Problem," *Management Science,* **23,** 490–498 (1977).
28. C. H. Jones, "Parametric Production Planning," *Management Science,* **13,** 843–866 (1967).
29. A. D. Flowers and S. E. Preston, "Work Force Scheduling with the Search Decision Rule," *OMEG,* **45,** 473–479 (1977).
30. W. B. Lee and B. M. Khumawala, "Simulation Testing of Aggregate Production Planning Models in an Implementation Methodology," *Management Science,* **20,** 903–911 (1974).
31. H. Hwang and C. N. Cha, "An Improved Version of the Production Switching Heuristic for the Aggregate Production Planning Problems," *International Journal of Production Research,* **33,** 2567–2577 (1995).
32. S. Eilon. "Five Approaches to Aggregate Production Planning," *AIIE Transactions,* **7,** 118–131 (1975).
33. A. C. Hax and D. Candea, *Production and Inventory Management,* Prentice-Hall, Englewood Cliffs, NJ, 1984.

34. P. J. Rondeau and L. A. Littral, "Evolution of Manufacturing Planning and Control Systems: From Reorder Point to Enterprise Resource Planning," *Production and Inventory Management Journal,* 2nd Quarter, 1–7 (2001).
35. V. A. Mabert, A. Soni, and M. A. Venkataramanan, "Enterprise Resource Planning: Common Myths Versus Evolving Reality," *Business Horizons,* May–June, 69–76 (2001).
36. A. Clewett, D. Franklin, and A. McCown, *Network Resource Planning for SAP R/3, BAAN IV, and PEOPLESOFT: A Guide to Planning Enterprise Applications,* McGraw-Hill, New York, 1998.
37. S. French, *Sequencing and Scheduling: An Introduction to the Mathematics of the Job Shop,* Ellis Horwood Limited, Halsted Press, Chichester, England, 1982.
38. K. Baker, *Introduction to Sequencing and Scheduling,* Wiley, New York, 1974.
39. T. J. Hodgson and J. M. Moore, "A Technical Note to 'Sequencing n Jobs on One Machine to Minimize the Number of Tardy Jobs,'" *Management Science,* **17**(1), 102–109 (1968).
40. S. M. Johnson, "Optional Two- and Three-Stage Production Schedules with Setup Times Included," *Naval Research Logistics Quarterly,* **1**(1) (1954).
41. S. S. Panwalkar and W. Iskander, "A Survey of Scheduling Rules," *Operations Research,* **25,** 45–61 (1977).
42. J. Lorenz and D. Pooch, "Assembly Line Balancing," in *Production Handbook,* 4th ed., J. White (ed.), Wiley, New York, 1987, 3.176–3.189.
43. E. A. Elsayed and T. Boucher, *Analysis and Control of Production Systems,* 2nd ed., Prentice-Hall, Englewood Cliffs, NJ, 1994.
44. B. R. Sarker and H. Pan, "Designing a Mixed-Model, Open-station Assembly Line Using Mixed-integer Programming," *Journal of the Operational Research Society,* **52**(5), 545–558 (2001).
45. B. R. Sarker and H. Pan, "Design Configuration for a Closed station, Mixed-model Assembly Line: A Filing Cabinet Manufacturing System," *International Journal of Production Research,* **39**(10), 2251–2270 (2001).
46. N. Thomopoulos, "Mixed Model Line Balancing with Smoothed Station Assignments," *Management Science,* **16**(9), 593–603 (1970).
47. Y. Moden, *Toyota Production System: Practical Approach to Production Management,* Industrial Engineering and Management Press, Atlanta, GA, 1983.
48. Y. Moden, *Toyota Production System: An Integrated A1212roach to Just-In-Time,* 2nd ed., Industrial Engineering and Management Press, Atlanta, GA, 1993.
49. R. T. Sumichrast and R. S. Russell, "Evaluating Mixed-model Assembly Line Sequencing Heuristics for Just-in-time Production Systems," *Journal of Operation Management,* **9**(3), 371–386 (1990).
50. J. F. Bard, A. Shtub, and S. B. Joshi, "Sequencing Mixed-model Assembly Lines to Level Parts Usage and Minimize Line Length," *International of Production Research,* **32**(10), 2431–2454 (1994).
51. W. B. Helgeson and D. P. Birnie, "Assembly Line Balancing Using the Ranke Positional Weight Technique," *Journal of Industrial Engineering* (old name of *IIE Transactions*), **16**(6), 394–398 (1963).
52. C. L. Moodie and H. H. Young, "A heuristic method of assembly line balancing," *Journal of Industrial Engineering* (old name of *IIE Transactions*), **12**(6), 394–398 (1965).
53. B. R. Sarker and J. G. Shanthikumar, "A Generalized Approach for Serial or Parallel Line Balancing," *International Journal of Production Research,* **21**(1), 109–133 (1983).
54. S. Wang and B. R. Sarker, "A Single-stage Supply Chain System Controlled by Kanbans Under Just-in-time Philosophy, *Journal of the Operational Research Society,* **55**(5), 485–494 (2004).
55. G. R. Parija and B. R. Sarker, "Operations Planning in a Supply Chain System with Fixed-interval Deliveries to Multiple Customers," *IIE Transactions Special Issue on Manufacturing Logistics,* **31**(11), 1075–1082 (1999).
56. I. J. Chen and A. Paulraj, "Understanding Supply Chain Management: Critical Research and a Theoretical Framework," *International Journal of Production Research,* **42**(1), 131–163 (2004).

BIBLIOGRAPHY

Adam, E. A., Jr., and R. J. Ebert, *Production and 0perations Management,* 4th ed., Prentice-Hall, Englewood Cliffs, NJ, 1989.

Carter, P., S. Melnyk, and R. Handfield, "Identifying the Basic Process Strategies for Time Based Competition," *Production and Inventory Management Journal,* 1st Quarter, 65–70 (1995).

Eilon, S., "Aggregate Production Scheduling," in *Handbook of Industrial Engineering,* G. Salvendy, (ed.), Wiley Interscience, New York, 1982, pp. 11.3.1–11.3.23.

Fabrycky, W. J., P. M. Ghare, and P. E. Torgersen, *Applied Operations Research and Management Science,* Prentice-Hall, Englewood Cliffs, NJ, 1984.

Gaither, N., *Production and Operations Management,* 6th ed., The Dryden Press, 1992.

Lasdon, L. S., and R. C. Tedung, "An Efficient Algorithm for Multi-item Scheduling," *Operations Research,* **19,** 946–966 (1971).

Mellichamp, J. M., and R. M. Love, "Production Switching Heuristics for the Aggregate Planning Problem," *Management Science,* **24,** 1242–1251 (1978).

Nori, V. S., and B. R. Sarker, "Cyclic Scheduling for a Multi-product, Single-facility Production System Operating Under a Just-in-time Delivery Policy," *Journal of the Operational Research Society,* **47**(7), 930–935 (1996).

Sarker, B. R., and C. V. Balan, "Cell Formation with Operations Times for Even Distribution of Workloads," *International Journal of Production Research,* **34**(5), 1447–1468 (1996).

Schonberger, R. J., *Japanese Manufacturing Techniques,* Free Press, New York, 1983, pp. 219–245.

Scott, B., D. N. Burt, W. Copacino, C. Gopal, H. L. Lee, R. P. Lynch, and S. Morris, "Supply Chain Challenges: Building Relationships: A Conversation," *Harvard Business Review,* **81**(7), 64 (2003).

Volimann, T. E., W. L. Berry, and D. C. Whybark, *Manufacturing Planning and Control Systems,* 3rd ed., Irwin, Homewood, IL, 1992.

Wagner, A. M., *Principles of Management Science,* Prentice-Hall, New York, (1970).

CHAPTER 5
PRODUCTION PROCESSES AND EQUIPMENT

Magd E. Zohdi
Industrial Engineering Department
Louisiana State University
Baton Rouge, Louisiana

William E. Biles
Industrial Engineering Department
University of Louisville
Louisville, Kentucky

Dennis B. Webster
Industrial Engineering Department
Louisiana State University
Baton Rouge, Louisiana

1	**METAL-CUTTING PRINCIPLES**	174
2	**MACHINING POWER AND CUTTING FORCES**	178
3	**TOOL LIFE**	180
4	**METAL-CUTTING ECONOMICS**	182
	4.1 Cutting Speed for Minimum Cost (V_{min})	185
	4.2 Tool Life Minimum Cost (T_m)	185
	4.3 Cutting Speed for Maximum Production (V_{max})	185
	4.4 Tool Life for Maximum Production (T_{max})	185
5	**CUTTING-TOOL MATERIALS**	185
	5.1 Cutting-Tool Geometry	186
	5.2 Cutting Fluids	187
	5.3 Machinability	188
	5.4 Cutting Speeds and Feeds	188
6	**TURNING MACHINES**	188
	6.1 Lathe Size	191
	6.2 Break-Even (BE) Conditions	191
7	**DRILLING MACHINES**	192
	7.1 Accuracy of Drills	197
8	**MILLING PROCESSES**	200
9	**GEAR MANUFACTURING**	203
	9.1 Machining Methods	204
	9.2 Gear Finishing	205
10	**THREAD CUTTING AND FORMING**	206
	10.1 Internal Threads	207
	10.2 Thread Rolling	208
11	**BROACHING**	209
12	**SHAPING, PLANING, AND SLOTTING**	210
13	**SAWING, SHEARING, AND CUTTING OFF**	213
14	**MACHINING PLASTICS**	214
15	**GRINDING, ABRASIVE MACHINING, AND FINISHING**	215
	15.1 Abrasives	215
	15.2 Temperature	218
16	**NONTRADITIONAL MACHINING**	219
	16.1 Abrasive Flow Machining	220
	16.2 Abrasive Jet Machining	220
	16.3 Hydrodynamic Machining	220

16.4	Low-Stress Grinding	220		16.21	Electrical Discharge Grinding	235
16.5	Thermally Assisted Machining	222		16.22	Electrical Discharge Machining	236
16.6	Electromechanical Machining	222		16.23	Electrical Discharge Sawing	237
16.7	Total Form Machining	222		16.24	Electrical Discharge Wire Cutting (Traveling Wire)	237
16.8	Ultrasonic Machining	225		16.25	Laser-Beam Machining	238
16.9	Water-Jet Machining	227		16.26	Laser-Beam Torch	239
16.10	Electrochemical Deburring	228		16.27	Plasma-Beam Machining	240
16.11	Electrochemical Discharge Grinding	229		16.28	Chemical Machining: Chemical Milling, Chemical Blanking	241
16.12	Electrochemical Grinding	229		16.29	Electropolishing	241
16.13	Electrochemical Honing	229		16.30	Photochemical Machining	241
16.14	Electrochemical Machining	231		16.31	Thermochemical Machining	242
16.15	Electrochemical Polishing	232		16.32	Rapid Prototyping and Rapid Tooling	243
16.16	Electrochemical Sharpening	232				
16.17	Electrochemical Turning	233		**REFERENCES**		**243**
16.18	Electro-Stream	233		**BIBLIOGRAPHY**		**244**
16.19	Shaped-Tube Electrolytic Machining	234				
16.20	Electron-Beam Machining	235				

1 METAL-CUTTING PRINCIPLES

Material removal by chipping process began as early as 4000 BC, when the Egyptians used a rotating bowstring device to drill holes in stones. Scientific work developed starting about the mid-19th century. The basic chip-type machining operations are shown in Fig. 1.

Figure 2 shows a two-dimensional type of cutting in which the cutting edge is perpendicular to the cut. This is known as *orthogonal* cutting, as contrasted with the three-dimensional *oblique* cutting shown in Fig. 3. The main three cutting velocities are shown in Fig. 4. The metal-cutting factors are defined as follows:

α	rake angle
β	friction angle
γ	strain
λ	chip compression ratio, t_2/t_1
μ	coefficient of friction
ψ	tool angle
τ	shear stress
ϕ	shear angle
Ω	relief angle
A_o	cross section, wt_1
e_m	machine efficiency factor
f	feed rate ipr (in./revolution), ips (in./stroke), mm/rev (mm/revolution), or mm/stroke
f_t	feed rate (in./tooth, mm/tooth) for milling and broaching
F	feed rate, in./min (mm/s)
F_c	cutting force

1 Metal-Cutting Principles 175

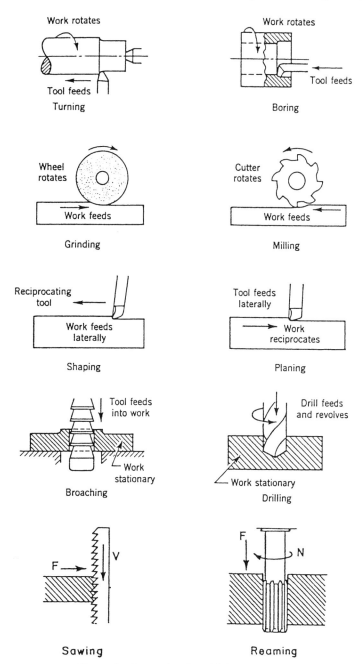

Figure 1 Conventional machining processes.

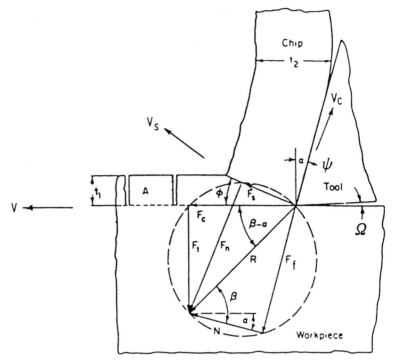

Figure 2 Mechanics of metal-cutting process.

F_f	friction force
F_n	normal force on shear plane
F_s	shear force
F_t	thrust force
HP_c	cutting horsepower
HP_g	gross horsepower
HP_u	unit horsepower
N	revolutions per minute
Q	rate of metal removal, in.^{3}/min
R	resultant force
T	tool life in minutes
t_1	depth of cut
t_2	chip thickness
V	cutting speed, ft/min
V_c	chip velocity
V_s	shear velocity

The *shear angle* ϕ controls the thickness of the chip and is given by

$$\tan \phi = \frac{\cos \alpha}{\lambda - \sin \alpha} \tag{1}$$

The *strain* γ that the material undergoes in shearing is given by

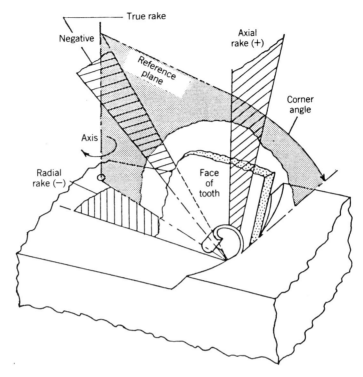

Figure 3 Oblique cutting.

$$\gamma = \cot \phi + \tan(\phi - \alpha)$$

The *coefficient of friction* μ on the face of the tool is

$$\mu = \frac{F_t + F_c \tan \alpha}{F_c - F_t \tan \alpha} \qquad (2)$$

The *friction force* F_t along the tool is given by

$$F_t = F_t \cos \alpha + F_c \sin \alpha$$

Cutting forces are usually measured with dynamometers and/or wattmeters. The shear stress τ in the shear plane is

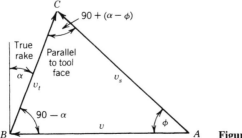

Figure 4 Cutting velocities.

$$\tau = \frac{F_c \sin\phi \cos\phi - F_t \sin^2\phi}{A}$$

The speed relationships are

$$\frac{V_c}{V} = \frac{\sin\phi}{\cos(\phi - \alpha)}$$

$$V_c = V/\lambda \tag{3}$$

2 MACHINING POWER AND CUTTING FORCES

Estimating the power required is useful when planning machining operations, optimizing existing ones, and specifying new machines. The power consumed in cutting is given by

$$\text{power} = F_c V \tag{4}$$

$$\text{HP}_c = \frac{F_c V}{33{,}000} \tag{5}$$

$$= Q\,\text{HP}_u \tag{6}$$

where F_c = cutting force, lb
  V = cutting speed, ft/min = $\pi DN/12$ (*rotating operations*)
  D = diameter, in.
  N = revolutions/min
  HP_u = specific power required to cut a material at a rate of 1 cu in./min
  Q = material removal rate, cu in./min

For SI units,

$$\text{Power} = F_c V \quad \text{watts} \tag{7}$$

$$= QW \quad \text{watts} \tag{8}$$

where F_c = cutting force, newtons
  V = m/s = $2\pi RN$
  W = specific power required to cut a material at a rate of 1 cu mm/s
  Q = material removal rate, cu mm/s

The specific energies for different materials, using sharp tools, are given in Table 1.

$$\text{Power} = F_c V = F_c 2\pi RN$$

$$= F_c R 2\pi N$$

$$= M 2\pi N \tag{9}$$

$$= \frac{MN}{63{,}025} \quad \text{HP} \tag{10}$$

where M = torque, in.–lbf
  N = revolutions per min

Table 1 Average Values of Energy per Unit Material Removal Rate

Material	BHN	$HP_c/in.^3$ per min	$W/mm^3/s$
Aluminum alloys	50–100	0.3	0.8
	100–150	0.4	1.1
Cast iron	125–190	0.5	1.6
	190–250	1.6	4.4
Carbon steels	150–200	1.1	3.0
	200–250	1.4	3.8
	250–350	1.6	4.4
Leaded steels	150–175	0.7	1.9
Alloy steels	180–250	1.6	4.4
	250–400	2.4	6.6
Stainless steels	135–275	1.5	4.1
Copper	125–140	1.0	2.7
Copper alloys	100–150	0.8	2.2
Leaded brass	60–120	0.7	1.9
Unleaded brass	50	1.0	2.7
Magnesium alloys	40–70	0.2	0.55
	70–160	0.4	1.1
Nickel alloys	100–350	2.0	5.5
Refractory alloys (tantalum, columbium, molybdenum)	210–230	2.0	5.5
Tungsten	320	3.0	8.0
Titanium alloys	250–375	1.3	3.5

Note. BHN, Brinnel hardness number.

In SI units,

$$= \frac{MN}{9549} \text{ kW} \tag{11}$$

where M = newton–meter
HP/cu in./min 2.73 = ? W/(cu mm/s)
$M = F_c R$ = power/$2\pi N$
$F_c = M/R$

$$\text{Gross power} = \text{Cutting power}/e_m \tag{13}$$

The cutting horsepowers for different machining operations are given below.
For turning, planing, and shaping,

$$HP_c = (HP_u)12CWVfd \tag{14}$$

For milling,

$$HP_c = (HP_u)CWFwd \tag{15}$$

For drilling,

$$HP_c = (HP_u)CW(N)f\left(\frac{\pi D^2}{4}\right) \tag{16}$$

For broaching,

$$HP_c = (HP_u)12CWVn_cwd_t \qquad (17)$$

where V = cutting speed, fpm
C = feed correction factor
f = feed, ipr (turning and drilling), ips (planing and shaping)
F = feed, ipm = $f \times N$
d = depth of cut, in.
d_t = maximum depth of cut per tooth, in.
n_c = number of teeth engaged in work
w = width of cut, in.
W = tool wear factor

Specific energy is affected by changes in feed rate. Table 2 gives feed correction factor (C). Cutting speed and depth of cut have no significant effect on power. Tool wear effect factor (W) is given in Table 3. The gross power is calculated by applying the overall efficiency factor (e_m).

3 TOOL LIFE

Tool life is a measure of the length of time a tool will cut satisfactorily, and may be measured in different ways. Tool wear, as in Fig. 5, is a measure of tool failure if it reaches a certain limit. These limits are usually 0.062 in. (1.58 mm) for high-speed tools and 0.030 in. (0.76 mm) for carbide tools. In some cases, the life is determined by surface finish deterioration and an increase in cutting forces. The cutting speed is the variable that has the greatest effect on tool life. The relationship between tool life and cutting speed is given by the Taylor equation:

$$VT^n = C \qquad (18)$$

where V = cutting speed, fpm (m/s)
T = tool life, min (s)
n = exponent depending on cutting condition
C = constant, the cutting speed for a tool life of 1 min

Table 2 Feed Correction (C) Factors for Turning, Milling, Drilling, Planing, and Shaping

Feed (ipr or ips)	mm/rev or mm/stroke	Factor
0.002	0.05	1.4
0.005	0.12	1.2
0.008	0.20	1.05
0.012	0.30	1.0
0.020	0.50	0.9
0.030	0.75	0.80
0.040	1.00	0.80
0.050	1.25	0.75

Table 3 Tool Wear Factors (W)

Type of Operations[a]	W
Turning	
Finish turning (light cuts)	1.10
Normal rough and semifinish turning	1.30
Extra-heavy-duty rough turning	1.60–2.00
Milling	
Slab milling	1.10
End milling	1.10
Light and medium face milling	1.10–1.25
Extra-heavy-duty face milling	1.30–1.60
Drilling	
Normal drilling	1.30
Drilling hard-to-machine materials and drilling with a very dull drill	1.50
Broaching	
Normal broaching	1.05–1.10
Heavy-duty surface broaching	1.20–1.30

[a] For all operations with sharp cutting tools.

Table 4 gives the approximate ranges for the exponent n. Taylor's equation is equivalent to

$$\log V = C - n \log T \tag{19}$$

which when plotted on log–log paper gives a straight line, as shown in Fig. 6.

Equation (20) incorporates the size of cut:

$$K = V T^n f^{n_1} d^{n_2} \tag{20}$$

Average values for $n_1 = 0.5$–0.8
$n_2 = 0.2$–0.4

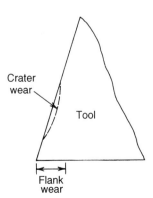

Figure 5 Types of tool wear.

Table 4 Average Values of n

Tool Material	Work Material	n
HSS (18-4-1)	Steel	0.15
	C.I.	0.25
	Light metals	0.40
Cemented carbide	Steel	0.30
	C.I.	0.25
Sintered carbide	Steel	0.50
Ceramics	Steel	0.70

Equation (21) incorporates the hardness of the workpiece:

$$K = VT^n f^{n1} d^{n2} (\text{BHN})^{1.25} \tag{21}$$

4 METAL-CUTTING ECONOMICS

The efficiency of machine tools increases as cutting speeds increase, but tool life is reduced. The main objective of metal-cutting economics is to achieve the optimum conditions, that is, the minimum cost while considering the principal individual costs: machining cost, tool cost, tool-changing cost, and handling cost. Figure 7 shows the relationships among these four factors.

$$\text{Machining cost} = C_o t_m \tag{22}$$

where C_o = operating cost per minute, which is equal to the machine operator's rate plus appropriate overhead
t_m = machine time in minutes, which is equal to $L/(fN)$, where L is the axial length of cut

$$\text{Tool cost per operation} = C_t \frac{t_m}{T} \tag{23}$$

where C_t = tool cost per cutting edge
T = tool life, which is equal to $(C/V)^{1/n}$

$$\text{Tool changing cost} = C_o t_c (t_m/T) \tag{24}$$

where t_c = tool changing time, min

$$\text{Handling cost} = C_o t_h$$

where t_h = handling time, min

The average unit cost C_u will be equal to

$$C_u = C_o t_m + \frac{t_m}{T}(C_t + C_o t_c) + C_o t_h \tag{25}$$

4 Metal-Cutting Economics 183

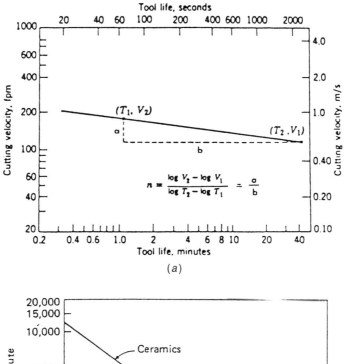

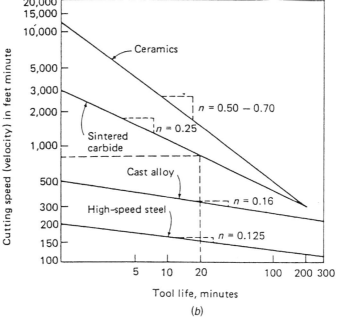

Figure 6 Cutting speed/tool life relationship.

184 Production Processes and Equipment

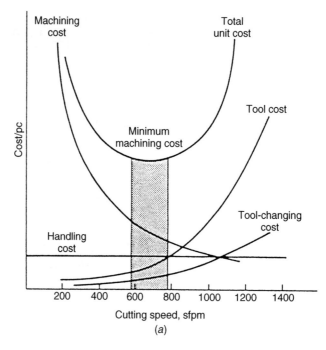

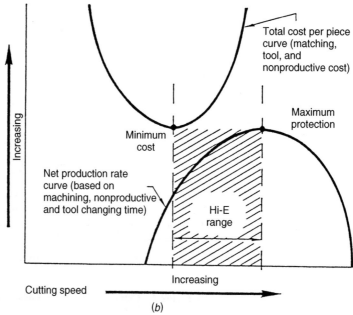

Figure 7 Cost factors.

4.1 Cutting Speed for Minimum Cost (V_{min})

Differentiating the costs with respect to cutting speed and setting the results equal to zero will result in V_{min}:

$$V_{min} = \frac{C}{(1/n - 1)[(C_o t + C_t)/C_o]^n} \tag{26}$$

4.2 Tool Life Minimum Cost (T_m)

Since the constant C is the same in Taylor's equation and Eq. (23), and if V corresponds to V_{min}, then the tool life that corresponds to the cutting speed for minimum cost is

$$T_{min} = \left(\frac{1}{n} - 1\right)\left(\frac{C_o t_c + C_t}{C_o}\right) \tag{27}$$

4.3 Cutting Speed for Maximum Production (V_{max})

This speed can be determined from Eq. (26) for the cutting speed for minimum cost by assuming that the tool cost is negligible, that is, by setting $C_1 = 0$:

$$V_{max} = \frac{C}{[(1/n - 1)\, t_c]^n} \tag{28}$$

4.4 Tool Life for Maximum Production (T_{max})

By analogy to Taylor's equation, the tool life that corresponds to the maximum production rate is given by

$$T_{max} = \left(\frac{1}{n} - 1\right) t_c \tag{29}$$

5 CUTTING-TOOL MATERIALS

The desirable properties for any tool material include the ability to resist softening at high temperature, which is known as red hardness; a low coefficient of friction; wear resistance; sufficient toughness and shock resistance to avoid fracture; and inertness with respect to workpiece material.

The principal materials used for cutting tools are carbon steels, cast nonferrous alloys, carbides, ceramic tools or oxides, and diamonds.

1. *High-carbon steels* contain (0.8–1.2%) carbon. These steels have good hardening ability, and with proper heat treatment hold a sharp cutting edge where excessive abrasion and high heat are absent. Because these tools lose hardness at around 600°F (315°C), they are not suitable for high speeds and heavy-duty work.
2. *High-speed steels* (HSS) are high in alloy contents such as tungsten, chromium, vanadium, molybdenum, and cobalt. High-speed steels have excellent hardenability and will retain a keen cutting edge to temperatures around 1200°F (650°C).
3. *Cast nonferrous alloys* contain principally chromium, cobalt, and tungsten, with smaller percentages of one or more carbide-forming elements, such as tantalum,

molybdenum, or boron. Cast-alloy tools can maintain good cutting edges at temperatures up to 1700°F (935°C) and can be used at twice the cutting speed as HSS and still maintain the same feed. Cast alloys are not as tough as HSS and have less shock resistance.

4. *Carbides* are made by powder-metallurgy techniques. The metal powders used are tungsten carbide (WC), cobalt (Co), titanium carbide (TiC), and tantalum carbide (TaC) in different ratios. Carbide will maintain a keen cutting edge at temperatures over 2200°F (1210°C) and can be used at speeds two or three times those of cast alloy tools.

5. *Coated tools,* cutting tools, and inserts are coated by titanium nitride (TiN), titanium carbide (TiC), titanium carbonitride (TiCN), aluminum oxide (Al_2O_3), and diamond. Cutting speeds can be increased by 50% due to coating.

6. *Ceramic or oxide tool* inserts are made from aluminum oxide (Al_2O_3) grains with minor additions of titanium, magnesium, or chromium oxide by powder-metallurgy techniques. These inserts have an extremely high abrasion resistance and compressive strength, lack affinity for metals being cut, resistance to cratering and heat conductivity. They are harder than cemented carbides but lack impact toughness. The ceramic tool softening point is above 2000°F (1090°C) and these tools can be used at high speeds (1500–2000 ft/min) with large depth of cut. Ceramic tools have tremendous potential because they are composed of materials that are abundant in the earth's crust. Optimum cutting conditions can be achieved by applying negative rank angles (5–7°), rigid tool mountings, and rigid machine tools.

7. *Cubic boron nitride* (CBN) is the hardest material presently available, next to diamond. CBN is suitable for machining hardened ferrous and high-temperature alloys. Metal removal rates up to 20 times those of carbide cutting tools were achieved.

8. *Single-crystal diamonds* are used for light cuts at high speeds of 1000–5000 fpm to achieve good surface finish and dimensional accuracy. They are used also for hard materials difficult to cut with other tool material.

9. *Polycrystalline diamond* cutting tools consist of fine diamond crystals, natural or synthetic, that are bonded together under high pressure and temperature. They are suitable for machining nonferrous metals and nonmetallic materials.

5.1 Cutting-Tool Geometry

The shape and position of the tool relative to the workpiece have a very important effect in metal cutting. There are six single-point tool angles critical to the machining process. These can be divided into three groups.

Rake angles affect the direction of chip flow, the characteristics of chip formation, and tool life. Positive rake angles reduce the cutting forces and direct the chip flow away from the material. Negative rake angles increase cutting forces but provide greater strength, as is recommended for hard materials.

Relief angles avoid excessive friction between the tool and workpiece and allow better access of coolant to tool–work interface.

The *side cutting-edge angle* allows the full load of the cut to be built up gradually. The *end cutting-edge angle* allows sufficient clearance so that the surface of the tool behind the cutting point will not rub over the work surface.

The purpose of the *nose radiuses* is to give a smooth surface finish and to increase the tool life by increasing the strength of the cutting edge. The elements of the single-point tool are written in the following order: back rake angle, side rake angle, end relief angle, side relief angle, end cutting-edge angle, side cutting-edge angle, and nose radius. Figure 8 shows the basic tool geometry.

Cutting tools used in various machining operations often appear to be very different from the single-point tool in Figure 8. Often they have several cutting edges, as in the case of drills, broaches, saws, and milling cutters. Simple analysis will show that such tools are comprised of a number of single-point cutting edges arranged so as to cut simultaneously or sequentially.

5.2 Cutting Fluids

The major roles of the cutting fluids—liquids or gases—are

1. Removal of the heat friction and deformation
2. Reduction of friction among chip, tool, and workpiece
3. Washing away chips
4. Reduction of possible corrosion on both workpiece and machine
5. Prevention of built-up edges

Cutting fluids work as coolants and lubricants. Cutting fluids applied depend primarily on the kind of material being used and the type of operation. The four major types of cutting fluids are

1. Soluble oil emulsions with water-to-oil ratios of 20:1 to 80:1
2. Oils

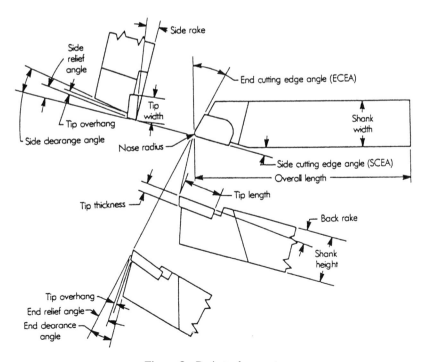

Figure 8 Basic tool geometry.

3. Chemicals and synthetics
4. Air

At low cutting speeds (40 ft/min and below), oils are highly recommended, especially in tapping, reaming, and gear and thread machining. Cutting fluids with the maximum specific heat, such as soluble oil emulsions, are recommended at high speeds.

5.3 Machinability

Machinability refers to a system for rating materials on the basis of their relative ability to be machined easily, long tool life, low cutting forces, and acceptable surface finish. Additives such as lead, manganese sulfide, or sodium sulfide with percentages less than 3% can improve the machinability of steel and copper-based alloys, such as brass and bronze. In aluminum alloys, additions up to 1–3% of zinc and magnesium improve their machinability.

5.4 Cutting Speeds and Feeds

Cutting speed is expressed in feet per minute (m/s) and is the relative surface speed between the cutting tool and the workpiece. It may be expressed by the simple formula CS = πDN /12 fpm in., where D is the diameter of the workpiece in inches in case of turning or the diameter of the cutting tool in case of drilling, reaming, boring, and milling, and N is the revolutions per minute. If D is given in millimeters, the cutting speed is CS = πDN/60,000 m/s.

Feed refers to the rate at which a cutting tool advances along or into the surface of the workpiece. For machines in which either the workpiece or the tool turns, feed is expressed in inches per revolution (ipr) (mm/rev). For reciprocating tools or workpieces, feed is expressed in inches per stroke (ips) (mm/stroke).

The recommended cutting speeds, and depth of cut that resulted from extensive research, for different combinations of tools and materials under different cutting conditions can be found in many references, including Society of Manufacturing Engineers (SME) publications such as *Tool and Manufacturing Engineers Handbook*[1]; *Machining Data Handbook*[2]; Metcut Research Associates, Inc.; *Journal of Manufacturing Engineers*; *Manufacturing Engineering Transactions*; *American Society for Metals (ASM) Handbook*[3]; *American Machinist's Handbook*[4]; *Machinery's Handbook*[5]; American Society of Mechanical Engineering (ASME) publications; Society of Automotive Engineers (SAE) Publications; and *International Journal of Machine Tool Design and Research*.

6 TURNING MACHINES

Turning is a machining process for generating external surfaces of revolution by the action of a cutting tool on a rotating workpiece, usually held in a lathe. Figure 9 shows some of the external operations that can be done on a lathe. When the same action is applied to internal surfaces of revolution, the process is termed *boring*. Operations that can be performed on a lathe are turning, facing, drilling, reaming, boring, chamfering, taping, grinding, threading, tapping, and knurling.

The primary factors involved in turning are speed, feed, depth of cut, and tool geometry. Figure 10 shows the tool geometry along with the feed (f) and depth of cut (d). The cutting speed (CS) is the surface speed in feet per minute (fpm) or meters per second (m/s). The

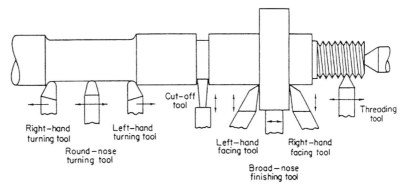

Figure 9 Common lathe operations.

feed (f) is expressed in inches of tool advance per revolution of the spindle (ipr) or (mm/rev). The depth of cut (d) is expressed in inches. Table 5 gives some of the recommended speeds while using HSS tools and carbides for the case of finishing and rough machining. The cutting speed (fpm) is calculated by

$$\text{CS} = \frac{\pi DN}{12} \quad \text{fpm} \tag{30}$$

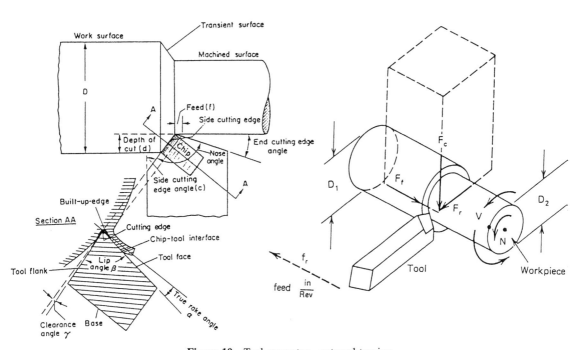

Figure 10 Tool geometry—external turning.

Table 5 Typical Cutting Speeds ft/min (m/s)

Material	High-Speed Steel		Carbide	
	Finish[a]	Rough[b]	Finish[a]	Rough[b]
Free cutting steels, 1112, 1315	250–350 (1.3–1.8)	80–160 (0.4–0.8)	600–750 (3.0–3.8)	350–500 (1.8–2.5)
Carbon steels, 1010, 1025	225–300 (1.1–1.5)	80–130 (0.4–0.6)	550–700 (2.8–3.5)	300–450 (1.5–2.3)
Medium steels, 1030, 1050	200–300 (1.0–1.5)	70–120 (0.4–0.6)	450–600 (2.3–3.0)	250–400 (1.3–2.0)
Nickel steels, 2330	200–300 (1.0–1.5)	70–110 (0.4–0.6)	425–550 (2.1–2.8)	225–350 (1.1–1.8)
Chromium nickel, 3120, 5140	150–200 (0.8–1.0)	60–80 (0.3–0.4)	325–425 (1.7–2.1)	175–300 (0.9–1.5)
Soft gray cast iron	120–150 (0.6–0.8)	80–100 (0.4–0.5)	350–450 (1.8–2.3)	200–300 (1.0–1.5)
Brass, normal	275–350 (1.4–1.8)	150–225 (0.8–1.1)	600–700 (3.0–3.5)	400–600 (2.0–3.0)
Aluminum	225–350 (1.1–1.8)	100–150 (0.5–0.8)	450–700 (2.3–3.5)	200–350 (1.0–1.8)
Plastics	300–500 (1.5–2.5)	100–200 (0.5–1.0)	400–650 (2.0–3.3)	150–300 (0.8–1.5)

[a] Cut depth, 0.015–0.10 in. (0.38–2.54 mm); feed 0.005–0.015 ipr (0.13–0.38 mm/rev).
[b] Cut depth, 0.20–0.40 in. (5.0–10.0 mm); feed, 0.030–0.060 ipr (0.75–1.5 mm/rev).

where D = workpiece diameter, in.
N = spindle revolutions per minute

For SI units,

$$\text{CS} = \frac{\pi D N}{1000} \quad \text{m/s} \tag{31}$$

where D is in mm
N is in revolutions per second

The tool advancing rate is $F = f \times N$ ipm (mm/s). The machining time (T_1) required to turn a workpiece of length L in. (mm) is calculated from

$$T_1 = \frac{L}{F} \quad \text{min (s)} \tag{32}$$

The machining time (T_2) required to face a workpiece of diameter D is given by

$$T_2 = \frac{D/2}{F} \quad \text{min (s)} \tag{33}$$

The rate of metal removal (MRR) (Q) is given by

$$Q = 12fd\text{CS} \quad \text{in.}^3/\text{min} \tag{34}$$

$$\text{Power} = Q\text{HP}_u \quad \text{HP} \tag{35}$$

$$\text{Power} = \text{Torque } 2\pi N$$

$$= \frac{\text{Torque} \times N}{63{,}025} \quad \text{HP} \tag{36}$$

where torque is in in.–lbf

For SI units,

$$\text{Power} = \frac{\text{Torque} \times N}{9549} \quad \text{kW} \tag{37}$$

where torque is in newton–meter and N in rev/min
$$\text{torque} = F_c \times R$$

$$F_c = \frac{\text{Torque}}{R} \tag{38}$$

where R = radius of workpiece

To convert to SI units,

$$\text{HP} \times 746 = ? \text{ watt (W)}$$

$$f \text{ (lb)} \times 4.448 = ? \text{ newtons}$$

$$\text{torque (in.–lb)} \times 0.11298 = ? \text{ newton–meter (Nm)}$$

$$\text{HP/(cu in./min)} \times 2.73 = ? \text{ W/(cu mm/s)}$$

$$\text{ft/min} \times 0.00508 = ? \text{ m/s}$$

$$\text{in.}^3 \times 16{,}390 = ? \text{ mm}^3$$

Alignment charts were developed for determining metal removal rate and motor power in turning. Figures 11 and 12 show the method of using these charts either for English or metric units. The unit power (P) is the adjusted unit power with respect to turning conditions and machine efficiency.

6.1 Lathe Size

The size of a lathe is specified in terms of the diameter of the work it will swing and the workpiece length it can accommodate. The main types of lathes are engine, turret, single-spindle automatic, automatic screw machine, multispindle automatic, multistation machines, boring, vertical, and tracer. The level of automation can range from semiautomatic to tape-controlled machining centers.

6.2 Break-Even (BE) Conditions

The selection of a specific machine for the production of a required quantity q must be done in a way to achieve minimum cost per unit produced. The incremental setup cost is given

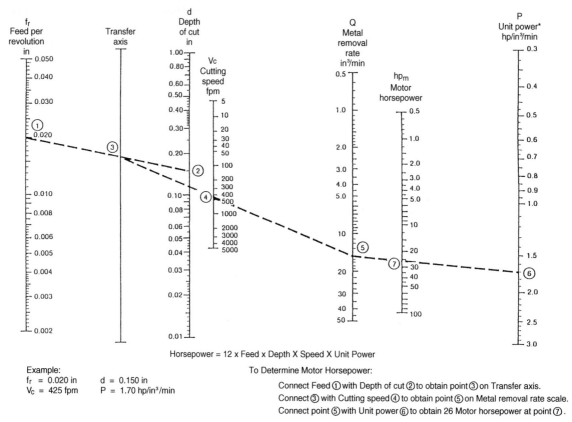

Figure 11 Alignment chart for determining metal removal rate and motor horsepower in turning—English units.

by ΔC_t, C_1 is the machining cost per unit on the first machine, and C_2 is the machining cost for the second machine, the break-even point will be calculated as follows:

$$BE = \Delta C \sqrt{(C_1 - C_2)}$$

7 DRILLING MACHINES

Drills are used as the basic method of producing holes in a wide variety of materials. Figure 13 indicates the nomenclature of a standard twist drill and its comparison with a single-point tool. Knowledge of the thrust force and torque developed in the drilling process is important for design consideration. Figure 14 shows the forces developed during the drilling process. From the force diagram, the thrust force must be greater than $2P_y + P_y^1$ to include the friction on the sides and to be able to penetrate in the metal. The torque required is equal to $P_2 X$. It is reported in the *Tool and Manufacturing Engineers Handbook*[1] that the following relations reasonably estimate the torque and thrust requirements of sharp twist drills of various sizes and designs.

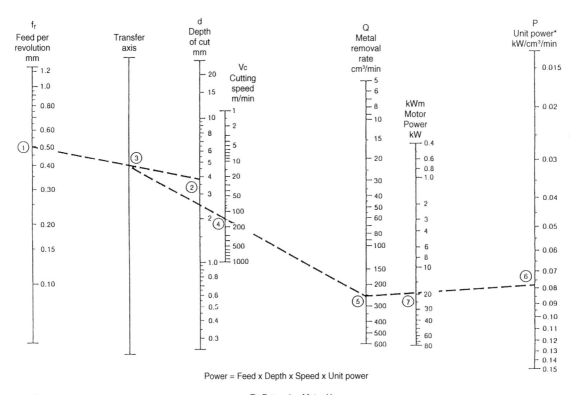

Power = Feed × Depth × Speed × Unit power

Example:
f_r = 0.5 mm   d = 3.8 mm
V_C = 130 m/min   P = 0.077 kW/cm³/min

To Determine Motor Horsepower:

Connect Feed ① with Depth of cut ② to obtain point ③ on Transfer axis.
Connect ③ with Cutting speed ④ to obtain point ⑤ on Metal removal rate scale.
Connect point ⑤ with Unit power ⑥ to obtain 19.02 kW at Motor, point ⑦.

Figure 12 Alignment chart for determining metal removal rate and motor power in turning—metric units.

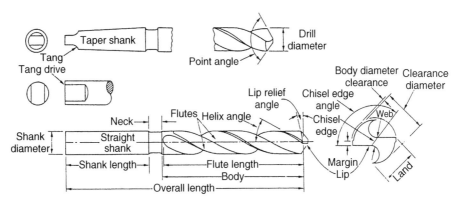

Figure 13 Drill geometry.

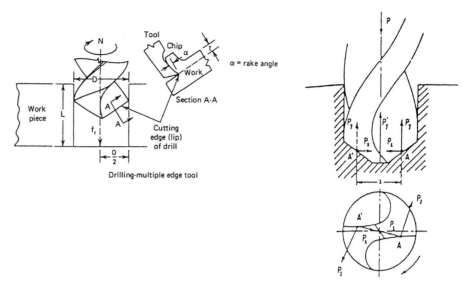

Figure 14 Thrust forces and torque in drilling operation.

$$\text{Torque:} \quad M = Kf^{0.8}d^{1.8}A \quad \text{in.–lbf} \tag{39}$$

$$\text{Thrust:} \quad T = 2Kf^{0.8}d^{0.8}B + kd^2E \quad \text{lb} \tag{40}$$

The thrust force has a large effect upon the required strength, rigidity, and accuracy, but the power required to feed the tool axially is very small.

$$\text{Cutting power:} \quad \text{HP} = \frac{MN}{63{,}025} \tag{41}$$

where K = work–material constant
f = drill feed, ipr
d = drill diameter, in.
A, B, E = design constants
N = drill speed, rpm

Tables 6 and 7 give the constants used with the previous equations. Cutting speed at the surface is usually taken as 80% of turning speeds and is given by

$$\text{CS} = \frac{\pi d N}{12} \quad \text{fpm}$$

Force in cutting direction:

$$F_c = \frac{33{,}000\,\text{HP}}{\text{CS}} \quad \text{lb} \tag{42}$$

For SI units,

$$\text{CS} = \frac{\pi d_1 N}{60{,}000} \quad \text{m/s} \tag{43}$$

Table 6 Work-Material Constants for Calculating Torque and Thrust (National Twist Drill)

Work Material	K
Steel, 200 BHN	24,000
Steel, 300 BHN	31,000
Steel, 400 BHN	34,000
Most aluminum alloys	7,000
Most magnesium alloys	4,000
Most brasses	14,000
Leaded brass	7,000
Cast iron, 65 Bhn	15,000
Free-machining mild steel, resulfurized	18,000
Austenitic stainless steel (type 316)	34,000

c = chisel-edge length, in.

d = drill diameter, in.

w = web thickness, in.

d_1 = drill diameter, in mm

Unit HP (hp/in.^{3}/min) × 2.73 = ? unit power (kW/cm^3/s)

$$kW = \frac{MN}{9549} \qquad (44)$$

M = torque   Nm

For drills of regular proportion the ratio c/d is = 0.18 and c = 1.15w, approximately.

Table 7 Torque and Thrust Constants Based on Ratios c/d or w/d (National Twist Drill)

c/d	w/d	Torque Constant A	Thrust Constant B	Thrust Constant E
0.03	0.025	1.000	1.100	0.001
0.05	0.045	1.005	1.140	0.003
0.08	0.070	1.015	1.200	0.006
0.10	0.085	1.020	1.235	0.010
0.13	0.110	1.040	1.270	0.017
0.15	0.130	1.080	1.310	0.022
0.18	0.155	1.085	1.355	0.030
0.20	0.175	1.105	1.380	0.040
0.25	0.220	1.155	1.445	0.065
0.30	0.260	1.235	1.500	0.090
0.35	0.300	1.310	1.575	0.120
0.40	0.350	1.395	1.620	0.160

It is a common practice to feed drills at a rate that is proportional to the drill diameter in accordance with

$$f = \frac{d}{65} \qquad (45)$$

For holes that are longer than $3d$, feed should be reduced. Also feeds and speeds should be adjusted due to differences in relative chip volume, material structure, cutting fluid effectiveness, depth of hole, and conditions of drill and machine. The advancing rate is

$$F = f \times N \quad \text{ipm} \qquad (46)$$

The recommended feeds are given in Table 8.

The time T required to drill a hole of depth h is given by

$$T = \frac{h + 0.3d}{F} \quad \text{min} \qquad (47)$$

The extra distance of $0.3d$ is approximately equal to the distance from the tip to the effective diameter of the tool. The rate of metal removal in case of blind holes is given by

$$Q = \left(\frac{\pi d^2}{4}\right) F \quad \text{in.}^3/\text{min} \qquad (48)$$

When torque is unknown, the horsepower requirement can be calculated by

$$\text{HP}_c = Q \times C \times W \times (\text{HP}_u) \quad \text{hp}$$

C, W, HP_u are given in previous sections.

$$\text{Power} = \text{HP}_c \times 396{,}000 \quad \text{in.–lb/min} \qquad (49)$$

$$\text{Torque} = \frac{\text{Power}}{2\pi N} \quad \text{in–lbf}$$

$$F_c = \frac{\text{Torque}}{R} \quad \text{lb}$$

Along the cutting edge of the drill, the cutting speed is reduced toward the center as the diameter is reduced. The cutting speed is actually zero at the center. To avoid the region of very low speed and to reduce high thrust forces that might affect the alignment of the finished hole, a pilot hole is usually drilled before drilling holes of medium and large sizes. For the case of drilling with a pilot hole

Table 8 Recommended Feeds for Drills

Diameter		Feed	
(in.)	(mm)	(ipr)	(mm/rev)
Under $\tfrac{1}{8}$	3.2	0.001–0.002	0.03–0.05
$\tfrac{1}{3}$–$\tfrac{1}{4}$	3.2–6.4	0.002–0.004	0.05–0.10
$\tfrac{1}{4}$–$\tfrac{1}{2}$	6.4–12.7	0.004–0.007	0.10–0.18
$\tfrac{1}{2}$–1	12.7–25.4	0.007–0.015	0.18–0.38
Over 1	25.4	0.015–0.025	0.38–0.64

$$Q = \frac{\pi}{4}(d^2 - d_p^2)F$$

$$= \frac{\pi}{4}(d + d_p)(d - d_p)F \quad \text{in.}^3/\text{min} \tag{50}$$

Due to the elimination of the effects of the chisel-edge region, the equations for torque and thrust can be estimated as follows:

$$M_p = M\left[\frac{1 - (d_p/d)^2}{(1 + d_1/d)^{0.2}}\right] \tag{51}$$

$$T_p = T\left[\frac{1 - d_1/d}{(1 + d_1/d)^{0.2}}\right] \tag{52}$$

where d_p = pilot hole diameter

Alignment charts were developed for determining motor power in drilling. Figures 15 and 16 show the use of these charts either for English or metric units. The unit power* (P) is the adjusted unit power with respect to drilling conditions and machine efficiency.

For English units,

$$\text{HP}_m = \frac{\pi D^2}{4} \times f \times N \times P^*$$

$$N = \frac{12V}{\pi D}$$

As

$$\text{HP}_m = \frac{\pi D^2}{4} \times f \times \frac{12V}{\pi D} \times P^*$$

$$= 3D \times f \times V \times P^*$$

For metric units,

$$\text{HP}_m = \frac{\pi D^2}{4 \times 100} \times \frac{f}{10} \times N \times P$$

P^* in kW/cm^3/min

$$N = \frac{1000V}{\pi D}$$

$$\text{HP}_m = \frac{\pi D^2}{4 \times 100} \times \frac{f}{10} \times \frac{100V}{\pi D} \times P^*$$

$$= 0.25D \times f \times V \times P^*$$

7.1 Accuracy of Drills

The accuracy of holes drilled with a two-fluted twist drill is influenced by many factors, including the accuracy of the drill point; the size of the drill, the chisel edge, and the jigs used; the workpiece material; the cutting fluid used; the rigidity and accuracy of the machine

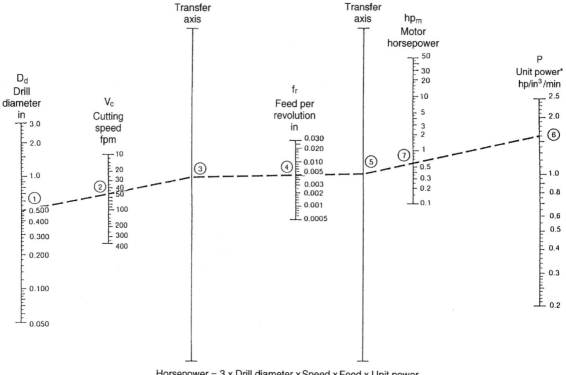

Figure 15 Alignment chart for determining motor horsepower in drilling—English units.

Example:
$D_d = 0.500$ in
$V_c = 50$ fpm
$f_r = 0.005$ in
$P = 1.6$ hp/in^3/min

To Determine Motor Horsepower:
Connect Drill diameter ① with Cutting speed ② to obtain point ③ on Transfer axis. Connect ③ with Feed ④ to obtain point ⑤ on Transfer axis. Connect ⑤ with Unit power ⑥ to obtain 0.60 Motor horsepower, point ⑦.

used; and the cutting speed. Usually, when drilling most materials, the diameter of the drilled holes will be oversize. Table 9 provided the results of tests reported by The Metal Cutting Tool Institute for holes drilled in steel and cast iron.

Gun drills differ from conventional drills in that they are usually made with a single flute. A hole provides a passageway for pressurized coolant, which serves as a means of both keeping the cutting edge cool and flushing out the chips, especially in deep cuts.

Spade drills (Fig. 17) are made by inserting a spade-shaped blade into a shank. Some advantages of spade drills are (1) efficiency in making holes up to 15 in. in diameter; (2) low cost, since only the insert is replaced; (3) deep hole drilling; and (4) easiness of chip breaking on removal.

Trepanning is a machining process for producing a circular hole, groove, disk, cylinder, or tube from solid stock. The process is accomplished by a tool containing one or more cutters, usually single-point, revolving around a center. The advantages of trepanning are (1) the central core left is solid material, not chips, which can be used in later work; and (2) the power required to produce a given hole diameter is highly reduced because only the annulus is actually cut.

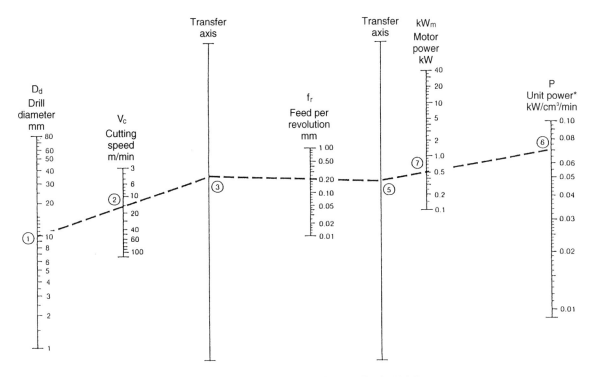

Figure 16 Alignment chart for determining motor power in drilling—metric units.

Table 9 Oversize Diameters in Drilling

Drill Diameter (in.)	Amount Oversize (in.)		
	Average Max.	Mean	Average Min.
1/16	0.002	0.0015	0.001
1/8	0.0045	0.003	0.001
1/4	0.0065	0.004	0.0025
1/2	0.008	0.005	0.003
3/4	0.008	0.005	0.003
1	0.009	0.007	0.004

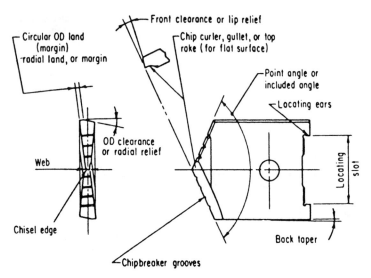

Figure 17 Spade-drill blade elements.

Reaming, boring, counterboring, centering and countersinking, spotfacing, tapping, and chamfering processes can be done on drills. Microdrilling and submicrodrilling achieve holes in the range of 0.000025–0.20 in. in diameter.

Drilling machines are usually classified in the following manner:

1. Bench: plain or sensitive
2. Upright: single-spindle or turret
3. Radial
4. Gang
5. Multispindle
6. Deep-hole: vertical or horizontal
7. Transfer

8 MILLING PROCESSES

The milling machines use a rotary multitooth cutter that can be designed to mill flat or irregularly shaped surfaces, cut gears, generate helical shapes, drill, bore, or do slotting work. Milling machines are classified broadly as vertical or horizontal. Figure 18 shows some of the operations that are done on both types.

Feed in milling (F) is specified in inches per minute, but it is determined from the amount each tooth can remove or feed per tooth (f_t). The feed in./min is calculated from

$$F = f_t \times n \times N \quad \text{in./min} \tag{53}$$

where n = number of teeth in cutter
N = rpm

Table 10 gives the recommended f_t for carbides and HSS tools. The cutting speed CS is calculated as follows:

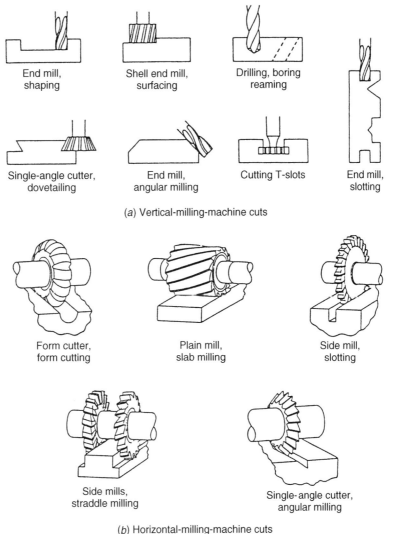

Figure 18 Applications of (*a*) vertical; (*b*) horizontal milling machines.

$$\text{CS} = \frac{\pi DN}{12} \quad \text{fpm}$$

where D = tool diameter, in.

Table 11 gives the recommended cutting speeds while using HSS and carbide-tipped tools. The relationship between cutter rotation and feed direction is shown in Fig. 19. In climb milling or down milling, the chips are cut to maximum thickness at initial engagement and decrease to zero thickness at the end of engagement. In conventional or up milling, the reverse occurs. Because of the initial impact, climb milling requires rigid machines with backlash eliminators.

Table 10 Recommended Feed per Tooth for Milling Steel with Carbide and HSS Cutters

	Feed per Tooth	
Type of Milling	Carbides	HSS
Face	0.008–0.015	0.010
Side or straddle	0.008–0.012	0.006
Slab	0.008–0.012	0008
Slotting	0.006–0.010	0.006
Slitting saw	0.003–0.006	0.003

The material removal rate (MRR) is $Q = F \times w \times d$, where w is the width of cut and d is the depth of cut. The horsepower required for milling is given by

$$HP_c = HP_u \times Q$$

Machine horsepower is determined by

$$HP_m = \frac{HP_c}{\text{Eff.}} + HP_i \qquad (54)$$

where HP_i = idle horsepower

Table 11 Table of Cutting Speeds (sfpm)–Milling

	HSS Tools		Carbide-tipped Tools	
Work Material	Rough Mill	Finish Mill	Rough Mill	Finish Mill
Cast iron	50–60	80–110	180–200	350–400
Semisteel	40–50	65–90	140–160	250–300
Malleable iron	80–100	110–130	250–300	400–500
Cast steel	45–60	70–90	150–180	200–250
Copper	100–150	150–200	600	1000
Brass	200–300	200–300	600–1000	600–1000
Bronze	100–150	150–180	600	1000
Aluminum	400	700	800	1000
Magnesium	600–800	1000–1500	1000–1500	1000–5000
SAE steels				
1020 (coarse feed)	60–80	60–80	300	300
1020 (fine feed)	100–120	100–120	450	450
1035	75–90	90–120	250	250
X-1315	175–200	175–200	400–500	400–500
1050	60–80	100	200	200
2315	90–110	90–110	300	300
3150	50–60	70–90	200	200
4150	40–50	70–90	200	200
4340	40–50	60–70	200	200
Stainless steel	60–80	100–120	240–300	240–300
Titanium	30–70		200–350	

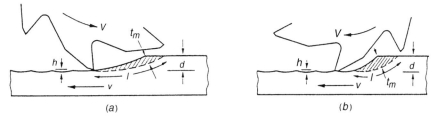

Figure 19 Cutting action in up-and-down milling.

Alignment charts were developed for determining metal removal rate (MRR) and motor power in face milling. Figures 21 and 22 show the method of using these charts either for English or metric units.

The time required for milling is equal to distance required to be traveled by the cutter to complete the cut (L_1) divided by the feed rate F. L_1 is equal to the length of cut (L) plus cutter approach A and the overtravel OT. The machining time T is calculated from

$$T = \frac{L + A + \text{OT}}{F} \quad \text{min} \tag{55}$$

OT depends on the specific milling operation.

The milling machines are designed according to the longitudinal table travel. Milling machines are built in different types, including

1. Column-and-knee: vertical, horizontal, universal, and ram
2. Bed-type, multispindle
3. Planer
4. Special, turret, profilers, and duplicators
5. Numerically controlled

9 GEAR MANUFACTURING

Gears are made by various methods, such as machining, rolling, extrusion, blanking, powder metallurgy, casting, or forging. Machining still is the unsurpassed method of producing gears

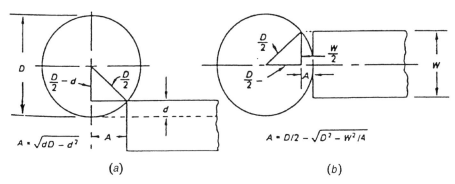

Figure 20 Allowance for approach in (*a*) plain or slot milling; (*b*) face milling.

204 Production Processes and Equipment

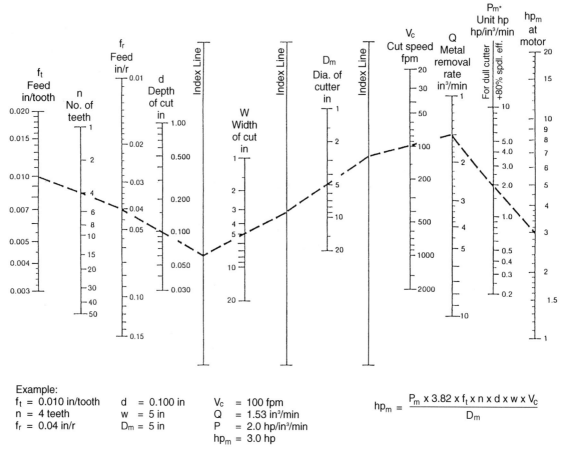

Example:
f_t = 0.010 in/tooth   d = 0.100 in V_c = 100 fpm
n = 4 teeth   w = 5 in   Q = 1.53 in³/min
f_r = 0.04 in/r D_m = 5 in   P = 2.0 hp/in³/min
   hp_m = 3.0 hp

$$hp_m = \frac{P_m \times 3.82 \times f_t \times n \times d \times w \times V_c}{D_m}$$

Figure 21 Alignment chart for determining metal removal rate and motor horsepower in face milling—English units.

of all types and sizes with high accuracy. Roll forming can be used only on ductile materials; however, it has been highly developed and widely adopted in recent years. Casting, powder metallurgy, extruding, rolling, grinding, molding, and stamping techniques are used commercially in gear production.

9.1 Machining Methods

There are three basic methods for machining gears: form cutting, template machining, and the generating process.

Form cutting uses the principle illustrated in Fig. 23. The equipment and cutters required are relatively simple, and standard machines, usually milling, are often used. Theoretically, there should be different-shaped cutters for each size of gear for a given pitch, as there is a slight change in the curvature of the involute. However, one cutter can be used for several gears having different numbers of teeth without much sacrifice in their operating action. The eight standard involute cutters are listed in Table 12. On the milling machine, the index or dividing head is used to rotate the gear blank through a certain number of degrees after each

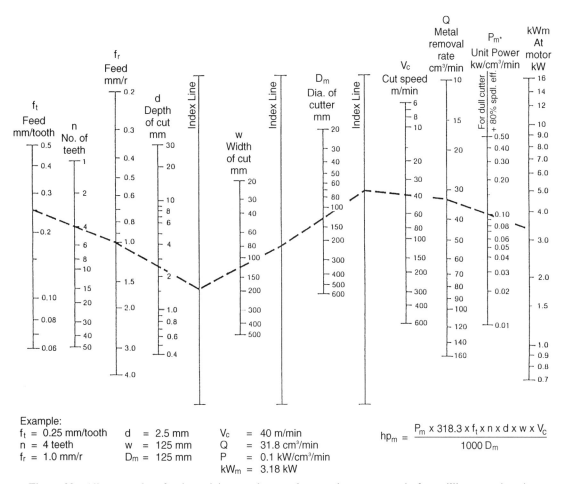

Figure 22 Alignment chart for determining metal removal rate and motor power in face milling—metric units.

cut. The rule to use is: turns of index handle = 40/N, where N is the number of teeth. Form cutting is usually slow.

Template machining utilizes a simple, single-point cutting tool that is guided by a template. However, the equipment is specialized, and the method is seldom used except for making large-bevel gears.

The *generating process* is used to produce most high-quality gears. This process is based on the principle that any two involute gears, or any gear and a rack, of the same diametral pitch will mesh together. Applying this principle, one of the gears (or the rack) is made into a cutter by proper sharpening and is used to cut into a mating gear blank and thus generate teeth on the blank. Gear shapers (pinion or rack), gear-hobbing machines, and bevel-gear generating machines are good examples of the gear generating machines.

9.2 Gear Finishing

To operate efficiently and have satisfactory life, gears must have accurate tooth profile and smooth and hard faces. Gears are usually produced from relatively soft blanks and are sub-

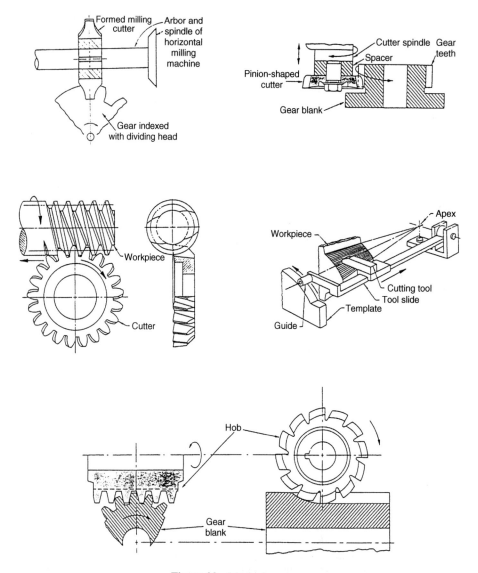

Figure 23 Machining gears.

sequently heat-treated to obtain greater hardness, if it is required. Such heat treatment usually results in some slight distortion and surface roughness. *Grinding and lapping* are used to obtain very accurate teeth on hardened gears. Gear-shaving and burnishing methods are used in gear finishing. Burnishing is limited to unhardened gears.

10 THREAD CUTTING AND FORMING

Three basic methods are used for the manufacturing of threads; *cutting, rolling,* and *casting.* Die casting and molding of plastics are good examples of casting. The largest number of

Table 12 Standard Gear Cutters

Cutter Number	Gear Tooth Range
1	135 teeth to rack
2	55–34
3	35–54
4	26–34
5	21–25
6	17–20
7	14–16
8	12–13

threads are made by rolling, even though it is restricted to standardized and simple parts, and ductile materials. Large numbers of threads are cut by the following methods:

1. Turning
2. Dies: manual or automatic (external)
3. Milling
4. Grinding (external)
5. Threading machines (external)
6. Taps (internal)

10.1 Internal Threads

In most cases, the hole that must be made before an internal thread is tapped is produced by drilling. The hole size determines the depth of the thread, the forces required for tapping, and the tap life. In most applications, a drill size is selected that will result in a thread having about 75% of full thread depth. This practice makes tapping much easier, increases the tap's life, and only slightly reduces the resulting strength. Table 13 gives the drill sizes used to produce 75% thread depth for several sizes of UNC threads. The feed of a tap depends on the lead of the screw and is equal to 1/lead ipr.

Cutting speeds depend on many factors, such as

1. Material hardness
2. Depth of cut
3. Thread profile
4. Tooth depth
5. Hole depth
6. Fineness of pitch
7. Cutting fluid

Cutting speeds can range from lead 3 ft/min (1 m/min) for high-strength steels to 150 ft/min (45 m/min) for aluminum alloys. Long-lead screws with different configurations can be cut successfully on milling machines, as in Fig. 24. The feed per tooth is given by the following equation:

Table 13 Recommended Tap-Drill Sizes for Standard Screw-Thread Pitches (American National Coarse-Thread Series)

Number or Diameter	Threads per Inch	Outside Diameter of Screw	Tap Drill Sizes	Decimal Equivalent of Drill
6	32	0.138	36	0.1065
8	32	0.164	29	0.1360
10	24	0.190	25	0.1495
12	24	0.216	16	0.1770
1/4	20	0.250	7	0.2010
3/8	16	0.375	5/16	0.3125
1/2	13	0.500	27/64	0.4219
3/4	10	0.750	21/32	0.6562
1	8	1.000	7/8	0.875

$$f_t = \frac{\pi d S}{nN} \qquad (56)$$

where d = diameter of thread
n = number of teeth in cutter
N = rpm of cutter
S = rpm of work

10.2 Thread Rolling

In thread rolling, the metal on the cylindrical blank is cold-forged under considerable pressure by either rotating cylindrical dies or reciprocating flat dies. The advantages of thread rolling include improved strength, smooth surface finish, less material used (~19%), and high production rate. The limitations are that blank tolerance must be close, it is economical only for large quantities, it is limited to external threads, and it is applicable only for ductile materials, less than Rockwell C37.

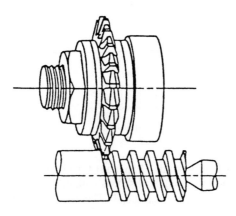

Figure 24 Single-thread milling cutter.

11 BROACHING

Broaching is unique in that it is the only one of the basic machining processes in which the feed of the cutting edges is built into the tool. The machined surface is always the inverse of the profile of the broach. The process is usually completed in a single, linear stroke. A broach is composed of a series of single-point cutting edges projecting from a rigid bar, with successive edges protruding farther from the axis of the bar. Figure 25 illustrates the parts and nomenclature of the broach. Most broaching machines are driven hydraulically and are of the pull or push type.

The maximum force an internal pull broach can withstand without damage is given by

$$P = \frac{A_y F_y}{s} \quad \text{lb} \tag{57}$$

where A_y = minimum tool selection, in.²
F_y = tensile yield strength of tool steel, psi
s = safety factor

The maximum push force is determined by the minimum tool diameter (D_y), the length of the broach (L), and the minimum compressive yield strength (F_y). The ratio L/D_y should be less than 25 so that the tool will not bend under load. The maximum allowable pushing force is given by

$$P = \frac{A_y F_y}{s} \quad \text{lb} \tag{58}$$

where F_y is minimum compressive yield strength.

If L/D_y ratio is greater than 25 (long broach), the *Tool and Manufacturing Engineers Handbook* gives the following formula:

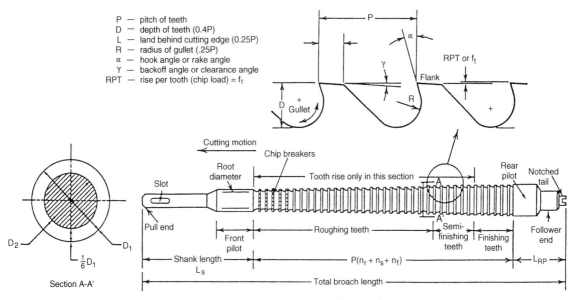

Figure 25 Standard broach part and nomenclature.

$$P = \frac{5.6 \times 10^7 D_r^4}{sL^2} \quad \text{lb} \tag{59}$$

D_r and L are given in inches.

Alignment charts were developed for determining metal removal rate (MRR) and motor power in surface broaching. Figures 26 and 27 show the application of these charts for either English or metric units.

Broaching speeds are relatively low, seldom exceeding 50 fpm, but, because a surface is usually completed in one stroke, the productivity is high.

12 SHAPING, PLANING, AND SLOTTING

The shaping and planing operations generate surfaces with a single-point tool by a combination of a reciprocating motion along one axis and a feed motion normal to that axis (Fig. 28). Slots and limited inclined surfaces can also be produced. In shaping, the tool is mounted on a reciprocating ram and the table is fed at each stroke of the ram. Planers handle large, heavy workpieces. In planing, the workpiece reciprocates and the feed increment is provided by moving the tool at each reciprocation. To reduce the lost time on the return stroke, they are provided with a quick-return mechanism. For mechanically driven shapers, the ratio of cutting time to return stroke averages 3:2, and for hydraulic shapers the ratio is 2:1. The average cutting speed may be determined by the following formula:

$$CS = \frac{LN}{12C} \quad \text{fpm} \tag{60}$$

where N = strokes per minute
L = stroke length, in.
C = cutting time ratio, cutting time divided by total time

For mechanically driven shapers, the cutting speed reduces to

$$CS = \frac{LN}{7.2} \quad \text{fpm} \tag{61}$$

or

$$CS = \frac{L_1 N}{600} \quad \text{m/min} \tag{62}$$

where L_1 is the stroke length in millimeters. For hydraulically driven shapers,

$$CS = \frac{LN}{8} \quad \text{fpm} \tag{63}$$

or

$$CS = \frac{L_1 N}{666.7} \quad \text{m/min} \tag{64}$$

The time T required to machine a workpiece of width W (in.) is calculated by

12 Shaping, Planing, and Slotting 211

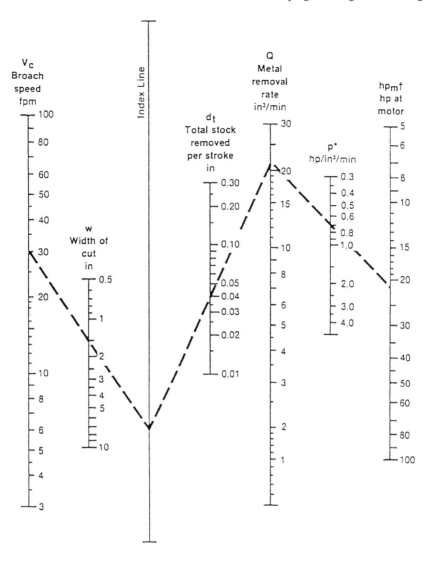

Example:
Material: Cast iron — HSS tools
Chipload 0.005 in/tooth
V_c = 30 fpm   w = 1.5 in
d_t = 0.040 in   Q = 22 in³/min
P = 0.7 hp/in³/min hp_m = 22 hp

$Q = 12\, V_c \times w \times d_t \text{ in}^3/\text{min}$

$hp_m = \dfrac{Q \times P}{E} = \dfrac{Q \times P}{0.7}$

Figure 26 Alignment chart for determining metal removal rate and motor horsepower in surface broaching with high-speed steel broaching tools—English units.

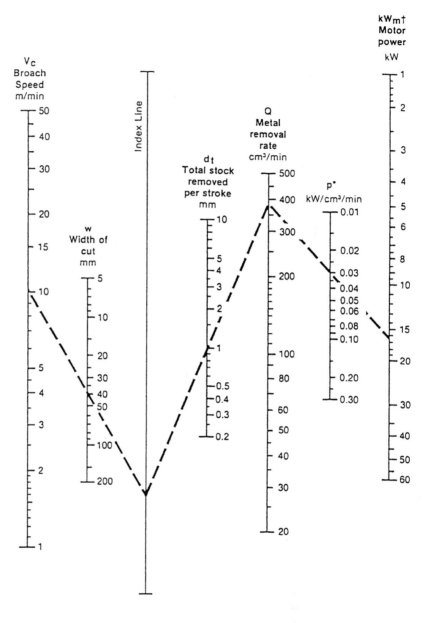

Figure 27 Alignment chart for determining metal removal rate and motor power in surface broaching with high-speed steel broaching tools—metric units.

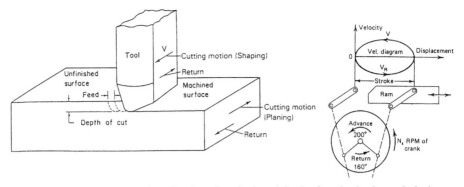

Figure 28 Basic relationships of tool motion, feed, and depth of cut in shaping and planing.

$$T = \frac{W}{N \times f} \quad \text{min} \tag{65}$$

where f = feed, in. per stroke

The number of strokes (S) required to complete a job is then

$$S = \frac{W}{f} \tag{66}$$

The power required can be approximated by

$$\text{HP}_c = Kdf(\text{CS}) \tag{67}$$

where d = depth of cut, in.
 CS = cutting speed, fpm
 K = cutting constant, for medium cast iron, 3; free-cutting steel, 6; and bronze, 1.5

or

$$\text{HP}_c = 12f \times d \times \text{CS} \times \text{HP}_\mu$$

$$F_c = \frac{33{,}000\text{HP}_c}{\text{CS}}$$

13 SAWING, SHEARING, AND CUTTING OFF

Saws are among the most common of machine tools, even though the surfaces they produce often require further finishing operations. Saws have two general areas of applications: contouring and cutting off. There are three basic types of saws: hacksaw, circular, and band saw.

The *reciprocating power hacksaw* machines can be classified as either positive or uniform-pressure feeds. Most of the new machines are equipped with a quick-return action to reduce idle time.

The machining time required to cut a workpiece of width W in. is calculated as follows:

$$T = \frac{W}{fN} \quad \text{min} \tag{68}$$

where F = feed, in./stroke
N = number of strokes per min

Circular saws are made of three types: metal saws, steel friction disks, and abrasive disks. Solid metal saws are limited in size, not exceeding 16 in. in diameter. Large circular saws have either replaceable inserted teeth or segmented-type blades. The machining time required to cut a workpiece of width W in. is calculated as follows:

$$T = \frac{W}{f_t nN} \quad \text{min} \tag{69}$$

where f_t = feed per tooth
n = number of teeth
N = rpm

Steel friction disks operate at high peripheral speeds ranging from 18,000 to 25,000 fpm (90 to 125 m/sec). The heat of friction quickly softens a path through the part. The disk, which is sometimes provided with teeth or notches, pulls and ejects the softened metal. About 0.5 min are required to cut through a 24-in. I-beam.

Abrasive disks are mainly aluminum oxide grains or silicon carbide grains bonded together. They will cut ferrous or nonferrous metals. The finish and accuracy is better than steel friction blades, but they are limited in size compared to steel friction blades.

Band saw blades are of the continuous type. Band sawing can be used for cutting and contouring. Band-sawing machines operate with speeds that range from 50–1500 fpm. The time required to cut a workpiece of width W in. can be calculated as follows:

$$T = \frac{W}{12 f_t nV} \quad \text{min} \tag{70}$$

where f_t = feed, in. per tooth
n = number of teeth per in.
V = cutting speed, fpm

Cutting can also be achieved by band-friction cutting blades with a surface speed up to 15,000 fpm. Other band tools include band filing, diamond bands, abrasive bands, spiral bands, and special-purpose bands.

14 MACHINING PLASTICS

Most plastics are readily formed, but some machining may be required. Plastic's properties vary widely. The general characteristics that affect their machinability are discussed below.

First, all plastics are poor heat conductors. Consequently, little of the heat that results from chip formation will be conducted away through the material or carried away in the chips. As a result, cutting tools run very hot and may fail more rapidly than when cutting metal. Carbide tools frequently are more economical to use than HSS tools if cuts are of moderately long duration or if high-speed cutting is to be done.

Second, because considerable heat and high temperatures do develop at the point of cutting, thermoplastics tend to soften, swell, and bind or clog the cutting tool. Thermosetting plastics give less trouble in this regard.

Third, cutting tools should be kept very sharp at all times. Drilling is best done by means of straight-flute drills or by "dubbing" the cutting edge of a regular twist drill to produce a zero rake angle. Rotary files and burrs, saws, and milling cutters should be run at high speeds to improve cooling, but with feed carefully adjusted to avoid jamming the gullets. In some cases, coolants can be used advantageously if they do not discolor the plastic or cause gumming. Water, soluble oil and water, and weak solutions of sodium silicate in water are used. In turning and milling plastics, diamond tools provide the best accuracy, surface finish, and uniformity of finish. Surface speeds of 500–600 fpm with feeds of 0.002–0.005 in. are typical.

Fourth, filled and laminated plastics usually are quite abrasive and may produce a fine dust that may be a health hazard.

15 GRINDING, ABRASIVE MACHINING, AND FINISHING

Abrasive machining is the basic process in which chips are removed by very small edges of abrasive particles, usually synthetic. In many cases, the abrasive particles are bonded into wheels of different shapes and sizes. When wheels are used mainly to produce accurate dimensions and smooth surfaces, the process is called *grinding*. When the primary objective is rapid metal removal to obtain a desired shape or approximate dimensions, it is termed *abrasive machining*. When fine abrasive particles are used to produce very smooth surfaces and to improve the metallurgical structure of the surface, the process is called *finishing*.

15.1 Abrasives

Aluminum oxide (Al_2O_3), usually synthetic, performs best on carbon and alloy steels, annealed malleable iron, hard bronze, and similar metals. Al_2O_3 wheels are not used in grinding very hard materials, such as tungsten carbide, because the grains will get dull prior to fracture. Common trade names for aluminum oxide abrasives are *Alundum* and *Aloxite*.

Silicon carbide (SiC), usually synthetic, crystals are very hard, being about 9.5 on the Moh's scale, where diamond hardness is 10. SiC crystals are brittle, which limits their use. Silicon carbide wheels are recommended for materials of low tensile strength, such as cast iron, brass, stone, rubber, leather, and cemented carbides.

Cubic boron nitride (CBN) is the second-hardest natural or manmade substance. It is good for grinding hard and tough-hardened tool-and-die steels.

Diamonds may be classified as natural or synthetic. Commercial diamonds are now manufactured in high, medium, and low impact strength.

Grain Size

To have uniform cutting action, abrasive grains are graded into various sizes, indicated by the numbers 4–600. The number indicates the number of openings per linear inch in a standard screen through which most of the particles of a particular size would pass. Grain sizes 4–24 are termed coarse; 30–60, medium; and 70–600, fine. Fine grains produce smoother surfaces than coarse ones but cannot remove as much metal.

Bonding materials have the following effects on the grinding process: (1) they determine the strength of the wheel and its maximum speed; (2) they determine whether the wheel is rigid or flexible; and (3) they determine the force available to pry the particles loose. If only a small force is needed to release the grains, the wheel is said to be soft. Hard wheels are recommended for soft materials and soft wheels for hard materials. The bonding materials used are vitrified, silicate, rubber, resinoid, shellac, and oxychloride.

Structure or Grain Spacing

Structure relates to the spacing of the abrasive grain. Soft, ductile materials require a wide spacing to accommodate the relatively large chips. A fine finish requires a wheel with a close spacing. Figure 29 shows the standard system of grinding wheels as adopted by the American National Standards Institute.

Speeds

Wheel speed depends on the wheel type, bonding material, and operating conditions. Wheel speeds range between 4500 and 18,000 sfpm (22.86 and 27.9 m/s). 5500 sfpm (27.9 m/s) is generally recommended as best for all disk-grinding operations. Work speeds depend on type of material, grinding operation, and machine rigidity. Work speeds range between 15 and 200 fpm.

Feeds

Cross feed depends on the width of grinding wheel. For rough grinding, the range is one-half to three-quarters of the width of the wheel. Finer feed is required for finishing, and it ranges between one-tenth and one-third of the width of the wheel. A cross feed between 0.125 and 0.250 in. is generally recommended.

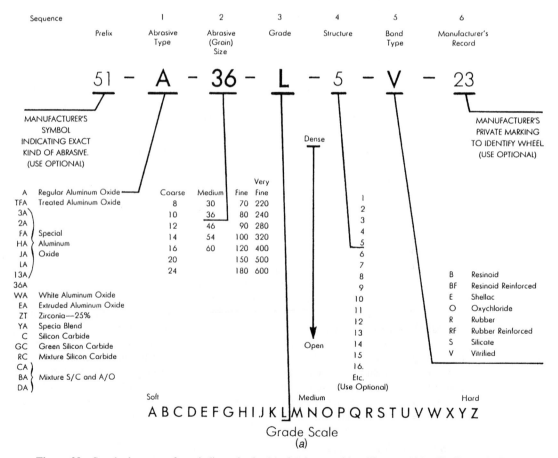

Figure 29 Standard systems for grinding wheels. (*a*) aluminum oxide, silicon carbide; (*b*) diamond, CBN.

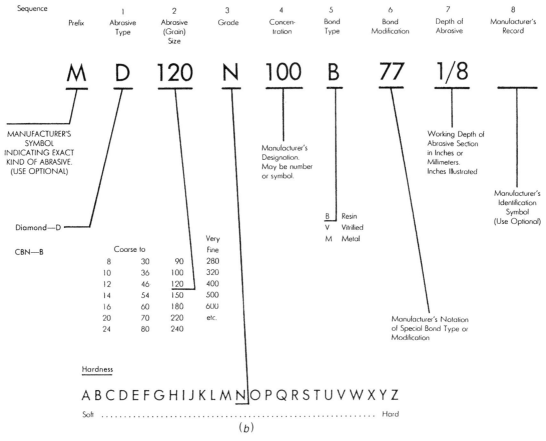

Figure 29 (*Continued*)

Depth of Cut

Rough-grinding conditions will dictate the maximum depth of cut. In the finishing operation, the depth of cut is usually small, 0.0002–0.001 in. (0.005–0.025 mm). Good surface finish and close tolerance can be achieved by "sparking out" or letting the wheel run over the workpiece without increasing the depth of cut till sparks die out. The *grinding ratio* (*G*-ratio) refers to the ratio of the cubic inches of stock removed to the cubic inches of grinding wheel worn away. *G*-ratio is important in calculating grinding and abrasive machining cost, which may be calculated by the following formula:

$$C = \frac{C_a}{G} + \frac{L}{tq} \tag{71}$$

where C = specific cost of removing a cu in. of material
C_a = cost of abrasive, \$/in.^{3}
G = grinding ratio
L = labor and overhead charge, \$/hr
q = machining rate, in.^{3}/hr
t = fraction of time the wheel is in contact with workpiece

Power Requirement

$$\text{Power} = (u)(\text{MRR}) = F_c \times R \times 2\pi N$$

$$\text{MRR} = \text{material removal rate} = d \times w \times v$$

where d = depth of cut
w = width of cut
v = work speed
u = specific energy for surface grinding. Table 14 gives the approximate specific energy requirement for certain metals.
R = radius of wheel
N = rev/unit time

15.2 Temperature

Temperature rise affects the surface properties and causes residual stresses on the workpiece. It is related to process variables by the following relation:

$$\text{Temperature rise} \propto D^{1/4} d^{3/4} \left(\frac{V}{v}\right)^{1/2} \tag{72}$$

where D = wheel diameter
V = wheel speed

Grinding Fluids

Grinding fluids are water-based emulsions for general guiding and oils for thread and gear grinding. Advantages include

1. Machining hard materials > RC50.
2. Fine surface finish, 10–80 μin. (0.25–2 μm).
3. Accurate dimensions and close tolerances I.0002 in. (I.005 mm) can be easily achieved.
4. Grinding pressure is light.

Machines

Grinding and abrasive machines include

1. Surface grinders, reciprocating or rotating table
2. Cylindrical grinders, work between centers, centerless, crankshaft, thread and gear form work, and internal and other special applications

Table 14 Approximate Specific Energy Required for Surface Grinding

Workpiece Material	Hardness	HP (in.³/min)	W/(mm³/s)
Aluminum	150 HB	3–10	8–27
Steel	(110–220) HB	6–24	16–66
Cast iron	(140–250) HB	5–22	14–60
Titanium alloy	300 HB	6–20	16–55
Tool steel	62–67 HRC	7–30	19–82

3. Jig grinders
4. Tool and cutter grinders
5. Snagging, foundry rough work
6. Cutting off and profiling
7. Abrasive grinding, belt, disk and loose grit
8. Mass media, barrel tumbling, and vibratory

Ultrasonic Machining

In ultrasonic machining, material is removed from the workpiece by microchipping or erosion through high-velocity bombardment by abrasive particles, in the form of a slurry, through the action of an ultrasonic transducer. It is used for machining hard and brittle materials and can produce very small and accurate holes 0.015 in. (0.4 mm).

Surface Finishing

Finishing processes produce an extra-fine surface finish; in addition, tool marks are removed and very close tolerances are achieved. Some of these processes follow.

Honing is a low-velocity abrading process. It uses fine abrasive stones to remove very small amounts of metals usually left from previous grinding processes. The amount of metal removed is usually less than 0.005 in. (0.13 mm). Because of low cutting speeds, heat and pressure are minimized, resulting in excellent sizing and metallurgical control.

Lapping is an abrasive surface-finishing process wherein fine abrasive particles are charged in some sort of a vehicle, such as grease, oil, or water, and are embedded into a soft material, called a *lap*. Metal laps must be softer than the work and are usually made of close-grained gray cast iron. Other materials, such as steel, copper, and wood, are used where cast iron is not suitable. As the charged lap is rubbed against a surface, small amounts of material are removed from the harder surface. The amount of material removed is usually less than 0.001 in. (0.03 mm).

Superfinishing is a surface-improving process that removes undesirable fragmentation, leaving a base of solid crystalline metal. It uses fine abrasive stones, like honing, but differs in the type of motion. Very rapid, short strokes, very light pressure, and low-viscosity lubricant–coolant are used in superfinishing. It is essentially a finishing process and not a dimensional one, and can be superimposed on other finishing operations.

Buffing

Buffing wheels are made from a variety of soft materials. The most widely used is muslin, but flannel, canvas, sisal, and heavy paper are used for special applications. Buffing is usually divided into two operations: cutting down and coloring. The first is used to smooth the surface and the second to produce a high luster. The abrasives used are extremely fine powders of aluminum oxide, tripoli (an amorphous silicon), crushed flint or quartz, silicon carbide, and red rouge (iron oxide). Buffing speeds range between 6000 and 12,000 fpm.

Electropolishing is the reverse of electroplating; that is, the work is the anode instead of the cathode and metal is removed rather than added. The electrolyte attacks projections on the workpiece surface at a higher rate, thus producing a smooth surface.

16 NONTRADITIONAL MACHINING

Nontraditional, or nonconventional, machining processes are material-removal processes that have recently emerged or are new to the user. They have been grouped for discussion here according to their primary energy mode; that is, mechanical, electrical, thermal, or chemical, as shown in Table 15.

Nontraditional processes provide manufacturing engineers with additional choices or alternatives to be applied where conventional processes are not satisfactory, such as when

- Shapes and dimensions are complex or very small
- Hardness of material is very high (>400 HB)
- Tolerances are tight and very fine surface finish is desired
- Temperature rise and residual stresses must be avoided
- Cost and production time must be reduced

Figure 30 and Table 16 demonstrate the relationships among the conventional and the nontraditional machining processes with respect to surface roughness, dimensional tolerance, and metal-removal rate. The *Machinery Handbook*[6] is an excellent reference for nontraditional machining processes, values, ranges, and limitations.

16.1 Abrasive Flow Machining

Abrasive flow machining (AFM) is the removal of material by a viscous, abrasive medium flowing, under pressure, through or across a workpiece. Figure 31 contains a schematic presentation of the AFM process. Generally, the putty-like medium is extruded through or over the workpiece with motion usually in both directions. Aluminum oxide, silicon carbide, boron carbide, or diamond abrasives are used. The movement of the abrasive matrix erodes away burrs and sharp corners and polishes the part.

16.2 Abrasive Jet Machining

Abrasive jet machining (AJM) is the removal of material through the action of a focused, high-velocity stream of fine grit or powder-loaded gas. The gas should be dry, clean, and under modest pressure. Figure 32 shows a schematic of the AJM process. The mixing chamber sometimes uses a vibrator to promote a uniform flow of grit. The hard nozzle is directed close to the workpiece at a slight angle.

16.3 Hydrodynamic Machining

Hydrodynamic machining (HDM) removes material by the stroking of high-velocity fluid against the workpiece. The jet of fluid is propelled at speeds up to Mach 3. Figure 33 shows a schematic of the HDM operation.

16.4 Low-Stress Grinding

Low-stress grinding (LSG) is an abrasive material-removal process that leaves a low-magnitude, generally compressive residual stress in the surface of the workpiece. Figure 34 shows a schematic of the LSG process. The thermal effects from conventional grinding can produce high tensile stress in the workpiece surface. The process parameter guidelines can be applied to any of the grinding modes: surface, cylindrical, centerless, internal, and so on.

Table 15 Current Commercially Available Nontraditional Material Removal Processes

	Mechanical		Electrical		Thermal		Chemical
AFM	Abrasive flow machining	ECD	Electrochemical deburring	EBM	Electron-beam machining	CHM	Chemical machining: chemical milling, chemical blanking
AJM	Abrasive jet machining	ECDG	Electrochemical discharge grinding	EDG	Electrical discharge grinding		
HDM	Hydrodynamic machining						
LSG	Low-stress grinding	ECG	Electrochemical grinding	EDM	Electrical discharge machining	ELP	Electropolish
RUM	Rotary ultrasonic machining	ECH	Electrochemical honing			PCM	Photochemical machining
		ECM	Electrochemical machining	EDS	Electrical discharge sawing	TCM	Thermochemical machining (or TEM, thermal energy method)
TAM	Thermally assisted machining	ECP	Electrochemical polishing	EDWC	Electrical discharge wire cutting		
		ECS	Electrochemical sharpening				
TFM	Total form machining	ECT	Electrochemical turning	LBM	Laser-beam machining		
USM	Ultrasonic machining	ES	Electro-stream™	LBT	Laser-beam torch		
WJM	Water-jet machining	STEM™	Shaped tube electrolytic machining	PBM	Plasma-beam machining		

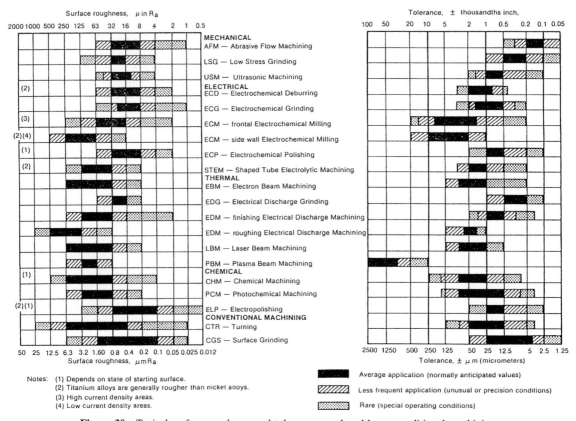

Figure 30 Typical surface roughness and tolerances produced by nontraditional machining.

16.5 Thermally Assisted Machining

Thermally assisted machining (TAM) is the addition of significant amounts of heat to the workpiece immediately prior to single-point cutting so that the material is softened but the strength of the tool bit is unimpaired (Fig. 35). While resistive heating and induction heating offer possibilities, the plasma arc has a core temperature of 14,500°F (8000°C) and a surface temperature of 6500°F (3600°C). The torch can produce 2000°F (1100°C) in the workpiece in approximately one-quarter revolution of the workpiece between the point of application of the torch and the cutting tool.

16.6 Electromechanical Machining

Electromechanical machining (EMM) is a process in which the metal removal is effected in a conventional manner except that the workpiece is electrochemically polarized. When the applied voltage and the electrolytic solution are controlled, the surface of the workpiece can be changed to achieve the characteristics suitable for the machining operation.

16.7 Total Form Machining

Total form machining (TFM) is a process in which an abrasive master abrades its full three-dimensional shape into the workpiece by the application of force while a full-circle, orbiting

Table 16 Material Removal Rates and Dimensional Tolerances

Process	Maximum Rate of Material Removal in.³/min cm³/min	Typical Power Consumption hp/in.³/min kW/cm³/min	Cutting Speed fpm m/min	Penetration Rate per Minute in. mm	Accuracy ± Attainable in. mm	Accuracy ± At Maximum Material Removal Rate in. mm	Typical Machine Input hp kW
Conventional turning	200 / 3300	1 / 0.046	250 / 76	— / —	0.0002 / 0.005	0.005 / 0.13	30 / 22
Conventional grinding	50 / 820	10 / 0.46	10 / 3	— / —	0.0001 / 0.0025	0.002 / 0.05	25 / 20
CHM	30 / 490	— / —	— / —	0.001 / 0.025	0.0005 / 0.013	0.003 / 0.075	— / —
PBM	10 / 164	20 / 0.91	50 / 15	10 / 254	0.02 / 0.5	0.1 / 2.54	200 / 150
ECG	2 / 33	2 / 0.019	0.25 / 0.08	— / —	0.0002 / 0.005	0.0025 / 0.063	4 / 3
ECM	1 / 16.4	160 / 7.28	— / —	0.5 / 12.7	0.0005 / 0.013	0.006 / 0.15	200 / 150
EDM	0.3 / 4.9	40 / 1.82	— / —	0.5 / 12.7	0.00015 / 0.004	0.002 / 0.05	15 / 11
USM	0.05 / 0.82	200 / 9.10	— / —	0.02 / 0.50	0.0002 / 0.005	0.0015 / 0.040	15 / 11
EBM	0.0005 / 0.0082	10,000 / 455	200 / 60	6 / 150	0.0002 / 0.005	0.002 / 0.050	10 / 7.5
LBM	0.0003 / 0.0049	60,000 / 2,731	— / —	4 / 102	0.0005 / 0.013	0.005 / 0.13	20 / 15

224 Production Processes and Equipment

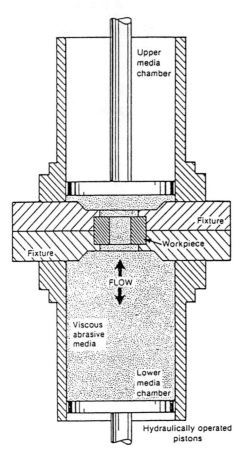

Figure 31 Abrasive flow machining.

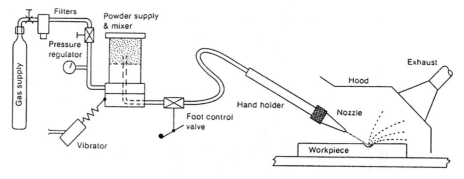

Figure 32 Abrasive jet machining.

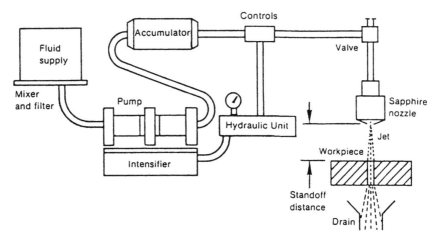

Figure 33 Hydrodynamic machining.

motion is applied to the workpiece via the worktable (Fig. 36). The cutting master is advanced into the work until the desired depth of cut is achieved. Uniformity of cutting is promoted by the fluid that continuously transports the abraded particles out of the working gap. Adjustment of the orbiting cam drive controls the precision of the overcut from the cutting master. Cutting action takes place simultaneously over the full surface of abrasive contact.

16.8 Ultrasonic Machining

Ultrasonic machining (USM) is the removal of material by the abrading action of a grit-loaded liquid slurry circulating between the workpiece and a tool vibrating perpendicular to

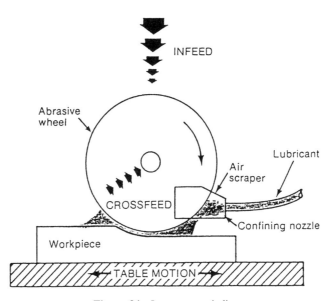

Figure 34 Low-stress grinding.

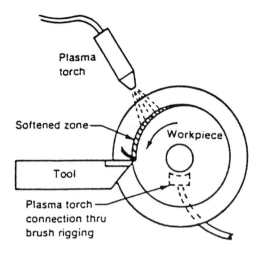

Figure 35 Thermally assisted machining.

the workface at a frequency above the audible range (Fig. 37). A high-frequency power source activates a stack of magnetostrictive material, which produces a low-amplitude vibration of the toolholder. This motion is transmitted under light pressure to the slurry, which abrades the workpiece into a conjugate image of the tool form. A constant flow of slurry (usually cooled) is necessary to carry away the chips from the workface. The process is sometimes called *ultrasonic abrasive machining* (UAM) or *impact machining*.

A prime variation of USM is the addition of ultrasonic vibration to a rotating tool—usually a diamond-plated drill. *Rotary ultrasonic machining* (RUM) substantially increases

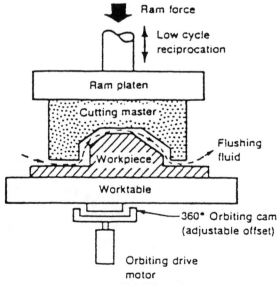

Figure 36 Total form machining.

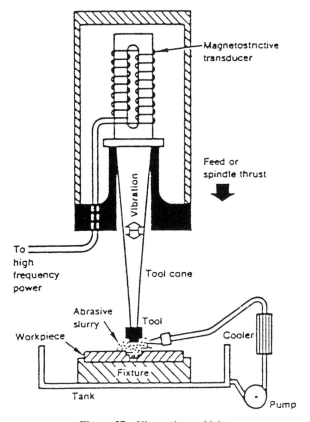

Figure 37 Ultrasonic machining.

the drilling efficiency. A piezoelectric device built into the rotating head provides the needed vibration. Milling, drilling, turning, threading, and grinding-type operations are performed with RUM.

16.9 Water-Jet Machining

Water-jet machining (WJM) is low-pressure hydrodynamic machining. The pressure range for WJM is an order of magnitude below that used in HDM. There are two versions of WJM: one for mining, tunneling, and large-pipe cleaning that operates in the region from 250 to 1000 psi (1.7 to 6.9 MPa); and one for smaller parts and production shop situations that uses pressures below 250 psi (1.7 MPa).

The first version, or high-pressure range, is characterized by use of a pumped water supply with hoses and nozzles that generally are hand-directed. In the second version, more production-oriented and controlled equipment, such as that shown in Fig. 38, is involved. In some instances, abrasives are added to the fluid flow to promote rapid cutting. Single or multiple-nozzle approaches to the workpiece depend on the size and number of parts per load. The principle is that WJM is high-volume, not high-pressure.

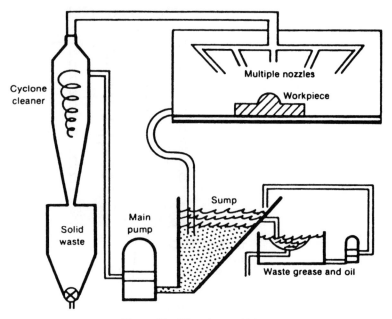

Figure 38 Water-jet machining.

16.10 Electrochemical Deburring

Electrochemical deburring (ECD) is a special version of ECM (Fig. 39). ECD was developed to remove burrs and fins or to round sharp corners. Anodic dissolution occurs on the workpiece burrs in the presence of a closely placed cathodic tool whose configuration matches the burred edge. Normally, only a small portion of the cathode is electrically exposed, so a maximum concentration of the electrolytic action is attained. The electrolyte flow usually is arranged to carry away any burrs that may break loose from the workpiece during the cycle.

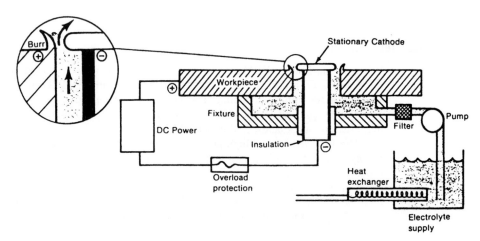

Figure 39 Electrochemical deburring.

Voltages are low, current densities are high, electrolyte flow rate is modest, and electrolyte types are similar to those used for ECM. The electrode (tool) is stationary, so equipment is simpler than that used for ECM. Cycle time is short for deburring. Longer cycle time produces a natural radiusing action.

16.11 Electrochemical Discharge Grinding

Electrochemical discharge grinding (ECDG) combines the features of both electrochemical and electrical discharge methods of material removal (Fig. 40). ECDG has the arrangement and electrolytes of electrochemical grinding (ECG), but uses a graphite wheel without abrasive grains. The random spark discharge is generated through the insulating oxide film on the workpiece by the power generated in an ac source or by a pulsating dc source. The principal material removal comes from the electrolytic action of the low-level dc voltages. The spark discharges erode the anodic films to allow the electrolytic action to continue.

16.12 Electrochemical Grinding

Electrochemical grinding (ECG) is a special form of electrochemical machining in which the conductive workpiece material is dissolved by anodic action, and any resulting films are removed by a rotating, conductive, abrasive wheel (Fig. 41). The abrasive grains protruding from the wheel form the insulating electrical gap between the wheel and the workpiece. This gap must be filled with electrolyte at all times. The conductive wheel uses conventional abrasives—aluminum oxide (because it is nonconductive) or diamond (for intricate shapes)—but lasts substantially longer than wheels used in conventional grinding. The reason for this is that the bulk of material removal (95–98%) occurs by deplating, while only a small amount (2–5%) occurs by abrasive mechanical action. Maximum wheel contact arc lengths are about ¾–1 in. (19–25 mm) to prevent overheating the electrolyte. The fastest material removal is obtained by using the highest attainable current densities without boiling the electrolyte. The corrosive salts used as electrolytes should be filtered and flow rate should be controlled for the best process control.

16.13 Electrochemical Honing

Electrochemical honing (ECH) is the removal of material by anodic dissolution combined with mechanical abrasion from a rotating and reciprocating abrasive stone (carried on a

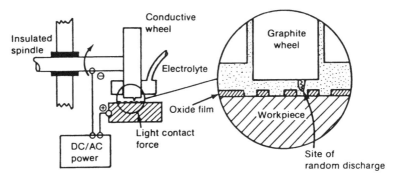

Figure 40 Electrochemical discharge grinding.

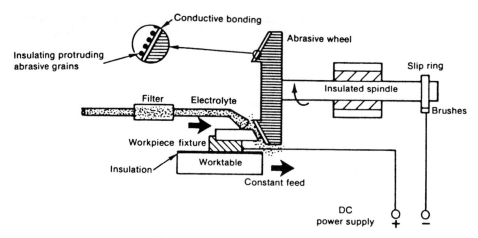

Figure 41 Electrochemical grinding.

spindle, which is the cathode) separated from the workpiece by a rapidly flowing electrolyte (Fig. 42). The principal material removal action comes from electrolytic dissolution. The abrasive stones are used to maintain size and to clean the surfaces to expose fresh metal to the electrolyte action. The small electrical gap is maintained by the nonconducting stones that are bonded to the expandable arbor with cement. The cement must be compatible with the electrolyte and the low dc voltage. The mechanical honing action uses materials, speeds, and pressures typical of conventional honing.

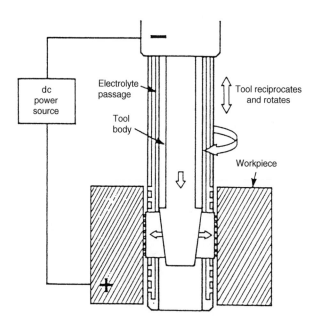

Figure 42 Electrochemical honing.

16.14 Electrochemical Machining

Electrochemical machining (ECM) is the removal of electrically conductive material by anodic dissolution in a rapidly flowing electrolyte, which separates the workpiece from a shaped electrode (Fig. 43). The filtered electrolyte is pumped under pressure and at controlled temperature to bring a controlled-conductivity fluid into the narrow gap of the cutting area. The shape imposed on the workpiece is nearly a mirror or conjugate image of the shape of the cathodic electrode. The electrode is advanced into the workpiece at a constant feed rate that exactly matches the rate of dissolution of the work material. Electrochemical machining is basically the reverse of electroplating.

Calculation of Metal Removal and Feed Rates in ECM

$$\text{Current } I = \frac{V}{R} \quad \text{amp}$$

$$\text{Resistance } R = \frac{g \times r}{A}$$

where g = length of gap (cm)
r = electrolyte resistivity
A = area of current path (cm^2)
V = voltage
R = resistance

$$\text{Current density } S = \frac{I}{A} = \frac{V}{r \times g} \quad \text{amp/cm}^2$$

The amount of material deposited or dissolved is proportional to the quantity of electricity passed (current × time):

$$\text{Amount of material} = C \times I \times t$$

where C = constant
t = time, sec

The amount removed or deposited by one faraday (96,500 coulombs = 96,500 amp–sec) is 1 gram-equivalent weight (G).

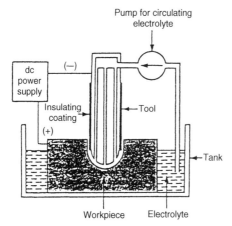

Figure 43 Electrochemical machining.

$$G = \frac{N}{n} \quad \text{(for 1 faraday)}$$

where N = atomic weight
n = valence

$$\text{Volume of metal removed} = \frac{I \times t}{96{,}500} \times \frac{N}{n} \times \frac{1}{d} \times h$$

where d = density, g/cm^3
h = current efficiency

$$\text{Specific removal rate } s = \frac{N}{n} \times \frac{1}{96{,}500} \times h \quad \text{cm}^3/\text{amp-sec}$$

$$\text{Cathode feed rate } F = S \times s \quad \text{cm/sec.}$$

16.15 Electrochemical Polishing

Electrochemical polishing (ECP) is a special form of electrochemical machining arranged for cutting or polishing a workpiece (Fig. 44). Polishing parameters are similar in range to those for cutting, but without the feed motion. ECP generally uses a larger gap and a lower current density than does ECM. This requires modestly higher voltages. (In contrast, electropolishing (ELP) uses still lower current densities, lower electrolyte flow, and more remote electrodes.)

16.16 Electrochemical Sharpening

Electrochemical sharpening (ECS) is a special form of electrochemical machining arranged to accomplish sharpening or polishing by hand (Fig. 45). A portable power pack and electrolyte reservoir supply a finger-held electrode with a small current and flow. The fixed gap incorporated on the several styles of shaped electrodes controls the flow rate. A suction tube picks up the used electrolyte for recirculation after filtration.

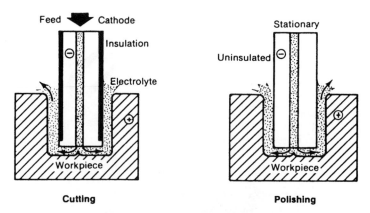

Figure 44 Electrochemical polishing.

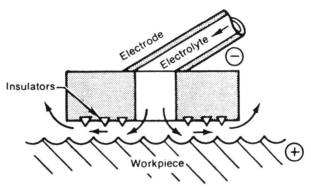

Figure 45 Electrochemical sharpening.

16.17 Electrochemical Turning

Electrochemical turning (ECT) is a special form of electrochemical machining designed to accommodate rotating workpieces (Fig. 46). The rotation provides additional accuracy but complicates the equipment with the method of introducing the high currents to the rotating part. Electrolyte control may also be complicated because rotating seals are needed to direct the flow properly. Otherwise, the parameters and considerations of electrochemical machining apply equally to the turning mode.

16.18 Electro-Stream

Electro-stream (ES) is a special version of electrochemical machining adapted for drilling very small holes using high voltages and acid electrolytes (see Fig. 47). The voltages are more than 10 times those employed in ECM or STEM, so special provisions for containment and protection are required. The tool is a drawn-glass nozzle, 0.001–0.002 in. smaller than

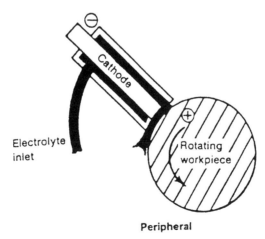

Figure 46 Electrochemical turning.

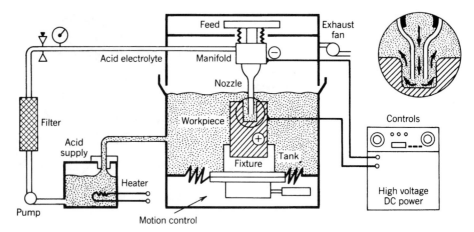

Figure 47 Electro-stream.

the desired hole size. An electrode inside the nozzle or the manifold ensures electrical contact with the acid. Multiple-hole drilling is achieved successfully by ES.

16.19 Shaped-Tube Electrolytic Machining

Shaped-tube electrolytic machining (STEM) is a specialized ECM technique for "drilling" small, deep holes by using acid electrolytes (Fig. 48). Acid is used so that the dissolved metal will go into the solution rather than form a sludge, as is the case with the salt-type electrolytes of ECM. The electrode is a carefully straightened acid-resistant metal tube. The tube is coated with a film of enamel-type insulation. The acid is pressure-fed through the tube and returns via a narrow gap between the tube insulation and the hole wall. The electrode is fed into the workpiece at a rate exactly equal to the rate at which the workpiece

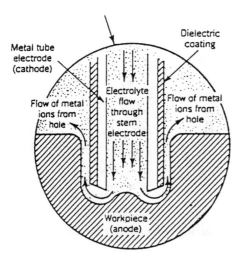

Figure 48 Shaped-tube electrolytic machining.

material is dissolved. Multiple electrodes, even of varying diameters or shapes, may be used simultaneously. A solution of sulfuric acid is frequently used as the electrolyte when machining nickel alloys. The electrolyte is heated and filtered, and flow monitors control the pressure. Tooling is frequently made of plastics, ceramics, or titanium alloys to withstand the electrified hot acid.

16.20 Electron-Beam Machining

Electron-beam machining (EBM) removes material by melting and vaporizing the workpiece at the point of impingement of a focused stream of high-velocity electrons (Fig. 49). To eliminate scattering of the beam of electrons by contact with gas molecules, the work is done in a high-vacuum chamber. Electrons emanate from a triode electron-beam gun and are accelerated to three-fourths the speed of light at the anode. The collision of the electrons with the workpiece immediately translates their kinetic energy into thermal energy. The low-inertia beam can be simply controlled by electromagnetic fields. Magnetic lenses focus the electron beam on the workpiece, where a 0.001-in. (0.025-mm) diameter spot can attain an energy density of up to 10^9 W/in.^{2} (1.55×10^8 W/cm^2) to melt and vaporize any material. The extremely fast response time of the beam is an excellent companion for three-dimensional computer control of beam deflection, beam focus, beam intensity, and workpiece motion.

16.21 Electrical Discharge Grinding

Electrical discharge grinding (EDG) is the removal of a conductive material by rapid, repetitive spark discharges between a rotating tool and the workpiece, which are separated by a flowing dielectric fluid (Fig. 50). (EDG is similar to EDM except that the electrode is in the form of a grinding wheel and the current is usually lower.) The spark gap is servocontrolled. The insulated wheel and the worktable are connected to the dc pulse generator.

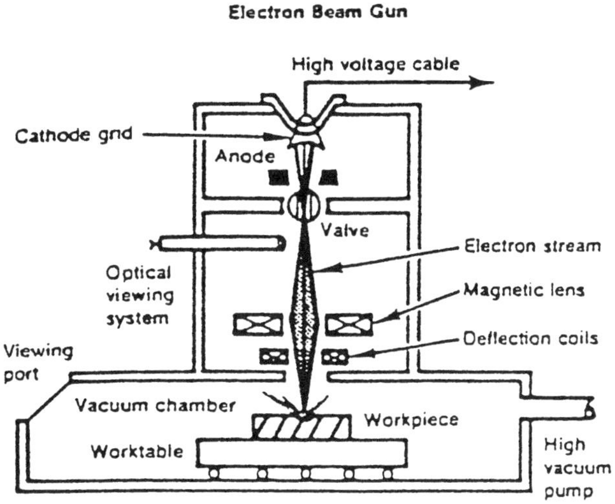

Figure 49 Electron-beam machining.

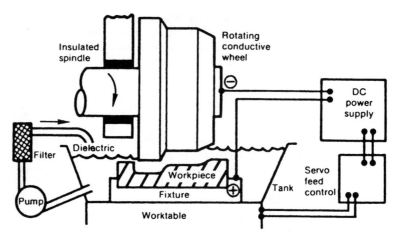

Figure 50 Electrical discharge grinding.

Higher currents produce faster cutting, rougher finishes, and deeper heat-affected zones in the workpiece.

16.22 Electrical Discharge Machining

Electrical discharge machining (EDM) removes electrically conductive material by means of rapid, repetitive spark discharges from a pulsating dc power supply with dielectric flowing between the workpiece and the tool (Fig. 51). The cutting tool (electrode) is made of electrically conductive material, usually carbon. The shaped tool is fed into the workpiece under servocontrol. A spark discharge then breaks down the dielectric fluid. The frequency and energy per spark are set and controlled with a dc power source. The servocontrol maintains a constant gap between the tool and the workpiece while advancing the electrode. The

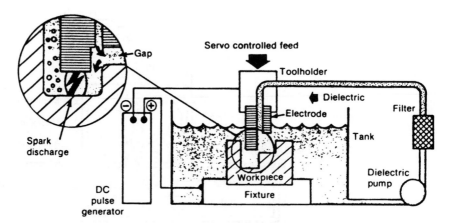

Figure 51 Electrical discharge machining.

dielectric oil cools and flushes out the vaporized and condensed material while reestablishing insulation in the gap. Material removal rate ranges from 16 to 245 cm^3/h. EDM is suitable for cutting materials regardless of their hardness or toughness. Round or irregular-shaped holes 0.002 in. (0.05 mm) diameter can be produced with L/D ratio of 20:1. Narrow slots as small as 0.002–0.010 in. (0.05–0.25 mm) wide are cut by EDM.

16.23 Electrical Discharge Sawing

Electrical discharge sawing (EDS) is a variation of electrical discharge machining (EDM) that combines the motion of either a band saw or a circular disk saw with electrical erosion of the workpiece (Fig. 52). The rapid-moving, untoothed, thin, special steel band or disk is guided into the workpiece by carbide-faced inserts. A kerf only 0.002–0.005 in. (0.050–0.13 mm) wider than the blade or disk is formed as they are fed into the workpiece. Water is used as a cooling quenchant for the tool, swarf, and workpiece. Circular cutting is usually performed under water, thereby reducing noise and fumes. While the work is power-fed into the band (or the disk into the work), it is not subjected to appreciable forces because the arc does the cutting, so fixturing can be minimal.

16.24 Electrical Discharge Wire Cutting (Traveling Wire)

Electrical discharge wire cutting (EDWC) is a special form of electrical discharge machining wherein the electrode is a continuously moving conductive wire (Fig. 53). EDWC is often called *traveling wire* EDM. A small-diameter tension wire, 0.001–0.012 in. (0.03–0.30 mm), is guided to produce a straight, narrow-kerf size 0.003–0.015 in. (0.075–0.375 mm). Usually, a programmed or numerically controlled motion guides the cutting, while the width of the kerf is maintained by the wire size and discharge controls. The dielectric is oil or deionized water carried into the gap by motion of the wire. Wire EDM is able to cut plates as thick as 12 in. (300 mm) and issued for making dies from hard metals. The wire travels with speed in the range of 6–300 in./min (0.15–8 mm/min). A typical cutting rate is 1 in.^{2} (645 mm^2) of cross-sectional area per hour.

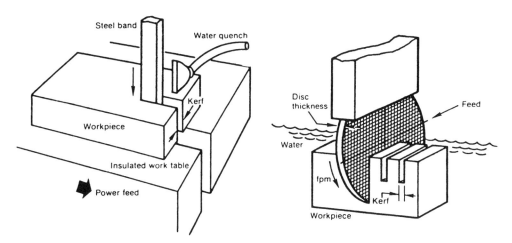

Figure 52 Electrical discharge sawing.

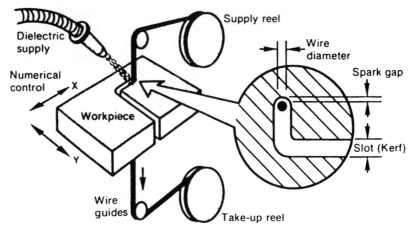

Figure 53 Electrical discharge wire cutting.

16.25 Laser-Beam Machining

Laser-beam machining (LBM) removes material by melting, ablating, and vaporizing the workpiece at the point of impingement of a highly focused beam of coherent monochromatic light (Fig. 54). Laser is an acronym for "light amplification by stimulated emission of radiation." The electromagnetic radiation operates at wavelengths from the visible to the infrared. The principal lasers used for material removal are the Nd:glass (neodymium:glass), the Nd:YAG (neodymium:yttrium–aluminum–garnet), the ruby and the carbon dioxide (CO_2). The last is a gas laser (most frequently used as a torch with an assisting gas—see LBT, laser-beam torch), while others are solid-state lasing materials.

For pulsed operation, the power supply produces short, intense bursts of electricity into the flash lamps, which concentrate their light flux on the lasing material. The resulting energy

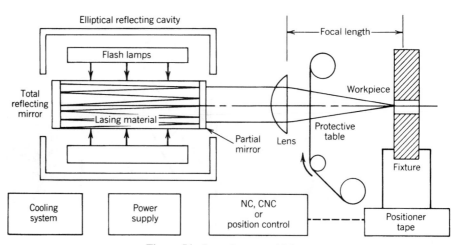

Figure 54 Laser-beam machining.

from the excited atoms is released at a characteristic, constant frequency. The monochromatic light is amplified during successive reflections from the mirrors. The thoroughly collimated light exits through the partially reflecting mirror to the lens, which focuses it on or just below the surface of the workpiece. The small beam divergence, high peak power, and single frequency provide excellent, small-diameter spots of light with energy densities up to 3×10^{10} W/in.^{2} (4.6×10^9 W/cm^2), which can sublime almost any material. Cutting requires energy densities of 10^7–10^9 W/in.^{2} (1.55×10^6–1.55×10^8 W/cm^2), at which rate the thermal capacity of most materials cannot conduct energy into the body of the workpiece fast enough to prevent melting and vaporization. Some lasers can instantaneously produce 41,000°C (74,000°F). Holes of 0.001 in. (0.025 mm), with depth-to-diameter 50 to 1 are typically produced in various materials by LBM.

16.26 Laser-Beam Torch

Laser-beam torch (LBT) is a process in which material is removed by the simultaneous focusing of a laser beam and a gas stream on the workpiece (see Fig. 55). A continuous-wave (CW) laser or a pulsed laser with more than 100 pulses per second is focused on or slightly below the surface of the workpiece, and the absorbed energy causes localized melting. An oxygen gas stream promotes an exothermic reaction and purges the molten material from the cut. Argon or nitrogen gas is sometimes used to purge the molten material while also protecting the workpiece.

Argon or nitrogen gas is often used when organic or ceramic materials are being cut. Close control of the spot size and the focus on the workpiece surface is required for uniform cutting. The type of gas used has only a modest effect on laser penetrating ability. Typically,

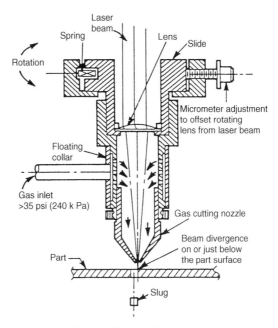

Figure 55 Laser-beam torch.

short laser pulses with high peak power are used for cutting and welding. The CO_2 laser is the laser most often used for cutting. Thin materials are cut at high rates, $1/8$–$3/8$ in. (3.2–9.5 mm) thickness is a practical limit.

16.27 Plasma-Beam Machining

Plasma-beam machining (PBM) removes material by using a superheated stream of electrically ionized gas (Fig. 56). The 20,000–50,000°F (11,000–28,000°C) plasma is created inside a water-cooled nozzle by electrically ionizing a suitable gas, such as nitrogen, hydrogen, or argon, or mixtures of these gases. Since the process does not rely on the heat of combustion between the gas and the workpiece material, it can be used on almost any conductive metal. Generally, the arc is transferred to the workpiece, which is made electrically positive. The plasma—a mixture of free electrons, positively charged ions, and neutral atoms—is initiated in a confined, gas-filled chamber by a high-frequency spark. The high-voltage dc power sustains the arc, which exits from the nozzle at near-sonic velocity. The high-velocity gases blow away the molten metal "chips." Dual-flow torches use a secondary gas or water shield to assist in blowing the molten metal out of the kerf, giving a cleaner cut. PBM is sometimes called *plasma-arc cutting* (PAC). PBM can cut plates up to 6.0 in. (152 mm) thick. Kerf width can be as small as 0.06 in. (1.52 mm) in cutting thin plates.

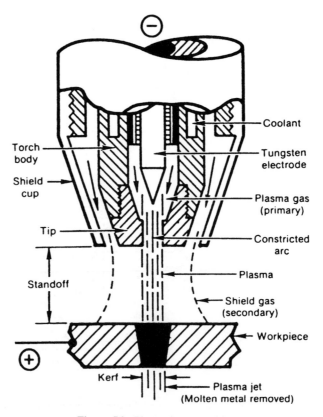

Figure 56 Plasma-beam machining.

16.28 Chemical Machining: Chemical Milling, Chemical Blanking

Chemical machining (CHM) is the controlled dissolution of a workpiece material by contact with a strong chemical reagent (Fig. 57). The thoroughly cleaned workpiece is covered with a strippable, chemically resistant mask. Areas where chemical action is desired are outlined on the workpiece with the use of a template and then stripped off the mask. The workpiece is then submerged in the chemical reagent to remove material simultaneously from all exposed surfaces. The solution should be stirred or the workpiece should be agitated for more effective and more uniform action. Increasing the temperatures will also expedite the action. The machined workpiece is then washed and rinsed, and the remaining mask is removed. Multiple parts can be maintained simultaneously in the same tank. A wide variety of metals can be chemically machined; however, the practical limitations for depth of cut are 0.25–0.5 in. (6.0–12.0 mm) and typical etching rate is 0.001 in./min (0.025 mm/min).

In chemical blanking, the material is removed by chemical dissolution instead of shearing. The operation is applicable to production of complex shapes in thin sheets of metal.

16.29 Electropolishing

Electropolishing (ELP) is a specialized form of chemical machining that uses an electrical deplating action to enhance the chemical action (Fig. 58). The chemical action from the concentrated heavy acids does most of the work, while the electrical action smooths or polishes the irregularities. A metal cathode is connected to a low-voltage, low-amperage dc power source and is installed in the chemical bath near the workpiece. Usually, the cathode is not shaped or conformed to the surface being polished. The cutting action takes place over the entire exposed surface; therefore, a good flow of heated, fresh chemicals is needed in the cutting area to secure uniform finishes. The cutting action will concentrate first on burrs, fins, and sharp corners. Masking, similar to that used with CHM, prevents cutting in unwanted areas. Typical roughness values range from 4 to 32 μin. (0.1 to 0.8 μm).

16.30 Photochemical Machining

Photochemical machining (PCM) is a variation of CHM where the chemically resistant mask is applied to the workpiece by a photographic technique (Fig. 59). A photographic negative, often a reduced image of an oversize master print (up to 100×), is applied to the workpiece and developed. Precise registry of duplicate negatives on each side of the sheet is essential

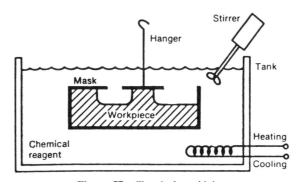

Figure 57 Chemical machining.

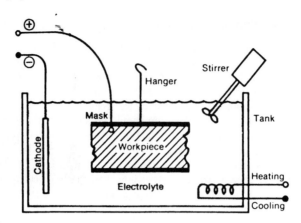

Figure 58 Electropolishing.

for accurately blanked parts. Immersion or spray etching is used to remove the exposed material. The chemicals used must be active on the workpiece, but inactive against the photoresistant mask. The use of PCM is limited to thin materials—up to $1/16$ in. (1.5 mm).

16.31 Thermochemical Machining

Thermochemical machining (TCM) removes the workpiece material—usually only burrs and fins—by exposure of the workpiece to hot, corrosive gases. The process is sometimes called *combustion machining, thermal deburring,* or *thermal energy method* (TEM). The workpiece is exposed for a very short time to extremely hot gases, which are formed by detonating an explosive mixture. The ignition of the explosive—usually hydrogen or natural gas and oxygen—creates a transient thermal wave that vaporizes the burrs and fins. The main body of

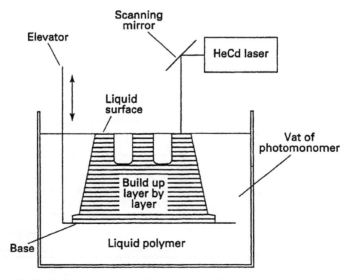

Figure 59 Rapid Prototyping using laser to photopolymerize the liquid photopolymer.

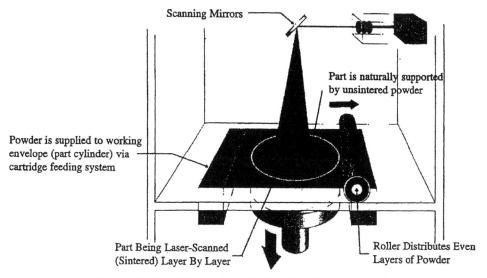

Figure 60 Rapid Prototyping using Sintering Process (powder).

the workpiece remains unaffected and relatively cool because of its low surface-to-mass ratio and the shortness of the exposure to high temperatures.

16.32 Rapid Prototyping and Rapid Tooling

In the past, when making a prototype, a full-scale model of a product, the designed part would have then machined or sculptured from wood, plastic, metal, or other solid materials. Now there is rapid prototyping (Fig. 59), also called desktop manufacturing, a process by which a solid physical model of a product is made directly from a three-dimensional CAD drawing.

Rapid prototyping entails several different consolidation techniques and steps: resin curing, deposition, solidification, and finishing. The conceptual design is viewed in its entirety and at different angles on the monitor through a three-dimensional CAD system. The part is then sliced into horizontal planes from 0.004 to 0.008 in. (0.10 to 0.20 mm). Then a helium:cadmium (He:Cd) laser beam passes over the liquid photopolymer resin. The ultraviolet (UV) photons harden the photosensitive resin. The part is lowered only one layer thickness. The recoater blade sweeps over the previously hardened surface, applying a thin, even coat of resin. Upon completion, a high-intensity broadband or continuum ultraviolet radiation is used to cure the mold. Large parts can be produced in sections, and then the sections are welded together.

Other techniques, such as selective laser sintering (SLS) (Fig. 60), use a thin layer of heat-fusable powder that has been evenly deposited by a roller. A CO_2 laser, controlled by a CAD program, heats the powder to just below the melting point and fuses it only along the programmed path.

REFERENCES

1. Society of Manufacturing Engineers, *Tool and Manufacturing Engineers Handbook,* Vol. 1, *Machining,* McGraw-Hill, New York, 1985.

2. *Machining Data Handbook,* 3rd ed., Machinability Data Center, Cincinnati, OH, 1980.
3. *Metals Handbook,* 8th ed., Vol. 3, Machining American Society for Metals, Metals Park, OH, 1985.
4. R. LeGrand (ed.), *American Machinist's Handbook,* 3rd ed., McGraw-Hill, New York, 1973.
5. *Machinery's Handbook,* 21st ed., Industrial Press, New York, 1979.
6. *Machinery Handbook,* Vol. 2, Machinability Data Center, Department of Defense, Cincinnati, OH, 1983.
7. K. G. Swift and J. D. Booker, *Process Selection,* Arnold, London, 1977.
8. C. Sommer, *Non-traditional Machining Handbook,* Advance Publishing, Houston, TX, 2000.

BIBLIOGRAPHY

Amstead, B. H., P. F. Ostwald, and M. L. Begeman, *Manufacturing Processes,* 8th ed., Wiley, New York, 1988.
ASM Handbook, Vol. 16: *Machining,* ASM International, Materials Park, OH, 1995.
Astakhov, V. P., *Metal Cutting Mechanics,* CRC Press, Boca Raton, FL, 1998.
Brown, J., *Advanced Machining Technology Handbook,* McGraw-Hill, New York, 1998.
Charles, J. A., F. Crane, and J. Furness, *Selection and Use of Engineering Materials,* 3rd ed., Butterworth-Heinemann, England, 1997.
DeGarmo, E. P., J. T. Black, and R. A. Kohser, *Material and Processes in Manufacturing,* 9th ed., Wiley, New York, 2003.
Doyle, L. E., G. F. Schrader, and M. B. Singer, *Manufacturing Processes and Materials for Engineers,* 3rd ed., Prentice Hall, Englewood Cliffs, NJ, 1985.
El Walkil, S. D., *Processes and Design for Manufacturing,* 2nd ed., PWS, Boston, MA, 1998.
Groover, M. P., *Automation, Production Systems and Computer-Integrated Manufacturing,* 2nd ed., Prentice Hall, Upper Saddle River, NJ, 2001.
Hamrock, B. J., B. Jacobson, and S. R. Schmid, *Fundamentals of Machine Elements,* McGraw-Hill, Boston, MA, 1999.
Jain, V. K., and P. C. Pandey, *Theory and Practice of Electro-chemical Machining,* Wiley, New York, 1993.
Kalpakjia, S., and S. R. Schmid, *Manufacturing Processes for Engineering Materials,* Prentice Hall, Englewood Cliffs, NJ, 2003.
Kronenberg, M., *Machining Science and Application,* Pergamon, London, 1966.
Lindberg, R. A., *Processes and Materials of Manufacture,* 2nd ed., Allyn & Bacon, Boston, MA, 1977.
McGeough, J. A., *Advanced Methods of Machining,* Wolters Kluwer Co., Dordrecht, The Netherlands, 1988.
Metal Cutting Tool Handbook, 7th ed., Industrial Press, Cleveland, OH, 1989.
Moore, H. D., and D. R. Kibbey, *Manufacturing Materials and Processes,* 3rd ed., Wiley, New York, 1982.
Niebel, B. W., and A. B. Draper, *Product Design and Process Engineering,* McGraw-Hill, New York, 1974.
Schey, J. A., *Introduction to Manufacturing Processes,* McGraw-Hill, New York, 1977.
Shaw, M. C., *Metal Cutting Principles,* Oxford University Press, Oxford, UK, 1984.
Sommer, C., *Non-traditional Machining Handbook,* Advance Publishing, Houston, TX, 2000.
Trent, E. M., and P. K. Wright, *Metal Cutting,* 4th ed., Butterworth Heinmann, 1999.
Walsh, R. A., *Machining and Metalworking Handbook,* McGraw-Hill, New York, 1994.
Waters, F., *Fundamentals of Manufacturing for Engineers,* UCL Press, University College, London, 1996.
Webster, J. A., *Abrasive Processes Theory, Technology, and Pracitce,* Dekker, New York, 1996.
Zohdi, M. E., "Statistical Analysis, Estimation and Optimization in the Grinding Process," *ASME Transactions,* 1973, Paper No. 73-DET-3.

CHAPTER 6
METAL FORMING, SHAPING, AND CASTING

Magd E. Zohdi
Industrial Engineering Department
Louisiana State University
Baton Rouge, Louisiana

William E. Biles
Industrial Engineering Department
University of Louisville
Louisville, Kentucky

1	**INTRODUCTION**	245	4.3	Permanent-Mold Casting	272
			4.4	Plaster-Mold Casting	274
2	**HOT-WORKING PROCESSES**	246	4.5	Investment Casting	274
	2.1 Classification of Hot-Working Processes	247	5	**PLASTIC-MOLDING PROCESSES**	275
	2.2 Rolling	247		5.1 Injection Molding	275
	2.3 Forging	249		5.2 Coinjection Molding	275
	2.4 Extrusion	251		5.3 Rotomolding	275
	2.5 Drawing	253		5.4 Expandable-Bead Molding	275
	2.6 Spinning	256		5.5 Extruding	276
	2.7 Pipe Welding	257		5.6 Blow Molding	276
	2.8 Piercing	257		5.7 Thermoforming	276
				5.8 Reinforced-Plastic Molding	276
3	**COLD-WORKING PROCESSES**	258		5.9 Forged-Plastic Parts	276
	3.1 Classification of Cold-Working Operations	258	6	**POWDER METALLURGY**	277
	3.2 Squeezing Processes	259		6.1 Properties of P/M Products	277
	3.3 Bending	260			
	3.4 Shearing	263	7	**SURFACE TREATMENT**	278
	3.5 Drawing	265		7.1 Cleaning	278
				7.2 Coatings	281
4	**METAL CASTING AND MOLDING PROCESSES**	268		7.3 Chemical Conversions	283
	4.1 Sand Casting	268		**BIBLIOGRAPHY**	284
	4.2 Centrifugal Casting	269			

1 INTRODUCTION

Metal-forming processes use a remarkable property of metals—their ability to flow plastically in the solid state without concurrent deterioration of properties. Moreover, by simply moving the metal to the desired shape, there is little or no waste. Figure 1 shows some of the metal-forming processes. Metal-forming processes are classified into two categories: hot-working processes and cold-working processes.

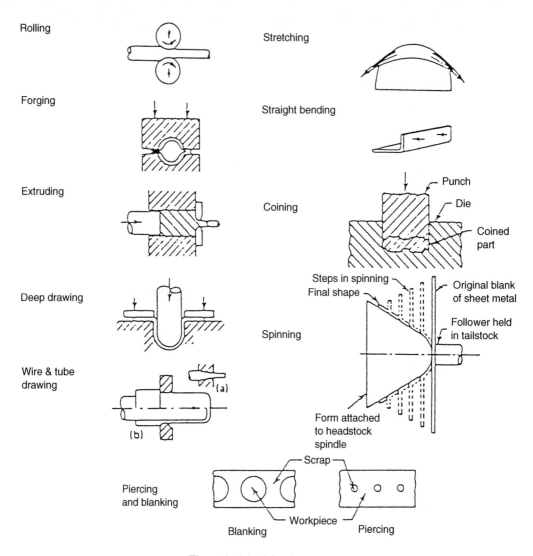

Figure 1 Metal-forming processes.

2 HOT-WORKING PROCESSES

Hot working is defined as the plastic deformation of metals above their recrystallization temperature. Here it is important to note that the crystallization temperature varies greatly with different materials. Lead and tin are hot worked at room temperature, while steels require temperatures of 2000°F (1100°C). Hot working does not necessarily imply high absolute temperatures.

Hot working can produce the following improvements:

  1. Production of randomly oriented, spherical-shaped grain structure, which results in a net increase not only in the strength but also in ductility and toughness.

2. The reorientation of inclusions or impurity material in metal. The impurity material often distorts and flows along with the metal.

This material, however, does not recrystallize with the base metal and often produces a fiber structure. Such a structure clearly has directional properties, being stronger in one direction than in another. Moreover, an impurity originally oriented so as to aid crack movement through the metal is often reoriented into a "crack-arrestor" configuration perpendicular to crack propagation.

2.1 Classification of Hot-Working Processes

The most obvious reason for the popularity of hot working is that it provides an attractive means of forming a desired shape. Some of the hot-working processes that are of major importance in modern manufacturing are

1. Rolling
2. Forging
3. Extrusion and upsetting
4. Drawing
5. Spinning
6. Pipe welding
7. Piercing

2.2 Rolling

Hot rolling (Fig. 2) consists of passing heated metal between two rolls that revolve in opposite directions, the space between the rolls being somewhat less than the thickness of the entering metal. Many finished parts, such as hot-rolled structural shapes, are completed entirely by hot rolling. More often, however, hot-rolled products, such as sheets, plates, bars, and strips, serve as input material for other processes, such as cold forming or machining.

In hot rolling, as in all hot working, it is very important that the metal be heated uniformly throughout to the proper temperature, a procedure known as *soaking*. If the temperature is not uniform, the subsequent deformation will also be nonuniform, the hotter exterior flowing in preference to the cooler and, therefore, stronger, interior. Cracking, tearing, and associated problems may result.

Isothermal Rolling

The ordinary rolling of some high-strength metals, such as titanium and stainless steels, particularly in thicknesses below about 0.150 in. (3.8 mm), is difficult because the heat in the sheet is transferred rapidly to the cold and much more massive rolls. This has been overcome by isothermal rolling. Localized heating is accomplished in the area of deformation by the passage of a large electrical current between the rolls, through the sheet. Reductions up to 90% per roll have been achieved. The process usually is restricted to widths below 2 in. (50 mm).

The rolling strip contact length is given by

$$L \simeq \sqrt{R(h_0 - h)}$$

where R = roll radius
h_0 = original strip thickness
h = reduced thickness

The roll-force F is calculated by

$$F = LwY_{avg} \tag{1}$$

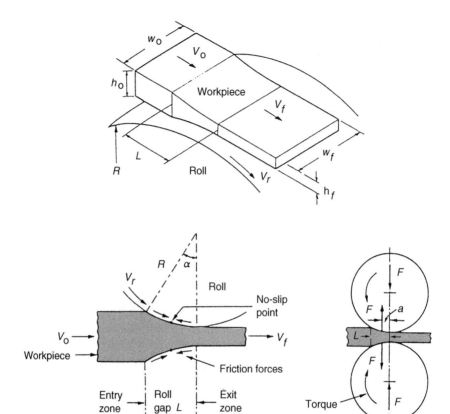

Figure 2 Hot rolling.

where w = width
Y_{avg} = average true stress

Figure 3 gives the true stress for different material at the true stress ϵ. The true stress ϵ is given by

$$\epsilon = \ln\left(\frac{h_0}{h}\right)$$

$$\text{Power/Roll} = \frac{2\pi FLN}{60,000} \quad \text{kW} \tag{2}$$

where F = newtons
L = meters
N = rev per min

or

$$\text{Power} = \frac{2\pi FLN}{33,000} \quad \text{hp} \tag{3}$$

where F = lb
L = ft

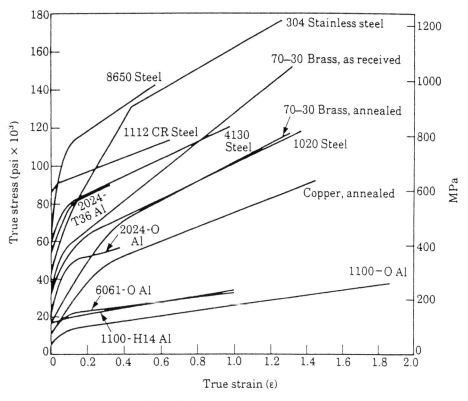

Figure 3 True stress–true strain curves.

2.3 Forging

Forging is the plastic working of metal by means of localized compressive forces exerted by manual or power hammers, presses, or special forging machines.

Various types of forging have been developed to provide great flexibility, making it economically possible to forge a single piece or to mass produce thousands of identical parts. The metal may be

1. Drawn out, increasing its length and decreasing its cross section
2. Upset, increasing the cross section and decreasing the length, or
3. Squeezed in closed impression dies to produce multidirectional flow

The state of stress in the work is primarily uniaxial or multiaxial compression.
The common forging processes are

1. Open-die hammer
2. Impression-die drop forging
3. Press forging
4. Upset forging
5. Roll forging
6. Swaging

Open-Die Hammer Forging

Open-die forging, (Fig. 4) does not confine the flow of metal, the hammer and anvil often being completely flat. The desired shape is obtained by manipulating the workpiece between blows. Specially shaped tools or a slightly shaped die between the workpiece and the hammer or anvil are used to aid in shaping sections (round, concave, or convex), making holes, or performing cutoff operations.

The force F required for an open-die forging operation on a solid cylindrical piece can be calculated by

$$F = Y_f \pi r^2 \left(1 + \frac{2\mu r}{3h}\right) \qquad (4)$$

where Y_f = flow stress at the specific ϵ [$\epsilon = \ln(h_0/h)$]
μ = coefficient of friction
r and h = radius and height of workpiece

Impression-Die Drop Forging

In impression-die or closed-die drop forging (Fig. 5), the heated metal is placed in the lower cavity of the die and struck one or more blows with the upper die. This hammering causes the metal to flow so as to fill the die cavity. Excess metal is squeezed out between the die faces along the periphery of the cavity to form a flash. When forging is completed, the flash is trimmed off by means of a trimming die.

The forging force F required for impression-die forging can be estimated by

$$F = KY_f A \qquad (5)$$

where K = multiplying factor (4–12) depending on the complexity of the shape
Y_f = flow stress at forging temperature
A = projected area, including flash

Press Forging

Press forging employs a slow-squeezing action that penetrates throughout the metal and produces a uniform metal flow. In hammer or impact forging, metal flow is a response to the energy in the hammer–workpiece collision. If all the energy can be dissipated through flow of the surface layers of metal and absorption by the press foundation, the interior regions of the workpiece can go undeformed. Therefore, when the forging of large sections is required, press forging must be employed.

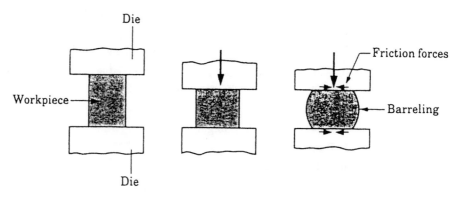

Figure 4 Open-die hammer forging.

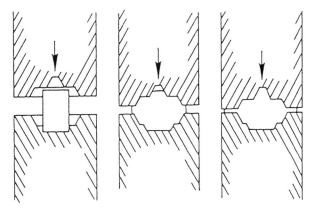

Figure 5 Impression-die drop forging.

Upset Forging
Upset forging involves increasing the diameter of the end or central portion of a bar of metal by compressing its length. Upset-forging machines are used to forge heads on bolts and other fasteners, valves, couplings, and many other small components.

Roll Forging
Roll forging, in which round or flat bar stock is reduced in thickness and increased in length, is used to produce such components as axles, tapered levers, and leaf springs.

Swaging
Swaging involves hammering or forcing a tube or rod into a confining die to reduce its diameter, the die often playing the role of the hammer. Repeated blows cause the metal to flow inward and take the internal form of the die.

2.4 Extrusion

In the extrusion process (Fig. 6), metal is compressively forced to flow through a suitably shaped die to form a product with reduced cross section. Although it may be performed

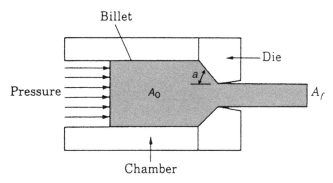

Figure 6 Extrusion process.

either hot or cold, hot extrusion is employed for many metals to reduce the forces required, to eliminate cold-working effects, and to reduce directional properties. The stress state within the material is triaxial compression.

Lead, copper, aluminum, and magnesium, and alloys of these metals, are commonly extruded, taking advantage of the relatively low yield strengths and extrusion temperatures. Steel is more difficult to extrude. Yield strengths are high and the metal has a tendency to weld to the walls of the die and confining chamber under the conditions of high temperature and pressures. With the development and use of phosphate-based and molten glass lubricants, substantial quantities of hot steel extrusions are now produced. These lubricants adhere to the billet and prevent metal-to-metal contact throughout the process.

Almost any cross-section shape can be extruded from the nonferrous metals. Hollow shapes can be extruded by several methods. For tubular products, the stationary or moving mandrel process is often employed. For more complex internal cavities, a spider mandrel or torpedo die is used. Obviously, the cost for hollow extrusions is considerably greater than for solid ones, but a wide variety of shapes can be produced that cannot be made by any other process.

The extrusion force F can be estimated from the formula

$$F = A_0 k \ln \left(\frac{A_0}{A}\right) \tag{6}$$

where k = extrusion constant depends on material and temperature (see Fig. 7)
A_0 = billet area
A_f = finished extruded area

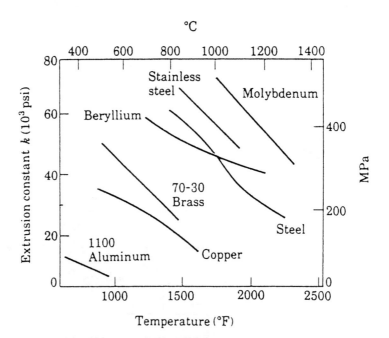

Figure 7 Extrusion constant k.

2.5 Drawing

Drawing (Fig. 8) is a process for forming sheet metal between an edge-opposing punch and a die (draw ring) to produce a cup, cone, box, or shell-like part. The work metal is bent over and wrapped around the punch nose. At the same time, the outer portions of the blank move rapidly toward the center of the blank until they flow over the die radius as the blank is drawn into the die cavity by the punch. The radial movement of the metal increases the blank thickness as the metal moves toward the die radius; as the metal flows over the die radius, this thickness decreases because of the tension in the shell wall between the punch nose and the die radius and (in some instances) because of the clearance between the punch and the die.

The force (load) required for drawing a round cup is expressed by the following empirical equation:

$$L = \pi dt S \left(\frac{D}{d} - k \right) \tag{7}$$

where L = press load, lb
d = cup diameter, in.
D = blank diameter, in.
t = work-metal thickness, in.
S = tensile strength, lbs/in.^{2}
k = a constant that takes into account frictional and bending forces, usually 0.6–0.7

The force (load) required for drawing a rectangular cup can be calculated from the following equation:

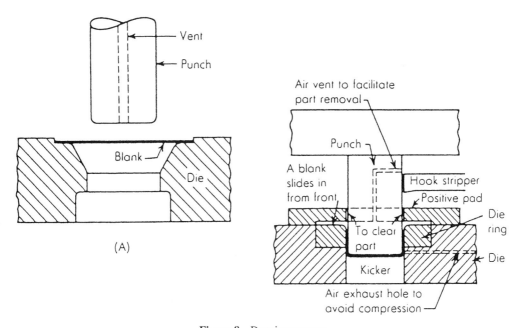

Figure 8 Drawing process.

$$L = tS(2\pi Rk_A + lk_B) \tag{8}$$

where L = press load, lb
 t = work-metal thickness, in.
 S = tensile strength, lb/in.^{2}
 R = corner radius of the cup, in.
 l = the sum of the lengths of straight sections of the sides, in.
k_A and k_B = constants

Values for k_A range from 0.5 (for a shallow cup) to 2.0 (for a cup of depth five to six times the corner radius). Values for k_B range from 0.2 (for easy draw radius, ample clearance, and no blankholding force) and 0.3 (for similar free flow and normal blankholding force of about $L/3$) to a maximum of 1.0 (for metal clamped too tightly to flow).

Figure 9 can be used as a general guide for computing maximum drawing load for a round shell. These relations are based on a free draw with sufficient clearance so that there is no ironing, using a maximum reduction of 50%. The nomograph gives the load required to fracture the cup (1 ton = 8.9 kN).

Blank Diameters

The following equations may be used to calculate the blank size for cylindrical shells of relatively thin metal. The ratio of the shell diameter to the corner radius (d/r) can affect the blank diameter and should be taken into consideration. When d/r is 20 or more,

$$D = \sqrt{d^2 + 4dh} \tag{9}$$

When d/r is between 15 and 20,

$$D = \sqrt{d^2 + 4dh - 0.5r} \tag{10}$$

When d/r is between 10 and 15,

$$D = \sqrt{d^2 + 4dh - r} \tag{11}$$

When d/r is below 10,

$$D = \sqrt{(d - 2r)^2 + 4d(h - r) + 2\pi r(d - 0.7r)} \tag{12}$$

where D = blank diameter
 d = shell diameter
 h = shell height
 r = corner radius

The above equations are based on the assumption that the surface area of the blank is equal to the surface area of the finished shell.

In cases where the shell wall is to be ironed thinner than the shell bottom, the volume of metal in the blank must equal the volume of the metal in the finished shell. Where the wall-thickness reduction is considerable, as in brass shell cases, the final blank size is developed by trial. A tentative blank size for an ironed shell can be obtained from the equation

$$D = \sqrt{d^2 + 4dh\frac{t}{T}} \tag{13}$$

where t = wall thickness
 T = bottom thickness

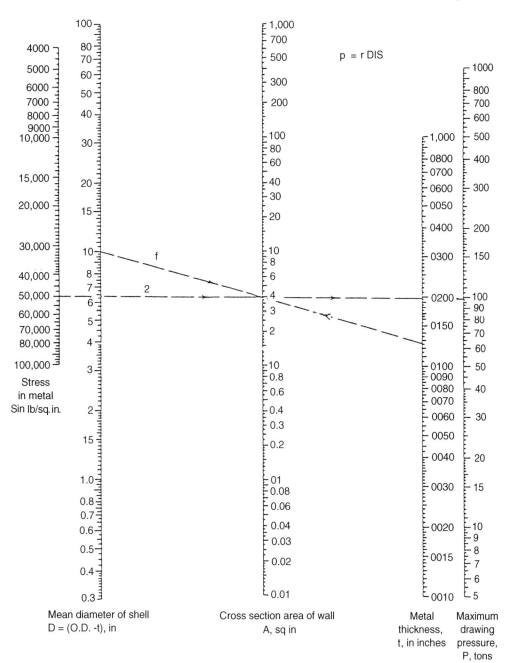

Figure 9 Nomograph for estimating drawing pressures.

2.6 Spinning

Spinning is a method of forming sheet metal or tubing into seamless hollow cylinders, cones, hemispheres, or other circular shapes by a combination of rotation and force. On the basis of techniques used, applications, and results obtainable, the method may be divided into two categories: *manual spinning* (with or without mechanical assistance to increase the force) and *power spinning*.

Manual spinning entails no appreciable thinning of metal. The operation ordinarily done in a lathe consists of pressing a tool against a circular metal blank that is rotated by the headstock.

Power spinning is also known as *shear spinning* because in this method metal is intentionally thinned, by shear forces. In power spinning, forces as great as 400 tons are used.

The application of shear spinning to conical shapes is shown schematically in Fig. 10. The metal deformation is such that forming is in accordance with the sine law, which states that the wall thickness of the starting blank and that of the finished workpiece are related as

$$t_2 = t_1(\sin \alpha) \tag{14}$$

where t_1 = the thickness of the starting blank
t_2 = the thickness of the spun workpiece
α = one-half the apex angle of the cone

Tube Spinning

Tube spinning is a rotary-point method of extruding metal, much like cone spinning, except that the sine law does not apply. Because the half-angle of a cylinder is zero, tube spinning follows a purely volumetric rule, depending on the practical limits of deformation that the metal can stand without intermediate annealing.

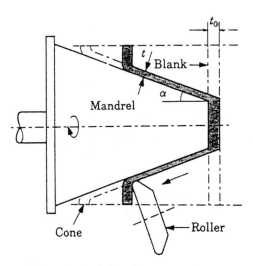

Figure 10 Setup and dimensional relations for one-operation power spinning of a cone.

2.7 Pipe Welding

Large quantities of small-diameter steel pipe are produced by two processes that involve hot forming of metal strip and welding of its edges through utilization of the heat contained in the metal. Both of these processes, *butt welding* and *lap welding* of pipe, utilize steel in the form of skelp—long and narrow strips of the desired thickness. Because the skelp has been previously hot rolled and the welding process produces further compressive working and recrystallization, pipe welding by these processes is uniform in quality.

In the butt-welded pipe process, the skelp is unwound from a continuous coil and is heated to forging temperatures as it passes through a furnace. Upon leaving the furnace, it is pulled through forming rolls that shape it into a cylinder. The pressure exerted between the edges of the skelp as it passes through the rolls is sufficient to upset the metal and weld the edges together. Additional sets of rollers size and shape the pipe. Normal pipe diameters range from $\frac{1}{8}$ to 3 in. (3 to 75 mm).

The lap-welding process for making pipe differs from butt welding in that the skelp has beveled edges and a mandrel is used in conjunction with a set of rollers to make the weld. The process is used primarily for larger sizes of pipe, from about 2 to 14 in. (50 to 400 mm) in diameter.

2.8 Piercing

Thick-walled and seamless tubing is made by the piercing process. A heated, round billet, with its leading end center-punched, is pushed longitudinally between two large, convex-tapered rolls that revolve in the same direction, their axes being inclined at opposite angles of about 6° from the axis of the billet. The clearance between the rolls is somewhat less than the diameter of the billet. As the billet is caught by the rolls and rotated, their inclination causes the billet to be drawn forward into them. The reduced clearance between the rolls forces the rotating billet to deform into an elliptical shape. To rotate with an elliptical cross section, the metal must undergo shear about the major axis, which causes a crack to open. As the crack opens, the billet is forced over a pointed mandrel that enlarges and shapes the opening, forming a seamless tube (Fig. 11).

This procedure applies to seamless tubes up to 6 in. (150 mm) in diameter. Larger tubes up to 14 in. (355 mm) in diameter are given a second operation on piercing rolls. To produce sizes up to 24 in. (610 mm) in diameter, reheated, double-pierced tubes are processed on a

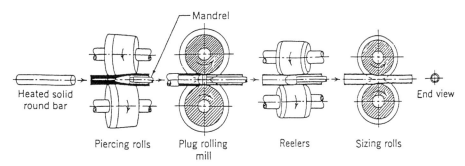

Figure 11 Principal steps in the manufacture of seamless tubing.

rotary rolling mill, and are finally completed by reelers and sizing rolls, as described in the single-piercing process.

3 COLD-WORKING PROCESSES

Cold working is the plastic deformation of metals below the recrystallization temperature. In most cases of manufacturing, such cold forming is done at room temperature. In some cases, however, the working may be done at elevated temperatures that will provide increased ductility and reduced strength, but will be below the recrystallization temperature.

When compared to hot working, cold-working processes have certain distinct advantages:

1. No heating required
2. Better surface finish obtained
3. Superior dimension control
4. Better reproducibility and interchangeability of parts
5. Improved strength properties
6. Directional properties can be imparted
7. Contamination problems minimized

Some disadvantages associated with cold-working processes include

1. Higher forces required for deformation
2. Heavier and more powerful equipment required
3. Less ductility available
4. Metal surfaces must be clean and scale-free
5. Strain hardening occurs (may require intermediate anneals)
6. Imparted directional properties may be detrimental
7. May produce undesirable residual stresses

3.1 Classification of Cold-Working Operations

The major cold-working operations can be classified basically under the headings of squeezing, bending, shearing, and drawing, as follows:

Squeezing	Bending	Shearing	Drawing
1. Rolling	1. Angle	1. Shearing	1. Bar and tube drawing
2. Swaging	2. Roll	Slitting	2. Wire drawing
3. Cold forging	3. Roll forming	2. Blanking	3. Spinning
4. Sizing	4. Drawing	3. Piercing	4. Embossing
5. Extrusion	5. Seaming	Lancing	5. Stretch forming
6. Riveting	6. Flanging	Perforating	6. Shell drawing
7. Staking	7. Straightening	4. Notching	7. Ironing
8. Coining		Nibbling	8. High-energy-rate forming
9. Peening		5. Shaving	
10. Burnishing		6. Trimming	
11. Die hobbing		7. Cutoff	
12. Thread rolling		8. Dinking	

3.2 Squeezing Processes

Most of the cold-working squeezing processes have identical hot-working counterparts or are extensions of them. The primary reasons for deforming cold rather than hot are to obtain better dimensional accuracy and surface finish. In many cases, the equipment is basically the same, except that it must be more powerful.

Cold Rolling
Cold rolling accounts for by far the greatest tonnage of cold-worked products. Sheets, strip, bars, and rods are cold-rolled to obtain products that have smooth surfaces and accurate dimensions.

Swaging
Swaging basically is a process for reducing the diameter, tapering, or pointing round bars or tubes by external hammering. A useful extension of the process involves the formation of internal cavities. A shaped mandrel is inserted inside a tube and the tube is then collapsed around it by swaging (Fig. 12).

Cold Forging
Extremely large quantities of products are made by cold forging, in which the metal is squeezed into a die cavity that imparts the desired shape. Cold heading is used for making enlarged sections on the ends of rod or wire, such as the heads on bolts, nails, rivets, and other fasteners.

Sizing
Sizing involves squeezing areas of forgings or ductile castings to a desired thickness. It is used principally on basses and flats, with only enough deformation to bring the region to a desired dimension.

Extrusion
This process is often called *impact extrusion* and was first used only with the low-strength ductile metals, such as lead, tin, and aluminum, for producing such items as collapsible tubes for toothpaste, medications, and so forth; small "cans" such as are used for shielding in electronics and electrical apparatus; and larger cans for food and beverages. In recent years, cold extrusion has been used for forming mild steel parts, often being combined with cold heading.

Another type of cold extrusion, known as *hydrostatic extrusion,* used high fluid pressure to extrude a billet through a die, either into atmospheric pressure or into a lower-pressure chamber. The pressure-to-pressure process makes possible the extrusion of relatively brittle

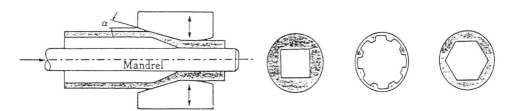

Figure 12 Cross sections of tubes produced by swaging on shaped mandrels. Rifling (spiral grooves) in small gun barrels can be made by this process.

materials, such as molybdenum, beryllium, and tungsten. Billet-chamber friction is eliminated, billet-die lubrication is enhanced by the pressure, and the surrounding pressurized atmosphere suppresses crack initiation and growth.

Riveting
In riveting, a head is formed on the shank end of a fastener to provide a permanent method of joining sheets or plates of metal together. Although riveting usually is done hot in structural work, in manufacturing it almost always is done cold.

Staking
Staking is a commonly used cold-working method for permanently fastening two parts together where one protrudes through a hole in the other. A shaped punch is driven into one of the pieces, deforming the metal sufficiently to squeeze it outward.

Coining
Coining involves cold working by means of positive displacement punch while the metal is completely confined within a set of dies.

Peening
Peening involves striking the surface repeated blows by impelled shot or a round-nose tool. The highly localized blows deform and tend to stretch the metal surface. Because the surface deformation is resisted by the metal underneath, the result is a surface layer under residual compression. This condition is highly favorable to resist cracking under fatigue conditions, such as repeated bending, because the compressive stresses are subtractive from the applied tensile loads. For this reason, shafting, crankshafts, gear teeth, and other cyclic-loaded components are frequently peened.

Burnishing
Burnishing involves rubbing a smooth, hard object under considerable pressure over the minute surface protrusions that are formed on a metal surface during machining or shearing, thereby reducing their depth and sharpness through plastic flow.

Hobbing
Hobbing is a cold-working process that is used to form cavities in various types of dies, such as those used for molding plastics. A male hob is made with the contour of the part that ultimately will be formed by the die. After the hob is hardened, it is slowly pressed into an annealed die block by means of hydraulic press until the desired impression is produced.

Thread Rolling
Threads can be rolled in any material sufficiently plastic to withstand the forces of cold working without disintegration. Threads can be rolled by flat or roller dies.

3.3 Bending

Bending is the uniform straining of material, usually flat sheet or strip metal, around a straight axis that lies in the neutral plane and normal to the lengthwise direction of the sheet or strip. Metal flow takes place within the plastic range of the metal, so that the bend retains a permanent set after removal of the applied stress. The inner surface of the bend is in compression; the outer surface is in tension.

Terms used in bending are defined and illustrated in Fig. 13. The neutral axis is the plane area in bent metal where all strains are zero.

Bend Allowances

Since bent metal is longer after bending, its increased length, generally of concern to the product designer, may also have to be considered by the die designer if the length tolerance of the bent part is critical. The length of bent metal may be calculated from the equation

$$B = \frac{A}{360} \times 2\pi(R_i + Kt) \tag{15}$$

where B = bend allowance, in. (mm) (along neutral axis)
  A = bend angle, deg
  R_i = inside radius of bend, in. (mm)
  t = metal thickness, in. (mm)
  K = 0.33 when R_i is less than $2t$ and is 0.50 when R_i is more than $2t$

Bending Methods

Two bending methods are commonly made use of in press tools. Metal sheet or strip, supported by a V block (Fig. 14), is forced by a wedge-shaped punch into the block. Edge bending (Fig. 14) is cantilever loading of a beam. The bending punch (1) forces the metal against the supporting die (2).

Bending Force

The force required for V bending is

$$P = \frac{KLSt^2}{W} \tag{16}$$

where P = bending force, tons (for metric usage, multiply number of tons by 8.896 to obtain kilonewtons)
  K = die opening factor: 1.20 for a die opening of 16 times metal thickness, 1.33 for an opening of eight times metal thickness
  L = length of part, in.
  S = ultimate tensile strength, tons/in.^{2}
  W = width of V or U die, in.
  t = metal thickness, in.

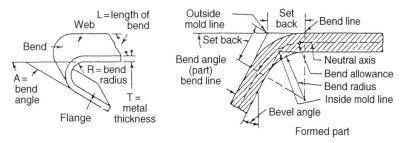

Figure 13 Bend terms.

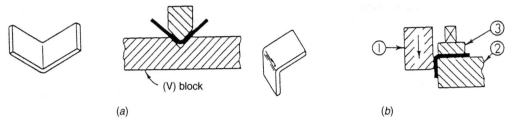

Figure 14 Bending methods. (*a*) V bending; (*b*) edge bending.

For U bending (channel bending), pressures will be approximately twice those required. For U bending, edge bending is required about one-half those needed for V bending. Table 1 gives the ultimate strength = S for various materials.

Several factors must be considered when designing parts that are to be made by bending. Of primary importance is the minimum radius that can be bent successfully without metal cracking. This, of course, is related to the ductility of the metal.

Angle Bending
Angle bends up to 150° in the sheet metal under about 1/16 in. (1.5 mm) in thickness may be made in a bar folder. Heavier sheet metal and more complex bends in thinner sheets are made on a press brake.

Roll Bending
Plates, heavy sheets, and rolled shapes can be bent to a desired curvature on forming rolls. These usually have three rolls in the form of a pyramid, with the two lower rolls being driven and the upper roll adjustable to control the degree of curvature. Supports can be swung clear to permit removal of a closed shape from the rolls. Bending rolls are available in a wide range of sizes, some being capable of bending plate up to 6 in. (150 mm) thick.

Cold-Roll Forming
This process involves the progressive bending of metal strip as it passes through a series of forming rolls. A wide variety of moldings, channeling, and other shapes can be formed on machines that produce up to 10,000 ft (3000 m) of product per day.

Seaming
Seaming is used to join ends of sheet metal to form containers such as cans, pails, and drums. The seams are formed by a series of small rollers on seaming machines that range

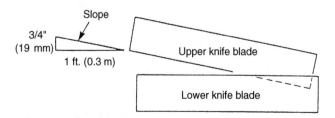

Figure 15 The rake is the angular slope formed by the cutting edges of the upper and lower knives.

Table 1 Ultimate Strength

Metal	(ton/in.²)
Aluminum and alloys	6.5–38.0
Brass	19.0–38.0
Bronze	31.5–47.0
Copper	16.0–25.0
Steel	22.0–40.0
Tin	1.1–1.4
Zinc	9.7–13.5

from small hand-operated types to large automatic units capable of producing hundreds of seams per minute in the mass production of cans.

Flanging
Flanges can be rolled on sheet metal in essentially the same manner as seaming is done. In many cases, however, the forming of flanges and seams involves drawing, since localized bending occurs on a curved axis.

Straightening
Straightening or flattening has as its objective the opposite of bending and often is done before other cold-forming operations to ensure that flat or straight material is available. Two different techniques are quite common. *Roll straightening* or *roller leveling* involves a series of reverse bends. The rod, sheet, or wire is passed through a series of rolls having decreased offsets from a straight line. These bend the metal back and forth in all directions, stressing it slightly beyond its previous elastic limit and thereby removing all previous permanent set.

Sheet may also be straightened by a process called *stretcher leveling*. The sheets are grabbed mechanically at each end and stretched slightly beyond the elastic limit to remove previous stresses and thus produce the desired flatness.

3.4 Shearing

Shearing is the mechanical cutting of materials in sheet or plate form without the formation of chips or use of burning or melting. When the two cutting blades are straight, the process is called *shearing*. Other processes, in which the shearing blades are in the form of curved edges or punches and dies, are called by other names, such as *blanking, piercing, notching, shaving,* and *trimming*. These all are basically shearing operations, however.

The required shear force can be calculated as

$$F = \left(\frac{S \times P \times t^2 \times 12}{R}\right)\left(1 - \frac{P}{2}\right) \tag{17}$$

where F = shear force, lb
S = shear strength (stress), psi
P = penetration of knife into material, %
t = thickness of material, in.
R = rake of the knife blade, in./ft (Fig. 13)

For SI units, the force is multiplied by 4.448 to obtain newtons (N). Table 2 gives the values of P and S for various materials.

Table 2 Values of Percent Penetration and Shear Strength for Various Materials

Material	Percent Penetration	Shear Strength, psi (MPa)
Lead alloys	50	3500 (24.1)–6000 (41.3)
Tin alloys	40	5000 (34.5)–10,000 (69)
Aluminum alloys	60	8000 (55.2)–45,000 (310)
Titanium alloys	10	60,000 (413)–70,000 (482)
Zinc	50	14,000 (96.5)
Cold worked	25	19,000 (131)
Magnesium alloys	50	17,000 (117)–30,000 (207)
Copper	55	22,000 (151.7)
Cold worked	30	28,000 (193)
Brass	50	32,000 (220.6)
Cold worked	30	52,000 (358.5)
Tobin bronze	25	36,000 (248.2)
Cold worked	17	42,000 (289.6)
Steel, 0.10C	50	35,000 (241.3)
Cold worked	38	43,000 (296.5)
Steel, 0.40C	27	62,000 (427.5)
Cold worked	17	78,000 (537.8)
Steel, 0.80C	15	97,000 (668.8)
Cold worked	5	127,000 (875.6)
Steel, 1.00C	10	115,000 (792.9)
Cold worked	2	150,000 (1034.2)
Silicon steel	30	65,000 (448.2)
Stainless steel	30	57,000 (363)–128,000 (882)
Nickel	55	35,000 (241.3)

Blanking
A blank is a shape cut from flat or preformed stock. Ordinarily, a blank serves as a starting workpiece for a formed part; less often, it is a desired end product.

Calculation of the forces and the work involved in blanking gives average figures that are applicable only when (1) the correct shear strength for the material is used, and (2) the die is sharp and the punch is in good condition, has correct clearance, and is functioning properly.

The total load on the press, or the press capacity required to do a particular job, is the sum of the cutting force and other forces acting at the same time, such as the blankholding force exerted by a die cushion.

Cutting Force: Square-End Punches and Dies
When punch and die surfaces are flat and at right angles to the motion of the punch, the cutting force can be found by multiplying the area of the cut section by the shear strength of the work material:

$$L = Stl \tag{18}$$

where L = load on the press, lb (cutting force)
S = shear strength of the stock, psi
t = stock thickness, in.
l = the length or perimeter of cut, in.

Piercing

Piercing is a shearing operation wherein the shearing blades take the form of closed, curved lines on the edges of a punch and die. Piercing is basically the same as blanking except that the piece punched out is the scrap and the remainder of the strip becomes the desired workpiece.

Lancing

Lancing is a piercing operation that may take the form of a slit in the metal or an actual hole. The purpose of lancing is to permit adjacent metal to flow more readily in subsequent forming operations.

Perforating

Perforating consists of piercing a large number of closely spaced holes.

Notching

Notching is essentially the same as piercing except that the edge of the sheet of metal forms a portion of the periphery of the piece that is punched out. It is used to form notches of any desired shape along the edge of a sheet.

Nibbling

Nibbling is a variation of notching in which a special machine makes a series of overlapping notches, each farther into the sheet of metal.

Shaving

Shaving is a finished operation in which a very small amount of metal is sheared away around the edge of a blanked part. Its primary use is to obtain greater dimensional accuracy, but it also may be used to obtain a square of smoother edge.

Trimming

Trimming is used to remove the excess metal that remains after a drawing, forging, or casting operation. It is essentially the same as blanking.

Cutoff

A cutoff operation is one in which a stamping is removed from a strip of stock by means of a punch and die. The cutoff punch and die cut across the entire width of the strip. Frequently, an irregularly shaped cutoff operation may simultaneously give the workpiece all or part of the desired shape.

Dinking

Dinking is a modified shearing operation that is used to blank shapes from low-strength materials, primarily rubber, fiber, and cloth.

3.5 Drawing

Cold Drawing

Cold drawing is a term that can refer to two somewhat different operations. If the stock is in the form of sheet metal, cold drawing is the forming of parts wherein plastic flow occurs over a curved axis. This is one of the most important of all cold-working operations because a wide range of parts, from small caps to large automobile body tops and fenders, can be drawn in a few seconds each. Cold drawing is similar to hot drawing, but the higher deformation forces, thinner metal, limited ductility, and closer dimensional tolerance create some distinctive problems.

If the stock is wire, rod, or tubing, cold drawing refers to the process of reducing the cross section of the material by pulling it through a die, a sort of tensile equivalent to extrusion.

Cold Spinning
Cold spinning is similar to hot spinning, discussed above.

Stretch Forming
In stretch forming, only a single male form block is required. The sheet of metal is gripped by two or more sets of jaws that stretch it and wrap it around the form block as the latter raises upward. Various combinations of stretching, wrapping, and upward motion of the blocks are used, depending on the shape of the part.

Shell or Deep Drawing
The drawing of closed cylindrical or rectangular containers, or a variation of these shapes, with a depth frequently greater than the narrower dimension of their opening, is one of the most important and widely used manufacturing processes. Because the process had its earliest uses in manufacturing artillery shells and cartridge cases, it is sometimes called *shell drawing*. When the depth of the drawn part is less than the diameter, or minimum surface dimension, of the blank, the process is considered to be *shallow drawing*. If the depth is greater than the diameter, it is considered to be *deep drawing*.

The design of complex parts that are to be drawn has been aided considerably by computer techniques, but is far from being completely and successfully solved. Consequently, such design still involves a mix of science, experience, empirical data, and actual experimentation. The body of known information is quite substantial, however, and is being used with outstanding results.

Forming with Rubber or Fluid Pressure
Several methods of forming use rubber or fluid pressure (Fig. 16) to obtain the desired information and thereby eliminate either the male or female member of the die set. Blanks of sheet metal are placed on top of form blocks, which usually are made of wood. The upper ram, which contains a pad of rubber 8–10 in. (200–250 mm) thick in a steel container, then descends. The rubber pad is confined and transmits force to the metal, causing it to bend to

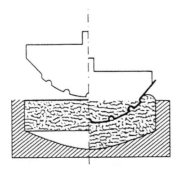

Figure 16 Form with rubber.

the desired shape. Since no female die is used and form blocks replace the male die, die cost is quite low.

The hydroform process or "rubber bag forming" replaces the rubber pad with a flexible diaphragm backed by controlled hydraulic pressure. Deeper parts can be formed with truly uniform fluid pressure.

The bulging oil or rubber is used for applying an internal bulging force to expand a metal blank or tube outward against a female mold or die, thereby eliminating the necessity for a complicated, multiple-piece male die member.

Ironing

Ironing is the name given to the process of thinning the walls of a drawn cylinder by passing it between a punch and a die where the separation is less than the original wall thickness. The walls are elongated and thinned while the base remains unchanged. The most common example of an ironed product is the thin-walled all-aluminum beverage can.

Embossing

Embossing is a method for producing lettering or other designs in thin sheet metal. Basically, it is a very shallow drawing operation, usually in open dies, with the depth of the draw being from one to three times the thickness of the metal.

High-Energy-Rate Forming

A number of methods have been developed for forming metals through the release and application of large amounts of energy in a very short interval (Fig. 17). These processes are called *high-energy-rate-forming processes* (HERF). Many metals tend to deform more readily under the ultrarapid rates of load application used in these processes, a phenomenon apparently related to the relative rates of load application and the movement of dislocations through the metal. As a consequence, HERF makes it possible to form large workpieces and difficult-to-form metals with less expensive equipment and tooling than would otherwise be required.

The high energy-release rates are obtained by five methods:

1. Underwater explosions
2. Underwater spark discharge (electrohydraulic techniques)

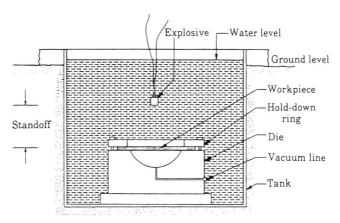

Figure 17 High-energy-rate forming.

3. Pneumatic–mechanical means
   4. Internal combustion of gaseous mixtures
   5. Rapidly formed magnetic fields (electromagnetic techniques)

4 METAL CASTING AND MOLDING PROCESSES

Casting provides a versatility and flexibility that have maintained casting position as a primary production method for machine elements. Casting processes are divided according to the specific type of molding method used in casting, as follows:

1. Sand
2. Centrifugal
3. Permanent
4. Die
5. Plaster-mold
6. Investment

4.1 Sand Casting

Sand casting consists basically of pouring molten metal into appropriate cavities formed in a sand mold (Fig. 18). The sand may be natural, synthetic, or an artificially blended material.

Molds
The two common types of sand molds are the *dry sand mold* and the *green sand mold*. In the dry sand mold, the mold is dried thoroughly prior to closing and pouring, while the green sand mold is used without any preliminary drying. Because the dry sand mold is more firm and resistant to collapse than the green sand mold, core pieces for molds are usually made in this way. Cores are placed in mold cavities to form the interior surfaces of castings.

Patterns
To produce a mold for a conventional sand cast part, it is necessary to make a pattern of the part. Patterns are made from wood or metal to suit a particular design, with allowances to compensate for such factors as natural metal shrinkage and contraction characteristics. These and other effects, such as mold resistance, distortion, casting design, and mold design,

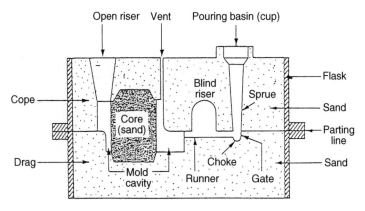

Figure 18 Sectional view of casting mold.

which are not entirely within the range of accurate prediction, generally make it necessary to adjust the pattern to produce castings of the required dimensions.

Access to the mold cavity for entry of the molten metal is provided by sprues, runners, and gates.

Shrinkage

Allowances must be made on patterns to counteract the contraction in size as the metal cools. The amount of shrinkage is dependent on the design of the coating, type of metal used, solidification temperature, and mold resistance. Table 3 gives average shrinkage allowance values used in sand casting. Smaller values apply generally to large or cored castings of intricate design. Larger values apply to small to medium simple castings designed with unrestrained shrinkage.

Machining

Allowances are required in many cases because of unavoidable surface impurities, warpage, and surface variations. Average machining allowances are given in Table 4. Good practice dictates use of minimum section thickness compatible with the design. The normal minimum section recommended for various metals is shown in Table 5.

4.2 Centrifugal Casting

Centrifugal casting consists of having a sand, metal, or ceramic mold that is rotated at high speeds. When the molten metal is poured into the mold, it is thrown against the mold wall, where it remains until it cools and solidifies. The process is increasingly being used for such products as cast-iron pipes, cylinder liners, gun barrels, pressure vessels, brake drums, gears, and flywheels. The metals used include almost all castable alloys. Most dental tooth caps are made by a combined lost-wax process and centrifugal casting.

Advantages and Limitations

Because of the relatively fast cooling time, centrifugal castings have a fine grain size. There is a tendency for the lighter nonmetallic inclusion, slag particles, and dross to segregate toward the inner radius of the castings (Fig. 19), where it can be easily removed by machining. Owing to the high purity of the outer skin, centrifugally cast pipes have a high

Table 3 Pattern Shrinkage Allowance (in./ft)

Metal	Shrinkage
Aluminum alloys	1/10–5/32
Beryllium copper	1/8–5/32
Copper alloys	3/16–7/32
Everdur	3/16
Gray irons	1/8
Hastelloy alloys	1/4
Magnesium alloys	1/8–11/64
Malleable irons	1/16–3/16
Meehanite	1/10–5/32
Nickel and nickel alloys	1/4
Steel	1/8–1/4
White irons	3/16–1/4

Table 4 Machining Allowances for Sand Castings (in.)

Metal	Casting Size	Finish Allowance
Cast irons	Up to 12 in.	3/32
	13–24 in.	1/8
	25–42 in.	3/16
	43–60 in.	1/4
	61–80 in.	5/16
	81–120 in.	3/8
Cast steels	Up to 12 in.	1/8
	13–24 in.	3/16
	25–42 in.	5/16
	43–60 in.	3/8
	61–80 in.	7/16
	81–120 in.	1/2
Malleable irons	Up to 8 in.	1/16
	9–12 in.	3/32
	13–24 in.	1/8
	25–36 in.	3/16
Nonferrous metals	Up to 12 in.	1/16
	13–24 in.	1/8
	25–36 in.	5/32

resistance to atmospheric corrosion. Figure 19 shows a schematic sketch of how a pipe would be centrifugally cast in a horizontal mold. Parts that have diameters exceeding their length are produced by vertical-axis casting (see Fig. 20).

If the centrifugal force is too low or too great, abnormalities will develop. Most horizontal castings are spun so that the force developed is about 65g. Vertically cast parts force is about 90–100g.

The centrifugal force (CF) is calculated from

$$\mathrm{CF} = \frac{mv^2}{r} \text{ lb}$$

$$m = \text{Mass} = \frac{W}{g} = \frac{\text{Weight, lb}}{\text{Acceleration of gravity (ft/s)}^2} = \frac{W}{32.2}$$

Table 5 Minimum Sections for Sand Castings (in.)

Metal	Section
Aluminum alloys	3/16
Copper alloys	3/32
Gray irons	1/8
Magnesium alloys	5/32
Malleable irons	1/8
Steels	1/4
White irons	1/8

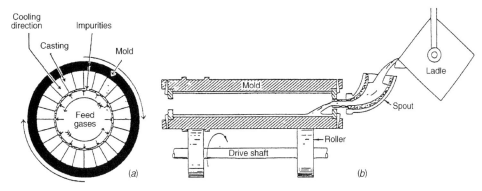

Figure 19 The principle of centrifugal casting is to produce the high-grade metal by throwing the heavier metal outward and forcing the impurities to congregate inward (*a*). Shown at (*b*) is a schematic of how a horizontal-bond centrifugal casting is made.

where v = velocity, ft/s = $r \times w$
r = radius, ft = $\tfrac{1}{2}D$
w = angular velocity, rad/s
w = $2\pi/60 \times$ rpm
D = inside diameter, ft

The number of g's is

$$g = CF/W$$

Hence,

$$g\text{'s} = \frac{1}{W} \times \left[\frac{W}{32.2 \times r} \left(\frac{r \times 2\pi}{60} \right)^2 \right]$$

$$= r \times 3.41 \times 10^{-4} \text{ rpm}^2$$

$$= 1.7 \times 10^{-4} \times D \times (\text{rpm})^2$$

The spinning speed for horizontal-axis molds may be found in English units from the equation

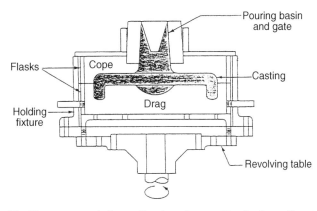

Figure 20 Floor-type vertical centrifugal casting machine for large-diameter parts.

$$N = \sqrt{(\text{Number of g's}) \times \frac{70{,}500}{D}}$$

where N = rpm
D = inside diameter of mold, ft

4.3 Permanent-Mold Casting

As demand for quality castings in production quantities increased, the attractive possibilities of metal molds brought about the development of the permanent-mold process. Although not as flexible regarding design as sand casting, metal-mold casting made possible the continuous production of quantities of casting from a single mold as compared to batch production of individual sand molds.

Metal Molds and Cores

In permanent-mold casting, both metal molds and cores are used, the metal being poured into the mold cavity with the usual gravity head as in sand casting. Molds are normally made of dense iron or meehanite, large cores of cast iron, and small or collapsible cores of alloy steel. All necessary sprues, runners, gates, and risers must be machined into the mold, and the mold cavity itself is made with the usual metal-shrinkage allowances. The mold is usually composed of one, two, or more parts, which may swing or slide for rapid operation. Whereas in sand casting the longest dimension is always placed in a horizontal position, in permanent-mold casting the longest dimension of a part is normally placed in a vertical position.

Production Quantities

Wherever quantities are in the range of 500 pieces or more, permanent-mold casting becomes competitive in cost with sand casting, and if the design is simple, runs as small as 200 pieces are often economical. Production runs of 1000 pieces or more will generally produce a favorable cost difference. High rates of production are possible, and multiple-cavity dies with as many as 16 cavities can be used. In casting gray iron in multiple molds, as many as 50,000 castings per cavity are common with small parts. With larger parts of gray iron, weighing from 12–15 lb, single-cavity molds normally yield 2000–3000 pieces per mold on an average. Up to 100,000 parts per cavity or more are not uncommon with nonferrous metals, magnesium providing the longest die life. Low-pressure permanent mold casting is economical for quantities up to 40,000 pieces (Fig. 21).

Die Casting

Die casting may be classified as a permanent-mold casting system; however, it differs from the process just described in that the molten metal is forced into the mold or die under high pressure (1000–30,000 psi [6.89–206.8 MPa]). The metal solidifies rapidly (within a fraction of a second) because the die is water-cooled. Upon solidification, the die is opened and ejector pins automatically knock the casting out of the die. If the parts are small, several of them may be made at one time in what is termed a *multicavity die.*

There are two main types of machines used: the hot-chamber and the cold-chamber types.

Hot-Chamber Die Casting. In the hot-chamber machine, the metal is kept in a heated holding pot. As the plunger descends, the required amount of alloy is automatically forced into the die. As the piston retracts, the cylinder is again filled with the right amount of molten metal.

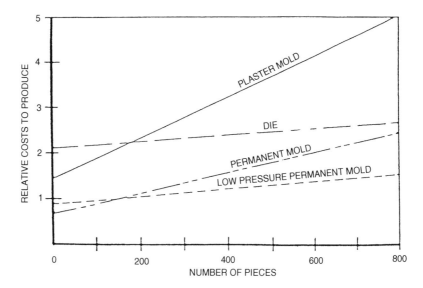

Figure 21 Cost comparison of various casting systems.

Metals such as aluminum, magnesium, and copper tend to alloy with the steel plunger and cannot be used in the hot chamber.

Cold-Chamber Die Casting. This process gets its name from the fact that the metal is ladled into the cold chamber for each shot. This procedure is necessary to keep the molten-metal contact time with the steel cylinder to a minimum. Iron pickup is prevented, as is freezing of the plunger in the cylinder.

Advantages and Limitations
Die-casting machines can produce large quantities of parts with close tolerances and smooth surfaces. The size is limited only by the capacity of the machine. Most die castings are limited to about 75 lb (34 kg) of zinc; 65 lb (30 kg) of aluminum; and 44 lb (20 kg) of magnesium. Die castings can provide thinner sections than any other casting process. Wall thickness as thin as 0.015 in. (0.38 mm) can be achieved with aluminum in small items. However, a more common range on larger sizes will be 0.105–0.180 in. (2.67–4.57 mm).

Some difficulty is experienced in getting sound castings in the larger capacities. Gases tend to be entrapped, which results in low strength and annoying leaks. Of course, one way to reduce metal sections without sacrificing strength is to design in ribs and bosses. Another approach to the porosity problem has been to operate the machine under vacuum. This process is now being developed.

The surface quality is dependent on that of the mold. Parts made from new or repolished dies may have a surface roughness of 24 μin. (0.61 μm). The high surface finish available means that, in most cases, coatings such as chromeplating, anodizing, and painting may be applied directly. More recently, decorative finishes of texture, as obtained by photoetching, have been applied. The technique has been used to simulate woodgrain finishes, as well as textile and leather finishes, and to obtain checkering and crosshatching.

4.4 Plaster-Mold Casting

In general, the various methods of plaster-mold casting are similar. The plaster, also known as *gypsum* or *calcium sulfate,* is mixed dry with other elements, such as talc, sand, asbestos, and sodium silicate. To this mix is added a controlled amount of water to provide the desired permeability in the mold. The slurry that results is heated and delivered through a hose to the flasks, all surfaces of which have been sprayed with a parting compound. The plaster slurry readily fills in and around the most minute details in the highly polished brass patterns. Following filling, the molds are subjected to a short period of vibration and the slurry sets in 5–10 min.

Molds
Molds are extracted from the flask with a vacuum head, following which drying is completed in a continuous oven. Copes and drags are then assembled, with cores when required, and the castings are poured. Upon solidification, the plaster is broken away and any cores used are washed out with a high-pressure jet of water.

4.5 Investment Casting

Casting processes in which the pattern is used only once are variously referred to as *lost-wax* or *precision-casting* processes. They involve making a pattern of the desired form out of wax or plastic (usually polystyrene). The expendable pattern may be made by pressing the wax into a split mold or by the use of an injection-molding machine. The patterns may be gated together so that several parts can be made at once. A metal flask is placed around the assembled patterns and a refractory mold slurry is poured in to support the patterns and form the cavities. A vibrating table equipped with a vacuum pump is used to eliminate all the air from the mold. Formerly, the standard procedure was to dip the patterns in the slurry several times until a coat was built up. This is called the *investment process.* After the mold material has set and dried, the pattern material is melted and allowed to run out of the mold.

The completed flasks are heated slowly to dry the mold and to melt out the wax, plastic, or whatever pattern material was used. When the molds have reached a temperature of 100°F (37.8°C), they are ready for pouring. Vacuum may be applied to the flasks to ensure complete filling of the mold cavities.

When the metal has cooled, the investment material is removed by vibrating hammers or by tumbling. As with other castings, the gates and risers are cut off and ground down.

Ceramic Process
The ceramic process is somewhat similar to the investment-casting in that a creamy ceramic slurry is poured over a pattern. In this case, however, the pattern, made out of plastic, plaster, wood, metal, or rubber, is reusable. The slurry hardens on the pattern almost immediately and becomes a strong green ceramic of the consistency of vulcanized rubber. It is lifted off the pattern while it is still in the rubberlike phase. The mold is ignited with a torch to burn off the volatile portion of the mix. It is then put in a furnace and baked at 1800°F (982°C), resulting in a rigid refractory mold. The mold can be poured while still hot.

Full-Mold Casting
Full-mold casting may be considered a cross between conventional sand casting and the investment technique of using lost wax. In this case, instead of a conventional pattern of wood, metals, or plaster, a polystyrene foam or styrofoam is used. The pattern is left in the mold and is vaporized by the molten metal as it rises in the mold during pouring. Before

molding, the pattern is usually coated with a zirconite wash in an alcohol vehicle. The wash produces a relatively tough skin separating the metal from the sand during pouring and cooling. Conventional foundry sand is used in backing up the mold.

5 PLASTIC-MOLDING PROCESSES

Plastic molding is similar in many ways to metal molding. For most molding operations, plastics are heated to a liquid or a semifluid state and are formed in a mold under pressure. Some of the most common molding processes are discussed below.

5.1 Injection Molding

The largest quantity of plastic parts is made by injection molding. Plastic compound is fed in powdered or granular form from a hopper through metering and melting stages and then injected into a mold. After a brief cooling period, the mold is opened and the solidified part is ejected.

5.2 Coinjection Molding

Coinjection molding makes it possible to mold articles with a solid skin of one thermoplastic and a core of another thermoplastic. The skin material is usually solid while the core material contains blowing agents.

The basic process may be one-, two-, or three-channel technology. In one-channel technology, the two melts are injected into the mold, one after the other. The skin material cools and adheres to the colder surface; a dense skin is formed under proper parameter settings. The thickness of the skin can be controlled by adjustment of injection speed, stock temperature, mold temperature, and flow compatibility of the two melts.

In two- and three-channel techniques, both plastic melts may be introduced simultaneously. This allows for better control of wall thickness of the skin, especially in gate areas on both sides of the part.

Injection-Molded Carbon-Fiber Composites
By mixing carbon or glass fibers in injection-molded plastic parts, they can be made lightweight yet stiffer than steel.

5.3 Rotomolding

In rotational molding, the product is formed inside a closed mold that is rotated about two axes as heat is applied. Liquid or powdered thermoplastic or thermosetting plastic is poured into the mold, either manually or automatically.

5.4 Expandable-Bead Molding

The expandable-bead process consists of placing small beads of polystyrene along with a small amount of blowing agent in a tumbling container. The polystyrene beads soften under heat, which allows a blowing agent to expand them. When the beads reach a given size, depending on the density required, they are quickly cooled. This solidifies the polystyrene in its larger foamed size. The expanded beads are then placed in a mold until it is completely filled. The entrance port is then closed and steam is injected, resoftening the beads and fusing them together. After cooling, the finished, expanded part is removed from the mold.

5.5 Extruding

Plastic extrusion is similar to metal extrusion in that a hot material (plastic melt) is forced through a die having an opening shaped to produce a desired cross section. Depending on the material used, the barrel is heated anywhere from 250 to 600°F (121 to 316°C) to transform the thermoplastic from a solid to a melt. At the end of the extruder barrel is a screen pack for filtering and building back pressure. A breaker plate serves to hold the screen pack in place and straighten the helical flow as it comes off the screen.

5.6 Blow Molding

Blow molding is used extensively to make bottles and other lightweight, hollow plastic parts. Two methods are used: injection blow molding and extrusion blow molding.

Injection blow molding is used primarily for small containers. The parison (molten-plastic pipe) or tube is formed by the injection of plasticized material around a hollow mandrel. While the material is still molten and still on the mandrel, it is transferred into the blowing mold where air is used to inflate it. Accurate threads may be formed at the neck.

In extrusion-type blow molding, parison is inflated under relatively low pressure inside a split-metal mold. The die closes, pinching the end and closing the top around the mandrel. Air enters through the mandrel and inflates the tube until the plastic contacts the cold wall, where it solidifies. The mold opens, the bottle is ejected, and the tailpiece falls off.

5.7 Thermoforming

Thermoforming refers to heating a sheet of plastic material until it becomes soft and pliable and then forming it either under vacuum, by air pressure, or between matching mold halves.

5.8 Reinforced-Plastic Molding

Reinforced plastics generally refers to polymers that have been reinforced with glass fibers. Other materials used are asbestos, sisal, synthetic fibers such as nylon and polyvinyl chloride, and cotton fibers. High-strength composites using graphite fibers are now commercially available with moduli of 50,000,000 psi (344,700,000 MPa) and tensile strengths of about 300,000 psi (2,068,000 MPa). They are as strong as or stronger than the best alloy steels and are lighter than aluminum.

5.9 Forged-Plastic Parts

The forging of plastic materials is a relatively new process. It was developed to shape materials that are difficult or impossible to mold and is used as a low-cost solution for small production runs.

The forging operation starts with a blank or billet of the required shape and volume for the finished part. The blank is heated to a preselected temperature and transferred to the forging dies, which are closed to deform the work material and fill the die cavity. The dies are kept in the closed position for a definite period of time, usually 15–60 sec. When the dies are opened, the finished forging is removed. Since forging involves deformation of the work material in a heated and softened condition, the process is applicable only to thermoplastics.

6 POWDER METALLURGY

In powder metallurgy (P/M), fine metal powders are pressed into a desired shape, usually in a metal die and under high pressure, and the compacted powder is then heated (sintered), with a protective atmosphere. The density of sintered compacts may be increased by repressing. Repressing is also performed to improve the dimensional accuracy, either concurrently or subsequently, for a period of time at a temperature below the melting point of the major constituent. P/M has a number of distinct advantages that account for its rapid growth in recent years, including (1) no material is wasted, (2) usually no machining is required, (3) only semiskilled labor is required, and (4) some unique properties can be obtained, such as controlled degrees of porosity and built-in lubrication.

A crude form of powder metallurgy appears to have existed in Egypt as early as 3000 BC, using particles of sponge iron. In the 19th century, P/M was used for producing platinum and tungsten wires. However, its first significant use related to general manufacturing was in Germany, following World War I, for making tungsten carbide cutting-tool tips. Since 1945 the process has been highly developed, and large quantities of a wide variety of P/M products are made annually, many of which could not be made by any other process. Most are under 2 in. (50.8 mm) in size, but many are larger, some weighing up to 50 lb (22.7 kg) and measuring up to 20 in. (508 mm).

Powder metallurgy normally consists of four basic steps:

1. Producing a fine metallic powder
2. Mixing and preparing the powder for use
3. Pressing the powder into the desired shape
4. Heating (sintering) the shape at an elevated temperature

Other operations can be added to obtain special results.

The pressing and sintering operations are of special importance. The pressing and repressing greatly affect the density of the product, which has a direct relationship to the strength properties. Sintering strips contaminants from the surface of the powder particles, permitting diffusion bonding to occur and resulting in a single piece of material. Sintering usually is done in a controlled, inert atmosphere, but sometimes it is done by the discharge of spark through the powder while it is under compaction in the mold.

6.1 Properties of P/M Products

Because the strength properties of powder metallurgy products depend on so many variables—type and size of powder, pressing pressure, sintering temperature, finishing treatments, and so on—it is difficult to give generalized information. In general, the strength properties of products that are made from pure metals (unalloyed) are about the same as those made from the same wrought metals. As alloying elements are added, the resulting strength properties of P/M products fall below those of wrought products by varying, but usually substantial, amounts. The ductility usually is markedly less, as might be expected because of the lower density. However, tensile strengths of 40,000–50,000 psi (275.8–344.8 MPa) are common, and strengths above 100,000 psi (689.5 MPa) can be obtained. As larger presses

278 Metal Forming, Shaping, and Casting

and forging combined with P/M preforms are used, to provide greater density, the strength properties of P/M materials will more nearly equal those of wrought materials. Coining can also be used to increase the strength properties of P/M products and to improve their dimensional accuracy.

7 SURFACE TREATMENT

Products that have been completed to their proper shape and size frequently require some type of surface finishing to enable them to satisfactorily fulfill their function. In some cases, it is necessary to improve the physical properties of the surface material for resistance to penetration or abrasion.

Surface finishing may sometimes become an intermediate step in processing. For instance, cleaning and polishing are usually essential before any kind of plating process. Another important need for surface finishing is for corrosion protection in a variety of environments. The type of protection provided will depend largely on the anticipated exposure, with due consideration to the material being protected and the economic factors involved.

Satisfying the above objectives necessitates the use of many surface-finishing methods that involve chemical change of the surface; mechanical work affecting surface properties, cleaning by a variety of methods; and the application of protective coatings organic and metallic.

7.1 Cleaning

Few, if any, shaping and sizing processes produce products that are usable without some type of cleaning unless special precautions are taken. Figure 22 indicates some of the cleaning methods available. Some cleaning methods provide multiple benefits. Cleaning and finish improvements are often combined. Probably of even greater importance is the combination of corrosion protection with finish improvement, although corrosion protection is more often a second step that involves coating an already cleaned surface with some other material or chemical conversion.

Liquid and Vapor Baths

Liquid and Vapor Solvents. The most widely used cleaning methods make use of a cleaning medium in liquid or vapor form. These methods depend on a solvent or chemical action between the surface contaminants and the cleaning material.

Petroleum Solvents. Among the more common cleaning jobs required is the removal of grease and oil deposited during manufacturing or intentionally coated on the work to provide protection. One of the most efficient ways to remove this material is by use of solvents that dissolve the grease and oil but have no effect on the base metal. Petroleum derivatives, such as Stoddard solvent and kerosene, are common for this purpose, but, since they introduce some danger of fire, chlorinated solvents, such as trichlorethylene, that are free of this fault are sometimes substituted.

Conditioned Water. One of the most economical cleaning materials is water. However, it is seldom used alone, even if the contaminant is fully water soluble, because the impurity of the water itself may contaminate the work surface. Depending on its use, water is treated with various acids and alkalies to suit the job being performed.

7 Surface Treatment **279**

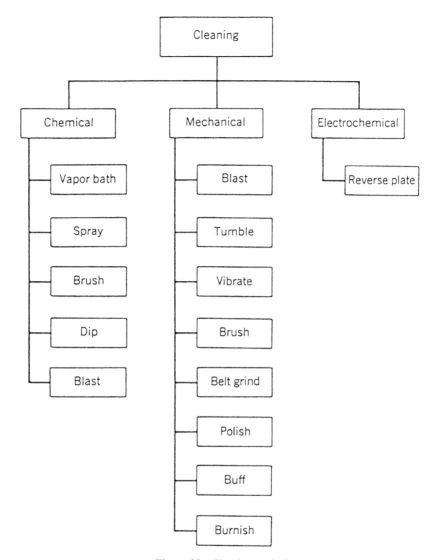

Figure 22 Cleaning methods.

Pickling. Water containing sulfuric acid in a concentration from about 10–25% and at a temperature of approximately 149°F (65°C) is commonly used in a process called *pickling* for removal of surface oxides or scale or iron and steel.

Mechanical Work Frequently Combined with Chemical Action. Spraying, brushing, and dipping methods are also used with liquid cleaners. In nearly all cases, mechanical work to cause surface film breakdown and particle movement is combined with chemical and solvent action. The mechanical work may be agitation of the product, as in dipping, movement of the cleaning agent, as in spraying, or use of a third element, as in rubbing brushing. In some applications, sonic or ultrasonic vibrations are applied to either the solution or the workpieces to speed the cleaning action. Chemical activity is increased with higher temperatures and optimum concentration of the cleaning agent, both of which must in some cases be controlled closely for efficient action.

Blasting

The term *blasting* is used to refer to all those cleaning methods in which the cleaning medium is accelerated to high velocity and impinged against the surface to be cleaned. The high velocity may be provided by air or water directed through a nozzle or by mechanical means with a revolving slinger. The cleaning agent may be either dry or wet solid media, such as sand, abrasive, steel grit, or shot, or may be liquid or vapor solvents combined with abrasive material. In addition to cleaning, solid particles can improve finish and surface properties of the material on which they are used. Blasting tends to increase the surface area and thus set up compressive stresses that may cause a warping of thin sections, but in other cases, it may be very beneficial in reducing the likelihood of fatigue failure. When used for the latter purpose, the process is more commonly known as *shot peening*.

Water Slurries. Liquid or vaporized solvents may, by themselves, be blasted against a surface for high-speed cleaning of oil and grease films with both chemical and mechanical action. Water containing rust-inhibiting chemicals may carry, in suspension, fine abrasive particles that provide a grinding cutting-type action for finish improvement along with cleaning. The blasting method using this medium is commonly known as *liquid honing*.

Abrasive Barrel Finishing

Barrel finishing, rolling, tumbling, and *rattling* are terms used to describe similar operations that consist of packing parts together with some cleaning media in a cylinder or drum, which can be rotated to cause movement among them. The media may be abrasive (either fine or coarse); metal stars, slugs, or balls; stones; wood chips; sawdust; or cereals. The work may be done wet or dry, depending on the materials being worked with, the kind of surface finish desired, and the kind of equipment available.

Wire Brushing

A number of cleaning operations can be quickly and easily performed by use of a high-speed rotating wire brush. In addition to cleaning, the contact rubbing of the wire ends across the work surface produce surface improvement by a burnishing-type action. Sharp edges and burrs can be removed.

Abrasive Belt Finishing

Continuous fabric belts coated with abrasive can be driven in several kinds of machines to provide a straight-line cutting motion for grinding, smoothing, and polishing work surfaces. Plane surfaces are the most common surfaces worked on with fabric belts.

Polishing

The term *polishing* may be interpreted to mean any nonprecision procedure providing a glossy surface, but is most commonly used to refer to a surface-finishing process using a flexible abrasive wheel. The wheels may be constructed of felt or rubber with an abrasive band, of multiple coated abrasive discs, of leaves of coated abrasive, of felt or fabric to which loose abrasive is added as needed, or of abrasives in a rubber matrix.

Buffing

About the only difference between buffing and polishing is that, for buffing, a fine abrasive carried in wax or a similar substance is charged on the surface of a flexible level.

Electropolishing

If a workpiece is suspended in an electrolyte and connected to the anode in an electrical circuit, it will supply metal to the electrolyte in a reverse plating process. Material will be removed faster from the high spots of the surface than from the depressions and will thereby increase the average smoothness. The cost of the process is prohibitive for very rough surfaces because larger amounts of metal must be removed to improve surface finish than would be necessary for the same degree of improvement by mechanical polishing. Electropolishing is economical only for improving a surface that is already good or for polishing complex and irregular shapes, the surfaces of which are not accessible to mechanical polishing and buffing equipment.

7.2 Coatings

Many products, particularly those exposed to view and those subject to change by the environment with which they are in contact, need some type of coating for improved appearance or for protection from chemical attack. The need for corrosion protection for maintenance and appearance is important. In addition to change of appearance, loss of actual material, change of dimensions, and decrease of strength, corrosion may be the cause of eventual loss of service or failure of a product. Material that must carry loads in structural applications, especially when the loads are cyclic in nature, may fail with fatigue if corrosion is allowed to take place. Corrosion occurs more readily in highly stressed material, where it attacks grain boundaries in such a way as to form points of stress concentration that may be nuclei for fatigue failure.

Harness and wear resistance, however, can be provided on a surface by plating with hard metals. Chromium plating of gages and other parts subject to abrasion is frequently used to increase their wear life. Coatings of plastic material and asphaltic mixtures are sometimes placed on surfaces to provide sound-deadening. The additional benefit of protection from corrosion is usually acquired at the same time.

Plastics of many kinds, mostly of the thermoplastic type because they are easier to apply and also easier to remove later if necessary, are used for mechanical protection. Highly polished material may be coated with plastic, which may be stripped off later, to prevent abrasion and scratches during processing. It is common practice to coat newly sharpened cutting edges of tools by dipping them in thermoplastic material to provide mechanical protection during handling and storage.

Organic Coatings

Organic coatings are used to provide pleasing colors, to smooth surfaces, to provide uniformity in both color and texture, and to act as a protective film for control of corrosion. Organic resin coatings do not ordinarily supply any chemical-inhibiting qualities. Instead, they merely provide a separating film between the surface to be protected and the corrosive environment. The important properties, therefore, are continuity, permeability, and adhesion characteristics.

Paints, Varnishes, and Enamels

Paints. Painting is a generic term that has come to mean the application of almost any kind of organic coating by any method. Because of this interpretation, it is also used generally to describe a broad class of products. As originally defined and as used most at present, paint is a mixture of pigment in a drying oil. The oil serves as a carrier for the pigment and in addition creates a tough continuous film as it dries. Drying oils, one of the common ones of which is linseed oil, become solid when large surface areas are exposed to air. Drying starts with a chemical reaction of oxidation. Nonreversible polymerization accompanies oxidation to complete the change from liquid to solid.

Varnish. Varnish is a combination of natural or synthetic resins and drying oil, sometimes containing volatile solvents as well. The material dries by a chemical reaction in the drying oil to a clear or slightly amber-colored film.

Enamel. Enamel is a mixture of pigment in varnish. The resins in the varnish cause the material to dry to a smoother, harder, and glossier surface than is produced by ordinary paints. Some enamels are made with thermosetting resins that must be baked for complete dryness. These baking enamels provide a toughness and durability not usually available with ordinary paints and enamels.

Lacquers

The term *lacquer* is used to refer to finishes consisting of thermoplastic materials dissolved in fast-drying solvents. One common combination is cellulose nitrate dissolved in butyl acetate. Present-day lacquers are strictly air-drying and form films very quickly after being applied, usually by spraying. No chemical change occurs during the hardening of lacquers; consequently, the dry film can be redissolved in the thinner. Cellulose acetate is used in place of cellulose nitrate in some lacquers because it is nonflammable. Vinyls, chlorinated hydrocarbons, acrylics, and other synthetic thermoplastic resins are also used in the manufacture of lacquers.

Vitreous Enamels

Vitreous, or porcelain, enamel is actually a thin layer of glass fused onto the surface of a metal, usually steel or iron. Shattered glass, ball milled in a fine particle size, is called *frit*. Frit is mixed with clay, water, and metal oxides, which produce the desired color, to form a thin slurry called *slip*. This is applied to the prepared metal surface by dipping or spraying and, after drying, is fired at approximately 1470°F (800°C) to fuse the material to the metal surface.

Metallizing

Metal spraying, or metallizing, is a process in which metal wire or powder is fed into an oxyacetylene heating flame and then, after melting, is carried by high-velocity air to be impinged against the work surface. The small droplets adhere to the surface and bond together to build up a coating.

Vacuum Metallizing

Some metals can be deposited in very thin films, usually for reflective or decorative purposes, as a vapor deposit. The metal is vaporized in a high-vacuum chamber containing the parts to be coated. The metal vapor condenses on the exposed surfaces in a thin film that follows the surface pattern. The process is cheap for coating small parts, considering the time element only, but the cost of special equipment needed is relatively high.

Aluminum is the most used metal for deposit by this method and is used frequently for decorating or producing a mirror surface on plastics. The thin films usually require mechanical protection by covering with lacquer or some other coating material.

Hot-Dip Plating

Several metals, mainly zinc, tin, and lead, are applied to steel for corrosion protection by a hot-dip process. Steel in sheet, rod, pipe, or fabricated form, properly cleansed and fluxed, is immersed in molten plating metal. As the work is withdrawn, the molten metal that adheres solidifies to form a protective coat. In some of the large mills, the application is made continuously to coil stock that is fed through the necessary baths and even finally inspected before being recoiled or cut into sheets.

Electroplating
Coatings of many metals can be deposited on other metals, and on nonmetals when suitably prepared, by electroplating. The objectives of plating are to provide protection against corrosion, to improve appearance, to establish wear- and abrasion-resistant surfaces, to add material for dimensional increase, and to serve as an intermediate step of multiple coating. Some of the most common metals deposited in this way are copper, nickel, cadmium, zinc, tin, silver, and gold. The majority are used to provide some kind of corrosion protection but appearance also plays a strong part in their use.

Temporary Corrosion Protection
It is not uncommon in industry for periods of time, sometimes quite long periods, to elapse between manufacture, assembly, shipment, and use of parts. Unless a new processing schedule can be worked out, about the only cure for the problem is corrosion protection suitable for the storage time and exposure. The coatings used are usually nondrying organic materials, called *shushing compounds,* that can be removed easily. The two principal types of compounds used for this purpose are petroleum-based materials, varying from extremely light oils to semisolids, and thermoplastics. The most common method of application of shushing compounds for small parts is by dipping. Larger parts that cannot be handled easily may be sprayed, brushed, or flow coated with the compound.

7.3 Chemical Conversions

A relatively simple and often fully satisfactory method for protection from corrosion is by conversion of some of the surface material to a chemical composition that resists from the environment. These converted metal surfaces consist of relatively thin (seldom more than 0.001 in. (0.025 mm) thick) inorganic films that are formed by chemical reaction with the base material. One important feature of the conversion process is that the coatings have little effect on the product dimensions.

Anodizing
Aluminum, magnesium, and zinc can be treated electrically in a suitable electrolyte to produce a corrosion-resistant oxide coating. The metal being treated is connected to the anode in the circuit, which provides the name *anodizing* for the process. Aluminum is commonly treated by anodizing that produces an oxide film thicker than, but similar to, that formed naturally with exposure to air. Anodizing of zinc has very limited use. The coating produced on magnesium is not as protective as that formed on aluminum, but does provide some protective value and substantially increases protection when used in combination with paint coatings.

Chromate Coatings
Zinc is usually considered to have relatively good corrosion resistance. This is true when the exposure is to normal outdoor atmosphere where a relatively thin corrosion film forms. Contact with either highly aerated water films or immersion in stagnant water containing little oxygen causes uneven corrosion and pitting. The corrosion products of zinc are less

dense than the base material, so that heavy corrosion not only destroys the product appearance, but also may cause malfunction by binding moving parts. Corrosion of zinc can be substantially slowed by the production of chromium salts on its surface. The corrosion resistance of magnesium alloys can be increased by immersion of anodic treatment in acid baths containing dichromates. Chromate treatment of both zinc and magnesium improves corrosion resistance, but is used also to improve adhesion of paint.

Phosphate Coatings
Phosphate coatings, used mostly on steel, result from a chemical reaction of phosphoric acid with the metal to form a nonmetallic coating that is essentially phosphoric salts. The coating is produced by immersing small items or spraying large items with the phosphating solution. Phosphate surfaces may be used alone for corrosion resistance, but their most common application is as a base for paint coatings. Two of the most common application methods are called *parkerizing* and *bonderizing.*

Chemical Oxide Coatings
A number of proprietary blacking processes, used mainly on steel, produce attractive black oxide coatings. Most of the processes involve the immersing of steel in a caustic soda solution, heated to about 300°F (150°C) and made strongly oxidizing by the addition of nitrites or nitrates. Corrosion resistance is rather poor unless improved by application of oil, lacquer, or wax. As in the case of most of the other chemical-conversion procedures, this procedure also finds use as a base for paint finishes.

BIBLIOGRAPHY

Abrasion-Resistant Cast Iron Handbook, American Foundry Society, 2000.
Altan, T., S.-I. Oh, and H. Gegel, *Metal Forming—Fundamentals and Application,* ASM International, Materials Park, OH, 1983.
ASM Handbook, Vol. 14: *Forming and Forging,* ASM International, Materials Park, OH, 1988.
Bhushan, B., and B. K. Gupta, *Handbook of Tribology: Materials, Coating, and Surface Treatment,* McGraw-Hill, New York, 1991.
Bhushan, B., *Modern Tribology Handbook,* CRC Press, Boca Raton, FL, 2001.
Clegg, A. J. *Precision Casting Processes,* Pergamon, New York, 1991.
Ginzburg, V. B., *High-Quality Steel Rolling: Theory and Practice,* Dekker, New York, 1993.
Hoffman, H., *Metal Forming Handbook,* Springer Verlag, 1998.
Hosford, W. F., and R. M. Caddell, *Metal Forming, Mechanics and Metallurgy,* 2nd ed., Prentice Hall, Englewood Cliffs, NJ, 1993.
Investment Casting Handbook, Investment Casting Institute, 1997.
Lange, K., *Handbook of Metal Forming,* McGraw-Hill, New York, 1985.
Lindsay, J. H., *Coatings and Coating Processes for Metals,* ASM International, Materials Park, OH, 1998.
Marciniak, Z., and J. L. Duncan, *The Mechanics of Sheet Metal Forming,* Edward Arnold, 1992.
Suchy, I., *Handbook of Die Design,* McGraw-Hill, New York, 1997.
Tool and Manufacturing Engineers Handbook, 4th ed., Vol. 2: *Forming,* Society of Manufacturing Engineers, Dearborn, MI, 1984.
Upton, B., *Pressure Die Casting,* Part 1: *Metals, Machines, Furnaces,* Pergamon, New York, 1982.
Wagoner, R. H., and J. L. Chenot, *Fundamentals of Metal Forming,* Wiley, New York, 1996.
Wagoner, R. H., and J. L. Chenot, *Metal Forming Analysis,* Cambridge University Press, Cambridge, UK, 2001.
Walton, C. F., and T. J. Opar, *Iron Castings Handbook,* 3rd ed., Iron Castings Society, 1981.

Wieser, P. P., *Steel Castings Handbook,* 6th ed., ASM International, Materials Park, OH, 1995.
Young, K. P., *Semi-Solid Processing,* Chapman & Hall, 1997.
Yu, K.-O., *Modeling for Casting and Solidification Processing,* Dekker, 2001.
Zohdi, M. E., "Statistical Analysis: Estimation and Optimization of Surface Finish," Proceedings of International Conference on Development of Production Systems, Copenhagen, Denmark, 1974.

CHAPTER 7
MECHANICAL FASTENERS

Murray J. Roblin
Chemical and Materials Engineering Department
California State Polytechnic University
Pomona, California

Updated by Anthony Luscher
Department of Mechanical Engineering
The Ohio State University
Columbus, Ohio

1	INTRODUCTION TO FASTENING AND JOINING	286		8.2 Torsional Stress Factor	302
	1.1 Assembly Features and Functions	287		8.3 Other Design Issues	303
	1.2 Some Examples of a Nesting Strategy	290	9	THEORETICAL BEHAVIOR OF THE JOINT UNDER TENSILE LOADS	304
	1.3 Three-Part Assembly	291		9.1 Critical External Load Required to Overcome Preload	305
2	INTRODUCTION TO FASTENING WITH BOLTS AND RIVETS	292		9.2 Very Large External Loads	307
3	BOLTED AND RIVETED JOINT TYPES	293	10	EVALUATION OF SLIP CHARACTERISTICS	308
4	EFFICIENCY	295	11	TURN-OF-NUT METHOD OF BOLT TIGHTENING	308
5	STRENGTH OF A SIMPLE LAP JOINT	295	12	TORQUE AND TURN TOGETHER	309
6	SAMPLE PROBLEM OF A COMPLEX BUTT JOINT (BEARING-TYPE CONNECTION)	296	13	ULTRASONIC MEASUREMENT OF BOLT STRETCH OR TENSION	310
	6.1 Preliminary Calculations	297	14	FATIGUE FAILURE AND DESIGN FOR CYCLICAL TENSION LOADS	312
7	FRICTION-TYPE CONNECTIONS	300			
8	UPPER LIMITS ON CLAMPING FORCE	302		REFERENCES	314
	8.1 Design-Allowable Bolt Stress and Assembly Stress Limits	302			

1 INTRODUCTION TO FASTENING AND JOINING

The study of fastening and joining is complex but worthy of study because of the economic, structural, reliability, safety, and structural efficiency benefits that can achieved. It is also an ever-changing field of study with advances constantly being made. As this document goes to press in 2005, Boeing is developing a commercial aircraft with will be a step increment in lower cost, comfort, and fuel efficiency. These goals could only be achieved by use of a full composite structure making extensive use of adhesives along precisely controlled bond lines. As a general reference for the study of assembly I recommend *Mechanical Assembly* by Daniel Whitney.[1]

The technical area of fastening and joining is a very wide area drawing from many engineering disciplines. It comprises a great deal of specialized knowledge within each joining and fastening methodology. A complete survey of all of these areas would require a series of texts and is beyond the scope of this book. Instead this text will focus on mechanical fastening via rivets and bolts. Throughout this chapter we use the term *fastening* to correspond to removable features and *joining* as the creation of a permanent connect.

Before these mechanical fastening systems are discussed in detail, a section on common assembly issues is presented. This section discusses how a creative consideration of assembly constraint can improve the design of mechanically fastened assemblies.

1.1 Assembly Features and Functions

To successfully join parts and form an assembly four primary tasks must be accomplished:

- Location of the parts relative to each other. Surfaces must contact to remove all degrees of freedom between the parts except those to be removed by the final locking features. Location is not complete until the part is fully constrained with respect to moments and translations.
- Transfer of service loads across the interfaces of the assembly. These are commonly the same features used to form location but need not be. They must, however, have sufficient strength and rigidity to transfer load.
- If necessary, part tolerance and manufacturing variability between the parts must be absorbed by use of shims or compliance features. This is not necessary in all assemblies.
- Addition of locking features such as bolts to finish the constraint of the parts to each other.

Although fastening can be done creatively in three-space this is not often done. The current situation is called by the author the *bolting paradigm* and can be characterized by

- Using assembly interfaces that are two dimensional and predominately planar.
- Having all bolts take loads in multiple directions (axial, shear, moment).
- Tolerancing is not a concern as long as the bolt can be assembled through the mating parts.

Figure 1 shows two examples. Another approach to joining parts, which can provide great benefit in loading and cost, is to use a more complex 3-dimensional assembly interface. In this strategy,

- Individual features take loads only in specific direction.
- Assemblies can be designed to be statically determinate or with various levels of overconstraint.
- Assembly features and not the fasteners determine tolerances.

To create such a three-dimensional fastening strategy, surfaces or features must be created to accomplish 3 assembly functions: locators, compliant features, and locks.

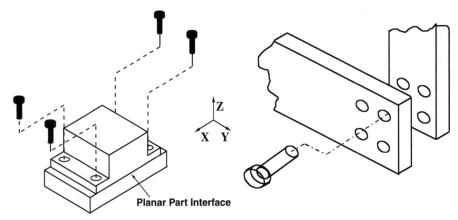

Figure 1 Examples of the 2-dimensional bolting paradigm.

Locators are features or mating surfaces that eliminate degrees of freedom between parts, transfer the service loads, and/or establish the major reference or datum planes or points that locate parts relative to each other. For two parts to remain together as an assembly and function, their relative location, alignment, and orientation must be fixed for all time.

Locators are available in many different geometries and topologies. They can be designed for very specific loading situations or for general situations. When designing with locators it is useful to consider the degrees of freedom eliminated by each feature or surface pair. Several features that can be machined, molded, or bent into a part are shown in Fig. 2.

To create this type of assembly, start the process by considering one part of the assembly as the reference frame and fixed to ground. Then assume that the other part (the mating part) starts out with all degrees of freedom available to it. Normally a part would have 6 possible degrees of motion in 3-space (for example 3 Cartesian coordinate translations and 3 rotations). Physical assembly is somewhat more difficult than this since it is possible to constrain an object in only one direction along an axis. The simplest example of this is a planar surface that constrains motion into the surface but not away from it. Therefore, both positive and negative directions need to be considered separately, leading to 12 possible degrees-of-motion that are available between two parts (6 Cartesian coordinate translations and 6 rotations).

For example, the "stop" feature constrains only motion that is inward normal to its large surface (in the positive x direction in Figure 2). It does not constrain motion in any

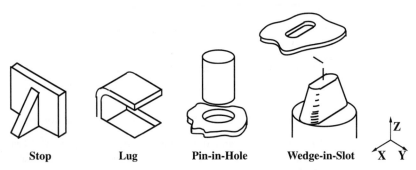

Figure 2 Several features that provide locational constraint between parts.

other direction or outwardly normal from its large surface. The "pin-in-hole" feature, on the other hand, constrains all planar motion normal to its centerline in both x and y directions. A "pin-in-hole feature, however, will allow in-plane rotation around the pin. The "wedge-in-slot" feature limits motion to being along the slot. In forming these locator definitions it is important to include surfaces on both parts since they both contribute to location functionality. As an example, defining a "pin" feature is not sufficient to determine the degrees of freedom which are eliminated. A pin-in-hole will remove different degrees of freedom than a pin-in-slot feature.

Note that several different physical features can be used to achieve identical locator functionality. Because these features are physically different they will differ in moldability, strength, and ability to absorb manufacturing variability. These secondary attributes should be considered after determining the degrees of freedom that need to be removed. In this way an assembly concept can progress from a very abstract level to a more physical level in a logical manner.

Figure 3 below shows another important aspect of using locators, which is their line of action in providing constraint. In Figure 3 assume that the rectangular box is being constrained against rotation by the 4 stop features shown surrounding it. In these graphics the gray arrows represent the normal to the constraint feature while the black arrow shows the force couple caused by the imposed moment. Note that in the top graphic the gray and black arrows are almost co-linear, while in the bottom graphic they are almost orthogonal.

Compliant features are designed to absorb any tolerance stack-up or misalignment between the data of mating parts. This is most often accomplished by the built-in compliance or flexibility of these features.

All parts are manufactured with some form of variability or tolerance. This variability occurs in many forms, including errors of size, location, orientation, and form. Tolerance stack-up can (a) produce a gap between the parts resulting in an undesirable rattle or loose-

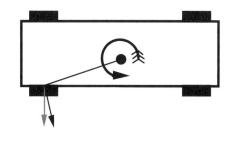

Good Locator Placement

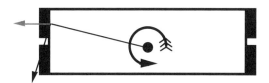

Poor Locator Placement

Figure 3 The importance of line-of-action in the use of locators.

ness, or, (b) produce an unintended interference leading to high stresses in the part and high assembly forces. The feature group "compliant features" eliminates any resultant gap or interferences between the data of mating parts by building in a certain amount of flexibility or compliance between the parts. Compliant features function in one of the following two ways:

- *Elastically:* These features are designed so that, under all tolerance stack-ups, a guaranteed minimum amount of preload is maintained in the system. The left graphic in Figure 4 shows a common feature used in the sheetmetal industry. The spring finger's undeformed height is above the surrounding surface. When parts are assembled the finger elastically deforms but maintains pressure between the parts.
- *Inelastically:* These features are designed so that, during initial assembly, they permanently deform to eliminate any gaps between parts. The right graphic in Figure 4 shows a crush rib that is often used in plastic parts. In this design a steel shaft needs to have a tight fit against a polymer housing. The first time that the shaft is inserted the rib is fractured to the exact diameter needed. Another example from metals is a crush washer, which is used to deform material and provide a sealing surface.

The compliant feature classification is important since it represents an inexpensive method of compensating for variability versus the very high cost of tightening tolerances in both the base and mating parts.

Locks are features or devices used to provide the final attachment between two parts. Locks can be any of a wide spectrum of fasteners, including, bolts, rivets, or snap-fits. It is important to note that whatever lock feature or fastener is chosen, its kinematic role in the overall structure is the same.

1.2 Some Examples of a Nesting Strategy

The use of the above feature types can best be shown by an example. Figure 5 is a schematic of 2 ways of attaching a transformer base to a sheetmetal panel. The graphic on the left shows a conventional way of fastening the structure by use of 4 fasteners. This approach, however, has the disadvantage of requiring 4 fasteners, and also can lead to tolerance issues between the 4 screw holes. The concept on the right shows an approach that uses the selective constraint of locating features. The sheetmetal is bent to form a 3-sided support structure that is kinematically equivalent to the lug feature shown in Fig. 2. They constrain the vertical direction as well as motion toward the back surface. By the use of 3 sets of these features the only allowable motion is shown in Figure 5. More importantly, only one locking feature

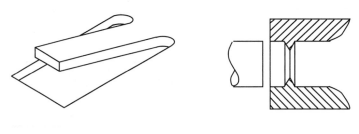

Cantilever Spring Feature   **Crush Rib Feature**

Figure 4   An example of an elastic and an inelastic compliance feature.

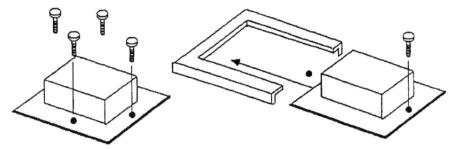

Figure 5 Transformer attached to a sheetmetal panel.

is required and it needs to provide constraint only in the opposite direction of the assembly motion. Tolerance windows can be opened and are set by the dimensions of the locators. If rattle needs to be eliminated, compliance features could have been added.

Figure 6 shows another example. It is the attachment of a lens to a bulb housing. This example uses 3 assembly feature types and is from an automotive interior. Most of the constraint between the parts is provided by the mating of the track and tab features on the housing and lens respectively. These two features remove all degrees of freedom except motion along and away from the assembly direction. The back of the lens hitting the front of the housing removes motion along the assembly direction. Motion away from the assembly motion is removed by the locking device, which in this case, is a snap-fit. Since rattle is a concern is an automotive environment, 4 spring tabs are the most cost-effective solution to providing some preload between the parts.

1.3 Three-Part Assembly

In many cases fastening efficiencies can be improved by the creation of what the author calls three-part assemblies. In this assembly strategy a part can be sandwiched between two other

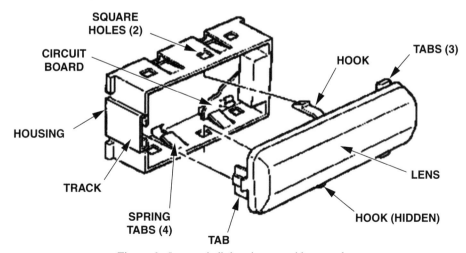

Figure 6 Lens to bulb housing assembly example.

parts that are fastened together. In the left-hand graphic below, the handle is attached to the top part while fasteners on the top part connect it to the bottom part. Consider the graphic on the right side below. A single set of fasteners is used to connect the handle, top, and bottom parts together. The top part is completely constrained and no additional fasteners are needed. Other examples of multipart assemblies include gears in a gear housing. Often several gears are in parallel bores. They are all constrained together when the two housings are fastened together.

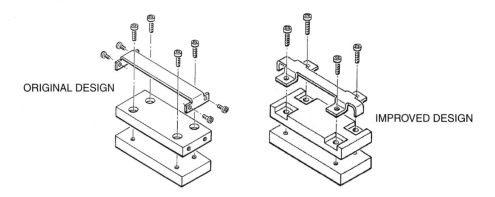

ORIGINAL DESIGN

IMPROVED DESIGN

2 INTRODUCTION TO FASTENING WITH BOLTS AND RIVETS

Two or more components may need to be fastened in such a way that they may be taken apart during the service life of the part. In these cases, the assembly must be fastened mechanically. Other reasons for choosing mechanical fastening over welding could be

1. Ease of part replacement, repair, or maintenance
2. Ease of manufacture
3. Designs requiring movable joints
4. Designs requiring adjustable joints

The most common mechanical fastening methods are bolts (threaded fasteners), and rivets.

To join two members by bolting or riveting requires holes to be drilled in the parts to accommodate the rivets or bolts. These holes reduce the load-carrying cross-sectional area of the members to be joined. Because this reduction in area is at least 10–15%, the load-carrying capacity of the bolted structure is reduced, which must be accounted for in the design. Alternatively, when one inserts bolts into the holes, only the cross section of the bolt or rivet supports the load. In this case, the reduction in the strength of the joint is reduced even further than 15%.

Even more critical are the method and care taken in drilling the holes. When one drills a hole in metal, not only is the cross-sectional area reduced, but the hole itself introduces stress risers and/or flaws on the surface of the holes that may substantially endanger the structure. First, the hole places the newly created surface in tension, and if any defects are created as a result of drilling, they must be accounted for in a quantitative way. Unfortunately, it is very difficult to obtain definitive information on the inside of a hole that would allow characterization of the introduced defect.

The only current solution is to make certain that the hole is properly prepared which means not only drilling or subpunching to the proper size, but also *reaming* the surface of the hole. To be absolutely certain that the hole is not a problem, one needs to put the surface of the hole in residual compression by expanding it slightly with an expansion tool or by pressing the bolt, which is just slightly larger than the hole. This method causes the hole to expand during insertion, creating a hole whose surface is in residual compression. While there are fasteners designed to do this, it is not clear that all of the small surface cracks of the hole have been removed to prevent flaws or stress risers from existing in the finished product.

Using bolts and rivets in an assembly can also provide an ideal location for water to enter the crevices between the two joined parts. This trapped water, under conditions where chlorides and sodium exist, can cause *crevice corrosion,* which is a serious problem if encountered.

Obviously, in making the holes as perfect as possible, you increase the cost of a bolted and/or riveted joint significantly, which makes welding or adhesive joining a more attractive option. Of course, as will be shown below, welding and joining have their own set of problems that can degrade the joint strength.

The analysis of the strength of a bolted or riveted joint involves many indeterminate factors resulting in inexact solutions. However, by making certain simplifying assumptions, we can obtain solutions that are conservative, acceptable, and practical. We discuss two types of solutions: *bearing-type connections,* which use ordinary or unfinished bolts or rivets, and *friction-type connections,* which use high-strength bolts. Today, economy and efficiency are obtained by using high-strength bolts for field connections together with welding in the shop. With the advent of lighter-weight welding power supplies, the use of field welding combined with shop welding is finding increasing favor.

While riveted joints do show residual clamping forces (even in cold-driven rivets), the clamping forces in the rivet is difficult to control, is not as great as that developed by high-strength bolts, and cannot be relied upon. Hot driven rivets have fallen out of favor due to the cost and safety issues involved. Most commercial and bridge structures are now designed using bolts. Studies have shown that the holes are almost completely filled for short rivets. As the grip length is increased, the clearances between rivet and plate material tend to increase.

3 BOLTED AND RIVETED JOINT TYPES

There are two types of riveted and bolted joints: *lap joints* and *butt joints.* See Figs. 7 and 8 for lap and butt joints, respectively. Note that there can be one or more rows of connectors, as shown in Fig. 8*a* and *b*.

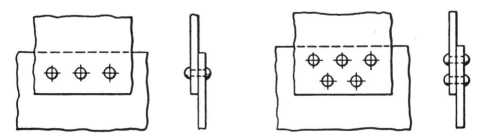

Figure 7 Lap joints. Connectors are shown as rivets only for convenience.

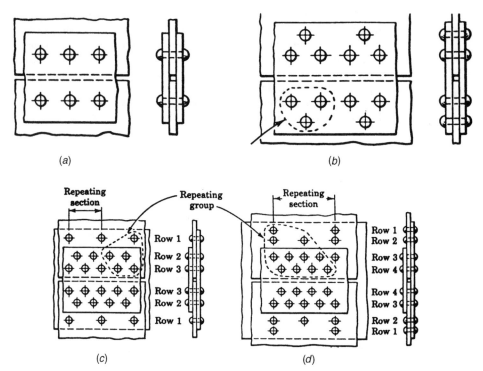

Figure 8 Butt joints: (*a*) single-row; (*b*) double-row; (*c*) triple-row (pressure-type); (*d*) quadruple row (pressure-type).

In a butt joint, plates are butted together and joined by two cover plates connected to each of the main plates. (Rarely, only one cover plate is used to reduce the cost of the joint.) The number of rows of connectors that fasten the cover plate to each main plate identifies the joint—single row, double row, and so on. See Fig. 8.

Frequently the outer cover plate is narrower than the inner cover plate, as in Fig. 8c and d, the outer plate being wide enough to include only the row in which the connectors are most closely spaced. This is called a *pressure joint* because caulking along the edge of the outer cover plate to prevent leakage is more effective for this type of joint.

The spacing between the connectors in a given row is called the *pitch*. When the spacing varies in different rows, as in Fig. 8d, the smallest spacing is called the *short pitch*, the next smallest the *intermediate pitch*, and the greatest the *long pitch*. The spacing between consecutive rows of connectors is called the *back pitch*. When the connectors (rivets or bolts) in consecutive rows are staggered, the distance between their centers is the *diagonal pitch*.

In determining the strength of a joint, computations are usually made for the length of a joint corresponding to a repeating pattern of connectors. The length of the repeating pattern, called the *repeating section*, is equal to the long pitch.

To clarify how many connectors belong in a repeating section, see Fig. 8c, which shows that there are five connectors effective in each half of the triple row—that is, two half connectors in row 1, two whole connectors in row 2, and one whole and two half connectors in row 3. Similarly, there are 11 connectors effective in each half of the repeating section in Fig. 8d.

When rivets are used in joints, the holes are usually drilled or, punched, and reamed out to a diameter of $\frac{1}{16}$ in. (1.5 mm) larger than the nominal rivet size. The rivet is assumed to be driven so tightly that it fills the hole completely. Therefore, in calculations the diameter of the hole is used because the rivet fills the hole. This is not true for a bolt unless it is very highly torqued. In this case, a different approach needs to be taken, as delineated later in this chapter.

The fastener patterns shown in Fig. 8b–d are designed with multiple rows in order to spread the load out among the different fasteners. The number of fasteners in each row is different and is determined by the concept of elastic matching. In this concept, the goal is to carry the same load in each fastener to the greatest extent possible.

Consider the three-row riveted joint as shown in Figure 8c. Notice that this is a statically indeterminate structure and the only way to solve for forces is to consider the stiffness of each load path. Each row of rivets forms a different load path and has a different stiffness due to the different distances from the center of the joint. Row 3, for example, has the shortest distance between its counterpart on each side of the joint. Because of this, the load path for row 3 is the stiffest of the 3 load paths. Row 1, on the other hand, has the longest load path, the greatest length of plate that can deform, and therefore the lowest stiffness of the three.

When solving statically indeterminate structures, load is distributed as the ratio of relative stiffness. As an example, if a load has three paths, the path with the highest stiffness will take the highest load. In the three-rowed joint, row 3 takes the greatest load since it is the stiffest. Row 2 has an intermediate stiffness, and row 1 has the lowest stiffness. In order to equalize the load on each fastener, the row with the highest load (row 3) should have the greatest number of fasteners, and row 1 should have the least. With an exact calculation of stiffness the loads can be fine tuned for joint efficiency.

4 EFFICIENCY

Efficiency compares the strength of a joint to that of a continuous solid plate as follows:

$$\text{Efficiency} = \frac{\text{Strength of the joint}}{\text{Strength of solid plate}}$$

5 STRENGTH OF A SIMPLE LAP JOINT

For bearing-type connections using rivets or ordinary bolts, we first consider failure in the bolt or rivet itself in shear. This leads to the following equation:

$$P_s = A_s \tau$$

Using the diameter of the bolt/rivet, this can be rewritten as

$$P_s = A_s \tau = \frac{\pi d^2 \tau}{4}$$

where P_s = the load
A_s = shear area of one connector
d = diameter of connector and/or hole

For the above example, friction is neglected. Figure 9 shows the shearing of a single connector.

Another possible type of failure is caused by tearing the main plate. Figure 10 demonstrates this phenomenon.

The above failure occurs on a section through the connector hole because this region has reduced tearing resistance. If p is the width of the plate or the length of a repeating section, the resisting area is the product of the net width of the plate $(p - d)$ times the thickness t. The failure load in tension therefore is

$$P_{\text{tension}} = A_t \sigma_t = (p - d)t(\sigma_t)$$

A third type of failure, called a *bearing failure,* is shown in Fig. 11 For this case, the edge of the plate yields and the bolt hole is enlarged into a slot. Actually, the stress that the connector bears against the edges of the hole varies from zero at the edges of the hole to the maximum value at the center of the bolt or rivet. However, common practice assumes the stress as uniformly distributed over the projected area of the hole. See Fig. 11.

The failure load in the bearing area can be expressed by

$$P_b = A_b \sigma_b = (td)\sigma_b$$

Other types of failure are possible but will not occur in a properly designed joint. These are tearing of the edge of the plate back of the connector hole (Fig. 12a) or a shear failure behind the connector hole (Fig. 12b) or a combination of both. Failures of this type will not occur when the distance from the edge of the plate is greater than 2 times the diameter of the bolt/rivet.

6 SAMPLE PROBLEM OF A COMPLEX BUTT JOINT (BEARING-TYPE CONNECTION)

The strength of a bearing-type connection is limited by the capacity of the rivets or ordinary bolts to transmit load between the plates or by the tearing resistance of the plates themselves, depending on which is smaller. The calculations are divided as follows:

1. Preliminary calculations to determine the load that can be transmitted by one rivet or bolt in shear or bearing *neglecting friction* between the plates
2. Calculations to determine which mode of failure is most likely

A repeating section 180 mm long of a riveted triple row butt joint of the pressure type is illustrated in Fig. 13. The rivet hole diameter $d = 20.5$ mm, the thickness of the main plate $t = 14$ mm, and the thickness of each cover plate $t = 10$ mm. The ultimate stresses in shear, bearing, and tension are respectively $\tau = 300$ MPa, $\sigma_b = 650$ MPa, and $\sigma_t = 400$ MPa. Using a factor of safety of 5, determine the strength of a repeating section, the effi-

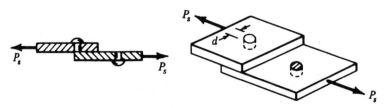

Figure 9 Shear failure.

6 Sample Problem of a Complex Butt Joint (Bearing-Type Connection)

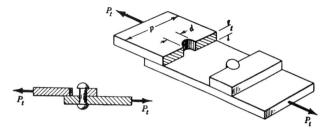

Figure 10 Tear of plate at section through connector hole. $P_t = A_t \sigma_t = (p - d)t\sigma_t$.

ciency of the joint, and the maximum internal pressure that can be carried in a 1.5 m diameter boiler where this joint is the longitudinal seam.

Solution: The use of ultimate stresses will determine the ultimate load, which is then divided by the factor of safety (in this case 5) to determine the safe working load. An alternative but preferable procedure is to use allowable stresses to determine the safe working load directly, which involves smaller numbers. Thus, dividing the ultimate stressed by 5, we find that the allowable stresses in shear, bearing, and tension, respectively, are $\tau = 300/5 = 60$ MPa, $\sigma_b = 650/5 - 130$ MPa, and $\sigma_t = 400/5 = 80$ MPa. The ratio of the shear strength τ to the tensile strength σ of a rivet is about 0.75.

6.1 Preliminary Calculations

To single shear one rivet,

$$P_s = \frac{\pi d^2}{4} \tau = \frac{\pi}{4}(20.5 \times 10^{-3})^2 (60 \times 10^6) = 19.8 \text{ kN}$$

As shown in the bottom of Fig. 13, to move the main plates the rivets must be sheared in two places. To double shear one rivet,

$$P_s = 2 \times 19.8 = 39.6 \text{ kN}$$

To have bearing failure in one rivet in the main plate,

$$P_b = (td)\sigma_b = (14.0 \times 10^{-3})(20.5 \times 10^{-3})(130 \times 10^6) = 37.3 \text{ kN}$$

To crush one rivet in one cover plate,

$$P'_b = (t'd)\sigma_b = (10 \times 10^{-3})(20.5 \times 10^{-3})(130 \times 10^6) = 26.7 \text{ kN}$$

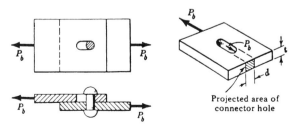

Figure 11 Exaggerated bearing deformation of upper plate. $P_b = A_b \sigma_b = (td)\sigma_b$.

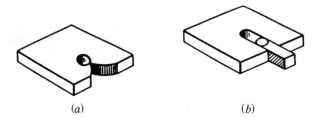

(a) (b)

Figure 12 Possible types of failure if connector hole is too close to edge of plate: (a) tear out; (b) shear behind connector.

Rivet capacity solution: The strength of a single rivet in row 1 in a repeating section is determined by the lowest value of the load that will single shear the rivet, crush it in the main plate, or crush it in one of the cover plates. Based on the values in the preceding calculations, this value is 19.8 kN per rivet.

The strength of each of the two rivets in row 2 depends on the lowest value required to double shear the rivet, crush it the main plate, or crush it in both cover plates. From the above preliminary calculations, this value is 37.3 kN per rivet or 2 × 37.3 + 74.6 kN for both rivets in row 2.

Each of the two rivets in the repeating section in row 3 transmits the load between the main plate and the cover plate in the same manner as those in row 2; hence for row 3, the strength = 74.6 kN.

The total rivet capacity is the sum of the rivet strengths in all rows (rows 1, 2, 3), as follows:

$$P_{total} = 19.8 + 74.6 + 74.6 = 169.0 \text{ kN}$$

Tearing capacity: The external load applied to the joint acts directly to tear the main plate at row 1, and the failure would be similar to Fig. 10. This is calculated as follows:

$$P_{tearing} = (p - d)\sigma_t = [(180 \times 10^{-3}) - (20.5 \times 10^{-3})](14 \times 10^{-3})(80 \times 10^6) = 178.6 \text{ kN}$$

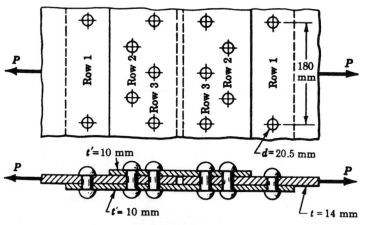

Figure 13

The external load applied does not act directly to tear the main plate at row 2 because part of the load is absorbed or transmitted by the rivet in row 1. Hence, if the main plate is to tear at row 2, the external load must be the sum of the tearing resistance of the main plate at row 2 plus the load transmitted by the rivet in row 1. See Figs. 14 and 15.

Thus,

$$P_{tearing2} = (p - 2d)t\sigma_t + \text{Rivet strength in row 1}$$
$$= [(180 \times 10^{-3}) - 2(20.5 \times 10^{-3})](14 \times 10^{-3})(80 \times 10^6)$$
$$+ 19.8 \times 10^3 = 175.5 \text{ kN}$$

Similarly, the external load required to tear the main plate at row 3 must include the rivet resistance in rows 1 and 2 or

$$P_3 = [(180 \times 10^{-3}) - 2(20.5 \times 10^{-3})](14 \times 10^{-3})(80 \times 10^6) + (19.8 \times 10^3) + (74.6 \times 10^3)$$
$$= 250.1 \text{ kN}$$

It is obvious that this computation need not be made because the tearing resistance of the main plates at rows 2 and 3 is equal, thus giving a larger value.

At row 3, the tearing resistance of the cover plates is resisted by the tensile strength of the reduced section of that row. The tensile strength of one cover plate is

$$P_c = [(180 \times 10^{-3}) - 2(20.5 \times 10^{-3})](10 \times 10^{-3})(80 \times 10^6) = 111.2 \text{ kN}$$

In an ordinary butt joint, the tensile capacity of both cover plates is twice this value. In a pressure joint, however, where one cover plate is shorter than the other, the load capacity of the shorter plate must be compared with the rivet load transmitted to it. In this example, the upper cover plate transmits the rivet load of four rivets in single shear, or $4 \times 19.8 = 79.2$ kN, which is less than its tear capacity of 111.2 kN. Hence, the load capacity of both cover plates becomes

$$P_c = 79.2 + 111.2 = 190.4 \text{ kN}$$

determined by rivet shear in the upper plate and by tension at row 3 in the lower plate. Thus, the safe load is the lowest of these several values = 169.0 kN, which is the rivet strength in shear.

$$\text{Efficiency} = \frac{\text{Safe load}}{\text{Strength of solid plate}} = \frac{169 \times 10^3}{(180 \times 10^3)(14 \times 10^{-3})(80 \times 10^6)} = 83.8\%$$

In this discussion, we have neglected friction and assumed that the rivets or bolts only act as pins in the structure or joint—in essence like spot welds spaced in the same way as the rivets or bolts are spaced.

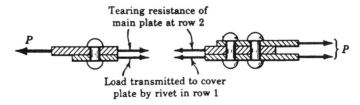

Figure 14

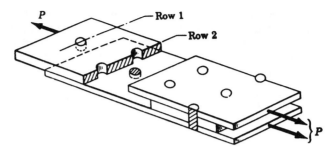

Figure 15 Failure by shear of rivet in row 1 plus tear of main plate in row 2.

7 FRICTION-TYPE CONNECTIONS

In friction-type connections, high-strength bolts of various grades are used and are tightened to high tensile stresses, thereby causing a large residual compression force between the plates. Tightening of the bolts to a predetermined initial tension is usually done using a calibrated torque wrench or by turn-of-the nut methods.

If done properly (as will be discussed later), transverse shear loads are now transferred by the friction between the plates and not by shear and the bearing of the bolt, as described in the previous sections. Heretofore, even though the bolts are not subject to shear, design codes, as a matter of convenience, specified an allowable shearing stress to be applied over the cross-sectional area of the bolt. Thus, friction-type joints were analyzed by the same procedures used for bearing-type joints and the frictional forces that existed, were taken as an extra factor of safety. In the ASME code, the "allowable stresses" listed in several places are not intended to limit assembly stresses in the bolts. These allowables are intended to force flange designers to overdesign the joint to use more and/or larger bolts and thicker flange members than they might otherwise be inclined to use.

Structurally, a bolt serves one of two purposes: it can act as a pin to keep two or more members from slipping relative to each other, or it can act as a heavy spring to clamp two or more pieces together.

In the vast majority of applications, the bolt is used as a clamp and, as such, it must be tightened properly. When we tighten a bolt by turning the head or the nut, we will stretch the bolt initially in the elastic region. More tightening past the elastic limit will cause the bolt to deform plastically. In either case, the bolt elongates and the plates deform in the opposite direction (equal compressive stresses in the materials being joined). In this way, you really have a spring system as shown (with substantial exaggeration) in Fig. 16.

The tensile stress introduced into the fastener during this initial tightening process results in a tension force within the fastener, which in turn creates the clamping force on the joint. This initial clamping force is called the *preload*. Preloading a fastener properly is a major challenge but is critical for fatigue in bolted systems.

When a bolt is loaded in a tensile testing machine, we generate a tension versus a change in length curve, as shown in Fig. 17. The initial straight line portion of the elastic curve is called the *elastic region*. Loading and unloading a bolt within this range of tension never results in a permanent deformation of the bolt because elastic deformation is recoverable. The upper limit of this straight section of the curve ends at the *elastic limit*. Loading beyond or above this limit results in *plastic deformation* of the bolt, which is not recoverable; thus, the bolt has a permanent set (it is longer than it was originally even though the load is completely removed). At the *yield point,* the bolt has a specific amount of permanent

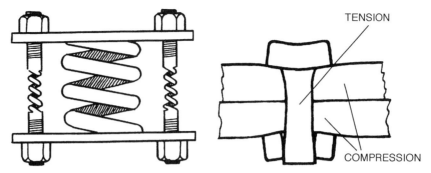

Figure 16 When analyzing the behavior of a bolted joint, pretend the members are a large spring being compressed (clamped) by a group of smaller springs (bolts). When tightened, these springs distort somewhat as shown but grossly exaggerated on the right.

plastic deformation, normally defined as 0.2 or 0.5% of the initial length. Permanent plastic deformation will increase up until the *ultimate tensile strength* (UTS), which is the maximum tension that can be carried in the bolt. The UTS is always greater than the yield stress, sometimes as much as twice yield. The final point on the curve is the *failure* or *rupture stress*, where the bolt breaks under the applied load.

If we load the bolt well into the plastic region of its curve and then remove the load, it will behave as shown in Fig. 18, returning to the zero load point along a line parallel to the original elastic line but offset by the amount of plastic strain the bolt has set.

On reloading the bolt below the previous load but above the original yield point, the behavior of the bolt will follow this new offset stress strain line and the bolt will behave elastically well beyond the original load that caused plastic deformation in the first place. The difference between the original yield strength of the material and the new yield strength is a function of the work hardening that occurred by taking it past the original yield strength on the first cycle. By following the above procedure, we have made the bolt stronger, at least as far as static loads are concerned.

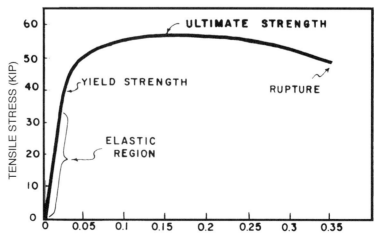

Figure 17 Engineering stress–strain curve (typical).

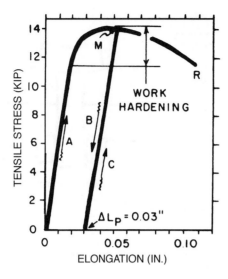

Figure 18 Elastic curve for a ⅜–16 × 4 socket-head cap screw loaded (A) to point M well past the yield strength and then unloaded (B) to give permanent deformation $L_p = 0.03$ in. If reloaded, it will follow path (C).

This is not wise practice, however, for more brittle materials can suffer a loss of strength by such treatments. Loss of strength in ASTM A490 bolts, because of repeated cycling past the yield (under water and wind loads), has been publicly cited as a contributing factor in the 1979 collapse of the roof on the Kemper Auditorium in Kansas City.

8 UPPER LIMITS ON CLAMPING FORCE

8.1 Design-Allowable Bolt Stress and Assembly Stress Limits

We need to follow the limits placed on bolt stresses by codes, company policies, and standard practices. Both structural steel and pressure vessel codes define maximum design allowable stresses for bolts. To distinguish between maximum design stress and the maximum stress that may be allowed in the fastener during assembly, we need to look at the design safety factor. These two will differ—that is, maximum design allowables will differ if a factor of safety is involved. For structural steels, bolts are frequently tightened well past the yield strength even though the design allowables are only 35–58% of yield. Pressure vessel bolts are commonly tightened to twice the design allowable. Aerospace, auto, and other industries may impose stringent limits on design stresses rather than on actual stresses to force the designer to use more or larger bolts.

8.2 Torsional Stress Factor

If the bolts are to be tightened by turning the nut or the head, they will experience a torsion stress as well as a tensile stress during assembly. If tightened to the yield stress, they will yield under this combination. If we plan to tighten to or near the yield stress, we must reduce the maximum tensile stresses allowed by a "torquing factor." If using as received steel on steel bolts, then a reduction in the allowable tensile stress of 10% is reasonable. If the fasteners are to be lubricated, use 5%.

8.3 Other Design Issues

- *Flange rotation.* Excessive bolt load can rotate raised face flanges so much that the ID of the gasket is unloaded, opening a leak path. The threat of rotation, therefore, can place an upper limit on planned or specified clamping forces.
- *Gasket crush.* Excessive preload can so compress a gasket that it will not be able to recover when the internal pressure or a thermal cycle partially unloads it. Contact the gasket manufacturer for upper limits. Note that these will be a function of the service temperature.
- *Stress cracking.* Stress cracking is encouraged by excessive tension in the bolts, particularly if service loads exceed 50% of the yield stress at least for low alloy quenched and tempered steels.
- *Combined loads.* These loads include weight, inertial affects, thermal effects, pressure, shock, earthquake loading, and so on. Both static and dynamic loads must be estimated. Load intensifiers such as prying and eccentricity should be acknowledged if present. Joint diagrams can be used (see later section) to add external loads and preloading. The parts designed must be able to withstand *worst case combinations of these pre- and service loads.*

Figure 19 shows the residual stresses in a group of 90 2¼-12 × 29 4330 studs that were tightened by stretching them 79% of their yield stress with a hydraulic tensioner. The studs and nuts were not new but had been tightened several times before these data were taken. Relaxation varied from 5% to 43% of the initial tension applied in these apparently identical studs. In many cases, similar scatter in relaxation also occurs after torquing.

Charts of this sort can be constructed on the basis of individual bolts or multibolt joints. Limits can be defined in terms of force, stress, yield instead of UTS, or even assembly torque.

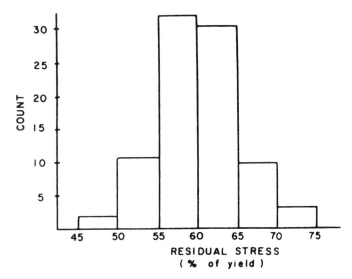

Figure 19 Residual stress as a percentage of yield strength, following removal of tension. Studs were all tensioned to 79% of yield. Torqued to 500 lb–ft.

9 THEORETICAL BEHAVIOR OF THE JOINT UNDER TENSILE LOADS

In this section, we examine the way a joint responds when exposed to the external loads it has been designed to support. This will be done by examining the elastic behavior of the joint. When we tighten a bolt on a flange, the bolt is placed in tension and it gets longer. The joint compresses in the vicinity of the bolt.

We need to plot separate elastic curves for the bolt and joint members by plotting the force in each of the two vertical axes and the deformation of each (elongation in the bolt and compression in the joint) on the horizontal axes. See Fig. 20.

Three things should be noted:

1. Typically the slope (K_B) of the bolts elastic curve is only ⅓ to ⅕ of the slope (K_J) of the joints elastic curve; i.e., the stiffness of the bolt is only ⅓ to ⅕ that of the joint.
2. The clamping force exerted by the bolt on the joint is opposed by an equal and opposite force exerted by the joint members on the bolt. (The bolt wants to shrink back to its original length and the joint wants to expand to its original thickness.)
3. If we continue to tighten the bolt, it or the joint will ultimately start to yield plastically, as suggested by the dotted lines. In future diagrams, we will operate only in the elastic region of each curve.

Rotscher first demonstrated what is called a *joint diagram* (Fig. 21). In Fig. 21, the tensile force in the bolt is called the *preload* in the bolt and is equal and opposite to the compressive force in the joint. If we apply an additional tension force to the bolt, this added load partially relieves the load on the joint, allowing it (if enough load is applied) to return to its original thickness while the bolt gets longer. Note that the increase in the length of the bolt is equal to the increase in thickness (reduction in compression) in the joint. In other words, the *joint expands to follow the nut as the bolt lengthens.*

Because the stiffness of the bolt is only ⅓ to ⅕ that of the joint, for an equal change in the strain, the change in load in the bolt must be only ⅓ to ⅕ of the change in the load in the joint. This is shown in Fig. 22.

The external tension load (L_x) required to produce this change of force and strain in the bolt and joint members is equal to the increase in the force on the bolt (ΔF_B) plus the reduction of force in the joint (ΔF_J):

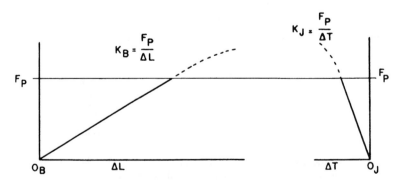

Figure 20 Elastic curves for bolt and joint members.

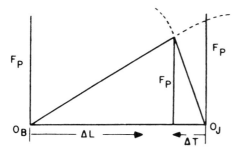

Figure 21 The elastic curves for bolt and joint can be combined to construct a joint diagram. O_B is the reference point for bolt length at zero stress. O_J is the reference point for joint thickness at zero stress.

$$L_X = \Delta F_B + \Delta F_J$$

The above relationship is demonstrated in Fig. 23.

Any external tension load, no matter how small, will be partially absorbed in replacing the force in the bolt (ΔF_B), and partially absorbed in replacing the reduction of force that the joint originally exerted on the bolt (ΔF_J). The force of the joint on the bolt plus the external load equal the new total tension force in the bolt—which is greater than the previous total—but the change in bolt force is less than the external load applied to the bolt. See Fig. 23, which recaps all of this. This is extremely important, because it is a way to move external loads around the bolt. In the case of alternating loads the fatigue life of the bolts can be greatly increased. That the bolt sees only a part of the external load, and that the amount it sees is dependent on the stiffness ratio, between the bold and the joint, have many implications for joint design, joint failure, measurement of residual preloads, and so on.

We can change the joint stiffness between the bolt and the joint by making the bolt much stiffer (i.e., a bolt with a larger diameter). The new joint diagram resulting from this change is shown in Fig. 24. Note that the bolt now absorbs a larger percentage of the same external load.

9.1 Critical External Load Required to Overcome Preload

If we keep adding external load to the original joint, we reach a point where the joint members are fully unloaded, as in Fig. 25. This is the critical external load, which is not equal to the original preload in the bolt but is often equal to the preload for several reasons.

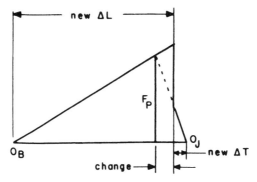

Figure 22 When an external tension load is applied, the bolt gets longer and joint compression is reduced. The change in deformation in the bolt equals the change in deformation in the joint.

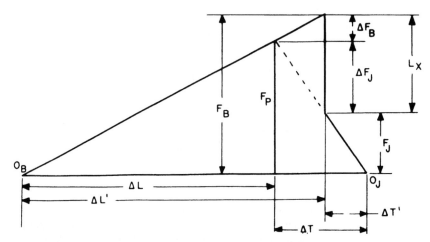

Figure 23 Summary diagram. F_P = initial preload; F_B = present bolt load; F_J = present joint load; L_X = external tension load applied to the bolt.

1. In many joints, the bolt has a low spring rate compared to the joint members. This is advantageous since the greatest percentage of external load is carried by the clamped members and not the bolt. Creation of a joint with a high ratio of K_J/K_B can be difficult. As an example, sheet metal joints are extremely thin and therefore have a low value of K_J. There are several strategies for lowering the stiffness of bolts in critical applications, including using hollow bolts as well as bolts with a reduced diameter in the middle.

2. Joints almost always relax after first tightening with relaxations of 10–20% of the initial preload being not uncommon. There are three main sources of relaxation in the joint. First the torsional load in the bolt is unstable and usually relaxes over time. Second, the nut is usually made of a more ductile material than the bolt and is designed to even out load among the threads. This is accomplished by a small movement of the threads relative to each other. Third, all of the mating surfaces that are

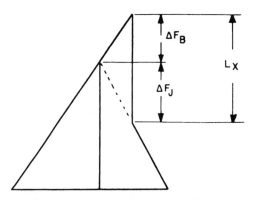

Figure 24 Joint diagram when stiffness of the bolt nearly equals that of joint.

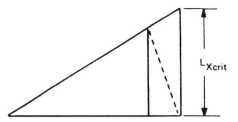
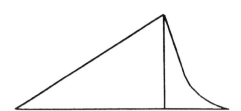

Figure 25 A critical external load (L_{xcrit}) fully unloads the joint (but not the bolt). No load sharing exists.

Figure 26 The spring rate of the joint is frequently nonlinear for small deflections.

compressed by the bolt contribute to relaxation. Entrapped particles as well as the high points of surface imperfections are crushed. If a bolt has ⅕ the stiffness of the joint, then the critical external load to free the joint members is 20% greater than the residual preload in the bolt when the load is applied. Therefore, the difference between the critical external load and the present preload is just about equal and opposite to the loss in preload caused by bolt relaxation. In other words, the critical external load equals the original preload before bolt relaxation.

9.2 Very Large External Loads

Any additional external load we add beyond the critical point will all be absorbed by the bolt. Although it is usually ignored in joint calculations, there is another curve we should be aware of. The compressive spring rate of many joint members is not a constant. A more accurate joint diagram would show this. See Fig. 26.

For joint diagrams, as shown in Fig. 23, we can make the following calculations where

F_P = initial preload (lb, N)
L_X = external tension load (lb, N)
ΔF_B = change in load in bolt (lb, N)
ΔF_J = change in load in joint (lb, N)
$\Delta L, \Delta L'$ = elongation of the bolt before and after application of the external load (in., mm)
$\Delta T, \Delta T'$ = compression of joint members before and after application of the external load (in., mm)
$L_{Xcritical}$ = external load required to completely unload the joint (lb, N) (not shown in diagram)

The stiffness (spring constants) of the bolt and joint are defined as follows:

$$\text{for the bolt } K_B = \frac{F_P}{\Delta L} \quad \text{for the joint } K_J = \frac{F_P}{\Delta T}$$

by manipulation $\Delta F_B = \dfrac{K_B}{K_B + K_J} \times L_X$ until joint separation, after which

$$\Delta F_B = \Delta L_X \text{ and } L_{Xcritical} = F_P \left\{ 1 + \frac{K_B}{K_J} \right\}$$

10 EVALUATION OF SLIP CHARACTERISTICS

A slip-resistant joint is one that has a low probability of slip at any time during the life of the structure. In this type of joint, the external applied load usually acts in a plane perpendicular to the bolt axis. The load is completely transmitted by frictional forces acting on the contact area of the plates fastened by the bolts. This frictional resistance is dependent on (1) the bolt preload and (2) the slip resistance of the fraying surfaces.

Slip-resistant joints are often used in connections subjected to stress reversals, severe stress fluctuations, or in any situation wherein slippage of the structure into a "bearing mode" would produce intolerable geometric changes. A slip load of a simple tension splice is given by

$$P_{\text{slip}} = k_s m \sum_{l=1}^{n} T_i$$

where k_s = slip coefficient
m = number of slip planes
$\sum_{l=1}^{n} T_i$ = the sum of the bolt tensions

If the bolt tension is equal in all bolts, then

$$p_{\text{slip}} = k_s\, m\, n\, T_i$$

where n = the number of bolts in the joint

The slip coefficient K_s varies from joint to joint, depending on the type of steel, different surface treatments, and different surface conditions, and along with the clamping force T_i shows considerable variation from its mean value. The slip coefficient K_s can only be determined experimentally, but some values are now available, as shown in Table 1.

11 TURN-OF-NUT METHOD OF BOLT TIGHTENING

To overcome the variability of torque control, efforts a more reliable tightening procedure is called the turn-of-nut method. (This is a strain-control method.) Initially it was believed that one turn from the snug position was the key, but because of out-of-flatness, thread imperfections, and dirt accumulation, it was difficult to determine the hand-tight position (the starting point—from the snug position). Current practice is as follows: run the nut up to a snug position using an impact wrench rather than the finger-tight condition (elongations are still within the elastic range). From the snug position, turn the nut in accordance with Table 2, provided by the RCSC specification.

Nut rotation is relative to bolt, regardless of the element (nut or bolt) being turned. For bolts installed by ⅔ turn and less, the tolerance should be ±30°; for bolts installed by ⅔ turn and more, the tolerance should be ±45°. All material within the grip of the bolt must be steel.

No research work has been performed by the council to establish the turn-of-nut procedure when bolt length exceeds 12 diameters. Therefore, the required rotation must be determined by actual tests in a suitable tension device simulating the actual conditions.

When bolts pass through a sloping interface greater than 1:20, a beveled washer is required to compensate for the lack of parallelism. As noted in Table 2, bolts require additional nut rotation to ensure that tightening will achieve the required minimum preload.

Table 1 Summary of Slip Coefficients

Type of Steel	Treatment	Average	Standard Deviation	Number of Tests
A7, A36, A440	Clean mill scale	0.32	0.06	180
A7, A36, A440, Fe37, Fe.52	Clean mill scale	0.33	0.07	327
A588	Clean mill scale	0.23	0.03	31
Fe37	Grit blasted	0.49	0.07	167
A36, Fe37, Fe52	Grit blasted	0.51	0.09	186
A514	Grit blasted	0.33	0.04	17
A36, Fe37	Grit blasted, exposed	0.53	0.06	51
A36, Fe37, Fe52	Grit blasted, exposed	0.54	0.06	83
A7, A36, A514, A572	Sand blasted	0.52	0.09	106
A36, Fe37	Hot-dip galvanized	0.18	0.04	27
A7, A36	Semipolished	0.28	0.04	12
A36	Vinyl wash	0.28	0.02	15
	Cold zinc plated	0.30	—	3
	Metallized	0.48	—	2
	Galvanized and sand blasted	0.34	—	1
	Sand blasted treated with linseed oil (exposed)	0.26	0.01	3
	Red lead paint	0.06	—	6

12 TORQUE AND TURN TOGETHER

Measuring of torque and turn at the same time can improve our control over preload. The final variation in preload in a large number of bolts is closer to ±5% than the 25–30% if we used torque or turn control alone. For this reason the torque–turn method is widely used today, especially in structural steel applications.

In this procedure, the nut is first snugged with a torque that is expected to stretch the fastener to a minimum of 75% of its ultimate strength. The nut is then turned (half a turn) or the like, which stretches the bolt well past its yield point. See Fig. 27.

This torque–turn method cannot be used on brittle bolts, but only on ductile bolts having long plastic regions. Therefore, it is limited to A325 fasteners used in structural steel work. Furthermore, it should never be used unless you can predict the working loads that the bolt

Table 2 Nut Rotation from Snug-Tight Condition

Bolt Length (as measured from underside of head to extreme end of point)	Both Faces Normal to Bolt Axis	One Face Normal to Bolt Axis and Other Face Sloped Not More Than 1:20 (bevel washer not used)	Both Faces Sloped Not More Than 1:20 from Normal to Bolt Axis (bevel washers not used)
Up to and including 4 diameters	⅓ turn	½ turn	⅔ turn
Over 4 diameters but not exceeding 8 diameters	½ turn	⅔ turn	⅚ turn
Over 8 diameters but not exceeding 12 diameters	⅔ turn	⅚ turn	1 turn

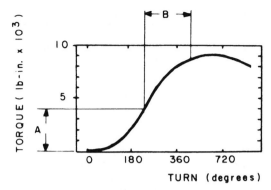

Figure 27 In turn-of-nut techniques, the nut is first tightened with an approximate torque (A) and then further tightened with a measured turn (B).

will see in service. Anything that loads the bolts above the original tension will create additional plastic deformation in the bolt. If the overloads are high enough, the bolt will break.

A number of knowledgeable companies have developed manual torque–turn procedures that they call "turn of the nut" but that do not involve tightening the fasteners past the yield point. Experience shows that some of these systems provide additional accuracy over turn or torque alone.

Other methods have also been developed to control the amount of tension produced in bolts during assembly, namely *stretch* and *tension control*.[2] All of these methods have drawbacks and limitations, but each is good enough for many applications. However, in more and more applications, we need to find a better way to control bolt tension and/or clamping forces. Fortunately, that better way is emerging, namely *ultrasonic measurement of bolt stretch or tension*.

13 ULTRASONIC MEASUREMENT OF BOLT STRETCH OR TENSION

Ultrasonic techniques, while not in common use, allow us to get past dozens of the variables that affect the results we achieve with torque and/or torque and turn control.

The basic concepts are simple. The two most common systems are *pulse-echo* and *transit time* instruments. In both, a small acoustic transducer is placed against one end of the bolt being tested. See Fig. 28. An electronic instrument delivers a voltage pulse to the transducer, which emits a very brief burst of ultrasound that passes down the bolt, echoes off the far end, and returns to the transducer. An electronic instrument measures precisely the amount of time required for the sound to make its round trip in the bolt.

As the bolt is tightened, the amount of time required for the round trip increases for two reasons:

1. The bolt stretches as it is tightened, so the path length increases.
2. The average velocity of sound within the bolt decreases because the average stress level in the bolt has increased.

Both of these changes are linear functions of the preload in the fastener, so that the total change in transit time is also a linear function of preload.

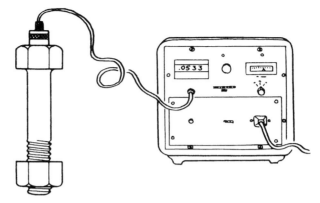

Figure 28 An acoustic transducer is held against one end of the fastener to measure the fastener's change in length as it is tightened.

The instrument is designed to measure the change in transit time that occurs during tightening and to report the results as

1. A change in length of the fastener
2. A change in the stress level within the threaded region of the fastener
3. A change in tension within the fastener

Using such an instrument is relatively easy. A drop of coupling fluid is placed on one end of the fastener to reduce the acoustic impedance between the transducer and the bolt. The transducer is placed on the puddle of fluid and held against the bolt, mechanically or magnetically. The instrument is zeroed for this particular bolt (because each bolt will have a slightly different acoustic length). If you wish to measure residual preload, or relaxation, or external loads at some later date, you record the length of the fastener at zero load at this time. Next the bolt is tightened. If the transducer can remain in place during tightening, the instrument will show you the buildup of stretch or tension in the bolt. If it must be removed, it is placed on the bolt after tightening to show the results achieved by torque, turn, or tension.

If, at some later date, you wish to measure the present tension, you dial in the original length of *that* bolt into the instrument and place the transducer back on the bolt. The instrument will then show you the difference in length or stress that now exists in the bolt.

Ultrasonic equipment is used primarily in applications involving relatively few bolts in critically important joints or quality control audits. Operator training in the use of this equipment is necessary and is a low-cost alternative to strain-gaged bolts in all sorts of studies. The author knows of at least one application where ultrasonic measurement is built into a production work cell. It is used to tighten the crankshaft main bearings to very precise values.

These instruments are new to the field, so you must be certain to find out from the manufacturers exactly what the equipment will or will not do as well as precise information needed for use or equipment calibration. Training is essential not only for the person ordering the equipment, but for all who will use it in the field or laboratory. Proper calibration is essential. If the equipment can only measure transit time, you must tell it how to interpret transit time data *for your application.*

14 FATIGUE FAILURE AND DESIGN FOR CYCLICAL TENSION LOADS

A fastener subjected to repeated cyclical tension loads can suddenly break. These failures are generally catastrophic in kind, even if the loads are well below the yield strength of the material.

Three essential conditions are necessary for a fatigue failure: cyclical tensile loads; stress levels above the endurance limit of the material; and a stress concentration region (such as a sharp corner, a hole, a surface scratch or other mark on the surface of the part, corrosion pits, an inclusion and/or a flaw in the material). Essentially no part is completely free of these types of defects unless great care has been taken to remove them.

The sequence of events leading up to a fatigue failure is as follows:

1. Crack inititation begins after about 90% of the total fatigue life (total number of cycles) has occurred. This crack always starts on the surface of the part.
2. The crack begins to grow with each half-cycle of tension stress, leaving beach marks on the part.
3. Growth of the crack continues until the now-reduced cross section is unable to support the load, at which time the part fails catastrophically.

A bolt is a very poor shape for good fatigue resistance. Although the average stress levels in the body may be well below the endurance limit, stress levels in the stress concentration points, such as thread roots, head to body fillets, and so on can be well over the endurance limit. One thing we can do to reduce or eliminate a fatigue problem is to attempt to overcome one or more of the three essential conditions without which failure would not occur. In general, most of the steps are intended to reduce stress levels, reduce stress concentrations, and/or reduce the load excursions seen by the bolt. The following are additional suggestions for reducing the chance of bolt fatigue.

Rolled Threads
Rolling provides a smoother thread finish than cutting and thus lowers the stress concentrations found at the root of the thread. In addition to overcoming the notch effect of cut threads, rolling induces compressive stresses on the surface rolled. This compressive "preload" must be overcome by tension forces before the roots will be in net tension. A given tension load on the bolt, therefore, will result in a smaller tension excursion at this critical point. Rolling the threads is best done after heat treating the bolt, but it is more difficult. Rolling before heat treatment is possible on larger-diameter bolts.

Fillets
Use bolts with generous fillets between the head and the shank. An elliptical fillet is better than a circular one and the larger the radius the better. Prestressing the fillet is wise (akin to thread rolling).

Perpendicularity
If the face of the nut, the underside of the bolt head, and/or joint surfaces are not perpendicular to thread axes and bolt holes, the fatigue life of the bolt can be seriously affected. For example, a 2° error reduces the fatigue life by 79%.[3]

Overlapping Stress Concentrations
Thread run-out should not coincide with a joint interface (where shear loads exist) and there should be at least two full bolt threads above and below the nut because bolts normally see stress concentrations at (1) thread run-out; (2) first threads to engage the nut, and head-to-shank fillets.

Thread Run-Out

The run-out of the thread should be gradual rather than abrupt. Some people suggest a maximum of 15° to minimize stress concentrations.

Thread Stress Distribution

Most of the tension in a conventional bolt is supported by the first two or three nut threads. Anything that increases the number of active threads will reduce the stress concentration and increase the fatigue life. Some of the possibilities are

1. Using so-called "tension nuts," which create nearly uniform stress in all threads.
2. Modifying the nut pitch so that it is slightly different than the pitch of the bolt, i.e., thread of nut 11.85 threads/in. used with a bolt having 12 threads/in.
3. Using a nut slightly softer than the bolt (this is the usual case); however, select still softer nuts if you can stand the loss in proof load capability.
4. Using a jam nut, which improves thread stress distribution by preloading the threads in a direction opposite to that of the final load.
5. Tapering the threads slightly. This can distribute the stresses more uniformly and increase the fatigue life. The taper is 15°.

Bending

Reduce bending by using a spherical washer because nut angularity hurts fatigue life.

Corrosion

Anything that can be done to reduce corrosion will reduce the possibilities of crack initiation and/or crack growth and will extend fatigue life. Corrosion can be more rapid at points of high stress concentration, which is also the point where fatigue failure is most prevalent. Fatigue and corrosion aid each other and it is difficult to tell which mechanism initiated or resulted in a failure.

Surface Conditions

Any surface treatment that reduces the number and size of incipient cracks will improve fatigue life significantly, so that polishing of the surface will greatly improve the fatigue life of a part. This is particularly important for punched or drilled holes, which can be improved by reaming and expanding to put the surface in residual compression. Shot peening of bolts or any surface smooths out sharp discontinuites and puts the surface in residual compression. Handling of bolts in such a way as not to ding one against the other is also important.

Reduce Load Excursions

It is necessary to identify the maximum safe preload that your joint can stand by estimating fastener strength, joint strength, and external loads. Also do whatever is required to minimize the bolt-to-joint stiffness ratio so that most of the excursion and external load will be seen by the joint and not the bolt. Use long, thin bolts even if it means using more bolts. Eliminate gaskets and/or use stiffer gaskets.

While there are methods available for estimating the endurance limit of a bolt, it is best to base your calculations on actual fatigue tests of the products you are going to use or your own experience with those products.

For the design criteria for fatigue loading of slip resistant joints, see Refs. 2 and 3.

REFERENCES

1. D. Whitney, *Mechanical Assembly,* Oxford University Press, New York, 2004.
2. J. H. Bickford, *An Introduction to the Design and Behavior of Bolted Joints,* 2nd ed., Marcel Dekker, New York, 1990.
3. G. L. Kulak, J. W. Fisher, and J. H. A. Struik, *Guide to Design Criteria for Bolted and Riveted Joints,* Wiley, New York, 1987.
4. *SPS Fastener Facts,* Standard Pressed Steel Co., Jenkintown, PA, Section IV-C-4.
5. G. Linnert, *Welding, Metallurgy, Carbon and Alloy Steels,* Vol. 4, American Welding Society, Miami, FL, 1994, Chap. 7.
6. N. Yurioka, "Weldability of Modern High Strength Steels," in *First US–Japan Symposium on Advances in Welding Metallurgy,* American Welding Society, Miami, FL, 1990, pp. 79–100.

CHAPTER 8
STATISTICAL QUALITY CONTROL

Magd E. Zohdi
Department of Industrial Engineering and Manufacturing
Louisiana State University
Baton Rouge, Louisiana

1	MEASUREMENTS AND QUALITY CONTROL	315	6	CONTROL CHARTS FOR ATTRIBUTES	322
				6.1 The p and np Charts	322
2	DIMENSION AND TOLERANCE	315		6.2 The c and u Charts	323
3	QUALITY CONTROL	316	7	ACCEPTANCE SAMPLING	325
	3.1 $\bar{X}$, R, and σ Charts	316		7.1 Double Sampling	325
				7.2 Multiple and Sequential Sampling	326
4	INTERRELATIONSHIP OF TOLERANCES OF ASSEMBLED PRODUCTS	321	8	DEFENSE DEPARTMENT ACCEPTANCE SAMPLING BY VARIABLES	326
5	OPERATION CHARACTERISTIC CURVE (OC)	322		BIBLIOGRAPHY	326

1 MEASUREMENTS AND QUALITY CONTROL

The metric and English measuring systems are the two measuring systems commonly used throughout the world. The metric system is universally used in most scientific applications, but, for manufacturing in the United States, has been limited to a few specialties, mostly items that are related in some way to products manufactured abroad.

2 DIMENSION AND TOLERANCE

In dimensioning a drawing, the numbers placed in the dimension lines are only approximate and do not represent any degree of accuracy unless so stated by the designer. To specify the degree of accuracy, it is necessary to add tolerance figures to the dimension. Tolerance is the amount of variation permitted in the part or the total variation allowed in a given dimension.

Dimensions given close tolerances mean that the part must fit properly with some other part. Both must be given tolerances in keeping with the allowance desired, the manufacturing processes available, and the minimum cost of production and assembly that will maximize profit. Generally speaking, the cost of a part goes up as the tolerance is decreased.

Allowance, which is sometimes confused with tolerance, has an altogether different meaning. It is the minimum clearance space intended between mating parts and represents the condition of tightest permissible fit.

3 QUALITY CONTROL

When parts must be inspected in large numbers, 100% inspection of each part is not only slow and costly, but does not eliminate all of the defective pieces. Mass inspection tends to be careless; operators become fatigued; and inspection gages become worn or out of adjustment more frequently. The risk of passing defective parts is variable and of unknown magnitude, whereas, in a planned sampling procedure, the risk can be calculated. Many products, such as bulbs, cannot be 100% inspected, since any final test made on one results in the destruction of the product. Inspection is costly and nothing is added to a product that has been produced to specifications.

Quality control enables an inspector to sample the parts being produced in a mathematical manner and to determine whether or not the entire stream of production is acceptable, provided that the company is willing to allow up to a certain known number of defective parts. This number of acceptable defectives is usually taken as 3 out of 1000 parts produced. Other values might be used.

3.1 $\overline{X}$, R, and σ Charts

To use quality techniques in inspection, the following steps must be taken (see Table 1).

1. Sample the stream of products by taking m samples, each of size n.
2. Measure the desired dimension in the sample, mainly the central tendency.
3. Calculate the deviations of the dimensions.
4. Construct a control chart.
5. Plot succeeding data on the control chart.

The arithmetic mean of the set of n units is the main measure of central tendency. The symbol $\overline{X}$ is used to designate the arithmetic mean of the sample and may be expressed in algebraic terms as

$$\overline{X}_i = (X_1 + X_2 + X_3 + \cdots + X_n)/n \tag{1}$$

where X_1, X_2, X_3, etc. represent the specific dimensions in question. The most useful measure of dispersion of a set of numbers is the standard deviation σ. It is defined as the root-mean-square deviation of the observed numbers from their arithmetic mean. The standard deviation σ is expressed in algebraic terms as

Table 1 Computational Format for Determining $\overline{X}$, R, and σ

Sample Number	Sample Values	Mean $\overline{X}$	Range R	Standard Deviation σ'
1	$X_{11}, X_{12}, \ldots, X_{1n}$	$\overline{X}_1$	R_1	σ'_1
2	$X_{21}, X_{22}, \ldots, X_{2n}$	$\overline{X}_2$	R_2	σ'_2
.	...	.	.	.
.	...	.	.	.
.	...	.	.	.
m	$X_{m1}, X_{m2}, \ldots, X_{mn}$	$\overline{X}_m$	R_m	σ'_m

$$\sigma_i = \sqrt{\frac{(X_1 - \overline{X})^2 + (X_2 - \overline{X})^2 + \cdots + (X_n - \overline{X})^2}{n}} \qquad (2)$$

Another important measure of dispersion, used particularly in control charts, is the range R. The range is the difference between the largest observed value and the smallest observed in a specific sample.

$$R = X_i(\max) - X_i(\min) \qquad (3)$$

Even though the distribution of the X values in the universe can be of any shape, the distribution of the $\overline{X}$ values tends to be close to the normal distribution. The larger the sample size and the more nearly normal the universe, the closer will the frequency distribution of the average $\overline{X}$'s approach the normal curve, as in Fig. 1.

According to the statistical theory (the Central Limit Theory), in the long run, the average of the $\overline{X}$ values will be the same as μ, the average of the universe. And in the long run, the standard deviation of the frequency distribution $\overline{X}$ values, $\sigma_{\bar{x}}$, will be given by

$$\sigma_{\bar{x}} = \frac{\sigma}{\sqrt{n}} \qquad (4)$$

where σ is the standard deviation of the universe. To construct the control limits, the following steps are taken:

1. Calculate the average of the average $\overline{\overline{X}}$ as follows:

$$\overline{\overline{X}} = \sum_1^m \overline{X}_i/m \qquad i = 1, 2, \ldots, m \qquad (5)$$

2. Calculate the average deviation, $\overline{\sigma}$ where

$$\overline{\sigma} = \sum_1^m \sigma'_i/m \qquad i = 1, 2, \ldots, m \qquad (6)$$

Statistical theory predicts the relationship between $\overline{\sigma}$ and $\sigma_{\bar{x}}$. The relationship for the $3\sigma_{\bar{x}}$ limits or the 99.73% limits is

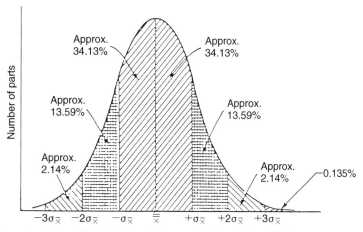

Figure 1 Normal distribution and percentage of parts that will fall within σ limits.

$$A_1\bar{\sigma} = 3\sigma_{\bar{x}} \qquad (7)$$

This means that control limits are set so that only 0.27% of the produced units will fall outside the limits. The value of $3\sigma_{\bar{x}}$ is an arbitrary limit that has found acceptance in industry.

The value of A_1 calculated by probability theory is dependent on the sample size and is given in Table 2. The formula for 3σ control limits using this factor is

$$\text{CL}(\bar{X}) = \bar{\bar{X}} \pm A_1\bar{\sigma} \qquad (8)$$

Once the control chart (Fig. 2) has been established, data ($\bar{X}_i$'s) that result from samples of the same size n are recorded on it. It becomes a record of the variation of the inspected dimensions over a period of time. The data plotted should fall in random fashion between the control limits 99.73% of the time if a stable pattern of variation exists.

So long as the points fall between the control lines, no adjustments or changes in the process are necessary. If five to seven consecutive points fall on one side of the mean, the process should be checked. When points fall outside of the control lines, the cause must be located and corrected immediately.

Table 2 Factors for $\bar{X}$, R, σ, and X Control Charts

Sample Size n	Factors for $\bar{X}$ Chart		Factors for R Chart		Factors for σ' Chart		Factors for X Chart		$\sigma = \bar{R}/d_2$
	From $\bar{R}$ A_2	From $\bar{\sigma}$ A_1	Lower D_3	Upper D_4	Lower B_3	Upper B_4	From $\bar{R}$ E_2	From $\bar{\sigma}$ E_1	d_2
2	1.880	3.759	0	3.268	0	3.267	2.660	5.318	1.128
3	1.023	2.394	0	2.574	0	2.568	1.772	4.146	1.693
4	0.729	1.880	0	2.282	0	2.266	1.457	3.760	2.059
5	0.577	1.596	0	2.114	0	2.089	1.290	3.568	2.326
6	0.483	1.410	0	2.004	0.030	1.970	1.184	3.454	2.539
7	0.419	1.277	0.076	1.924	0.118	1.882	1.109	3.378	2.704
8	0.373	1.175	0.136	1.864	0.185	1.815	1.054	3.323	2.847
9	0.337	1.094	0.184	1.816	0.239	1.761	1.011	3.283	2.970
10	0.308	1.028	0.223	1.777	0.284	1.716	0.975	3.251	3.078
11	0.285	0.973	0.256	1.744	0.321	1.679	0.946	3.226	3.173
12	0.266	0.925	0.284	1.717	0.354	1.646	0.921	3.205	3.258
13	0.249	0.884	0.308	1.692	0.382	1.618	0.899	3.188	3.336
14	0.235	0.848	0.329	1.671	0.406	1.594	0.881	3.174	3.407
15	0.223	0.817	0.348	1.652	0.428	1.572	0.864	3.161	3.472
16	0.212	0.788	0.364	1.636	0.448	1.552	0.848	3.152	3.532
17	0.203	0.762	0.380	1.621	0.466	1.534	0.830	3.145	3.588
18	0.194	0.738	0.393	1.608	0.482	1.518	0.820	3.137	3.640
19	0.187	0.717	0.404	1.597	0.497	1.503	0.810	3.130	3.687
20	0.180	0.698	0.414	1.586	0.510	1.490	0.805	3.122	3.735
21	0.173	0.680	0.425	1.575	0.523	1.477	0.792	3.114	3.778
22	0.167	0.662	0.434	1.566	0.534	1.466	0.783	3.105	3.819
23	0.162	0.647	0.443	1.557	0.545	1.455	0.776	3.099	3.858
24	0.157	0.632	0.451	1.548	0.555	1.445	0.769	3.096	3.895
25	0.153	0.619	0.459	1.540	0.565	1.435	0.765	3.095	3.931

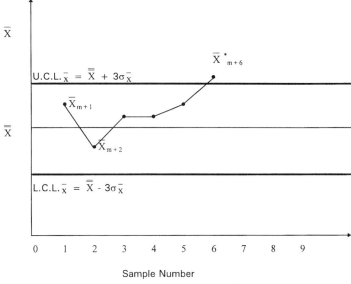

Figure 2 Control chart $\bar{X}$.

Statistical theory also gives the expected relationship between $\bar{R}$ ($\Sigma R_i/m$) and $\sigma_{\bar{x}}$. The relationship for the $3\sigma_{\bar{x}}$ limits is

$$A_2 \bar{R} = 3\sigma_{\bar{x}} \tag{9}$$

The values for A_2 calculated by probability theory, for different sample sizes, are given in Table 2.

The formula for 3σ control limits using this factor is

$$\text{CL}(\bar{X}) = \bar{\bar{X}} \pm A_2 \bar{R} \tag{10}$$

In control chart work, the ease of calculating R is usually much more important than any slight theoretical advantage that might come from the use of σ. However, in some cases where the measurements are costly and it is necessary that the inferences from a limited number of tests be as reliable as possible, the extra cost of calculating σ is justified. It should be noted that, because Fig. 2 shows the averages rather than individual values, it would have been misleading to indicate the tolerance limits on this chart. It is the individual article that has to meet the tolerances, not the average of a sample. Tolerance limits should be compared to the machine capability limits. Capability limits are the limits on a single unit and can be calculated by

$$\text{Capability limits} = \bar{\bar{X}} \pm 3\sigma \tag{11}$$

$$\sigma = \bar{R}/d_2$$

Since $\sigma' = \sqrt{n}\,\sigma_{\bar{x}}$, the capability limits can be given by

$$\text{Capability limits } (X) = \bar{\bar{X}} \pm 3\sqrt{n}\,\sigma_{\bar{x}} \tag{12}$$

$$= \bar{\bar{X}} \pm E_1 \sigma \tag{13}$$

$$= \bar{\bar{X}} \pm E_2 \bar{R} \tag{14}$$

The values for d_2, E_1, and E_2 calculated by probability theory, for different sample sizes, are given in Table 2.

Figure 3 shows the relationships among the control limits, the capability limits, and assumed tolerance limits for a machine that is capable of producing the product with this specified tolerance. Capability limits indicate that the production facility can produce 99.73% of its products within these limits. If the specified tolerance limits are greater than the capability limits, the production facility is capable of meeting the production requirement.

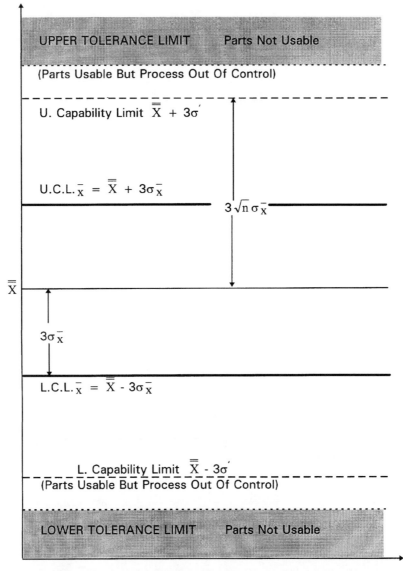

Figure 3 Control, capability, and tolerance (specification limits).

If the specified tolerance limits are tighter than the capability limits, a certain percentage of the production will not be usable and 100% inspection will be required to detect the products outside the tolerance limits.

To detect changes in the dispersion of the process, the R and σ charts are often employed with $\overline{X}$ and X charts.

The upper and lower control limits for the R chart are specified as

$$\text{UCL}(R) = D_4 \overline{R} \tag{15}$$

$$\text{LCL}(R) = D_3 \overline{R} \tag{16}$$

Figure 4 shows the $\overline{R}$ chart for samples of size 5.

The upper and lower control for the T chart are specified as

$$\text{UCL}(\sigma) = B_4 \overline{\sigma} \tag{17}$$

$$\text{LCL}(\sigma) = B_3 \overline{\sigma} \tag{18}$$

The values for D_3, D_4, B_3, and B_4 calculated by probability theory, for different sample sizes, are given in Table 2.

4 INTERRELATIONSHIP OF TOLERANCES OF ASSEMBLED PRODUCTS

Mathematical statistics states that the dimension on an assembled product may be the sum of the dimensions of the several parts that make up the product. It states also that the standard deviation of the sum of any number of independent variables is the square root of the sum of the squares of the standard deviations of the independent variables. So if

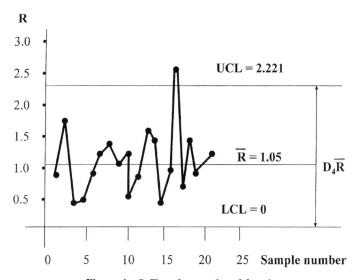

Figure 4 R Chart for samples of 5 each.

$$X = X_1 \pm X_2 \pm \cdots \pm X_n \tag{19}$$

$$\overline{X} = \overline{X}_1 \pm \overline{X}_2 \pm \cdots \pm \overline{X}_n \tag{20}$$

$$\sigma(X) = \sqrt{(\sigma_1)^2 + (\sigma_2)^2 + \cdots + (\sigma_n)^2} \tag{21}$$

Whenever it is reasonable to assume that the tolerance ranges of the parts are proportional to their respective σ' values, such tolerance ranges may be combined by taking the square root of the sum of the squares:

$$T = \sqrt{T_1^2 + T_2^2 + T_3^2 + \cdots + T_n^2} \tag{22}$$

5 OPERATION CHARACTERISTIC CURVE (OC)

Control charts detect changes in a pattern of variation. If the chart indicates that a change has occurred when it has not, Type I error occurs. If three-sigma limits are used, the probability of making a Type I error is approximately 0.0027.

The probability of the chart indicating no change, when in fact it has, is the probability of making a Type II error. The operation characteristic curves are designed to indicate the probability of making a Type II error. An OC curve for an $\overline{X}$ chart of three-sigma limits is illustrated in Fig. 5.

6 CONTROL CHARTS FOR ATTRIBUTES

Testing may yield only one of two defined classes: within or outside certain limits, acceptable or defective, working or idle. In such a classification system, the proportion of units falling in one class may be monitored with a p chart.

In other cases, observation may yield a multivalued, but still discrete, classification system. In such case, the number of discrete observations, such as events, objects, states, or occurrences, may be monitored by a c chart.

6.1 The p and np Charts

When sampled items are tested and placed into one of two defined classes, the proportion of units falling into one class p is described by the binomial distribution. The mean and standard deviation are given as

$$\mu = np$$

$$\sigma = \sqrt{np(1 - p)}$$

Dividing by the sample size n, the parameters are expressing as proportions. These statistics can be expressed as

$$\overline{p} = \frac{\text{Total number in the class}}{\text{Total number of observations}} \tag{23}$$

$$s_p = \sqrt{\frac{\overline{p}(1 - \overline{p})}{n}} \tag{24}$$

The control limits are either set at two-sigma limits with Type I error as 0.0456 or at three-sigma limits with Type I error as 0.0027. The control limits for the p chart with two-sigma limits (Fig. 6) are defined as

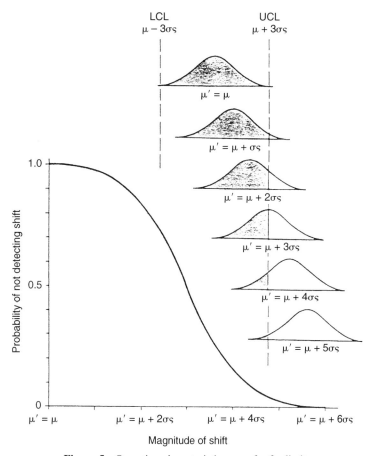

Figure 5 Operating characteristic curve for 3σ limit.

$$\text{CL}(p) = \bar{p} \pm 2S_p \tag{25}$$

However, if subgroup size is constant, the chart for actual numbers of rejects np or pn may be used. The appropriate model for three-sigma control limits on an np chart is

$$\text{CL}(np) = n\bar{p} \pm 3\sqrt{n\bar{p}(1-\bar{p})} \tag{26}$$

6.2 The c and u Charts

The random variable process that provides numerical data that are recorded as a number c rather than a proportion p is described by the Poisson distribution. The mean and the variance of the Poisson distribution are equal and expressed as $\mu = \sigma^2 = np$. The Poisson distribution is applicable in any situation when n and p cannot be determined separately, but their product np can be established. The mean and variable can be estimated as

$$\bar{c} = S_c^2 = \frac{\sum_1^m C_i}{m} = \frac{\sum_1^m (np)_i}{m} \tag{27}$$

The control limits (Fig. 7) are defined as

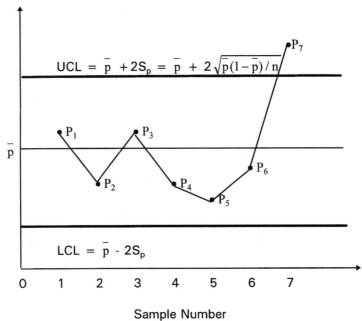

Figure 6 P charts.

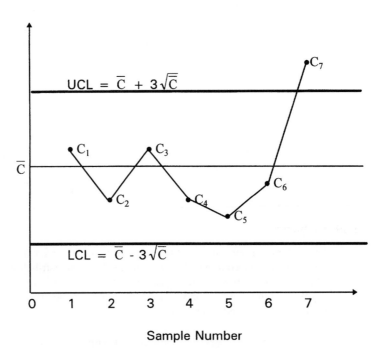

Figure 7 C charts.

$$\text{CL}(c) = \overline{C} \pm 3S_c \tag{28}$$

If there is change in the area of opportunity for occurrence of a nonconformity from subgroup to subgroup, such as number of units inspected or the lengths of wires checked, the conventional c chart showing only the total number of nonconformities is not applicable. To create some standard measure of the area of opportunity, the nonconformities per unit (c/n) or u is used as the control statistic. The control limits are

$$\text{CL}(u)\overline{u} \pm 3\frac{\sqrt{\overline{u}}}{\sqrt{n_i}} \tag{29}$$

$$\text{where } \overline{u} = \frac{\Sigma C_i}{\Sigma n_i} = \frac{\text{Total nonconformities found}}{\text{Total units inspected}}$$

$c = nu$ is Poisson-distributed, u is not

7 ACCEPTANCE SAMPLING

The objective of acceptance sampling is to determine whether a quantity of the output of a process is acceptable according to some criterion of quality. A sample from the lot is inspected and the lot is accepted or rejected in accordance with the findings of the sample.

Acceptance sampling plans call for the random selection of sample of size n from a lot containing N items. The lot is accepted if the number of defectives found in the sample are $\leq c$, the acceptance number. A rejected lot can either be returned to the producer, nonrectifying inspections, or it can be retained and subjected to a 100% screening process, rectifying inspection plan improves the outgoing quality. A second attribute-inspection plan might use two samples before requiring the acceptance or rejection of a lot. A third plan might use multiple samples or a sequential sampling process in evaluating a lot. Under rectifying inspection programs, the average outgoing quality level (AOQ), the average inspection lot (I), and the average outgoing quality limit (AOQL) can be predicted for varying levels of incoming fraction defective p.

Assuming that all lots arriving contain the same proportion of defectives p, and that rejected lots will be subjected to 100% inspection, AOQ and I are given below:

$$\text{AOQ} = \frac{P_a p(N - n)}{N - pn - (1 - P_a)p(Nn)} \tag{30}$$

$$I = n + (1 - P_a)(N - n) \tag{31}$$

The average outgoing quality (AOQ) increases as the proportion defective in incoming lots increases until it reaches a maximum value and then starts to decrease. This maximum value is referred to as the average outgoing quality limit (AOQL). The hypergeometric distribution is the appropriate distribution to calculate the probability of acceptance P_a; however, the Poisson distribution is used as an approximation.

Nonrectifying inspection program does not significantly improve the quality level of the lots inspected.

7.1 Double Sampling

Double sampling involves the possibility of putting off the decision on the lot until a second sample has been taken. A lot may be accepted at once if the first sample is good enough or rejected at once if the first sample is bad enough. If the first sample is neither, the decision is based on the evidence of the first and second samples combined.

The symbols used in double sampling are

N = lot size

n_1 = first sample

c_1 = acceptance number for first sample

n_2 = second sample

c_2 = acceptance number of the two samples combined

Computer programs are used to calculate the OC curves; acceptance after the first sample, rejection after the first sample, acceptance after the second sample, and rejection after the second sample. The average sample number (ASN) in double sampling is given by

$$\text{ASN} = [P_a(n_1) + P_r(n_1)]n_1 + [P_a(n_2) + P_r(n_2)](n_1 + n_2) \qquad (32)$$

7.2 Multiple and Sequential Sampling

In multiple sampling, three or more samples of a stated size are permitted and the decision on acceptance or rejection is revealed after a stated number of samples. In sequential sampling, item-by-item inspection, a decision is possible after each item has been inspected and when there is no specified limit on the total number of units to be inspected. OC curves are developed through computer programs. The advantage of using double sampling, multiple sampling, or sequential sampling is to reach the appropriate decision with fewer items inspected.

8 DEFENSE DEPARTMENT ACCEPTANCE SAMPLING BY VARIABLES

MIL-STD-105 A, B, C, D, and then ABC-STD-105, are based on the acceptance quality level (AQL) concept. The plans contain single, double, or multiple sampling, depending on the lot size and AQL and the probability of acceptance at this level P_a. Criteria for shifting to tightened inspection, requalification for normal inspection, and reduced inspection are listed in the tables associated with plan.

MIL-STD-414 plans were developed to reduce inspection lots by using sample sizes compared to MIL-STD-105. They are similar, as both procedures and tables are based on the concept of AQL; lot-by-lot acceptance inspection; both provide for normal, tightened, or reduced inspection; sample sizes are greatly influenced by lot size; several inspection levels are available; and all plans are identified by sample size code letter. MIL-STD-414 could be applied either with a single specification limit, L or U, or with two specification limits. Known-sigma plans included in the standard were designated as having "variability known." Unknown-sigma plans were designated as having "variability unknown." In the latter-type plans, it was possible to use either the standard deviation method or the range method in estimating the lot variability.

BIBLIOGRAPHY

Aft, L. S., *Fundamentals of Industrial Quality Control,* 3rd ed., Addison–Wesley, Menlo Park, CA, 1988.

ASTM Manual on Presentation of Data and Control Chart Analysis, Special Technical Pub. 15D, American Society for Testing and Materials, Philadelphia, PA, 1976.

Bibliography

Bhushan, B. (ed.), *Modern Tribology Handbook,* CRC Press, Boca Raton, FL, 2001.

Clements, R., *Quality ITQM/ISO 9000,* Prentice-Hall, Englewood Cliffs, NJ, 1995.

Control Chart Method of Controlling Quality During Production, ANSI Standard 21.3-1975, American National Standards Institute, New York, 1975.

Devore, J. L., *Probability and Statistics for Engineering and the Sciences,* Duxburg Press, New York, 1995.

Dodge, H. F., *A General Procedure for Sampling Inspection by Attributes—Based on the AQL Concept,* Technical Report No. 10, The Statistics Center, Rutgers—The State University, New Brunswick, NJ, 1959.

Drake, P. J., *Dimension and Tolerancing Handbook,* McGraw-Hill, New York, 1999.

Duncan, A. J., *Quality Control and Industrial Statistics,* 5th ed., Richard D. Irwin, Homewood, IL, 1986.

Feigenbaum, A. V., *Total Quality Control—Engineering and Mangement,* 3rd ed., McGraw-Hill, New York, 1991.

Grant, E. L., and R. S. Leavenworth, *Statistical Quality Control,* 7th ed., McGraw-Hill, New York, 1996. (Software included)

Juran, J. M., and F. M. Gryna, Jr., *Quality Control Handbook,* McGraw-Hill, New York, 1988.

Juran, J. M., and F. M. Gryna, Jr., *Quality Planning and Analysis,* 4th ed., McGraw-Hill, New York, 2000.

Lamprecht, J., *Implementing the ISO 9000 Series,* Marcel Dekker, New York, 1995.

Military Standard 105E, Sampling Procedures and Tables for Inspection by Attributes, Superintendent of Documents, Government Printing Office, Washington, DC, 1969.

Military Standard 414, Sampling Procedures and Tables for Inspection by Variables for Percent Defective, Superintendent of Documents, Government Printing Office, Washington, DC, 1957.

Military Standard 690-B, Failure Rate Sampling Plans and Procedures, Superintendent of Documents, Government Printing Office, Washington, DC, 1969.

Military Standard 781-C, Reliability Design Qualification and Production Acceptance Tests, Exponential Distribution, Superintendent of Documents, Government Printing Office, Washington, DC, 1977.

Military Standard 1235B, Single and Multi-Level Continuous Sampling Procedures and Tables for Inspection by Attributes, Superintendent of Documents, Government Printing Office, Washington, DC, 1981.

Montgomery, D. C., and G. C. Runger, *Introduction to Statistical Quality Control,* Wiley, New York, 2001.

Rabbit, J., and P. Bergh, *The ISO 9000 Book,* ME, Dearborn, MI, 1993.

Society of Manufacturing Engineers, *Quality Control and Assembly,* 4th ed., Dearborn, MI, 1994.

Supply and Logistic Handbook—Inspection H 105. Administration of Sampling Procedures for Acceptance Inspection, Superintendent of Documents, Government Printing Office, Washington, DC, 1954.

Walpole, R. E., R. H. Myers, S. L. Myers, and K. Ye, *Probability and Statistics for Engineers and Scientists,* Prentice Hall, Englewood Cliffs, NJ, 2002.

Zohdi, M. E., *Manufacturing Processes Quality Evaluation and Testing,* International Conference, Operations Research, January 1976.

CHAPTER 9
COMPUTER-INTEGRATED MANUFACTURING

William E. Biles
Department of Industrial Engineering
University of Louisville
Louisville, Kentucky

Magd E. Zohdi
Department of Industrial and Manufacturing Engineering
Louisiana State University
Baton Rouge, Louisiana

1 INTRODUCTION 328

2 DEFINITIONS AND CLASSIFICATONS 329
 2.1 Automation 329
 2.2 Production Operations 330
 2.3 Production Plants 331
 2.4 Models for Production Operations 332

3 NUMERICAL-CONTROL MANUFACTURING SYSTEMS 334
 3.1 Numerical Control 334
 3.2 The Coordinate System 334
 3.3 Selection of Parts for NC Machining 335
 3.4 CAD/CAM Part Programming 336
 3.5 Programming by Scanning and Digitizing 336
 3.6 Adaptive Control 336
 3.7 Machinability Data Prediction 337

4 INDUSTRIAL ROBOTS 337
 4.1 Definition 337
 4.2 Robot Configurations 338
 4.3 Robot Control and Programming 340
 4.4 Robot Applications 340

5 COMPUTERS IN MANUFACTURING 340
 5.1 Hierarchical Computer Control 341
 5.2 CNC and DNC Systems 342
 5.3 The Manufacturing Cell 342
 5.4 Flexible Manufacturing Systems 343

6 GROUP TECHNOLOGY 343
 6.1 Part Family Formation 343
 6.2 Parts Classification and Coding 344
 6.3 Production Flow Analysis 345
 6.4 Types of Machine Cell Designs 347
 6.5 Computer-Aided Process Planning 347

BIBLIOGRAPHY 348

1 INTRODUCTION

Modern manufacturing systems are advanced automation systems that use computers as an integral part of their control. Computers are a vital part of automated manufacturing. They control stand- alone manufacturing systems, such as various machine tools, welders, laser-beam cutters, robots, and automatic assembly machines. They control production lines and are beginning to take over control of the entire factory. The computer-integrated-manufacturing system (CIMS) is a reality in the modern industrial society. As illustrated in Fig. 1, CIMS combines computer-aided design (CAD), computer-aided manufacturing (CAM), computer-aided inspection (CAI), and computer-aided production planning (CAPP), along with automated material handling. This chapter focuses on computer-aided manufac-

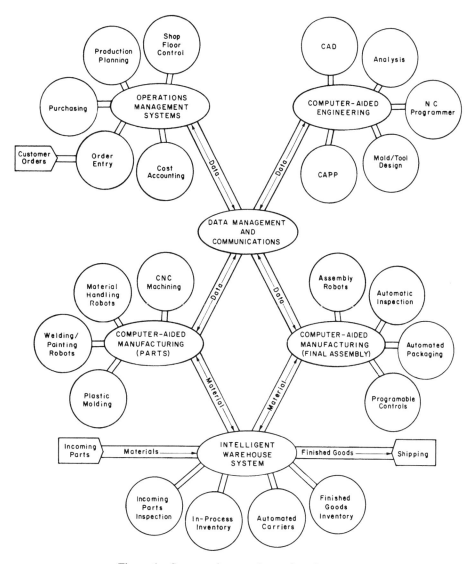

Figure 1 Computer-integrated manufacturing system.

turing for both parts fabrication and assembly, as shown in Fig. 1. It treats numerical-control (NC) machining, robotics, and group technology. It shows how to integrate these functions with automated material storage and handling to form a CIM system.

2 DEFINITIONS AND CLASSIFICATIONS

2.1 Automation

Automation is a relatively new word, having been coined in the 1930s as a substitute for the word *automatization,* which referred to the introduction of automatic controls in manufacturing. Automation implies the performance of a task without human assistance. Manufacturing processes are classified as manual, semiautomatic, or automatic, depending on the extent of human involvement in the ongoing operation of the process.

The primary reasons for automating a manufacturing process are to

1. Reduce the cost of the manufactured product, through savings in both material and labor
2. Improve the quality of the manufactured product by eliminating errors and reducing the variability in product quality
3. Increase the rate of production
4. Reduce the lead time for the manufactured product, thus providing better service for customers
5. Make the workplace safer

The economic reality of the marketplace has provided the incentive for industry to automate its manufacturing processes. In Japan and in Europe, the shortage of skilled labor sparked the drive toward automation. In the United States, stern competition from Japanese and European manufacturers, in terms of both product cost and product quality, has necessitated automation. Whatever the reasons, a strong movement toward automated manufacturing processes is being witnessed throughout the industrial nations of the world.

2.2 Production Operations

Production is a transformation process in which raw materials are converted into the goods demanded in the marketplace. Labor, machines, tools, and energy are applied to materials at each of a sequence of steps that bring the materials closer to a marketable final state. These individual steps are called *production operations*.

There are three basic types of industries involved in transforming raw materials into marketable products:

1. *Basic producers*. These transform natural resources into raw materials for use in manufacturing industry—for example, iron ore to steel ingot in a steel mill.
2. *Converters*. These take the output of basic producers and transform the raw materials into various industrial products—for example, steel ingot is converted into sheet metal.
3. *Fabricators*. These fabricate and assemble final products—for example, sheet metal is fabricated into body panels and assembled with other components into an automobile.

The concept of a computer-integrated-manufacturing system as depicted in Fig. 1 applies specifically to a "fabricator" type of industry. It is the "fabricator" industry that we focus on in this chapter.

The steps involved in creating a product are known as the "manufacturing cycle." In general, the following functions will be performed within a firm engaged in manufacturing a product:

1. *Sales and marketing*. The order to produce an item stems either from customer orders or from production orders based on product demand forecasts.
2. *Product design and engineering*. For proprietary products, the manufacturer is responsible for development and design, including component drawings, specifications, and bill of materials.

2 Definitions and Classifications 331

3. *Manufacturing engineering.* Ensuring manufacturability of product designs, process planning, design of tools, jigs, and fixtures, and "troubleshooting" the manufacturing process.
4. *Industrial engineering.* Determining work methods and time standards for each production operation.
5. *Production planning and control.* Determining the master production schedule, engaging in material requirements planning, operations scheduling, dispatching job orders, and expediting work schedules.
6. *Manufacturing.* Performing the operations that transform raw materials into finished goods.
7. *Material handling.* Transporting raw materials, in-process components, and finished goods between operations.
8. *Quality control.* Ensuring the quality of raw materials, in-process components, and finished goods.
9. *Shipping and receiving.* Sending shipments of finished goods to customers, or accepting shipments of raw materials, parts, and components from suppliers.
10. *Inventory control.* Maintaining supplies of raw materials, in-process items, and finished goods so as to provide timely availability of these items when needed.

Thus, the task of organizing and coordinating the activities of a company engaged in the manufacturing enterprise is complex. The field of industrial engineering is devoted to such activities.

2.3 Production Plants

There are several ways to classify production facilities. One way is to refer to the volume or rate of production. Another is to refer to the type of plant layout. Actually, these two classification schemes are related, as will be pointed out.

In terms of the volume of production, there are three types of manufacturing plants:

1. *Job shop production.* Commonly used to meet specific customer orders; great variety of work; production equipment must be flexible and general purpose; high skill level among workforce—for example, aircraft manufacturing.
2. *Batch production.* Manufacture of product in medium lot sizes; lots produced only once at regular intervals; general-purpose equipment, with some specialty tooling—for example, household appliances, lawn mowers.
3. *Mass production.* Continuous specialized manufacture of identical products; high production rates; dedicated equipment; lower labor skills than in a job shop or batch manufacturing—for example, automotive engine blocks.

In terms of the arrangement of production resources, there are three types of plant layouts. These include

1. *Fixed-position layout.* The item is placed in a specific location and labor and equipment are brought to the site. Job shops often employ this type of plant layout.
2. *Process layout.* Production machines are arranged in groups according to the general type of manufacturing process; forklifts and hand trucks are used to move materials from one work center to the next. Batch production is most often performed in process layouts.

332   Computer-Integrated Manufacturing

    3. *Product-flow layout.* Machines are arranged along a line or in a *U* or *S* configuration, with conveyors transporting work parts from one station to the next; the product is progressively fabricated as it flows through the succession of workstations. Mass production is usually conducted in a product-flow layout.

2.4   Models for Production Operations

In this section, we examine three types of models by which we can examine production operations, including graphical models, manufacturing process models, and mathematical models of production activity. Process-flow charts depict the sequence of operations, storages, transportations, inspections, and delays encountered by a workpart of assembly during processing. As illustrated in Fig. 2, a process-flow chart gives no representation of the layout or physical dimensions of a process, but focuses on the succession of steps seen by the product. It is useful in analyzing the efficiency of the process, in terms of the proportion of time spent in transformation operations as opposed to transportations, storages, and delays.

    The manufacturing-process model gives a graphical depiction of the relationship among the several entities that comprise the process. It is an input–output model. Its inputs are raw materials, equipment (machine tools), tooling and fixtures, energy, and labor. Its outputs are completed workpieces, scrap, and waste. These are shown in Fig. 3. Also shown in this figure are the controls that are applied to the process to optimize the utilization of the inputs in producing completed workpieces, or in maximizing the production of completed workpieces at a given set of values describing the inputs.

    Mathematical models of production activity quantify the elements incorporated into the process-flow chart. We distinguish between operation elements, which are involved whenever the work part is on the machine and correspond to the circles in the process-flow chart, and nonoperation elements, which include storages, transportations, delays, and inspections. Letting T_o represent operation time per machine, T_{no} the nonoperation time associated with each operation, and n_m the number of machines or operations through which each part must be processed, then the total time required to process the part through the plant (called the manufacturing lead time, T_l) is

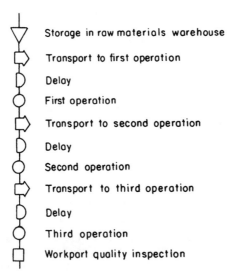

Figure 2   Flow process chart for a sample workpart.

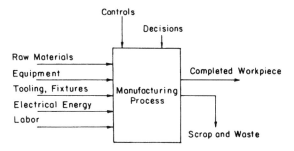

Figure 3 General input–output model of the manufacturing process.

$$T_l = n_m(T_o + T_{no})$$

If there is a batch of p parts,

$$T_l = n_m(pT_o + T_{no})$$

If a setup of duration T_{su} is required for each batch,

$$T_l = n_m(T_{su} + pT_o + T_{no})$$

The total batch time per machine, T_b, is given by

$$T_b = T_{su} + pT_o$$

The average production time T_a per part is therefore

$$T_a = \frac{T_{su} + pT_o}{p}$$

The average production rate for each machine is

$$R_a = 1/T_a$$

As an example, a part requires six operations (machines) through the machine shop. The part is produced in batches of 100. A setup of 2.5 hr is needed. Average operation time per machine is 4.0 min. Average nonoperation time is 3.0 hr. Thus,

$$n_m = 6 \text{ machines}$$
$$p = 100 \text{ parts}$$
$$T_{su} = 2.5 \text{ hr}$$
$$T_o = 4/60 \text{ hr}$$
$$T_{no} = 3.0 \text{ hr}$$

Therefore, the total manufacturing lead time for this batch of parts is

$$T_l = 6[2.5 + 100(0.06667) + 3.0] = 73.0 \text{ hr}$$

If the shop operates on a 40-hr week, almost two weeks are needed to complete the order.

3 NUMERICAL-CONTROL MANUFACTURING SYSTEMS

3.1 Numerical Control

The most commonly accepted definition of numerical control (NC) is that given by the Electronic Industries Association (EIA): a system in which motions are controlled by the direct insertion of numerical data at some point. The system must automatically interpret at least some portion of these data.

The numerical control system consists of five basic, interrelated components:

1. Data input devices
2. Machine control unit
3. Machine tool or other controlled equipment
4. Servo-drives for each axis of motion
5. Feedback devices for each axis of motion

The major components of a typical NC machine tool system are shown in Fig. 4.

The programmed codes that the machine control unit (MCU) can read may be perforated tape or punched tape, magnetic tape, tabulating cards, or signals directly from computer logic or some computer peripherals, such as disk or drum storage. Direct computer control (DCC) is the most recent development, and one that affords the help of a computer in developing a part program.

3.2 The Coordinate System

The Cartesian coordinate system is the basic system in NC control. The three primary linear motions for an NC machine are given as X, Y, and Z. Letters A, B, and C indicate the three rotational axes, as in Fig. 5.

NC machine tools are commonly classified as being either point-to-point or continuous path. The simplest form of NC is the point-to-point machine tool used for operations such as drilling, tapping, boring, punching, spot welding, or other operations that can be completed at a fixed coordinate position with respect to the workpiece. The tool does not contact the workpiece until the desired coordinate position has been reached; consequently, the exact path by which this position is reached is not important.

With continuous-path (contouring) NC systems, there is contact between the workpiece and the tool as the relative movements are made. Continuous-path NC systems are used

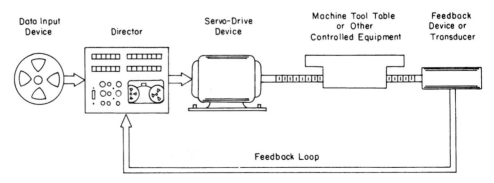

Figure 4 Simplified numerical control system.

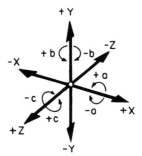

Figure 5 An example of typical axis nomenclature for machine tools.

primarily for milling and turning operations that can profile and sculpture workpieces. Other NC continuous-path operations include flame cutting, sawing, grinding, and welding, and even operations such as the application of adhesives. We should note that continuous-path systems can be programmed to perform point-to-point operations, although the reverse (while technically possible) is infrequently done.

3.3 Selection of Parts for NC Machining

Parts selection for NC should be based on an economic evaluation, including scheduling and machine availability. Economic considerations affecting NC part selection including alternative methods, tooling, machine loadings, manual versus computer-assisted part programming, and other applicable factors.

Thus, NC should be used only where it is more economical or does the work better or faster, or where it is more accurate than other methods. The selection of parts to be assigned to NC has a significant effect on its payoff. The following guidelines, which may be used for parts selection, describe those parts for which NC may be applicable.

1. Parts that require *substantial tooling costs* in relation to the total manufacturing costs by conventional methods
2. Parts that require *long setup times* compared to the machine run time in conventional machining
3. Parts that are machined in *small or variable lots*
4. A *wide diversity of parts* requiring frequent changes of machine setup and a large tooling inventory if conventionally machined
5. Parts that are *produced at intermittent times* because demand for them is cyclic
6. Parts that have *complex configurations* requiring close tolerances and intricate relationships
7. Parts that have *mathematically defined complex contours*
8. Parts that require *repeatability* from part to part and lot to lot
9. *Very expensive* parts where human error would be very costly and increasingly so as the part nears completion
10. *High-priority* parts where lead time and flow time are serious considerations
11. Parts with *anticipated design changes*
12. Parts that involve a *large number of operations* or *machine setups*
13. Parts where *non-uniform cutting conditions* are required

14. Parts that require 100% *inspection* or require measuring many checkpoints, resulting in high inspection costs
15. *Family of parts*
16. *Mirror-image parts*
17. *New parts* for which conventional tooling does not already exist
18. Parts that are suitable for *maximum machining* on NC machine tools

3.4 CAD/CAM Part Programming

Computer-aided design (CAD) consists of using computer software to produce drawings of parts or products. These drawings provide the dimensions and specifications needed by the machinist to produce the part or product. Some well-known CAD software products include *AutoCAD, Cadkey,* and *Mastercam.*

Computer-aided manufacturing (CAM) involves the use of software by NC programmers to create programs to be read by a CNC machine in order to manufacture a desired shape or surface. The end product of this effort is an NC program stored on disk, usually in the form of G codes, that when loaded into a CNC machine and executed will move a cutting tool along the programmed path to create the desired shape. If the CAM software has the means of creating geometry, as opposed to importing the geometry from a CAD system, it is called *CAD/CAM.* CAD/CAM software, such as Mastercam, is capable of producing instructions for a variety of machines, including lathes, mills, drilling and tapping machines, and wire electrostatic discharge machining (EDM) processes.

3.5 Programming by Scanning and Digitizing

Programming may be done directly from a drawing, model, pattern, or template by digitizing or scanning. An optical reticle or other suitable viewing device connected to an arm is placed over the drawing. Transducers will identify the location and translate it either to a tape puncher or other suitable programming equipment. Digitizing is used in operations such as sheet-metal punching and hole drilling. A scanner enables an operator to program complex free-form shapes by manually moving a tracer over the contour of a model or premachined part. Data obtained through the tracer movements are converted into tape by a minicomputer. Digitizing and scanning units have the capability of editing, modifying, or revising the basic data gathered.

3.6 Adaptive Control

Optimization processes have been developed to improve the operational characteristics of NC machine-tool systems. Two distinct methods of optimization are adaptive control and machinability data prediction. Although both techniques have been developed for metal-cutting operations, adaptive control finds application in other technological fields.

The adaptive control (AC) system is an evolutionary outgrowth of numerical control. AC optimizes an NC process by sensing and logically evaluating variables that are not controlled by position and velocity feedback loops. Essentially, an adaptive control system monitors process variables, such as cutting forces, tool temperatures, or motor torque, and alters the NC commands so that optimal metal removal or safety conditions are maintained.

A typical NC configuration (Fig. 6*a*) monitors position and velocity output of the servo system, using feedback data to compensate for errors between command response. The AC

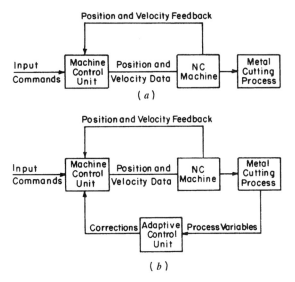

Figure 6 Schematic diagrams for conventional and adaptive NC systems.

feedback loop (Fig. 6b) provides sensory information on other process variables, such as workpiece–tool air gaps, material property variations, wear, cutting depth variations, or tool deflection. This information is determined by techniques such as monitoring forces on the cutting tool, motor torque variations, or tool–workpiece temperatures. The data are processed by an adaptive controller that converts the process information into feedback data to be incorporated into the machine control unit output.

3.7 Machinability Data Prediction

The specification of suitable feeds and speeds is essentially in conventional and NC cutting operations. Machinability data are used to aid in the selection of metal-cutting parameters based on the machining operation, the tool and workpiece material, and one or more production criteria. Techniques used to select machinability data for conventional machines have two important drawbacks in relation to NC applications: data are generally presented in a tabular form that requires manual interpolation, checkout, and subsequent revisions; and tests on the machine tool are required to find optimum conditions.

Specialized machinability data systems have been developed for NC application to reduce the need for machinability data testing and to decrease expensive NC machining time. Part programming time is also reduced when machinability information is readily available.

A typical process schematic showing the relationship between machinability data and NC process flow is illustrated in Fig. 7.

4 INDUSTRIAL ROBOTS

4.1 Definition

As defined by the Robot Institute of America, "a robot is a reprogrammable, multifunctional manipulator designed to handle material, parts, tools or specialized devices through variable

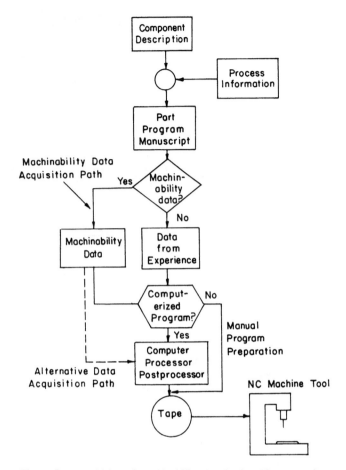

Figure 7 Acquisition of machinability data in the NC process flow.

programmed motions for the performance of a variety of tasks." Robots have the following components:

1. *Manipulator.* The mechanical unit or "arm" that performs the actual work of the robot, consisting of mechanical linkages and joints with actuators to drive the mechanism directly through gears, chains, or ball screws.
2. *Feedback devices.* Transducers that sense the positions of various linkages or joints and transmit this information to the controller.
3. *Controller.* Computer used to initiate and terminate motion, store data for position and sequence, and interface with the system in which the robot operates.
4. *Power supply.* Electric, pneumatic, and hydraulic power systems used to provide and regulate the energy needed for the manipulator's actuators.

4.2 Robot Configurations

Industrial robots have one of three mechanical configurations, as illustrated in Fig. 8. Cylindrical coordinate robots have a work envelope that is composed of a portion of a cylinder.

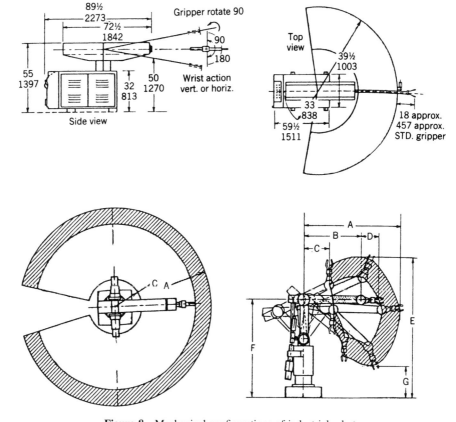

Figure 8 Mechanical configurations of industrial robots.

Spherical coordinate robots have a work envelope that is a portion of a sphere. Jointed-arm robots have a work envelope that approximates a portion of a sphere. There are six motions or degrees of freedom in the design of a robot—three arm and body motions and three wrist movements.

Arm and body motions:

1. *Vertical traverse*—an up-and-down motion of the arm
2. *Radial traverse*—an in-and-out motion of the arm
3. *Rotational traverse*—rotation about the vertical axis (right or left swivel of the robot body)

Wrist motions:

4. *Wrist swivel*—rotation of the wrist
5. *Wrist bend*—up-and-down movement of the wrist
6. *Wrist yaw*—right or left swivel of the wrist

The mechanical hand movement, usually opening and closing, is not considered one of the basic degrees of freedom of the robot.

4.3 Robot Control and Programming

Robots can also be classified according to type of control. Point-to-point robot systems are controlled from one programmed point in the robot's control to the next point. These robots are characterized by high load capacity, large working range, and relative ease of programming. They are suitable for pick-and-place, material handling, and machine loading tasks.

Contouring robots, on the other hand, possess the capacity to follow a closely spaced locus of points that describe a smooth, continuous path. The control of the path requires a large memory to store the locus of points. Continuous-path robots are therefore more expensive than point-to-point robots, but they can be used in such applications as seam welding, flame cutting, and adhesive beading.

There are three principal systems for programming robots:

1. *Manual method.* Used in older, simpler robots, the program is set up by fixing stops, setting switches, and so on.
2. *Walk-through.* The programmer "teaches" the robot by actually moving the hand through a sequence of motions or positions, which are recorded in the memory of the computer.
3. *Lead-through.* The programmer drives the robot through a sequence of motions or positions using a console or teach pendant. Each move is recorded in the robot's memory.

4.4 Robot Applications

A current directory of robot applications in manufacturing includes the following:

1. Material handling
2. Machine loading and unloading
3. Die casting
4. Investment casting
5. Forging and heat treating
6. Plastic molding
7. Spray painting and electroplating
8. Welding (spot welding and seam welding)
9. Inspection
10. Assembly

Research and development efforts are under way to provide robots with sensory perception, including voice programming, vision and "feel." These capabilities will no doubt greatly expand the inventory of robot applications in manufacturing.

5 COMPUTERS IN MANUFACTURING

Flexible manufacturing systems combined with automatic assembly and product inspection, on the one hand, and integrated CAD/CAM systems, on the other hand, are the basic components of the computer-integrated manufacturing system. The overall control of such systems is predicated on hierarchical computer control, such as illustrated in Fig. 9.

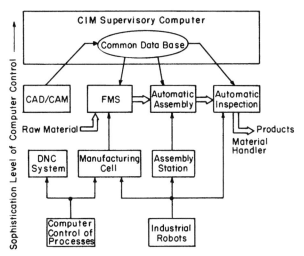

Figure 9 Hierarchical computer control in manufacturing.

5.1 Hierarchical Computer Control

The lowest level of the hierarchical computer control structure illustrated in Fig. 9 contains stand-alone computer control systems of manufacturing processes and industrial robots. The computer control of processes includes all types of CNC machine tools, welders, electrochemical machining (ECM), electrical discharge machining (EDM), and laser-cutting machines.

When a set of NC or CNC machine tools is placed under the direct control of a single computer, the resulting system is known as a *direct-numerical-control* (DNC) system. DNC systems can produce several different categories of parts or products, perhaps unrelated to one another. When several CNC machines and one or more robots are organized into a system for the production of a single part or family of parts, the resulting system is called a *manufacturing cell*. The distinction between DNC systems and a manufacturing cell is that in DNC systems the same computer receives data from and issues instructions to several separate machines, whereas in manufacturing cells the computer coordinates the movements of several machines and robots working in concert. The computer receives "completion of job" signals from the machines and issues instructions to the robot to unload the machines and change their tools. The software includes strategies for handling machine breakdowns, tool wear, and other special situations.

The operation of several manufacturing cells can be coordinated by a central computer in conjunction with an automated material-handling system. This is the next level of control in the hierarchical structure and is known as a *flexible manufacturing system* (FMS). The FMS receives incoming workpieces and processes them into finished parts, completely under computer control.

The parts fabricated in the FMS are then routed on a transfer system to automatic assembly stations, where they are assembled into subassemblies or final product. These assembly stations can also incorporate robots for performing assembly operations. The subassemblies and final product may also be tested at automatic inspection stations.

As shown in Fig. 9, FMS, automatic assembly, and automatic inspection are integrated with CAD/CAM systems to minimize production lead time. These four functions are coordinated by means of the highest level of control in the hierarchical structure-computer-integrated-manufacturing (CIM) systems. The level of control is often called *supervisory computer control*.

The increase in productivity associated with CIM systems will not come from a speedup of machining operations, but rather from minimizing the direct labor employed in the plant. Substantial savings will also be realized from reduced inventories, with reductions in the range of 80–90%.

5.2 CNC and DNC Systems

The distinguishing feature of a CNC system is a dedicated computer, usually a microcomputer, associated with a single machine tool, such as a milling machine or a lathe. Programming the machine tools is managed through punched or magnetic tape, or directly from a keyboard. DNC is another step beyond CNC, in that a number of CNC machines, ranging from a few to as many as 100, are connected directly to a remote computer. NC programs are downloaded directly to the CNC machine, which then processes a prescribed number of parts.

5.3 The Manufacturing Cell

The concept of a manufacturing cell is based on the notion of cellular manufacturing, wherein a group of machines served by one or more robots manufactures one part or one part family. Figure 10 depicts a typical manufacturing cell consisting of a CNC lathe, a CNC milling machine, a CNC drill, open conveyor to bring workparts into the cell, another to remove completed parts from the cell, and a robot to serve all these components. Each manufacturing cell is self-contained and self-regulating. The cell is usually made up of 10 or fewer machines. Those cells that are not completely automated are usually staffed with fewer personnel than machines, with each operator trained to handle several machines or processes.

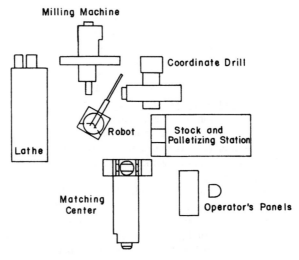

Figure 10 A typical manufacturing cell.

5.4 Flexible Manufacturing Systems

Flexible manufacturing systems (FMS) combine many different automation technologies into a single production system. These include NC and CNC machine tools, automatic material handling between machines, computer control over the operation of the material handling system and machine tools, and group technology principles. Unlike the manufacturing cell, which is typically dedicated to the production of a single parts family, the FMS is capable of processing a variety of part types simultaneously under NC control at the various workstations.

Human labor is used to perform the following functions to support the operation of the FMS:

- Load raw workparts into the system
- Unload finished workparts from the system
- Change tools and tool settings
- Equipment maintenance and repair

Robots can be used to replace human labor in certain areas of these functions, particularly those involving material or tool handling. Figure 11 illustrates a sample FMS layout.

6 GROUP TECHNOLOGY

Group technology is a manufacturing philosophy in which similar parts are identified and grouped together to take advantage of similarities in design and/or manufacture. Similar parts are grouped into part families. For example, a factory that produces as many as 10,000 different part numbers can group most of these parts into as few as 50 distinct part families. Since the processing of each family would be similar, the production of part families in dedicated manufacturing cells facilitates workflow. Thus, group technology results in efficiencies in both product design and process design.

6.1 Part Family Formation

The key to gaining efficiency in group-technology-based manufacturing is the formation of part families. A part family is a collection of parts that are similar either due to geometric features such as size and shape or because similar processing steps are required in their

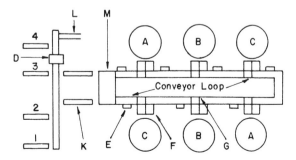

Figure 11 A flexible manufacturing system.

manufacture. Parts within a family are different, but are sufficiently similar in their design attributes (geometric size and shape) and/or manufacturing attributes (the sequence of processing steps required to make the part) to justify their identification as members of the same part family.

The biggest problem in initiating a group-technology-based manufacturing system is that of grouping parts into families. Three methods for accomplishing this grouping are

1. *Visual inspection.* This method involves looking at the part, a photograph, or a drawing and placing the part in a group with similar parts. It is generally regarded as the most time-consuming and least accurate of the available methods.
2. *Parts classification and coding.* This method involves examining the individual design and/or manufacturing attributes of each part, assigning a code number to the part on the basis of these attributes, and grouping similar code numbers into families. This is the most commonly used procedure for forming part families.
3. *Production flow analysis.* This method makes use of the information contained on the routing sheets describing the sequence of processing steps involved in producing the part, rather than part drawings. Workparts with similar or identical processing sequences are grouped into a part family.

6.2 Parts Classification and Coding

As previously stated, parts classification and coding is the most frequently applied method for forming part families. Such a system is useful in both design and manufacture. In particular, parts coding and classification, and the resulting coding system, provide a basis for interfacing CAD and CAM in CIM systems. Parts classification systems fall into one of three categories:

1. Systems based on part design attributes:
   Basic external shape
   Basic internal shape
   Length/diameter ratio
   Material type
   Part function
   Major dimensions
   Minor dimensions
   Tolerances
   Surface finish
2. Systems based on part manufacturing attributes:
   Primary process
   Minor processes
   Major dimensions
   Length/diameter ratio
   Surface finish
   Machine tool
   Operation sequence
   Production time
   Batch size
   Annual production requirement
   Fixtures needed
   Cutting tools
3. Systems based on a combination of design and manufacturing attributes.

The part code consists of a sequence of numerical digits that identify the part's design and manufacturing attributes. There are two basic structures for organizing this sequence of digits:

1. Hierarchical structures in which the interpretation of each succeeding digit depends on the value of the immediately preceding digit
2. Chain structures in which the interpretation of each digit in the sequence is position-wise fixed

The Opitz system is perhaps the best known coding system used in parts classification and coding. The code structure is

12345 6789 ABCD

The first nine digits constitute the basic code that conveys both design and manufacturing data. The first five digits, 12345, are called the *form code* and give the primary design attributes of the part. The next four digits, 6789, constitute the *supplementary code* and indicate some of the manufacturing attributes of the part. The next four digits, ABCD, are called the *secondary code* and are used to indicate the production operations of type and sequence. Figure 12 gives the basic structure for the Opitz coding system. Note that digit 1 establishes two primary categories of parts, rotational and non-rotational, among nine separate part classes.

The MICLASS (Metal Institute Classification System) was developed by the Netherlands Organization for Applied Scientific Research to help automate and standardize a number of design, manufacturing, and management functions. MICLASS codes range from 12 to 30 digits, with the first 12 constituting a universal code that can be applied to any part. The remaining 18 digits can be made specific to any company or industry. The organization of the first 12 digits is as follows:

1st digit	main shape
2nd and 3rd digits	shape elements
4th digit	position of shape elements
5th and 6th digits	main dimensions
7th digit	dimension ratio
8th digit	auxiliary dimension
9th and 10th digits	tolerance codes
11th and 12th digits	material codes

MICLASS allows computer-interactive parts coding, in which the user responds to a series of questions asked by the computer. The number of questions asked depends on the complexity of the part and ranges from as few as 7 to more than 30, with an average of about 15.

6.3 Production Flow Analysis

Production flow analysis (PFA) is a method for identifying part families and associated grouping of machine tools. PFA is used to analyze the operations sequence of machine routing for the parts produced in a shop. It groups parts that have similar sequences and routings into a part family. PFA then establishes machine cells for the producing part families. The PFA procedure consists of the following steps:

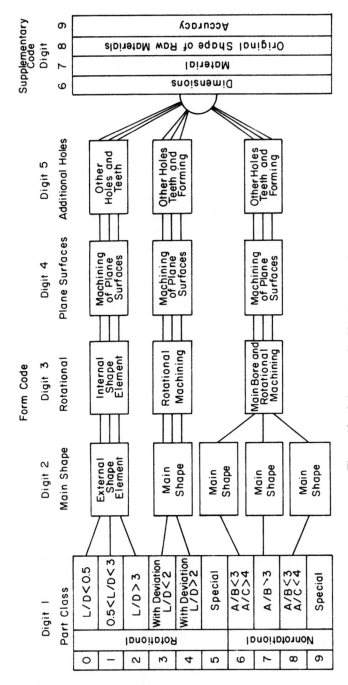

Figure 12 Opitz parts classification and coding system.

1. Data collection, gathering part numbers and machine routings for each part produced in the shop
2. Sorting process routings into "packs" according to similarity
3. Constructing a PFA chart, such as depicted in Fig. 13, that shows the process sequence (in terms of machine code numbers) for each pack (denoted by a letter)
4. Analysis of the PFA chart in an attempt to identify similar packs. This is done by rearranging the data on the original PFA chart into a new pattern that groups packs having similar sequences. Figure 14 shows the rearranged PFA chart. The machines grouped together within the blocks in this figure form logical machine cells for producing the resulting part family.

6.4 Types of Machine Cell Designs

The organization of machines into cells, whether based on parts classification and coding or PFA, follows one of three general patterns:

1. Single-machine cell
2. Group-machine layout
3. Flow-line cell layout

The single-machine pattern can be used for workparts whose attributes allow them to be produced using a single process. For example, a family composed of 40 different machine bolts can be produced on a single turret lathe.

The group-machine layout was illustrated in Fig. 13. The cell contains the necessary grouping of machine tools and fixtures for processing all parts in a given family, but material handling between machines is not fixed. The flow-line cell design likewise contains all machine tools and fixtures needed to produce a family of parts, but these are arranged in a fixed sequence with conveyors providing the flow of parts through the cell.

6.5 Computer-Aided Process Planning

Computer-aided process planning (CAPP) involves the use of a computer to automatically generate the operation sequence (routing sheet) based on information about the workpart. CAPP systems require some form of classification and coding system, together with standard process plans for specific part families. The flow of information in a CAPP system is initiated

Machine \ Part No.	1	2	3	4	5	6	7	8	9	10	11	12	13	14	15	16	17	18	19	20
Lathe	x	x		x	x		x	x	x		x	x		x	x		x	x	x	x
Milling Mach. I	x	x	x		x	x	x		x		x		x	x		x				x
Milling Mach. II			x	x				x		x		x	x		x		x	x	x	
Drilling Mach.	x	x	x	x		x	x	x		x	x	x	x		x	x	x			x
Grinding Mach.	x	x	x	x		x			x			x	x		x			x		x

Figure 13 PFA chart.

Part No. / Machine	1	2	20	7	11	14	9	5	4	18	12	8	17	15	19	3	13	6	16	10
Lathe	x	x	x	x	x	x	x	x												
Milling Mach. I	x	x	x	x	x	x	x	x												
Drilling Mach.	x	x	x	x	x	x														
Grinding Mach.	x	x	x				x													
Lathe									x	x	x	x	x	x	x					
Milling Mach. II									x	x	x	x	x	x	x					
Drilling Mach.									x	x	x	x	x							
Grinding Mach.									x	x	x			x						
Milling Mach. I																x	x	x	x	
Milling Mach. II																x	x			x
Drilling Mach.																x	x	x	x	x
Grilling Mach.																x	x	x		

Figure 14 Rearranged PFA chart.

by having the user enter the part code for the workpart to be processed. The CAPP program then searches the part family matrix file to determine if a match exists. If so, the standard machine routing and the standard operation sequence are extracted from the computer file. If no such match exists, the user must then search the file for similar code numbers and manually prepare machine routings and operation sequences for dissimilar segments. Once this process has been completed, the new information becomes part of the master file so that the CAPP system generates an ever-growing data file.

BIBLIOGRAPHY

Groover, M. P., *Automation, Production Systems and Computer-Integrated Manufacturing,* Prentice-Hall, Upper Saddler River, NJ, 2001.

Groover, M. P., M. Weiss, R. N. Nagel, and N. G. Odrey, *Industrial Robotics: Technology, Programming, and Applications,* McGraw-Hill, New York, 1986.

Boothroyd, G., P. Dewhurst, and W. Knight, *Product Design for Manufacture and Assembly,* Marcel Dekker, New York, 1994.

Buzacott, J. A., and D. D. Yao, "Flexible Manufacturing Systems: A Review of Analytical Models," *Management Science,* **32,** 890–895 (1986).

Chow, W. M., *Assembly Line Design,* Marcel Dekker, New York, 1990.

Moodie, C., R. Uzsoy, and Y. Yih, *Manufacturing Cells: A Systems Engineering View,* Taylor & Francis, London, 1995.

Opitz, H., and H. P. Wiendohl, "Group Technology and Manufacturing Systems for Medium Quantity Production," *International Journal of Production Research,* **9,** 181–203 (1971).

CHAPTER 10
MATERIAL HANDLING

William E. Biles
John S. Usher
Department of Industrial Engineering
University of Louisville
Louisville, Kentucky

Magd E. Zohdi
Department of Industrial and Manufacturing Engineering
Louisiana State University
Baton Rouge, Louisiana

1	INTRODUCTION	349
2	BULK MATERIAL HANDLING	350
	2.1 Conveying of Bulk Solids	350
	2.2 Screw Conveyors	351
	2.3 Belt Conveyors	351
	2.4 Bucket Elevators	352
	2.5 Vibrating or Oscillating Conveyors	355
	2.6 Continuous-Flow Conveyors	356
	2.7 Pneumatic Conveyors	356
3	BULK MATERIALS STORAGE	360
	3.1 Storage Piles	360
	3.2 Storage Bins, Silos, and Hoppers	360
	3.3 Flow-Assisting Devices and Feeders	361
	3.4 Packaging of Bulk Materials	361
	3.5 Transportation of Bulk Materials	363
4	UNIT MATERIAL HANDLING	365
	4.1 Introduction	365
	4.2 Analysis of Systems for Material Handling	365
	4.3 Identifying and Defining the Problem	366
	4.4 Collecting Data	366
	4.5 Unitizing Loads	372
5	MATERIAL-HANDLING EQUIPMENT CONSIDERATIONS AND EXAMPLES	375
	5.1 Developing the Plan	375
	5.2 Conveyors	376
	5.3 Hoists, Cranes, and Monorails	382
	5.4 Industrial Trucks	385
	5.5 Automated Guided Vehicle Systems	387
	5.6 Automated Storage and Retrieval Systems	390
	5.7 Carousel Systems	391
	5.8 Shelving, Bin, Drawer, and Rack Storage	393
6	IMPLEMENTING THE SOLUTION	394

1 INTRODUCTION

Material handling is defined by the Materials Handling Institute (MHI), www.mhia.org, as the movement, storage, control, and protection of materials and products throughout the process of their manufacture, distribution, consumption, and disposal. The five commonly recognized aspects of material handling are

1. *Motion.* Parts, materials, and finished products that must be moved from one location to another should be moved in an efficient manner and at minimum cost.
2. *Time.* Materials must be where they are needed at the moment they are needed.

3. *Place.* Materials must be in the proper location and positioned for use.
4. *Quantity.* The rate of demand varies between the steps of processing operations. Materials must be continually delivered to, or removed from, operations in the correct weights, volumes, or numbers of items required.
5. *Space.* Storage space, and its efficient utilization, is a key factor in the overall cost of an operation or process.

The science and engineering of material handling is generally classified into two categories, depending on the form of the material handled. *Bulk solids handling* involves the movement and storage of solids that are flowable, such as fine, free-flowing materials (e.g., wheat flour or sand), pelletized materials (e.g., soybeans or soap flakes), or lumpy materials (e.g., coal or wood bark). *Unit handling* refers to the movement and storage of items that have been formed into unit loads. A *unit load* is a single item, a number of items, or bulk material that is arranged or restrained so that the load can be stored, picked up, and moved between two locations as a single mass. The handling of liquids and gases is usually considered to be in the domain of fluid mechanics, whereas the movement and storage of containers of liquid or gaseous material properly comes within the domain of unit material handling.

2 BULK MATERIAL HANDLING

The handling of bulk solids involves four main areas: (1) conveying, (2) storage, (3) packaging, and (4) transportation.

2.1 Conveying of Bulk Solids

The selection of the proper equipment for conveying bulk solids depends on a number of interrelated factors. First, alternative types of conveyors must be evaluated and the correct model and size must be chosen. Because standardized equipment designs and complete engineering data are available for many types of conveyors, their performance can be accurately predicted when they are used with materials having well-known conveying characteristics. Some of the primary factors involved in conveyor equipment selection are as follows:

1. *Capacity requirement.* The rate at which material must be transported (e.g., tons per hour). For instance, belt conveyors can be manufactured in relatively large sizes, operate at high speeds, and deliver large weights and volumes of material economically. On the other hand, screw conveyors can become very cumbersome in large sizes, and cannot be operated at high speeds without severe abrasion problems.
2. *Length of travel.* The distance material must be moved from origin to destination. For instance, belt conveyors can span miles, whereas pneumatic and vibrating conveyors are limited to hundreds of feet.
3. *Lift.* The vertical distance material must be transported. Vertical bucket elevators are commonly applied in those cases in which the angle of inclination exceeds 30°.
4. *Material characteristics.* The chemical and physical properties of the bulk solids to be transported, particularly flowability.
5. *Processing requirements.* The treatment material incurs during transport, such as heating, mixing, and drying.

6. *Life expectancy.* The period of performance before equipment must be replaced; typically, the economic life of the equipment.
7. *Comparative costs.* The installed first cost and annual operating costs of competing conveyor systems must be evaluated in order to select the most cost-effective configuration.

Table 1 lists various types of conveyor equipment for certain common industrial functions. Table 2 provides information on the various types of conveyor equipment used with materials having certain characteristics.

The choice of the conveyor itself is not the only task involved in selecting a conveyor system. Conveyor drives, motors, and auxiliary equipment must also be chosen. Conveyor drives comprise from 10 to 30% of the total cost of the conveyor system. Fixed-speed drives and adjustable speed drives are available, depending on whether changes in conveyor speed are needed during the course of normal operation. Motors for conveyor drives are generally three-phase, 60-cycle, 220-V units; 220/440-V units; 550-V units; or four-wire, 208-V units. Also available are 240-V and 480-V ratings. Auxiliary equipment includes such items as braking or arresting devices on vertical elevators to prevent reversal of travel, torque-limiting devices or electrical controls to limit power to the drive motor, and cleaners on belt conveyors.

2.2 Screw Conveyors

A screw conveyor consists of a helical shaft mount within a pipe or trough. Power may be transmitted through the helix, or in the case of a fully enclosed pipe conveyor through the pipe itself. Material is forced through the channel formed between the helix and the pipe or trough. Screw conveyors are generally limited to rates of flow of about 10,000 ft^3/hr. Figure 1 shows a chute-fed screw conveyor, one of several types in common use. Table 3 gives capacities and loading conditions for screw conveyors on the basis of material classifications.

2.3 Belt Conveyors

Belt conveyors are widely used in industry. They can traverse distances up to several miles at speeds up to 1000 ft/min and can handle thousands of tons of material per hour. Belt

Table 1 Types of Conveyor Equipment and Their Functions

Function	Conveyor Type
Conveying materials horizontally	Apron, belt, continuous flow, drag flight, screw, vibrating, bucket, pivoted bucket, air
Conveying materials up or down an incline	Apron, belt, continuous flow, flight, screw, skip hoist, air
Elevating materials	Bucket elevator, continuous flow, skip hoist, air
Handling materials over a combination horizontal and vertical path	Continuous flow, gravity-discharge bucket, pivoted bucket, air
Distributing materials to or collecting materials from bins, bunkers, etc.	Belt, flight, screw, continuous flow, gravity-discharge bucket, pivoted bucket, air
Removing materials from railcars, trucks, etc.	Car dumper, grain-car unloader, car shaker, power shovel, air

Table 2 Material Characteristics and Feeder Type

Material Characteristics	Feeder Type
Fine, free-flowing materials	Bar flight, belt, oscillating or vibrating, rotary vane, screw
Nonabrasive and granular materials, materials with some lumps	Apron, bar flight, belt, oscillating or vibrating, reciprocating, rotary plate, screw
Materials difficult to handle because of being hot, abrasive, lumpy, or stringy	Apron, bar flight, belt, oscillating or vibrating, reciprocating
Heavy, lumpy, or abrasive materials similar to pit-run stone and ore	Apron, oscillating or vibrating, reciprocating

conveyors are generally placed horizontally or at slopes ranging from 10 to 20°, with a maximum incline of 30°. Direction changes can occur readily in the vertical plane of the belt path, but horizontal direction changes must be managed through such devices as connecting chutes and slides between different sections of belt conveyor.

Belt-conveyor design depends largely on the nature of the material to be handled. Particle-size distribution and chemical composition of the material dictate selection of the width of the belt and the type of belt. For instance, oily substances generally rule out the use of natural rubber belts. Conveyor-belt capacity requirements are based on peak load rather than average load. Operating conditions that affect belt-conveyor design include climate, surroundings, and period of continuous service. For instance, continuous service operation will require higher-quality components than will intermittent service, which allows more frequent maintenance. Belt width and speed depend on the bulk density of the material and lump size. The horsepower to drive the belt is a function of the following factors:

1. Power to drive an empty belt
2. Power to move the load against the friction of the rotating parts
3. Power to elevate and lower the load
4. Power to overcome inertia in placing material in motion
5. Power to operate a belt-driven tripper

Table 4 provides typical data for estimating belt-conveyor and design requirements. Figure 2 illustrates a typical belt-conveyor loading arrangement.

2.4 Bucket Elevators

Bucket elevators are used for vertical transport of bulk solid materials. They are available in a wide range of capacities and may operate in the open or totally enclosed. They tend to

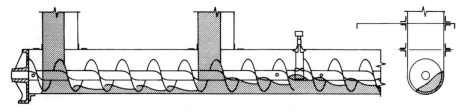

Figure 1 Chute-fed screw conveyor.

Table 3 Capacity and Loading Conditions for Screw Conveyors

Capacity tons/hr	ft³/hr	Diam. of Flights (in.)	Diam. of Pipe (in.)	Diam. of Shafts (in.)	Hanger Centers (ft)	Max. Size Lumps All Lumps	Max. Size Lumps Lumps 20–25%	Max. Size Lumps Lumps 10% or Less	Speed (rpm)	Max. Torque Capacity (in.-lb)	Feed Section Diam. (in.)	Hp at Motor 15 ft Max. Length	30 ft Max. Length	45 ft Max. Length	60 ft Max. Length	75 ft Max. Length	Max. hp Capacity at Speed Listed
5	200	9	2½	2	10	¾	1½	2¼	40	7,600	6	0.43	0.85	1.27	1.69	2.11	4.8
10	400	10	2½	2	10	¾	1½	2½	55	7,600	9	0.85	1.69	2.25	3.00	3.75	6.6
15	600	10	2½	2	10	¾	1½	2½	80	7,600	9	1.27	2.25	3.38	3.94	4.93	9.6
		12	2½	2	12	1	2	3	45	7,600	10	1.27	2.25	3.38	3.94	4.93	5.4
		12	3½	3						16,400		1.27	2.25	3.38	3.94	4.93	11.7
20	800	12	2½	2	12	1	2	3	60	7,600	10	1.69	3.00	3.94	4.87	5.63	7.2
			3½	3						16,400		1.69	3.00	3.94	4.87	5.63	15.6
25	1000	12	2½	2	12	1	2	3	75	7,600	10	2.12	3.75	4.93	5.63	6.55	9.0
			3½	3						16,400		2.12	3.75	4.93	5.63	6.55	19.5
		14	3½	3		1¼	2½	3½	45	16,400	12	2.12	3.75	4.93	5.63	6.55	11.7
30	1200	14	3½	3	12	1¼	2½	3½	55	16,400	12	2.25	3.94	5.05	6.75	7.50	14.3
35	1400	14	3½	3	12	1¼	2½	3½	65	16,400	12	2.62	4.58	5.90	7.00	8.75	16.9
40	1600	16	3½	3	12	1½	3	4	50	16,400	14	3.00	4.50	6.75	8.00	10.00	13.0

Table 4 Data for Estimating Belt Conveyor Design Requirements

Belt Width (in.)	Cross-Sectional Area of Load (ft²)	Belt Speed		Belt Plies		Max. Size Lump (in.)		Belt Speed (ft/min)	50 lb/ft³ Material			100 lb/ft³ Material			Add hp for Tripper
		Normal Operating Speed (ft/min)	Max. Advisable Speed (ft/min)	Min.	Max.	Sized Material 80% Under	Unsized Material Not Over 20%		Capacity (tons/hr)	hp 10-ft Lift	hp 100-ft Centers	Capacity (tons/hr)	hp 10-ft Lift	hp 100-ft Centers	
14	0.11	200	300	3	5	2	3	100	16	0.17	0.22	32	0.34	0.44	1.00
								200	32	0.34	0.44	64	0.68	0.88	
								300	48	0.52	0.66	96	1.04	1.32	
16	0.14	200	300	3	5	2½	4	100	22	0.23	0.28	44	0.46	0.56	1.25
								200	44	0.45	0.56	88	0.90	1.12	
								300	66	0.68	0.84	132	1.36	1.68	
18	0.18	250	350	4	6	3	5	100	27	0.29	0.35	54	0.58	0.7	1.50
								250	67	0.71	0.88	134	1.42	1.76	
								350	95	1.00	1.21	190	2.00	2.42	
20	0.22	250	350	4	6	3½	6	100	33	0.35	0.42	66	0.70	0.84	1.60
								250	82	0.86	1.03	164	1.72	2.06	
								350	115	1.22	1.45	230	2.44	2.9	
24	0.33	300	400	4	7	4½	8	100	49	0.51	0.51	98	1.02	1.02	1.75
								300	147	1.53	1.52	294	3.06	3.04	
								400	196	2.04	2.02	392	4.08	4.04	
30	0.53	300	450	4	8	7	12	100	79	0.80	0.75	158	1.60	1.5	2.50
								300	237	2.40	2.25	474	4.80	4.5	
								450	355	3.60	3.37	710	7.20	6.74	
36	0.78	400	600	4	9	8	15	100	115	1.22	0.80	230	2.44	1.59	3.53
								400	460	4.87	3.18	920	9.74	6.36	
								600	690	7.30	4.76	1380	14.6	9.52	
42	1.09	400	600	4	10	10	18	100	165	1.75	1.14	330	3.50	2.28	4.79
								400	660	7.00	4.56	1320	14.0	9.12	
								600	990	11.6	6.84	1980	23.2	13.68	
48	1.46	400	600	4	12	12	21	100	220	2.33	1.52	440	4.66	3.04	6.42
								400	880	9.35	6.07	1760	18.7	12.14	
								600	1320	14.0	9.10	2640	28.0	18.2	
54	1.90	450	600	6	12	14	24	100	285	3.02	1.97	570	6.04	3.94	10.56
								450	1282	13.6	8.85	2564	27.2	17.7	
								600	1710	18.1	11.82	3420	36.2	23.6	
60	2.40	450	600	6	13	16	28	100	360	3.82	2.49	720	7.64	4.98	
								450	1620	17.2	11.20	3240	34.4	22.4	
								600	2160	22.9	14.95	4320	45.8	29.9	

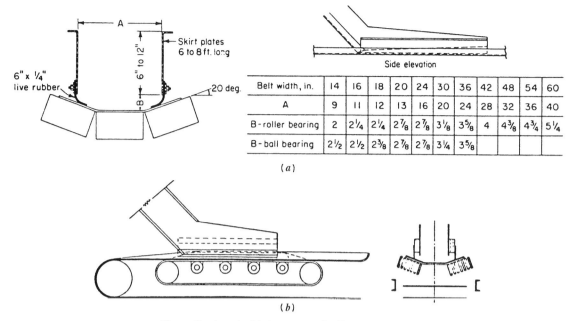

Figure 2 A typical belt conveyor loading arrangement.

be acquired in highly standardized units, although specifically engineered equipment can be obtained for use with special materials, unusual operating conditions, or high capacities. Figure 3 shows a common type of bucket elevator, the spaced-bucket centrifugal-discharge elevator. Other types include spaced-bucket positive-discharge elevators, V-bucket elevators, continuous-bucket elevators, and super-capacity continuous-bucket elevators. The latter handle high tonnages and are usually operated at an incline to improve loading and discharge conditions.

Bucket elevator horsepower requirements can be calculated for space-bucket elevators by multiplying the desired capacity (tons per hour) by the lift and dividing by 500. Table 5 gives bucket elevator specifications for spaced-bucket, centrifugal-discharge elevators.

2.5 Vibrating or Oscillating Conveyors

Vibrating conveyors are usually directional-throw devices that consist of a spring-supported horizontal pan or trough vibrated by an attached arm or rotating weight. The motion imparted to the material particles abruptly tosses them upward and forward so that the material travels in the desired direction. The conveyor returns to a reference position, which gives rise to the term *oscillating conveyor*. The capacity of the vibrating conveyor is determined by the magnitude and frequency of trough displacement, angle of throw, and slope of the trough, and the ability of the material to receive and transmit through its mass the directional "throw" of the trough. Classifications of vibrating conveyors include (1) mechanical, (2) electrical, and (3) pneumatic and hydraulic vibrating conveyors. Capacities of vibrating conveyors are very broad, ranging from a few ounces or grams for laboratory-scale equipment to thousands of tons for heavy industrial applications. Figure 4 depicts a leaf-spring mechanical vibrating conveyor, and provides a selection chart for this conveyor.

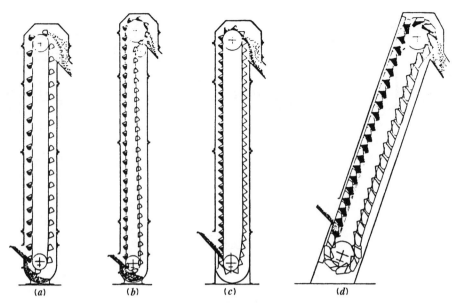

Figure 3 Bucket elevators.

2.6 Continuous-Flow Conveyors

The continuous-flow conveyor is a totally enclosed unit that operates on the principle of pulling a surface transversely through a mass of bulk solids material, such that it pulls along with it a cross section of material that is greater than the surface of the material itself. Figure 5 illustrates a typical configuration for a continuous-flow conveyor. Three common types of continuous flow conveyors are (1) closed-belt conveyors, (2) flight conveyors, and (3) apron conveyors. These conveyors employ a chain-supported transport device, which drags through a totally enclosed boxlike tunnel.

2.7 Pneumatic Conveyors

Pneumatic conveyors operate on the principle of transporting bulk solids suspended in a stream of air over vertical and horizontal distances ranging from a few inches or centimeters to hundreds of feet or meters. Materials in the form of fine powders are especially suited to this means of conveyance, although particle sizes up to a centimeter in diameter can be effectively transported pneumatically. Materials with bulk densities from one to more than 100 lb/ft^3 can be transported through pneumatic conveyors.

The capacity of a pneumatic conveying system depends on such factors as the bulk density of the product, energy within the conveying system, and the length and diameter of the conveyor.

There are four basic types of pneumatic conveyor systems: (1) pressure, (2) vacuum, (3) combination pressure and vacuum, and (4) fluidizing. In pressure systems, the bulk solids material is charged into an air stream operated at higher-than-atmospheric pressures, such that the velocity of the air stream maintains the solid particles in suspension until it reaches the separating vessel, usually an air filter or cyclone separator. Vacuum systems operate in much the same way, except that the pressure of the system is kept lower than atmospheric

Table 5 Bucket Elevator Specifications

Size of Bucket (in.)[a]	Elevator Centers (ft)	Capacity (tons/hr) Material Weighing 100 lb/ft³[b]	Size Lumps Handled (in.)[c]	Bucket Speed (ft/min)	rpm Head Shaft	Horsepower[b] Required at Head Shaft	Additional Horsepower[b] per Foot for Intermediate Lengths	Bucket Spacing (in.)	Shaft Diameter (in.) Head	Shaft Diameter (in.) Tail	Diameter of Pulleys (in.) Head	Diameter of Pulleys (in.) Tail	Belt Width (in.)
6 × 4 × 4¼	25	14	¾	225	43	1.0	0.02	12	1¹⁵⁄₁₆	1¹¹⁄₁₆	20	14	7
	50	14	¾	225	43	1.6	0.02	12	1¹⁵⁄₁₆	1¹¹⁄₁₆	20	14	7
	75	14	¾	225	43	2.1	0.02	12	1¹⁵⁄₁₆	1¹¹⁄₁₆	20	14	7
8 × 5 × 5½	25	27	1	225	43	1.6	0.04	14	1¹⁵⁄₁₆	1¹¹⁄₁₆	20	14	9
	50	30	1	260	41	3.5	0.05	14	1¹⁵⁄₁₆	1¹¹⁄₁₆	24	14	9
	75	30	1	260	41	4.8	0.05	14	2⁷⁄₁₆	1¹¹⁄₁₆	24	14	9
10 × 6 × 6¼	25	45	1¼	225	43	3.0	0.063	16	1¹⁵⁄₁₆	1¹⁵⁄₁₆	20	16	11
	50	52	1¼	260	41	5.2	0.07	16	2⁷⁄₁₆	1¹⁵⁄₁₆	24	16	11
	75	52	1¼	260	41	7.2	0.07	16	2¹⁵⁄₁₆	1¹⁵⁄₁₆	24	16	11
12 × 7 × 7¼	25	75	1½	260	41	4.7	0.1	18	2⁷⁄₁₆	1¹⁵⁄₁₆	24	18	13
	50	84	1½	300	38	8.9	0.115	18	2¹⁵⁄₁₆	1¹⁵⁄₁₆	30	18	13
	75	84	1½	300	38	11.7	0.115	18	3⁷⁄₁₆	2⁷⁄₁₆	30	18	13
14 × 7 × 7¼	25	100	1¾	300	38	7.3	0.14	18	2¹⁵⁄₁₆	2⁷⁄₁₆	30	18	15
	50	100	1¾	300	38	11.0	0.14	18	3⁷⁄₁₆	2⁷⁄₁₆	30	18	15
	75	100	1¾	300	38	14.3	0.14	18	3⁷⁄₁₆	2⁷⁄₁₆	30	18	15
16 × 8 × 8½	25	150	2	300	38	8.5	0.165	18	2¹⁵⁄₁₆	2⁷⁄₁₆	30	20	18
	50	150	2	300	38	12.6	0.165	18	3⁷⁄₁₆	2⁷⁄₁₆	30	20	18
	75	150	2	300	38	16.7	0.165	18	3¹⁵⁄₁₆	2⁷⁄₁₆	30	20	18

[a] Size of buckets given: width × projection × depth.
[b] Capacities and horsepowers given for materials weighing 100 lb/ft³. For materials of other weights, capacity and horsepower will vary in direct proportion. For example, an elevator handling coal weighing 50 lb/ft³ will have half the capacity and will require approximately half the horsepower listed above.
[c] If volume of lumps averages less than 15% of total volume, lumps of twice the size listed may be handled.

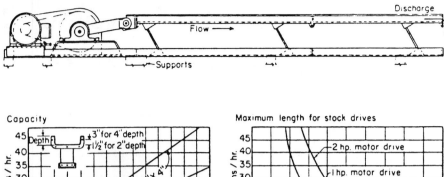
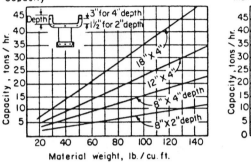
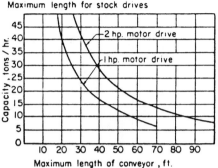

Figure 4 Leaf-spring mechanical vibrating conveyor.

pressure. Pressure–vacuum systems combine the best features of these two techniques, with a separator and a positive-displacement blower placed between the vacuum "charge" side of the system and the pressure "discharge" side. One of the most common applications of pressure–vacuum systems is with the combined bulk vehicle (e.g., hopper car) unloading and transporting to bulk storage. Fluidizing systems operate on the principle of passing air through a porous membrane, which forms the bottom of the conveyor, thus giving finely divided, non-free-flowing bulk solids the characteristics of free-flowing material. This technique, commonly employed in transporting bulk solids over short distances (e.g., from a storage bin to the charge point to a pneumatic conveyor), has the advantage of reducing the volume of conveying air needed, thereby reducing power requirements. Figure 6 illustrates these four types of pneumatic conveyor systems.

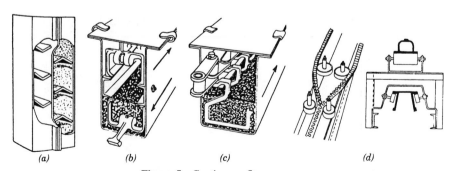

Figure 5 Continuous-flow conveyor.

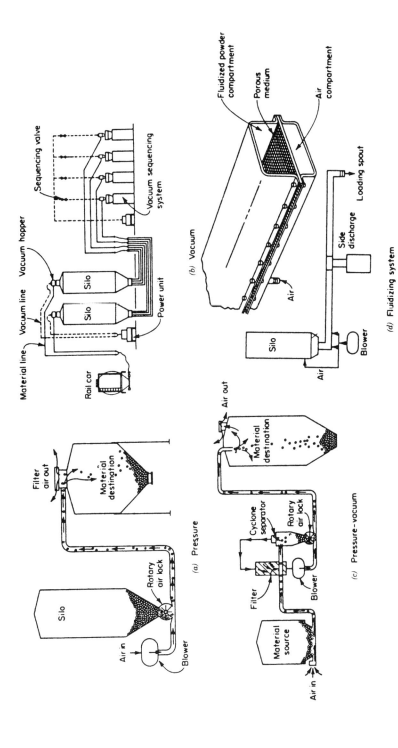

Figure 6 Four types of pneumatic corveyor systems.

3 BULK MATERIALS STORAGE

3.1 Storage Piles

Open-yard storage is a commonplace approach to the storage of bulk solids. Belt conveyors are most often used to transport to and from such a storage area. Cranes, front-end loaders, and draglines are commonly used at the storage site. Enclosed storage piles are employed where the bulk solids materials can erode or dissolve in rainwater, as in the case of salt for use on icy roads. The necessary equipment for one such application, the circular storage facility, is (1) feed conveyor, (2) central support column, (3) stacker, (4) reclaimer, (5) reclaim conveyor, and (6) the building or dome cover.

3.2 Storage Bins, Silos, and Hoppers

A typical storage vessel for bulk solids materials consists of two components—a bin and a hopper. The bin is the upper section of the vessel and has vertical sides. The hopper is the lower part of the vessel, connecting the bin and the outlet, and must have at least one sloping side. The hopper serves as the means by which the stored material flows to the outlet channel. Flow is induced by opening the outlet port and using a feeder device to move the material, which drops through the outlet port.

If all material stored in the bin moves whenever material is removed from the outlet port, *mass flow* is said to prevail. However, if only a portion of the material moves, the condition is called *funnel flow*. Figure 7 illustrates these two conditions.

Many flow problems in storage bins can be reduced by taking the physical characteristics of the bulk material into account. Particle size, moisture content, temperature, age, and oil content of the stored material affect flowability. Flow-assisting devices and feeders are usually needed to overcome flow problems in storage bins.

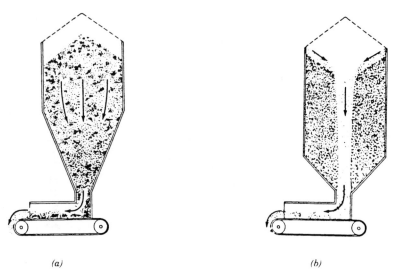

Figure 7 Mass flow (*a*) and funnel flow (*b*) in storage bins.

3.3 Flow-Assisting Devices and Feeders

To handle those situations in which bin design alone does not produce the desired flow characteristics, flow-assisting devices are available. Vibrating hoppers are one of the most important types of flow-assisting devices. These devices fall into two categories: *gyrating devices,* in which vibration is applied perpendicular to the flow channel; and *whirlpool devices,* which apply a twisting motion and a lifting motion to the material, thereby disrupting any bridges that might tend to form. Screw feeders are used to assist in bin unloading by removing material from the hopper opening.

3.4 Packaging of Bulk Materials

Bulk materials are often transported and marketed in containers, such as bags, boxes, and drums. Packaged solids lend themselves to material handling by means of unit material handling.

Bags

Paper, plastic, and cloth bags are common types of containers for bulk solids materials. Multiwall paper bags are made from several plies of kraft paper. Bag designs include valve and open-mouth designs. Valve-type bags are stitched or glued at both ends prior to filling, and are filled through a valve opening at one corner of the bag. Open-mouth bags are sealed at one end during manufacture, and at the open end after filling. Valve bags more readily lend themselves to automated filling than open-mouth bags, yielding higher packing rates.

Bag size is determined by the weight or volume of material to be packed and its bulk density. Three sets of dimensions must be established in bag sizing:

1. *Tube*—outside length and width of the bag tube before closures are fabricated
2. *Finished face*—length, width, and thickness of the bag after fabrication
3. *Filled face*—length, width, and thickness of the bag after filling and closure

Figure 8 shows the important dimensions of multiwall paper bags, and Table 6 gives their relationships to tube, finished face, and filled face dimensions.

Boxes

Bulk boxes are fabricated from corrugated kraft paper. They are used to store and ship bulk solid materials in quantities ranging from 50 lb to several hundred pounds. A single-wall corrugated kraft board consists of an outside liner, a corrugated medium, and an inside liner. A double-wall board has two corrugated mediums sandwiched between three liners. The

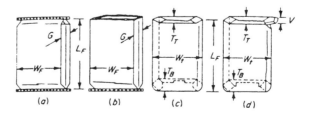

Figure 8   Dimensions of multiwall paper bags.

Table 6 Dimensions of Multiwall Paper Bags

Bag Type	Tube Dimensions	Finished-Face Dimensions	Filled-Face Dimensions	Valve Dimensions
Sewn open-mouth	Width = $W_t = W_f + G_f$ Length = $L_t = L_f$	Width = $W_f = W_t - G_f$ Length = $L_f = L_t$ Gusset = G_f	Width = $W_F = W_f + \frac{1}{2}$ in. Length = $L_F = L_f - 0.67 G_f$ Thickness = $G_F = G_f + \frac{1}{2}$ in.	
Sewn valve	Width = $W_t = W_f + G_f$ Length = $L_t = L_f$	Width = $W_f = W_t - G_f$ Length = $L_f = L_t$ Gusset = G_f	Width = $W_F = W_f + 1$ in. Length = $L_F = L_f - 0.67 G_f$ Thickness = $G_F = G_f + 1$ in.	Width = $V = G_f \pm \frac{1}{2}$ in.
Pasted valve	Width = $W_t = W_f$ Length = L_t	Width = $W_f = W_t$ Length = $L_f = L_t - (T_T + T_B)/2 - 1$ Thickness at top = T_T Thickness at bottom = T_B	Width = $W_F = W_f - T_T + 1$ in. Length = $L_F = L_f - T_T + 1$ in. Thickness = $T_F = T_T + \frac{1}{2}$ in.	Width = $V = T_T \begin{cases} +0 \text{ in.} \\ -1 \text{ in.} \end{cases}$

specifications for bulk boxes depend on the service requirements; 600 lb/in.^{2} is common for loads up to 1000 lb, and 200 lb/in.^{2} for 100-lb loads. Bulk boxes have the advantages of reclosing and of efficient use of storage and shipping space, called *cube.* Disadvantages include the space needed for storage of unfilled boxes and limited reusability. Figure 9 shows important characteristics of bulk boxes.

Folding cartons are used for shipping bulk solids contained in individual bottles, bags, or folding boxes. Cartons are of less sturdy construction than bulk boxes, because the contents can assist in supporting vertically imposed loads.

3.5 Transportation of Bulk Materials

The term *transportation of bulk materials* refers to the movement of raw materials, fuels, and bulk products by land, sea, and air. A useful definition of a bulk shipment is any unit greater than 4000 lb or 40 ft^3. The most common bulk carriers are railroad hopper cars, highway hopper trucks, portable bulk bins, barges, and ships. Factors affecting the choice

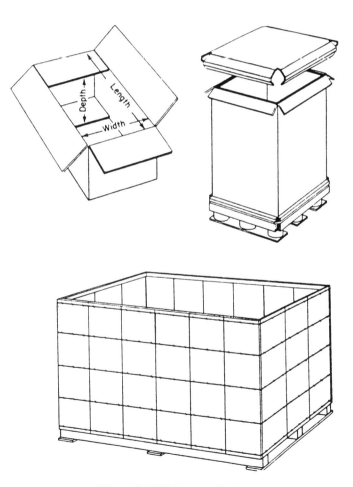

Figure 9 Bulk boxes and cartons.

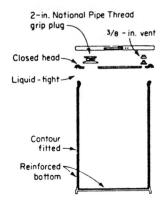

Drum type	Outside dimensions	
	Dia., in.	Height, in.
55-gal. lever top	21	40 3/4
55-gal. lever top	23 1/2	30 3/4
55-gal. lever top	22	34 3/4
41-gal. lever top	20 1/2	30 1/4
30-gal. lever top	19	26 1/4
6.28-cu. ft. rectangular	17 5/8*	37 1/2
55-gal. liquid	22	37 1/2
30-gal. liquid	19	28
55-gal. fiber	20 3/8	40 3/4
30-gal. fiber	17 3/8	30 3/4

* Side dimension, square

Figure 10 Storage drums.

of transportation include the characteristics of material size of shipment, available transportation routes from source to destination (e.g., highway, rail, water), and the time available for shipment.

Railroad Hopper Cars
Railroad hopper cars are of three basic designs:

1. Covered, with bottom-unloading ports
2. Open, with bottom-unloading ports
3. Open, without unloading ports

Gravity, pressure differential, and fluidizing unloading systems are available with railroad hopper cars. Loading of hopper cars can be done with most types of conveyors: belt, screw, pneumatic, and so on. Unloading of bottom-unloading hopper cars can be managed by constructing a special dumping pit beneath the tracks with screw or belt takeaway conveyors.

Hopper Trucks
Hopper trucks are used for highway transportation of bulk solids materials. The most common types include (1) closed type with a pneumatic conveyor unloading system and (2) the open dump truck. With the first type, a truck can discharge its cargo directly into a storage silo. The shipment weights carried by trucks depend on state highway load limits, usually from 75,000 to 125,000 lb.

4 UNIT MATERIAL HANDLING

4.1 Introduction

Unit material handling involves the movement and storage of unit loads, as defined in Section 1. Examples include automobile body components, engine blocks, bottles, cans, bags, pallets of boxes, bins of loose parts, and so on. As the previous definition implies, the word *unit* refers to the single entity that is handled. That entity can consist of a single item or numerous items that have been unitized for purposes of movement and storage.

This section discusses some of the procedures employed in material-handling system design, and describes various categories, with examples, of material-handling equipment used in handling unit loads.

4.2 Analysis of Systems for Material Handling

Material handling is an indispensable element in most production and distribution systems. Yet, while material handling is generally considered to add nothing to the value of the materials and products that flow through the system, it does add to their cost. In fact, it has been estimated that 30–60% of the end-price of a product is related to the cost of material handling. Therefore, it is essential that material handling systems be designed and operated as efficiently and cost-effectively as possible.

The following steps can be used in analyzing production systems and solving the inherent material-handling problems:

1. Identify and define the problem(s).
2. Collect relevant data.

3. Develop a plan.
4. Implement the solution.

Unfortunately, when most engineers perceive that a material-handling problem exists, they skip directly to step 4; that is, they begin looking for material-handling equipment that will address the symptoms of the problem without looking for the underlying root causes of the problem, which may be uncovered by execution of all four steps listed above.

Thus, the following sections explain how to organize a study and provide some tools to use in an analysis of a material-handling system according to this four-step procedure.

4.3 Identifying and Defining the Problem

For a new facility, the best way to begin the process of identifying and defining the problems is to become thoroughly familiar with all of the products to be produced by the facility, their design and component parts, and whether the component parts are to be made in the facility or purchased from vendors. Then, one must be thoroughly knowledgeable about the processes required to produce each part and product to be made in the facility. One must also be cognizant of the production schedules for each part and product to be produced; that is, parts or products produced per shift, day, week, month, year, and so on. Finally, one must be intimately familiar with the layout of the facility in which production will take place; not just the area layout, but the volume (or cubic space) available for handling materials throughout the facility.

Ideally, the persons or teams responsible for the design of material-handling systems for a new facility will be included and involved from the initial product design stage through process design, schedule design, and layout design. Such involvement in a truly concurrent engineering approach will contribute greatly to the efficient and effective handling of materials when the facility becomes operational.

In an existing facility, the best way to begin the process of identifying and defining the problems is to tour the facility, looking for material-handling aspects of the various processes observed. It is a good idea to take along a checklist, such as that shown in Fig. 11. Another useful guide is the Material Handling Institute (MHI) list of "The Ten Principles of Material Handling," as given in Fig. 12.

Once the problem has been identified, its scope must be defined. For example, if most of the difficulties are found in one area of the plant, such as shipping and receiving, the study can be focused there. Are the difficulties due to lack of space? Or is part of the problem due to poor training of personnel in shipping and receiving? In defining the problem, it is necessary to answer the basic questions normally asked by journalists: Who? what? when? where? why?

4.4 Collecting Data

In attempting to answer the journalistic questions above, all relevant data must be collected and analyzed. At a minimum, the data collection and analysis must be concerned with the products to be produced in the facility, the processes (fabrication, assembly, and so on) used to produce each product, the schedule to be met in producing the products, and the facility layout (three-dimensional space allocation) supporting the production processes.

Some useful data can be obtained by interviewing management, supervisors, operators, vendors, and competitors, by consulting available technical and sales literature, and through personal observation. However, most useful data are acquired by systematically charting the

Material Handling Checklist

☐ Is the material handling equipment more than 10 years old?
☐ Do you use a wide variety of makes and models which require a high spare parts inventory?
☐ Are equipment breakdowns the result of poor preventive maintenance?
☐ Do the lift trucks go too far for servicing?
☐ Are there excessive employee accidents due to manual handling of materials?
☐ Are materials weighing more than 50 pounds handled manually?
☐ Are there many handling tasks that require 2 or more employees?
☐ Are skilled employees wasting time handling materials?
☐ Does material become congested at any point?
☐ Is production work delayed due to poorly scheduled delivery and removal of materials?
☐ Is high storage space being wasted?
☐ Are high demurrage charges experienced?
☐ Is material being damaged during handling?
☐ Do shop trucks operate empty more than 20% of the time?
☐ Does the plant have an excessive number of rehandling points?
☐ Is power equipment used on jobs that could be handled by gravity?
☐ Are too many pieces of equipment being used, because their scope of activity is confined?
☐ Are many handling operations unnecessary?
☐ Are single pieces being handled where unit loads could be used?
☐ Are floors and ramps dirty and in need of repair?
☐ Is handling equipment being overloaded?
☐ Is there unnecessary transfer of material from one container to another?
☐ Are inadequate storage areas hampering efficient scheduling of movement?
☐ Is it difficult to analyze the system because there is no detailed flow chart?
☐ Are indirect labor costs too high?

Figure 11 Material-handling checklist.

flows of materials and the movements that take place within the plant. Various graphical techniques are used to record and analyze this information.

An assembly chart, shown in Fig. 13, is used to illustrate the composition of the product, the relationship among its component parts, and the sequence in which components are assembled.

The operations process chart, shown in Fig. 14, provides an even more detailed depiction of material flow patterns, including sequences of production and assembly operations. It begins to afford an idea of the relative space requirements for the process.

The flow process chart, illustrated in Fig. 15, tabulates the steps involved in a process, using a set of standard symbols adopted by the American Society of Mechanical Engineers (ASME). Shown at the top of the chart, these five symbols allow one to ascribe a specific status to an item at each step in processing. The leftmost column in the flow process chart lists the identifiable activities comprising the process, in sequential order. In the next column, one of the five standard symbols is selected to identify the activity as an operation, transportation, inspection, delay, or storage. The remaining columns permit the recording of more detailed information.

Note that in the flow process chart in Fig. 16, for each step recorded as a "transport," a distance (in feet) is recorded. Also, in some of the leftmost columns associated with a transport activity, the type of material handling equipment used to make the move is recorded—for example, "fork lift." However, material-handling equipment could be used for any of the activities shown in this chart. For example, automated storage and retrieval systems (AS/RSs) can be used to store materials, accumulating conveyors can be used to queue materials during a delay in processing, or conveyors can be configured as a moving assembly line so that operations can be performed on the product while it is being transported through the facility.

1. Planning principle. All material handling should be the result of a deliberate plan where the needs, performance objectives, and functional specification of the proposed methods are completely defined at the outset.
2. Standardization Principle. Material handling methods, equipment, controls, and software should be standardized within the limits of achieving overall performance objectives and without sacrificing needed flexibility, modularity, and throughput.
3. Work Principle. Material handling work should be minimized without sacrificing productivity or the level of service required of the operation.
4. Ergonomic Principle. Human capabilities and limitations must be recognized and respected in the design of material handling tasks and equipment to ensure safe and effective operations.
5. Unit Load Principle. Unit loads shall be appropriately sized and configured in a way that achieves the material flow and inventory objectives at each stage in the supply chain.
6. Space Utilization. Effective and efficient use must be made of all available space.
7. System Principle. Material movement and storage activities should be fully integrated to form a coordinated, operational system, which spans receiving, inspection, storage, production, assembly, packaging, unitizing, order selection, shipping, transportation, and the handling of returns.
8. Automation Principle. Material handling operations should be mechanized and/or automated where feasible to improve operational efficiency, increase responsiveness, improve consistency and predictability, decrease operating costs, and eliminate repetitive or potentially unsafe manual labor.
9. Environmental Principle. Environmental impact and energy consumption should be considered as criteria when designing or selecting alternative equipment and material handling systems.
10. Life-cycle Cost Principle. A thorough economic analysis should account for the entire life cycle of all material handling equipment and resulting systems.

Figure 12 Ten principles of material handling.

In the columns grouped under the heading *possibilities,* opportunities for improvement or simplification of each activity can be noted.

The flow diagram, depicted in Fig. 16, provides a graphical record of the sequence of activities required in the production process, superimposed upon an area layout of a facility. This graphical technique uses the ASME standard symbol set and augments the flow process chart.

The "from–to" chart, illustrated in Fig. 17, provides a matrix representation of the required number of material moves (unit loads) in the production process. A separate from–to chart can also be constructed that contains the distances materials must be moved between activities in the production process. Of course, such a chart will be tied to a specific facility layout and usually contains assumptions about the material-handling equipment to be used in making the required moves.

The activity relationship chart, shown in Fig. 18, can be used to record qualitative information regarding the flow of materials between activities or departments in a facility. Read like a highway mileage table in a typical road atlas, which indicates the distances between pairs of cities, the activity relationship chart allows the analyst to record a qualitative relationship that should exist between each pair of activities or departments in a facility layout. The relationships recorded in this chart show the importance that each pair of activities be located at varying degrees of closeness to each other (using an alphabetic symbol) and the reason for the assignment of that rating (using a numeric symbol). Together these

4 Unit Material Handling   **369**

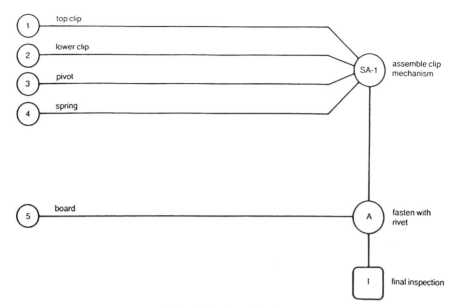

Figure 13 Assembly chart.

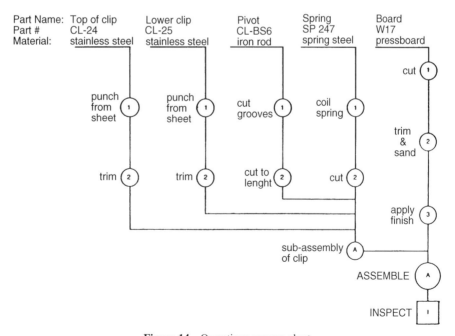

Figure 14 Operations process chart.

370 Material Handling

Symbol	Name	Results
○	Operation	Produces, prepares, and accomplishes
◇	Transportation	Moves
□	Inspection	Verifies
D	Delay	Interfere, waits
▽	Storage	Keeps, retains

SUMMARY						JOB		ANALYSIS	
	PRESENT		PROPOSED		DIFERENCE	Manufacture of a tissue box	QUESTION EACH DETAIL	WHAT? WHY? WHERE?	WHEN? WHO? HOW?
	NO.	TIME	NO.	TIME	NO.	TIME			
○ OPERATIONS	5								
◇ TRANSPORTATIONS	9					☐ OPERATOR	DATE		
□ INSPECTIONS	1					☒ MATERIAL	NUMBER		
D DELAYS	2					CHART BEGINS Receiving (raw materials)	PAGE 1	OF 1	
▽ STORAGES	3					CHART ENDS Shipping (finished product)			
Distance Traveled		1485 FT.		FT.		FT.	CHARTED BY T.P.C.		

DETAILS OF (PRESENT/PROPOSED) METHOD	OPERATION	TRANSPORT	INSPECTION	DELAY	STORAGE	DISTANCE IN FEET	QUANTITY	TIME	ELIMINATE	COMBINE	SEQUENCE	PLACE	PERSON	IMPROVE	SAFER?	$ SAVED?	NOTES
1. Receive raw materials	○	◇	□	D	▽	50											
2. Inspect	○	◇	▣	D	▽												
3. Move by fork lift	○	◇	□	D	▽	40											
4. Store	○	◇	□	D	▽												
5. Move by fork lift	○	◇	□	D	▽	45											
6. Set up and print	⊘	◇	□	D	▽												
7. Moved by printer	○	◇	□	D	▽	120											
8. Stack at end of printer	⊘	◇	□	D	▽												
9. Move to stripping	○	◇	□	D	▽	165											
10. Delay	○	◇	□	D	▽												
11. Being stripped	⊘	◇	□	D	▽												
12. Move to temp. storage	○	◇	□	D	▽	150											
13. Storage	○	◇	□	D	▽												
14. Move to folders	○	◇	□	D	▽	200											
15. Delay	○	◇	□	D	▽												
16. Set up, fold, glue	⊘	◇	□	D	▽												
17. Mechanically moved	○	◇	□	D	▽	90											
18. Stack, count, crate	⊘	◇	□	D	▽												
19. Move by fork lift	○	◇	□	D	▽	525											
20. Storage	○	◇	□	D	▽												

Figure 15 Flow process chart.

4 Unit Material Handling **371**

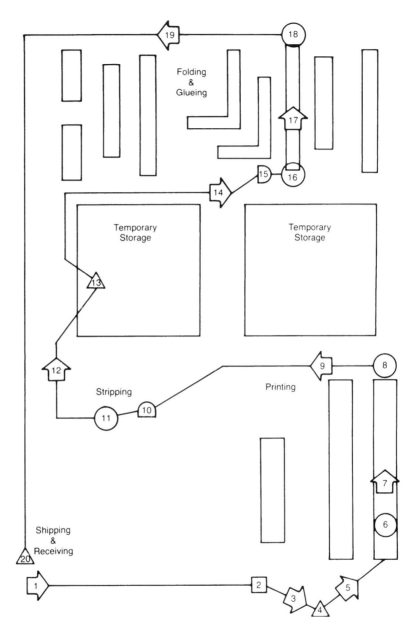

Figure 16 Flow diagram.

To \ From	A	B	C	D	E	F	G
A		13	37	69	85	57	33
B	13		21	53	69	41	17
C	37	21		29	45	17	41
D	69	53	29		13	33	57
E	85	69	45	13		49	73
F	57	41	17	33	49		21
G	33	17	41	57	73	21	

Figure 17 "From–to" chart.

charting techniques provide the analyst extensive, qualitative data about the layout to support a production process. This is very useful from the standpoint of designing a material handling system.

4.5 Unitizing Loads

Principle 7 of the MHI Ten Principles of Material Handling (Fig. 12) is the *unit size principle,* also known as the *unit load principle,* which states, " Increase the quantity, size, or weight of unit loads or flow rate." The idea behind this principle is that if materials are consolidated into large quantities or sizes, fewer moves of this material will have to be made to meet needs of the production processes. Therefore, less time will be required to move the unitized material than that required to move the same quantity of non-unitized material. So, unitizing materials usually results in low-cost, efficient material-handling practices.

The decision to unitize is really a design decision in itself, as illustrated in Fig. 19. Unitization can consist of individual pieces through unit packs, inner packs, shipping cartons, tiers on pallets, pallet loads, containers of pallets, truckloads, and so on. The material-handling system must then be designed to accommodate the level of unitized parts at each step of the production process.

As shown in Fig. 19, once products or components have been unitized into shipping cartons, further consolidation may easily be achieved by placing the cartons on a pallet, slip sheet, or some other load-support medium for layers (or tiers) of cartons comprising the unit

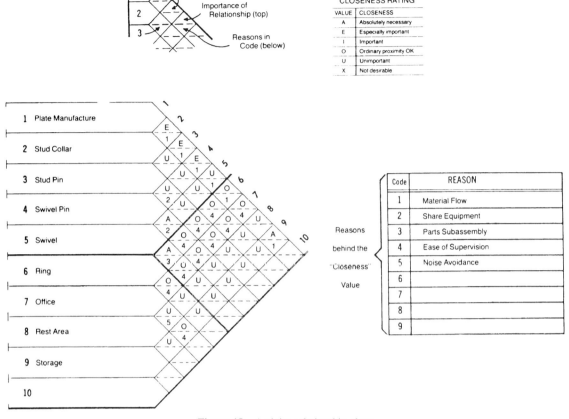

Figure 18 Activity relationship chart.

load. Since the unit load principle requires the maximum utilization of the area on the pallet surface, another design problem is to devise a carton stacking pattern that achieves this objective. Examples of pallet loading patterns that can achieve optimal surface utilization are illustrated in Fig. 20. Charts of such patterns are available from the U.S. Government (General Services Administration). There are also a number of providers of computer software programs for personal computers that generate pallet-loading patterns.

Highly automated palletizer machines as well as palletizing robots are available that can be programmed to form unit loads in any desired configuration. Depending on the dimensions of the cartons to be palletized, and the resulting optimal loading pattern selected, the palletized load may be inherently stable due to overlapping of cartons in successive tiers; for example, the various pinwheel patterns shown in Fig. 20.

However, other pallet-loading patterns may be unstable, such as the block pattern in Fig. 21, particularly when cartons are stacked several tiers high. In such instances, the loads may be stabilized by stretch-wrapping the entire pallet load with plastic film, or by placing bands around the individual tiers. The wrapping or banding operations themselves can be automated by use of equipment that exists in the market today.

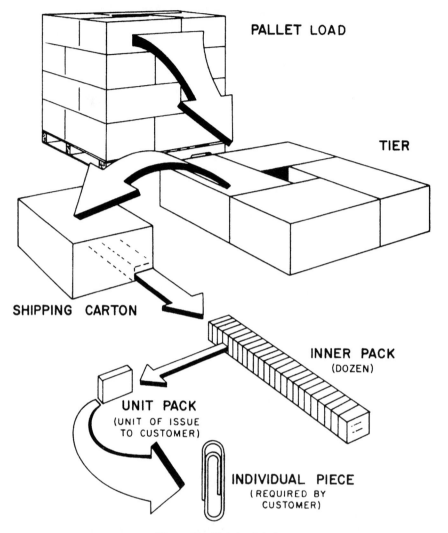

Figure 19 Unit load design.

Once the unit load has been formed, there are only four basic ways it can be handled while being moved. These are illustrated in Fig. 22 and consist of the following:

1. Support the load from below.
2. Support or grasp the load from above.
3. Squeeze opposing sides of the load.
4. Pierce the load.

These handling methods are implemented individually, or in combination, by commercially available material-handling equipment types.

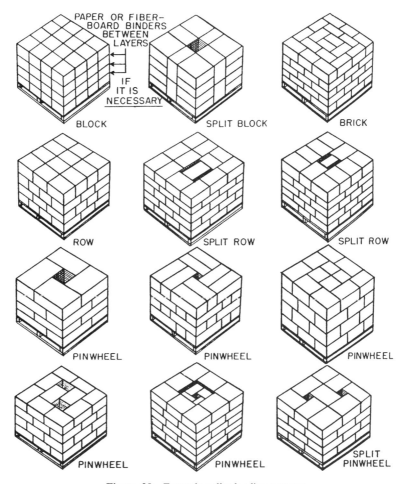

Figure 20 Example pallet-loading patterns.

5 MATERIAL-HANDLING EQUIPMENT CONSIDERATIONS AND EXAMPLES

5.1 Developing the Plan

Once the material-handling problem has been identified and the relevant data have been collected and analyzed, the next step in the design process is to develop a plan for solving the problem. This usually involves the design and/or selection of appropriate types, sizes, and capacities of material-handling equipment. To properly select material handling equipment, it must be realized that in most cases, the solution to the problem does not consist merely of selecting a particular piece of hardware, such as a section of conveyor. Rather, handling should be viewed as part of an overall system, with all activities interrelated and meshing together. Only on this basis can the best overall type of equipment or system be planned.

The following sections provide examples of some of the more common types of unit load material handling and storage equipment used in production facilities.

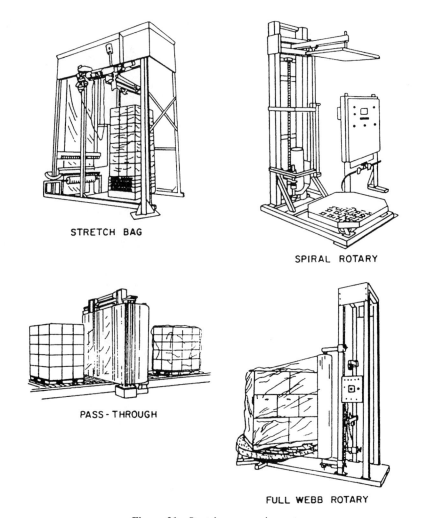

Figure 21 Stretch wrap equipment.

5.2 Conveyors

Conveyors are generally used to transport materials long distances over fixed paths. Their function may be solely the movement of items from one location in a process or facility to another point, or they may move items through various stages of receiving, processing, assembly, finishing, inspection, packaging, sortation, and shipping.

Conveyors used in material handing are of two basic types:

1. *Gravity conveyors,* including chutes, slides, and gravity wheel or roller conveyors that essentially exploit the use of gravity to move items from a point at a relatively high elevation to another point at a lower elevation. As listed in Fig. 12, MHI Principle 5 indicates that one should maximize the use of gravity in designing material-handling systems.

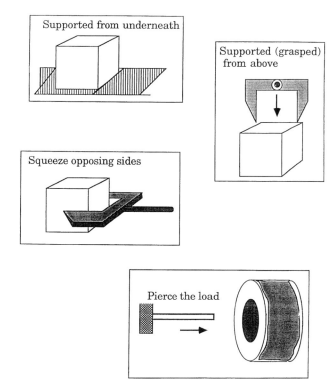

Figure 22 Ways to handle a load.

2. *Powered conveyors,* which generally use electric motors to drive belts, chains, or rollers in a variety of in-floor, floor-mounted, or overhead configurations.

In general, conveyors are employed in unit material handling when

1. Loads are uniform.
2. Materials move continuously.
3. Routes do not vary.
4. Load is constant.
5. Movement rate is relatively fixed.
6. Cross traffic can be bypassed.
7. Path is relatively fixed.
8. Movement is point-to-point.
9. Automatic counting, sorting, weighing, or dispatching is needed.
10. In-process storage is required.
11. In-process inspection is required.
12. Production pacing is necessary.
13. Process control is required.

378 Material Handling

14. Controlled flow is needed.
15. Materials are handled at extreme temperatures, or other adverse conditions.
16. Handling is required in a hazardous area.
17. Hazardous materials are handled.
18. Machines are integrated into a system.
19. Robots are integrated into a system.
20. Materials are moved between workplaces.
21. Manual handling and/or lifting is undesirable.
22. Changes in production volume or pace are needed.
23. Visual surveillance of a production process is required.
24. Floor space can be saved by utilizing overhead space.
25. Flexibility is required to meet changes in production processes.
26. Integration between computer-aided design and computer-aided manufacturing is required.

This section further details essential information on four main classes of conveyors used in unit material handling:

1. Gravity conveyors
2. Powered conveyors
3. Chain-driven conveyors
4. Power-and-free conveyors

Gravity Conveyors
Gravity conveyors exploit gravity to move material without the use of other forms of energy. Chutes, skate wheel conveyors, and roller conveyors are the most common forms of gravity conveyors. Figure 23 illustrates wheel and roller conveyors. Advantages of gravity conveyors

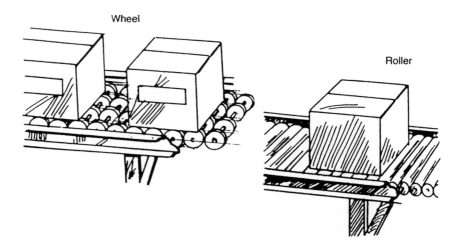

Figure 23 Gravity conveyors.

are low cost, relatively low maintenance, and negligible breakdown rate. The main requirement for using gravity conveyors is the ability to provide the necessary gradient in the system configuration at the point at which gravity units are placed.

Powered Conveyors

The two principal types of powered conveyors are belt conveyors and roller conveyors, as shown in Fig. 24. Electric motors provide the energy to drive the belt or rollers on these conveyors.

Belt conveyors are used either in the horizontal plane or with inclines up to 30°. They can range in length from a few feet to hundreds of feet, and are usually unidirectional. Changes in direction must be managed through the use of connecting chutes or diverters to conveyors running in another direction.

Roller conveyors are used for heavier loads than can be moved with belt conveyors, and are generally of sturdier construction. When used as accumulating conveyors, roller conveyors can also be used to provide spacing between items. Inclines are possible to about 10°; declines of about 15° are possible.

Powered conveyors should be operated at about 65 ft/min (about 1 mile/hr).

Chain-driven Conveyors

Chain conveyors are those in which closed-loop systems of chain, usually driven by electric motors, are used to pull items or carts along a specified path. The three principal types of chain-driven conveyors used in unit material handling are flight conveyors, overhead towlines and monorails, and in-floor towlines. Figure 25 illustrates an overhead towline type of chain-driven conveyor.

Flight conveyors consist of one or more endless strands of chain with spaced transverse flights or scrapers attached, which push the material along through a trough. Used primarily in bulk material handling, its primary function in unit material handling includes movement of cans or bottles in food canning and bottling. A flight conveyor is generally limited to speeds of up to 120 ft/min.

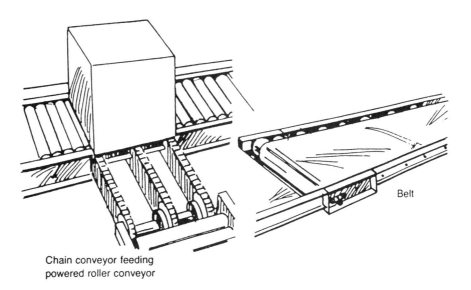

Figure 24 Belt and roller conveyors.

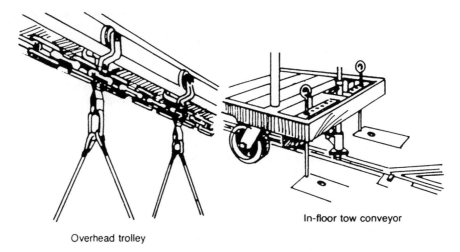

Figure 25 Chain-driven conveyors.

In-floor towlines consist of chain tracks mounted in the floor. A cart is pulled along the track by attaching a pin-type handle to the chain. In-floor towlines are capable of greater speeds than overhead towlines and have a smoother pickup action. They are difficult to maintain and lack flexibility for rerouting.

Overhead towlines consists of a track mounted 8–9 ft above the floor. Carts on the floor are attached to the chain, which moves through the overhead track. Overhead towlines free the floor for other uses, and are less expensive and more flexible than in-floor towlines.

Power-and-Free Conveyors
Power-and-free conveyors are a combination of powered trolley conveyors and unpowered monorail-type conveyors. Two sets of tracks are used, one positioned above the other. The upper track carries the powered trolley or monorail-type conveyor, which is chain-driven. The lower track is the free, unpowered monorail. Load-carrying free trolleys are engaged by pushers attached to the powered trolley conveyors. Load trolleys can be switched to and from adjacent unpowered free tracks.

Interconnections on power-and-free conveyors may be manually or automatically controlled. Track switches may divert trolleys from "power" to "free" tracks. Speeds may vary from one "power" section to another, and programmable logic controllers (PLC) or computers can be used to control power-and-free conveyors.

Power-and-free conveyors, shown in Fig. 26, are relatively expensive and are costly to relocate.

Sortation Conveyors
Sortation conveyors are used for merging different product streams, identifying units for later operations, inducting units into a different device, and/or separating units so they can be conveyed to specific destinations. *Direct diverters,* as shown in Fig. 27 are stationary or movable arms that deflect, push, or pull a product to desired destination. Since they do not come in contact with the conveyor, they can be used with almost any flat surface conveyor. They are usually hydraulically or pneumatically operated, but also can be motor driven. They are generally a simple and low cost means of sortation.

Figure 26 Power-and-free conveyor.

Pop-up diverters, shown in Fig. 28, consist of one or more sets of powered rollers, wheels, or chains that "pop-up" to lift and divert the unit off of the conveyor at an angle. The device is then lowered so that nondiverted units can pass by. These are generally capable of sorting only flat-bottomed items that can make good contact with the diverter.

Sliding shoe sorters, sometimes referred to as "slat sorters," use a series of diverter shoes that to push product off the conveyor, as shown in Fig. 29. The shoes move from side to side through the slats in the conveyor in order to divert the product to either side. These systems can divert products of different sizes by simply assigning a different number of shoes to a given product. They offer gentle and gradual handling of products and, as such, are widely used in the parcel-handling industry.

Tilt-tray sorters, shown in Fig. 30, are individual carriers conveyed in a continuous loop, that tilt at the appropriate time so that product can slide off the carrier to its destination. These are often used for small to medium size items with high throughput requirements.

Cross-belt sorters, shown in Fig. 31 are generally designed as either a continuous loop, where a small number of carriers are tied togehter, or in a train style (asynchronous) layout, where a small number of carriers are tied together with potential for several trains running the track simultaneously. Each carrier is equipped with small belt conveyor, called the cell,

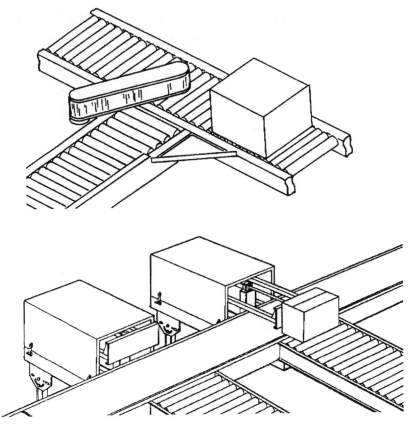

Figure 27 Direct diverters.

that is mounted perpendicular to direction of travel of loop and discharges product at the appropriate destination.

A general guide for selecting various types of sortation conveyors can be found in Fig. 32. More equipment information, picture banks, and selection guidelines can be found at the College-Industry Council for Material Handling Education web site: *www.mhia.org/cicmhe*.

5.3 Hoists, Cranes, and Monorails

Overhead material handling system components (e.g., tracks, carriers/trolleys, hoists, monorails, and cranes), can make effective use of otherwise unused overhead space to move materials in a facility. This can free up valuable floor space for other uses than material handling, reduce floor-based traffic, and reduce handling time by employing "crow-fly" paths between activities or departments.

Hoists, cranes, and monorails are used for a variety of overhead handling tasks. A hoist is a device for lifting and/or lowering a load and typically consists of an electric or pneumatically powered motor; a lifting medium, such as a chain, cable, or rope; a drum for reeling the chain, cable, or rope; and a handling device at the end of the lifting medium, such as a hook, scissor clamp mechanism, grapple, and so on. Hoists may be manually operated or automatically controlled by PLC or computer.

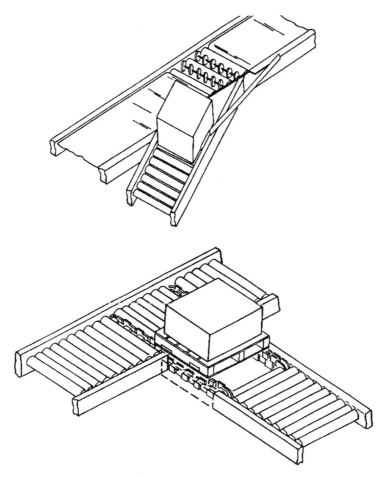

Figure 28 Pop-up diverters.

A monorail is a single-beam overhead track that provides a horizontal single path or route for a load as it is moved through a facility. The lower flange of the rail serves as a runway for a trolley-mounted hoist. A monorail system is used to move, store, and queue material overhead.

Through the use of switches, turntables, and other path-changing devices, an overhead monorail can be made to follow multiple predetermined paths, carrying a series of trolleys through various stations in processing or assembly. A chain-driver overhead monorail is very similar to the overhead towline in its configuration, except that it generally carries uniformly spaced trolleys overhead instead of pulling carts along the floor.

However, newer monorail technology has led to the development of individually powered and controllable trolleys that travel on the monorail (see Fig. 33). These devices are termed *automated electrified monorails* (AEM). The speed of the individually powered AEM vehicles can be changed en route and can function in nonsynchronous or flexible production environments.

Monorails can be made to dip down at specific points to deliver items to machines or other processing stations.

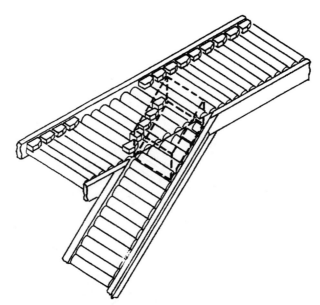

Figure 29 Sliding shoe sorter.

A crane also involves a hoist mounted on a trolley. Frequently, the trolley may be transported, as in the case of the bridge cranes shown in Fig. 34. Cranes may be manually, electrically, or pneumatically powered.

A jib crane has a horizontal beam on which a hoist trolley rides. The beam is cantilevered from a vertical support mast about which the beam can rotate or pivot (see the wall bracket-type jib crane in Fig. 34). This rotation permits the jib crane a broad range of coverage within the cylindrical work envelope described by the degrees of freedom of the beam, hoist, and mast.

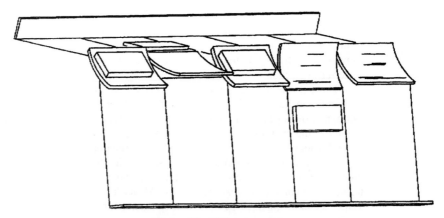

Figure 30 Tilting tray sorter.

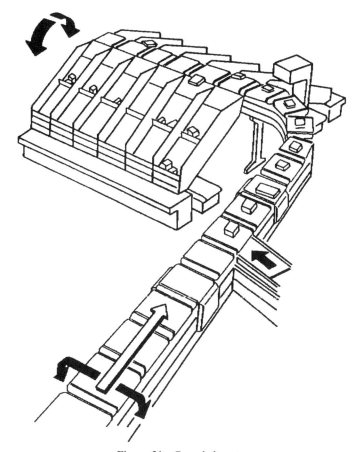

Figure 31 Cross-belt sorter.

5.4 Industrial Trucks

Industrial trucks provide flexible handling of materials along variable (or random) flow paths. The two main categories of industrial trucks are hand trucks and powered trucks, illustrated by the examples in Fig. 35.

Four-wheeled and multiple-wheeled carts and trucks include dollies, platform trucks, and skid platforms equipped with jacks. Hand-operated lift trucks include types equipped with hand-actuated hydraulic cylinders, and others having mechanical-lever systems.

Perhaps the most familiar type of powered truck is the forklift, which uses a pair of forks—capable of variable spacing—riding on a vertical mast to engage, lift, lower, and move loads. Lift trucks may be manually propelled or powered by electric motors, gasoline, liquified propane, or diesel-fueled engines. With some models, the operator walks behind the truck. On others, he or she rides on the truck, in either a standing or sitting position. Figure 36 depicts several types of forklift trucks.

Lift trucks are very effective in lifting, stacking, and unloading materials from storage racks, highway vehicles, railroad cars, and other equipment. Some lift trucks are designed

386 Material Handling

Type of Sortation	Maximum Sorts per Minute	Typical Load Range (lb)	Minimum Distance Between Spurs	Impact on Load	Relative Initial Cost	Typical Repair Cost	Package Orientation Maintained
Manual	15–25	1–75	Pkg. width +3 in.	Gentle	Lowest	Lowest	Yes
Deflector	20–60	1–50	3–5 ft	Med.–rough	Low	Low	No
Push/pull	30–70	1–150	Pkg. width +6 in.	Med.	Low–med	Low–med	No
Wheel transfer	5–10	10–150	1 ft	Gentle–med	Low–med	Low	No
Roller transfer	15–20	10–300	1 ft	Gentle–med	Med	Low–med	No
V-belt transfer	15–20	1–200	2–3 ft	Gentle–med	Med–high	Med–high	No
Roller diverter	50–120	10–500	4–5 ft	Gentle	Med–high	Med	Yes
Wheel diverter	65–150	3–300	4–5 ft	Gentle	Med	Med	Yes
V-belt diverter	65–120	1–250	4–5 ft	Gentle	Med–high	Med–high	Yes
Tilt tray	65–200	1–250	1 ft	Med.–rough	High	Med–high	No
Tilt slat	65–200	1–300	12 ft	Gentle–med	High	High	Yes
Sliding shoe	50–150	1–200	4–5 ft	Gentle	High	Med–high	Yes
Diverter	20–70	1–150	Pkg. width +6 in.	Med	Low–med	Low–med	No
Pop-up belt & chain	30–120	1–250	1 ft	Med.–rough	High	Med–high	No

Figure 32 Sortation selection guidelines.

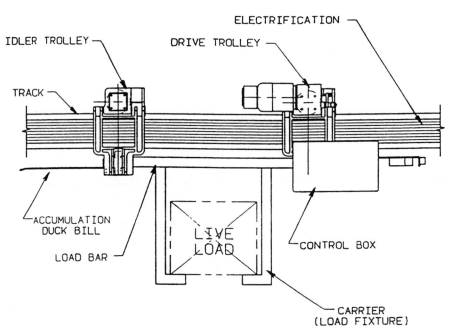

Figure 33 Automated electrified monorail.

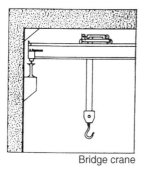 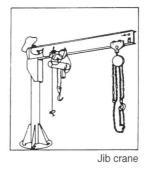 

Figure 34 Cranes.

for general-purpose use, while others are designed for specific tasks, such as narrow-aisle or high-rack handling.

5.5 Automated Guided Vehicle Systems

An automated guided vehicle system (AGVS) has similar uses as an industrial truck-based material-handling system. However, as implied by their name, the vehicles in an AGVS are under automatic control and do not require operators to guide them. In general, the vehicles in an AGVS are battery-powered, driverless, and capable of being automatically routed between, and positioned at, selected pickup or dropoff stations strategically located within a facility. Most of the vehicles in industrial use today are transporters of unit loads. However, when properly equipped, AGVs can provide a number of other functions, such as serving as automated storage devices or assembly platforms.

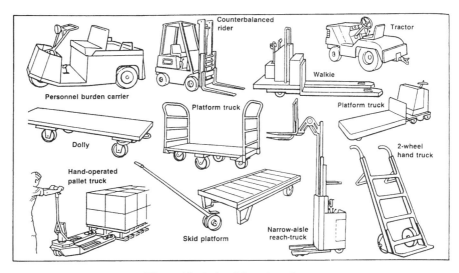

Figure 35 Industrial truck equipment.

388 Material Handling

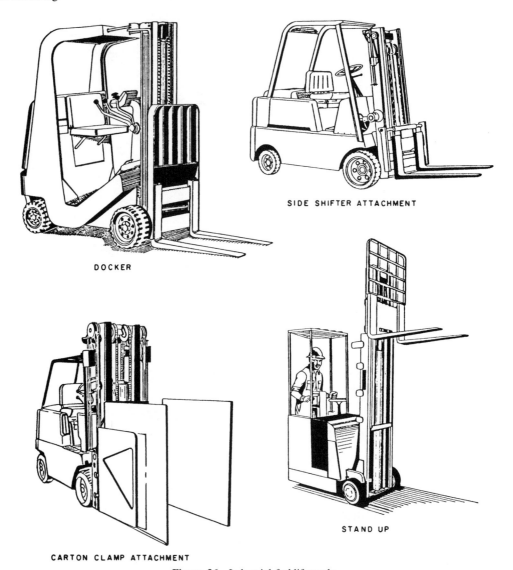

Figure 36 Industrial forklift trucks.

The four commonly recognized operating environments for AGVSs are distribution warehouses, manufacturing storerooms and delivery systems, flexible manufacturing systems, and assembly systems. Vehicles are guided by inductive-loop wires embedded in the floor of a facility, a chemical stripe painted on the floor, or laser-based navigation systems. All vehicular motion, as well as load pickup and delivery interfaces, are under computer control.

Examples of typical unit load AGVs are shown in Fig. 37 equipped with various types of load-handling decks that can be used.

5 Material-Handling Equipment Considerations and Examples 389

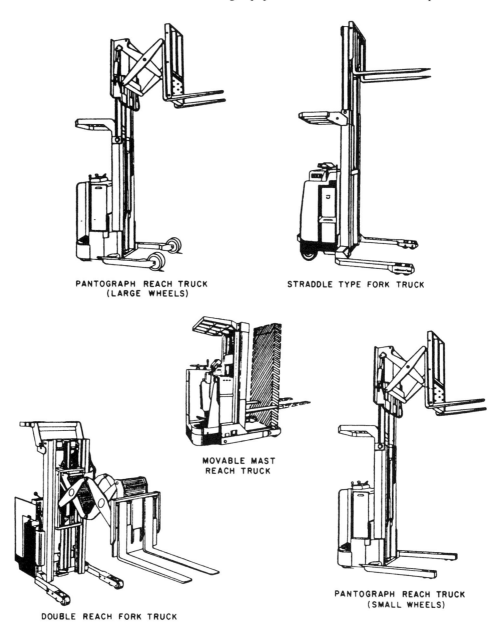

Figure 36 (*Continued*)

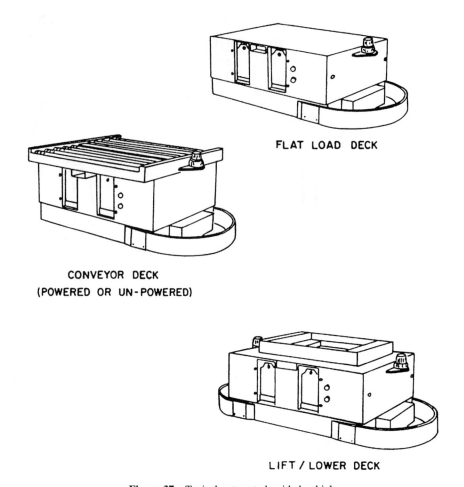

Figure 37 Typical automated guided vehicles.

5.6 Automated Storage and Retrieval Systems

An automated storage and retrieval system (AS/RS) consists of a set of racks or shelves arrayed along either side of an aisle through which a machine travels that is equipped with devices for storing or retrieving unit loads from the rack or shelf locations. As illustrated in Fig. 38, the AS/RS machine resembles a vertically oriented bridge crane (mast) with one end riding on a rail mounted on the floor and the other end physically connected to a rail or channel at the top of the rack structure. The shuttle mechanism travels vertically along the mast as it, in turn, travels horizontally through the aisle. In this manner, it carries a unit load from an input station to the storage location in the rack structure, then extends into the rack to place the load. The procedure is reversed for a retrieval operation; that is, the empty shuttle is positioned at the correct rack location by the mast, then it is extended to withdraw the load from storage and transport it to the output station, usually located at the end of the aisle.

The AS/RS machines can have people on board to control the storage/retrieval operations, or they can be completely controlled by a computer. The objective in using AS/RSs

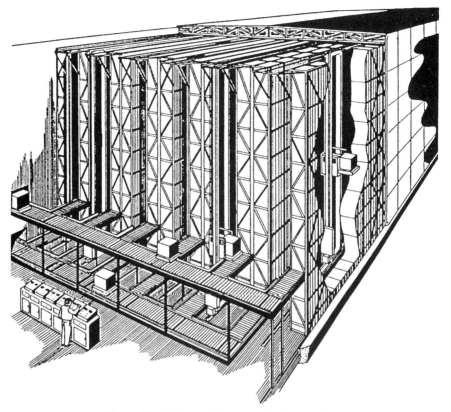

Figure 38 Automated storage and retrieval system.

is to achieve very dense storage of unit loads while simultaneously exercising very tight control of the inventory stored in these systems.

AS/RSs which store palletized unit loads can be 100 feet or more high and hundreds of feet deep. However, these systems can also be of much smaller dimensions and store small parts in standard-sized drawers. Such systems are called *miniload AS/R systems.* An example is shown in Fig. 39.

In a typical miniload AS/RS, the machine retrieves the proper coded bins from specified storage locations and brings them to an operator station. Each bin can be divided into a number of sections, and the total weight of parts contained in each bin typically ranges from 200 to 750 lb. While the operator is selecting items from one bin at the operator station, the machine is returning a previously accessed bin to storage. The system can be operator-directed through a keyboard entry terminal, or it may be operated under complete computer control.

5.7 Carousel Systems

A carousel is a series of linked bin sections mounted on either an oval horizontal track (horizontal carousel) or an oval vertical track (vertical carousel). A horizontal carousel is illustrated in Fig. 40. When activated, usually by an operator, the bins revolve in whichever

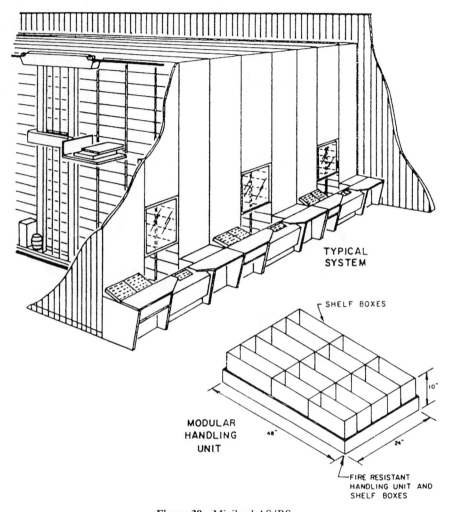

Figure 39 Miniload AS/RS.

direction will require the minimum travel distance to bring the desired bin to the operator. The operator then either picks or puts away stock into the bin selected.

Some of the standard applications for carousels are

- Picking less than full case lots of small items for customer or dealer orders
- Storing small parts or subassemblies on the shop floor or in stockrooms
- Storing tools, maintenance parts, or other items that require limited access or security
- Storing work-in-process kits for assembly operations
- Storing documents, tapes, films, manuals, blueprints, etc.
- Storage and accumulation of parts between processing operations
- Storage during electrical burn-in of products by mounting a continuous oval electrified track on top of the carousel

5 Material-Handling Equipment Considerations and Examples   **393**

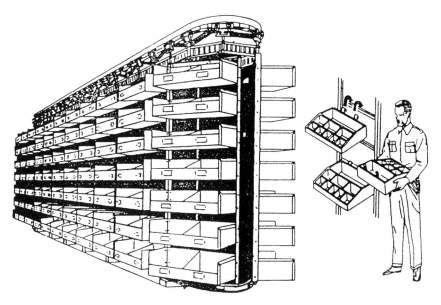

Figure 40   Horizontal carousel.

Since carousels bring the bins to the operator, multiple carousel units can be placed adjacent to each other (≤ 18 in.), with no aisles. Carousels can also be multitiered, with the various tiers capable of rotating in directions and at speeds independent of the other tiers. These designs result in very dense but easily accessible storage systems.

Carousels can be of almost any length, from a minimum of 10 feet to over 100 feet. However, most are in the 30- to 50-foot range so as to minimize bin access times. Carousels may be arranged so that one operator can pick or put away items in bins located on different systems or tiers while the other systems or tiers are positioning the next bin for access.

5.8 Shelving, Bin, Drawer, and Rack Storage

Shelving is used to economically store small, hand-stackable items that are generally not suited to mechanized handling and storage due to their handling characteristics, activity, or quantity. Standard shelving units are limited to about seven feet in height, but mezzanines can be used to achieve multiple storage levels and density.

Bin storage is, in most instances, identical in application to shelf storage, but is generally used for small items that do not require the width of a conventional shelf module. Bin storage usually represents a small part of the total storage system in terms of physical space, but it may represent a significant portion of the total storage in terms of the item positions used or stock keeping units (SKUs).

Modular drawer cabinets provide the advantages of increased security and density of storage over shelving and bin storage. As illustrated in Fig. 41, drawers provide the operator a clear view of all parts stored in them when pulled out. They can be configured to hold a large number of different SKU parts by partitioning the drawer volume into separate storage cells with dividers. This provides high storage density, good organization, and efficient utilization of storage space for small parts in applications such as tool cribs, maintenance shops, and parts supply rooms.

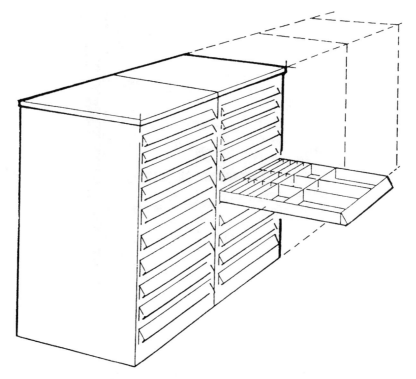

Figure 41 Modular cabinet drawer storage.

A pallet rack, as illustrated in Fig. 42, is a framework designed primarily to facilitate the storage of unit loads. This framework consists of upright columns and horizontal members for supporting the loads, and diagonal bracing for stability. The structural members may be bolted into place or lock-fitted.

Standard pallet racks can also be equipped with shelf panel inserts to facilitate their use for storage of binnable or shelvable materials. They may be loaded or unloaded by forklift trucks, by AS/R machines, or by hand. They may be fixed into position or made to slide along a track for denser storage. The primary purpose in using rack storage is to provide a highly organized unit material storage system that facilitates highly efficient operations in either manufacturing or distribution.

6 IMPLEMENTING THE SOLUTION

After the best system for solving the material handling problem has been designed, it is recommended that computer simulation be used to test the design before implementation. Although somewhat expensive to build and time-consuming to use, a valid simulation model can effectively test the overall operation of the material handling system as designed. It can identify potential bottleneck flows or choke points, isolate other costly design errors, determine efficient labor distribution, and evaluate various operating conditions that can be encountered. In other words, simulation enables the material-handling system designer to look into the future and get a realistic idea of how the system will operate before proceeding to cost justification—or before the millwrights start bolting the wrong equipment together.

6 Implementing the Solution   395

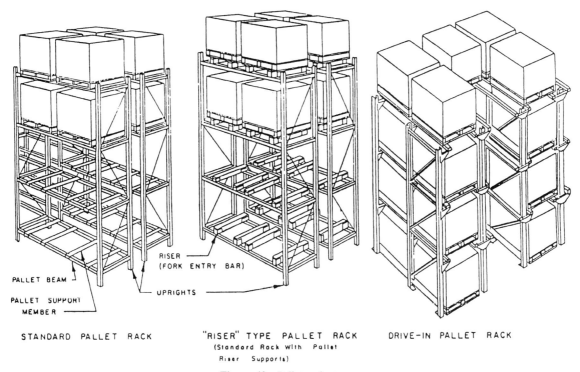

Figure 42 Pallet rack storage.

The final step is to implement the solution. Once total system costs—initial costs, recurring costs and salvage costs—have been calculated, an engineering economic analysis should be done to justify the investments required. Then the justification must be presented to, and approved by, appropriate managers. Once approval is obtained, a carefully prepared, written bid specification called a *request for quotation* (RFQ) is typically sent to several qualified vendors or contractors. Competing bids or proposals submitted by the vendors or contractors must then be evaluated carefully to ascertain whether they all are quoting on the same type and grade of equipment and components.

Each step of the equipment-acquisition process must be closely monitored to ensure that any construction is accomplished in a correct and timely manner and that equipment-installation procedures are faithfully followed. Once the completed facility is operational, it should be fully tested before final acceptance from the vendors and contractors. Operating personnel must be fully trained to use systems installed in the new facility.

CHAPTER 11
COATINGS AND SURFACE ENGINEERING: PHYSICAL VAPOR DEPOSITION

Allan Matthews
Department of Engineering Materials
Sheffield University
Sheffield, United Kingdom

Suzanne L. Rohde
Department of Mechanical Engineering
The University of Nebraska
Lincoln, Nebraska

1	INTRODUCTION	396	4	PROCESS DETAILS	400
				4.1 Evaporative Processes	400
2	GLOW DISCHARGE PLASMA	397		4.2 Sputter Deposition Processes	402
				4.3 Beam Processes	410
3	FILM FORMATION AND GROWTH	400		REFERENCES	411

1 INTRODUCTION

The term *physical vapor deposition* (PVD) is used to describe processes in which at least one of the coating species is atomised from a solid source within a coating chamber, to then condense on a substrate, forming a film. It is different from *chemical vapor deposition* (CVD), as that process utilizes gaseous reagents as the source of coating material. Also, PVD is typically carried out under low-pressure vacuum conditions, whereas CVD can be performed over a wide range of operating pressures, from high vacuum to atmospheric.

PVD processes can be categorized according to the means of atomizing the source material; the main division, as shown in Fig. 1, is between evaporative methods and sputtering. Evaporation, as the name suggests, involves the thermal vaporization of the source, whereas sputtering is a kinetically controlled mechanism in which the source material or "target" is bombarded with gas atoms, which then transfer momentum to atoms in the target, leading to the ejection of coating atoms. The sputtering and evaporation techniques both originated at about the same time. The first sputtering experiments were reported by Grove[1] in 1852 and the first reports of evaporation were by Faraday[2] in 1857.

Although PVD was originally used as a means of depositing elemental metallic coatings, it has been increasingly used for alloy and ceramic deposition. In the latter case this can be achieved by using a ceramic source or by using a metal source and inletting a reactive gas such as oxygen, nitrogen, or methane, for example, to produce oxides, nitrides, or carbides. Similarly, different means now exist to deposit multielement alloy or even composite metal/ceramic films. Furthermore, the process has been transformed by the addition of a plasma within the deposition chamber, which provides control of film nucleation and growth kinetics to allow the production of coatings with previously unachievable properties. In the first part

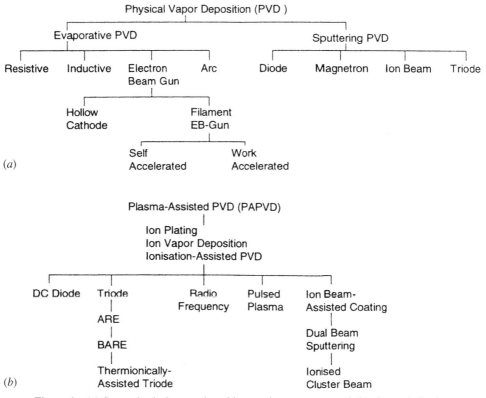

Figure 1 (a) Some physical vapor deposition coating processes, and (b) plasma derivatives

of this chapter we therefore discuss some basic plasma principles and then outline some aspects of the influence of plasma bombardment on coating morphology. Then the main evaporative and sputtering processes are discussed, followed by a description of some hybrid techniques.

2 GLOW DISCHARGE PLASMA

As mentioned above, plasma assistance is now used routinely in PVD processes. Usually the substrate is the cathode in such systems and is bombarded by ions prior to and during film growth. To explain the mechanisms occurring, it is convenient to consider a simple dc diode argon plasma, shown schematically in Fig. 2.[3] The "negative glow" is the visible plasma, which is a partially ionized gas with equal numbers of positive and negative charges (ions and electrons). The plasma is thus virtually field-free; most of the voltage in the discharge is dropped across the cathode sheath, which is also called the cathode fall region. Positive ions are thus accelerated toward the cathode. However, they may undergo charge exchange collisions with neutrals as they traverse the sheath region. This means that both accelerated neutrals and ions will arrive at the cathode with a range of energies. Davis and Vanderslice[4] showed how the energy spectra can be calculated for a dc diode discharge, and Fancey and Matthews[3,5] have indicated how these results will apply in plasma-assisted (PA)

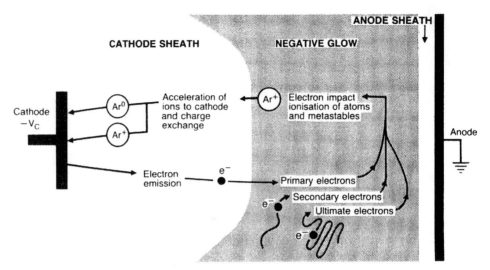

Figure 2 Schematic of an argon dc diode glow discharge

PVD systems. A key parameter is the ratio between L (the cathode fall distance) and λ (the mean free path for charge exchange). As we shall see, modern PAPVD systems utilize ionization enhancement devices, which reduce L and thus L/λ, which results in few collisions in the sheath and most of the ions arriving with the full acceleration energy. This gives the benefit that more will thus have the energy required to create the surface nucleation and growth effects necessary for optimised coatings. Some of these energy requirements are listed in Table 1.

The original PAPVD system was patented by Berghaus[6] in 1938, but it was not until Mattox[7] coined the term ion plating in the early 1960s that the potential of the process was recognized. These researchers used the dc diode configuration. Enhancing this system not only means that the ions will not undergo energy-reducing collisions, it also means that there will be more ions available. Several measures can be used to define the actual level of ionization. For example, the term "ionization efficiency" has been used to define the percentage ratio between the ions arriving at the substrate surface during deposition and the total bombardment.[3,5,8,9] Usually the latter parameter is derived from the background chamber pressure. This, however, assumes that the arrival rate of depositing species is very much lower than the arrival rate of "background" gas atoms. Others have taken a rather different view, and have chosen to use the ratio between the ion current and the metal species arrival rate as the determining parameter.[10–12] This latter approach was first used by researchers who modeled growth using computer routines to predict the coating structure.[13,14] Both of these approaches can be criticized to some extent. The former becomes increasingly invalid for higher deposition rates and lower chamber pressures, while the latter does not take into account the background chamber pressure (which is known to have a very dominant influence on coating structure). Notwithstanding these deficiencies, both of these measures of process effectiveness confirm that the most important single goal in plasma-assisted PVD is to achieve an adequate level of ionisation, and different processes ensure this occurs by a variety of means, as discussed in the following sections.

Table 1 Important Mechanisms in Plasma-Assisted PVD (From Ref. 3)

Mechanism	Description	Effects	Energy Requirements
Ionization	Electrons emitted from substrate accelerate across substrate (cathode) sheath to gain sufficient energy for electron impact ionization in negative glow region.	Maintains the discharge in a diode configuration; provides means of ionization in addition to any enhancement applied to the process.	For argon and many other gases (e.g., nitrogen), maximum collision cross section for electron impact ionization is typically 70–100 eV.
Substrate surface contamination	Desorption of adsorbed impurities on the substrate surface prior to deposition.	Prevents, for example, contaminants reducing adhesion between coating and substrate.	Several eV.
Adatom mobility	Removal/surface diffusion of adsorbed atoms on the growing coating surface.	Promotes coating densification.	Several eV.
Atomic displacement	Atoms in the substrate and coating are displaced from their normal sites, creating lattice defects.	Can lead to, for example, intermixing of substrate and coating atoms. Increased defect densities promote rapid interdiffusion. Improves coating–substrate adhesion. May also increase coating desification.	Thresholds are in the region of 20–50 eV.
Sputtering	Substrate and subsequent growing coating is sputtered.	Can be used to clean the substrate prior to deposition. Promotes atomic mixing, which can improve coating–substrate adhesion and coating densification.	Thresholds (argon bombarding metals) are in the region of 15–35 eV.
Entrapment	Support gas (argon) is incorporated in coating during deposition.	Gas atoms may cluster within the coating to form bubbles. May be detrimental to coating properties.	Argon entrapment probability is believed to be very low, below 100 eV.

Source. Reference 3.

3 FILM FORMATION AND GROWTH

As Jehn[15] has pointed out, a number of film nucleation and growth models have been published. However, these typically apply only for certain idealized conditions, such as a single-phase substrate surface, no alloy or compound formation, and deposition under ultra-high vacuum conditions. Such models are not strictly applicable for ceramic deposition, especially under plasma bombardment conditions. It is thus perhaps more instructive to consider here those models that rely on empirical observation to describe the nature of film growth morphology under different conditions.

The most widely reported of these models is the Thornton structure zone diagram[16] (Fig. 3), which was an extension on the earlier zone model of Movchan and Demchishin.[17] The zone models are based on the observation that PVD coatings deposited under high-pressure (i.e., medium-vacuum) conditions tend to be porous and columnar. This is known as a zone 1 structure. At higher temperatures and/or lower pressures the coatings become more fine grained, densely packed, and fibrous. Thornton called this a zone T structure. At higher temperatures the films exhibit a dense columnar morphology known as zone 2. At even higher temperatures the films exhibit an equiaxed grain structure (zone 3) similar to a recrystallized solid. Under nonbombardment conditions zone 2 morphologies would normally occur only at substrate temperatures above about 50% of the melting temperature of the coating material (i.e., over 1200 K for many ceramics). The remarkable effect of ion bombardment is that it can induce such structures at substrate temperatures of less than one tenth of the coating material melting temperature.

4 PROCESS DETAILS

4.1 Evaporative Processes

Although there are several ways by which the metal can be evaporated in PAPVD, as given in Fig. 1, the electron beam and arc techniques dominate. These are considered separately here.

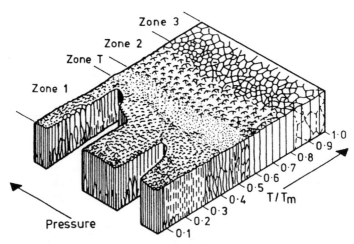

Figure 3 The Thornton zone diagram. T (K) is the substrate temperature and T_m (K) is the melting point of the coating material.

In electron beam evaporation the electron source is most often a heated filament;[8,9,18,19] alternatively, the "hollow cathode" effect can be utilized to produce a collimated or diffuse beam of electrons from a confining tube[20-23] (Fig. 4). In either case the resultant beam may be magnetically focused onto a crucible to achieve evaporation and this crucible may itself be biased positively to permit greater control of the power input (and also the plasma conditions in the deposition chamber)[24-26] (Fig. 5). Many systems that utilize electron beam evaporation adopt configurations in which the electron energy is relatively low (e.g., tens of eV) with a high current flow. Additionally, the beam is frequently directed through the deposition chamber volume. Such arrangements ensure that the electron beam significantly enhances the ionization. In some cases the electrons may be used to directly preheat the components to be coated, by biasing them positively prior to the initiation of the precoating and deposition plasmas.[24,25] Additional control of the plasma during deposition may be achieved by the incorporation of magnetic confinement, or by features such as an additional positive electrode[27,28] and/or electron emission source[29,30] (Fig. 6).

Such modifications are bound to induce inhomogeneity into the plasma, and this can have a detrimental influence on the uniformity of the resultant coatings.[31,32] However, the impact of such effects can be minimised by appropriately locating the enhancement device relative to the vapor source.[3,5]

An advantage of systems that incorporate ionization enhancement that is independent of the vapor source is that more effective plasma heating can be achieved prior to deposition. Also, independent enhancement provides the possibility to achieve controlled plasma diffusion processes (such as ion/plasma nitriding or carburizing),[33-35] which can be particularly useful as precoating surface treatments.[8,36,37]

An alternative evaporative process is arc evaporation, which has the advantage that it generates copious quantities of energetic electrons and achieves a high degree of metal ionization—both desirable characteristics for the optimization of PAPVD systems.[38-41] Figures 7 and 8[42] illustrate the arc evaporation process. Martin[42] has described this as follows. The emission site of a discrete cathode spot is active for a short period, extinguishes, and then reestablishes close to the original site. The emission site is a source of electrons and atoms of the source material, which are subsequently ionized above the arc site. The ions can either flow back toward the cathode or be accelerated outward toward anodic surfaces,

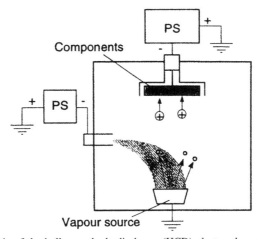

Figure 4 Schematic of the hollow cathode discharge (HCD) electron beam (EB) PVD process.

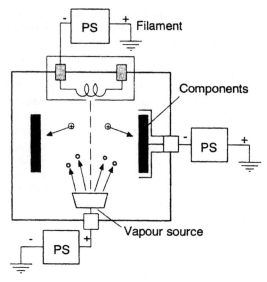

Figure 5 Schematic of the low-voltage thermionic electron beam (EB) PVD process.

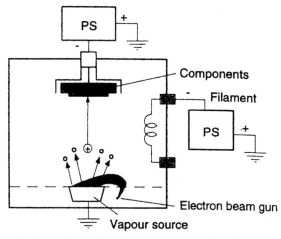

Figure 6 Schematic of the thermionically assisted triode PVD process.

together with the electrons. In the case of titanium the degree of ionisation of the metal can be over 80% and typical ion energies are 40–100 eV.

Certain designs of arc sources can also be run as magnetron sputtering sources.[43,44] It has been found that if the arc source is run during the sputter-cleaning phase, then coating adhesion can be improved.[45–47]

4.2 Sputter Deposition Processes

The simplest sputtering plasma is that of dc diode glow discharge, and thus will provide the model in which the basics of the sputtering processes are discussed. In diode sputtering there

4 Process Details   403

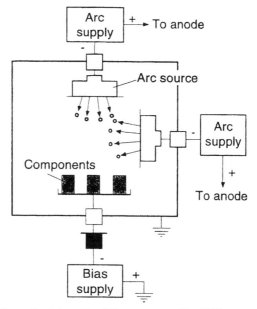

Figure 7 Schematic of the arc evaporation PVD process.

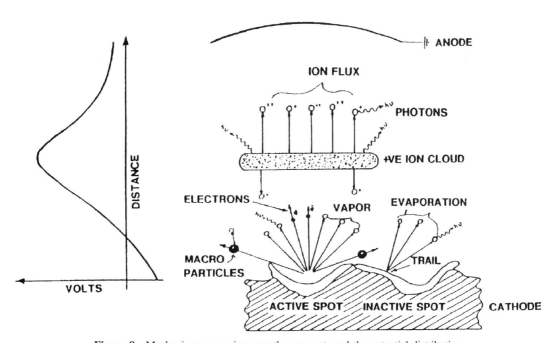

Figure 8 Mechanisms occurring near the arc spot, and the potential distribution.

are only two electrodes, a positive anode and a negatively biased cathode. If the potential applied between these two electrodes is constant over time it is termed a dc diode discharge. However, if this potential changes with time it may be either a pulsed-diode discharge (unipolar or bipolar), which has only recently come into use for sputtering applications, or if the frequency is sufficiently high and reverses in polarity, it is a radio frequency (rf) diode discharge. Each of these has relative advantages and disadvantages, particularly in terms of ceramic coatings. However, all diode plasmas have certain similar mechanisms; in particular, the previous comments with regard to ions traversing the sheath and acquiring sufficient energy to permit kinetically induced mechanisms will apply.

dc Diode Sputtering

Although dc diode sputtering was used in the 19th century to deposit thin metallic films on mirrors, it was not until reliable radio-frequency sputtering was developed in the 1970s that it was possible to sputter insulating materials. Since many ceramic thin-film materials are good insulators, most early applications of sputtering were thus limited to the deposition of metallic coatings. DC diode sputtering would therefore not have been widely used in the deposition of ceramic thin films if the technology necessary for high-rate-reactive sputtering from a metallic target had not been developed in the early 1980s.[48,49]

Provided the target material is sufficiently conductive to avoid the buildup of charge on its surface, the target can be effectively sputtered using simple dc diode geometry. Typically, in dc diode sputtering a potential of about 1000 V is applied between the cathode and anode, and gas pressures in the range from a few mtorr up to 100 mtorr are used to generate a plasma. The substrates are immersed in the discharge, which expands to fill the chamber. The substrates can be heated, cooled, held at ground potential, electrically isolated (i.e., floated), or biased relative to the plasma, each of which affects the properties of the resulting thin films. In general, substrates in a dc diode discharge are subject to a significant amount of electron heating, making this process unsuitable for many semiconductor applications. However, in the case of ceramic thin films for engineering applications this additional heating can be advantageous.

rf Diode Sputtering

The use of an oscillating power source to generate a sputtering plasma offers several advantages over dc methods. The main one is that when the frequency of oscillation is greater than about 50 kHz, it is no longer necessary for both electrodes to be conductive because the electrode can be coupled through an impedance[50] and will take up a negative dc offset voltage due to the greater mobility of electrons compared to ions in the reversing field. The coupled electrode must be much smaller than the direct electrode to effectively sputter only the insulating (coupled) electrode; this is usually accomplished by utilizing the grounded chamber walls as the other electrode. An impedance matching network is integrated into the circuit between the rf generator and the load to introduce the inductance necessary to form a resonant circuit.

An additional benefit of using rf frequencies above 50 kHz is that the electrons have a longer residence in the negative glow region and also have sufficient energy to directly ionize the gas atoms; hence, the number of electrons required to sustain the discharge is substantially reduced.[51,52] This, in turn, means that lower sputtering pressures can be used, reducing the risk of film contamination. The most commonly used frequencies are 13.56 and 27 MHz; these are the FCC-specified frequencies for medical and industrial use.[50] The applications of rf sputtering are quite varied and include deposition of metals, metallic alloys, oxides, nitrides, and carbides.[53–56]

A number of comprehensive reviews and discussions of rf discharges are available in the literature.[50,57–60] In general, the primary advantages of rf sputtering are

- Ability to sputter insulators as well as almost any other material
- Operation at lower pressures

Unfortunately, the deposition rates in rf sputtering are often limited by the low thermal conductivity of the insulating target materials which can lead to the formation of "hot spots" on the target; the hot spots generate stresses that may cause fracture of the brittle target materials. For this reason, it may be preferable to deposit insulating films reactively from a metal source. Although compound materials can be readily sputtered in rf discharges, the resulting films may not be representative of the initial target composition.

Pulsed Power Sputtering

The recent availability of sophisticated electrical supplies capable of providing controlled pulsed power waveforms with excellent arc control has opened up new opportunities in sputter deposition.[61–63] In particular, pulse power has proved highly effective in the reactive sputtering of oxide coatings.

Sproul[64] describes the use of bipolar pulsed power for sputtering of metal targets in oxygen as follows. The polarity of the target power is switched from negative to positive, and during the positive pulse any charging of oxide layers formed on the target surface is discharged when electrons are attracted to the positive surface. During the negative pulse, ions are attracted to the target surface and sputtering takes place initially from all surfaces on the target, even those that have formed a compound, since the charge on that surface has been neutralised during the positive pulse.

Bipolar pulse power can be either symmetric or asymmetric, depending on the relative positive and negative maximum voltages.[65] Pulsed dc power is said to be symmetric if these voltages are equal. With modern power supplies it is possible to control the on and off time for the pulses.[65,66]

The successful use of symmetric bipolar pulsed dc power has been reported for the deposition of oxide coatings from two magnetron targets. These targets, mounted side by side, are both connected to the same bipolar pulsed dc power supply. One dc power lead is connected to one target, and the other to the second target. With this system, one of the sputter targets is briefly the anode, while the other is the cathode and this continuously swaps, with reversals in polarity. Sputtering from the cathode surface during the negative pulse keeps the target surface clean, and when it switches to act as an anode it is not covered by an oxide. As pointed out by Sproul, this procedure avoids the disappearing anode problem, which can occur in pulsed dc sputtering of oxides when all surfaces in the chamber become covered with an insulating oxide.

Asymmetric bipolar pulsed dc has unequal pulse heights and will be applied to one target. Usually, the negative pulse voltage is greater than the positive one, and in this case there is usually no off time between pulses. The width of the positive pulse is typically 10–20% of the negative one, thus a high proportion of the cycle is spent in sputtering mode, providing deposition rates that are near to those for nonreactive dc sputtering of metals, provided that pulsed dc power is combined with effective partial pressure control. For example, Ref. 67 reports reactive sputtering of aluminium in oxygen at 76% of the metal rate under non-reactive conditions. Typical frequencies used are 20–200 kHz. Sproul[64] states that the frequency selected depends on the material being deposited. Whereas no arcing occurs for titanium dioxide deposition at a pulse frequency of 30 kHz, it takes 50–70 kHz to prevent arcing in aluminium oxide deposition.

Triode Sputtering

Triode sputtering uses a thermionic cathode separate from the sputtering target to sustain the plasma. The target electrode then extracts ions from the plasma. This additional electron source, typically either a simple biased conductor or a thermionic electron emitter, provides a means of sustaining the discharge that is independent of the secondary electron generation at the cathode. Thus, the discharge may be maintained at pressures as low as 10^{-3} Pa (10^{-5} torr) and at discharge voltages as low as 40 V.[68] By varying the emission of the electron source, the discharge current can be varied independently of the sputtering voltage, allowing high ion densities at the target and substrate while maintaining a low discharge potential.

Triode sputtering, both dc and rf, has been used successfully to deposit films of a great variety of materials for semiconductor, wear-resistant, optical, and other coating applications.[68–70] The primary advantages of triode sputtering are[70]

- Lower discharge pressures
- Lower discharge voltages
- Higher deposition rates
- Independent control of the plasma density and the bombardment conditions of the sputtering target

The major weaknesses of triodes are that they

- Are often more complicated to use
- Can increase film contamination from the electron source
- Are difficult to scale up for industrial processing
- May not be suitable in temperature-sensitive and reactive processes because of the electron source

Magnetron Sputtering

Magnetron sputtering has already been mentioned in regard to pulsed dc power. It is a variant on the diode sputtering process, in which a magnetic field is used to trap electrons in the vicinity of the target and to spiral them round a "racetrack," thereby increasing the degree of ionization occurring and therefore the sputtering rate. Probably the greatest research advances in PAPVD in recent years have occurred around the magnetron sputtering process. The benefits of magnetron sputtering have been known for several decades (i.e., in terms of increasing the deposition rate). However, until very recently sputtering was found to be less effective as a means of producing hard ceramic coatings on tools and components than electron beam and arc evaporation. This was because the levels of ionization achieved were considerably less than those possible with the enhanced plasmas, which are characteristic of these other methods.

The most important step in improving the competitiveness of magnetron sputtering was reported by Window and Savvides[71,72] in 1986, when they described their studies on the unbalanced magnetron (UBM) effect. They found that the ion flux to the substrate could be considerably increased by running magnetron cathodes in "unbalanced" mode (UM). They identified three main magnetic arrangements (Fig. 9). In the type I configuration all the field lines originate from the central magnet, with some not passing into the outer magnet. In the intermediate case all of the field lines starting on the central pole go to the outer pole. In the type II configuration, all of the field lines originate in the outer magnet, with some not passing to the central pole. Following the work of Window and Savvides, a number of researchers began building upon these concepts.[70] In the late 1980s Sproul and his co-workers researched multicathode high-rate-reactive sputtering systems.[73,74] They studied different magnetic configurations, principally in the dual-cathode arrangement shown schematically in Fig. 10. They found that by strengthening the outer magnets of a magnetron cathode with

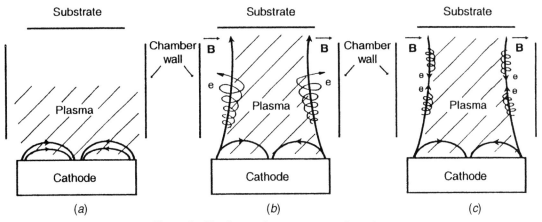

Figure 9 The three main magnetron configurations.

NdFeB and arranging two of these magnetrons in an opposed closed field" configuration (i.e., with opposite poles facing each other), the substrate bias current could be increased to well over 5 mA/cm^2 at a pressure of 5 mtorr. Under these conditions they were able to deposit hard well-adhered TiN coatings.

Another person utilizing the opposed magnetron configuration at that time was Tominaga.[75] He presented an arrangement whereby the permanent magnetic field could be in-

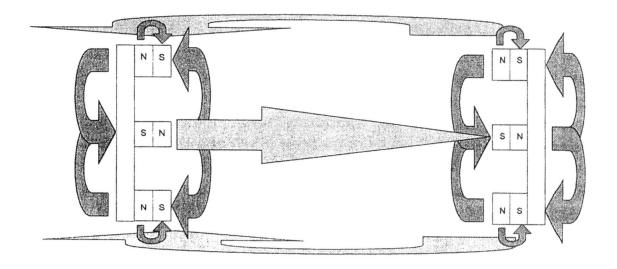

Magnetic field lines

Figure 10 Schematic of the opposed unbalanced magnetron configuration.

creased by external electromagnetic coils. He stated that this allowed the trapping of electrons by the closed field arrangement to increase ionization or to allow lower working gas pressure with the same level of ionization as achievable with conventional magnetron sputtering.

Another group that has researched confined and unbalanced magnetron sputtering is that led by Musil and Kadlec,[76,77] who in early collaboration with Munz investigated a circular planer unbalanced magnetron arrangement surrounded by two magnetic coils and a set of permanent magnets. They described this as a multipolar magnetic plasma confinement (MMPC), and cite high ion currents at substrates, even when located at large distances from the magnetron.

Howson and co-workers have also studied unbalanced magnetrons, especially for high-power, large-area applications,[78,79] and thus demonstrated how additional anodes or electro-

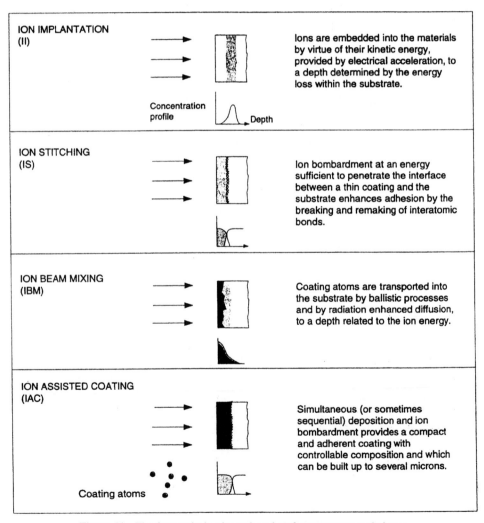

Figure 11 The four main ion beam-based surface treatment techniques.

magnets can be used to control the ion current to the substrate. This may be seen as an extension of early work by Morrison,[80] who demonstrated that a "magnetically hidden" anode placed in a magnetron sputtering system could double the plasma density and provide a more uniform plasma throughout the chamber.[81] Hofman et al. have described an improved magnetron sputtering system, which combines the magnetic anode effect and the opposed closed-field unbalanced magnetron.[82] Leyendecker et al. utilize a further modification on this theme.[83]

A recent innovation to the magnetron sputtering technique has been described by Rossnagel and Hopwood.[84] They placed a rf coil between the target and the substrate, to increase the degree of ionization of the coating species. In the case of aluminium he achieved ionization levels of up to 80%. Although the technique was developed for metal films for use in metallization of small aspect ratio semiconductor structures, it has been shown that reactive deposition of ceramics can be made more effectively by this method. In particular the use of this approach in the reactive deposition of crystalline alumina has been achieved at much lower temperatures than previously reported.[85,86]

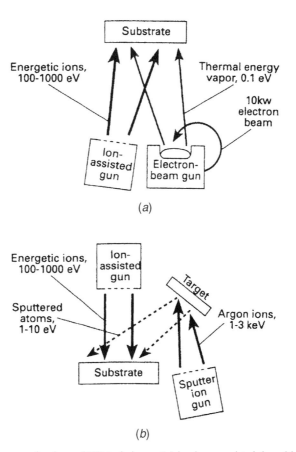

Figure 12 Two common ion beam PVD techniques: (*a*) ion beam-assisted deposition (IBAD), and (*b*) dual ion beam sputtering (DIBS).

4.3 Beam Processes

A third group of processes that can be included under the PVD heading are those that utilize ion or laser beams, either to influence the growing film or to produce the coating flux. Dearnaley[87] has cited the four ion beam processes, shown in Fig. 11[88] as representing the effects that ion beams can have on surfaces. The ion-assisted coating (IAC) or ion beam-assisted deposition (IBAD) methods compare most closely with the plasma-based PVD processes described earlier. Hubler and Hirvonen[89] cite the two geometries shown in Fig. 12 as being the most common of these. According to these authors, what distinguishes IBAD from the other PVD methods is that the source of vapour and the source of energetic ions are separated into two distinct pieces of hardware. In many plasma-based PVD processes the evaporant flux and the ion flux are derived by extraction from the plasma, whereas in IBAD they can be controlled independently. The other major difference between the plasma techniques and IBAD is the chamber pressure used. Typically PAPVD processes are carried out at pressures of several millitorrs, whereas IBAD techniques must usually operate in a near collision-free pressure regime, better than 10^{-5} torr. This means that IBAD is often restricted to line-of-sight applications. Ceramics such as ZrO_2,[90] TiO_2,[91] Al_2O_3,[92] Si_3N_4,[93] AlN,[94] TiN,[95] BN,[96] and TiC[97] have been deposited by IBAD methods.

There is also considerable interest in processes to deposit diamond-like carbon (DLC) films using IBAD methods, or even by direct gas deposition from an ion or plasma source.[98–100] A further variation on this theme is the filtered-arc source[101–103] that utilizes deflection coil to eliminate macrodroplets and produce a beam of ionized atoms, which can be carbon (for DLC films) or a metal such as titanium for reactive deposition of a ceramic.

Another important beam technique in PVD is pulsed laser deposition. According to Voevodin and Donley[104] this is especially appropriate for DLC coatings, as the controlled

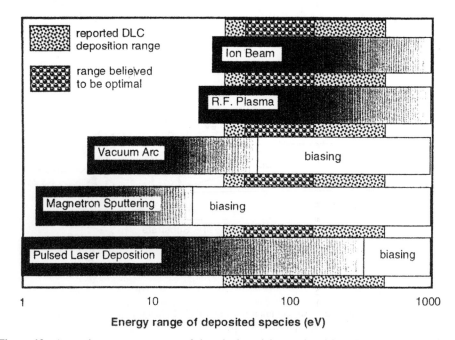

Figure 13 Approximate energy ranges of deposited particles produced by selected PVD techniques. Energy regions corresponding to DLC formation are also shown.

energy range achievable for the arriving species is wider than other techniques (Fig. 13). Also flux control and species control is said to be excellent, for example nonhydrogenated DLC can readily been produced. Several ceramics have also been deposited by this method.[105,106]

REFERENCES

1. W. R. Grove, *Philos. Trans. R. Soc. London A,* **142,** 87 (1852).
2. M. Faraday, *Philos. Trans.,* **147,** 145 (1857).
3. K. S. Fancey and A. Matthews, in *Advanced Surface Coatings,* D. S. Rickerby and A. Matthews (eds.), Blackie, London, 1991.
4. W. D. Davis and T. A. Vanderslice, *Phys. Rev.,* **131,** 219 (1963).
5. K. S. Fancey and A. Matthews, *IEE Trans. Plasma Sci.,* **18,** 869 (1990).
6. B. Berghaus, UK Patent 510993, 1938.
7. D. M. Mattox, *Electrochem. Technol.,* **2,** 95 (1964).
8. A. Matthews, *J. Vac. Sci. Technol.,* **A3**(6), 2354 (1985).
9. A. Matthews, in *Physics and Chemistry of Protective Coatings,* W. D. Sproul, J. E. Greene, and J. A. Thornton (eds.), AIP Proceedings 149 (1986).
10. I. Petrov, F. Adibi, J. E. Green, L. Hultman, and J.-E. Sundgren, *Appl. Phys. Lett.,* **63,** 36 (1993).
11. G. Hakansson, L. Hultman, J.-E. Sundgren, J. E. Greene, and W.-D. Munz, *Surf. Coat. Technol.,* **48,** 51 (1991).
12. F. Abidi, I. Petrov, J. E. Greene, L. Hultman, and J.-E. Sundgren, *J. Appl. Phys.,* **73,** 8580 (1993).
13. K.-H. Muller, *Phys. Rev. B,* **35,** 7906 (1987).
14. K.-H. Muller, *J. Vac. Sci. Technol.,* **A5,** 2161 (1987).
15. H. A. Jehn, in *Advanced Techniques for Surface Engineering,* W. Gissler and H. A. Jehn (eds.), Kluwer Academic, New York, 1992.
16. J. A, Thornton, *J. Vac. Sci. Technol.,* **11,** 666 (1974).
17. B. A. Movchan and A. V. Demchishin, *Fiz. Metal. Metalloved.,* **28,** 653 (1969).
18. Y. Enomoto and K. Matsubara, *J. Vac. Sci. Technol.,* **12,** 827 (1975).
19. H. K. Pulker, *Coatings on Glass,* Elsevier, Amsterdam, 1984.
20. J. R. Morley and H. R. Smith, *J. Vac. Sci. Technol.,* **9,** 1377 (1972).
21. S. Komiya and K. Tsuruska, *J. Vac. Sci. Technol.,* **13,** 520 (1976).
22. C. T. Wan, D. L. Chambers, and D. C. Carmichael, in *Proceedings of the Fourth International Conference on Vacuum Metallurgy,* Tokyo, Japan (Iron and Steel Institute of Japan), 1973, p. 23.
23. S. Komiyas and K. Tsuruokak, *J. Vac. Sci. Technol.,* **13,** 520–524 (1976).
24. E. Moll, in W. Gissler and H. Jehn (eds.), *Advanced Techniques for Surface Engineering,* Kluwer, Dordrecht, 1992.
25. E. Moll and H. Daxinger, U.S. Patent 4 197 175, 1980.
26. R. Buhl, H. K. Pulker, and E. Moll, *Thin Solid Films,* **80,** 265–270 (1981).
27. M. Koboyashi and Y. Doi, *Thin Solid Films,* **54,** 17 (1978).
28. R. F. Bunshah, *Thin Solid Films,* **80,** 255 (1981).
29. A. Matthews and D. G. Teer, *Thin Solid Films,* **72,** 541 (1980).
30. G. A. Baum, Dow Chemical Co., Report No. RFP-686 UC-25, Colorado, 1967.
31. K. S. Fancey, M. Williams, A. Leyland, and A. Matthews, *Vacuum,* **43,** 235 (1992).
32. K. S. Fancey and A. Matthews, *Thin Solid Films,* **193–194,** 171 (1990).
33. A. Leyland, K. S. Fancey, and A. Matthews, *Surf. Eng.,* **7,** 207 (1991).
34. P. R. Stevenson, A. Leyland, M. A. Parkin, and A. Matthews, *Surf. Coat. Technol.,* **63,** 135–143 (1994).
35. A. Leyland, D. B. Lewis, P. R. Stevenson, and A. Matthews, *Surf. Coat. Technol.,* **62,** 608–617 (1993).
36. D. B. Lewis, A. Leyland, P. R. Stevenson, J. Cawley, and A. Matthews, *Surf. Coat. Technol.,* **60,** 416–423 (1993).

37. A. Matthews and A. Leyland, *Surf. Coat. Technol.,* **71,** 88–92 (1995).
38. P. A. Lindfors, W. M. Mularie, and G. K. Wehner, *Surf. Coat. Technol.,* **29,** 275 (1986).
39. R. L. Boxman and S. Goldsmith, *Surf. Coat. Technol.,* **33,** 152 (1987).
40. P. J. Martin, D. R. McKenzie, P. P. Netterfield, P. Swift, S. W. Filipczuk, K. J. Muller, C. G. Pacey, and B. James, *Thin Solid Films,* **153,** 91 (1987).
41. D. M. Sanders, D. B. Boercker, and S. Falabella, *IEEE Trans. Plasma Sci.,* **18,** 883 (1990).
42. P. H. Martin, Lecture Summaries of the Workshop on Current Topics in Advanced PVD Technology for Optical and Wear Resistant Applications, San Diego, April 1991.
43. P. A. Robinson and A. Matthews, *Surf. Coat. Technol.,* **43–44,** 288 (1990).
44. Hauzer Holding BV, Eur. Patent Appl. PCT/EP90 01032 (1990).
45. W.-D. Munz, F. J. M. Hauzer, D. Schulze, and B. Buil, *Surf. Coat. Technol.,* **49,** 161 (1991).
46. W.-D. Munz, D. Schulze, and F. J. M. Hauzer, *Surf. Coat. Technol.,* **50,** 169 (1992).
47. W.-D. Munz, K. Vannisselroy, R. Tietma, T. Turkmans, and G. Keiren, *Surf. Coat. Technol.,* **58,** 205 (1993).
48. W. D. Sproul, *Thin Solid Films,* **107,** 141–147 (1983).
49. W. D. Sproul and J. A. Tomashek, United States Patent No. 4 428 811 (1984).
50. R. J. Hill, (ed.), *Physical Vapor Deposition,* Temescal, Livermore, CA, 1986.
51. J. L. Vossen and J. J. Cuomo, "Glow Discharge Sputter Deposition," in *Thin Film Processes,* J. L. Vossen and W. Kern (eds.), Academic Press, New York, 1978, pp. 12–73.
52. J. A. Thornton, "Coating Deposition by Sputtering," in *Deposition Technologies for Thin Films and Coatings,* R. F. Bunshah (ed.), Noyes, Park Ridge, NJ, 1982, pp. 170–243.
53. B. Chapman, *Sputtering, Glow Discharge Process,* Wiley, New York, 1980, pp. 177–296.
54. J.-E. Sundgren, B.-O. Johansson, and S. E. Karlsson, *Thin Solid Films,* **105,** 353–366 (1983).
55. J.-E. Sundgren, B.-O. Johansson, S.-E. Karlsson, and H. T. G. Hentzell, *Thin Solid Films,* **105,** 367–384 (1983).
56. J.-E. Sundgren, B.-O. Johansson, S.-E. Karlsson, and H. T. G. Hentzell, *Thin Solid Films,* **105,** 385–393 (1983).
57. J. L. Cecchi, "Introduction to Plasma Concepts and Discharge Configurations," in *Handbook of Plasma Processing Technology,* S. M. Rossnagel, J. J. Cuomo, and W. D. Westwood (eds.), Noyes, Park Ridge, NJ, 1990, p. 14,069.
58. J. S. Logan, "RF Diode Sputter Etching and Deposition," in *Handbook of Plasma Processing,* S. M. Rossnage, J. J. Cuomo, and W. D. Westwood (eds.), Noyes, Park Ridge, NJ, 1990, pp. 140–159.
59. B. Chapman, "RF Discharges," in *Glow Discharge Processes,* Wiley, New York, pp. 139–175, 1980.
60. G. N. Jackson, *Thin Solid Films,* **5,** 209–246 (1970).
61. S. Schiller, K. Goedicke, V. Kirchoff, and T. Kopte, *Proceedings of the 38th Annual Technical Conference of the Society of Vacuum Coaters, Chicago, April 1995,* SVC, Albuquerque, 1995.
62. P. Frach, Chr. Gottfried, H. Bastzsch, and K. Goedicke, *Proceedings of the Int. Conf. on Metallurgical Coatings and Thin Films,* San Diego, April 1996.
63. S. Schiller, K. Goedicke, J. Reschke, V. Kirchhoff, S. Schneider, and F. Milde, *Surf. Coat. Technol.,* **61,** 331–227 (1993).
64. W. D. Sproul, *Proceedings of the 39th Annual Technical Conference of the Society of Vacuum Coaters, Philadelphia, May 1996,* SVC, Albuquerque, NM, 1996.
65. R. A. Scholl, *Proceedings of the 39th Annual Technical Conference of the Society of Vacuum Coaters, Philadelphia, May 1996,* SVC, Albuquerque, NM, 1996.
66. J. C. Sellers, *Proceedings of the 39th Annual Technical Conference of the Society of Vacuum Coaters, Philadelphia, May 1996,* SVC, Albuquerque, NM, 1996.
67. J. M. Schneider, A. A. Voevodin, M. S. Wong, W. D. Sproul, and A. Matthews, *Surf. Coat. Technol.,* **96,** 262 (1997).
68. S. L. Rohde, S. A. Barnett, and C.-H. Choi, *J. Vac. Sci. Technol.,* **7**(3), 2273–2279 (1989).
69. G. Mah, C. W. Nordin, and V. F. Fuller, *J. Vac. Sci. Technol.,* **11**(1), 371–373 (1974).
70. S. L. Rohde, in *ASM Handbook,* Vol. 5: *Surface Engineering,* ASM International, Metals Park, OH, 1994.
71. B. Window and N. Savvides, *J. Vac. Sci. Technol.,* **A4**(3), 196 (1986).

72. N. Savvides and B. Window, *J. Vac. Sci. Technol.,* **A4**(3), 504 (1986).
73. W. D. Sproul, P. J. Rudnik, M. E. Graham, and S. L. Rohde, *Surf. Coat. Technol.,* **43/44,** 270 (1990).
74. S. L. Rohde, I. Petrov, W. D. Sproul, S. A. Barnett, P. J. Rudnik, and M. E. Graham, *Thin Solid Films,* **193/194,** 117 (1990).
75. K. Tominaga, *Vacuum,* **41,** 1154 (1990).
76. S. Kadlec, J. Musil, V. Valvoda, W.-D. Munz, H. Pewtersei, and J. Schroeder, *Vacuum,* **41**(7–9), 2233 (1990).
77. S. Kadlec, J. Musil, and W.-D. Munz, *J. Vac. Sci. Technol.,* **A8**(3), 1318 (1990).
78. R. P. Howson, H. A. J'Afer, and A. Spencer, *Thin Solid Films,* **193/194,** 127 (1990).
79. R. P. Howson and H. A. J'Afer, *J. Vac. Sci. Technol.,* **A10**(4), 1784 (1992).
80. C. F. Morrison, United States Patent No. 4351472 (1982).
81. C. F. Morrison, R. P. Welty, *Anodic Plasma Generation in Magnetron Sputtering,* Boulder Colorado, Vac-Tec Systems, Boulder, CO, 1982.
82. D. Hofman, S. Beisswanger, and A. Feuerstein, *Surf. Coat. Technol.,* **49,** 330 (1991).
83. T. Leyendecker, O. Lemmer, S. Esser, and J. Ebberink, *Surf. Coat. Technol.,* **48,** 175 (1991).
84. S. M. Rossnagel and J. Hopwood, *J. Vac. Sci. Technol.,* **12,** 449 (1994).
85. J. M. Schneider, W. D. Sproul, and A. Matthews, *Surf. Coat. Technol.,* **98,** 1473 (1998).
86. J. M. Schneider, W. D. Sproul, A. A. Voevodin, and A. Matthews, *J. Vac. Sci. Technol.,* **A15**(3), 1084 (1997).
87. G. Dearnaley, in *Advanced Surface Coatings: A Handbook of Surface Engineering,* D. S. Rickerby and A. Matthews, Blackic, London, 1991.
88. K. O. Legg, J. K. Cochran, H. Solnik-Legg, and X. L. Mann, *Nucl. Inst. Meth. Res.,* **137/138,** 535 (1985).
89. G. K. Hubler and J. K. Hirvonen, *ASM Handbook,* Vol. 5: *Surface Engineering,* ASM International, Metals Park, OH, 1994.
90. P. J. Martin, R. P. Netterfield, and W. G. Sainty, *J. Appl. Phys.,* **55,** 235 (1984).
91. P. J. Martin, H. A. Macleod, R. P. Netterfield, C. G. Pacey, and W. G. Sainty, *Appl. Opt.,* **22,** 178 (1983).
92. F. L. Williams, R. D. Jacobson, J. R. McNeil, G. J. Exarhos, and J. J. McNally, *J. Vac. Sci. Technol.,* **6,** 2020 (1988).
93. G. K. Hubler, C. A. Carosella, P. G. Burkhalter, R. K. Feitag, C. M. Cotell, and W. D. Coleman, *Nucl. Instrum. Methods Phys. Res. B.,* **59/60,** 268 (1991).
94. J. D. Targove, L. J. Lingg, J. P. Lecham, C. K. Hwangbo, H. A. Macleod, J. A. Leavitt, and L. C. McIntyre, Jr. in *Materials Modification and Growth Using Ion Beams,* U. J. Gibson, et al. (eds.), *Proc. Mater. Res. Soc. Symp.,* **94,** 311 (1987).
95. R. A. Kant, S. A. Dillich, B. D. Sartwell, and J. A. Sprague, *Proc. Mater. Res. Soc. Symp.,* **128,** 427 (1989).
96. W. G. Sainty, P. J. Martin, R. P. Netterfield, D. R. McKenzie, D. J. H. Cockayne, and D. M. Dwarte, *J. Appl. Phys.,* **64,** 3980 (1988).
97. S. Pimbert-Michaux, C. Chabrol, M. F. Denanot, and J. Delafond, *Mater. Sci. Eng. A,* **115,** 209 (1989).
98. A. Dehbi-Alaoui, A. Matthews, and J. Franks, *Surf. Coat. Technol.,* **47,** 722 (1991).
99. A. A. Voevodin, J. M. Schneider, P. Stevenson, and A. Matthews, *Vacuum,* **46,** 299 (1995).
100. J. C. Angus, P. Koidl, and S. Domitz, in *Plasma Deposited Thin Films,* J. Mart and F. Jansen (eds.), CRC, Boca Raton, FL, 1986.
101. P. J. Martin, R. P. Netterfield, A. Bendavid, and T. J. Kinder, *Surf. Coat. Technol.,* **54/55,** 136 (1992).
102. V. N. Zhitomirsky, R. L. Boxman, and S. Goldsmith, *Surf. Coat. Technol.,* **68/69,** 146 (1994).
103. D. R. McKenzie, D. Muller, and B. A. Pailthorpe, *Phys. Rev. Lett.,* **67,** 773 (1991).
104. A. A. Voevodin and M. S. Donley, *Surf. Coat. Technol.,* **82,** 199 (1996).
105. S. V. Prasad, J. S. Zabinski, and N. T. McDevitt, *STLE Tribology Transactions,* **38,** 57 (1995).
106. A. A. Voevodin, M. A. Capano, A. J. Safriet, M. S. Donley, and J. S. Zabinski, *Appl. Phys. Lett.,* **69,** 188 (1996).

CHAPTER 12
PRODUCT DESIGN AND MANUFACTURING PROCESSES FOR SUSTAINABILITY

I. S. Jawahir
P. C. Wanigarathne
X. Wang
College of Engineering
University of Kentucky
Lexington, Kentucky

1 INTRODUCTION		**414**
1.1	General Background on Sustainable Products and Processes	414
1.2	Projected Visionary Manufacturing Challenges	416
1.3	Significance of Sustainable Project Design and Manufacture	416
2	**NEED FOR SUSTAINABILITY SCIENCE AND ITS APPLICATIONS IN PRODUCT DESIGN AND MANUFACTURE**	**420**
3	**PRODUCT DESIGN FOR SUSTAINABILITY**	**421**
3.1	Measurement of Product Sustainability	422
3.2	The Impact of Multi-Life-Cycles and Perpetual Life Products	423
3.3	Product Sustainability Assessment	424
3.4	Product Sustainability Index (PSI)	425
4	**PROCESSES FOR SUSTAINABILITY**	**426**
4.1	Selection of Sustainability Measures for Manufacturing Operations	426
5	**CASE STUDY**	**431**
5.1	Assessment of Process Sustainability for Product Manufacture in Machining Operations	431
5.2	Performance Measures Contributing to Product Sustainability in Machining	434
5.3	Optimized Operating Parameters for Sustainable Machining Process	435
5.4	Assessment of Machining Process Sustainability	438
6	**FUTURE DIRECTIONS**	**439**
	REFERENCES	**439**

1 INTRODUCTION

1.1 General Background on Sustainable Products and Processes

Sustainability studies in general have so far been focused on environmental, social, and economical aspects, including public health, welfare, and environment over their full commercial cycle, defined as the period from the extraction of raw materials to final disposition.[1] Sustainability requirements are based on the utilization of available, and the generation of new, resources for the needs of future generations. Sustainable material flow on our planet has been known to exist for over 3.85 billion years, and using the nature's simple framework in terms of *cyclic, solar,* and *safe means* has been shown to offer the most efficient products for sustainability.[2,3] It is also generally known that sustainable products are fully compatible with nature throughout their entire life cycle. Designing and manufacturing sustainable products is a major, high-profile challenge to the industry as it involves highly complex, interdisciplinary approaches and solutions. Most research and applications so far, however, have heavily focused on environmental sustainability. Sustainable products are shown to increase corporate profits while enhancing society as a whole, because they are cheaper to make, have fewer regulatory constraints and less liability, can be introduced to the market quicker, and are preferred by the public.[4] By designing a product with environmental parameters in mind, companies can increase profits by reducing material input costs, by extending product life cycles by giving them second and third life spans, or by appealing to a specific consumer base.[5] Recent effort on designing for environment includes the development of a customized software tool for determining the economic and environmental effects of "end-of-life" product disassembly process.[6]

In recent years, several sustainability product standards have emerged. Figure 1 shows a partial list of such standards.[7–18] While most standards are based on environmental benefits, some standards such as the Forest Stewardship Council Certified Wood Standards, the Sustainable Textile Standards, or the Global Sullivan Principles deal with social and economic criteria as well. The Institute for Market Transformation to Sustainability (MTS) has also recently produced a manual for standard practice for sustainable products economic benefits.[19] This profusion of competing standards may well become an obstacle to the management of product sustainability in the market place, leading to confusion among consumers and manufacturers alike. What is called for is the development of a sustainability management system that creates clear accountability methods across industries and market segments, and that determines not only *substantive* elements (e.g., "how much CO_2 was emitted in making the product?"), but also process elements (including the manufacturing systems and operations involved).

The idea of recycling, reuse, and remanufacturing has in recent times emerged with sound, innovative, and viable engineered materials, manufacturing processes, and systems to provide multiple life-cycle products. This is now becoming a reality in selected application areas of product manufacture. The old concept of "from cradle to grave" is now transforming into "from cradle to cradle,"[20] and this is a very powerful and growing concept in the manufacturing world, which takes its natural course to mature. Added to this is the awareness and the need for eco-efficiency and the environmental concerns often associated with minimum toxic emissions into the air, soil, and water; production of minimum amounts of useless waste; and minimum energy consumption at all levels. Finally, a future sustainability management system needs to identify how the public can be educated about sustainability, so that market incentives are created to persuade producers to follow more rigorous, evolving sustainability standards. Only at that point can a sustainability program be counted as successful.

Since the 1990s, environmental and energy factors have become an increasingly important consideration in design and manufacturing processes due to more stringent regulations promulgated by local, state, and federal governments as well as professional organizations in the United States and other industrial countries. The pressure on industry from the government as well as consumer sector has demanded new initiatives in environmentally benign design and manufacturing.[21] In the government sector, the Environmental Protection Agency (EPA) and the Department of Energy (DoE) have been the two leaders in these initiatives. EPA has initiated several promotional programs, such as Design for the Environment program, Product Stewardship program, and Sustainable Industries Partnership program, working with individual industry sectors to compare and improve the performance, human health, environmental risks, and costs of existing and alternative products, processes, and practices.[22] EPA has also worked with selected industry sectors such as metal casting, metal finishing, shipbuilding and ship repair, and specialty-batch chemical industries to develop voluntary, multimedia performance improvement partnerships. Similarly, DoE has launched a Sustainable Design Program, which focuses on the systematic consideration, during the design process, of an activity, project, product, or facility's life-cycle impacts on the sustainable use of environmental and energy resources.[23,24] Recently, DoE has also sponsored a series of new vision workshops and conferences, producing the Remanufacturing Vision Statement—2020 and Roadmaps, encouraging the industry groups to work together in strategic relationships for producing more efficient production methods utilizing life-cycle considerations.[25,26]

The big three automotive companies, DaimlerChrysler, Ford, and General Motors, have been fierce competitors in the marketplace, but they have worked together on shared technological and environmental concerns under the umbrella of the United States Council for Automotive Research (USCAR), formed in 1992 by the three companies. USCAR has sought specific technologies in recycling, reuse, and recovery of auto parts, batteries, lightweight materials, engines, and other power sources, as well as safety and emission reduction, sharing the results of joint projects with member companies.[27]

1.2 Projected Visionary Manufacturing Challenges

The National Research Council (NRC) report on "Visionary Manufacturing Challenges for 2020" identifies six grand challenges for the future: Concurrent Manufacturing, Integration of Human and Technical Resources, Conversion of Information to Knowledge, Environmental Compatibility, Reconfigurable Enterprises, and Innovative Processes.[28] Five of the ten most important strategic technology areas identified by the NRC report for meeting the above six grand challenges involve sustainability applications for products and processes:

- *Waste-free processes.* Manufacturing processes that minimize waste and energy consumption
- *New materials processes.* Innovative processes for designing and manufacturing new materials and components
- *Enterprise modeling and simulation.* System synthesis, modeling, and simulation for all manufacturing operations
- *Improved design methodologies.* Products and process design methods that address a broad range of product requirements
- *Education and training.* New educational and training methods that enable the rapid assimilation of knowledge

1.3 Significance of Sustainable Product Design and Manufacture

Figure 1 shows the exponential increase in shareholder value when the *innovation-based sustainability* concepts are implemented against the traditional cost-cutting, substitution-based growth.[29] The business benefits of sustainability are built on the basis of 3Rs: Reduce, Reuse, and Recycle. A market-driven "logic of sustainability" is now emerging based on the growing expectations of stakeholders on performance. A compelling case for market transformation from short-term profit focus to innovation-based stakeholder management methods has been proposed in a well-documented book by Chris Laszlo.[30] This covers five major logics of sustainability:

	Logo	Program	Website	1	2	3
PRODUCT SPECIFIC STANDARDS	FSC	Forest Stewardship Council Certified Wood	http://www.fscoax.org http://www.certifiedwood.org	X	X	X
	CLEAN CAR CAMPAIGN	Clean Vehicles	http://www.cleancarcampaign.org/standard.html http://www.environmentaldefense.org/greencar	X		
		Certified Organic Products Labeling	http://www.ota.com	X		
		Certified Green e Power	http://www.green-e.org	X		
	LEED	U. S. Green Building Council LEED Rating System	http://www.usgbc.org	X		
		Salmon Friendly Products	http://www.sustainableproducts.com/susproddcf2.html#Salmonm	X		
	Certified Cleaner and Greener	Cleaner and Greener^{sm} Certification	http://www.cleanerandgreener.org	X		
OVERALL STANDARDS	the NATURAL STEP	Natural Step System Conditions	http://www.NaturalStep.org	X		
		Nordic Swan Ecolabel	www.ecolabel.no/ecolabel/english/about.html	X	X	X
	GREEN SEAL	Green Seal Product Standards	http://www.greenseal.org	X		
		Global Reporting Initiative (GRI) Sustainability Reporting Guidelines (2000) Social Equity Performance Indicators	http://www.sustainableproducts.com/susproddef2.html#Performance_Indicators		X	
	SUSTAINABLE PRODUCTS	Life Cycle Assessment (LCA)	http://www.sustainableproducts.com/susproddef.html	X		
	MTS	Sustainable Textile Standard		X	X	X

Figure 1 Partial list of currently available sustainable products standards.

1. Scientific (e.g., human-induced global climate change)
2. Regulatory (e.g., Title I of the Clean Air Act, amended in the United States in 1990)
3. Political (e.g., agenda of the green parties in Europe)
4. Moral, based on values and principles
5. Market, focusing on the shareholder value implications of stakeholder value.

The global challenge of sustainability may be restated as follows: address the needs of a growing, developing global population without depleting our natural resources and without ruining the environment with our wastes. Fundamental knowledge must be developed and new innovative technologies established to meet this need. Engineers must move beyond their traditional considerations of functionality, cost, performance, and time-to-market, to consider also sustainability. Engineers must begin thinking in terms of minimizing energy consumption, waste-free manufacturing processes, reduced material utilization, and resource recovery following the end of product use—all under the umbrella of a total life cycle view. Of course, all this must be done with involvement of stakeholders, and the development of innovative technologies, tools, and methods.

Simply designing a green (environmentally friendly) product does not guarantee sustainable development for the following reasons: (a) a product cannot be green if the public does not buy it—business economics and marketing are critical for product acceptance, and (b) a green product is often just focused on the "use" stage of the product life cycle, with environmental burdens shifted to other life-cycle stages—sustainability requires a comprehensive, multi-life-cycle view. Certainly, industrialized countries have made some improvement in terms of being green with their use of materials, but waste generation continues to increase. As much as 75% of material resources used to manufacture goods are returned to the environment as wastes within a year.[31] This wasting of potential resources is disconcerting now, but over the next 50 years, as the demand for resources increases tenfold and total waste increases by a comparable amount, this resource wasting could be viewed as tragic.

Countries around the world, especially in Western Europe and Japan, recognize that a concerted effort is needed to meet the global challenge of sustainability. The governments and manufacturers in these regions are well ahead of the United States. in addressing the sustainability challenge through the development of energy/material-efficient technologies/products, low-impact manufacturing (value creation) processes, and post-use (value recovery) operations. The European Union has established the Waste Electrical and Electronic Equipment (WEEE) Directive to manage the recovery and post-use handling of these products.[32] While mandated recovery rates can be met economically by material recycling at present, remanufacturing and reuse are developing into very competitive alternatives. Another European Union directive calls for the value recovery of end-of-life vehicles (ELVs) and their components, with 85% of the vehicle (by weight) to be reused or recycled by 2015.[33] If a company exports its products to the EU, it must conform to these directives. Japan is enacting regulations that closely follow those of the EU. Manufacturers in both the EU and Japan have begun to redesign their products to accommodate recycling.[34,35] The Sustainable Mobility Project, a sector project of World Business Council for Sustainable Development (WBCSD), includes participation from 12 major auto/energy companies globally. The project deals with developing a vision for sustainable mobility 30 years from now and identifying the pathways to get there.[36,37] Each year, approximately 15 million cars and trucks reach the end of their useful life in the United States. Currently, about 75% of a car is profitably recovered and recycled because the majority of it is metal that gets remelted. The balance of materials, which amounts to 2.7–4.5 million tons per year of shredder residue, goes to the landfill.[38] It is very clear that U.S. manufacturers lag far behind their overseas competitors in this regard.

As noted above, regulatory drivers are currently forcing European and Japanese companies to develop innovative products, processes, and systems to remain competitive. Many in U.S. industry believe that making products and processes more environmentally friendly will increase costs. This can be the case if environmental improvement is achieved through increased control efforts, more expensive materials, etc. However, if improved sustainability is achieved through product and process innovations, then in addition to environmental benefits, cost, quality, productivity, and other improvements will also result. Through innovation, discarded products and manufacturing waste streams can be recovered and reengineered into valuable feed streams, producing benefits for the society, the environment, and U.S. industry. The United States is in danger of losing market share to its overseas competitors because it is not subject to the same drivers for change. It has been shown that manufacturing is responsible for much of the waste produced by the U.S. economy. In terms of energy usage, about 70% of the energy consumed in the industrial sector is used to provide heat and power for manufacturing [39]. Much of the heat and power required within industry is due simply to material acquisition and processing. Through new technology and innovative products and processes, utilizing previously processed materials for example, these energy requirements can be drastically reduced.

A significant effort has been undertaken by various groups from a range of disciplines to characterize, define, and formulate different forms and means of sustainable development. Continued progress in sustainable development heavily depends on sustained growth, primarily focusing on three major contributing areas of sustainability: environment, economy, and society (see Fig. 2). A relatively less-known and significantly impacting element of sustainability is sustainable manufacture, which includes sustainable products, processes, and

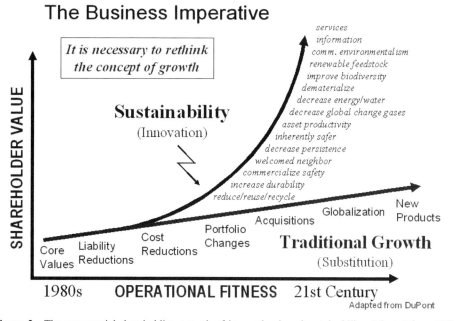

Figure 2 The exponential shareholding growth of innovation-based sustainability. Adapted from [29].

systems in its core. The understanding of the integral role of these three functional elements of sustainability in product manufacture is important to develop quantitative predictive models for sustainable product design and manufacture. This integral role of sustainable manufacture, with its three major functional elements (innovative product development—value design, manufacturing processes—value creation, and value recovery), all contributing to the sustained growth through the economic sustainability component, has been discussed[40] (Fig. 3).

2 NEED FOR SUSTAINABILITY SCIENCE AND ITS APPLICATIONS IN PRODUCT DESIGN AND MANUFACTURE

Sustainable development is now understood to encompass the full range of economic, environmental, and societal issues (often referred to as the "triple bottom line") that define the overall quality of life. These issues are inherently interconnected, and healthy survival requires engineered systems that support an enhanced quality of life and the recognition of this interconnectivity. Recent work, with details of integration requirements and sustainability indicators, shows that the sustainability science and engineering are emerging as a meta-discipline.[41,42] We are already beginning to see the consequences of engineered systems that are inconsistent with the general philosophy of sustainability. Because of our indiscriminate release of global warming gases, the recent EPA report on Global Climate Change forecasts

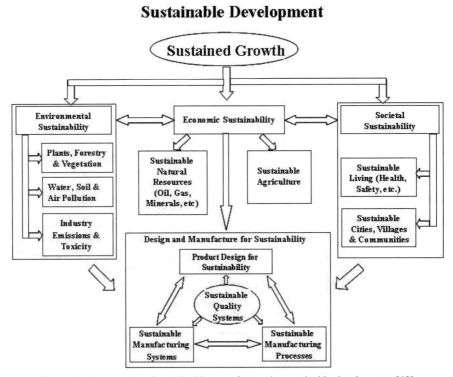

Figure 3 Integral role of sustainable manufacture in sustainable development [40].

some alarming changes in the earth's temperature, with concomitant increases in the sea level of 1 meter by 2100.[43] Obviously, fundamental changes are needed in engineered systems to reverse this trend.

The application of basic sustainability principles in product design and manufacture will serve as a catalyst for sustainable products to emerge in the marketplace. While the sustainable products make a direct contribution to economic sustainability, it also significantly contributes to environmental and societal sustainability. Building sustainability in manufactured products is a great challenge to the manufacturing world. The basic premise here is that, using the product sustainability principles comprehensively, all manufactured products can be designed, manufactured, assembled, used, and serviced/maintained/upgraded, and at the end of its life-cycle, these products can also be effectively disassembled, recycled, reused/remanufactured, and allowed to go through another cycle, and more. This multi-life-cycle approach and the associated need for product sustainability principles bring out an enormous technological challenge for the future. A cursory look at what would be required shows a long list of things to be performed; for example,

1. Known theories will be utilized while new theories emerge for sustainable product design.
2. Effective manufacturing processes with improved/enhanced sustainability applications will be developed and implemented.
3. Sustainable manufacturing systems will be developed to provide the overall infrastructure for sustainable product manufacture.

3 PRODUCT DESIGN FOR SUSTAINABILITY

Manufactured products can be broadly classified as consumer products, industrial products, aerospace products, biomedical products, pharmaceutical products, etc. Figure 4 shows some samples of manufactured products made from metals (steels, aluminum, hard alloys, plastics, polymers, composites, etc.), using a range of manufacturing processes. These products have well-defined functionalities and projected life cycles. Only a few of these products can be and are currently recycled, and very little progress has been made in using the recycled material effectively for remanufacturing other products. The fundamental question here is how to evaluate the product sustainability.

Understanding the need to design products beyond one life cycle has in recent times virtually forced the product designers to consider "end-of-life" status associated with product disassembly, recycling, recovery, refurbishment, and reuse. End-of-life options can be evaluated based on the concept of sustainability to achieve an optimum mix of economic and environmental benefits. Early work on product design for disassembly set the direction for research,[44] followed by disassembly analysis for electronic products.[45,46] Recovery methods[47] and models for materials separation methods[48,49] have been shown. End-of-life analysis for product recycling focuses primarily on the single life-cycle model.[50–54] More recent work shows that automotive end-of-life vehicle recycling deals with complex issues of post-shred technologies.[55] Also, screening of shredder residues and advanced separation mechanisms have been developed.[56,57] Significant work has been reported on recycling of plastics and metals.[58] Design guidelines have been developed for robust design for recyclability.[59] The application of some of the concepts developed in information theory for recycling of materials has been shown in a recent work through the measure of entropy.[60]

Eco-efficient and biodesign products for sustainability have been urged in recent times.[61,62] Environmental requirements were considered in sustainable product development

Figure 4 Samples of manufactured products [40].

using adapted quality function deployment (QFD) methods and environmental performance indicator (EPI) for several industrial cases,[63] followed by a material grouping method for product life-cycle assessment.[64] A further extension of this work includes estimation of life-cycle cost and health impact in injection-molded parts.[65] Software development for environmentally conscious product design includes BEES (Building for Environmental and Economic Sustainability) by the National Institute for Standards and Technology (NIST)[66] and Design for Environment Software.[6] Recently, by extending the previously developed sustainability target method (STM) to the product's end-of-life stage, analytical expressions were derived for the effectiveness of material recycling and reuse and were also corrected with the product's performance.[67]

3.1 Measurement of Product Sustainability

Quantification of product sustainability becomes essential for comprehensive understanding of the "sustainability content" in a manufactured product. The societal appreciation, need, and even the demand for such sustainability rating would become apparent with increasing awareness and the user value of all manufactured products more like the food labeling, energy efficiency requirements in appliances, and fuel efficiency rating in automotive vehicles.

Almost all previous research deals with a product's environmental performance and its associated economic and societal effects largely intuitively, and much of the qualitative descriptions offered are all, with the possible exception of a few recent efforts, difficult to measure and quantify. Thus, these analyses mostly remain nonanalytical and less scientific in terms of the need for quantitative modeling of product sustainability. The complex nature of the *systems property* of the term "product sustainability" seems to have limited the development of a science base for sustainability. Moreover, the partial treatment and acceptance of the apparent overall effects of several sustainability-contributing measures in relatively simplistic environmental, economic and societal impact categories has virtually masked the influence of other contributing factors such as product's functionality, manufacturability, and reusability with multiple life cycles. Consideration of a total and comprehensive evaluation of product sustainability will always provide much cheaper consumer costs, over the entire life cycle of the product, while the initial product cost could be slightly higher. This benefit is compounded when a multiple life-cycle approach, as seen in the next section, is adopted. The overall economic benefits and the technological advances involving greater functionality and quality enhancement are far too much to outscore with the current practice. The technological and societal impact is great for undertaking such an innovative approach to define the scientific principles of the overall product sustainability.

Sustainability science is more than a reality and is inevitable. Sound theories and models involving the application of basic scientific principles for each contributing factor are yet to emerge, but the momentum for this is growing worldwide. The available wide range of manufactured products from the consumer electronics, automotive, aerospace, biomedical, pharmaceutical sectors, etc., need to be evaluated for product sustainability and the associated economic and societal benefits.

3.2 The Impact of Multi-Life-Cycles and Perpetual Life Products

Figure 5 shows the various life-cycle stages for multi-life-cycle products leading toward the eventual "perpetual-life products." Even with innovative technology development, sustainability cannot be achieved in the absence of an engaged society. In the near future companies are envisioned to assume responsibility for the total product life cycle; but the consumer will remain responsible for preserving value during use and ensuring that post-use value

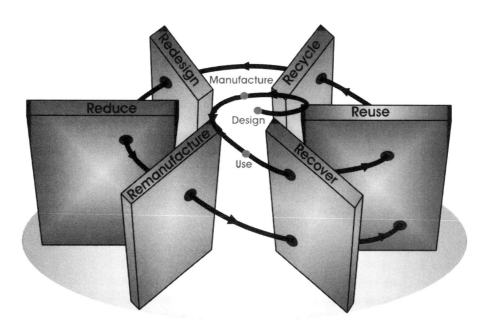

Figure 5 Generic multi-life-cycle products leading toward perpetual life products.

enters the recovery stream. Education and knowledge transfer will play an important role in communicating systemic changes, such as ecolabeling and the establishment of product reclamation centers. Engineers within industry must be educated to use innovative tools and methods for design/decision-making and to apply novel processes for manufacturing, value recovery, recycling, reuse, and remanufacturing. A diverse cadre of future engineers must receive a broad education prior to entering the workforce and have an awareness of a wide variety of issues, such as public opinion, environmental indicators, and life-cycle design.

The impact of products on sustainability does not start and end with manufacturing. The material flow in the product life cycle includes all activities associated with the product, including design, material acquisition and manufacturing processes (value creation), and use (value retention) and post-use (value recovery) processes (e.g., reuse, remanufacturing, and recycling) as illustrated in Fig. 6.[68] Distribution and take-back logistics are other important elements of the product life cycle. Obviously, some waste associated with the product life cycle is inevitable, and this waste is lost value and is associated with inefficiencies in the cycle. Increased use of innovative value recovery processes during the life cycle represents an underutilized potential business opportunity and means to be more sustainable.

3.3 Product Sustainability Assessment

While the concept of product sustainability continues to grow and become more compelling, the assessment of product sustainability has become difficult and challenging. There are no universally acceptable measurement methods for product sustainability as yet. This is largely due to the difficulty in quantifying and assessing some of the integral elements of product sustainability such as the societal and ethical aspects of sustainability. Also, the effects of the (social) use of products, in intended and unintended ways, are different from their material and production aspects, further complicating such an assessment. In 1987, the Brundtland Commission defined sustainability as "meeting the needs of present without compromising the ability of future generations to meet their own needs."[69] This rather ambiguous definition has very much limited the establishment of meaningful goals and measurable metrics for sustainability. Consideration of the key aspects of business performance subsequently extended this definition to include the effects of economic, environmental, and societal factors, each providing several categories of sustainable product indicators.[70] The initial product rating system developed by VDI[71] was subsequently modified to include variable weighting factors for products.[72]

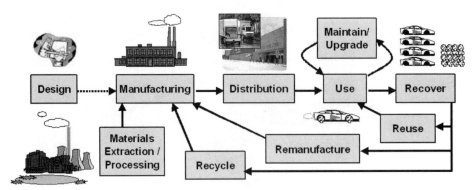

Figure 6 The material flow in the product life cycle for sustainability [68].

The ongoing work at the University of Kentucky (Lexington) within the Collaborative Research Institute for Sustainable Products (CRISP) involves a multidisciplinary approach for formulating the product sustainability level. A group of design, manufacturing/industrial, and materials engineers along with social scientists, economists, and marketing specialists are actively participating in a large program to establish the basic scientific principles for developing a product sustainability rating system. This includes the development of a science-based product sustainability index (PSI).[40]

3.4 Product Sustainability Index (PSI)

The PSI will represent the "level of sustainability" built in a product by taking into account of the following six major contributing factors:

1. Product's environmental impact
   - Life-cycle factor (including product's useful life span)
   - Environmental effect (including toxicity, emissions, etc.)
   - Ecological balance and efficiency
   - Regional and global impact (e.g., CO_2 emission, ozone depletion, etc.)
2. Product's societal impact
   - Operational safety
   - Health and wellness effects
   - Ethical responsibility
   - Social impact (quality of life, peace of mind, etc.)
3. Product's functionality
   - Service life/durability
   - Modularity
   - Ease of use
   - Maintainability/serviceability (including unitized manufacture and assembly effects)
   - Upgradability
   - Ergonomics
   - Reliability
   - Functional effectiveness
4. Product's resource utilization and economy
   - Energy efficiency/power consumption
   - Use of renewable source of energy
   - Material utilization
   - Purchase/market value
   - Installation and training cost
   - Operational cost (labor cost, capital cost, etc.)

5. Product's manufacturability
   - Manufacturing methods
   - Assembly
   - Packaging
   - Transportation
   - Storage
6. Product's recyclability/remanufacturability
   - Disassembly
   - Recyclability
   - Disposability
   - Remanufacturing/reusability

Quantifiable and measurable means can be developed and for each factor within each group, and then be combined to produce a single rating for each group. This rating can be on a percentage basis on a 0–10 scale, with 10 being the best. Each product will be required to comply with appropriate ratings for all groups. Standards will be developed to establish an "acceptable" level of rating for each group. While the rating of each group contributes to the product's sustainability, the composite rating will represent the overall "sustainability index" of a product, the product sustainability index (PSI). This overall product sustainability can be expressed in terms of a percentage level, on a 0–10 scale, or on a letter grade, such as A, B, C. Variations of these implementation methods of PSI are shown in Fig. 7.

4 PROCESSES FOR SUSTAINABILITY

The primary focus in identifying and defining the various contributing elements and sub-elements of manufacturing process sustainability is to establish a unified, standard scientific methodology to evaluate the degree of sustainability of a given manufacturing process. This evaluation can be performed irrespective of the product life-cycle issues, recycling, remanufacturability, etc., of the manufactured product. The overall goal of the new international standards, ISO 14001, is to support environmental protection and prevention of pollution in balance with socioeconomic needs.[73] Requirements of sustainable manufacturing covering recycling, and decision-making aspects such as supply chain, quality initiatives, environmental costing, and life-cycle assessment are well covered in a recent handbook on environmentally conscious manufacturing.[74] An early attempt to develop a sustainable process index was based on the assumption that in a truly sustainable society the basis of economy is the sustainable flow of solar energy.[75] More recent work predicts manufacturing wastes and energy for sustainable processes through a customized software system.[76] A modeling effort for the impact of fuel economy regulations on design decisions in automotive manufacturing was presented in a recent paper.[77]

4.1 Selection of Sustainability Measures for Manufacturing Operations

Manufacturing processes are numerous and, depending on the product being manufactured, the method of manufacture, and their key characteristics, these processes differ very widely. This makes the identification of the factors/elements involved in process sustainability and the demarcation of their boundaries complex. For example, the production process of a

4 Processes for Sustainability 427

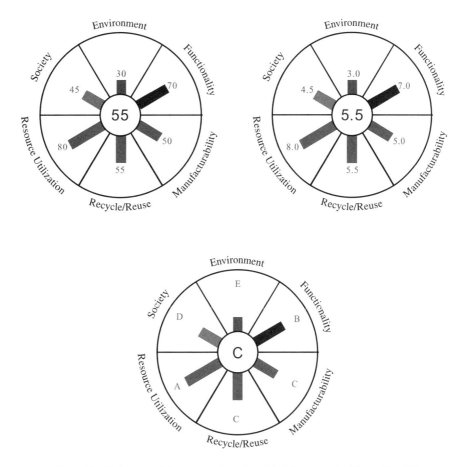

Figure 7 Variations of the proposed product label for a sustainable product [40].

simple bolt involves a few clearly defined production stages: bolt design, tool/work material selection, metal removal/forming, finishing, packaging, transporting, storage, dispatching, etc. It is difficult to consider all these stages in evaluating the manufacturing process sustainability, even though these processes either directly or indirectly contribute to the manufacturing process sustainability. Also, the processing cost largely depends on the method used to produce the part/component and the work material selected. In a never-ending effort to minimize the manufacturing costs, the industrial organizations are struggling to maintain the product quality, operator and machine safety, and power consumption. If the processing includes the use of coolants or lubricants or the emission of toxic materials or harmful chemicals, this then poses environmental, safety, and personnel health problems. In general, among the various influencing factors, the following six factors can be regarded as significant to make a manufacturing process sustainable:

- Energy consumption
- Manufacturing costs
- Environmental impact

- Operational safety
- Personnel health
- Waste management

Figure 8 shows these interacting parameters. The selection and primary consideration of these six parameters at this preliminary stage of sustainability evaluation do not restrict the inclusion of other likely significant parameters. This may include parameters such as the product's functionality requirements, which affect the decision making process and are related to machining costs and energy consumption, but would be expected to hold a secondary effect on the process sustainability. The product's functionality aspect is more closely related to the sustainability of a product. Marketing strategies and the initial capital equipment investment can also indirectly affect the sustainability of a machining process, but these are not included in the present analysis.

All six selected parameters have different expectation levels, as shown in Table 1. But there is an obvious fact that all these factors cannot achieve their best levels due to technological and cost implications. Also, there exist strong interactions among these factors, often requiring a trade-off. Thus, only an optimized solution would be practical, and this would involve combinations of minimum and maximum levels attainable within the constraints imposed. The attainable level is again very relative and case-specific. Measurement and quantification of the effect of contributing factors shown in Table 1 pose a significant technical challenge for use in an optimization system.

Energy Consumption

During manufacturing operations the power consumption level can be observed and evaluated against the theoretical values to calculate the efficiency of the power usage during the operation. Significant work has been done in this area to monitor the power consumption rate and to evaluate energy efficiency. Energy savings in manufacturing processes is a most needed sustainability factor, which needs to be considered for the entire operational duration of the machine, with significant overall savings in the long run. For any manufacturing operation, the energy consumed can be measured in real time. If the same task/operation is performed at two different machine shops or on two different machines, the power consumption may vary, due to the differences in the machines and the conditions used in man-

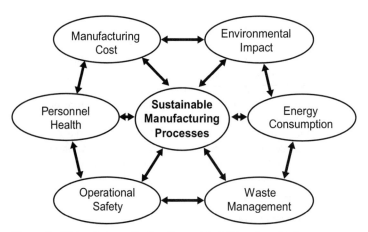

Figure 8 Major factors affecting the sustainability of machining processes.

Table 1 Measurable Sustainability Factors in Machining Processes and Their Desired Levels

Measurement Factor	Desired level
Energy consumption	Minimum
Environmental friendliness	Maximum
Machining costs	Minimum
Operational safety	Maximum
Personnel health	Maximum
Waste reduction	Maximum

Source: Wanigarathne et al. [84].

ufacturing processes. Notably, the application of proper coolants and lubricants, the selection of cutting tool inserts, cutting conditions, and cutting tool-work material combinations, and facilitating improved tribological conditions can all reduce the power consumption, typically in a machining process. Also, the functional features built in a machine tool design contribute toward energy savings in a machining operation. For instance, the horizontal movement of the turret of a lathe may use less energy than a vertically slanted turret in machining centers, as more power is used to keep the location locked. This may amount to several kilowatt-hours of energy, in real value.

Hence, it is clear that setting a standard for power consumption is relative and complex in the industry. Use of minimum energy is, however, most desirable from the perspective of the global energy standards/requirements. In the case of machining, there is an attainable minimum energy level for every machining operation. The power used in the real operation can be compared against this in assessing the amount of excess energy utilized. Depending on the proximity of the two values, one can determine the relative efficiency of the power/energy consumption, and then take measures to improve the process by reducing the gap. After these modifications and improvements, the specific process can be rated for sustainable use of energy. In sustainability assessment of energy/power consumption, it is generally anticipated that the preferred source of energy is environmentally friendly—solar or from a renewable source. If the renewable sources are available in abundance and are used in industry widely, the source of energy factor can be included in the process sustainability rating system.

Manufacturing Costs

Manufacturing cost involves a range of expenditures starting from the process planning activity until the part is dispatched to the next workstation, including the idling time and queuing time. Within the context of manufacturing process sustainability assessment, our interest is only on the manufacturing costs involved in and during the manufacturing operation time, including the cost of tooling. For example, in a machining operation, the material removal rate depends on the selected cutting conditions and the capabilities of the machine tools and cutting tools used. The criterion for selecting appropriate machine tools and cutting tools would generally facilitate a cost-effective machining operation. Numerous software tools are available for optimizing the machining cost through the use of proper cutting conditions. Recently, a new technique for multipass dry turning and milling operations has been developed. This technique employs analytical, experimental, and hybrid methods for performance-based machining optimization and cutting tool selection based on tool-life criterion involving minimum cost.[78,79] In addition, there are several other direct and indirect cost factors coming from the environmental effects and operator's health and safety aspects.

The cost components for recycling and reusability of consumables such as the coolants also need to be considered in the overall machining cost along with the tooling cost contributions to the machining process.

Environmental Impact
Basic factors contributing to environmental pollution, such as emissions from metal working fluids, metallic dust, and use of toxic, combustible, or explosive materials, contribute to this factor. Compliance with the U.S. Environmental Protection Agency (EPA),[80] Occupational Safety and Health Administration (OSHA),[81] and National Institute for Occupational Safety and Health (NIOSH)[82] regulations is required. Measurable parameters have been defined and are continually updated. The ISO 14000 series of standards[83] has been designed to help enterprises meet and improve their environmental management system needs. The management system includes setting of goals and priorities, assignment of responsibility for accomplishing these goals, measuring and reporting of results, and external verification of claims. Since the standards have been designed as voluntary, the decision to implement will be a business decision. The motivation may come from the need to better manage compliance with environmental regulations, from the search for process efficiencies, from customer requirements, from community or environmental campaign group pressures, or simply from the desire to be good corporate citizens. The machining process sustainability rating will eventually force the industries to impose and show progress at every stage of the production process.

Operational Safety
The amount of unsafe human interaction during a manufacturing operation and the ergonomic design of the human interface are in focus for this category. Also, compliance with the regulatory safety requirements is made mandatory. Statistical data on safety violations and the associated corrective measures that are quantifiable are usually being reviewed and updated.

In general, safety aspects in relation to a manufacturing process can be divided into two broad categories: personnel safety and work safety. Safety of the operator and the occupants of the manufacturing station are considered paramount to the work safety. The amount of human interaction during a manufacturing operation and the safety precautions provided against the foreseeable accidents will be the primary focus in evaluating the operational safety as a sustainability parameter. The ergonomic design of the human interface with the work environment will be important in safety evaluations. The compliance to and the proper implementation of regulatory safety requirements will also be considered in assessing the personnel safety factor.

Personnel Health
Assessment of the personnel health element contributing to the machining process sustainability is based on the compliance with the regulatory requirements, imposed on industries by governmental and regulatory enforcement units such as EPA,[80] OSHA,[81] and NIOSH[82] on emissions and waste from machining operations and their impact on directly exposed labor. One of the most prominent ways machine operators are affected by is exposure to the mist and vapors of metalworking fluids, as most metalworking fluids used as coolants and lubricants in machining operations contain large amounts of chemicals added to "enhance" the machining performance. Over time, the fluid containers become an ideal environment for the growth of harmful bacteria. There are a few ways to avoid this problem, but it appears that none of these methods is in practice due to inadequacy of knowledge and implementation issues.

Waste Management
Recycling and the disposal of all types of manufacturing wastes, during and after the manufacturing process is complete, are accounted for in this category. Scientific principles are still emerging with powerful techniques such as lean principles being applied, in quantifiable terms. Zero waste generation with no emissions into the environment is the ideal condition to be expected for products and processes, although it is technologically not feasible as yet. However, efforts to find means to reduce or eliminate wastes are continuing. For example, some cutting fluids can be forced to degrade biologically before being disposed of. The same technology can be used to control the growth of harmful bacteria in the cutting fluid containers as well as on the waste chip dumpers to make them biologically safe to handle and dispose of.

5 CASE STUDY

5.1 Assessment of Process Sustainability for Product Manufacture in Machining Operations

This case study provides a description of how the sustainability measures can be selected along with an approximate method for modeling these measures for optimizing machining processes for maximum sustainability.[84] The machining processes used herein are to be considered a generic case for manufacturing processes.

Energy Consumption
Figure 9 shows a comparison of cutting force at varying coolant flow rates, ranging from no-coolant applied dry machining to flood cooling involving a large amount of coolant, in machining of automotive aluminum alloy A390 work material. This range includes three typical coolant flow rates of minimum quantity lubrication (MQL) conditions generally known as "near-dry" machining. As seen, the measurable cutting force component can serve as a direct indicator of the power/energy consumption rate. The optimum power consumed seems to lie beyond the largest coolant flow rate tested for near-dry machining (i.e., 60 mL/hr). Obviously, the flood cooling method, despite the demonstrated major benefits of increased tool life, seems to show the highest power consumption for the test conditions used in the experimental work.

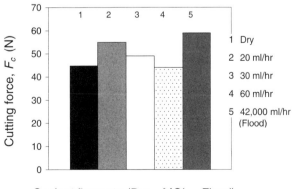

Figure 9 Cutting force variation when using different coolant flow rates [84].

Figure 10 shows a comparison of cutting force generated in machining of 4140 steel for three different conditions (flood cooling, oil-based MQL, and dry machining.)[84] A fairly consistent trend is seen for this operation. In contrast, in machining of aluminum A390 at a lower cutting speed of 300 m/min, a reverse trend is observed, with dry machining consuming the largest energy as seen in Fig. 11. Comparison of Figs. 9 and 11 shows that cutting speed plays a major role in power consumption rate with respect to cooling rates, given all other conditions remaining constant. Complex relationships are also expected for varying tool geometry, work–tool material combinations, etc., thus making it essential to model the machining process for optimal cooling conditions to provide minimum power consumption. This technical challenge calls for a need to have full and comprehensive knowledge on all major influencing variables for the entire range of machining. Analytical modeling effort must continue despite significant experimental difficulties involved. The sustainability contribution made by power consumption rate is too great to ignore, particularly when the environmental concerns are also addressed by reduced coolant applications and dry machining methods. Experimental work has shown that in continued machining over time, oil-based MQL-assisted machining operations produced a slight decrease in cutting force, while the flood cooling method generated a steadily increasing trend for the cutting force.[84,85]

Machining Cost

Optimal use of machines and tooling, including jigs and fixtures, can provide reduced manufacturing costs. Limited analytical and empirical models are available for this evaluation, and accurate calculations are highly complex and customized applications would be necessary. Developing comprehensive analytical models to account for the overall machining cost is feasible as significant generic and rudimentary calculation methods are already in existence.

Environmental Impact

The environmental impact due to machining is an important contributing factor for machining sustainability evaluation. Basic factors affecting environmental pollution, such as emissions from metal working fluids, metallic dust, machining of dangerous material (combustible or explosive), and amount of disposed untreated wastes, are among those considered under this category. The use of metalworking fluids in machining operations creates enormous health

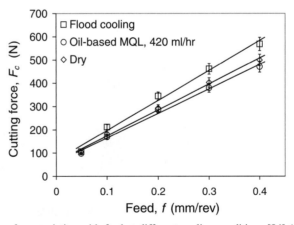

Figure 10 Cutting force variation with feed at different cooling conditions [84] (AISI 4140 Steel).

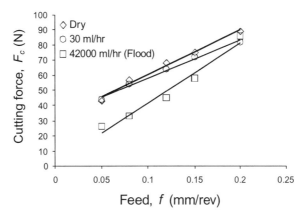

Figure 11 Cutting force variation with feed at different cooling conditions [84] (Aluminum A390).

and environmental problems. Keeping the metalworking fluid tanks clean with no seepage to the environment and without harmful bacterial growth, etc., are some of the important aspects to consider in the evaluation of environmental friendliness. All measures designed to minimize the environmental pollution, such as the use of fume hoods and the treatment of metalworking fluids, must be given credit in the final assessment. Adherence and compliance to prevent emissions as vapor or mist, as regulated by EPA[80] and OSHA,[81] are essential, and in consistent implementation of practices, as per regulations, the machining environment needs to be inspected and certified.

Operational Safety
Some examples of this category are auto power doors, safety fences and guards, safety display boards, safety training, facilities to safe interactions with machines, methods of lifting and handling of work, mandatory requirement to wear safety glasses, hats, and coats, availability of fire safety equipment, and first aid facilities in house. In addition to this routine inspection for operational safety and regular safety, specific training programs promoting safety precautions will be given due credit. Also, the routine maintenance of machines and availability of onsite fume detectors are considered desirable in assessing the operational safety measures for sustainability in machining operations.

Personnel Health
Compliance with the regulatory requirements according to EPA, NIOSH, and OSHA on emissions from manufacturing operations and their impact on directly exposed labor can be the basis for this category of assessment. For example the use of dry machining techniques or near-dry machining techniques can largely avoid and/or reduce the problems of mist generation and metalworking fluid handling, typically encountered in flood cooling. Such measures will be assessed superior over traditional flood cooling methods in the personnel health factor assessment. In addition to those factors such as compliance with regulations regarding space per machine and man-count in the factory, safety precautions in handling of explosives or radioactive materials in machining will be considered significant, too.

Waste Reduction and Management
Different types of wastes resulting from the machining process have to be treated, disposed of, and managed properly. Cost-effective and energy-efficient recycling of metal chips, de-

bris, contaminated coolants, etc., would contribute to sustainable machining process. Breaking chips into small, manageable sizes and shapes becomes a basic requirement for automated machining and for disposing of chips for recycling and/or reuse. Significant work has been done on chip breaking and control in metal machining[86–89] and a new method developed for quantifying the chip form in terms of its size and shape[90] has been in wide use in practical applications. A computer-based predictive planning system developed for chip-form predictions in turning operations includes the effects of chip-groove configurations, tool geometry, and work material properties on chip-forms/chip breakability.[91] More recent efforts to identify the variations of chip-forms in machining with varying cooling conditions have shown a promising opportunity for MQL-assisted machining processes, as shown in Fig. 12, where at relatively high feeds the size and shapes of the chips produced are more suitable for chip disposal for recycling and reuse.[85]

5.2 Performance Measures Contributing to Product Sustainability in Machining

In designing a product for machining and in the subsequent process planning operations, an important consideration is assuring the product's functionality and specification requirements in terms of surface finish, part accuracy, etc. The anticipated surface roughness value and the related surface integrity issues contribute to the service life of a machined product. Maintaining low tool-wear rates leading to increased tool life is essential in process modeling for surface roughness and surface integrity in machining.

Tool-Life Evaluation in Dry and Near-Dry Machining and Tool Insert Recycling and Reuse Issues for Sustainability

Significant efforts have been made to improve tool-life predictions in coated grooved tools, which provide the required increased tool life in dry machining with coated grooved tools.[92,93] More recent work on predictive modeling of tool life extended this work to include the effects of mist applications in near-dry machining, thus offering predictability of tool life in sustainable manufacturing.[94,95]

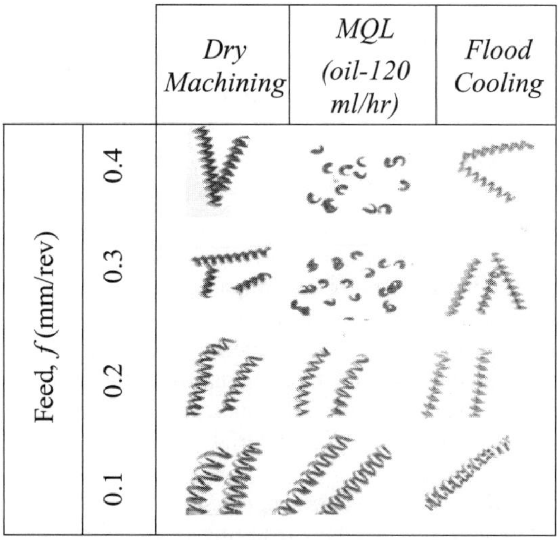

Figure 12 Chip-form variation with different cooling methods [84].

The life of a cutting tool insert can be as short as a few minutes, and no more than an hour in most cases. Tool inserts are made from wear-resistant, hard materials and are coated with specialized hard materials. Once a tool insert is used up, with all its effective cutting edges worn, it is a normal practice to replace it. In large companies involved in a range of machining operations, the weight of tool inserts discarded per year may easily pass few hundred tones. Hence, new technologies need to be developed for tool insert recycling and/ or reusing, given the large amount of waste in the form of worn tool inserts. Recent progress on tool recoating efforts, showing the technological feasibility with performance improvements and economic benefits, is encouraging.[96,97]

Surface Roughness and Surface Integrity Analysis for Sustainability
Comparison of surface roughness produced in machining under a range of cooling conditions (from dry to flood cooling) shows that flood cooling produces the least desirable surface roughness in turning operations of automotive alloys such as AISI 4140 steel, despite the popular belief that coolants would provide better surface roughness (Fig. 13). However, in machining of aluminum alloy A390, the trend is reversed, as seen in Fig. 14, where dry machining produces a rougher surface.

In addition, the surface integrity is affected greatly by the residual stress formation in machining processes. Material behavior at a cutting tool with finite edge roundness has been modeled using a thermo-elasto-viscoplastic finite element method to study the influence of sequential cuts, cutting conditions, etc., on the residual stress induced by cutting.[98,99] This work led to the conclusion that material fracture (or material damage) in machining is an important phenomenon to understand the actual material behavior on a finished surface and the surface integrity, both directly influencing the product sustainability.

5.3 Optimized Operating Parameters for Sustainable Machining Processes

Figure 15 illustrates a method for selecting optimal cutting conditions in the rough pass of multipass dry turning operations providing sustainability benefits. Thick lines represent constraints of surface roughness, tool life, material removal rate, and chip form/chip breakability. Thin lines are the contours of the objective function in the optimization method. Points A and B represent different optimized results of cutting conditions subject to different initial

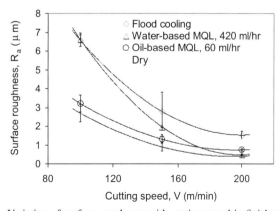

Figure 13 Variation of surface roughness with cutting speed in finish turning [84].

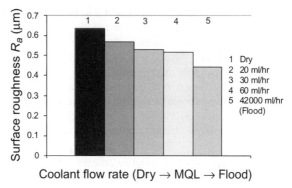

Figure 14 Surface roughness variation when using different cooling methods [84].

requirement of the total depth of cut. A comprehensive criterion including major machining performance measures is used in the optimization process.

In the optimization method described here, an additional sustainability criterion will be considered. The user is able to control the optimization process by configuring and assigning weighting factors for both machining performance measures and sustainability measures. The total objective function will combine all machining performance and sustainability measures prevalent in the given manufacturing process. The aim of the optimization process is to make a trade-off among these measures, and therefore to provide the optimal combinations of operating parameters and to propose ways of enhancing and improving sustainability level. Figure 16 shows a flow chart of the proposed predictive models and optimization method for sustainability assessment in machining processes. This shows that three of the six key sustainability parameters can be modeled using analytical and numerical techniques because of the deterministic nature of these parameters, while modeling of the other three parameters would require nondeterministic means such as fuzzy logic. The resulting hybrid sustainability

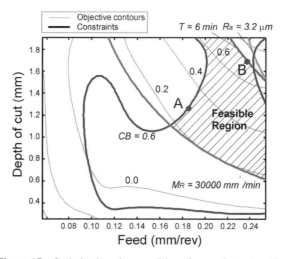

Figure 15 Optimized cutting conditions for rough turning [84].

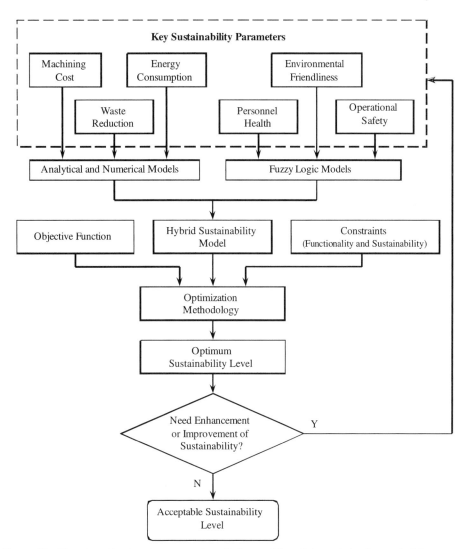

Figure 16 Flowchart showing the proposed predictive models and the optimization method for process sustainability assessment of machining processes [84].

model for machining processes along with the objective function and the relevant constraints, including functional constraints such as relevant machining performance measures and sustainability constraints, can be used in the optimization module to provide the optimum sustainability level of the given machining process representing the actual, overall sustainability level in the form of a sustainability index. With a subsequent decision making process, the process sustainability can be either improved using a feedback loop, or be accepted as is. The optimization problem is formulated as follows:

Maximize $U = U$ (MC, EC, WR, PH, OS, EF)

W. r. t.   Cutting conditions and shop floor data

Subject to   Functional constraints:

$$R_a \leq R'_a, F \leq F', T \geq T', M_R \geq M'_R, \text{CB} \geq \text{CB}'$$

Cutting condition constraints:

$$V_{\min} \leq V \leq V_{\max}, f_{(\min)} \leq f \leq f_{(\max)}, d_{(\min)} \leq d \leq d_{(\max)}, \text{etc.}$$

Sustainability constraints:

$$\text{MC} \leq \text{MC}', \text{EC} \leq \text{EC}', \text{WR} \geq \text{WR}', \text{PH} \geq \text{PH}', \text{OS} \geq \text{OS}', \text{EF} \geq \text{EF}'$$

where MC is machining cost, EC is energy consumption, WR is waste reduction, PH is personnel health, OS is operational safety, EF is environmental friendliness, R_a is surface roughness, F is cutting force, T is tool-life, M_R is material removal rate, CB is chip breakability, V is cutting speed, f is feed rate, and d is depth of cut.

5.4 Assessment of Machining Process Sustainability

The basic driving force in sustainability studies as it applies to manufacturing processes is the recent effort to develop a manufacturing process sustainability index. The idea in developing this practically implementable concept is to isolate the manufacturing process from the global picture of sustainability, and develop it up to the "level of acceptance" for common practice in industry. This can be achieved in different stages. First, in the characterization stage, the most important measures of the rating system for the manufacturing process sustainability must be identified and established through literature, in-house/field surveys, and appropriate experimental work. Shown in Fig. 16 are some of the key parameters that can be considered. These observations and the existing modeling capabilities will then be used to model the impact of the manufacturing process on the contributing major sustainability parameters. A hybrid modeling technique involving analytical and numerical methods, coupled with empirical data and artificial intelligence techniques, must be developed to scientifically quantify the influence of each parameter. Then, the modeled production process can be optimized to achieve desired level of sustainability with respect to constraints imposed by all involved variables. These optimized results can then be used to modify the existing processes and enhance the manufacturing performance with respect to the main factors considered. Finally, the optimized results can be used in defining the sustainability rating for the specific manufacturing process. In establishing the final sustainability rating for the selected process weighing factors can also be used to bring out focused evaluation and to serve the customized application.

User friendliness and communication efficiency are among the two most-needed features of the proposed new sustainability assessment system for machining processes. Two proposed methods can be employed: explicit and implicit. A symbolic representation of the proposed sustainability assessment method is shown in Fig. 17.

The explicit grading method uses a spider chart axis for each factor selected. The spider chart is then divided into five different transforming regions, and each region is represented by a color. On this spider chart axis the relevant rating can be marked very clearly and the rating value can be indicated next to the point, on demand. The points closest to the outer periphery are the highest ranking categories, while the points closer to the origin are considered to be worst with respect to the expected sustainability rating and these areas need to be improved for enhanced sustainability.

To implicitly show the level of sustainability, a color-coding system—five shades of green—is used in the background. The darker the color, the further the factor is away from the expected level and when the color turns bright green it reaches the maximum. Colors are assigned with the darkest closer to the origin and the brightest closer to the outer periphery of the spider chart.

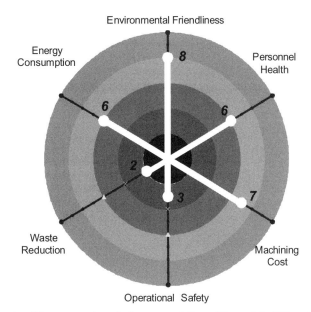

Figure 17 Example of the proposed symbolic representation of the sustainability rating system for six contributing factors [84].

The proposed process sustainability assessment method will heavily involve science-based sustainability principles for product design and manufacture. The overall sustainability level of the machining process will be established through a new sustainability index to be developed comprehensively using the modeling and optimization method shown in Fig. 16.

6 FUTURE DIRECTIONS

Continued trends in sustainability applications for products and processes indicate the need for identifying relevant sustainability metrics and for developing science-based methodologies for quantification of these factors.[40,84,100,101] Achieving global sustainability is a major challenge and this requires international cooperative research and applications.[102]

REFERENCES

1. The Institute for Market Transformation to Sustainability (MTS), Sustainable Products Corporation, Washington, DC, 2004, http://MTS.sustainableproducts.com
2. E. Datschefki, *Sustainable Products: Using Nature's Cyclic/Solar/Safe Protocol for Design, Manufacturing and Procurement,* BioThinking International, E-Monograph, 2002.
3. E. Datschefki, "Cyclic, Solar, Safe—BioDesign's Solution Requirements for Sustainability," *J. Sustainable Product Design,* 42–51 (1999).
4. The Institute for Market Transformation to Sustainability (MTS), "Suatainable and Green Profits Are Starting to Drive the Economy," Sustainable Products Corporation, Washington, DC, http://MTS.sustainableproducts.com
5. Green Product Design at 1-2, BSR Ed. Fund.
6. W. Knight and M. Curtis, "Design for Environment Software Development," *Journal of Sustainable Product Design,* 36–44 (1999).

7. Forest Stewardship Council Certified Wood, http://www.fscoax.org, http://www.certifiedwood.org
8. Clean Vehicles, http://www.cleancarcampaign.org/standard.html, http://www.environmentaldefense.org/greencar
9. Certified Organic Products Labeling, http://www.ota.com
10. Certified Green e-Power, http://www.green-e.org
11. U.S. Green Building Council LEED Rating System, http://www.usgbc.org
12. Salmon Friendly Products, http://www.sustainableproducts.com/susproddef2.html#Salmonm
13. Cleaner and Greener Certification, http://www.cleanerandgreener.org
14. Natural Step System Conditions, http://www.NaturalStep.org
15. Nordic Swan Ecolabel, www.ecolabel.no/ecolabel/english/about.html
16. Green Seal Product Standards, http://www.greenseal.org
17. Global Reporting Initiative (GRI) Sustainability Reporting Guidelines, Social Equity Performance Indicators (2002), http://www.sustainableproducts.com/susproddef2.html#Performance_Indicators
18. Life Cycle Assessment (LCA), http://www.sustainableproducts.com/susproddef.html
19. The Institute for Market Transformation to Sustainability (MTS), *Standard Practice for Sustainable Products Economic Benefits—Including Buildings and Vehicles,* January 2003.
20. W. McDonough and M. Braungart, *Cradle to Cradle,* North Point Press, New York, 2002.
21. Gutowski, T. G., et al., "WTEC Panel on Environmentally Benign Manufacturing—Final Report," ITRI-WTEC, Loyola College in Maryland., ISBN 1-883712-61-0, April 2001.
22. EPA Report, "Living the Vision—Accomplishments of National Metal Finishing Strategic Goal Program," EPA 240-R-00-007, January 2001.
23. DoE web site, http://www.doe.gov/.
24. K. L. Peterson and J. A. Dorsey, "Roadmap for Integrating Sustainable Design into Site-Level Operations," DoE Report, PNNL 13183, March 2000 (DoE, 1998).
25. DoE, "Visioning Workshop for Remanufacturing," Office of Industrial Technologies, Department of Energy, September 11, 1998. http://www.reman.rit.edu/reference/pubs/2020vision/DOCVisionStatement.htm
26. D. Bauer and S. Siddhaye, "Environmentally Benign Manufacturing Technologies: Draft Summary of the Roadmaps for U.S. Industries," International Technology Research Institute, Loyola College in Maryland, 1999. http://www.itri.loyola.edu/ebm/usws/welcome.htm.
27. USCAR web site, http://www.uscar.org/
28. NRC Report, *Visionary Manufacturing Challenges for 2020,* National Academy of Sciences, National Academy Press, Washington, DC, 1998.
29. World Resource Institute, http://www.wri.org
30. C. Laszlo, *The Sustainable Company: How to Create Lasting Value Through Social and Environmental Performance,* Island Press, Washington, DC, 2003.
31. J. C. Crittenden, "Activities Report: October 1998–September 1999," National Center for Clean Industrial and Treatment Technologies (CenCITT), Houghton, MI, April 2000.
32. European Union, "Directive of the European Parliament and of the Council on Waste Electrical and Electronic Equipment" (WEEE), 2002/95/EC, Brussels. (ELVs by 2015—85%)
33. http://www.plastics-in-elv.org/Regulatory_Comprehensive.htm
34. http://www.jama.org/library/position092303.htm
35. http://europa.eu.int/comm/environment/waste/elv_index.htm
36. http://www.wbcsd.org/templates/TemplateWBCSD5/layout.asp?MenuID=1
37. http://wbcsdmobility.org
38. E. J. Daniels, J. A. Carpenter, Jr., C. Duranceau, M. Fisher, C. Wheeler, and G. Wislow, "Sustainable End-of-Life Vehicle Recycling: R & D Collaboration Between Industry and the U.S. DOE," *JOM, A Publication of the Minerals, Metals & Materials Society,* 28–32 (August 2004).
39. DoE web site, http://www.doe.gov/.
40. I. S. Jawahir and P. C. Wanigarathne, "New Challenges in Developing Science-Based Sustainability Principles for Next Generation Product Design and Manufacture," Keynote Paper, in *Proc TMT 2004,* S. Ekinovic et al. (eds.), DOM STAMPE zenica Neum, Bosnia and Herzegovina, September 2004, pp. 1–10.

41. J. W. Sutherland, V. Kumar, J. C. Crittenden, M. H. Durfee, J. K. Gershenson, H. Gorman, D. R. Hokanson, N. J. Hutzler, D. J. Michalek, J. R. Mihelcic, D. R. Shonnard, B. D. Solomon, and S. Sorby, "An Education Program in Support of a Sustainable Future," *Proc. of ASME/IMECE, MED*, Vol. 14, 2003, pp. 611–618, (IMECE2003-43422), appeared on CD-ROM.
42. J. R. Mihelcic, J. C. Crittenden, M. J. Small, D. R. Shonnard, D. R. Hokanson, Q. Zhang, H. Chen, S. A. Sorby, V. U. James, J. W. Sutherland, and J. L. Schnoor, "Sustainability Science and Engineering: The Emergence of a New Meta-discipline," *Environmental Science & Technology,* **37**(23), 5314–5324 (2003).
43. EPA web site, http://www.epa.gov/ (2004).
44. F. Jovane, L. Alting, A. Armillotta, W. Eversheim, K. Feldmann, G. Seliger, and N. Roth, "A Key Issue in Product Life Cycle: Disassembly," *Annals of CIRP,* **42**(2), pp. 651–658 (1993).
45. H. C. Zhang, T. C. Kuo, H. Lu, and S. H. Huang, "Environmentally Conscious Design and Manufacturing: A State-of-the-art Survey," *Journal of Manufacturing Systems,* **16**(5), pp. 352–371 (1997).
46. S. Yu, K. Jin, H. C. Zhang, F. Ling, and D. Barnes, "A Decision Making Model for Materials Management of End-of-Life Electronic Products," *Journal of Manufacturing Systems,* **19**(2), 94–107 (2000).
47. G. Seliger and H. Perlewitz, "Disassembly Factories for the Recovery of Resources in Product and Material Cycles," *Proceedings of Japanese Society of Precision Engineering,* Sapporo, Japan, 1998.
48. M. S. Sodhi, J. Young, and W. A. Knight, "Modeling Material Separation Processes in Bulk Recycling," *International Journal of Production Research,* **37**(10), 2239–2252 (1997).
49. M. Sodhi and Knight, W., "Product Design for Disassembly and Bulk Recycling," *Annals of the CIRP,* **47**(1), 115–118 (1998).
50. A. Gesing, "Assuring the Continued Recycling of Light Metals in End-of-Life Vehicles: A Global Perspective," *JOM,* 18–27 (August 2004).
51. A. van Schaik, M. A. Reuter, and K. Heiskanen, "The Influence of Particle Size Reduction and Liberation on the Recycling Rate of End-of-Life Vehicles," *Minerals Engineering,* **17**(2), 331–347 (2004).
52. E. J. Daniels, "Automotive Materials Recycling: A Status Report of U.S. DOE and Industry Collaboration," in *Ecomaterials and Ecoprocesses,* H. Mostaghaci (ed.), CIM, Montreal, Canada, 2003, pp. 389–402.
53. M. Sodhi and B. Reimer, "Models for Recycling Electronics End-of-Life Products," *OR-SPEKTRUM,* Vol. 23, 2001, pp. 97–115.
54. C. M. Rose, K. Ishii, and K. Masui, "How Product Characteristics Determine End-of-Life Strategies," *Proceedings of the 1998 IEEE International Symposium on Electronics and the Environment,* Oak Brook, Illinois, 1998, pp. 322–327.
55. A. van Schaik and M. A. Reuter, "The Optimization of End-of-life Vehicles Recycling in the European Union", *JOM, A Publication of the Minerals, Metals & Materials Society,* 39–42 (August 2004).
56. V. Sendijarevec et al., "Screening Study to Evaluate Shredder Residue Materials," Annual SAE Congress, SAE Paper no. 2004-01-0468, 2004, Detroit, MI, 2004.
57. G. R. Winslow et al., "Advanced Separation of Plastics from Shredder Residue," Annual SAE Congress, SAE Paper no. 2004-01-0469, 2004, Detroit, MI, 2004
58. H. Antrekowitsch and F. Prior, "Recycling of Metals and Plastics from Electronic Scrap," *Proceedings of Global Conference on Sustainable Product Development and Life Cycle Engineering,* Berlin, Germany, 2004, pp. 121–126.
59. B. H. Lee, S. Rhee, and K. Ishii, "Robust Design for Recyclability Using Demanufacturing Complexity Metrics," *Proceedings of the 1997 ASME Design Engineering Technical Conference and Computers in Engineering Conference,* Sacramento, California, no: 97-DET/DFM-4345, 1997.
60. T. G. Gutowski and J. B. Dahmus, "Material Entropy and Product Recycling," *Proc. Global Conference on Sustainable Product Development and Life Cycle Engineering,* Berlin, 2004, pp.135–138.
61. M. Frei, "Eco-effective Product Design: The Contribution of Environmental Management in Designing Sustainable Products," *Journal of Sustainable Product Design,* 16–25 (October 1998).
62. E. Datschefski, "Cyclic, Solar, Safe—BioDesign's Solution Requirements for Sustainability," *Journal of Sustainable Product Design,* 42–51 (January 1999).

63. H. Kaebernick, S. Kara, and M. Sun, "Sustainable Product Development and Manufacturing by Considering Environmental Requirements," *Robotics and Computer Integrated Manufacturing,* 461–468 (2003).
64. M. Sun, C. J. Rydh, and H. Kaebernick, "Material Grouping for Simplified Life Cycle Assessment," *Journal of Sustainable Product Design,* 45–58 (2004).
65. S. Kara, J. Hanafi, S. Manmek, and H. Kaebernick, "Life Cycle Cost and Health Impact Estimation of Injection Moulded Parts," *Proc. Global Conference on Sustainable Product Development and Life Cycle Engineering,* Berlin, 2004, pp. 73–77.
66. http://www.bfrl.nist.gov/oae/software/bees.html
67. D. A. Dickinson and R. J. Caudill, "Sustainable Product and Material End-of-life Management: An Approach for Evaluating Alternatives," *Proc. IEEE,* 153–158 (2003).
68. J. W. Sutherland, K. Gunter, D. Allen, B. Bras, T. Gutowski, C. Murphy, T. Piwonka, P. Sheng, D. Thurston, E. Wolff, "A Global Perspective on the Environmental Challenges Facing the Automotive Industry: State-of-the-Art and Directions for the Future," *International Journal of Vehicle Design,* **35**(1/2), 86–102, (2004).
69. *Our Common Future: From One Earth to One World,* World Commission on Environment and Development, Bruntdland commission report, Oxford University Press, Oxford, 1987, pp. 22–23 IV.
70. J. Fiksel, J. McDaniel and D. Spitzley, "Measuring Product Sustainability", *Journal of Sustainable Product Design,* 7–16 (July 1998).
71. VDI 2225 Sheet 3, "Design Engineering Methodics, Engineering Design and Optimum Cost, Valuation of Costs" (in German), Berlin, Beuth, 1998.
72. M. Voβ and H. Birkhofer, "How Much Ecology Does Your Company Want?—A Technique for Assessing Product Concepts Based on variable Weighting Factors," *Proc. Global Conference on Sustainable Product Development and Life Cycle Engineering,* Berlin, 2004, pp. 301–304.
73. ISO 14001: *Environmental Management,* 1996.
74. *Handbook of Environmentally Conscious Manufacturing,* N. Madu (ed.), Kluwer Academic Publishers, 2001.
75. C. Krotscheck and M. Naradoslowsky, "The Sustainable Process Index: A New Dimension in Echological Evaluation," *Echological Engineering,* **6,** 241–258 (1996).
76. K. R. Haapala, K. N. Khadke, and J. W. Sutherland, "Predicting Manufacturing Waste and Energy for Sustainable Product Development via WE-Fab Software," *Proc. Global Conference on Sustainable Product Development and Life Cycle Engineering,* Berlin, 2004, pp. 243–250.
77. J. J. Michalek, P. Y. Papalambros, and S. J. Skerlos, "Analytical Framework for the Evaluation of Government Policy and Sustainable Design: Automotive Fuel Economy Example", *Proc. Global Conference on Sustainable Product Development and Life Cycle Engineering,* Berlin, 2004, pp. 273–280.
78. X. Wang, Z. J. Da, A. K. Balaji, and I. S. Jawahir, "Performance-Based Optimal Selection of Cutting Conditions and Cutting Tools in Multi-Pass Turning Operations Using Genetic Algorithms," *International Journal of Production Research,* **40**(9), 2053–2065 (2002).
79. X. Wang and I. S. Jawahir, "Web-based Optimization of Milling Operations for the Selection of Cutting Conditions using Genetic Algorithms," *Journal of Engineering Manufacture—Proc. Inst. Mech. Engineers, United Kingdom,* **218,** 647–655 (2004).
80. http://www.epa.gov/oaintrnt/content/kc_brochure.pdf, A Case Study of the Kansas City Science & Technology, May 2003.
81. Final Report of the OSHA Metalworking Fluids Standards Advisory Committee, 1999.
82. http://www.cdc.gov/niosh/homepage.html
83. ISO 14000.
84. P. C. Wanigarathne, J. Liew, X. Wang, O. W. Dillon, Jr., and I. S. Jawahir, "Assessment of Process Sustainability for Product Manufacture in Machining Operations," *Proc. Global Conference on Sustainable Product Development and Life Cycle Engineering,* Berlin, 2004, pp. 305–312.
85. P. C. Wanigarathne, K. C. Ee, and I. S. Jawahir, "Near-Dry Machining for Environmentally Benign Manufacturing: A Comparison of Machining Performance with Flood Cooling and Dry Machining," *Proc. 2nd International Conference on Design and Manufacture for Sustainable Development,* Cambridge, United Kingdom, September 2003, pp. 39–48.

86. I. S. Jawahir and C. A. van Luttervelt, "Recent Developments in Chip Control Research and Applications," *Annals of the CIRP,* **42**(2), 659–693 (1993).
87. X. D. Fang, J. Fei, and I. S. Jawahir, "A Hybrid Algorithm for Predicting Chip-form/Chip Breakability in Machining," *International Journal of Machine Tools & Manufacture,* **36**(10), 1093–1107 (1996).
88. R. Ghosh, O. W. Dillon, and I. S. Jawahir, "An Investigation of 3-D Curled Chip in Machining, Part 1: A Mechanics-Based Analytical Model," *Journal of Machining Science & Technology,* **2**(1), 91–116 (1998).
89. R. Ghosh, O. W. Dillon, and I. S. Jawahir, "An Investigation of 3-D Curled Chip in Machining, Part 2: Simulation and Validation Using FE Techniques," *Journal of Machining Science & Technology,* **2**(1), 117–145 (1998).
90. X. D. Fang and I. S. Jawahir, "Predicting Total Machining Performance in Finish Turning Using Integrated Fuzzy-Set Models of the Machinability Parameters," *International Journal of Production Research,* **32**(4), 833–849 (1994).
91. Z. J. Da, M. Lin, R. Ghosh, and I. S. Jawahir, " Development of an Intelligent Technique for Predictive Assessment of Chip Breaking in Computer-aided Process Planning of Machining Operations," *Proc. Int. Manufacturing and Materials Processing* (*IMMP*), Gold Coast, Australia, July 1997, pp. 1311–1320.
92. I. S. Jawahir, P. X. Li, R. Ghosh, and E. L. Exner, "A New Parametric Approach for the Assessment of Comprehensive Tool-wear in Coated Grooved Tools," *Annals of the CIRP,* **44**(1), pp. 49–54 (1995).
93. K. C. Ee, A. K. Balaji, P. X. Li, and I. S. Jawahir, "Force Decomposition Model for Tool-wear in Turning with Grooved Cutting Tools," *Wear,* **249,** 985–994 (2002).
94. P. Marksberry, *An Assessment of Tool-life Performance in Near-Dry Machining of Automotive Steel Components for Sustainable Manufacturing,* Thesis, University of Kentucky, Lexington, 2004, 489 pages.
95. P. Marksberry and I. S. Jawahir, "A Comprehensive Tool-life Performance Model in Near-dry Machining for Sustainable Manufacturing," *Journal of Wear,* accepted to be published in 2005.
96. M. Bromark, R. Cahlin, P. Hadenqvist, S. Hogmark, G. Hakansson, and G. Hansson, "Influence of Recoating on the Mechanical and Tribological Performance of TiN-coated HSS," *Surface and Coating Technology,* **76–77,** 481–486 (1995).
97. K. D. Bouzakis, G. Skordaris, S. Hadjiyiannis, A. Asimakopoulos, J. Mirisidis, N. Michailidis, G. Erkens, R. Kremer, F. Klocke, and M. Kleinjans "A Nanoindendation-based Determination of Internal Stress Alterations in PVD Films and their Cemented Carbides Substrates Induced by Recoating Procedures and their Effect on the Cutting Performance," *Thin Solid Films,* **447–448,** 264–271 (2004).
98. K. C. Ee, O. W. Dillon, Jr., and I. S. Jawahir, "Finite Element Modeling of Residual Stresses in Machining Induced by Cutting Tool with a Finite Edge Radius," *Proc. 6th CIRP Int. Workshop on Modeling of Machining Operations,* McMaster University, Hamilton, ON, Canada, May 2003, pp. 101–112.
99. K. C. Ee, O. W. Dillon, Jr. and I. S. Jawahir, "An Analysis of the Effects of Chip-groove Geometry on Residual Stress Formation in Machining using Finite Element Methods," *Proc. 7th CIRP Int. Workshop on Modeling of Machining Operations,* ENSAM, Cluny, France, May 2004, pp. 267–274.
100. S. K. Sikdar, "Sustainable Development and Sustainability Metrics," *AIChE Journal,* **49**(8), 1928–1932 (2003).
101. D. Tanzil, G. Ma, and B. R. Beloff, "Sustainability Metrix," *Proc. 11th Int. Conf. on Greening of Industry Network—Innovating for Sustainability,* October 2003.
102. G. Seliger, "Global Sustainability—A Future Scenario," *Proc. Global Conference on Sustainable Product Development and Life Cycle Engineering,* Berlin, 2004, pp. 29–35.

PART 2
MANAGEMENT, FINANCE, QUALITY, LAW, AND RESEARCH

CHAPTER 13
MANAGING PROJECTS IN ENGINEERING ORGANIZATIONS USING INTERORGANIZATIONAL TEAMS

Karen L. Higgins
NAVAIR Weapons Division
China Lake, California

Joseph A. Maciariello
Peter F. Drucker and Masatoshi Ito Graduate School of Management
Claremont Graduate University
Claremont, California

1	INTRODUCTION	447
2	CURRENT BUSINESS ENVIRONMENT AND UPDATE TO PROJECT MANAGEMENT CHAPTER	448
3	IMPLICATIONS OF THE CURRENT BUSINESS ENVIRONMENT	449
4	RESEARCH-BASED MODEL FOR SUCCESSFUL MANAGEMENT OF INTERDEPENDENT ORGANIZATIONS AND PROJECT TEAMS	450
	4.1 Interorganizational Teams	450
	4.2 Leading Interorganizational Project Teams	450
	4.3 Management Coordination Systems	451
	4.4 Components of an MCS for Interorganizational Teams	452
	4.5 Avoiding Failure and Achieving Success	466
5	EXAMPLES OF PROJECT SUCCESS AND FAILURE IN TWO DEFENSE PROJECTS	470
	5.1 Example: Failed Project Management	470
	5.2 Example: Successful Project Management	473
6	PRACTICAL APPLICATIONS FOR LEADING INTERORGANIZATIONAL TEAMS	476
7	SUMMARY: LEADING INTERORGANIZATIONAL ENGINEERING TEAMS	477
	APPENDIX: CORPEDIA INTERNET MODULES: AN APPLIED FRAMEWORK FOR PROJECT MANAGEMENT	478
	REFERENCES	479
	ANNOTATED BIBLIOGRAPHY: PROJECT MANAGEMENT REFERENCE MATERIAL	480

1 INTRODUCTION

Project management is foundational to running a complex business, whether that business is design and development of new products or production of existing ones. Projects by nature have two essential characteristics: each has a beginning and an end and each produces a unique product—one that plows new ground.[1] To manage projects even in the simplest environment a successful project manager must possess a combination of expertise, experience, technical and process insight, and people skills. Project management is at its core the ability to

- Articulate requirements
- Organize tasks
- Motivate people toward common goals
- Acquire and manage resources

It often requires political savvy to balance the demands of cost, schedule, and performance. A project manager must have highly honed communication skills to set expectations, describe goals, and define tasks.

2 CURRENT BUSINESS ENVIRONMENT AND UPDATE TO PROJECT MANAGEMENT CHAPTER

Basic project management skills as noted above are important to the success of any project. However, the current business environment adds another level to the project management challenge. As we look at the world around us, we see a magnitude and rate of change that can be overwhelming. Explosions in information technology, globalization of industries, and tremendous dynamics characterize this environment. Most all of our actions illustrate our highly interconnected world.

To accommodate this more complex environment, the material presented in this chapter has been updated from the second edition of the *Mechanical Engineers' Handbook*. It has transitioned from a focus on "The Management Control of Projects" in single organizations to "Managing Projects in Engineering Organizations Using Interorganizational Teams" in this third edition. Chapter 67 in the second edition of the handbook described techniques used by project managers and placed these techniques within an overall systems framework of management. While the chapter in the second edition was an expansion toward integration of techniques and management, the third edition is completely new. It reflects the sea of changes to project management that have occurred over the past decade. These changes have been a response to the expansion of technological knowledge, reflected, for example, in the four volumes of the third edition of the *Mechanical Engineers' Handbook*. This expansion and splintering of technical knowledge makes it much more difficult for an engineering organization to maintain every technical specialty it requires at a scale that is efficient and effective. Simultaneously, information technology has significantly reduced the cost of communication among partner organizations, thus facilitating desirable interorganization alliances and partnerships.

Using this more complex and dynamic business environment as a backdrop, the updated chapter addresses some of the challenges, successes, and failures of project management for organizations that are more interdependent. This material builds on an existing body of knowledge of project management. However, the chapter does not focus primarily on generic tools or on tools other authors have developed so well. Rather, this chapter concentrates primarily on the lessons to be learned that go beyond the basic tools and skills needed for project management. Basic tools are discussed extensively in the Annotated Bibliography and in the Appendix.

This chapter presents an integrated way of looking at project management. The material will be useful for both new and experienced project managers who are eager to adapt to the ever-changing environment and the interdependent world. The lessons described will help project managers create the right internal and external environments for managing successful projects using interdependent, interorganizational teams. It will allow the project manager to appreciate the cause and effect relationships of key components in these environments.

The chapter presents material in five parts:

1. A discussion of implications that the current business environment has on project management
2. A research-based system model that gives a success-oriented perspective for interdependent organizations and teams
3. Real-life examples and lessons learned
4. A set of imperatives applicable to project management
5. A sampling of "how to" references for traditional project management approaches and tools. These samples are listed and described in the Annotated Bibliography and the Appendix to this chapter

The model presented in this chapter is based on actual project management experiences in high-technology engineering organizations in both commercial and defense industries. These organizations were highly interdependent and used interorganizational teams. The real-life examples depict successes and failures and illustrate the dynamic model so that the project manager can develop an intuitive feel for successful project management, particularly for projects using teams of people from different organizations who work together to create a product.

3 IMPLICATIONS OF THE CURRENT BUSINESS ENVIRONMENT

Successful project managers in these organizations must expand their skill sets to accommodate the effects of the interdependence that surrounds our organizations. With technology advances in this information age, new products today range from the large and tangible, like an aircraft engine, to the nearly invisible software program. Because of their size and complexity, products today may also contain more than one "project"; thus, coordination among projects is also required to develop a product. As a result, project management today presents additional challenges of complexity, uncertainty, and interdependence. Project managers must keep pace with this change. The greater productivity and efficiencies that we are currently experiencing in our organizations reflect the results of the adaptation and transition of our management practices to the changes and dynamics of the new environment.

A few implications for the project manager of this changing and complicated environment include

- *More collaboration with other organizations.* More complex products require many different skills that may not all be found in the same organization. Project managers must manage projects that span organizational boundaries and lead teams of people who may be from different organizations; that is, they may lead interorganizational teams.
- *More emphasis on influence derived from relationships rather than authority derived from position.* When working with interorganizational teams, the project manager does not have full formal authority over those who work in different organizations.
- *More emphasis on informal and lasting relationships based on trust.* While this aspect has always been an important complement to establishing formal agreements, the

changing nature of tasks and the dynamic environment requires greater flexibility and adaptability than can be achieved through formal contracts and agreements.

4 RESEARCH-BASED MODEL FOR SUCCESSFUL MANAGEMENT OF INTERDEPENDENT ORGANIZATIONS AND PROJECT TEAMS

This section introduces a dynamic systems model for managing projects using interorganizational teams. The model considers an organization's internal interactions as well as its interactions with the environment and with external organizations whose members are part of an interorganizational project team. It provides project leaders and organizational executives a framework they can use to effectively collaborate in engineering organizations in today's interdependent environment. Leaders will learn successful actions to pursue and failure modes to avoid. Real case examples are used to illustrate the utility of the model.

4.1 Interorganizational Teams

An interorganizational project team consists of a web of interdependent organizations established for the purpose of achieving objectives that are agreeable to partner organizations. These teams lack clear organizational boundaries. The boundaries that do exist are fluid. Effective collaboration relies on resource exchanges, information exchanges, lateral interpersonal relationships, and trust. Working relationships cut across intra- and interorganizational boundaries. The participants must be led as well as managed, thus project managers not only must have management skills, but also must be leaders of team members.

4.2 Leading Interorganizational Project Teams

The model for leading projects presented in this chapter is informed by the theoretical and empirical research carried out by Higgins[2] at Texas Instruments (TI). The formal research project examined 6 interorganizational teams within 2 business units at TI. The 6 interorganizational teams included participants from 23 external partner organizations.

Subsequent analysis and executive practice has led to the design of a general model for leading interorganizational project teams. A detailed understanding of this model, lessons learned from real-life examples analyzed using the model, and four imperatives derived from the model will help managers of engineering-based organizations to successfully lead project teams.

When designing a management control system for a traditional organization, we use mechanisms such as command authority, responsibility centers, budgets, performance monitoring, and rewards and punishments to motivate goal-directed behavior. Leaders in organizations with interorganizational teams do not have direct authority over employees and leaders in partner organizations, but these leaders do have influence. They can empower people and coordinate their behavior even though they cannot "command or control." The leaders have to think about management in a different way. They have to think more in terms of how all the pieces fit together. To coordinate this whole collaborative organization, they must use more of a systems orientation to produce a product or service with which customers are happy and to achieve other organizational and interorganizational goals. We turn now to explain how to lead complex collaboration in these organizations.

4.3 Management Coordination Systems

Systems thinking forms the basis of a model for a management coordination system (MCS) to accomplish the goals of an organization. This model is also applicable to project teams and to programs that span multiple projects involving multiple organizations. This MCS model depicts the components and interactions in an organization that are essential to achieving success. Project leaders and organizational executives can use this model to address challenges and to increase success. Achieving goal alignment is fundamental to success; thus, we begin with a description of the goal alignment process.

Achieving Goal Alignment
Management control in a simple organization is a matter of selecting the right people and aligning their goals with those of the organization. Goal alignment occurs when individuals are motivated to achieve goals and objectives that are consistent with those of the organization. Even in the simplest organization there are at least two levels of goals: individual goals and organizational goals. Complexity is increased somewhat when an organization is responsible for multiple projects. In this case project goals add a third level. The goal formulation process can be highly effective in motivating people if the goals of individuals and those of the organization coincide or overlap. If people are motivated they tend to be more productive and the organization is more likely to fulfill its mission. Yet even in this simple organization it is difficult to achieve goal alignment. Individuals and the organization have multiple goals. One must understand individual motivations and provide rewards that will motivate individuals to achieve the goals of the organization.

Goal alignment is an even greater challenge on interorganizational teams. One of Higgins' interorganizational teams at Texas Instruments exemplifies the challenges of aligning the goals among organizations. This team was involved with another company in a joint venture. The primary focus of one of the companies in the venture was to earn the highest return possible on its investment. The other company's main objective was to deliver a high-quality product, and although they also wanted to earn a satisfactory return, the primary goals of the two organizations didn't match. There was inherent conflict in the goals of the joint venture. Conflict arose because it was difficult to reconcile ongoing decisions with differing organizational goals.

When one overlays each individual within each organization onto an interorganizational team, the complexity increases even more. Within TI, for example, individuals working on some of the interorganizational project teams were primarily motivated by technical excellence. They joined TI because they wanted to use and improve their technical skills and they received satisfaction from doing excellent technical work. On other teams, TIers were motivated by financial rewards. So, even motivations of personnel within TI were different.

Project teams in large organizations that have multiple project teams may operate under rules and motivations that are different from one another. One team's goal, for instance, might be to complete a report by the next day, and another team might need to get an integrated product tested. They both may need to use the same analysis equipment. However, the leader on the first team might not really care about the needs of the second team and might use the equipment without concern for these needs. When teams internal to each organization share the same resources, each team wishes to have sufficient resources to meet its schedule; this often leads to conflict within the same organization. Now, overlay the conflicts experienced between internal teams onto interorganizational teams and the problem of achieving goal alignment becomes even more challenging.

When project teams cut across organizations, it becomes very difficult to simultaneously motivate individuals, motivate teams, and align goals within an organization and between organizations. Alignment involves multiple layers of interactions (i.e., a system of systems). When leading these project teams and organizations, one must understand the dynamics within and among organizations and must collaborate well to successfully achieve all of these goals.

The project leader's role is critical and primary to achieving alignment. Leaders can take a number of steps to align individual, team, organizational, and interorganizational goals. The remainder of this section describes components of a general MCS that can be used by leaders to achieve both alignment and successful collaboration in project teams.

4.4 Components of an MCS for Interorganizational Teams

Table 1 lists the components of an MCS for managing interorganizational teams. The basic building blocks of this system are *primary management systems* that are designed to achieve *performance* objectives while satisfying employees and customers. Other components, such as *external environment, internal environment,* and *external organizations* are influenced by and influence these basic building blocks.

Primary management systems include individuals, a leadership system, and formal and informal management systems. Within organizations that are successful in managing interorganizational project teams there is no longer a pyramid with a CEO at the top saying, "This is our strategy and this is what we're going to do." Rather there is a mix or system of leaders at all different levels who ask, "how can we combine our talents to accomplish our goals?" On large programs, project team leaders often interact with project leaders from their own and from other organizations to meet goals. Leaders also design formal and influence informal management systems to motivate individuals and achieve performance objectives.

Next is the external environment. Each organization or project interacts with its external environment. Some of the dynamics experienced come from this environment. Leaders often find themselves asking questions like: "What should I do if my project requirements change, or if technology has advanced, or if there's no budget among customers for my product, or if an organization represented on our team goes out of business?" These kinds of external events are constantly churning and present major challenges for all project leaders and executives.

Then there is an internal environment within an organization, and on teams. The internal environment includes a number of soft variables, like trust and relationships. Information processing is also part of this internal environment and reflects how effectively members

Table 1 Management Coordination System Components

Basic building blocks
    Primary management systems
        Individuals
        Leadership system
        Formal and informal management systems
    Performance
External environment
Internal environment
    Trust/commitment/conflict/adaptability
    Information processing
    Goal alignment
Interorganizational influences

within an organization and among organizations share information. The effort required to achieve goal alignment is another part of this internal environment.

Next, there are influences from partners. These influences are referred to as interorganizational influences. Finally, these components or interconnected parts of an organization have cause-and-effect relationships with each other.

Figure 1 is a diagram of many of the interactions among the components defined in Table 1. We will describe the basic building blocks and the components of this diagram, and their interactions, step-by-step, so that you can see how executives can lead collaborative work using management coordination systems. You will also see by all the arrows and flows in Fig. 1 that there are a number of implied dynamics in this system.

Management Systems

Management systems are components that leaders can influence to achieve success in an organization. Their influence ranges from recognizing and leveraging an individual's intrinsic motivation to shaping the culture within the informal management system over time to directly and immediately changing elements in the formal management system.

The Individual's Intrinsic Motivation. We begin our explanation of the management systems in Fig. 1 very simply, with the individual and with intrinsic motivation of individuals. What is intrinsic motivation? Intrinsic motivation consists of the internal factors that individuals bring to an organization. Internal factors define who we are and provide the impetus for our

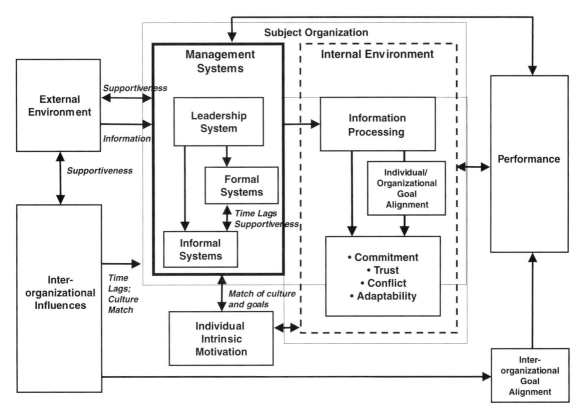

Figure 1 Management coordination system for managing interorganizational projects.

actions. Doing well in the world, serving others, feeling good about what we do, the desire for achievement, and acting with integrity are examples of internal factors that motivate individuals. Intrinsic motivation of individuals is a "self-control" system. Because it is such a powerful force in motivation, organizations should seek individuals whose internal motives and values match the goals and values of the organization and project.

Executives, in their hiring activities, can select associates with attributes they desire. Furthermore, many attributes can be enhanced by a leader's actions, oftentimes simply by setting an example or giving feedback. An executive may also enhance the intrinsic motivation of individuals by devising programs to improve technical skills, interpersonal skills, and knowledge of specific challenges facing the team and organization. These intrinsic characteristics of individuals influence commitment to work as well as commitment to team members and commitment to the organization. When these intrinsic characteristics match the goals and objectives of a team or organization, they contribute to its success.

The Leadership System. Leadership is the most important of the management systems since it has significant influence upon formal and informal systems and therefore on performance. Leadership is defined as the actions, values, style, and characteristics that enhance the ability of executives of business units and projects and of external organizations to operate in an informal, interactive network of organizations.

Leadership is considered a system because in a boundaryless organization, unlike in hierarchical organizations, leadership includes both organizational and interorganizational leadership. Leadership in a boundaryless organization is not merely a person at the top of the hierarchy setting the vision and values for the rest of the organization or project. Rather, it is a system of people in various leadership positions both inside and outside of the organization who are responsible for the direction of the organization and the interorganizational project team.

This system of leadership has to formulate as well as implement strategy. Implementation of strategy, in turn, requires decisions concerning the allocation of resources. In this process the leadership group—organization, organizational unit, and interorganizational team leaders—shares opinions and engages in dialogue and conflict over direction. The system of leadership is responsible for uniting and motivating the rest of the members, at all levels, toward the vision and goals of the organization and its interorganizational teams.

The leadership team thus should combine its forces and understand the influence it can have on overall performance. These leaders should ask themselves: "how by our values, style, actions, and other characteristics can we influence our organization toward success and how can we use our formal and informal management systems to create an environment that leads to success?"

The most important characteristics possessed by effective leaders in the TI study were

- Integrity
- Relational skills
- A track record of success
- The ability to

    Use influence rather than direct authority in relationships

    Empower others

    Understand the "big picture"

    Establish a vision and set goals

These characteristics, almost identical to those identified by Kotter,[3] emphasize *what a leader does, not merely what a leader is*. They complement the core skills described earlier for project management. What do the leaders value? Do they create trust? Do they empower people? Or do they just monitor every little thing—micromanage, in other words? Are they competent? What are their motives? and What kind of strategies do they pursue?

Formal Management System. Leaders actively design formal management systems in organizations. The formal management systems of an organization include organization and project structures, such as hierarchies and teams; formal monitoring systems and rewards; formal coordination mechanisms, such as electronic communications; training and indoctrination policies; and planning and control processes, including goal alignment and continuous improvement processes.

One example of interaction occurs between formal systems and intrinsic motivation. A formal reward based on achievement may reinforce feelings of self-worth and therefore appeal to and enhance intrinsic motivation. The Higgins study at Texas Instruments found that the most important formal systems for managing collaborative and interorganizational teams were

- The formal organizational structure for the interorganizational team
- The roles of the members of the interorganizational team
- Formal training in team building skills
- Training in management policies
- Planning and control processes including the goal alignment process (referred to as goal flowdown and catchball) and continuous improvement methods

Characteristics of the formal systems that led to ineffective performance were

- Overly loose or overly tight planning and control processes
- Formal coordination mechanisms that rely completely on rules and standard operating procedures
- Ineffective feedback of information from the environment regarding goal performance
- Lack of a direct link between performance and rewards
- Unclear or constantly changing organization structure and roles

It takes time to design formal systems such as the organization structure, a performance monitoring system, or a reward system. And, it takes time to change them. So leaders must be mindful that when they set a policy or a system in place, it is not quickly changed.

Informal Management Systems. Although the formal system is a powerful mechanism for coordination in an organization, the greater power comes from its interaction with the informal system. It is a framework upon which the informal system hangs. The informal organization refers to the emergent patterns of social life "that evolve within the framework of a formally established organization."[4] Informal management systems include factors that are intangible but nevertheless very real. They include culture or beliefs, interpersonal relationships, ad hoc roles, and informal rewards. Does an organization see itself as a dynamic, great organization? Does it have high integrity? How do individuals interact and what do they believe? Do individuals in the organization share the espoused values of the organization? For example, an employee may say, "I won't cheat anybody because honesty is a value of my organization. I will not cheat and furthermore the organization won't tolerate it."

Interpersonal relationships in the informal system may be fragile, and perhaps in some cases, extremely so. Think about interacting with people on a personal level. Suppose a person does something that irritates you. That can happen quickly, even in less than an hour.

Your reaction may be, "I'm not going to work with you anymore. You really upset me so I am not going to have anything to do with you." These are very real actions and although intangible, they're a critical part of the informal system within an organization and subsequently may change the nature of the working relationships in the organization.

Culture and beliefs in the informal system are also intangible, but may not change so quickly. The fieldwork at TI indicated that certain cultural values are important for the smooth functioning of interorganizational teams. These values, which were present at TI during the study period, are

- Ethics and integrity
- A technical problem-solving orientation
- A strong work ethic
- Respect for the individual
- The importance of teaming

Through his actions, a former president of TI set an example of strong ethics and integrity that has since become folklore in the company. In this example, his action was ingrained into the culture of the company and has been sustained. This man of high integrity didn't tolerate any kind of underhanded dealings. He received word that in trying to do business in a foreign country, employees there reported that a negotiator wanted TI to bribe him to get his business. He had said, "Hey, give me money and we'll do business with you." The president was upset by the request for a bribe, and even though the prospective business would have made a big difference to the bottom-line of the company and to future profitability, he said, "No, we don't do things like this." He personally got on a plane and brought the TI employee back home. The president meant what he said about bribes! And, the example he set became a defining act for the culture of the company.

So, you can see the effects of a leader's action. If a leader espouses high integrity and takes actions that reflect that integrity, people will believe in the leader, they will trust the leader, and say, "Okay, that is who we are; those are our values." But actions take a long time to get instilled into a culture. It happens over time. It takes longer than just signing a policy statement, or making a change in rewards, or changing another formal system.

Project teams tend to have their own subcultures as well. Though they may blend the stronger cultural aspects from the organizations of their members, these subcultures also evolve from formal systems, from interpersonal interactions and from actions of the project leaders.

Although actions may take a long time to influence an organization's culture, they are extremely powerful both in positive and negative ways. Leaders should understand the influence they have on the values and beliefs of their organization. This influence occurs by stating organizational values and then implementing them through consistent actions even if the actions are costly.

Interaction between Formal and Informal Systems. Another interaction in Fig. 1 is between the *formal system* and the *informal system*. The two systems are highly interlinked, with the informal system breathing life into the formal system skeleton, giving the organization personality and soul. Remember that the informal system includes the beliefs and values of an organization, the network of informal relationships; and the informal recognition and reward system.

For example, assume that an organization's *formal system* rewards getting projects done on time, and at any cost. The performance and reward plan might stipulate that a bonus will be given to team members if they complete a priority project by a given date. Assume further that an espoused value of the organization stipulates, "we don't take *shortcuts*, we produce excellent products" and this is what people believe the values of the organization to be. If you continue to reward the behavior, "performance at any cost," what do you think that will do to the values of the organization? How can individuals believe the espoused values if the organization rewards something else?

A real conflict between performance and values often occurs with "superstars." These people are high performers, but they may be very wearing on their teammates. They simply want to get the job done without placing the same value on "teamwork" that the organization does, without supporting other teammates or minimizing dysfunctional conflicts. Organizations have these superstars who say, "I don't care about you and your work because you don't help me with my project. If I step on your toes by achieving my goals, I don't really care. I'm going to get it done at any cost." Sometimes an organization rewards these individuals because they do get things done.

One-high technology organization decided, "Yes, it's great to have people who get things done at all cost." But they found that the heaviest cost of this behavior was long-term; as they continually rewarded these people, others noticed and said, "Wait a minute. You're saying, on one hand, you want us to be a team and you want us to work together and yet you're letting this person go off and be so different just because he gets things done. That's not fair!"

The solution is for the leaders to act in a way that is consistent with espoused values at every opportunity. Take the case of the superstar who doesn't support the values of the company. If a leader sees people operating contrary to organizational values, they must let it be known that they will not tolerate this behavior. Some people get the point and change their behavior or leave. Others may need to be dismissed. However, in asking a star performer to leave the organization, leaders do not want to send a signal that outstanding performance is unwanted. Rather, they have to make it very clear why the individual is leaving. It isn't because the star performer excelled that he was let go, because the organization does reward outstanding performance. Rather, it is because he did not perform *within the culture;* he did not accept the organization's beliefs.

Time Lags and Long-Term Effects. You can see from this example that if a formal system rewards "getting things done at all cost," but the informal system emphasizes teamwork, there's a conflict between the formal and the informal systems. They are not supportive of one another. But the conflict normally doesn't hurt the organization immediately. Changes in the formal system, such as in the reward system, don't impact the informal system right away. *There is a time lag. Espoused values* are *stated beliefs* in the informal system, *actions* reflect true *values,* and if, over time, actions do not match beliefs, the espoused values are not believed, creating a disconnect between formal systems and espoused values in the informal system.

Espoused values are embedded in the culture of an organization through *two-way interactions* between formal and informal systems. Interactions occur and individuals watch what happens and measure actions against what was said, continually analyzing actions in light of values. Over time, actions become the beliefs that are part of the culture. Prior to that time, however, the very divergence of espoused values and actions can affect many contributors to success. This divergence is critical because it not only reduces employee satisfaction, but may also undermine the trust individuals have for their leaders.

Through their actions, executives in our superstar example have stated their real beliefs: "We are not going to emphasize getting the job done at all costs." In that case, individuals have to (1) share the values of the organization and (2) perform. If they don't adhere to the values, they may be candidates for being released even if they perform well. So to produce internal consistency between informal and formal systems, and maintain high integrity, leaders must practice what they preach. As some have noted, trust is built when leaders "walk their talk."

Feedback. But we are not finished. The last and linking element among these primary management systems is feedback, as shown in Fig. 1. If the team does well and meets its goals, adjustments to the MCS may not be necessary. If the team fails to meet its goals, a leader might say, "Wait a minute, something is wrong here. As the leader of the organization I may want to change some things. Maybe we should reorganize. Or perhaps examine our values. Or look to see if our people are as productive as they should be or if they are in some way demoralized. Are they getting what they need from this organization?"

The feedback loop is where we start seeing the importance of adopting a *systems orientation*. We recognize that some components and elements in the model affect other pieces and there is a feedback of results onto the rest of the system. The feedback process continues indefinitely following the precepts of management control theory.[5]

Performance

The last building block in the model in Fig. 1 is performance. Performance is measured by assessing team, organizational, and interorganizational success with regards to project goals. Overall objectives of the primary management systems are to influence performance. To get a better understanding of their influence, let's summarize how they work together. First, the leadership system should be in place. Then one attempts to get the intrinsic motivation of individuals lined up with the organization. Next, the job is to get informal and formal systems in synch, at least to the point where they are not counterproductive to each other and they are in line with leadership values that promote success. Then strategic thinking is carried out, plans are established and implemented. Finally, if all goes well, performance goals for all stakeholders (team members, organizations, and customers) are met.

External Environment

Variables external to the organization are other components of the model that may significantly impact the performance of the organization. Performance, in turn, can affect these external variables, thus the bidirectional supportiveness arrow in Fig. 1. We consider below two external variables, environmental intensity and unique environmental variables. We also describe how leaders must adapt the MCS to the external environment.

Environmental Intensity. Some combination of three variables, *complexity, dynamics,* and *uncertainty,* characterize the environmental intensity in which an organization and its projects operate. The nature of the industry itself, including the customer and the market environment, and the interdependency of the organization with external organizations all influence these intensity variables.

The *number of sources* required to get the information needed to make decisions and the *range of knowledge required* to gain an understanding of these inputs are measures of *complexity*. The *rate of change* over one and five years in products, technology, legal environment, internal structures, number of employees, competition, and the customer base provides a way to assess *dynamics*. *Uncertainty* is assessed by the extent to which individuals in a business unit have *enough information on the environment* to make effective decisions and to predict the outcome of their decisions.

Differences in external environments may account for differences in ultimate success of a project. For example, in the Higgins study, major differences existed in the environments faced by TI and by the 23 external organizations on the 6 interorganizational project teams.

Differences in these environmental variables accounted for differences in their management coordination systems and, to a considerable extent, for differences in the performance of these teams when their MCSs did not appropriately accommodate the environment.

The external environment may be highly dynamic and move quickly. Technology, for example, changes very rapidly and significantly impacts affected organizations. An organization's MCS must be flexible and adaptable enough to adjust to the complexity, dynamics, and uncertainty of the external environment and to changes that are happening, whether they're rapid or slow. If changes are rapid, a highly bureaucratic organization, one that has rigid rules and procedures that it cannot deviate from, is unlikely to prosper and may not survive.

Unique Environmental Factors. Unique features in the environment include specific industry features, such as downsizing trends and availability of specific human, capital, and physical resources and budget cuts enforced upon an organizational unit and its projects. These factors can affect the business unit, including its performance on project goals. For example, an organization in an industry that is "downsizing" is beset by extremely demoralizing influences on internal relationships. These external influences may impede the effectiveness of the projects of an organization.

Adapting Primary Management Systems to the External Environment. A way to acquire flexibility and adapt to a highly intense external environment is to create an organic organization structure, one that permits supportiveness between the environment and the primary management systems. An organic structure is just that; it's a living thing, it's a dynamic thing. It almost breathes; it changes; it grows; it morphs into other things. An organic organization is very much based on the informal management system. It relies on relationships, trust and interactions among people.

Supportiveness between the external environment and the organization means that if the external environment changes rapidly, the internal environment must be able to adapt to those changes quickly. If an organization lacks informal relationships that are required for adaptation to a rapidly changing environment, the leader can develop them—but it takes time.

A leader may create situations that strengthen an informal system. Rewarding constructive informal behavior when it occurs can do this. For example, if someone has a good idea and has carried it out informally, the leader could say, "Great idea, let's do it. It isn't in our rule book, but do it because it makes sense." That's how a leader can encourage development of the informal system. It takes a deliberate thought process, but one builds trust and says, "Well, okay, you can go out on a limb a little bit. Go ahead and, oh, by the way, if you want to, talk to so and so."

A leader can also encourage informal relationships and networking by chartering a formal team to solve a specific problem with expectations that informal relationships will form. This technique has worked effectively in many organizations where individuals from different parts of the same organization or from different organizations have been placed on a team. These leaders travel extensively, visiting each other's sites. As they work together, they interact socially and break down barriers of mistrust, learn each other's skills, and build close ties. Team members are then more likely to call upon each other when they need support. This approach builds a productive set of relationships that did not exist before executives set up the formal team. Informal relations make the organization more adaptive and better able to cope with environmental changes. Project teams are a perfect setting to encourage relationship building, since goals are common and individuals must work together to achieve these defined goals.

Responses to Environmental Conditions. Three of the six interorganizational teams at TI were engaged in commercial projects, and three were engaged in defense projects. The environment faced by teams in the defense unit differed from the environment of the commercial unit in ways that are common to organizations engaged in different types of knowledge work. First, the nature of the market and competition faced by each unit was completely different. *Complexity, uncertainty,* and *dynamics* were much higher in the commercial unit. Second, the goals of the defense teams were *predetermined* to a significant degree. The primary tasks in defense projects involved solving technical problems according to stipulated requirements. In contrast, the very goals themselves had to be determined in response to changing customer goals and requirements in the commercial unit.

These different environments influence other elements of the MCS and thereby affect success. First, the effects of a dynamic and uncertain environment can be very detrimental to the goal alignment process and therefore to success. Second, the dynamics of the environment can cause the formal and informal systems to go significantly out of synch with each other and with the requirements of the environment, thus hampering success. For example, in a fast-paced environment, the informal system may be dominant, especially if there are established relationships on which to build. However, since formal systems such as contracts cannot change as quickly, they may lag and actually act counter to informal agreements.

Figure 2 contrasts two different approaches to intense environmental conditions, a *proactive* approach and a *passive* approach. Executives in organizations who use a more passive approach—reacting to the environment—may experience lower success, especially as a result of goal alignment difficulties and mismatches. Executives who use a more proactive

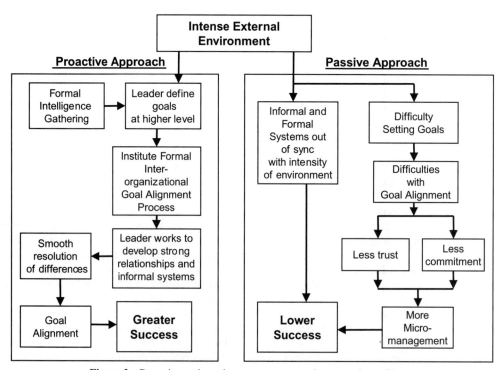

Figure 2 Proactive and passive responses to environmental conditions.

approach—gathering information on the environment, communicating that information throughout the organization, translating that information into overarching goals, and instituting formal goal alignment processes—should experience greater success. If the environment is intense, comparable to that faced by the commercial unit, or if it is less intense, as in the case of the defense unit, there are implications both for the appropriate style of leadership and for the design of formal and informal management systems of the organizational unit. A highly uncertain and dynamic environment will require more informal, relationship-based, mechanisms and high levels of trust.

By devising common goals and by championing the process of building team relationships, the leader can strengthen the informal systems of both the internal and interorganizational units. But even in a highly uncertain and dynamic environment, a complementary formal system is required to ensure that individuals know their roles and goals.

Internal Environment
The internal environment, as presented in Fig. 1, consists of a number of processes and variables indicative of how an organization works. It is in the operating environment where both formal and informal systems are used for processing information, whether via formal email systems or informal grapevines. Information processing is a product of the primary management systems. Goal alignment requires an information processing sequence. Individual, organizational, and interorganizational goal alignment are critical for developing commitment, trust, constructive conflict, and adaptability within the internal environment. Not only do the primary management systems affect information processing and goal alignment but they also interact to create an internal environment, working together to encourage or discourage trust, commitment, constructive conflict, and adaptability. So primary management systems have direct and indirect effects on performance.

Goal Alignment. Goal alignment appears as a component of the internal environment in Fig. 1 and involves information processing. The information processing sequence is accomplished through the interaction of leadership and formal and informal management systems. For example, the goal alignment sequence might start with leaders articulating their goals. Or, a leader might say, "These are my goals and I'm going to communicate them and get reactions from my people." A key issue is deciding on the actual communication steps necessary to carry out the information processing leading to goal alignment.

Texas Instruments carried out this sequence using a communication process called "catchball." Catchball starts with overarching goals—a big idealistic vision, which they promoted in all of their magazines and newspapers. Everybody knew these overarching goals. At first people would look at them, and say, "Okay, but I don't fit there. I don't see myself being networked into the TI world. I solder circuit boards. How do I fit?"

In response, TI's top executives gathered literally hundreds of supervisors over days and weeks and communicated these goals. Each of these supervisors went back and said, "This is how my organization fits." It was like throwing a ball: "Here you go; here are the goals, what are you going to do with them?" The supervisor responds: "I caught them. Now I'm going to share them with my people or with the next supervisory level down. Okay, here you go. What does it mean to your particular organization or project?"

Interviews at TI demonstrated that people at all levels knew the overarching goals as well as the goals of their project, if they were on a project team. Furthermore, they knew their own contribution to those goals. They were committed to these goals because they knew and accepted the goals and they trusted that the organization was concerned about the goals and that the goals were for their benefit. They were highly motivated to accomplish their goals.

So, this process of communication is extremely important. And the way TI did it was very effective because they involved everyone. Everyone had to write down in his or her own performance plan: "This is what I'm going to do and this is how it fits." The catchball process at TI illustrates an exemplary goal alignment process for an organization, its project teams, and its individuals.

Individuals have their own motivations, and each has different motivations: great work, achievement, a sense of belonging, integrity, technical challenge, money, and so on. The organization has profitability, customer satisfaction, quality, and employee satisfaction as goals. And project goals fall between individual and organizational goals. Thus, a key issue and a critical role for the leader is to align the goals of the individual with the goals of the organization. It requires knowing your people, knowing how to motivate them, and articulating overall goals well enough so that they can all know where they fit and can hold these goals as their own. It is important that individuals agree with the goals of the organization and be committed to accomplishing them.

Trust and Commitment. Trust and commitment in an organization's internal environment are important reflections of how primary management systems interact to achieve results. In addition to the primary management systems, goal alignment as implemented in the internal environment is also an important influence on trust and commitment. Goal alignment enhances commitment and trust, and commitment and trust enhance goal alignment (see Fig. 1). If individuals have the same or overlapping goals with the organization and know what these goals are, they are going to trust that they are working in synch with the organization. And trust is the lubricant for constructive interpersonal relationships. Trust suggests to the individual, "I know the organization isn't going to fire me unless I do badly, and then I'll know why I got fired. I know that they're going to try to motivate me and I believe what they say is true."

Trust is also enhanced by the direct actions of leaders. How does one develop trust? The way to reinforce trust is to communicate. Keep people informed of exactly what steps are being taken, what is known, and what is not known. And a covenant of trust is critical to solving recurring problems. Trust is imperative in leading people through crises and through times of great change. Having developed trust, the leader maintains it by communicating exactly what is known and exactly what is not known. When as a leader you do not know something, you simply say, "I don't know" and that fosters trust. Truth telling is a cornerstone of trust.

So, in all of these things a leader's actions matter. Leaders have the greatest impact on trust by having high integrity and by making sure that their actions actually promote that integrity.

Trust and the Dangers of Micromanagement. While there are many things that build trust, the practice of micromanagement, or excessive monitoring of performance of knowledge workers, diminishes trust and normally should be avoided. When leaders micromanage people they are essentially saying, "I don't trust you. I'm going to watch you carefully." It's especially demeaning to knowledge workers and potentially very dysfunctional to individual and organizational outcomes.

Figure 3 depicts a reinforcing process when micromanagement is present. Qualitative and quantitative analysis in the TI field study uncovered a detrimental reinforcing feedback process, particularly in TI's commercial business unit. This feedback process involved the use of a micromanaging leadership style in contrast to an empowering leadership style. The diagram in Fig. 3 represents a potential scenario in which a detrimental reinforcing feedback process may develop among four elements of the model in Fig. 3.

Figure 3 A detrimental reinforcing feedback process.

First, arrow 1 in Fig. 3 shows that a lack of organizational trust encourages a micromanaging leadership style. Second, a micromanaging style of dictating how things are to be done rather than empowering people to achieve a goal decreases goal consistency and goal acceptance. This in turn (arrow 2) negatively affects the goal alignment sequence. Third, lack of goal acceptance and goal alignment decreases organizational performance (arrow 3). Poor organizational performance intensifies the micromanaging style (arrow 4), which itself further reduces trust (arrow 6). Poor performance also directly diminishes trust (arrow 5).

To summarize, in Fig. 3 we see a detrimental, reinforcing feedback process linking micromanagement to lack of organizational trust to poor goal alignment and poor system performance. These relationships are entirely consistent with the internal environmental relationships shown in Fig. 1. Figure 3 illustrates three reinforcing feedback processes:

- Low organizational trust encourages micromanagement, and micromanagement reduces the level of trust (arrows 1 and 6).
- Poor goal performance encourages micromanagement tendencies (arrow 4), and micromanagement decreases goal performance by negatively affecting the goal alignment sequence (arrows 2 and 3).
- Poor goal performance reduces trust (arrow 5), and low trust reduces goal performance through its effect on tendencies to micromanage (arrow 1) and through its negative impact on the goal alignment sequence (arrows 2 and 3).

This scenario is found everyday in organizations. How often do we see managers immediately react to "bad news" by jumping in and trying to take over—micromanaging rather than allowing individuals to solve problems? And just as frequently, the individual will react to this management style—perhaps not immediately, but at a later time—by losing trust or by merely paying lip service to goals (thus diminishing goal alignment). The process of uncovering these three feedback processes is not intuitive but rather a result of systemic thinking and analysis. Yet a deep understanding of these processes is essential to effectively leading interorganizational team projects.

Constructive Conflict and Adaptability. To increase adaptability to change, and to become more flexible, a leader should create an environment that encourages people to learn. One way leaders facilitate learning is by encouraging constructive dissent. They open themselves to being challenged. One leader challenges people with, "You know what I don't like? It is to have people just sit there and nod their heads and say, yeah, yeah, yeah. They sit there and agree with everything I say and then go away and do something else, or offer dissenting opinions later."

If leaders create an environment that encourages people to open up and speak up, they will get wonderful ideas and learn things they may never have known. It is a trusting environment, one that tolerates failure and allows the sharing of ideas and constructive conflict. It allows the organization to act quickly and flexibly. An open environment fosters a learning kind of adaptability, a desirable internal variable, as seen in Fig. 1.

Constructive conflict comes about by setting up conditions whereby an individual doesn't feel afraid when challenging a leader, regardless of that leader's position or status. People get better, teams get better, and the organization gets better by creating an environment where constructive conflict is encouraged and where people are allowed to learn through their failures. When this environment is not encouraged, success diminishes.

Actions of the leaders, formal policies for dealing with conflict, an informal system that values trust and builds relationships all work together to create an internal environment that's very positive and one that permits adaptability. This type of internal environment facilitates adaptability to very intense external environments.

To this point in this chapter we have focused primarily on the individual, on project teams, on the organization, and on dynamics. We turn now to examine influences within an interorganizational team.

Interorganizational Influences

By adding other organizations and teams across organizations, we increase complexity and dynamics exponentially. Now individuals work on the same project but they may be from different organizations; the leader no longer has direct authority over individuals from these partner organizations. Numerous bilateral interactions occur among organizations on a team. Achieving goal alignment among partner organizations is essential to success. There is also a need for supportiveness between an interorganizational team and its environment. These additional components and complexities are represented prominently in Fig. 1.

There are several ways in which interorganizational project teams can influence internal business units of partner organizations:

- Leadership characteristics of external team leaders
- Teams' formal and informal management systems
- The internal environment of a team, including goal alignment
- Commitment and trust among organization's members on the team
- Team success

We will discuss effective methods of managing these interorganizational influences by using an example of excellent practices from the TI study. These practices were evident on the Joint Standoff Weapon interorganizational program team at Texas Instruments.

This team comprised a number of organizations working on the project. TI, of course, was one of them. The government was one of them, including government personnel from all over the country. The team also included multiple suppliers and other specialists. The two major leaders on the interorganizational team, TI and the government, started the project by telling all team partners, "We will spend time in team building because we believe it is critical to a successful project. This project is highly important so we have to establish trust in our team, to determine and align goals, to gain pride in our goals, and finally to commit to our goals."

Team members spent days together. Many complained, saying, "Why are you wasting this time? Let's just get on with it." But they spent days working on these issues. They worked at understanding what the project was all about; what were project goals; how they were going to operate; how they were going to resolve conflicts; and what were the ground rules.

Whenever a team member came to one organization to complain about someone from another organization, executives would say, "We do not want to hear this. You work it out because we are all working together. We are working on the same goal and the goal is . . . ." Leadership continued to make these very clear statements regarding what the team was about. As a result, there was no problem understanding what project goals were, or how the team was going to operate, or that leaders would not tolerate adversarial relationships on the team. The leadership on this team—and it was a system of leaders—was saying, "We are working together. We have linked arms and we won't tolerate any other kind of negative behavior." And they didn't. The result was the strongest, most successful team the researcher had witnessed. Top leaders took the time to establish the ground rules and to set the team in place. Then they empowered people saying, "These are our values," "These are our goals," and "This is how we will work together." So, the integrity, the abilities, the characteristics of the leadership members filtered into the rest of the team and extraordinary performance resulted.

Gaining interorganizational goal alignment is absolutely essential to the success of a project. It's more difficult, however, to accomplish this alignment. The difficulty arises because goals not only must be aligned within a single organization, they also must be aligned among organizations. This alignment process must reconcile goals among organizations and each may begin with very different goals.

It is also essential that organizations on an interorganizational team share values. Optimal performance depends on a match among the cultures of the organizations; or at least a match of the values that matter most to the success of the interorganizational team. If an organization with high integrity tries to team with a company with less than high integrity, the result is a values clash. The project team will not work effectively because trust cannot be present, nor would you know how to go about establishing trust when dealing with such a partner.

To summarize, although complex interactions exist within a single organization, the interactions are even more complex when you add other organizations to form an interorganizational team. Leading interorganizational teams poses the challenge of leading great complexity.

There is a lag in potential interorganizational influences, as shown in Fig. 1. If on an interorganizational team, leaders and members from one organization influence the leaders and members from another organization, then the values and practices found in this interorganizational team may over time permeate to each of the organizations themselves.

Complex dynamics result when collaboration occurs within and across organizations on an interorganizational team. The external environment influences the leadership system. Leadership can influence the formal system and then the informal system across organizations. Interactions of these systems create internal environments both in an organization and on teams whose members come from multiple organizations. Finally, there are time lags in the influence of one organization on the systems of partner organizations.

4.5 Avoiding Failure and Achieving Success

The interorganizational teams studied at TI achieved a considerable amount of success, but the defense teams were much more successful than the commercial teams. By examining the characteristics of project management we learn how to avoid mistakes and increase success.

Any one or combination of the variables in Fig. 1 may cause failures. The TI study did reveal some failures that are fairly common in engineering projects. These failures were somewhat different and more extensive in commercial projects. So we first examine the failures on commercial teams.

Failures on Commercial Teams

The environments faced by commercial teams were more dynamic than for defense teams and the challenges were greater. Although the commercial teams overall were successful, failures were evident in the primary management systems and were also influenced by external influences, system interactions, and the internal environment.

Management System Failures. There were three management system failures in the projects of the commercial unit: infrastructure and roles, planning and control processes, and leadership.

1. *Infrastructure and roles.* Goals were often not accomplished because of changes in the structure of teams and roles of team members. Lack of a clear organization structure and roles caused confusion. One individual when asked what prevented accomplishing organizational goals said, "We've accomplished a great deal in spite of that [absence of clear lines of roles], but we would have accomplished a great deal more if we had paid more attention to defining roles, responsibilities, and accountabilities of individuals."

2. *Planning and control processes.* Both *overly rigid* and *too loose* planning and control processes may decrease performance on team goals. It was widely noted in the commercial unit that "bureaucracy" and "procedural requirements" prevented individuals from accomplishing organizational goals. Conversely, excessively loose control processes, such as lack of formal contracts and stable requirements and lack of a formal process for understanding the external environment, diminish performance on team goals. One person noted, "I think it gets back to not having structure, that formal statement of work that we're used to on military programs . . . that caused . . . the goals to shift or appear to shift to a lot of individuals on the team."

3. *Leadership style.* Micromanaging was the leadership style for some teams in the commercial business unit. Micromanaging in the highly dynamic and uncertain environment negatively influences success and this was reflected in the relatively low performance on goals for commercial teams. This is consistent with the analysis presented in Fig. 3.

Failures Related to Interaction of Primary Management Systems. Interaction of primary management systems also created failure modes, in particular in goal setting, performance and rewards, and formal and informal systems.

1. *Leadership and goal setting.* Questionnaires and interviews indicated a sense of failure in setting common vision and goals. A number of individuals expressed a lack of internal consistency in the vision and goal-setting process. Here is how one person put it:

   > So constantly changing of your target, I guess that's an okay thing on the commercial side, because it's changing a lot, but it kind of goes back to [different leaders saying]: "This is the product we should develop" and according to how much influence they have over the developing engineers, that's the product they begin to develop.

   A lack of an overarching vision creates a lack of focus for the creative and technical efforts necessary to achieve goals. Again,

   > We're motivated to do something new and different in the industry.... We have more ideas than we have clear focus sometimes. ...We get defocused and off chasing something that prevents us from accomplishing the first goals we set for ourselves. ...Sometimes we just try to do too much."

   Having common organizational goals is motivating to individuals. But, the kind of goal instability witnessed on the commercial teams often results in lack of acceptance of organizational goals by individuals.

2. *Disconnect between performance and rewards.* Constantly changing goals on commercial teams created a "disconnect" between performance and rewards. Accomplishment of team goals was hindered by "a switching of what the goals are supposed to be, unclearness or constant change or putting a carrot out there and then taking it—not rewarding for the meeting of those goals." And the absence of feedback between performance and rewards decreases the ability of individuals to understand and attain organizational goals. Here is how one person described the disconnect: "Bonuses...are good, however, they're not very effective because they don't tell you why you got it so you don't know what to go redo in order to get it again."

3. *Low supportiveness between formal and informal systems.* Low supportiveness between formal and informal systems diminishes organizational performance. It was noted that lack of supportiveness between formal and informal organization structure hampered the team's performance on organizational goals. This is how the need for supportiveness was described:

   > You don't need a rigid structure, but you need some structure....differentiating between rigid and totally ad hoc has been difficult....somewhere in between is what works best. ...Particularly in a company like TI where you've got a lot of smart people and they are so smart that they can tie themselves up in debate for weeks and months and not do anything.

Effects of External Influences on Goal Performance. In addition to management system and leadership difficulties, a number of influences from outside the commercial business unit negatively affected team performance. They did so through (1) lack of goal clarity and stability, (2) a mismatch between the environment and the design of formal and informal management systems, and (3) resource constraints.

1. *The lack of goal clarity and stability.* The environment faced by the commercial unit was highly intense. The tasks of the commercial unit were determined by market requirements. Therefore, the primary tasks of commercial teams included defining market requirements and team goals. The environment faced by the defense business unit, on the other hand, was more stable since defense teams began from a statement

of work or from better-defined requirements. This external environment was a major difference faced by the two business units as noted by a member of the commercial unit.

> I think there is some unclarity [in goals] in that it is a much different step coming from the defense side to the commercial side. The defense side you have a product that you have to develop. On the commercial side, you are depending on market to tell you what you need to develop.

This dynamic and uncertain environment in the commercial unit reduced goal clarity and stability and affected performance negatively. The more dynamic environment faced by the commercial unit caused a significant mismatch between individual and organizational goals, thus leading to poorer performance on organizational goals relative to the defense unit.

In the commercial unit, goals must change and one must tolerate flexible goals while developing a feedback process to improve the ability to set goals. Here is how one team member described the desired flexibility:

> Depending on the business area, you have to be a little more tolerant of some mushy goals when you're trying to start something up. You want people to be aggressive and go after the lofty goals, but if they miss the mark you don't want the penalty to be so bad or the lack of reward to be so bad that it discourages them... You'd like for it to be enough for them to correct their goal-setting ability and maybe have a more realistic goal. I think you have to do that over a period of time.

2. *Mismatch to the environment.* The dynamics of the environment greatly affected the goal setting process in the commercial unit. This in turn negatively influenced performance on goals. In the presence of high environmental dynamics, an informal orientation that allows for greater flexibility in adapting to changes in the environment should increase success. The commercial unit appeared to require a more informal orientation in its management systems to successfully adapt to its relatively intense environment.

3. *Resource constraints.* Respondents in the commercial unit cited the lack of resource availability as the greatest barrier to the accomplishment of goals. To the unit and its teams these represent environmental constraints.

System Interaction Failures. Failures that result from the interaction among different parts of the model are referred to as system interaction failures. The ones that occurred in the commercial unit are precisely those described previously in Fig. 3, "The Dangers of Micromanagement." These involve the reinforcing feedback loop among Leadership Style [Management Systems]; Goal Alignment [Internal Environment]; Goal Performance [Performance]; and Organizational Trust [Internal Environment]. To understand these effects one must understand the interaction of variables throughout the system represented in Fig. 1.

Failures in the Internal Environment. The most serious problem in the internal environment of the commercial unit was a reported lack of trust. Some team members reported that managers in their organization were not open with team members. Members also complained about insufficient information, a variable related to trust. Finally, team members in the commercial unit did not believe they had sufficient information about the external environment and therefore they had difficulty adapting to the environment in a successful way.

Success and Failures in Defense Teams

Teams in the defense unit were considerably more successful than were those in the commercial unit. Defense teams experienced notable success in the Internal Environment and in

Intrinsic Motivation, while experiencing some failures in Management Systems and in the External Environment. There was no evidence that these failures reduced performance of defense teams.

Internal Environment Effects. There was strong evidence that the Formal Goal Flowdown Process and related Information Processing Capabilities positively influenced overall performance. The greater use of the Goal Flowdown Process in the defense unit and the presence of strong Information Processing Capabilities positively influenced performance. This positive finding, however, clearly indicates that inadequate goal flowdown and communication could also decrease goal performance:

> [What most influences success is] good communications overall, good flowdown of business goals, business problems, business occurrences so the people understand what's going on, they know better what to do, and they feel more motivated to do a good job, and they feel more involved overall.

Intrinsic Motivation. Qualitative data also related the "quality of people" in the organization to the successful performance of overall goals:

> You know the other thing that influences the success of the business unit is the quality of people you have...I think we've got just outstanding people ... I would say that that was probably the more significant reason for success and that is we had the experienced people with the right skill levels that we could do the job right the first time.

Management Systems Effects. The only potentially dysfunctional element in the management systems was a relatively high reliance on rules and standard operating procedures on defense teams when their environments indicated the need for more informal system elements. Therefore, the use of more formal coordination mechanisms had a detrimental effect even in the less uncertain and complex environments faced by defense teams.

External Environment Effects. External factors, on the other hand, had a large, often negative effect on goal performance. Thirty-seven percent of responses to questionnaires in the defense unit cited "resource" and "time" constraints as barriers to accomplishing goals.

Summary: Commercial and Defense Teams

Interorganizational teams in the defense business unit worked in a more stable environment than did commercial teams. Formal goal-setting processes and interpersonal relationships fostered good interorganizational goal alignment and were critical to the success of these teams.

The teams in the defense unit operated with more of a system's perspective than did teams in the commercial unit. They were able to recognize the positive and negative effects of differences within their leadership system, that is, in culture and leadership values among interorganization teams and their leaders. They understood the importance of trust and relationships to achieving success. At times, they were aware of the detrimental effects of strong formal coordination mechanisms that were often too strong for the environment in which they operated.

Because of their concentrated involvement in TI-sponsored training programs, teaming and goal alignment became "near-cultural" values of individuals on defense teams. They were also aware of the *proactive influence* their leaders had on internal elements and formal management systems. Given their systems' understanding of the management coordination system coupled with a more benign environment in which they operated, defense teams experienced greater success than did the commercial teams.

5 EXAMPLES OF PROJECT SUCCESS AND FAILURE IN TWO DEFENSE PROJECTS

Understanding core competencies of project management and becoming familiar with various tools are important cornerstones to successful project management. However, this understanding of basics and even an appreciation for the more subtle elements are not enough to ensure success. Given greater organizational interdependencies and the increasing use of interorganizational teams, it becomes imperative for project managers to grasp the cause-and-effect relationships when they use these tools and when they create a teaming environment.

This section begins by sharing two examples from which the reader can begin to see what caused success and what created failure. Both examples are taken from the defense industry and are based on actual program outcomes. The products involved are a highly complex aircraft, and development of operational mission software that is integrated on another aircraft. These examples, one of grandiose failure and one of substantial success, will use the MCS presented earlier as a framework to understand critical elements and important interrelationships for project management.

5.1 Example: Failed Project Management

> The A-12 never amounted to more than a mockup, but the consequences of this unfortunate program are going to affect the size and composition of naval aviation for years to come.[5]

The failure of this program is one that not only profoundly touched its own success, but reverberated to touch the entire industry. The A-12 information summarized below came from publicly available sources.[6–8]

Background
In 1983, the Deputy Secretary of Defense ordered the Navy to undertake a large development program for a stealthy, deep-strike aircraft, that is, one that could go far into threat-laden territory, accomplish its mission, and survive. In 1998, after almost seven years of litigation, Federal Claims Court ordered a multibillion dollar judgment in favor of two A-12 contractors. There are many reasons this huge program ended up in litigation but certain elements of project management were contributors to its abject failure. We will use this example, comparing it to the model discussed earlier, to illustrate lessons learned and instances where alternative actions could have increased the chances for success.

The A-12 aircraft program was a large, multifaceted program that combined multiple projects for its subsystems. Its complexity and size, even in the early design and prototype phase, involved multiple technical disciplines, numerous stakeholders, and participation by management and technical teams from both government and industry. The program required project management for the aircraft subsystems, as well as integration of these projects at a total product or program level. This example provides lessons learned for project managers—especially those who find themselves in a larger and more complex program environment that spans organizational boundaries.

So why did this program fail? Could it have succeeded in some way, even given the environmental constraints? What actions could you as a project manager have taken and what lessons can be learned from the managers who were involved?

Lessons Learned: Elements of Failure
Let us consider three major elements of the model presented earlier to better understand the lessons we can glean from this example and thus consider actions that may be under the

control of the project/program manager. First, the management systems must be complementary and must accommodate as well as facilitate adaptability to the external environment. Leadership in an interorganizational setting must be collaborative among organizations involved. Leadership characteristics and style must include high-integrity actions and the ability to empower others and to facilitate open and honest communication. The formal and informal systems are two synergistic components that can also contribute to success or failure. As noted earlier, the formal system refers to formal contracts, organization structures, roles, measurements and tools to report progress, and formal processes such as communication. On the informal side, characteristics of culture and informal communication are important.

Second, goals must be well understood, unambiguous and aligned among stakeholders and among those organizations involved. They must also support constraints in the external environment. Finally, the internal environment supports and influences the management systems. Highly influential in this internal environment are trust and communication. In this example, problems in all three of these elements contributed to failure of the program.

1. *Management systems* were not well used and were not conducive to accommodating the external environment or an interorganizational setting.
    a. *Leadership.* External environmental constraints, such as limited resources, coupled with pressures to succeed, appeared to influence leadership behavior in all organizations involved. These organizations did not openly acknowledge the magnitude of risk and performance status. Leadership actions that may have discounted risk also influenced the informal belief system for the same reason.
    b. *Informal systems.* Program culture and beliefs gave mixed messages and created an environment that discounted risk and did not facilitate open communication.
        i. *Beliefs* and motives for program. There was an emerging contest between the Air Force and Navy for the future of medium to long-range attack missions.[6] The A-12 was at the crux of this contest, i.e., the Navy saw the A-12 more as a path to retain this mission rather than as an answer to the deep-strike need. These circumstances set the stage for conflicting goals and could have driven an almost zealous passion to ensure its success—perhaps regardless of risk.
        ii. A highly *classified environment* isolated project managers and technical contributors and created a culture of secrecy and nonsharing of information.
    c. *Formal systems.* In an attempt to accommodate severe funding restrictions and pressures to succeed imposed by the external environment, formal systems in the organizations did not well support the goals and were often discounted.
        i. *Formal agreement and goals.* A formal contracting vehicle stipulated a fixed price or maximum amount that could be bid to force the program to be done within funding limits. However, this contracting approach, though it attempted to meet the resource limitations, did not appropriately address program goals in terms of performance or schedule.
        ii. *Formal strategy and competitors.* To encourage reduced cost, the desired strategy was to set up a competition among three industry teams. However, this strategy mostly failed; one team dropped out early, refusing to use the fixed-price contract, recognizing the inadequate funding and assessing technology as high risk. Another nearly dropped out, but went forward with huge misgivings.[6]

iii. *Formal processes and tools.* Success at meeting goals was not achieved from the earliest stages. However, although tools and processes such as schedule, cost, and performance tracking were in place, program management did not rely upon them. For example, a couple years after the program began, an inquiry indicated that the cost performance data "clearly indicated significant cost and schedule problems. . . . However, neither the contractors nor the Navy program manager relied on these data; instead, they used overly optimistic recovery plans and schedule assumptions. The inquiry concluded that government and contractor program managers lacked the objectivity to assess the situation and they disregarded financial analysts who surfaced the problems."[8]

2. *Goals.* Goals of the various stakeholders were conflicting and did not accommodate constraints of the external environment. Funding constraints dictated a need for stable and known requirements early on, whereas the high-technology nature of the system being built was not conducive to this need.

   a. *Navy goals.* The Navy thought it could meet the need for deep-strike operations by upgrading another aircraft, the A-6.[6] However, the Department of Defense directed that the Navy initiate the new program. Thus, the Navy's commitment and passion for the program to meet the deep-strike need were not as high as they could have been.

   b. *Customer requirements.* The performance requirements were not agreed on and evolved during the program. Fenster notes that even the basic requirement to counter a threat was changing: "Added to these issues were growing uncertainties about the threats such an aircraft might expect to encounter and the threats it was supposed to address."[6]

   c. *Technology and funding constraints.* The technology and funding constraints in the external environment did not match goals of the program. Stealth technology was immature and funding was inadequate. "At the time, the Navy did not want to embark on a stealth aircraft program—the technologies were by no means ready, and the funds needed to reach such a result were nowhere to be found."[6] Areas such as composite materials and advanced avionics components proved most challenging.[8]

3. *Internal environment.* The internal environment was not supportive of the characteristics required in an interorganizational setting or in a highly dynamic and complex business environment. Two characteristics were particularly deficient: communication and trust.

   a. *Communication* in an open, risk-free way was stifled.

      i. *Pressures to succeed.* Program execution and desire to have a product, coupled with the overwhelming pressures to succeed superseded open, risk-free communication.

      ii. *Classification.* Program development was done in a highly classified and compartmentalized manner to protect technology and to reduce external oversight. This type of classification makes normal communication, let alone communication of risk, exceedingly challenging and in some cases not possible. "Many observers believe that the A-12's demise was attributable in good part to the hiding of its existence, costs, and failures in secret accounts and programs"[7]

b. *Trust* was absent. In the case of the A-12, because of its sophistication and challenges, the relationship between industry and government team members "must be one of trust."[6]

   i. *Erosion.* Trust between industry and government team members, especially at program management level, had eroded. Additionally, some facts regarding the failed competitive strategy and the high risks were not generally communicated early enough to allow correction. Subsequent to the Defense Secretary's Major Aircraft Review, "the A-12 contractors revealed that the project faced serious engineering problems and a $2 billion cost overrun, which would delay the first flight by over a year, to the fall of 1991, and raised the unit cost substantially."[8]

   ii. *Micromanaging and blame.* When risks began to manifest themselves in schedule and cost overruns, there was much micromanaging and finger pointing among government and industry team members. The behavior, rather than collaborative, became confrontational without each organization taking responsibility for whatever problems they could influence.

5.2 Example: Successful Project Management

Background

The AV-8B Harrier is a sophisticated weapons-carrying aircraft capable of both horizontal and vertical takeoff. Its primary missions include close air support (CAS) and air-to-ground weapons delivery supporting ground forces from air-capable ships and austere sites for both Marines and Navy. Aircraft such as these have software programs called operational flight programs. These OFPs integrate sensors, weapons, and displays so that a pilot can fly the aircraft while accomplishing an operational mission.

In the early 1990s, the Marines began a program to integrate new subsystems while translating the software from a cumbersome assembly language into a high-order language (HOL). Participants in this program included international and U.S. defense contractors. A combination of government and industry personnel comprised the project team and worked together to design and test the new software. Both government and contractor facilities were used; Marine pilots performed developmental testing on a government open-air range.

The program had an abundance of challenges. Overall program management for the entire aircraft was located on the east coast, and project managers for the software development were located on the west coast, making communication more difficult. Use of a high-order language when combined with the sophistication of the new subsystems increased technical complexity of this effort. International involvement coupled with the tightness of funding created additional external constraints. At times, lack of promised funding caused hiccups, frustrations, and delays.

Additionally, software development processes were not robust, and program status tracking was not accurate enough to ensure timely information. Further, communication regarding issues and risk was done at lower levels; direct communication from project managers to program managers was infrequent and led to misinterpretation on occasion. Other problems outside the control of the program also surfaced when problems were found with aircraft engines. Test aircraft were put in a nonoperational status so that operational aircraft could use the remaining good engines. Thus, flight-testing of the new software could not be done in a timely manner. The schedule became tighter and required an extremely challenging 5–8 flight tests per week to finish. International partners were frustrated with lack of delivery and program management had lost faith in the technical teams. Team members from all organizations were demoralized and tired from working long hours with minimal success. Pressures were building.

Finally, a program review was held: the program schedule and requirements were rebaselined to better meet funding constraints. Partner organizations were brought into this rebaselining effort. Key managers were replaced and more discipline was put into the program. Today, this program is on schedule, has delivered tremendous capability, and is a model of management success. The software development project, in particular, was turned completely around. Robust processes were put into place, team members went from being demoralized and less productive to highly energetic and proud; project milestones were continually met.

So how did this success happen? The early challenges were daunting and the program had been in jeopardy of being canceled. Why hadn't this program failed like the A-12?

Lessons Learned: Elements of Success
Interestingly enough, the reasons for the success of this rebaselined program were nearly the exact opposite of the reasons for failure of the earlier A-12 example. Again we will build an understanding of these success modes using the MCS model described earlier. Management systems, goals, and the internal environment were key elements in the success of this project.

1. *Management systems* became strong, mutually supportive, and consistent with the external environment. They well accommodated an interorganizational structure.
    a. *Leadership.* Leadership was by far the most influential management system. The project manager for the OFP had an extraordinary systems orientation and intuitive understanding of the cause and effect of actions. Not only did this particular high-integrity leader implement formal systems, he also appreciated how linked the informal world of culture and relationships were, and he acted consistently on these beliefs. He supported and truly believed in the goals of the project and in the capabilities of all team members to accomplish the goals. His communication skills were extraordinary, up, down, and sideways. According to his supervisor, "The environment in which AV-8 software was developed was one of trust and communication across a team. The team was not afraid to take risk and trusted their leadership."
    b. *Informal systems*
        i. *Beliefs and culture.* Through his own beliefs, the project manager instilled a deep appreciation for the goals of the program into the team. His vision provided a path to success. With awareness and intent to change the culture, he created an environment that was risk tolerant, highly communicative, and embedded with trust. He translated his vision into values for the team culture: a belief in the end product and self-assurance. The culture also incorporated a readiness to change. A defining moment for this team may have been when leadership laid it on the line to the entire project team: if we don't succeed, this project won't succeed and we won't be here 5 years from now.
        ii. *Relationships.* Small teams were set up to accomplish various tasks. Leadership used training opportunities, such as classes on high-performance organizations, to build tight-knit teams. Integrated together and enabled by the open culture, close trusting relationships developed and allowed free and open communication of status and issues.

c. *Formal systems.* The formal systems were revamped to be more adaptive, while providing more rigorous processes, accountability, status reporting, and progress measurement.

   i. *Tracking systems.* Even though it was initially costly to incorporate, the use of more formal tracking systems resulted in a greater awareness of progress and issues. One such system, an earned value management system, was implemented to rigorously track tasks, schedules, and funding expended. This immediate awareness ensured timely resolution and communication to program management.

   ii. *Software development process.* The team began to use a more rigorous software development process (the Software Engineering Institute's Capabilities Maturity Model [CMM]) in earnest. This investment of time and money has since paid off manyfold. There are five levels of this model, with level 5 being the most effective. Graduating from one level to another means step increases in productivity. The AV-8B software team began at level 1 and moved quickly to level 4.

   In addition to using this CMM process to improve efficiency, the project manager required all individuals to take training in a PSP/TSP (Personal Software Process/Team Software Process) to account for their own productivity. This particular tool had high personal risk for team members associated with it since it required formally accounting for every minute of time spent during the day. Initially there was a fear that it would be used in a negative way to find inefficiencies that would affect a person's performance evaluation. However, leadership assured people it was only a tool for increasing productivity of the team, and because leadership had created an environment of trust, team members relaxed and used it to greatly improve. The combination of all these tools accelerated movement from level 2 to level 4 in 16 months—a three-fold decrease from the average 50-month timeline. Once these tools and processes were in place, the team witnessed significant drops in schedule variance (from 30 to 2.5%), decreased software defects from 2 per thousand lines of code to 1/2 per thousand lines of code, and a 34% increase in productivity to write and test software.

d. *Compatibility among management systems.* According to the supervisor, the project managers well understood the importance of the interaction between formal and informal systems. They thought that you must "grasp cultural implications, the 'soft side' while at the same time implementing the processes. If you don't do it all simultaneously, it won't work." This leadership attitude, along with the attention to ensuring that formal and informal systems complemented one another and were built together, was a primary reason for success.

2. *Goals* were well defined, measured, and achievable. Rebaselining of the program ensured that the goals and requirements matched allotted schedule and funding. The overarching goal for program success, punctuated by near failure of the program, helped align potentially disparate goals among team members and stakeholders. The goal-setting process took into consideration the requirements of all stakeholders, including international partners. Once set, leadership communicated these goals well and often, and used formal systems to track progress toward them. Team members could understand how their parts fit.

3. *Internal environment* was characterized by high levels of honest communication and feedback and exuded trust. As noted in the discussion of management systems, the

stage for open communication was set by a project manager who "walked his talk" and was constructive when risk-laden discussions occurred. Use of a personal productivity system in a constructive, rather than punitive way, was a foundation to this building up of trust.

6 PRACTICAL APPLICATIONS FOR LEADING INTERORGANIZATIONAL TEAMS

Leadership at all levels has a dominant influence on the success of interorganizational project teams. By considering the previous examples of success and failure, and using the management coordination system model as a framework, we can see that common themes emerge. In the A-12 example, interaction among management systems created a low trust, low collaborative environment that was incompatible with the external environment. Formal systems were perfunctory and real issues never surfaced. Goals were not aligned among stakeholders or even among team members. The program failed. Alternatively, in the AV-8B example, the interaction among management systems was supremely supportive, leadership had an intuitive understanding for the formal and informal components within the project, goals were aligned, and the internal environment reflected high trust complemented by effective interpersonal relationships.

We can distill these examples and the model into four imperatives that project managers, or any leader, can use when leading teams of people—particularly interorganizational teams. These four imperatives, provided in Table 2, put into actionable recommendations the lessons learned in this chapter.

The first imperative is that leaders should think in terms of a system of leadership. It's no longer just one person up at the top of the pyramid making all of the decisions. The environment is too complex. Decisions are too complex. One cannot lead by using a command and control structure. A system of interactive leadership is needed that permits rapid information processing and adaptation to changes in general and unique environmental forces. This system of leadership should be supported by good information flows throughout the team. It should maintain an environment of trust, interpersonal relationships, and collaboration using influence rather than control.

Leaders develop goals and the competencies that are necessary to achieve those goals. Then leaders shape their formal and informal management systems. They may directly shape their formal systems. They may seek to influence their informal systems. This proactive

Table 2 Four Imperatives for Leading Interorganizational Teams

A "System" of leadership
    Translates input from the external environment into strategy, goals, and competencies
    Shapes management systems to influence the internal environment, goal alignment, and performance
    Leverages intrinsic motivation of individuals

Goal alignment
    Shapes the internal environment
    Directly influences success

System interaction and supportiveness
    Complementary and dynamic interaction ensures adaptation and flexibility to external environment
    Leverages awareness of interdependence among external environment, management systems, internal environment, external organizations, and performance

Trust-based interpersonal relationships
    Acts as a lubricant for complementary and dynamic interaction

shaping can be employed to develop trust, commitment, constructive conflict, and adaptability. But this ability of the leader to proactively shape must be understood and used appropriately. A leader may also insure that information processing is continuous and focused on monitoring critical external and internal variables.

The second imperative is goal alignment. Goal alignment is one of the most critical elements in achieving goals. It directly influences performance. If there is any ambiguity about goals or about how to achieve them, the organization is bound to fail. If there is a lack of commitment to goals, either because they are not accepted or because individuals do not trust the organization, then goal alignment and performance will be impeded.

The third imperative is system interaction and supportiveness. The primary management systems should be supportive of one another and highly tuned to the environment. The primary management systems must facilitate an internal environment of trust, commitment, constructive conflict, learning, and adaptability. The core values of the organization will probably remain fairly stable, but they may change at the margin. Values will change with leadership. Values will change as reward systems change. What leaders must do is insure that these changes, whether in the informal system or in the formal system, are supportive of each other. This supportiveness will guard against such counterproductive behavior as "I'm going to reward you because you got your project done in two weeks and I don't care who you killed doing it" while the stated value is to be respectful of people. These kinds of conflicts create a tense environment and they reduce trust.

The fourth and final imperative is trust-based interpersonal relationships that must be a hallmark of interorganizational project teams. These organizations have to rely extensively on interactive, organic structures that depend on trust while facilitating the development of trust and allowing tasks to be accomplished quickly. If elaborate rules and regulations bind an organization, that organization will not be able to react quickly. Trust-based interpersonal relationships are critical among individuals in the organization, within the leadership system, and between organizations.

Powerful organizations result when leaders understand these four imperatives, understand how they interact together, and use them to lead complex projects. The leader has influence on each of these four. Barring any unforeseen problems in the external environment, organizations that are so led should succeed. The people in these organizations should be committed, have high levels of trust, and succeed. Furthermore, people should also be happy working in these organizations.

7 SUMMARY: LEADING INTERORGANIZATIONAL ENGINEERING TEAMS

We have now discussed the full model for leading project organizations. Our discussion covered some of the most important dynamics, basically the process of connecting the dots. It revealed that all of these systems are interconnected. There are lags, such as those between setting up a formal reward system and seeing it played out in the informal culture. There are also lags in management or a leader's actions in instilling a value system or in developing others who are empowering.

Numerous dynamics exist in the coordination of collaborative work. Collaborative work is always changing. It is never constant, nor can it be, because the external environment is always changing. In today's environment, one has to design adaptability into organizations, which means that leaders must encourage and rely on the informal management systems more than ever.

Developing trust within and between organizations is essential to success. How does a leader establish this trust when each organization is an autonomous entity with its own self-interest? How does one ensure that one does not have one organization on an interorganizational team trying to control or withhold information from partners? How does one establish trust in the private sector where top leaders each report to their own board, their own shareholders, and their own stakeholders? For these difficult issues, simplistic answers will not suffice.

So, in deciding to collaborate with another organization, a leader must first understand the values of potential partners. Before partnering, values should be aligned among organizations on a team. Integrity is the primary value that should be synchronized. Integrity means that leaders of partner organizations will live up to their commitments. Integrity means that leaders say what they mean and mean what they say. Trust is a by-product of integrity.

Once leaders and partners trust each other and form a strong leadership system based on integrity, the remaining ingredients for success in leading network organizations can be put into place to further enhance success. These ingredients include a formal goal-alignment process that leads to defining goals that are agreeable to all. Formal processes for conflict resolution should be established. Because not all conflicts among partners can be resolved, interorganizational teams should be dissolved when partners are unable to work out conflicts. Rules for disbanding teams should be established in advance.

Following the four imperatives, a successful organization or project in a collaborative venture will have a strong leadership system that proactively seeks to leverage individual motivation and shape the internal environment using informal and formal systems. It will have internal and interorganizational goal alignment; and its primary management systems will be supportive and facilitate trust-based interpersonal relationships.

APPENDIX: CORPEDIA INTERNET MODULES: AN APPLIED FRAMEWORK FOR PROJECT MANAGEMENT

Corpedia Education in Phoenix, Arizona developed ten Internet modules in partnership with the Project Management Institute (PMI). These modules describe, illustrate, and encourage application of the tools of Project Management contained in **PMBOK** (see Annotated Bibliography). The eleventh module of the series, "Knowledge Assessment," contains 200 questions and serves as a "final exam" for the series. You may subscribe to these online courses by contacting Corpedia Education, Phoenix, Arizona (*http://www.corpedia.com*).

Corpedia Online Module 2101: An Applied Framework for Project Management: Introduction. The first course in the series introduces basic ideas and concepts of project management and provides a roadmap to the next nine, more detailed courses. Ideas and concepts are taught through a series of explanations, interactions, and "action planning" question-and-answer sessions. This general format—explanations, interactions with the user, and action planning by the user—is repeated in each of nine more detailed courses in this series.

Corpedia Online Module 2102: Initiating Processes. The second course in the series describes the key components of the initiating processes within project management. The course defines the initiating components, processes, and pitfalls, providing a full understanding of what to expect at the beginning of a project. Special attention is given to the roles and responsibilities of the project manager, project stakeholders, key personnel, and vendors.

Corpedia Online Module 2111: Planning Processes (Part A). The third course in the series introduces some of the planning processes within project management. It illustrates which components comprise the planning processes and how the planning processes fit into the overall project management processes. In addition, the course elaborates on scope planning, scope definition, and risk management planning.

Corpedia Online Module 2112: Planning Processes (Part B). The fourth course continues discussion of planning processes. Topics covered in the course include quality planning, activity definition, activity sequencing, resource planning, and estimating activity durations. The course describes what to expect when defining quality standards, what types of activities to consider in project planning, how to sequence a project's activities, and methods for estimating multiple, concurrent activities.

Corpedia Online Module 2113: Planning Processes (Part C). This course continues a description of the planning processes. Topics covered include cost estimating, schedule development, risk identification, and qualitative analysis of risk. The course provides an opportunity to apply the concepts to the user's organization through interactive questions and drafting of action plans.

Corpedia Online Module 2114: Planning Processes–(Part D). This course continues a description on the planning processes. Some of the topics covered include quantitative analysis of risk, risk response planning, procurement planning, solicitation planning, and organizational planning. The course provides an opportunity to apply these concepts to the user's organization.

Corpedia Online Module 2115: Planning Processes (Part E). This course continues a description of planning processes. Topics include staff acquisition, cost budgeting, communications planning, and techniques for developing project plans.

Corpedia Online Module 2116: Executing Processes. This course describes the executing processes once project work begins and places these processes within the overall context of project management processes. Topics covered include execution of project plans, solicitation of bids, source selection, quality assurance, contract administration, information distribution, and team development.

Corpedia Online Module 2117: Controlling Processes. This course identifies the controlling processes within project management and their place within the context of project management. Topics covered include performance reporting, quality control, scope verification, integrated change control, scope change control, schedule control, cost control, and risk monitoring and control.

Corpedia Online Module 2118: Closing Processes. This course covers the closing processes for a project and includes activities and action plans that allow users to apply the concepts to their own organization. Topics covered include contract closeout and administrative closing processes.

REFERENCES

1. E. Verzuh, *The Fast Forward MBA in Project Management,* Wiley, New York, 1999, pp. 10–11.
2. K. L. Higgins, *Management Coordination System for an Interorganizational Network,* Ph.D. dissertation, Claremont Graduate University, Claremont, CA, 1997.
3. J. P. Kotter, *The Leadership Factor,* The Free Press, New York, 1988.
4. P. Blau and W. R. Scott, *Formal Organizations: A Comparative Approach,* Chandler, San Francisco, 1962, pp. 6–7.
5. J. A. Maciariello and C. J. Kirby, *Management Control Systems: Using Adaptive Systems to Attain Control,* Prentice-Hall, Englewood Cliffs, NJ, 1994.

6. H. L. Fenster, "The A-12 Legacy: It Wasn't an Airplant—It Was a Train Wreck," Naval Institute Press, U.S. Naval Institute, Annapolis, MD, February 1999, Vol. 125/2/1152, pp. 33–39.
7. M. Corbin, "Project on Government Oversight, Defense, Concerns and Questions: The A-12 Aircraft Financial Fiasco," October 2, 1996 (©The Project on Government Oversight 2003). *http://www.pogo.org/p/defense/da-961008-reform.html*
8. J. Pike, "Military Analysis Network, A-12 Avenger II," *http://www.fas.org/man/dod-101/sys/ac/a-12.htm;* last updated October 3, 2003.

ANNOTATED BIBLIOGRAPHY: PROJECT MANAGEMENT REFERENCE MATERIAL

The concept of organizing work to achieve a desired result is not new; many books have been written on management over the past 50 years. Project management, however, has its own unique challenges that are being better understood and addressed in the literature. Project management has become a management discipline of its own, supported by a plethora of reference material and even a "Project Management Institute." A sampling of this reference material is provided in this section so that both the novice and experienced project manager can better understand the basic skills and tools required for success. The references chosen are by no means comprehensive. Indeed, the reader has access to the greater body of work on project management that is referenced further in these materials.

The sample material is intended to cover the spectrum of project management from cut-and-dried use of traditional tools and time-proven quantitative approaches, to illustrative examples and case studies to basic qualitative and so-called "soft-science" concepts dealing with people, and finally to expansion of concepts and tools to today's complex and dynamic environment. Six sample references, a description of their content, and a guide for mastering the tools of project management are provided below.

1. Project Management Institute, *A Guide to the Project Management Body of Knowledge,* (**PMBOK® Guide**), Third Edition, Newtown Square, PA, 2004.

    PMBOK is an American National Standard Reference book. As a basic reference book and standard for the project management profession, it describes "generally recognized good practices" for successful project management. The Project Management Institute (founded in 1969) uses it to certify project management professionals and to accredit educational programs.

    This reference book covers project management definitions, project management processes, and nine project management knowledge areas. The nine knowledge areas are the management of Integration, Scope, Time, Cost, Quality, Human Resources, Communications, Risk, and Procurement. The reference also provides a further list of books and professional/technical organizations around the world for the advancement of project management. Information on additional resources can be accessed through the Project Management Institute website, *http://www.pmi.org*.

    The ten Internet modules described in the Appendix have been written by Twin Star Consulting Company, and developed for online training by Corpedia Education (*http://www.corpedia.com*) in partnership with the Project Management Institute.

2. Finding help on standard tools of project management in **PMBOK**.

    a. *How do I go about defining the scope of my project?* (**PMBOK,** Chapter 5)

    To plan the scope of your project you must understand the needs of your customer and how the project will satisfy those needs. Then you must engage in the process of elaborating those needs into a detailed statement of work and into a Work Breakdown Structure (WBS). The statement of work must be feasible and meet the requirements of your customer.

b. *How do I develop plans and keep track of the schedule for my project?* (**PMBOK,** Chapter 6)

The WBS provides the basis for carrying out detailed planning and sequencing of all the activities of the project. A network is sometimes used to depict these activities. A network includes activities, time durations, dependencies, and constraints. The list of activities, when performed, should result in fulfillment of the statement of work. Once the plan is developed in detail it is compared to resources required and those likely to become available. A schedule is put together to perform each portion of the plan according to required times.

c. *How do I develop a plan and budget to manage financial resources of my project?* (**PMBOK,** Chapter 7)

Using the detailed activity list and relationships among activities, cost estimates are prepared at the most detailed level of the WBS and summed to the top. This provides an estimate of expenditures necessary to complete the project. The expenditure plan should be developed before the project commences and used as a guide in contract negotiations. Once work begins, the plan of expenditures must be converted to budgets for each organization contributing to the project. Actual expenditures should be compared to the plan on a periodic basis. Adjustments should be made to work plans, financial plans, schedules and budgets to maintain financial control.

d. *How do I determine quality standards that are relevant to my project? How do I satisfy these standards?* (**PMBOK,** Chapter 8)

First, you must determine which quality standards are relevant to the project. And this must be done in agreement with the customer. The scope statement provides the reference point for determining the required "goodness" of project deliverables. Then, Cost of Quality may be measured and monitored. The Cost of Quality is the sum of all financial resources expended to achieve quality standards, including rework to correct the external cost of nonconformance to quality specifications experienced by the customer.

e. *What are the risks associated with the project and how should these risks be managed?* (**PMBOK,** Chapter 11)

There are many potential risks in the management of complex projects. These include the risks of nonperformance or performance risks; risks of not achieving quality specifications; risks of missing schedule requirements; and financial risks. Each risk should be identified and monitored systematically throughout the duration of the project. A checklist of potential risk areas may be used to assess the generic risks that apply with any risk factors unique to the project added to the checklist.

f. *How should I go about staffing my project over its entire life?* (**PMBOK,** Chapter 9)

The needs of a project during each of its phases determine the kind and quality of human resources required and therefore the staffing plan. The project manager must think through competency and experience requirements, the availability of appropriate personnel, and the mix of individuals who are likely to form a well-functioning team.

g. *How do I make sure all members of my project team have the information they require, when they require it, and in the form that they consider most useful.* (**PMBOK,** Chapter 10)

The project manager should determine what information technology, both hardware and software, is required to support the scope and other requirements of the project. In addition, the project manager must ensure that the content of information provided is appropriate to the needs of each team member, as well as customers and suppliers. Lastly, the information technology must facilitate effective communications among those team members who must cooperate with each other to effectively perform project work.

h. *What process should I put into place to acquire goods and services from vendors?* (**PMBOK,** Chapter 12)
The project manager requires a list of approved vendors along with the experiences of the project organization with each vendor. In addition, the project manager should understand the process of procurement in the project organization. The project manger should develop plans to ensure that vendors perform according to specifications, at the right level of quality, and on time.

i. *Finally, what are my plans for executing all of my plans and controlling progress so as to achieve the scope, quality, cost, and schedule requirements of the project?* (**PMBOK,** Chapter 4)
Project integration requires that project managers coordinate and integrate all the plans of the project. Integration involves making decisions on necessary trade-offs among key success variables of the project. Earned Value is an integrated measure that incorporates cost and schedule control information into one measure of performance. This measure can serve as a key integrated measure of performance on the project.

In addition to an Earned Value Management System, Change Control procedures should be initiated to ensure that alterations to project scope are fully tracked and documented along with approved changes in quality and schedule.

3. Verzuh, E., *The Fast Forward MBA in Project Management,* Wiley, New York, 1999. This book gives tips and descriptions of project management in simple-to-understand language. It uses real-world examples to illustrate concepts. The detailed planning model gives a simplified example of how to organize and manage a project. The project-management challenges it addresses include personnel, estimating cost and schedule, budgeting, authority and organization charts (including crossing organizational boundaries), accounting controls, and communication. A project manager's role is broken down into three functions: project definition, project planning, and project control. This book also provides examples of tools that can be used to manage projects, such as work breakdown structures, network diagrams, resource histograms, spreadsheets, Gantt charts, and work packages.

4. Frame, J. D., *Managing Projects in Organizations: How to Make the Best Use of Time, Techniques, and People,* revised edition, Josey-Bass, San Francisco, 1995.
This book, revised from its first edition in 1987, is an introduction to project management. It extrapolates traditional tools and approaches for use by the information-age knowledge worker. The book describes terms and different stages of project life cycle. It also focuses on two important lessons: "avoiding pitfalls and making things happen." Four sources of project failure addressed are organizational factors; poorly identified customer needs; inadequately specified project requirements; and poor planning and control.

The book's three parts provide basic how-to's in project management. Part one describes project context and includes people, teams, and organization concerns. It provides insights into organizational realities such as authority; organizational structures such as matrix forms; finding capable people, including desirable psychological profiles; management styles; team structure; and dynamics and rewards. Part two considers needs and requirements analysis, including cost and schedule. It emphasizes having clear need and articulating requirements well; sorting out needs and priorities of multiple customers; and shifting, overspecified or overly flexible requirements. Part three describes planning and control. It illustrates basic tools such as project plan, schedules and work-breakdown structures, Gantt charts, PERT/CPM (Critical Path Method) Networks, and various budgeting and resources tools. Sections on core competencies and additional reference material allow the reader to expand knowledge in project management.

5. Frame, J. D., *The New Project Management: Tools for an Age of Rapid Change, Complexity and Other Business Realities,* 2nd ed., Jossey Bass, San Francisco, 2002. Revised from its 1994 edition, this book provides more advanced insight into project management than Frame's other works. It goes beyond traditional approaches and provides a good basis for understanding implications of uncertainty, complexity, and dynamics in producing change in today's business environment. The first of two parts of this book addresses complexity and change and how to cope with them, including change management. This part also covers risk management, provides advice on how to identify all customers, and presents a more subtle concept of bridging the gap of understanding between customers and the technical team. The second part reviews the skill sets required by project manager in today's environment. Included are such concepts as political skills, building authority, creating a team spirit in a matrix environment, decision-making skills, assessing the quality of estimates, outsourcing, integrated cost and schedule controls, accountability, and metrics for performance. The second part also examines new scheduling techniques like "time-boxed" and "critical chain" scheduling. Finally, it describes the direction of project management and the evolving role of the project manager.

6. Kerzner, Harold, *Project Management Case Studies,* Wiley, Hoboken, NJ, 2003. One of few sources for project management case studies, this book provides a collection of cases describing dilemmas and actual circumstances that project managers may experience. It can help the reader make academic exercises come to life in easy-to-understand descriptions and dialog. The cases presented include commercial as well as government projects from a wide range of industries, including aerospace, banking, manufacturing, and electronics. The experiences described in the book illustrate issues with such formal systems as organizational structure, roles, and project management tools as well as with such informal systems as culture, relationships, and conflict. Risks and risk management are simply and clearly depicted, and the reader is often challenged to determine the effects of project manager actions.

CHAPTER 14
MANAGING PEOPLE

Hans J. Thamhain
Department of Management
Bentley College
Waltham, Massachusetts

1	**CHALLENGES OF MANAGING IN ENGINEERING AND TECHNOLOGY**	484	5.2 Power-Sharing and Dual Accountability	494
			5.3 The Important Role of Salary	494
2	**THE UNIQUE NATURE OF MANAGING ENGINEERING PERSONNEL**	485	6 **TEAMWORK IN ENGINEERING ORGANIZATIONS**	495
			6.1 Toward Self-Direction and Virtual Teams	496
3	**CHANGING ROLES OF MANAGERIAL LEADERSHIP**	486	6.2 Measuring Project Team Performance	497
4	**MOTIVATION AND ENGINEERING PERFORMANCE**	488	7 **BUILDING HIGH-PERFORMING TEAMS**	497
	4.1 Implications for Engineering Performance	489	7.1 How the Four-Stage Model of Team Development Works	499
	4.2 Motivation as a Function of Risks and Challenges	491	8 **RECOMMENDATIONS FOR EFFECTIVE ENGINEERING MANAGEMENT**	500
	4.3 Success as a Function of Motivation	492		
	4.4 Manage in the Range of High Motivation	492	9 **CONCLUDING REMARK**	503
5	**THE POWER SPECTRUM IN ENGINEERING AND TECHNOLOGY MANAGEMENT**	493	**REFERENCES**	503
			BIBLIOGRAPHY	504
	5.1 Motivation, Managerial Power, and Performance	493		

1 CHALLENGES OF MANAGING IN ENGINEERING AND TECHNOLOGY

People are one of the most critical assets affecting business performance in today's technology-based business environment.* Activities often cluster around projects with team efforts that span organizational lines involving a broad spectrum of personnel, support groups, subcontractors, vendors, partners, government agencies, and customer organizations. Effective linkages, cooperation, and alliances among various organizational functions are critical for proper communications, decision-making, and control. This requires sophisticated teamwork and, typical for many high-tech organizations, the ability to manage across func-

---

*For additional discussion on these challenges and issues of high-tech management see Armstrong[1], Barkema et al.,[2] Barner,[3] Dillon,[4] Gray & Larson,[5] and Thamhain.[6]

tional lines with little or no formal authority, dealing effectively with resource sharing, multiple reporting relationships, and accountabilities.

2 THE UNIQUE NATURE OF MANAGING ENGINEERING PERSONNEL

Managers in engineering and technology-based enterprises see themselves as different from those in less technical environments. Their work requires unique organizational structures, policies, and interaction of people. Their management style and leadership must not only be consistent with the nature of the work and the business process, but also be conducive to the special needs of the people and consistent with the unique culture of the technology-based organization and its values.

Specifically, six organizational subsystems make engineering management unique and different from other types of management and influence the type of leadership style appropriate for these situations: (1) work, (2) people, (3) work process, (4) managerial tools and techniques, (5) organizational culture, and (6) business environment, as graphically shown in Fig. 1.

- *Work.* Engineering-oriented work is by and large more complex, requiring special skills, equipment, tools, processes, and support systems. The unit of work is often a project, organized and executed by a multidisciplinary team. Cross-functional integration, progress measurements, and controlling the work toward desired results are usually more challenging with increasing technology, involving higher levels of creativity, risks, and uncertainties.

    MANAGERIAL IMPACT: Engineering work impacts organizational structure, work planning, work processes, personnel recruiting and advancement, skill development, management style, organizational culture, and business strategy.

- *People.* Because of the type of work and its challenges, engineering-oriented environments attract different people. On average, these people have highly specialized skills. They are better educated, self-motivated/directed, and require a minimum of supervision. They enjoy problem solving and find technical challenges motivating and intellectually stimulating. They often enjoy a sense of community and team spirit, while having little tolerance for personal conflict and organizational politics.

Figure 1 Business subsystems unique for technology-intensive organizations.

MANAGERIAL IMPACT: Because of the relationship between people and work issues, these two impact areas are similar. Organizational structure, work plans and processes, personnel recruiting and advancement, skill development, management style, and organizational culture are the primary areas impacted by people in technology.

- *Work process.* Because of the nature of technical work with its complexities, uncertainties, risks, and need for innovation, work processes are less sequential and centrally administered, but more team-based, self-directed, and agile, often structured for parallel, concurrent execution of the work.

    MANAGERIAL IMPACT: The work process design impacts primarily people issues, management style, and organizational culture. New organizational models and management methods, such as *stage-gate, concurrent engineering,* and *design–build processes* evolved together with the refinement of long-time concepts such as the *matrix, project,* and *product management.*

- *Managerial tools and techniques.* While many of the managerial tools and techniques are also used in other environments, the unique nature of engineering requires special tools and application. Spiral planning, stakeholder mapping, concurrent engineering, and integrated product developments are just a few examples of the specialized nature of tools needed to produce team-based solutions to complex problems.

    MANAGERIAL IMPACT: Managerial tools affect the people and the work process. Matching organizational culture with any of these tools is a great challenge. Stakeholder involvement during the tool selection, development, and implementation, and trade-offs among efficiency, speed, control, flexibility, creativity and risk are critical to the effective use of these tools and techniques.

- *Organizational culture.* The challenges of technology-driven environments create a unique organizational culture with their own norms, values, and work ethics. These cultures are more team oriented regarding decision making, work flow, performance evaluation, and work group management. Authority must often be earned and emerges within the work group as a result of credibility, trust, and respect, rather than organizational status and position. Rewards come to a considerable degree from satisfaction with the work and its surroundings, with recognition of accomplishments as important motivational factors for stimulating enthusiasm, cooperation, and innovation.

    MANAGERIAL IMPACT: Organizational culture has a strong influence on the people and work process. It impacts organizational systems from hiring practices to performance evaluation and reward systems, to organizational structure and management style.

- *Business environment.* Engineering-oriented businesses operate in an environment that is fast changing regarding market structure, suppliers, and regulations. Short product life cycles, intense global competition, low brand loyalty, low barriers to entry, and strong dependency on other technologies and support systems are typical for these business environments.

    MANAGERIAL IMPACT: The need for speed, agility, and efficiency impacts the work process design, supply chain, organizational structure, management methods, tools, and techniques, but also business strategy and competitive behavior that also includes cooperation and resource sharing via alliances, mergers, acquisitions, consortia, and joint ventures.

3 CHANGING ROLES OF MANAGERIAL LEADERSHIP

Managers must work effectively with people who are the heart and soul of an organization and are critical to the successful implementation of any plan, operational initiative, or specific project undertaking. The real value of this human asset is measured in terms of integrated skill sets, attitudes, ambitions, and compassions for the business. To develop an engineering team that has all of the right ingredients, matched with the needs of the enterprise, is not a simple task, but requires sophisticated leadership and skillful orchestration from the top. The mandate for managers is clear: they must weave together the best practices and programs for continuously developing their people toward highest possible performance. However, even the best practices and most sophisticated methods do not guarantee success. They must be carefully integrated with the business process, its culture and value system. The challenges are especially felt in today's technology organizations, which have become highly complex internally and externally, with a bewildering array of multifaceted activities, requiring sophisticated cross-functional cooperation, integration, and joint decision-making. Because of these dynamics, technology organizations are seldom structured along traditional functional lines, rather they operate as matrices or hybrid organizations with a great deal of power and resource sharing. In addition, lines of authority and responsibility blur among formal management functions, project personnel and other subject experts. This more empowered and self-directed work force needs a much higher bandwidth of skills to solve operational problems consistent with current and future market needs.

Because of these challenges, technology-based organizations must deal effectively with the following management issues:

1. *Manage technical work content.* Any job has technical content, such as electrical engineering, biochemistry, market research, or financial analysis. Especially in technology organizations, a lot of jobs have broad and complex technological context. The ability to manage the technical work, on an individual job level and collectively throughout the organization, is a *threshold competency,* critically important to the success of any business. Its managerial components relate to staffing, skill sets, professional development, support technologies, experiential learning, and in-depth management.

2. *Manage talent.* Businesses do not produce great results because of their equipment, buildings, and infrastructure. Rather, it is the people and their ideas and actions that bring the system alive. For many technology companies, talent is everything! The type of talent, along with its fit with the business needs and organizational culture, determines everything from idea generation to problem resolution and business results. Talent does not occur at random, nor should it be taken for granted. It needs to be searched out, attracted, developed, and maintained. An organization's personal policies and award systems must be consistent with the talent objectives. Losing a top talent is a sin! Companies like GE conduct postmortems on every top talent loss and hold their management accountable for those losses.

3. *Manage knowledge.* Technology companies are knowledge factories. In essence, they buy, trade, transfer, and sell knowledge. Their value lies increasingly in the collective knowledge that becomes the basis for creating new ideas, concepts, products, and services. The emphasis must be on orchestrated management of this collective knowledge. New products and services usually do not come from a single brilliant idea, but are the result of broad-based collaborative efforts throughout the organization. They are the result of an intricately connected, vast knowledge network with high interconnectivity and low cross-organizational impedance. Setting up effective support systems and managing the development, processing, filtering, sharing, and transferring knowledge toward value added is a very important and challenging task, which by and large involves people and sophisticated people skills.

4. *Manage information.* Similar to knowledge, information management has a strong human side, which often does not receive sufficient attention. Regardless of the avail-

5. *Manage communications.* Communication is the backbone of a firm's command and control structure. It is the catalyst to crucial integration of organizational efforts toward unified results. This is especially important in technology firms with their unconventional organizational structures and strong need for cross-functional coordination. Communication systems are much broader than information systems. They also include a wider range of human issues, such as interaction dynamics, power, politics, trust, respect, and credibility. While greatly supported by IT, some of the most powerful, effective, and expedient forms of communication are conducted in group meetings that include various forms of verbal and nonverbal information flows.

6. *Manage collaboration and commitment.* With the complexities of technological undertakings, broad-scale collaboration is often necessary to solve a problem or achieve a mission objective. This includes company internal as well as external forms of joint ventures, ranging from informal cross-functional agreements to multicompany consortia, co-developments, and joint ventures.

7. *Build supportive organizational environment.* Successful companies have cultures and environments that support their people. These companies provide visibility and recognition to their people. They also show the impact of these accomplishments on the company's mission and project-related objectives. This creates an ambiance where people are interested and excited about their work, which produces higher levels of ownership, cross-functional communications, cooperation, and some tolerance for risk and conflict.

8. *Ensure direction and leadership.* Managers themselves are change agents. Their concern for people, assistance to problem solving, and enthusiasm for the enterprise mission and objectives can foster a climate of high motivation, work involvement, commitment, open communications, and willingness to cooperate across the organization.

Engineering managers often describe their organizational environments as "unorthodox," with ambiguous authority and responsibility relations. They argue that such environments require broader management skill sets and more sophisticated leadership than traditional business situations. In this more open, dynamic business environment, enterprise performance is based to a large degree on teamwork. Yet, attention to individuals, their competence, accountability, commitment, and sense for self-direction, is crucial for organizations to function effectively. Such a team-centered management style is based on the thorough understanding of the motivational forces and their interaction with the enterprise environment.

4 MOTIVATION AND ENGINEERING PERFORMANCE

Understanding people is important in any management situation, but is critical in today's technology-based engineering organizations. To perform effectively, leaders must cross organizational lines and gain services from personnel not reporting directly to them. They must build multidisciplinary teams into cohesive groups and deal with a variety of networks, such as line departments, staff groups, team members, clients, and senior management, each having different cultures, interests, expectations, and charters. These engineering managers must relate socially as well as technically, and understand the culture and value system of the organization in which they work. The days of the manager who gets by with only technical expertise or pure administrative skills are gone.

What works best? Observations of best-in-class practices show consistently and measurably two important characteristics of high performers in technology organizations: (1) they enjoy work and are excited about the contributions they make to their company and society, and (2) they have strong needs, both professionally and personally, and use their professional work as vehicle to fulfill these needs. Sixteen professional needs are strongly associated with high-tech job performance. As summarized in Table 1, the fulfillment of these needs drives professional people to higher performance; conversely, the inability to fulfill these needs may become a barrier to individual performance and teamwork.* The rationale for this important correlation is found in the complex interaction of organizational and behavioral elements. Effective team management involves three primary components: (1) people skills, (2) organizational structure, and (3) management style. All three components are influenced by the specific task to be performed and the surrounding environment. That is, the degree of satisfaction of any of the needs is a function of (1) having the right mix of people with appropriate skills and traits, (2) organizing the people and resources according to the tasks to be performed, and (3) adopting the right leadership style.

4.1 Implications for Engineering Performance

The significance of assessing these motivational forces lies in several areas. *First,* Table 1 provides insight into the broad needs that technology-oriented professionals seem to have. These needs must be satisfied *continuously* before the people can reach high levels of performance.† Second, the list of needs also provides a model for *benchmarking,* as a framework for monitoring, defining, and assessing the needs of people on a team and ultimately building a work environment that is conducive to high performance.‡

Figure 2 illustrates the motivational process with the *inducement-contribution model.* The model shows that people are motivated to work toward personal and professional goals because they satisfy specific needs, such as needs for recognition, promotion, or pay increase. In the process of satisfying their needs, people make a contribution to the organization and its performance. If, on the other hand, employees cannot reach their goals and fulfill their needs through activities within the "primary organization" (i.e., employer), they may try working toward their goals through an outside organization. They may involve themselves in volunteer work or recreational activities. While this may help to satisfy the employees, their efforts and energy are directed within a secondary organization, which provides in-

---

*Several field studies have been conducted over the past 15 years, investigating needs and conditions most favorably associated with team performance in R&D and high-technology product environments. The findings are being used here to draw conclusions on professional needs and leadership style effectiveness. For details of these field studies see Thamhain[7-9] and Thamhain & Wilemon.[10-12]

†This is consistent with finding from other studies, which show that in technology-based environments a significant correlation exists between professional satisfaction and organizational performance. Some of the earlier studies on needs and performance in technology-intensive work environments go back to the 1970s (cf. Gemmill & Thamhain[11]). Later, Thamhain[12] found strong correlations between professional satisfaction of project team personnel and project performance toward high-tech product developments.

‡In most cases, managers have a lot more freedom in satisfying employee needs than they recognize. As an example, managers have many options in assigning "professionally interesting work." While top-down, the total work structure and organizational goals might be fixed and not negotiable, managers have a great deal of freedom and control over the way the work is distributed and assigned and brought into perspective.

Table 1 Sixteen Professional Needs Affecting Individual and Team Performance

1. *Interesting and challenging work.* This is an intrinsic motivator that satisfies professional esteem needs and helps to integrate personal goals with the objective of the organization.
2. *Professionally stimulating work environment.* Promotes professional involvement, creativity, and interdisciplinary support. Conducive to team building, effective communication, conflict resolution and commitment. Job satisfaction is a good benchmark for assessing fulfillment of this need.
3. *Professional growth.* Perceived via promotional opportunities, salary advances, skill building, and professional recognition.
4. *Overall leadership.* Leadership satisfies the need for clear direction and unified guidance of the work and its people. It involves dealing effectively with individual contributors, managers, and support groups. Leadership involves a wide spectrum of technical, administrative, and people skills.
5. *Tangible rewards.* A large spectrum of rewards is available to managers to satisfy extrinsic and intrinsic needs, such as salary, bonuses, incentives, promotions, recognition, praise, office decor, and educational opportunities.
6. *Job expertise.* People must have the knowledge and skills necessary to perform the required task. This includes the technical expertise, as well as the administrative and human skills needed for effective role performance.
7. *Assisting in problem solving.* Examples include facilitating solutions to technical, administrative, and personal problems. It is an important need, which, if not satisfied, often leads to frustration, conflict, and suboptimal job performance.
8. *Clearly defined objectives.* Goals, objectives, and outcomes of an effort must be clearly communicated to all affected personnel. Conflict can develop over ambiguities or missing information.
9. *Management control.* People like to see a reasonable degree of structure, direction, and control of their work, without being micromanaged. This is also important for effective team performance, innovation, and creativity toward established organizational goals.
10. *Job security.* This is one of the very fundamental needs that must be satisfied before people consider higher-order growth needs. Job security is "perceived" and has many facets, ranging from overall enterprise performance to anxieties over job skills or stakeholder satisfaction.
11. *Good interpersonal relations.* Components include trust, respect, credibility, low conflict, and collegial behavior, just to name a few. People must feel comfortable with their co-workers to interact effectively, share risks, and unify toward organizational objectives.
12. *Proper planning.* People need a roadmap to see where they are going. This is absolutely essential for the successful roll-out of engineering work. It is also a precondition for resource negotiations and commitment.
13. *Clear role definition.* This helps in defining the work team, minimizing role conflict and power struggles, and legitimizing authority and reporting relations.
14. *Open communications.* Helps to satisfy the need for a free flow of information both horizontally and vertically, keeping people informed and unified toward desired results.
15. *Minimum changes.* Although change is often inevitable, most people don't like change and see it as unnecessary, impeding creativity and performance. Advanced planning and proper communication can affect the personal perception and attitude toward change and minimize its negative impact.
16. *Senior management support.* Support is usually needed in many areas, such as resource provision, administrative tools and processes, project support from resource groups, and provision of necessary facilities and equipment. Management support is particularly crucial to larger, multifunctional, and complex undertakings.

4 Motivation and Engineering Performance

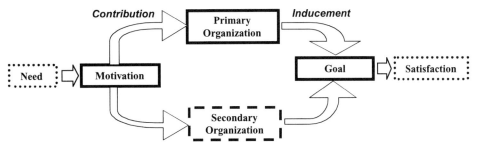

Figure 2 The inducement–induction model of motivation.

ducement and receives their contributions, which are being "leaked away" from the employer.

An employee's need satisfaction is as important to the healthy functioning of the organization as it is to the employee. Good managers are closely involved with the staff and their work. The manager should also identify any *unrealistic* goals and correct them, change situations that impede the attainment of realistic goals, and support people in reaching their goals and cheer them on, which is recognition. This will help to refuel the individual desire to reach the goal and keep the employees' energies channeled through the primary organization. The tools that help the manager to facilitate professional satisfaction are work sign-on, delegation, career counseling and development, job training and skill development, and effective managerial direction and leadership with proper recognition and visibility of individual and team accomplishments.

4.2 Motivation as a Function of Risks and Challenges

Additional insight into motivational drive and its dynamics can be gained by considering motivational strength as a function of the probability and desire to achieve the goal. We can push others, or ourselves, toward success or failure because of our mental predisposition called *self-fulfillment prophesies*. Our motivational drive and personal efforts increase or decrease relative to the likelihood of the expected outcome. Personal motivation toward reaching a goal changes with the probability of success (perception of doability) and challenge. Figure 3 expresses the relationship graphically. A person's motivation is very low if the probability of achieving the goal is very low or zero. As the probability of reaching the

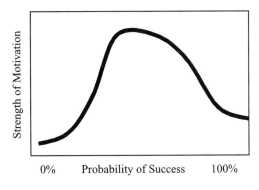

Figure 3 The Pygmalion effect, relating strength of motivation to the probability of success.

goal increases, so does the motivational strength. However, this increase continues only up to a certain level; when success is more or less assured, motivation decreases. This is an area where the work is often perceived as routine, uninteresting, and holding little potential for professional growth.

4.3 Success as a Function of Motivation

The saying "The harder you work, the luckier you get" expresses the effect of motivation in pragmatic terms. People who have a "can-do" attitude, who are confident and motivated, are more likely to succeed in their mission. Winning is in the attitude. This is the essence of the self-fulfillment prophecy. In fact, high-performing engineering teams often have a very positive image of their capabilities. As a result, they are more determined to produce the desired results, often against high risks and odds.

4.4 Manage in the Range of High Motivation

Applying motivational models to the high-technology workplace suggests that managers should ensure proper matching of people to their jobs. Further, managers must foster a work environment and direct personnel in a manner that promotes "can-do" attitudes and facilitates continuous assistance and guidance toward successful task completion. The process involves four primary issues: (1) work assignments, (2) team organization, (3) skill development, and (4) management. Specific suggestions for stimulation and sustaining motivation in technology-oriented professionals are summarized below:

1. *Work assignments*
   - Explain the assignment, its importance to the company, and the type of contributions expected.
   - Understand the employee's professional interests, desires, and ambitions and try to accommodate to them.
   - Understand the employee's limitations, anxieties, and fears. Often these factors are unjustly perceived and can be removed in a face-to-face discussion.
   - Develop the employee's interest in an assignment by pointing out its importance to the company and possible benefits to the employee.
   - Assure assistance where needed and share risks.
   - Show how to be successful. Develop a "can-do" attitude.
   - If possible, involve the employee in the definition phase of the work assignment, for instance via up-front planning, a feasibility analysis, needs assessment, or a bid proposal.

2. *Team organization*
   - Select the team members for each task or project carefully, assuring the necessary support skills and interpersonal compatibility.
   - Plan each engineering project properly to assure clear directions, objectives, and task charters.
   - Assure leadership within each task group.
   - Sign-on key personnel on a one-on-one basis according to the guidelines discusses in item 1.

3. *Skill development*
   - Plan the capabilities needed in your engineering department for the long-range. Direct your staffing and development activities accordingly.
   - Encourage people to keep abreast in their professional field.
   - Provide for on-the-job experimental training via selected work assignments and managerial guidance.
   - Provide the opportunity for some formal training via seminars, courses, conferences, and professional society activities.
   - Use career counseling sessions and performance reviews to help in guiding skill development and matching them with personal and organizational objectives.

4. *Management*
   - Develop interest in the work itself by showing its importance to the company and the potential for professional rewards and growth.
   - Promote project visibility, team spirit, and upper-management involvement.
   - Assign technically and managerially competent task leaders for each team, and provide top-down leadership for each project and for the engineering function as a whole.
   - Manage the quality of the work via regular task reviews and by staying involved with the project team, without infringing on their autonomy and accountability.
   - Plan your projects up front. Conduct a feasibility study and requirements analysis first. Assure the involvement of the key players during these early phases.
   - Break activities or projects into phases and define measurable milestones with specific results. Involve personnel in the definition phase. Obtain their commitment.
   - Try to detect and correct technical problems early in their development.
   - Foster a professionally stimulating work environment.
   - Unify the task team behind the overall objectives. Stimulate the sense of belonging and mutual interdependence.
   - Refuel the commitment and interest in the work by recognizing accomplishments frequently.
   - Assist in problem-solving and group decision-making.
   - Provide the proper resources.
   - Keep the visibility and priority for the project high. No interruptions.
   - Avoid threats. Deal with fear, anxieties, mistrust, and conflicts.
   - Facilitate skill development and technical competency.
   - Manage and lead.

5 THE POWER SPECTRUM IN ENGINEERING AND TECHNOLOGY MANAGEMENT

5.1 Motivation, Managerial Power, and Performance

Why do people comply with the requests or demands of others, such as their superiors? One reason is that they see the other people as being able affect their needs and wants. People comply with the requests of others if they perceive them to be able to influence specific outcomes. These outcomes could be desirable, such as a salary increase, or undesirable, such as a reprimand or demotion. In organizational environments the influence over others is referred to as *managerial power*. Managers, as well as anyone else, use this power to achieve interpersonal influence, which is leadership in its applied form. Therefore power is the force that when successfully activated motivates others toward desired results. In the next section, we will look specifically into the power spectrum that is available to leaders in a technology-oriented environment.

5.2 Power-Sharing and Dual Accountability

Engineering and technology managers must build multidisciplinary teams and successfully deal with a variety of interfaces across the enterprise, such as functional departments, staff groups, team members, clients, and senior management. In contrast to traditional management situations, where the *organization* provides the power needed to manage in the form of legitimate authority, reward, and punishment (style I), engineering managers must often earn part or all of their authority. This so-called system II management style* complements the organizationally derived power bases of authority, reward, and punishment with bases developed by the individual manager, relying largely on credibility, trust, and respect. Examples are technical and managerial expertise, friendship, work challenge, promotional opportunities, fund allocations, charisma, personal favor, project goal identification, recognition, and visibility. Effective engineering management combines both styles I and II.† The five commonly recognized bases of managerial power are summarized in Fig. 4.

5.3 The Important Role of Salary

Salary and other financial rewards play a very special role in the managerial power spectrum. Yet engineering and technology mangers do not always realize the relatively high importance

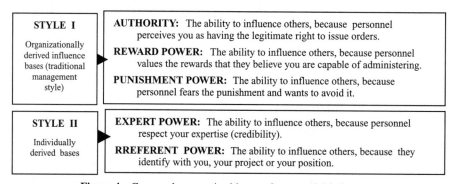

Figure 4 Commonly recognized bases of managerial influence.

---

* According to Likert[13] this knowledge and relationship-oriented style of leadership is also classified as style IV.

† Various research studies by Gemmill, Thamhain, and Wilemon provide an insight into the power spectrum available to project managers.[11,12,14] The original study was published by Gemmill and Thamhain.[11] This study used Kendall's rank-order correlation techniques to measure the association between the strength of a managerial power base and certain performance measures, such as project support, communications effectiveness, personal involvement and commitment, and overall management performance.

of salary as an influence base. Managers often argue that salary and other financial rewards fulfill only lower level needs, while recognition, pride, and accomplishments are the true motivators of engineers, scientists, and other technical professionals. This is true in general and is also supported by the writings in this chapter. However, the argument holds only if the personnel perceive a fair and adequate compensation. Otherwise, salary becomes a barrier to effective teamwork, a handicap for attracting and holding quality people, and a source of steady conflict. To illustrate, a person who is motivated to make an extra effort might indeed enjoy the praise and recognition that comes with the well-done job. The person may further infer that the job was important to the company. Receiving such praise on a regular basis will ensure the employee that he or she is making significant contributions to the enterprise and could also expect a better than average raise or even be in line for a promotion. Now, suppose a subsequent salary review results in a less than expected or zero increase. The employee would question the sincerity and value of any praise, recognition, or other intrinsic rewards received in the past and anticipated for the future. The employee might also feel confused, frustrated, angry, or manipulated. Obviously, this is *not* a situation that leads to long-range motivation, sustained personal drive, and high morale. Taken together, salary and other financial rewards are important bases of managerial power. They must be used judiciously in line with the employee's output, efforts, and contributions to the organization. All should be communicated to the employee and supported with well-earned recognition of the accomplishments.

6 TEAMWORK IN ENGINEERING ORGANIZATIONS

Virtually all managers recognize the critical importance of effective teamwork as a key determinant of project performance and success in today's competitive world. These managers also spend considerable resources on continuous improvement of their teams. When describing effective project teams that reach desired results on time and within budget, managers point at specific factors, which are graphically shown in Fig. 5. Building high-performing teams is the process of transforming the energies of individuals within a group into a unified team effort focused on desired results. Building and managing these engineering teams requires an understanding of the organization and its interfaces, authority, and

Figure 5 Characteristics of high-performing teams.

power structures. It also requires leadership, including a broad set of skills across a wide spectrum of technological, organizational, and interpersonal areas.

Managerial principle and practices have changed dramatically. For example, not too long ago, team leadership was considered a subset of project management, with its principles defined by "management science." Team leaders could ensure successful integration of their work by focusing on properly defining the work, timing, and resources, and by following established procedures for project tracking and control. Today, these factors have become threshold competencies, important, but unlikely to guarantee by themselves project success, especially for complex, technology-based undertakings. In addition to the technical issues, engineering managers must be able to deal with uncertainties and risks caused by technological, economic, political, social, and regulatory factors. Moreover, they must organize and manage their teams across organizational lines, dealing with resource sharing, multiple reporting relationships, and broadly based alliances.

6.1 Toward Self-Direction and Virtual Teams

Because of these complexities, traditional forms of hierarchical team structure and team leadership are seldom effective and are being replaced by self-directed, self-managed team concepts (Barner[3]; Thamhain and Wilemon[12]). Often the manager becomes a social architect who understands the interaction of organizational and behavioral variables, facilitates the work process, and provides overall project leadership for developing multidisciplinary task groups into unified teams and fostering a climate conducive to involvement, commitment, and conflict resolution.

Especially with the evolution of contemporary organizations, such as the matrix, traditional bureaucratic hierarchies have declined and horizontally oriented teams and work units have become increasingly important (Fisher[15]; Marshall[16]; Shonk[17]). Moreover, the team leader's role as supervisor has diminished in favor of *empowerment and self-direction* of the team (Table 2). In addition, advances of information technology, made it feasible and effective to link team members over the Internet or other media, creating a *virtual team* environment (Table 3). *Virtual teams* and *virtual project organizations* are powerful managerial tools, especially for companies with geographically dispersed project operations, and for linking contractors, customers, and regulators with the core of the project team.

Table 2 Self-Directed Teams

Definition: A group of people chartered with specific responsibilities for managing themselves and their work, with minimal reliance on group-external supervision, bureaucracy and control. Team structure, task responsibilities, work plans and team leadership often evolve based on needs and situational dynamics.

Benefits: Ability to handle complex assignments, requiring evolving and innovative solutions that cannot be easily directed via top-down supervision. Widely shared goals, values, information and risks. Flexibility toward needed changes. Capacity for conflict resolution, team building and self-development. Effective cross-functional communications and work integration. High degree of self-control, accountability, ownership, and commitment toward established objectives.

Challenges: A unified, mature team does not just happen, but must be carefully organized and developed by management. A high degree of self-motivation, and sufficient job, administrative and people skills must exist among the team members. Empowerment and self-control might lead to unintended results and consequences. *Self-directed* teams are not necessarily *self-managed;* they often require more sophisticated external guidance and leadership than conventionally structured teams.

Table 3 Virtual Teams

Definition: A group of project team members, linked via the Internet or media channels to each other and various project partners, such as contractors, customers, and regulators. Although physically separated, technology links these individuals so they can share information and operate as a unified project team. The number of elements in a virtual team and their permanency can vary, depending on need and feasibility. An example of a virtual team is a project review conducted among the team members, contractors and customer, over an Internet website.

Benefits: Ability to share information and communicate among team members and organizational entities of geographically dispersed projects. Ability to share and communicate information in a synchronous and asynchronous mode (application: communication across time-zones, holidays and shared work spaces). Creating unified visibility of project status and performance. Virtual teams, to some degree, bridge and neutralize the culture and value differences that exist among different task teams of a project organizations.

Challenges: The effectiveness of the virtual team depends on the team members' ability to work with the given technology. Information flow and access is not necessarily equal for all team members. Information may not be processed uniformly throughout the team. The virtual team concept does not fit the culture and value system of all members and organizations. Project tracking, performance assessment and managerial control of project activities is often very difficult. Risks, contingencies, and problems are difficult to detect and assess. Virtual organizations often do not provide effective methods for dealing with conflict, power, candor, feedback and resource issues. Because of the many limitations, more traditional team processes and communications are often needed to augment virtual teams.

6.2 Measuring Project Team Performance

"A castle is only as strong as the people who defend it." This Japanese proverb also applies to organizations and business processes. They are only as effective as their unified team efforts. However, team performance involves highly complex, interrelated sets of variables, including attitudes, personal preferences, and perceptions, which are difficult to measure and often impossible to quantify. In fact, only one in ten managers has a specific metric for actually measuring team performance (Thamhain[14]). This creates tough challenges, especially in technology-based organizations where much of the work is team-oriented.

In spite of these challenges, a framework for measuring team performance can be defined. Using the characteristics of high-performing teams summarized in Fig. 1, a metrics for measuring team performance can be developed. Table 4 provides such a metrics broken into four categories: (1) work and team structure, (2) communications and control, (3) team leadership, and (4) attitude and values. These broad measures can provide a model for measuring team performance and for developing the team or the organization.

The significance of establishing team performance measures lies in two areas. First, it offers some inside as to what an effective team environment looks like, providing the basic framework for team assessment, benchmarking and development. Second, it leads to a better understanding of how team characteristics affect team performance, providing building blocks for organization development, such as defining drivers and barriers to team performance and a framework for leadership style development.

7 BUILDING HIGH-PERFORMING TEAMS

Given the realities of today's business environment, its technical complexities, cross-functional dependencies, and incrementally evolving solutions, decision processes are often distributed throughout the team. Power and responsibility is shifting from managers to project

Table 4 Benchmarking Team Performance

Work and Team Structure
- Team participates in project definition, work plans evolve dynamically
- Team structure and responsibilities evolve and change as needed
- Broad information sharing
- Team leadership evolves based on expertise, trust, respect
- Minimal dependence on bureaucracy, procedures, politics

Communication and Control
- Effective cross-functional channels, linkages
- Ability to seek out and process information
- Effective group decision-making and consensus
- Clear sense of purpose and direction
- Self-control, accountability, and ownership
- Control is stimulated by visibility, recognition, accomplishments, autonomy

Team Leadership
- Minimal hierarchy in member status and position
- Internal team leadership based on situational expertise, trust, and need
- Clear management goals, direction, and support
- Inspires and encourages

Attitudes and Values
- Members are committed to established objectives and plans
- Shared goals, values, and project ownership
- High involvement, energy, work interest, need for achievement, pride self-motivated
- Capacity for conflict resolution and resource sharing
- Team building and self-development
- Risk sharing, mutual trust and support
- Innovative behavior
- Flexibility and willingness to change
- High morale and team spirit
- High commitment to established project goals
- Continuous improvement of work process, efficiency, quality
- Ability to stretch beyond agreed-on objectives

team members, who take higher levels of responsibility, authority, and control for specific project results. That is, teams can rarely be managed "top-down." Teams become *self-directed,* gradually replacing the more traditional, hierarchically structured organization. These emerging team processes rely strongly on group interaction, resource and power sharing, group decision-making, accountability, self-direction, and control. Organizing, building, and managing such self-directed teams requires a keen understanding of the organization and its processes. In addition, managers must realize the organizational dynamics involved during the various phases of the team development process. A four-stage model, originally developed by Hersey et al.[18] and graphically shown in Fig. 6, can be useful as a framework to analyze and develop the work group toward a fully integrated high-performing team. The four stages are labeled (1) *team formation,* (2) *team start-up,* (3) *partial integration,* and (4)

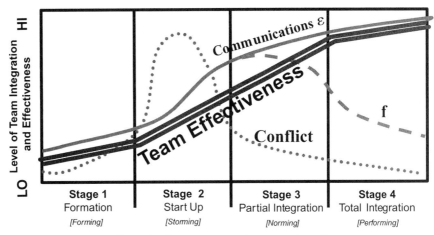

Figure 6 Four-stage team development model.

full integration. These stages are also known as *forming, storming, norming,* and *performing,* giving an indication of team behavior at each one of the stages.

7.1 How the Four-Stage Model of Team Development Works

To become fully integrated and unified in their values and skill sets, a work group needs to be skillfully nurtured and developed. Leaders must recognize the professional interests, anxieties, communication needs, and the challenges of their team members and anticipate them as the team goes through the various stages of integration. Moreover, team leaders must adapt their managerial style to the specific situation of each stage.* That is, team leaders must recognize what works best at each stage and what is most conducive to the team development process. Many of the problems that occur during the formation of the new project team or during its life cycle are normal and often predictable. However, they present barriers to effective team performance. The problems must be quickly identified and dealt with. Early stages, such as the *team formation* and *start-up,* usually require a predominately directive style of team leadership. Providing clear guidelines on the project mission its objectives and requirements and creating the necessary infrastructure and logistics support for the project team are critically important in helping the team to pass through the first two stages of their development quickly. During the third stage of *partial integration,* or *norming,* the team still needs a considerable amount of guidance and administrative support, as well as guidance in dealing with the inevitable human issues of conflict, power and politics, credibility, trust, respect, and the whole spectrum of professional career and development. This is the stage where a combination of *directive and participative leadership* will produce most favorable results. Finally, a team that reached the fully integrated stage by definition becomes "self-directed." That is, such a fully integrated, unified team can work effectively with a minimum degree of external supervision and administrative support. While at this

---

*The discussions on leadership style effectiveness is based on field research by H. J. Thamhain and D. L. Wilemon, published in "A High-Performing Engineering Project Team," in *The Human Side of Managing Innovation,* R. Katz (ed.), Oxford University Press, New York, 1997.

stage, the team often appears to have very little need for external managerial intervention, it requires highly sophisticated external leadership to maintain this delicate state of team effectiveness and focus. The situational leadership dynamics of teambuilding is graphically shown in the Project Life Cycle Model of Fig. 7.

8 RECOMMENDATIONS FOR EFFECTIVE ENGINEERING MANAGEMENT

The nature of engineering management, with its multidisciplinary technology requirements, ambiguous authority definition and complex operating environment, requires experience and sophistication for effective role performance. A number of suggestions may be helpful to increase the managerial effectiveness.

Understand Motivational Needs. Engineering and technology managers need to understand the interaction of organizational and behavioral elements to build an environment conducive to their personnel's motivational needs. This will enhance active participation and minimize dysfunctional conflict. The effective flow of communication is one of the major factors determining the quality of the organizational environment. Since the manager must build task and project teams at various organization level, it is important that key decisions are communicated properly to all task-related personnel. Regularly scheduled status review meetings can be an important vehicle for communicating and tracking project-related issues.

Adapt Leadership to the Situation. Because their environment is temporary and often untested, engineering and technology managers should seek a leadership style that allows them to adapt to the often conflicting demands existing within their organization, support departments, customers, and senior management. They must learn to "test" the expectations of

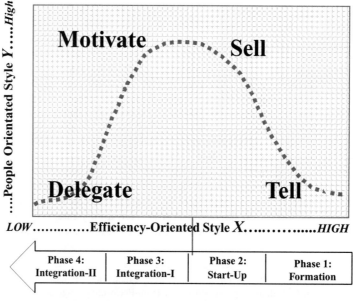

Figure 7 Situational leadership style versus team maturity.

others by observation and experimentation. Although it is difficult, they must be ready to alter their leadership style as demanded both by the specific tasks and by their participants.

Accommodate Professional Interests, Build Enthusiasm and Excitement. Technology managers should try to accommodate the professional interests and desires of supporting personnel when negotiating their tasks. Task effectiveness depends on how well the manager provides work challenges to motivate those individuals who provide support. Work challenge further helps to unify personal goals with the goals and objectives of the organization. Although the total work of an engineering department may be fixed, the manager has the flexibility of allocating task assignments among various contributors.

Build Technical Expertise. Technology managers should develop or maintain technical expertise in their fields. Without an understanding of the technology to be managed, they are unable to win the confidence of their team members, to build credibility with the customer community, to participate in search for solutions, or to lead a unified engineering/technology effort.

Plan Ahead. Effective planning early in the life cycle of a new technology program is highly recommended. Planning is a pervasive activity that leads to personnel involvement, understanding, and commitment. It helps to unify the task team, provides visibility, and minimizes future dysfunctional conflict.

Provide Role Model. Finally, engineering and technology managers can influence the work climate by their own actions. Their concern for project team members, their ability to integrate personal goals and needs of personnel with organizational goals, and their ability to create personal enthusiasm for the work itself can foster a climate that is high in motivation, work involvement, open communication, creativity, and engineering/technology performance.

Define Work Process and Team Structure. Successful engineering team management requires an infrastructure conducive to cross-functional teamwork and technology transfer. This includes properly defined interfaces, task responsibilities, reporting relations, communication channels, and work transfer protocols. The tools for systematically describing the work process and team structure come from the conventional project management system; they include (1) *project charter,* defining the mission and overall responsibilities of the project organization, including performance measures and key interfaces; (2) *project organization chart,* defining the major reporting and authority relationships; (3) *responsibility matrix* or *task roster;* (4) *project interface chart, such as the n-squared chart;* and (5) *job descriptions.*

Develop Organizational Interfaces and Communication Channels. Overall success of an engineering team depends on effective communications and cross-functional integration. Each task team should clearly understand its task inputs and outputs, interface personnel, and work transfer mechanism, which includes outside contractors and suppliers. In addition, to conventional technology, such as voice mail, e-mail, electronic bulletin boards, and conferencing, workspace design, regular meetings, reviews, and information sessions can facilitate the free flow of information, both horizontally and vertically. Moreover, team-based reward systems can help to facilitate cooperation with cross-functional partners. Quality function deployment (QFD) concepts, n-square charting, and well-defined phase-gate criteria can be useful tools for developing cross-functional linkages and promoting interdisciplinary cooperation and alliances.

Communicate Organizational Goals and Objectives. Management must communicate and update the organizational goals and project objectives. The relationship and contribution of individual work to the overall engineering project, the business plans, and the importance to the organizational mission must be clear to all team personnel. Senior management can help in unifying the team behind the project objectives by developing a "priority image," through their personal involvement and visible support.

Build a High-Performance Image. Building a favorable image for an ongoing project in terms of high priority, interesting work, importance to the organization, high visibility, and potential for professional rewards is crucial for attracting and holding high-quality people.

Create Proper Reward Systems. Personnel evaluation and reward systems should be designed to reflect the desired power equilibrium and authority/responsibility sharing needed for the concurrent engineering organization to function effectively. Creating a system and its metrics for reliably assessing performance in the engineering environment is a great challenge. However, several models, such as the Integrated Performance Index (Pillai et al.[19]), have been proposed and provide a potential starting point for customization. A QFD philosophy, where everyone recognizes the immediate "customer" for whom a task is performed helps to focus efforts on desired results and customer satisfaction.

Build Commitment. Managers should ensure team member commitment to their project plans, specific objectives, and results. If such commitments appear weak, managers should determine the reason for such lack of commitment and attempt to modify negative views. Anxieties and fear of the unknown are often a major reason for low commitment (Stum[20]). Conflict over technical, administrative, or personal issues and lack of interest in the work may be other reasons for lacking commitment.

Manage Conflict and Problems. Engineering activities are highly disruptive to the enterprise. Conflict is inevitable. Managers should focus their efforts on problem avoidance. That is, managers and team leaders, through experience, should recognize potential problems and conflicts at their onset, and deal with them before they become big and their resolutions consume a large amount of time and effort (Haque et al.[21]).

Conduct Team Building Sessions. A mixture of *focus-team* sessions, *brainstorming, experience exchanges,* and *social gatherings* can be powerful tools for developing the work group into an effective, fully integrated, and unified project team (Thamhain and Wilemon[12]). Intensive team building efforts may be especially needed during the formation stage of a new project team, but should be conducted throughout a project life cycle.

Foster a Culture of Continuous Support and Improvement. Successful engineering management focuses on people behavior and their roles within the organization. High-performance engineering organizations have cultures and support systems that demand broad participation in their organization development. It is important to establish support systems, such as discussion groups, action teams, and suggestion systems, to capture and leverage the lessons learned and to identify problems as part of a continuous improvement process. Tools such as the *project maturity model* and the *six sigma management process* can provide the framework and toolset for analyzing and fine tuning the team development and its management process.

Provide Proper Direction and Leadership. Engineering managers and team leaders can influence the attitude and commitment of their people by their own actions. Concern for the team members and enthusiasm for the project can foster a climate of high motivation, work involvement, open communications, and commitment.

9 CONCLUDING REMARK

Managerial leadership at all levels has significant impact on the work environment that ultimately affects engineering team performance. Effective managers understand the interaction of organizational and behavioral variables. They can foster a climate of active participation and minimal dysfunctional conflict. This requires carefully developed skills in leadership, administration, organization, and technical expertise. It further requires the ability to involve top management and ensure organizational visibility, resource availability, and overall support for the engineering work across the enterprise, its support functions, suppliers, sponsors, and partners.

REFERENCES

1. Armstrong, D., "Building Teams Across Boarders," *Executive Excellence,* 17(3), 10 (2000).
2. Barkema, H., J. Baum, and E. Mannix, "Management Challenges in a New Time," *Academy of Management Journal,* 45(5), 916–930 (2002).
3. Barner, R., "The New Millennium Workplace," *Engineering Management Review* (*IEEE*), 25(3), 114–119 (1997).
4. Dillon, P., "A Global Challenge," *Forbes Magazine,* 168 (Sep. 10), 73–80 (2001).
5. Gray, C., and E. Larson, *Project Management.* Irwin McGraw-Hill, New York, 2000.
6. Thamhain, H., "Criteria for Effective Leadership in Technology-oriented Project Teams," in *The Frontiers of Project Management Research,* Slevin, Cleland, and Pinto (eds.), Project Management Institute, Newton Square, PA, 259–270 (2002).
7. Thamhain, H., "Managing Engineers Effectively," *IEEE Transactions on Engineering Management,* 30(4), 231–237 (1983).
8. Thamhain, H., and D. Wilemon, "Building High-performance Engineering Project Teams," *IEEE Transactions on Engineering Management,* 34(1), 130–142 (1987).
9. Thamhain, H., "Leading Technology-based Project Teams," *Engineering Management Journal,* 16(2), 42–51 (2004).
11. Gemmill, G., and H. Thamhain, "Influence Styles of Project Managers: Some Project Performance Correlates," *Academy of Management Journal,* 17(2), 216–224 (1974).
12. Thamhain, H., and D. Wilemon, "Building Effective Teams for Complex Project Environments," *Technology Management,* 5(2), 203–212 (1999).
13. Likert, R., *New Patterns of Management,* McGraw-Hill, New York, 1967.
14. Thamhain, H., "Managing Innovative R&D Teams," *R&D Management,* 33(3), 297–312 (2003).
15. Fisher, K., *Leading Self-Directed Work Teams,* McGraw-Hill, New York, 1993.
16. Marshall, E., *Transforming the Way We Work,* AMACOM, New York, 1995.
17. Shonk, J. H., *Team-Based Organizations.* Irwin, Homewood, IL, 1996.
18. Hersey, P., K. Blanchard, and D. Johnson, *Management of Organizational Behavior: Leading Human Resources,* Prentice-Hall, Upper Saddle River, NJ, 2000.
19. Pillai, A., A. Joshi, and K. Raoi, "Performance Measurement of R&D Projects in a Multi-project, Concurrent Engineering Environment," *International Journal of Project Management,* 20(2), 165–172 (2002).
20. Stum, D., "Maslow Revisited: Building the Employee Commitment Pyramid," *Strategy and Leadership,* 29(4), 4–9 (2001).

21. Haque, B., K. Pawar, and R. Barson, "The Application of Business Process Modeling to Organizational Analysis of Concurrent Engineering Environments," *Technovation,* 23(2), 147–162 (2003).

BIBLIOGRAPHY

Bhatnager, A., "Great Teams," *The Academy of Management Executive,* 13(3), 50–63 (1999).

Jassawalla, A. R., and H. C. Sashittal, "Building Collaborate Cross-functional New Product Teams," *The Academy of Management Executive,* 13(3), 50–63 (1999).

Keller, R., "Cross-Functional Project Groups in Research and New Product Development," *Academy of Management Journal,* 44(3), 547–556 (2001).

Kruglianskas, I., and H. Thamhain, "Managing Technology-based Projects in Multinational Environments, *IEEE Transactions on Engineering Management,* 47(1), 55–64 (2000).

MacCormack, A., R. Verganti, and M. Iansiti, "Developing Products on Internet Time," *Management Science,* 47(1), 22–35 (2001); *Engineering Management Review,* 29(2), 90–104 (2001).

Shim, D., and M. Lee, "Upward Influence Styles of R&D Project Leaders," *IEEE Transactions on Engineering Management,* 48(4), 394–413 (2001).

Thamhain, H., "Managing Self-directed Teams Toward Innovative Results. *Engineering Management Journal,* 8(3), 31–39 (1996).

Thamhain, H., "Team Management," in *Project Management Handbook,* J. Knutson (ed.), Wiley, New York, 2001.

Thamhain, H., *Management of Technology,* Wiley, New York, 2005.

Williams, J., *Team Development for High-tech Project Managers.* Artech House, Norwood, MA, 2002.

CHAPTER 15

FINANCE AND THE ENGINEERING FUNCTION

William Brett
New York, New York

1	INTRODUCTION AND OUTLINE	505	4	PROFIT AND LOSS STATEMENT	516
	1.1 Needs of Owners, Investors, and Lenders	505		4.1 Financial Ratios	517
	1.2 Needs of Top Managers	505	5	CASH FLOW OR SOURCE AND APPLICATION OF FUNDS	518
	1.3 Needs of Middle Managers of Line Functions	507		5.1 Accelerated Depreciation	520
	1.4 Needs of Staff Groups (Product Planners, Engineers, Market Researchers)	508	6	EVALUATING RESULTS AND TAKING ACTION	523
	1.5 Needs of Accountants	508		6.1 Comparing Current Results with Budgets and Forecasts	523
2	A FINANCIAL MODEL	508		6.2 Identifying Problems and Solutions	525
3	BALANCE SHEET	509		6.3 Initiating Action	527
	3.1 Current Assets	510	7	FINANCIAL TOOLS FOR THE INDEPENDENT PROFESSIONAL ENGINEER	528
	3.2 Current Liabilities	511		7.1 Simple Record-Keeping	529
	3.3 Accrual Accounting	512		7.2 Getting the System Started	530
	3.4 Interest-Bearing Current Liabilities	512		7.3 Operating the System	530
	3.5 Net Working Capital	512	8	CONCLUSIONS	530
	3.6 Current Ratio	513			
	3.7 Fixed Assets	513			
	3.8 Total Capital	514			
	3.9 Second Year Comparison	515			

1 INTRODUCTION AND OUTLINE

Finance is fundamental; accounting is merely the set of procedures, techniques, and reports that make possible the effective execution of the finance function. Harold Geneen, the legendary chairman of International Telephone and Telegraph, included in his *Sayings of Chairman Hal,* "The worst thing a manager can do is run out of money." He meant it! The corporate function of Finance is that function which makes the decisions, or rather provides the recommendations to top management who really make the decisions, that prevent the enterprise from running out of money. Accounting gathers, organizes, and disseminates information that make it possible to make these decisions accurately and timely. In modern business, accounting performs many correlative functions, some in such detail and so esoteric as to appear to be an end in themselves.

The objectives of this chapter on finance and accounting are to describe

- How accounting systems work to provide information for top managers and owners
- How financial management is carried out

Additionally, this chapter provides a concise description of how an accounting system is constructed to provide for the needs of middle management and staff groups such as engineers and marketers.

The purposes and uses of accounting systems, data, and reports are quite different for different people and functions in the business community. The engineer needs to understand accounting principles and processes as they apply to his or her function and also to understand the way in which others of the enterprise view business and what their information needs are. The following are five major groups that have distinctly differing points of view and objectives:

- Owners, investors, lenders, and boards of directors
- Top managers
- Middle managers of line functions
- Staff groups such as product planners, engineers, and market researchers
- Accountants

1.1 Needs of Owners, Investors, and Lenders

The first group—owners, investors, and lenders—have as their primary concern the preservation and protection of the capital or the assets of the business. The Board of Directors represents the interest of the owners and can be considered to be the agents of the owners (stockholders). The board members provide continuing review of the performance of top management as well as approval or disapproval of policies and key investment decisions.

This entire group wants to be assured that the property of the business—fixed plant and equipment, inventories, etc.—is being conserved. Next, they want to be assured that there will be sufficient liquidity, which means only that there will be enough cash available to pay all the bills as they come due. Finally, they want to see evidence of some combination of regular payout or growth in value—a financial return such as regular dividends or indications that the enterprise is increasing in value. Increase in value may be evidenced by growth in sales and profits, by increases in the market value of the stock, or by increased value of the assets owned. If the dividend payout is small, the growth expectations will be large.

The information available to the owners is, at a minimum, that which is published for public companies—the balance sheet, cash flow, and profit and loss statement. Special reports and analyses are also provided when indicated.

1.2 Needs of Top Managers

The top managers must be sensitive to the needs and desires of the owners as expressed by the Board of Directors and of the bankers and other lenders so that all of the purposes and objectives of owners and lenders are also the objectives of top managers. Additionally, top management has the sole responsibility for

- Developing long-range strategic plans and objectives
- Approving short-range operating and financial plans
- Ensuring that results achieved are measuring up to plan
- Initiating broad-gauge corrective programs when results are not in conformance with objectives

Reports of financial results to this group must be in considerable detail and identified by major program, product, or operating unit in order to give insight sufficient to correct problems in time to prevent disasters. The degree of detail is determined by the management style of the top executive. Usually such reports are set up so that trouble points are automatically brought to the top executives' attention, and the detail is provided make it possible to delve into the problems.

In addition to the basic financial reports to the owners, directors and top managers need:

- Long-term projections
- One-year budgets
- Periodic comparison of budget to actual
- Unit or facility results
- Product line results
- Performance compared to standard cost

1.3 Needs of Middle Managers of Line Functions

For our present purposes we will consider only managers of the sales and the manufacturing groups and their needs for financial, sales, and cost information. The degree to which the chief executive shares information down the line varies greatly among companies, ranging from a highly secretive handling of all information to a belief that sharing all the facts of the business improves performance and involvement through greater participation. In the great bulk of publicly held, large corporations, with modern management, most of the financial information provided to top management is available to staff and middle management, either on routine basis or on request. There are additional data that are needed by lower-level line managers where adequate operational control calls for much greater detail than that which is routinely supplied to top executives.

The fundamental assignment of the line manager in manufacturing and sales is to execute the policies of top management. To do this effectively, the manager needs to monitor actions and evaluate results. In an accounting context this means the manufacturing manager, either by formal rules or by setting his or her personal rules of thumb, needs to

- Set production goals
- Set worker and machine productivity standards
- Set raw material consumption standards
- Set overhead cost goals
- Establish product cost standards
- Compare actual performance against goals
- Develop remedial action plans to correct deficiencies
- Monitor progress in correcting variances

The major accounting and control tools needed to carry out this mission include

- Production standards
- Departmental budgets
- Standard costs
- Sales and production projections
- Variance reports
- Special reports

It is important that the line manager understands the profit and loss picture in his or her area of control and that job performance is not merely measured against preset standards but that he or she is considered to be an important contributor to the entire organization. It is, then, important that managers understand the total commercial environment in which they are working, so that full disclosure of product profits is desirable. Such a philosophy requires that accounting records and reports be clear and straightforward, with the objective of exposing operating issues rather than being designed for a tax accountant or lawyer.

The top marketing executive must have a key role in the establishment of prices and the determination of the condition of the market so that he or she is a full partner in managing the enterprise for profits. He or she therefore needs to participate with the manufacturing executive in the development of budgets and longer-range financial plans. Thus the budget becomes a joint document of marketing and manufacturing, with both committed to its successful execution.

The marketing executive needs to be furnished with all of the information indicated above as appropriate for the manufacturing manager.

1.4 Needs of Staff Groups (Product Planners, Engineers, Market Researchers)

The major requirement of accounting information for staff is that it provide a way to measure the economic effect of proposed changes to the enterprise. For the engineer this may mean changes in equipment or tooling or redesign of the product as a most frequent kind of change that must be evaluated before funds can be committed.

Accounting records that show actual and standard costs by individual product and discrete operation are invaluable in determining the effect of change in design or process. If changes in product or process can result in changes in total unit sales or in price, the engineer needs to know those projected effects. His or her final projections of improved profits will then incorporate the total effect of engineering changes.

The accounting records need to be in sufficient detail that new financial projections can be made reliably, with different assumptions of product features, sales volume, cost, and price.

1.5 Needs of Accountants

The accounting system must satisfy the strategic, operational, and control requirements of the organization as outlined above, but it has other external demands that must be satisfied. The accountants have the obligation to maintain records and prepare reports to shareholders that are "in conformity with generally accepted accounting principles consistently applied." Therefore, traditional approaches are essential so that the outside auditor as well as the tax collector will understand the reports and find them acceptable. There seems to be little need to sacrifice the development of good, effective control information for operating executives in order to satisfy the requirements of the tax collector or the auditor. The needs are compatible.

The key financial reporting and accounting systems typically used by each group are explained next.

2 A FINANCIAL MODEL

A major concern of the owners or the Board of Directors and the lenders to the business must be to ensure the security of the assets of the business. The obvious way to do this in a small enterprise is occasionally to take a look. It is certainly appropriate for directors to visit facilities and places where inventories are housed to ensure that the assets really do exist, but this can only serve as a spot check and an activity comparable to a military inspection—everything looks very good when the troops know that the general is coming. The most useful and convenient way, as well as the most reliable way, to protect the assets is by careful study of financial records and a comparison with recent history to determine the trends in basic values within the business. A clear and consistent understanding of the condition of the assets of the business requires the existence of a uniform and acceptable system of accounting for them and for reporting their condition. The accounting balance sheet provides this.

In the remainder of this chapter, a set of examples based on the experience of one fictitious company is developed. The first element in the case study is the corporate balance sheet. From there the case moves back to the profit and loss and the cash flow statements. The case moves eventually back to the basic statements of expense and revenue to demonstrate how these records are used by the people managing the business—how these records enable them to make decisions concerning pricing, product mix, and investment in new plant and processes. The case will also show how these records help management to direct the business into growth patterns, a strengthened financial position, or increased payout to the owners.

The name of the fictitious company is the Commercial Construction Tool Company, Incorporated, and will be referred to as CCTCO throughout the remainder of this chapter. The company manufactures a precision hand tool, which is very useful in the positioning and nailing of various wooden structural members as well as sheathing in the construction of frame houses. The tool is a proprietary product on which the patents ran out some time ago; however, the company has had a reputation for quality and performance that has made it very difficult for competition to gain much headway. The tool has a reputation and prestige among users such that no apprentice carpenter would be without one. The product is sold through hardware distributors who supply lumber yards and independent retail hardware stores. About three years ago the company introduced a lighter weight and somewhat simplified model for use in the "do-it-yourself" market. Sales of the home-use model have been good and growing rapidly, and there is some concern that the HOMMODEL (home model) is cannibalizing sales of the COMMODEL (commercial model).

The company has one manufacturing facility and its general offices and sales offices are at the same location.

At the first directors' meeting after the year-end closing of the books the board is presented with the financial statements starting with the balance sheets for the beginning and end of the year. The principle of the balance sheet is that the enterprise has a net value to the owners (net worth) equal to the value of what is owned (the assets) less the amount owed to others (the liabilities).

3 BALANCE SHEET

When any business starts, the first financial statement is the balance sheet. In the case of CCTCO, the company was started many years ago to exploit the newly patented product. The beginning balance sheet was the result of setting up the initial financing. To get the enterprise started the original owners determined that $1000 (represents one million dollars, since in all of the exhibits and tables the last three zeros are deleted) was needed. The inventor and friends and associates put up $600 as the owners share—600,000 shares of common stock at a par value of $1 per share. Others, familiar with the product and the originators of the business, provided $400 represented by notes to be paid in 20 years—long-term debt. The original balance sheet was as shown below:

Assets		Liabilities and Net Worth	
		Liabilities	
Cash	1000	Long-term debt	400
		Net worth	–0–
		Capital stock	600
Total assets	1000	Total liabilities and net worth	1000

The first financial steps of the company were to purchase equipment and machinery for $640 and raw materials for $120. The equipment was sent COD, but the raw material was to be paid for in 30 days. Immediately the balance sheet became more complex. There were now current assets—cash and inventory of raw materials—as well as fixed assets—machinery. Current liabilities showed up now in the form of accounts payable-the money owed for the raw material. All this before anything was produced. Now the balance sheet had become

Assets		Liabilities and Net Worth	
		Liabilities	
Cash	360	Accounts payable	120
Inventories	120	Current liabilities	120
Current assets	480	Long-term debt	400
Fixed assets	640	Total liabilities	560
		Net worth	
		Capital stock	600
Total assets	1120	Total liabilities and net worth	1120

After a number of years of manufacturing and selling product the balance sheet became as shown in Table 1. This important financial report requires explanation.

Assets are generally of three varieties:

- *Current.* Usually liquid and will probably be turned over at least once each year.
- *Fixed.* Usually real estate and the tools of production, frequently termed plant, property, and equipment.
- *Intangible.* Assets without an intrinsic value, such as good will or development costs which are not written off as a current expense but are declared an asset until the development has been commercialized.

3.1 Current Assets

In CCTCO's balance sheet the first item to occur is cash, which the company tries to keep relatively low, sufficient only to handle the flow of checks. Any excess over that amount the treasurer applies to pay off short-term debt, which has been arranged with local banks at one-half of one percent over the prime rate.

Accounts receivable are trade invoices not yet paid. The terms offered by CCTCO are typical—2% 10 days net 30, which means that if the bill is paid by the customer within 10

Table 1 Commercial Construction Tool Co., Inc.

Balance Sheet	Beginning
Assets	
Current assets	
Cash	52
Accounts receivable	475
Inventories	941
Total current assets	1468
Fixed assets	
Gross plant and equipment	2021
Less reserve for depreciation	471
Net plant and equipment	1550
Total assets	3018
Liabilities	
Current liabilities	
Accounts payable	457
Short-term debt	565
Long-term debt becoming current	130
Total current liabilities	1152
Long-term liabilities	
Interest-bearing debt	843
Total liabilities	1995
Net worth	
Capital stock	100
Earned surplus	923
Total net worth	1023
Total liabilities and net worth	3018

days after receipt, he or she can take a 2% discount, otherwise the total amount is due within 30 days. Distributors in the hardware field are usually hard pressed for cash and are frequently slow payers. As a result, receivables are the equivalent of two and a half month's sales, tying up a significant amount of the company's capital.

Inventories are the major element of current assets and consist of purchased raw materials, primarily steel, paint, and purchased parts; work in process, which includes all material that has left the raw material inventory point but has not yet reached the stage of completion where it is ready to be shipped; and finished goods. In order to provide quick delivery service to customers, CCTCO finds it necessary to maintain inventories at the equivalent of about three months' shipments—normally about 25% of the annual cost of goods sold.

3.2 Current Liabilities

Skipping to the liability section of the report, in order to look at all the elements of the liquid segment of the balance sheet, we next evaluate the condition of current liabilities. This section is composed of two parts: interest-bearing debt and debt that carries no interest charge. The noninterest-bearing part is primarily accounts payable, which is an account parallel but opposite to accounts receivable. It consists of the trade obligations not yet paid for steel, paint, and parts as well as office supplies and other material purchases. Sometimes included in this category are estimates of taxes that have been accrued during the period but not yet paid as well as other services used but not yet billed or paid for.

3.3 Accrual Accounting

At this point it is useful to define the term "accrued" or "accrual" as opposed to "cash" basis accounting. Almost all individual, personal accounting is done on a cash basis, that is, for individual tax accounting, no transaction exists until the money changes hands—by either writing a check or paying cash. In commercial and industrial accounting the accrual system is normally used, in which the transaction is deemed to occur at the time of some other overt act. For example, a sale takes place when the goods are shipped against a bona fide order, even though money will not change hands for another month. Taxes are charged based on the pro rata share for the year even though they may not be paid until the subsequent year. Thus costs and revenues are charged when it is clear that they are in fact obligated. This tends to anticipate and level out income and costs and to reduce the effect of fluctuations resulting only from the random effect of the time at which payments are made. Business managers wish to eliminate, as far as possible, wide swings in financial results and accrual accounting assists in this, as well as providing a more clearly cause-related set of financial statements. It also complicates the art of accounting quite considerably.

3.4 Interest-Bearing Current Liabilities

Interest-bearing current obligations are of two types: short-term bank borrowings and that portion of long-term debt that must be paid during the current year. Most businesses, and particularly those with a seasonal variation in sales, find it necessary to borrow from banks on a regular basis. The fashion clothing industry needs to produce three or four complete new lines each year and must borrow from the banks to provide the cash to pay for labor and materials to produce the fall, winter, and spring lines. When the shipments have been made to the distributors and large retail chains and their invoices have been paid, the manufacturer can "get out of the banks," only to come back to finance the next season's line. Because CCTCO's sales have a significant summer bulge at the retail level, they must have heavy inventories in the early spring, which drop to a fairly low level in the fall. Bank borrowings are usually required in February through May, but CCTCO is normally out of the banks by year end, so that the year-end balance sheet has a sounder look than it would have in April. The item "short-term debt" of $565 consists of bank loans that had not been paid back by the year's end.

The second part of interest-bearing current liabilities is that part of the long-term debt that matures within 12 months, and will have to be paid within the 12-month period. Such obligations are typically bonds or long-term notes. These current maturities represent an immediate drain on the cash of the business and are therefore classed as a current liability. As CCTCO has an important bond issue with maturities taking place uniformly over a long period, it has long-term debt maturing in practically every year.

3.5 Net Working Capital

The total of current assets less current liabilities is known as "net working capital." Although it is not usually defined in the balance sheet, it is important in the financial management of a business because it represents a large part of the capitalization of an enterprise and because, to some degree, it is controllable in the short run.

In times of high interest rates and cash shortages, companies tend to take immediate steps to collect their outstanding bills much more quickly. They will carefully "age" their receivables, which means that an analysis showing receivables ranked by the length of time they have been unpaid will be made and particular pressure will be brought to bear on those invoices that have been outstanding for a long time. On the other hand, steps will be taken to slow the payment of obligations; discounts may be passed up if the need for cash is sufficiently pressing and a general slowing of payments will occur.

Considerable pressure will be exerted to reduce inventories in the three major categories of raw material, work in process, and finished goods as well as stocks of supplies. Annual inventory turns can sometimes be significantly improved. There are, however, irreducible minimums for net working capital, and going beyond those points may result in real damage to the business through reducing service, increasing delivery times, damaging credit ratings, and otherwise upsetting customer and supplier relationships.

The effect of reducing net working capital, in a moderate and constructive way, spreads through the financial structure of the enterprise. The need for borrowing is reduced and interest expense is thereby reduced and profits are increased. Also, another effect on the balance sheet further improves the financial position. As the total debt level is reduced and the net worth is increased, the ratio of debt to equity is reduced, thus improving the financial community's assessment of strength. An improved rating for borrowing purposes may result, making the company eligible for lower interest rates. Other aspects of this factor will be covered in more detail in the discussion of net worth and long term debt.

3.6   Current Ratio

The need to maintain the strength of another important analysis ratio puts additional resistance against the objective of holding net working capital to the minimum. Business owners feel the need to maintain a healthy "current ratio." In order to be in a position to pay current bills, the aggregate of cash, receivables, and inventories must be available in sufficient amount. One measure of the ability to pay current obligations is the ratio of current assets to current liabilities, the current ratio. In more conservative times and before the days of leverage, a ratio of 2.0 or even 3.0 was considered strong, an indication of financial stability. In times of high interest rates and with objectives of rapid growth, much lower ratios are acceptable and even desirable. CCTCO's ratio of 1.27 ($1468/$1152) is considered quite satisfactory.

3.7   Fixed Assets

Continuing the evaluation on the asset side of the balance sheet we find the three elements of fixed assets, that is, gross plant and equipment, reserve for depreciation, and net plant and equipment. Gross plant is the original cost of all the assets now owned and is a straightforward item. The concept of depreciation is one which is frequently misunderstood and partly because of the name "reserve for depreciation." The name seems to indicate that there is a reserve of cash, put away somewhere that can be used to replace the old equipment when necessary. This is not the case. Accountants have a very special meaning for the word reserve in this application. It means, to an accountant, the sum of the depreciation expense that has been applied over the life, up to now, of the asset.

When an asset, such as a machine, is purchased, it is assigned an estimated useful life in years. In a linear depreciation system, the value of the asset is reduced each year by the same percentage that one year is to its useful life. For example, an asset with a 12-year useful life would have an 8.33% annual depreciation rate (100 times the reciprocal of 12).

The critical reason for reducing the value each year is to reduce the profit by an amount equivalent to the degree to which equipment is transformed into product. With high income taxes, the depreciation rate is critical to ensuring that taxes are held to the legal minimum. When the profit and loss statement is covered, the effect on profits and cash flow as a result of using nonlinear, accelerated depreciation rates will be covered. The important point to understand is that the reserve for depreciation does not represent a reserve of cash but only an accounting artifice to show how much depreciation expense has been taken (charged against profits) so far and, by difference, to show the amount of depreciation expense that may be taken in the future.

The difference between gross plant and reserve for depreciation, net plant and equipment, is not necessarily the remaining market value of the equipment at all, but is the amount of depreciation expense that may be charged against profits in future years. The understanding of this principle of depreciation is critical to the later understanding of profits and cash flow.

3.8 Total Capital

Together, the remaining items (long-term debt and net worth) on the liability side of the balance sheet make up the basic investment in the business. In the beginning, the entrepreneurs looked for money to get the business started. It came from two sources, equity investors and lenders. The equity investors were given an ownership share in the business, with the right to a portion of whatever profits might be made or a pro rata share of the proceeds of liquidation, if that became necessary. The lenders were given the right to regular and prescribed interest payments and were promised repayment of principal on a scheduled basis. They were not to share in the profits, if any. A third source of capital became available as the enterprise prospered. Profits not paid out in dividends were reinvested in plant and equipment and working capital. Each of these sources has an official name:

Lenders:	Long-term debt
Equity investors:	Capital stock
Profits reinvested:	Earned surplus

In many cases the cash from equity investors is divided into two parts, the par value of the common shares issued, traditionally $1 each, and the difference between par and the actual proceeds from the sale of stock. For example, the sale of 1000 shares of par value $1 stock, for $8000 net of fees, would be expressed

Capital stock (1000 shares at $1 par):	$1000
Paid in surplus:	$7000
Total capitalization:	$8000

The final item on the balance sheet, earned surplus or retained earnings, represents the accumulated profits generated by the business which have not been paid out, but were reinvested.

Net worth is the total of capital stock and earned surplus and can also be defined as the difference, at the end of an accounting period, between the value of the assets, as stated on the corporate books, and the obligations of the business.

All of this is a simplified view of the balance sheet. In actual practice there are a number of other elements that may exist and take on great importance. These include preferred stocks, treasury stock, deferred income taxes, and goodwill. When any of these special situations occur, a particular review of the specific case is needed in order to understand the implications to the business and their effect on the financial condition of the enterprise.

3.9 Second Year Comparison

The balance sheet in Table 1 is a statement of condition. It tells the financial position of the company at the beginning of the period. At the end of the year the Board of Directors is presented two balance sheets—the condition of the business at the beginning and at the end of the period, as shown in Table 2. The Board is interested in the trends represented by the change in the balance sheet over a 1-year period.

Total assets have increased by $395 over the period—probably a good sign. Net worth or owners' equity has increased by $27, which is $368 less than the increase in assets. The money for the increase in assets comes from creating substantially more liabilities or obligations as well as the very small increase in the net worth. A look at the liabilities shows the following (note the errors from rounding that result from the use of computer models for financial statements):

	Increase
Accounts payable	$ 46
Short-term debt	36
Long-term debt	286
	$368

Table 2 Commercial Construction Tool Co., Inc.—Costs and Revenues, Bad Year—Actual

Balance Sheet	Beginning	Ending	Change
Assets			
Current assets			
Cash	52	62	10
Accounts receivable	475	573	98
Inventories	941	1000	59
Total current assets	1468	1635	167
Fixed assets			
Gross plant and equipment	2021	2521	500
Less reserve for depreciation	471	744	273
Net plant and equipment	1550	1777	227
Total assets	3018	3413	395
Liabilities			
Current liabilities			
Accounts payable	457	503	46
Short-term debt	565	600	36
Long-term debt becoming current	130	130	0
Total current liabilities	1152	1233	82
Long-term liabilities			
Interest-bearing debt	843	1129	286
Total liabilities	1995	2362	368
Net worth			
Capital stock	100	100	0
Earned surplus	923	950	27
Total net worth	1023	1050	27
Total liabilities and net worth	3018	3413	395

Changes in net working capital are evaluated to determine the efficiency in the use of cash and the soundness of the short-term position. No large changes that would raise significant questions have taken place. Current assets increased $167 and current liabilities by $82. These increases result from the fact that sales had increased, which had required higher inventories and receivables. The current ratio (current assets over current liabilities) had strengthened to 1.33 from 1.27 at the beginning of the period, indicating an improved ability to pay bills and probably increased borrowing power.

A major change in the left-hand (asset) side was the increase in fixed assets. Gross plant was up $500, nearly 25%, indicating an aggressive expansion or improvement program.

Net worth and earned surplus were up by $27, an important fact, sure to receive attention from the board.

To understand why the balance sheet had changed and to further evaluate the year's results, the directors needed a profit and loss statement and a cash flow statement.

4 PROFIT AND LOSS STATEMENT

The profit and loss statement (P&L) is probably the best understood and most used statement provided by accountants: It summarizes most of the important annual operating data and it acts as a bridge from one balance sheet to the next. It is a summary of transactions for the year—where the money came from and where most of it went. Table 3 is the P&L for CCTCO for the year.

For the sake of simplicity, net sales are shown as Sales. In many statements, particularly internal reports, gross sales are shown followed by returns and discounts to give a net sales figure. Cost of sales is a little more complex. Sales may be made from inventory or off the production line on special order. Stocks of finished goods or inventories are carried on the books at their cost of production. The formula for determining the cost of product shipped to customers is

Beginning inventory + Cost of production − Ending inventory = Cost of sales

Table 3 Commercial Construction Tool Co., Inc.—
Costs and Revenues, Bad Year—Actual

Profit and Loss Statement	($000)
Sales	4772
Cost of production	4097
Beginning inventory	941
Ending inventory	1000
Net change	59
Cost of sales	4038
Gross margin	734
Selling expense	177
Administrative	249
Operating profit	308
Interest	169
Profit before tax	138
Income tax	66
Net income	72

Additionally, CCTCO uses a standard burden rate system of applying overhead costs to production. The difference between the overhead charged to production at standard burden rates and the actual overhead costs for the period, in this case $62, is called unabsorbed burden and is added to the cost of production for the year, or it may be charged off as a period cost. The procedures for developing burden rates will be treated in more detail in a subsequent section.

Gross margin is the difference between sales dollars and the cost of manufacture. After deducting the costs of administrative overhead and selling expense, operating profit remains. Interest expense is part of the total cost of capital of the business and is therefore separate from operations. The last item, income tax, only occurs when there is a profit.

4.1 Financial Ratios

The combination of the P&L and the balance sheet makes it possible to calculate certain ratios that have great significance to investors. The ratios are shown in Table 4. The first and most commonly used as a measure of success is the return on sales. This is a valuable ratio to measure progress of a company from year to year, but is of less importance in comparing one company to another. A more useful ratio would be returns to value added. Value added is the difference between the cost of purchased raw materials and net sales, and represents the economic contribution of the enterprise. It is a concept used more extensively in Europe than the United States and is the basis of the Value Added Tax (VAT), quite common in Europe and at this writing being considered in the United States.

Return on assets begins to get closer to the real interest of the investor. It represents the degree to which assets are profitable, and would indicate, from an overall economic point of view, whether the enterprise was an economic and competitive application of production facilities.

A ratio even more interesting to the investor is the return on invested capital. Total assets, as was described earlier, are financed by three sources:

- Equity—made up of stock, that is, owners' investment and profits retained in the business
- Interest-bearing debt—composed of bonds, notes, and bank loans
- Current liabilities—composed of operating debts such as accounts payable and taxes payable, which do not require interest payments

Table 4 Commercial Construction Tool Co., Inc.—
Costs and Revenues, Bad Year—Actual

Financial Ratios	
Return on sales	1.51
Return on assets	2.24
Return on invested capital	3.56
Return on equity	6.94
Asset to sales ratio	0.67
Debt percent to debt plus equity	69
Average cost of capital	20.67

Because the current liabilities are normally more than offset by current assets, the economic return is well described by the return on total or invested capital, which is net profit after taxes divided by the sum of equity plus interest-bearing debt.

A rate of return percentage of great interest to the owner is the return on equity. This rate of return compared to the return on total capital represents the degree to which the investment is or can be leveraged. It is to the interest of the investor to maximize the return on his or her dollars invested, so, to the degree that money can be borrowed at interest rates well below the capacity of the business to provide a return, the total profits to the owners will increase. Return on equity is a function of the ratio of debt to debt plus equity (total capital) and is a measure of the leverage percentage in the business. It is to the advantage of the owners to increase this ratio in order to increase the return on equity up to the point that the investment community, including bankers, concludes that the company is excessively leveraged and is in unsound financial condition. At that point it becomes more difficult to borrow money and interest rates of willing lenders increase significantly. Fashions in leverage change depending on the business cycle. In boom times with low interest rates, highly leveraged enterprises are popular, but tend to fall into disfavor when times are tough.

A more direct measure of leverage is "debt percent to debt plus equity" or debt to total capital. The 69% for CCTCO indicates that lenders really "own" 69% of the company and investors only 31%.

Another ratio of interest to investors is the asset turnover or asset to sales ratio. If sales from a given asset base can be very high, the opportunity to achieve high profits appears enhanced. On the other hand, it is very difficult to change the asset to sales ratio very much without changing the basic business. Certain industries or businesses are characterized as being capital intensive, which means they have a high asset to sales ratio or a low asset turnover. It is fundamental to the integrated forest products industries that they have a high asset to sales ratio, typically one to one. The opposite extreme, for example, the bakery industry, may have a ratio of 0.3–0.35 and turn over assets about three times per year. Good management and very effective use of facilities coupled with low inventories can make the best industry performer 10% better than the average, but there is no conceivable way that the fundamental level can be dramatically and permanently changed.

The final figure in Table 4, that of average cost of capital, cannot be calculated from only the P&L and balance sheet. One component of the total cost of capital is the dividend payout, which is not included in either report. It was stated previously that the P&L shows where most of the money went—it does not include dividends and payments for new equipment and other capital goods. For this we need the cash flow, also known as the source and application of funds, shown in Table 5.

5 CASH FLOW OR SOURCE AND APPLICATION OF FUNDS

There are two sources of operating cash for any business: the net profits after tax and noncash expenses. In Table 5, the cash generated by the business is shown as $344, the sum of net profit and depreciation. This is actually the operating cash generated and does not include financing cash sources, which are also very important. These sources include loans, capital contributions, and the sale of stock and are included in the cash flow statement as well as in the balance sheet where they have already been reviewed in a previous section of this chapter.

It seems clear and not requiring further explanation that the net profit after tax represents money remaining at the end of the period, but the treatment of noncash expenses as a source

Table 5 Commercial Construction Tool Co., Inc.—Costs and Revenues, Bad Year—Actual

Source and Application of Funds	($000)
Net profit after tax	72
Depreciation expense	273
Cash generated	344
Increase in net working capital	
Change in cash	10
Change in receivables	98
Change in inventories	59
Change in payables	−46
Net change	121
Capital expenditures	500
Operating cash requirements	621
Operating cash flow	−276
Dividends	45
Net cash needs	−321
Increase in debt	321

of operating funds is less self-evident. Included in the cost of production and sales in the previous section were materials and labor and many indirect expenses such as rent and depreciation, which were included in the P&L in order to achieve two objectives:

- Do not overstate annual earnings.
- Do not pay more income taxes than the law requires.

In the section on fixed assets, when discussing the balance sheet, it was pointed out that the reserve for depreciation is not an amount of money set aside and available for spending. It is the total of the depreciation expense charged so far against a still existing asset. The example was a piece of equipment with a useful life of 12 years, the total value of which was reduced by 8.33% (the reciprocal of 12 times 100) each year. This accounting action is taken to reduce profits to a level that takes into consideration the decreasing value of equipment over time, to reduce taxes, and to avoid overstating the value of assets. Depreciation expense is not a cash expense—no check is written—it is an accounting convention. The cash profit to the business is therefore understated in the P&L statement because less money was spent for expenses than indicated. The overstatement is the amount of depreciation and other noncash expenses included in costs for the year.

In the P&L in Table 4, included in the cost of sales of $4038, is $273 of depreciation expense. If this noncash item were not included as an expense of doing business, profit before tax would be increased from $138 to $411. Taxes were calculated at a 48% rate, so the revised net profit after tax would be $214. This new net profit would also be cash generated from operations instead of the $344 actually generated ($72 profit plus $273 depreciation) when noncash expenses are included as costs. The reduction in cash available to the business resulting from ignoring depreciation is exactly equal to the increase in taxes paid on profits. The anomaly is that the business has more money left at the end of the year when profits are lower!

5.1 Accelerated Depreciation

This is a logical place to examine various kinds of depreciation systems. So far, only a straight-line approach has been considered—the example used was a 12-year life resulting in an 8.33% annual expense or writedown rate. Philosophical arguments have been developed to support a larger writedown in the early years and reducing the depreciation rate in later years. Some of the reasons advanced include

- A large loss in value is suffered when a machine becomes second hand.
- The usefulness and productivity of a machine is greater in the early years.
- Maintenance and repair costs of older machines are larger.
- The value of older machines does not change much from one year to the next.

The reason that accelerated systems have come into wide use is more practical than philosophical. With faster, early writedowns the business reduces its taxes now and defers them to a later date. Profits are reduced in the early years but cash flow is improved. There are two common methods of accelerating depreciation in the early years of a machine's life:

- Sum of the digits
- Double declining balance

Table 6 compares the annual depreciation expense for the two accelerated systems to the straightline approach. For these examples a salvage value of zero is assumed at the end of the period of useful life. At the time of asset retirement and sale, a capital gain or loss would be realized as compared to the residual, undepreciated value of the asset, or zero, if fully depreciated.

The methods of calculation are represented by the following equations and examples where:

N = number of years of useful life
A = year for which depreciation is calculated
P = original price of the asset
D_a = depreciation in year A
B = book value at year end

The equation for straight-line depreciation is

$$D_a = \frac{1}{n} \times P$$

$$B = P - (D_1 + D_2 + \cdots + D_a)$$

In the example with an asset costing $40,000 with an 8-year useful life:

$$D_a = \tfrac{1}{8} \times 40{,}000 = 0.125 \times 40{,}000 = 5000$$

To calculate depreciation by the sum of the years' digits method, use

$$D_a = [(N + 1 - A)/(N + N - 1 + N - 2 + \cdots + 1)] \times P$$

For the third year, for example,

$$D_3 = [(8 + 1 - 3)/(8 \times 7 + 6 + 5 + 4 + 3 + 2 + 1)] \times 40{,}000$$
$$D_3 = [(6)/(36)] \times 40{,}000 = 0.1667 \times 40{,}000 = 6667$$

Table 6 Accelerated Depreciation Methods[a]

Straight-Line Method[b]

Year	Rate	Depreciation Expense	Book Value, Year End
1	0.125	5000	35000
2	0.125	5000	30000
3	0.125	5000	25000
4	0.125	5000	20000
5	0.125	5000	15000
6	0.125	5000	10000
7	0.125	5000	5000
8	0.125	5000	0

Sum of the Years' Digits Method[c]

Year	Rate	Depreciation Expense	Book Value, Year End
1	0.2222222	8889	31111
2	0.1944444	7778	23333
3	0.1666667	6667	16667
4	0.1388889	5556	11111
5	0.1111111	4444	6667
6	0.0833333	3333	3333
7	0.0555556	2222	1111
8	0.0277778	1111	0

Double the Declining Balance Method[d]

Year	Rate	Depreciation Expense	Book Value, Year End
1	0.25	10000	30000
2	0.1875	7500	22500
3	0.140625	5625	16875
4	0.1054688	4219	12656
5	0.0791016	3164	9492
6	0.0593262	2373	7119
7	0.0444946	1780	5339
8	0.1334839	5339	0

[a] Basic assumptions: equipment life, 8 years; original price, $40,000; estimated salvage value, $0.
[b] Annual rate equation: one divided by the number of years times the original price.
[c] Annual rate equation: sum of the number of years divided into the years of life remaining.
[d] Annual rate equation: twice the straight-line rate times the book value at the end of the preceding year.

The depreciation rates shown in Table 6 under the double declining balance method are calculated to show a comparison of write-off rates between systems. The actual calculations are done quite differently:

$$D_a = \frac{2}{n} \times B_{a-1}$$

$$B_a = P - (D_1 + D_2 + \cdots + D_{a-1})$$

In the third year, then,

$$D_3 = \tfrac{2}{8} \times 22{,}500 = 5625$$

and

$$B_3 = 40{,}000 - (10{,}000 + 7500 + 5625) = 16{,}875$$

Note in Table 6 that the double declining balance method, as should be expected, if allowed to continue forever, never succeeds in writing off the entire value. The residue is completely written off in the final year of the asset's life. The sum of the years' digits is a straight line and provides for a full write-off at the end of the period.

Figure 1 depicts, graphically, the annual depreciation expense using the three methods.

In many cases, a company will succeed in attaining both the advantages to cash flow and tax minimization of accelerated depreciation as well as the maximizing of earnings by using straight-line depreciation. This is done by having one set of books for the tax collector and another for the shareholders and the investing public. This practice is an accepted approach and, where followed, is explained in the fine print of the annual report.

A number of special depreciation provisions and investment tax credit arrangements are available to companies from time to time. The provisions change as tax laws are revised either to encourage investment and growth or to plug tax loopholes, depending on which is politically popular at the time. The preceding explains the theory—applications vary considerably with changes in the law and differences in corporate objectives and philosophy.

The cash generated by the business has, as its first use, the satisfaction of the needs for working capital, that is, the needs for funds to finance increases in inventories, receivables,

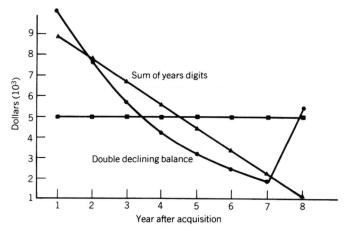

Figure 1 Comparison of depreciation methods. ($40,000 original price; 8-year life; and no residual salvage value.)

and cash in the bank. Each of these assets requires cash in order to provide them. Offsetting these uses of cash are the changes that may take place in the short-term debts of the enterprise and accounts payable. In Table 5, we see that $121 is required in increased net working capital, essentially all of which goes to provide for increased inventories and receivables needed to support sales increases.

The largest requirement for cash is the next item, that of capital expenditures, which has consumed $500 of the cash provided to the business. The total needs of the company for cash—the operating cash requirements—have risen to $621 compared to the cash generated of $344, and that is not the end of cash needs. The shareholders have become accustomed to a return on their investment—an annual cash dividend. The dividend is not considered part of operating cash flow nor is it a tax deductible expense as interest payments are. The dividend, added to the net operating cash flow of −$276, results in a borrowing requirement for the year of $321.

To summarize, the Board of Directors has been furnished a set of operating statements and financial ratios as shown in Tables 2–5. These ratios show a superficial picture of the economics of the enterprise from a financial viewpoint and present some issues and problems to the directors. The condition of the ratios and rates of return for CCTCO are of great concern to the directors and lead to some hard questions for management.

Why, when the total cost of capital, that is, interest plus dividends as a percentage of debt plus equity is 20.67%, is the return on total capital only 3.56%? Why does it take $0.67 worth of assets to provide $1 worth of sales in a year? Why is it that profit after tax is only 1.51% of sales? The board will not be pleased with performance and will want to know what can and will be done to improve. The banks will perhaps have concerns about further loans and shareholders or prospective shareholders will wonder about the price of the stock.

The answers to these questions require a level of cost and revenue information normally supplied to top management.

6 EVALUATING RESULTS AND TAKING ACTION

Corporate chief executives who allowed themselves to be as badly surprised by poor results at the end of the year as the chief executive of CCTCO would be unlikely to last long enough to take corrective action. However, the results at CCTCO can provide clear examples of the usefulness of accounting records in determining the cause of business problems and in pointing in the direction of practical solutions.

6.1 Comparing Current Results with Budgets and Forecasts

The first step of the chief executive at CCTCO was to compare actual results with those projected for the year. It had been the practice at CCTCO to prepare a comprehensive business plan and budget at the beginning of each year. Monthly and yearly, reports comparing actual with budget were made available to top officers of the company. Tables 7–10 show a comparison of the budgeted P&L, performance ratios, balance sheet, and cash flow for the year compared to the actual performance already reviewed by the board.

An examination of the budget/actual comparisons revealed many serious deviations from plan. Net worth and long-term debt were trouble spots. Profits were far from expected results, and cash flow was far below plan.

The president searched the reports for the underlying causes in order to focus his attention and questions on those corporate functions and executives that appeared to be responsible for the failures. He concluded that there were seven critical variances from the budget, which when understood, should eventually lead to the underlying real causes. They included

Table 7 Commercial Construction Tool Co., Inc.—P&L Statement ($000)

	Budget	Actual	Variance	Percent
Sales	5261	4772	−489	−9.29
Cost of production	3972	4097	−1	−0.03
Beginning inventory	941	941	0	0.00
Ending inventory	1007	1000	−7	−0.68
Net change	66	59	−7	−10.36
Cost of sales	3906	4038	−132	−3.37
Gross margin	1355	734	−621	−45.81
Selling expense	160	177	−17	−10.90
Administrative	231	249	−18	−7.88
Operating profit	964	308	−656	−68.08
Interest	154	169	−15	−9.93
Profit before tax	810	138	−671	−82.92
Income tax	389	66	322	82.92
Net income	421	72	−349	−82.92

Element	Variance	Percent
Sales	−489	−9.29
Cost of sales	−132	−3.37
Selling expense	−17	−10.90
Administrative expense	−18	−7.88
Interest	−15	−9.93
Net working capital	−68	−14.50

The president asked the VP Sales and the VP Manufacturing to report to him as to what had happened to cause these variances from plan and what corrective action could be taken. He instructed the Controller to provide all the cost and revenue analyses needed to arrive at answers.

Table 8 Commercial Construction Tool Co., Inc.—Financial Ratios

	Budget	Actual	Variance
Return on sales	8.00	1.51	−6.50
Return on assets	12.98	2.24	−10.74
Return on invested capital	20.47	3.56	−16.92
Return on equity	34.76	6.94	−27.82
Asset to sales ratio	0.62	0.67	−0.06
Debt percent to debt plus equity	60	69	−9.53
Average cost of capital	16.43	20.67	−4.24

Table 9 Commercial Construction Tool Co., Inc.—Variance Analysis, Balance Sheet ($000)

	Budget	Actual	Variance	Percent
Assets				
Current assets				
Cash	56	62	6	10.78
Accounts receivable	631	573	−59	−9.29
Inventories	1007	1000	−7	−0.68
Total current assets	1695	1635	−59	−3.51
Fixed assets				
Gross plant and equipment	2521	2521	0	0.00
Less reserve for depreciation	744	744	0	0.00
Net plant and equipment	1777	1777	0	0.00
Total assets	3472	3413	−59	−1.71
Liabilities				
Current liabilities				
Accounts payable	491	503	13	2.59
Short-term debt	604	600	−4	−0.68
Long-term debt becoming current	130	130	0	0.00
Total current liabilities	1225	1233	9	0.70
Long-term liabilities				
Interest-bearing debt	848	1129	281	33.17
Total liabilities	2073	2362	290	13.98
Net worth				
Capital stock	100	100	0	0.00
Earned surplus	1300	950	−349	−26.87
Total net worth	1400	1050	−349	−24.95
Total liabilities and net worth	3472	3413	−59	−1.71

In two weeks the three executives made a presentation to the president that provided a comprehensive understanding of the problems, recommended solutions to them, and a timetable to implement the program. The following is a summary of that report.

6.2 Identifying Problems and Solutions

Causes of Last Year's Results

The poor operating results of last year are caused almost entirely by a change in product mix from the previous year and not contemplated in the budget established 15 months ago. The introduction of the HOMMODEL nearly two years ago resulted in very few sales in the early months following its initial availability. However, early last year, sales accelerated dramatically, caught up with, and passed those of the COMMODEL. For a number of reasons this has had a poor effect on the financial structure of our company:

Table 10 Commercial Construction Tool Co., Inc.—Variance Analysis, Source and Application of Funds ($000)

	Budget	Actual	Variance	Percent
Net profit after tax	419	72	−347	−82.84
Depreciation expense	273	273	0	0.00
Cash generated	692	344	−347	−50.20
Increase in net working capital				
Change in cash	4	10	6	150.90
Change in receivables	156	98	−59	−37.53
Change in inventories	66	59	−7	−10.36
Change in payables	−34	−46	−13	37.85
Net change	193	121	−72	−37.40
Capital expenditures	500	500	0	0.00
Operating cash requirements	693	621	−72	−10.42
Operating cash flow	−1	−276	−275	27500.00
Dividends	45	45	0	0.00
Net cash needs	−46	−321	−275	593.31
Increase in debt	46	321	275	593.31

- Lack of experience on the new product has resulted in costs higher than standard.
- Standard margins are lower for the HOMMODEL.
- Travel and communications costs were high because of the new product introduction.
- Prices on the HOMMODEL were lower than standard because of special introductory dealer discounts and deals.
- Receivables increased because of providing initial stocking plans for new dealers handling the HOMMODEL.
- Higher interest expense resulted from higher debt—a direct result of cash flow shortfall.

The only significant variance unrelated to the new product was the fact that factory and office rents were raised during the year.

The following product mix table (Table 11) summarizes a number of accounting documents and shows the effect of product mix on profits.

Recommended Corrective Action

As the major problems are caused by the new product cannibalizing sales of the old COMMODEL, action is directed toward increasing margins on the HOMMODEL to nearly that of the COMMODEL and increasing the proportion of sales of the latter. This will be accomplished by simultaneously reducing unit cost and increasing selling price of the new product. The following program will be undertaken:

- Increase the unit price to 3.52 and eliminate deals and promotion pricing for a margin improvement of 0.34.
- Productivity improvements realized in the last two months of the year will reduce costs by 0.15 for the year.
- Proposed changes in material and finish will further reduce costs by 0.032.

Table 11 Product Line Comparison: Unit Volume, Price, and Costs

	Budget	Actual	Variance
Commodel			
Sales (1000s)	740	530	(210)
Unit price	4.203	4.280	0.077
Unit cost	3.0612	3.139	0.078
Unit margin	1.142	1.141	−0.001
Sales $	$3,110,220	$2,268,400	(841,820)
Cost $	$2,265,140	$1,663,670	($16,380)
Margin $	$845,080	$604,730	(240,350)
HOMMODEL			
Sales (1000s)	670	830	160
Unit price	3.210	3.016	(0.194)
Unit cost	2.449	2.932	(0.483)
Unit margin	0.761	0.084	(0.677)
Sales $	$2,150,700	$2,503,280	352,580
Cost $	$1,640,830	$2,433,560	792,730
Margin $	$509,870	$69,720	(440,150)
Total			
Sales $	$5,260,920	$4,771,680	(489,240)
Cost $	$3,905,970	$4,097,230	$776,350
Margin $	$1,354,950	$674,450	(680,500)
Selling expense	$160,000	$177,000	17,000
Administrative expense	$231,000	$249,000	18,000
Operating profit	$963,950	$248,450	($715,500)

These changes in price and cost will bring the standard margin of the HOMMODEL to 1.21, slightly more than that of the COMMODEL, thus eliminating any unfavorable effect of cannibalizing.

This report enabled the president to assure the board that the recommended steps would be taken and the year to come would provide better results.

6.3 Initiating Action

Following Board approval, the president asked the manufacturing manager, in conjunction with marketing, to prepare a five-year projection of operating results. The projection, as shown in Table 12, was prepared in a personal computer spreadsheet by the manufacturing manager and showed an increase in operating profit to just over $1,000,000 by the end of the five-year period.

The manufacturing manager was able to demonstrate the logic of his conclusions by showing the economic and operating assumptions on which the projections were based, as shown below:

Concerning the COMMODEL:

1. Unit sales will increase 1.5% annually.
2. Unit prices will increase at 2.5% annually, 0.5% less than the expected inflation rate of 3.0%.
3. Unit costs will increase at the same rate as prices.

Table 12 Product Line Comparison: Unit Volume, Price, and Costs

	Projections				
	Year 1	Year 2	Year 3	Year 4	Year 5
COMMODEL					
Sales (1000s)	700	717	735	754	773
Unit price	4.280	4.387	4.497	4.609	4.724
Unit cost	3.139	3.217	3.298	3.380	3.465
Unit margin	1.141	1.170	1.199	1.229	1.259
Sales $	$2,996,000	$3,070,900	$3,147,672	$3,226,364	$3,307,023
Cost $	$2,197,300	$2,252,233	$2,308,538	$2,366,252	$2,425,408
Margin $	$798,700	$818,668	$839,134	$860,113	$881,615
HOMMODEL					
Sales (1000s)	650	666	683	700	717
Unit price	3.520	3.608	3.698	3.791	3.885
Unit cost	2.750	2.819	2.889	2.961	3.035
Unit margin	0.770	0.789	0.809	0.829	0.850
Sales $	$2,288,000	$2,403,830	$2,525,524	$2,653,379	$2,787,706
Cost $	$1,787,500	$1,877,992	$1,973,066	$2,072,952	$2,177,895
Margin $	$500,500	$525,838	$552,458	$580,427	$609,811
Total					
Sales $	$5,284,000	$5,474,730	$5,673,196	$5,879,743	$6,094,729
Cost $	$3,984,800	$4,130,225	$4,281,604	$4,439,204	$4,603,303
Margin $	$1,299,200	$1,344,505	$1,391,593	$1,440,539	$1,491,426
Selling expense	$177,000	$182,310	$187,779	$193,413	$199,215
Administrative expense	$249,000	$256,470	$264,164	$272,089	$280,252
Operating profit	$873,200	$905,725	$939,649	$975,037	$1,011,959

Concerning the HOMMODEL:

1. Unit sales will increase at 4.0% annually.
2. Unit prices will increase at the same rate as for the COMMODEL, 2.5% annually.
3. Unit costs will increase at the same rate as prices.

Concerning expenses, both selling and administrative expenses will increase at 3.0% annually.

Using his model, the manufacturing manager was able to demonstrate to the board the reasonableness and the sensitivity of his projections. The cell formulas used in the spreadsheet are shown in Table 13.

7 FINANCIAL TOOLS FOR THE INDEPENDENT PROFESSIONAL ENGINEER

In the 1990s and for some years prior to that time, it became common for engineers to become independent consultants or "free lances." This was partly brought about by corporate downsizing and the tendency of companies to bring in part-time technical assistance for specific projects rather than to develop an in-house capability that was not needed at all times. One of the implications of this development is that the engineer needs to be able to account for his own expenses and income as a "business." This accounting must satisfy the requirement of the U.S. Internal Revenue Service and records need to be adequate to con-

Table 13 Cell Formulas Used for Projections

	Product Line Comparison: Unit Volume, Price, and Costs		
	Year 1	Year 2	Formula
COMMODEL			
Sales (1000s)	700	710	+C7*1.015
Unit price	4.280	4.387	+C8*1.025
Unit cost	3.139	3.217	+C9*1.025
Unit margin	1.141	1.170	+D8-D9
Sales $	$2,996,000	$3,116,963	+D7*D8*1000
Cost $	$2,197,300	$2,286,016	+D7*D9*1000
Margin $	$798,700	$830,948	+D11-D9*D7*1000
HOMMODEL			
Sales (1000s)	650	676	+C15*1.04
Unit price	3.520	3.608	+C16*1.025
Unit cost	2.750	2.819	+C17*1.025
Unit margin	0.770	0.789	+D16-D17
Sales $	$2,288,000	$2,439,008	+D15*D16*1000
Cost $	$1,787,500	$1,905,475	+D15*D17*1000
Margin $	$500,500	$533,533	+D19-D17*D15*1000
Total			
Sales $	$5,284,000	$5,555,971	+D19+D11
Cost $	$3,984,800	$4,191,491	+D20+D12
Margin $	$1,299,200	$1,364,481	+D21+D13
Selling expense	$177,000	$182,310	+C27*1.03
Administrative expense	$249,000	$256,470	+C28*1.03
Operating profit	$873,200	$925,701	+D25-D27-D28

In a like manner, relationship and cell formulae can be developed for year-by-year balance sheets and cash flows.

vince the IRS that tax submissions are accurate, that they satisfy the tax law, and that there is no fraud or indication of deception.

7.1 Simple Record-Keeping

With present home and business accounting software for the personal computer, the keeping of basic records can be made accurate, simple, and convincing to an IRS investigator and to the engineer's accountant.

The records of a private engineering practice should be *cash* rather than accrual and therefore can be based on bank and credit card transactions. Small cash transactions can be handled through a petty cash account that is replenished by check and that contains a journal of expenditures. A personal computer system can be set up that will automatically categorize each check written and even split a check into a number of categories, when necessary.

At the time of the publication of this edition the most popular program for personal finance was *QUICKEN*, but others are available and some banks will provide software and on-line access to a checking account. These systems make it possible to group and print out with full back-up and audit trail capability so that full quick disclosure is constantly available in a format that makes IRS audits become a matter solely of interpreting the law rather than tracking obscure expenditures or elements of income.

7.2 Getting the System Started

The first step should be to select an accountant. Although it is possible to maintain all needed records and prepare tax returns with computer software, the use of an accountant will probably save taxes through his knowledge of the law and is, for most engineers, essential. Following are some of the early decisions that should made with the accountant:

- Incorporation or not
- Computer needs
- Software needs
- Definition of categories or accounts
- Setting up bank accounts
- Level of accountant involvement

7.3 Operating the System

The basic approach to relatively painless small business accounting is that when a check is written or a deposit made, the transaction is entered in the computer at the time of the transaction and never again! As bills become due, the check is entered in and printed by the software or base; the funds are even transferred to the payees by the software. From that base, transaction lists, tabulations, and groupings are all done without writing or performing manual arithmetic. Cross columns *always* balance.

At the end of the fiscal year, the data can be transferred into a tax preparation program that will sort data and calculate the tax. At that time, the data can be transmitted to the accountant with a detailed, by category, listing of each transaction in hard copy or machine language or both. The accountant has very little number-crunching to do and accounting fees are minimal.

In the past, the problems of accounting for a business were a significant deterrent to freelancing. Sound, simple computer approaches eliminate that part of the terror of being on your own.

8 CONCLUSIONS

This chapter is intended to portray the principles of financial reporting without describing the underlying cost accounting systems needed to manage a business. These become so complex and are so varied that they are beyond the scope of this work.

The capacity to understand the meaning of financial reports and to make time projections based on historical reports coupled with sound assumptions for the future is frequently important to the engineer. Additionally, the ability to devise and administer a simple accounting system used to manage an engineering practice is, especially today, a useful skill.

The section is designed to provide a basis in these capabilities.

CHAPTER 16
DETAILED COST ESTIMATING

Rodney D. Stewart
Mobile Data Services
Huntsville, Alabama

1	THE ANATOMY OF A DETAILED ESTIMATE	532	9	TREATMENT OF RECURRING AND NONRECURRING ACTIVITIES	544
	1.1 Time, Skills, and Labor-Hours Required to Prepare an Estimate	533	10	WORK BREAKDOWN STRUCTURE INTERRELATIONSHIPS	544
2	DISCUSSION OF TYPES OF COSTS	535		10.1 Skill Matrix in a Work Breakdown Structure	545
	2.1 Initial Acquisition Costs	535		10.2 Organizational Relationships to a Work Breakdown Structure	545
	2.2 Fixed and Variable Costs	536			
	2.3 Recurring and Nonrecurring Costs	536	11	METHODS USED WITHIN THE DETAILED ESTIMATING PROCESS	545
	2.4 Direct and Indirect Costs	536		11.1 Detailed Resource Estimating	546
3	COLLECTING THE INGREDIENTS OF THE ESTIMATE	536		11.2 Direct Estimating	546
	3.1 Labor-Hours	536		11.3 Estimating by Analogy (Rules of Thumb)	546
	3.2 Materials and Subcontracts	538		11.4 Firm Quotes	546
	3.3 Labor Rates and Factors	538		11.5 Handbook Estimating	547
	3.4 Indirect Costs, Burden, and Overhead	538		11.6 The Learning Curve	547
	3.5 General and Administrative Costs	538		11.7 Labor-Loading Methods	548
	3.6 Fee, Profit, or Earnings	538		11.8 Statistical and Parametric Estimating as Inputs to Detailed Estimating	548
	3.7 Assembly of the Ingredients	539			
4	THE FIRST QUESTIONS TO ASK (AND WHY)	539	12	DEVELOPING A SCHEDULE	552
	4.1 What Is It?	540	13	TECHNIQUES USED IN SCHEDULE PLANNING	552
	4.2 What Does It Look Like?	540			
	4.3 When Is It to Be Available?	540	14	ESTIMATING ENGINEERING ACTIVITIES	552
	4.4 Who Will Do It?	540		14.1 Engineering Skill Levels	552
	4.5 Where Will It Be Done?	541		14.2 Design	553
5	THE ESTIMATE SKELETON: THE WORK BREAKDOWN STRUCTURE	541		14.3 Analysis	553
				14.4 Drafting	553
6	THE HIERARCHICAL RELATIONSHIP OF A DETAILED WORK BREAKDOWN STRUCTURE	541	15	MANUFACTURING/ PRODUCTION ENGINEERING	554
				15.1 Engineering Documentation	555
7	FUNCTIONAL ELEMENTS DESCRIBED	542	16	ESTIMATING MANUFACTURING/ PRODUCTION AND ASSEMBLY ACTIVITIES	555
8	PHYSICAL ELEMENTS DESCRIBED	543			

532 Detailed Cost Estimating

17	MANUFACTURING ACTIVITIES	556	21.5	Engineering Change Allowance	561
18	IN-PROCESS INSPECTION	558	21.6	Engineering Prototype Allowance	562
19	TESTING	558	21.7	Design Growth Allowance	562
20	COMPUTER SOFTWARE COST ESTIMATING	559	21.8	Cost Growth Allowance	562
21	LABOR ALLOWANCES	560	22	ESTIMATING SUPERVISION, DIRECT MANAGEMENT, AND OTHER DIRECT CHARGES	562
	21.1 Variance from Measured Labor-Hours	561	23	THE USE OF "FACTORS" IN DETAILED ESTIMATING	563
	21.2 Personal, Fatigue, and Delay (PFD) Time	561	24	CONCLUDING REMARKS	563
	21.3 Tooling and Equipment Maintenance	561			
	21.4 Normal Rework and Repair	561			

1 THE ANATOMY OF A DETAILED ESTIMATE

The detailed cost estimating process, like the manufacture of a product, is comprised of parallel and sequential steps that flow together and interact to culminate in a completed estimate. Figure 1 shows the anatomy of a detailed estimate. This figure depicts graphically how the various cost estimate ingredients are synthesized from the basic man-hour estimates and material quantity estimates. Man-hour estimates of each basic skill required to accomplish the job are combined with the labor rates for these basic skills to derive labor-dollar estimates. In the meantime, material quantities are estimated in terms of the units by which they are measured or purchased, and these material quantities are combined with their costs per unit to develop detailed direct material dollar estimates. Labor overhead or burden is

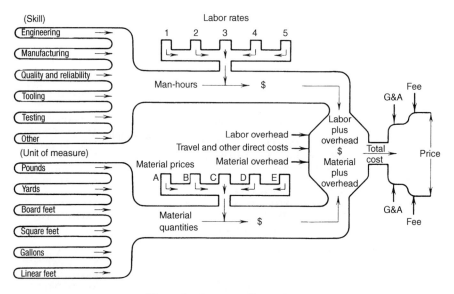

Figure 1 Anatomy of an estimate.

applied to direct material costs. Then travel costs and other direct costs are added to produce total costs; general and administrative expenses and fee or profit are added to derive the "price" of the final estimate.

The labor rates applied to the basic man-hour estimates are usually "composite" labor rates; that is, they represent an average of the rates within a given skill category. For example, the engineering skill may include draftsmen, designers, engineering assistants, junior engineers, engineers, and senior engineers. The number and titles of engineering skills vary widely from company to company, but the use of a composite labor rate for the engineering skill category is common practice. The composite labor rate is derived by multiplying the labor rate for each skill by the percentage of man-hours of that skill required to do a given task and adding the results. For example, if each of the six skills have the following labor rates and percentages, the composite labor rate is computed as follows:

Skill	Labor Rate ($/h)	Percentage in the Task
Draftsman	12.00	7
Designer	16.00	3
Engineering assistant	20.00	10
Junior engineer	26.00	20
Engineer	30.00	50
Senior engineer	36.00	10
Total		100

Composite labor rate − (0.07 × $12.00) + (0.03 × $16.00) + (0.10 × $20.00) + (0.20 × $26.00) + (0.50 × $30.00) + (0.10 × $36.00) = $27.12. Similar computations can be made to obtain the composite labor rate for skills within any of the other categories.

Another common practice is to establish separate overhead or burden pools for each skill category. These burden pools carry the peripheral costs that are related to and are a function of the labor-hours expended in that particular skill category. Assuming that the burden pool is established for each of the labor skills shown in Fig. 1, one can write an equation to depict the entire process. This equation is shown in Fig. 2. Thus far we have only considered a one-element cost estimate. The addition of multielement work activities or work outputs will greatly increase the number of mathematical computations, and it becomes readily evident that the anatomy of an estimate is so complex that computer techniques for computation are essential for all but the simplest estimate.

1.1 Time, Skills, and Labor-Hours Required to Prepare an Estimate

The resources (skills, calendar time, and labor-hours) required to prepare a cost estimate depend on a number of factors. One factor is the estimating method utilized. Another is the level of technology or state of the art involved in the job or task being estimated. A rule of thumb can be utilized to develop a rough idea of the estimating time required. The calendar time required to develop an accurate and credible estimate is usually about 8% of the calendar time required to accomplish a task involving existing technology and 18% for a task involving a high technology (i.e., nuclear plant construction, aerospace projects). These percentages are divided approximately as shown in Table 1.

534 Detailed Cost Estimating

$$T = \{[(E_H \times E_R) \times (1 + E_O)] + [(M_H \times M_R) \times (1 + M_O)] + [(TO_H \times TO_R) \times (1 + TO_O)] + [(Q_H \times Q_R) \times (1 + Q_O)] + [(TE_H + TE_R) \times (1 + TE_O)] + [(O_H \times O_R) \times (1 + O_O)] + S_D + S_O + [M_D \times (1 + M_{OH})] + T_D + C_D + OD_D\} \times \{GA + 1.00\} \times \{F + 1.00\}$$

(a)

$$T = \{(L1_H \times L1_R) \times (1 + L1_O)] + [(L2_H \times L2_R) \times (1 + L2_O) \cdots + [(LN_H \times LN_R \times (1 + LN_O)] + S_D + S_O + [M_D \times (1 + M_{OH})] + T_D + CD + OD_D\} \times \{1 + GA\} \times \{1 \times F\}$$

where $L1, L2, \ldots LN$ are various labor rate categories

(b)

Symbols:
- T = total cost
- E_H = engineering labor hours
- E_R = engineering composite labor rate in dollars per hour
- E_O = engineering overhead rate in decimal form (i.e., 1.15 = 115%)
- M_H = manufacturing labor hours
- M_R = manufacturing composite labor rate in dollars per hour
- M_O = manufacturing overhead rate in decimal form
- TO_H = tooling labor hours
- TO_R = tooling composite labor rate in dollars per hour
- TO_O = tooling overhead in decimal form
- Q_H = quality, reliability, and safety labor hours
- Q_R = quality, reliability, and safety composite labor rate in dollars per hour
- Q_O = quality, reliability, and safety overhead rate in decimal form
- TE_H = testing labor hours
- TE_R = testing composite labor rate in dollars per hour
- TE_O = testing overhead rate in decimal form
- O_H = other labor hours
- O_R = labor rate for other hours category in dollars per hour
- O_O = overhead rate for the hours category in decimal form
- S_D = major subcontract dollar
- S_O = other subcontract dollars
- M_D = material dollars
- M_{OH} = material overhead in decimal form (10% = 0.10)
- T_D = travel dollars
- C_D = computer dollars
- OD_D = other direct dollars
- GA = general and administrative expense in decimal form (25% = 0.25)
- F = fee in decimal form (0.10 = 10%)

Fig. 2 Generalized equation for cost estimating.

Note that the largest percentage of the required estimating time is for defining the output. This area is most important because it establishes a good basis for estimate credibility and accuracy, as well as making it easier for the estimator to develop supportable labor-hour and material estimates. These percentages also assume that the individuals who are going to perform the task or who have intimate working knowledge of the task are going to assist in estimate preparation. Hence the skill mix for estimating is very similar to the skill mix required for actually performing the task.

Labor-hours required for preparation of a cost estimate can be derived from these percentages by multiplying the task's calendar period in years by 2000 labor-hours per year,

Table 1 Estimating Time as a Percentage of Total Job Time

	Existing Technology (%)	High Technology (%)
Defining the output	4.6	14.6
Formulating the schedule and ground rules	1.2	1.2
Estimating materials and labor-hours	1.2	1.2
Estimating overhead, burden, and G&A	0.3	0.3
Estimating fee, profit, and earnings	0.3	0.3
Publishing the estimate	0.4	0.4
Total	8.0	18.0

multiplying the result by the percentage in Table 1, and then multiplying the result by 0.1 and by the number of personnel on the estimating team. Estimating team size is a matter of judgment and depends on the complexity of the task, but it is generally proportional to the skills required to perform the task (as mentioned). Examples of the application of these rules of thumb for determining the resources required to prepare a cost estimate follow:

1. A three-year, high-technology project involving 10 basic skills or disciplines would require the following number of labor-hours to estimate:

$$3 \times 2000 \times 0.18 \times 100 = 1080 \text{ labor-hours}$$

2. A six-month "existing-technology" project requiring five skills or disciplines would require $0.6 \times 2000 \times 0.08 \times 0.1 \times 5 = 48$ labor-hours to develop an estimate.

These relationships are drawn from the author's experience in preparing and participating in cost estimates and can be relied on to give you a general guideline in preparing for the estimating process. But remember that these are "rules of thumb," and exercise caution and discretion in their application.

2 DISCUSSION OF TYPES OF COSTS

Detailed estimating requires the understanding of and the distinction between initial acquisition costs, fixed and variable costs, recurring and nonrecurring costs, and direct and indirect costs. These distinctions are described in the material that follows.

2.1 Initial Acquisition Costs

Businesspersons, consumers, and government officials are becoming increasingly aware of the need to estimate accurately and to justify the initial acquisition cost of an item to be purchased, manufactured, or built. Initial acquisition costs usually refer to the total costs to procure, install, and put into operation a piece of equipment, a product, or a structure. Initial acquisition costs do not consider costs associated with the use and possession of the item. Individuals or businesses who purchase products now give serious consideration to maintenance, operation, depreciation, energy, insurance, storage, and disposal costs before purchasing or fabricating an item, whether it be an automobile, home appliance, suit of clothes, or industrial equipment. Initial acquisition costs include planning, estimating, designing, and/ or purchasing the components of the item; manufacturing, assembly, and inspection of the item; and installing and testing the item. Initial acquisition costs also include marketing, advertising, and markup of the price of the item as it flows through the distribution chain.

2.2 Fixed and Variable Costs

The costs of all four categories of productive outputs (processes, products, projects, and services) involve both fixed and variable costs. The relationship between fixed and variable costs depends on a number of factors, but it is principally related to the kind of output being estimated and the rate of output. Fixed cost is that group of costs involved in an ongoing activity whose total will remain relatively constant regardless of the quantity of output or the phase of the output cycle being estimated. Variable cost is the group of costs that vary in relationship to the rate of output. Therefore, where it is desirable to know the effect of output rate on costs, it is important to know the relationship between the two forms of cost as well as the magnitude of these costs. Fixed costs are meaningful only if they are considered at a given point in time, since inflation and escalation will provide a variable element to "fixed" costs. Fixed costs may only be truly fixed over a given range of outputs. Rental of floor space for a production machine is an example of a fixed cost, and its use of electrical power will be a variable cost.

2.3 Recurring and Nonrecurring Costs

Recurring costs are repetitive in nature and depend on continued output of a like kind. They are similar to variable costs because they depend on the quantity or magnitude of output. Nonrecurring costs are incurred to generate the very first item of output. It is important to separate recurring and nonrecurring costs if it is anticipated that the costs of continued or repeated production will be required at some future date.

2.4 Direct and Indirect Costs

As discussed earlier, direct costs are those that are attributable directly to the specific work activity or work output being estimated. Indirect costs are those that are spread across several projects and allocable on a percentage basis to each project. Table 2 is a matrix giving examples of these costs for various work outputs.

3 COLLECTING THE INGREDIENTS OF THE ESTIMATE

Before discussing the finer points of estimating, it is important to define the ingredients and to provide a preview of the techniques and methods utilized to collect these estimate ingredients.

3.1 Labor-Hours

Since the expenditure of labor-hours is the basic reason for the incurrence of costs, the estimating of labor-hours is the most important aspect of cost estimating. Labor-hours are estimated by four basic techniques: (1) use of methods, time, and measurement (MTM) techniques; (2) the labor-loading or staffing technique; (3) direct judgment of man-hours required; and (4) use of estimating handbooks. MTM methods are perhaps the most widespread methods of deriving labor-hour and skill estimates for industrial processes. These

Table 2 Examples of Costs for Various Outputs

	Process	Product	Project	Service
Initial acquisition costs	Plant construction costs	Manufacturing costs, marketing costs, and profit	Planning costs, design costs, manufacturing costs, test and checkout costs, and delivery costs	
Fixed costs	Plant maintenance costs	Plant maintenance costs	Planning costs and design costs	Building rental
Variable costs	Raw material costs	Labor costs	Manufacturing costs, test and checkout costs, and delivery costs	Labor costs
Recurring costs	Raw material costs	Labor and material costs	Manufacturing costs, test and checkout costs, and delivery costs	Labor costs
Nonrecurring costs	Plant construction costs	Plant construction costs	Planning costs and design costs	Initial capital equipment investment
Direct costs	Raw material	Manufacturing costs	Planning, design manufacturing, test and checkout and delivery costs	Labor and materials costs
Indirect costs	Energy costs	Marketing costs and profit	Energy costs	Energy costs

methods are available from and taught by the MTM Association for Standards and Research, located in Fair Lawn, New Jersey. The association is international in scope and has developed five generations of MTM systems for estimating all aspects of industrial, manufacturing, or machining operations. The MTM method subdivides operator motions into small increments that can be measured, and provides a means for combining the proper manual operations in a sequence to develop labor-hour requirements for accomplishing a job.

The labor-loading or staffing technique is perhaps the simplest and most widely used method for estimating the labor-hours required to accomplish a given job. In this method, the estimator envisions the job, the work location, and the equipment or machines required, and estimates the number of people and skills that would be needed to staff a particular operation. The estimate is usually expressed in terms of a number of people for a given number of days, weeks, or months. From this staffing level, the estimated on-the-job labor-hours required to accomplish a given task can be computed.

Another method closely related to this second method is the use of direct judgment of the number of labor-hours required. This judgment is usually made by an individual who has had direct hands-on experience in either performing or supervising a like task.

Finally, the use of handbooks is a widely utilized and accepted method of developing labor-hour estimates. Handbooks usually provide larger time increments than the MTM method and require a specific knowledge of the work content and operation being performed.

3.2 Materials and Subcontracts

Materials and subcontract dollars are estimated in three ways: (1) drawing "takeoffs" and handbooks, (2) dollar-per-pound relationships, and (3) direct quotations or bids. The most accurate way to estimate material costs is to calculate material quantities directly from a drawing or specification of the completed product. Using the quantities required for the number of items to be produced, the appropriate materials manufacturer's handbook, and an allowance for scrap or waste, one can accurately compute the material quantities and prices. Where detailed drawings of the item to be produced are not available, a dollar-per-pound relationship can be used to determine a rough order of magnitude cost. Firm quotations or bids for the materials or for the item to be subcontracted are better than any of the previously mentioned ways of developing a materials estimate because the supplier can be held to the bid.

3.3 Labor Rates and Factors

The labor rate, or number of dollars required per labor-hour, is the quantity that turns a labor-hour estimate into a cost estimate; therefore, the labor rate and any direct cost factors that are added to it are key elements of the cost estimate. Labor rates vary by skill, geographical location, calendar date, and the time of day or week applied. Overtime, shift premiums, and hazardous-duty pay are also added to hourly wages to develop the actual labor rate to be used in developing a cost estimate. Wage rate structures vary considerably, depending on union contract agreements. Once the labor rate is applied to the labor-hour estimate to develop a labor cost figure, other factors are commonly used to develop other direct cost allowances, such as travel costs and direct material costs.

3.4 Indirect Costs, Burden, and Overhead

Burden or overhead costs for engineering activities very often are as high as 100% of direct engineering labor costs, and manufacturing overheads go to 150% and beyond. A company that can keep its overhead from growing excessively, or a company that can successfully trim its overhead, can place itself in an advantageously competitive position. Since overhead more than doubles the cost of a work activity or work output, trimming the overhead has a significant effect on reducing overall costs.

3.5 General and Administrative Costs

General and administrative costs range up to 20% of total direct and indirect costs for large companies. General and administrative costs are added to direct and overhead costs and are recognized as a legitimate business expense.

3.6 Fee, Profit, or Earnings

The fee, profit, or earnings will depend on the amount of risk the company is taking in marketing the product, the market demand for the item, and the required return on the company's investment. This subject is one that deserves considerable attention by the cost estimator. Basically, the amount of profit depends on the astute business sense of the company's management. Few companies will settle for less than 10% profit, and many will not make an investment or enter into a venture unless they can see a 20 to 30% return on their investment.

3.7 Assembly of the Ingredients

Once resource estimates have been accumulated, the process of reviewing, compiling, organizing, and computing the estimate begins. This process is divided into two general subdivisions of work: (1) reviewing, compiling, and organizing the input resource data, and (2) computation of the costs based on desired or approved labor rates and factors. A common mistake made in developing cost estimates is the failure to perform properly the first of these work subdivisions. In the process of reviewing, compiling, and organizing the data, duplications in resource estimates are discovered and eliminated; omissions are located and remedied; overlapping or redundant effort is recognized and adjusted; and missing or improper rationale, backup data, or supporting data are identified, corrected, or supplied. A thorough review of the cost estimate input data by the estimator or estimating team, along with an adjustment and reconciliation process, will accomplish these objectives.

Computation of a cost estimate is mathematically simple since it involves only multiplication and addition. The number of computations can escalate rapidly, however, as the number of labor skills, fiscal years, and work breakdown structure elements are increased. One who works frequently in industrial engineering labor hour and material-based cost estimating will quickly come to the conclusion that some form of computer assistance is required.

With the basic ingredients and basic tools available, we are now ready to follow the steps required to develop a good detailed cost estimate. All steps are needed for any good cost estimate. The manner of accomplishing each step, and the depth of information needed and time expended on each step, will vary considerably, depending on what work activity or work output is being estimated. These steps are as follows:

1. Develop the work breakdown structure.
2. Schedule the work elements.
3. Retrieve and organize historical cost data.
4. Develop and use cost estimating relationships.
5. Develop and use production learning curves.
6. Identify skill categories, levels, and rates.
7. Develop labor-hour and material estimates.
8. Develop overhead and administrative costs.
9. Apply inflation and escalation factors.
10. Price (compute) the estimated costs.
11. Analyze, adjust, and support the estimate.
12. Publish, present, and use the estimate.

4 THE FIRST QUESTIONS TO ASK (AND WHY)

Whether you are estimating the cost of a process, product, or service, there are some basic questions you must ask to get started on a detailed cost estimate. These questions relate principally to the requirements, descriptions, location, and timing of the work.

540   Detailed Cost Estimating

4.1 What Is It?

A surprising number of detailed cost estimates fail to be accurate or credible because of a lack of specificity in describing the work that is being estimated. The objectives, ground rules, constraints, and requirements of the work must be spelled out in detail to form the basis for a good cost estimate. First, it is necessary to determine which of the four generic work outputs (process, product, project, or service) or combination of work outputs best describe the work being estimated. Then it is necessary to describe the work in as much detail as possible.

4.2 What Does It Look Like?

Work descriptions usually take the form of detailed specifications, sketches, drawings, materials lists, and parts lists. Weight, size, shape, material type, power, accuracy, resistance to environmental hazards, and quality are typical factors that are described in detail in a specification. Processes and services are usually defined by the required quality, accuracy, speed, consistency, or responsiveness of the work. Products and projects, on the other hand, usually require a preliminary or detailed design of the item or group of items being estimated. In general, more detailed designs will produce more accurate cost estimates. The principal reason for this is that as a design proceeds, better definitions and descriptions of all facets of this design unfold. The design process is an interactive one in which component or subsystem designs proceed in parallel; component or subsystem characteristics reflect on and affect one another to alter the configuration and perhaps even the performance of the end item. Another reason that a more detailed design results in a more accurate and credible cost estimate is that the amount of detail itself produces a greater awareness and visibility of potential inconsistencies, omissions, duplications, and overlaps.

4.3 When Is It to Be Available?

Production rate, production quantity, and timing of production initiation and completion are important ground rules to establish before starting a cost estimate. Factors such as raw material availability, labor skills required, and equipment utilization often force a work activity to conform to a specific time period. It is important to establish the optimum time schedule early in the estimating process, to establish key milestone dates, and to subdivide the overall work schedule into identifiable increments that can be placed on a calendar timescale. A work output schedule placed on a calendar time scale will provide the basic inputs needed to compute start-up costs, fiscal-year funding, and inflationary effects.

4.4 Who Will Do It?

The organization or organizations that are to perform an activity, as well as the skill categories and skill levels within these organizations, must be known or assumed to formulate a credible cost estimate. Given a competent organization with competent employees, another important aspect of developing a competitive cost estimate is the determination of the make or buy structure and the skill mix needs throughout the time period of a work activity. Judicious selection of the performers and wise time phasing of skill categories and skill levels can rapidly produce prosperity for any organization with a knowledge of its employees, its products, and its customers.

4.5 Where Will It Be Done?

Geographical factors have a strong influence on the credibility and competitive stature of a cost estimate. In addition to the wide variation in labor costs for various locations, material costs vary substantially from location to location, and transportation costs are entering even more heavily into the cost picture than in the past. The cost estimator must develop detailed ground rules and assumptions concerning location of the work, and then estimate costs accurately in keeping with all location-oriented factors.

5 THE ESTIMATE SKELETON: THE WORK BREAKDOWN STRUCTURE

The first step in developing a cost estimate of any type of work output is the development of a work breakdown structure. The work breakdown structure serves as a framework for collecting, accumulating, organizing, and computing the direct and directly related costs of a work activity or work output. It also can be and usually is utilized for managing and reporting resources and related costs throughout the lifetime of the work. There is considerable advantage in using the work breakdown structure and its accompanying task descriptions as the basis for scheduling, reporting, tracking, and organizing, as well as for initial costing. Hence it is important to devote considerable attention to this phase of the overall estimating process. A work breakdown structure is developed by subdividing a process, product, project, or service into its major work elements, then breaking the major work elements into subelements, and subelements into sub-subelements, and so on. There are usually 5 to 10 subelements under each major work element.

The purpose of developing the work breakdown structure is fivefold:

1. To provide a lower-level breakout of small tasks that are easy to identify, man-load, schedule, and estimate
2. To ensure that all required work elements are included in the work output
3. To reduce the possibility of overlap, duplication, or redundancy of tasks
4. To furnish a convenient hierarchical structure for the accumulation of resource estimates
5. To give greater overall visibility as well as depth of penetration into the makeup of any work activity

6 THE HIERARCHICAL RELATIONSHIP OF A DETAILED WORK BREAKDOWN STRUCTURE

A typical work breakdown structure is shown in Fig. 3. Note that the relationship resembles a hierarchy where each activity has a higher activity, parallel activities, and lower activities. A basic principle of work breakdown structures is that the resources or content of each work breakdown are made up of the sum of the resources or content of elements below it. No work element that has lower elements exceeds the sum of those lower elements in resource requirements. The bottommost elements are estimated at their own level and sum to higher levels. Many numbering systems are feasible and workable. The numbering system utilized here is one that has proved workable in a wide variety of situations.

One common mistake in using work breakdown structures is to try to input or allocate effort to every element, even those at a higher level. Keep in mind that this should not be

542 Detailed Cost Estimating

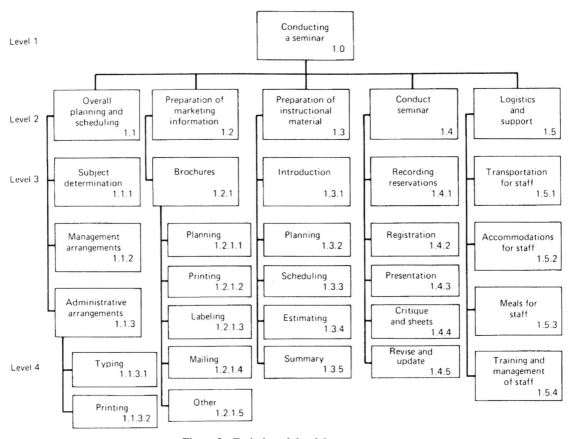

Figure 3 Typical work breakdown structure.

done because each block or work element contains only that effort included in those elements *below* it. If there are no elements below it, then it can contain resources. If there is need to add work activities or resources not included in a higher-level block, add an additional block below it to include the desired effort. Level 1 of a work breakdown structure is usually the top level, with lower levels numbered sequentially as shown. The "level" is usually equal to the number of digits in the work element block. For example, the block numbered 1.1.3.2 is in level 4 because it contains four digits.

7 FUNCTIONAL ELEMENTS DESCRIBED

When subdividing a work activity or work output into its elements, the major subdivisions can be either functional or physical elements. The second level in a work breakdown structure usually consists of a combination of functional and physical elements if a product or project is being estimated. For a process or service, all second-level activities could be functional. Functional elements of a production or project activity can include activities such as planning, project management, systems engineering and integration, testing, logistics, and operations. A process or service can include any of hundreds of functional elements. Typical examples of the widely dispersed functional elements that can be found in a work breakdown structure for a service are advising, assembling, binding, cleaning, fabricating, inspecting, packaging, painting, programming, receiving, testing, and welding.

8 PHYSICAL ELEMENTS DESCRIBED

The physical elements of a work output are the physical structures, hardware, products, or end items that are supplied to the consumer. These physical elements represent resources because they require labor and materials to produce. Hence they can and should be a basis for the work breakdown structure.

Figure 4 shows a typical work breakdown structure of just the physical elements of a well-known consumer product, the automobile. The figure shows how just one automobile company chose to subdivide the components of an automobile. For any given product or project, the number of ways that a work breakdown structure can be constructed are virtually unlimited. For example, the company could have included the carburetor and engine cooling system as part of the engine assembly (this might have been a more logical and workable arrangement since it is used in costing a mass-production operation). Note that the structure shows a level-3 breakout of the body and sheet metal element, and the door (a level-3 element) is subdivided into its level-4 components.

This *physical element* breakout demonstrates several important characteristics of a work breakdown structure. First, note that level 5 would be the individual component parts of each assembly or subassembly. It only took three subdivisions of the physical hardware to get

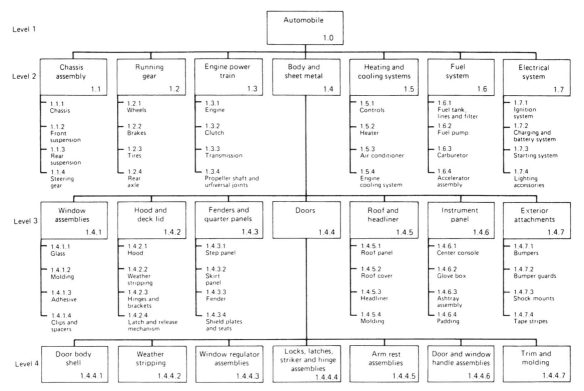

Figure 4 Work breakdown structure of an automobile.

down to a point where the next level breakout would be the individual parts. One can see rapidly that breaking down every level-2 element three more levels (down to level 5) would result in a very large work breakdown structure. Second, to convert this physical hardware breakout into a true work breakdown structure would require the addition of some functional activities. To provide the manpower as well as the materials required to procure, manufacture, assemble, test, and install the components of each block, it is necessary to add an "assembly," "fabrication," or "installation" activity block.

9 TREATMENT OF RECURRING AND NONRECURRING ACTIVITIES

Most work consists of both nonrecurring activities, or "one-of-a-kind" activities needed to produce an item or to provide a service, and recurring or repetitive activities that must be performed to provide more than one output unit. The resources requirements (labor-hours and materials) necessary to perform these nonrecurring and recurring activities reflect themselves in nonrecurring and recurring costs.

Although not all estimates require the separation of nonrecurring and recurring costs, it is often both convenient and necessary to separate costs because one may need to know what the costs are for an increased work output rate. Since work output rate principally affects the recurring costs, it is desirable to have these costs readily accessible and identifiable.

Separation of nonrecurring and recurring costs can be done in two ways that are compatible with the work breakdown structure concept. First, the two costs can be identified, separated, and accounted for within each work element. Resources for each task block would, then, include three sets of resource estimates: (1) nonrecurring costs, (2) recurring costs, and (3) total costs for that block. The second convenient method of cost separation is to start with identical work breakdown structures for both costs, and develop two separate cost estimates. A third estimate, which sums the two cost estimates into a total, can also use the same basic work breakdown structure. If there are elements unique to each cost category, they can be added to the appropriate work breakdown structure.

10 WORK BREAKDOWN STRUCTURE INTERRELATIONSHIPS

As shown in the automobile example, considerable flexibility exists concerning the placement of physical elements (the same is true with functional elements) in the work breakdown structure. Because of this, and because it is necessary to define clearly where one element leaves off and the other takes over, it is necessary to provide a detailed definition of what is included in each work activity block. In the automotive example, the rear axle unit could have been located and defined as part of the power train or as part of the chassis assembly rather than as part of the running gear. Where does the rear axle leave off and the power train begin? Is the differential or part of the differential included in the power train? These kinds of questions must be answered—and they usually are answered—before a detailed cost estimate is generated, in the form of a work breakdown structure dictionary. The dictionary describes exactly what is included in each work element and what is excluded; it defines where the interface is located between two work elements; and it defines where the assembly effort is located to assemble or install two interfacing units.

A good work breakdown structure dictionary will prevent many problems brought about by overlaps, duplications, and omissions, because detailed thought has been given to the interfaces and content of each work activity.

10.1 Skill Matrix in a Work Breakdown Structure

When constructing a work breakdown structure, keep in mind that each work element will be performed by a person or group of people using one or more skills. There are two important facets of the labor or work activity for each work element: skill mix and skill level. The skill mix is the proportion of each of several skill categories that will be used in performing the work. *Skill categories* vary widely and depend on the type of work being estimated. For a residential construction project, for example, typical skills would be bricklayer, building laborer, carpenter, electrician, painter, plasterer, or plumber. Other typical construction skills are structural steelworker, cement finisher, glazier, roofer, sheet metal worker, pipefitter, excavation equipment operator, and general construction laborer. Professional skill categories such as lawyers, doctors, financial officers, administrators, project managers, engineers, printers, writers, and so forth are called on to do a wide variety of direct-labor activities. Occasionally, skills will be assembled into several broad categories (such as engineering, manufacturing, tooling, testing, and quality assurance) that correspond to overhead or burden pools.

Skill level, on the other hand, depicts the experience or salary level of an individual working within a given skill category. For example, engineers are often subdivided into various categories such as principal engineers, senior engineers, engineers, associate engineers, junior engineers, and engineering technicians. The skilled trades are often subdivided into skill levels and given names that depict their skill level; for example, carpenters could be identified as master carpenters, journeymen, apprentices, and helpers. Because skill categories and skill levels are designated for performing work within each work element, it is not necessary to establish separate work elements for performance of each skill. A work breakdown structure for home construction would not have an element designated *carpentry,* because carpentry is a skill needed to perform one or more of the work elements (i.e., roof construction, wall construction).

10.2 Organizational Relationships to a Work Breakdown Structure

Frequently all or part of a work breakdown structure will have a direct counterpart in the performing organization. Although it is not necessary for the work breakdown structure to be directly correlatable to the organizational structure, it is often convenient to assign the responsibility for estimating and for performing a specific work element to a specific organizational segment. This practice helps to motivate the performer, since it assigns responsibility for an identifiable task, and it provides the manager greater assurance that each part of the work will be accomplished. In the planning and estimating process, early assignment of work elements to those who are going to be responsible for performing the work will motivate them to do a better job of estimating and will provide greater assurance of completion of the work within performance, schedule, and cost constraints, because the functional organizations have set their own goals. Job performance and accounting for work accomplished versus funds spent can also be accomplished more easily if an organizational element is held responsible for a specific work element in the work breakdown structure.

11 METHODS USED WITHIN THE DETAILED ESTIMATING PROCESS

The principal methods used *within* the detailed estimating process are detailed resource estimating, direct estimating, estimating by analogy, firm quotes, handbook estimating, and the parametric estimating technique mentioned earlier. These methods are described briefly in the following sections.

11.1 Detailed Resource Estimating

Detailed resource estimating involves the synthesis of a cost estimate from resource estimates made at the lowest possible level in the work breakdown structure. Detailed estimating presumes that a detailed design of the product or project is available and that a detailed manufacturing, assembly, testing, and delivery schedule is available for the work. This type of estimating assumes that skills, labor-hours, and materials can be identified for each work element through one or more of the methods that follow. A detailed estimate is usually developed through a synthesis of work element estimates developed by various methods.

11.2 Direct Estimating

A direct estimate is a judgmental estimate made in a "direct" method by an estimator or performer who is familiar with the task being estimated. The estimator will observe and study the task to be performed and then forecast resources in terms of labor-hours, materials, and/or dollars. For example, a direct estimate could be quoted as "so many dollars." Many expert estimators can size up and estimate a job with just a little familiarization. One estimator I know can take a fairly complex drawing and, within just a few hours, develop a rough order-of-magnitude estimate of the resources required to build the item. Direct estimating is a skill borne of experience in both estimating and in actually performing the "hands-on" work.

11.3 Estimating by Analogy (Rules of Thumb)

This method is similar to the direct estimating method in that considerable judgment is required, but an additional feature is the comparison with some existing or past task of similar description. The estimator collects resource information on a similar or analogous task and compares the task to be estimated with the similar or analogous activity. The estimator would say that "this task should take about twice the time (man-hours, dollars, materials, etc.) as the one used as a reference." This judgmental factor (a factor of 2) would then be multiplied by the resources used for the reference task to develop the estimate for the new task. A significant pitfall in this method of estimating is the potential inability of the estimator to identify subtle differences in the two work activities and, hence, to be estimating the cost of a system based on one that is really not similar or analogous.

11.4 Firm Quotes

One of the best methods of estimating the resources required to complete a work element or to perform a work activity is the development of a firm quotation by the supplier or vendor. The two keys to the development of a realistic quotation are (1) the solicitation of bids from at least three sources, and (2) the development of a detailed and well-planned request for quotation. Years of experience by many organizations in the field of procurement have indicated that three bids are optimum from the standpoint of achieving the most realistic and reasonable price at a reasonable expenditure of effort. The solicitation of at least three bids provides sufficient check and balance and furnishes bid prices and conditions for comparison, evaluation, and selection. A good request for quotation (RFQ) is essential, however, to evaluate the bids effectively. The RFQ should contain ground rules, schedules, delivery locations and conditions, evaluation criteria, and specifications for the work. The RFQ should also state and specify the format required for cost information. A well-prepared RFQ will result in a quotation or proposal that will be easily evaluated, verified, and compared with independent estimates.

11.5 Handbook Estimating

Handbooks, catalogs, and reference books containing information on virtually every conceivable type of product, part, supplies, equipment, raw material, and finished material are available in libraries and bookstores and directly from publishers. Many of these handbooks provide labor estimates for installation or operation, as well as the purchase costs of the item. Some catalogs either do not provide price lists or provide price lists as a separate insert to permit periodic updates of prices without changing the basic catalog description. Information services provide microfilmed cassettes and on-line databases for access to the descriptions and costs of thousands and even tens of thousands of items.

If you produce a large number of estimates, it may pay to subscribe to a microfilm catalog and handbook data access system or, at least, to develop your own library of databases, handbooks, and catalogs.

11.6 The Learning Curve

The learning curve is a mathematical and graphical representation of the reduction in time, resources, or costs either actually or theoretically encountered in the conduct of a repetitive human activity. The theory behind the learning curve is that successive identical operations will take less time, use fewer resources, or cost less than preceding operations. The term *learning* is used because it relates primarily to the improvement of mental or manual skills observed when an operation is repeated, but *learning* can also be achieved by a shop or organization through the use of improved equipment, purchasing, production, or management techniques. When the learning curve is used in applications other than those involving the feedback loop that brings improvement of an individual's work activities, it is more properly named by one or more of the following terms:

Productivity improvement curve	Production improvement curve
Manufacturing progress function	Production acceleration curve
Experience curve	Time reduction curve
Progress curve	Cost improvement curve
Improvement curve	

Learning curve theory is based on the concept that as the total quantity of units produced doubles, the hours required to produce the last unit of this doubled quantity will be reduced by a constant percentage. This means that the hours required to produce unit 2 will be a certain percentage less than the hours required to produce unit 1; the hours required to produce unit 4 will be the same percentage less than the hours required to produce unit 2; the hours required to produce unit 8 will be the same percentage less than unit 4; and this constant percentage of reduction will continue for doubled quantities as long as uninterrupted production of the same item continues. The complement of this constant percentage of reduction is commonly referred to as the *slope*. This means that if the constant percentage of reduction is 10%, the slope would be 90%. Table 3 gives an example of a learning curve with 90% slope when the number of hours required to produce the first unit is 100.

The reason for using the term *slope* in naming this reduction will be readily seen when the learning curve is plotted on coordinates with logarithmic scales on both the x and y axes (in this instance, the learning "curve" actually becomes a straight line). But first, let us plot the learning curve on conventional coordinates. You can see by the plot in Fig. 5 that it is truly a curve when plotted on conventional coordinates, and that the greater the production quantity, the smaller the incremental reduction in labor-hours required from unit to unit.

When the learning curve is plotted on log–log coordinates. as shown in Fig. 6, it becomes a straight line. The higher the slope, the flatter the line; the lower the slope, the steeper the line.

The effects of plotting curves on different slopes can be seen in Fig. 7, which shows the effects on labor-hour reductions of doubling the quantities produced 12 times. Formulas for the unit curve and the cumulative average curve are shown in Table 4.

Care should be taken in the use of the learning curve to avoid an overly optimistic (low) learning curve slope and to avoid using the curve for too few units in production. Most learning curve textbooks point out that this technique is credibly applicable only to operations that are done by hand (employ manual or physical operations) and that are highly repetitive.

11.7 Labor-Loading Methods

One of the most straightforward methods of estimating resources or labor-hours required to accomplish a task is the labor-loading or shop-loading method. This estimating technique is based on the fact that an experienced participant or manager of any activity can usually perceive, through judgment and knowledge of the activity being estimated, the number of individuals of various skills needed to accomplish a task. The shop-loading method is similar in that the estimator can usually predict what portion of an office or shop's capacity will be occupied by a given job. This percentage shop-loading factor can be used to compute labor-hours or resources if the total shop labor or total shop operation costs are known. Examples of the labor-loading and shop-loading methods based on 1896 labor-hours of on-the-job work per year are shown in Table 5.

11.8 Statistical and Parametric Estimating as Inputs to Detailed Estimating

Statistical and parametric estimating involves collecting and organizing historical information through mathematical techniques and relating this information to the work output that is

Table 3 Learning Curve Values

Cumulative Units	Hours per Unit	Percent Reduction
1	100.00	
2	90.00	10
4	81.00	10
8	72.90	10
16	65.61	10
32	59.05	10

11   Methods Used within the Detailed Estimating Process   **549**

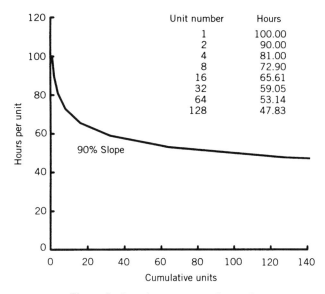

Figure 5   Learning curve on a linear plot.

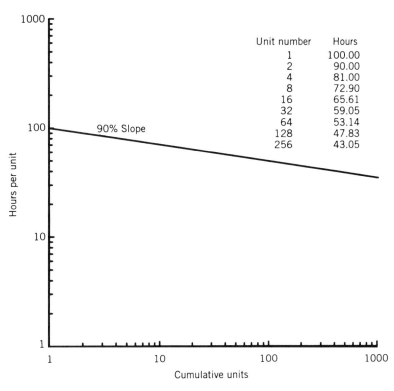

Figure 6   Learning curve on a log–log plot.

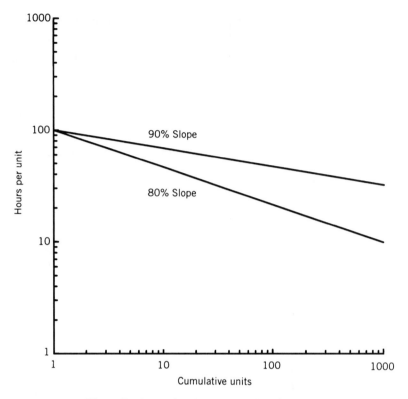

Figure 7 Comparison between two learning curves.

Table 4 Learning Curve Formulas

Unit curve $Y_x = KX^N$

where
- Y_x = number of direct labor-hours required to produce the Xth unit
- K = number of direct labor-hours required to produce the first unit
- x = number of units produced
- n = slope of curve expressed in positive hundredths (e.g., $n = 0.80$ for an 80% curve)
- $N = \dfrac{\log n}{\log 2}$

Cumulative average curve

$$V_x \approx \frac{K}{X(1+N)}[(x+0.5)^{(1+N)} - (0.5)^{(1-N)}]$$

where V_x = the cumulative average number of direct labor-hours required to produce x units

Table 5 Labor-Loading and Shop-Loading Methods

	Time Increment (Year)						
	1	2	3	4	5	6	7
Labor-loading method							
Engineers	1	1	1	2	1	0	0
Hours	1896	1896	1896	3792	1896	0	0
Technicians	3	4	4	6	2	1	0
Hours	5688	7584	7584	11,376	3792	1896	0
Draftsmen	0	0	1	3	6	4	2
Hours	0	0	1896	5688	11,376	7584	3792
Shop-loading method							
Electrical shop (5 workers)	10%	15%	50%	50%	5%	0%	0%
Hours	948	1422	4740	4740	474	0	0
Mechanical shop (10 workers)	5%	5%	10%	80%	60%	10%	5%
Hours	948	948	1896	15,168	11,376	1896	948

being estimated. There are a number of methods that can be used to correlate historical cost and manpower information; the choice depends principally on mathematical skills, imagination, and access to data. These mathematical and statistical techniques provide some analytical relationship between the product, project, or service being estimated and its physical characteristics. The format most commonly used for statistical and parametric estimating is the *estimating relationship,* which relates some physical characteristic of the work output (weight, power requirements, size, or volume) with the cost or labor-hours required to produce it. The most widely used estimating relationship is linear. That is, the mathematical equation representing the relationship is a linear equation, and the relationship can be depicted by a straight line when plotting on a graph with conventional linear coordinates for the x (horizontal) and y (vertical) axes. Other forms of estimating relationships can be derived based on curve-fitting techniques.

Estimating relationships have some advantages but certain distinct limitations. They have the advantage of providing a quick estimate even though very little is known about the work output except its physical characteristics. They correlate the present estimate with past history of resource utilization on similar items, and their use simplifies the estimating process. They require the use of statistical or mathematical skills rather than detailed estimating skills, which may be an advantage if detailed estimating skills are not available to the estimating organization.

On the other hand, because of their dependence on past (historical) data, they may erroneously indicate cost trends. Some products, such as mass-produced electronics, are providing more capability per pound, and lower costs per pound, volume, or component count every year. Basing electronics costs on past history may, therefore, result in noncompetitively high estimates. History should not be repeated if that history contains detrimental inefficiencies, duplications, unnecessary redundancies, rework, and overestimates. Often it is difficult to determine what part of historical data should be used to reflect future resource requirements accurately.

Finally, the parametric or statistical estimate, unless used at a very low level in the estimating process, does not provide in-depth visibility, and it does not permit determination of cost effects from subtle changes in schedule, performance, skill mix variations, or design

552 Detailed Cost Estimating

requirements. The way to use the statistical or parametric estimate most effectively is to subdivide the work into the smallest possible elements and then to use statistical or parametric methods to derive the resources required for these small elements.

12 DEVELOPING A SCHEDULE

Schedule elements are time-related groupings of work activities that are placed in sequence to accomplish an overall desired objective. Schedule elements for a process can be represented by very small (minutes, hours, or days) time periods. The scheduling of a process is represented by the time the raw material or raw materials take during each step to travel through the process. The schedule for manufacturing a product or delivery of a service is, likewise, a time flow of the various components or actions into a completed item or activity.

A project (the construction or development of a fairly large, complex, or multidisciplinary tangible work output) contains distinct schedule elements called *milestones*. These milestones are encountered in one form or another in almost all projects:

1. Study and analysis
2. Design
3. Procurement of raw materials and purchased parts
4. Fabrication or manufacturing of components and subsystems
5. Assembly of the components and subsystems
6. Testing of the combined system to qualify the unit for operation in its intended environment
7. Acceptance testing, preparation, packaging, shipping, and delivery of the item
8. Operation of the item

13 TECHNIQUES USED IN SCHEDULE PLANNING

There are a number of analytical techniques used in developing an overall schedule of a work activity that help to ensure the correct allocation and sequencing of schedule elements: precedence and dependency networks, arrow diagrams, critical path bar charts, and program evaluation and review techniques (PERT). These techniques use graphical and mathematical methods to develop the best schedule based on sequencing in such a way that each activity is performed only when the required predecessor activities are accomplished.

14 ESTIMATING ENGINEERING ACTIVITIES

Engineering activities include the design, drafting, analysis, and redesign activities required to produce an end item. Costing of engineering activities is usually based on labor-loading and staffing resource estimates.

14.1 Engineering Skill Levels

The National Society of Professional Engineers has developed position descriptions and recommended annual salaries for nine levels of engineers. These skills levels are broad enough in description to cover a wide variety of engineering activities. The principal activities performed by engineers are described in the following paragraphs.

14.2 Design

The design activity for any enterprise includes conceptual design, preliminary design, final design, and design changes. The design engineer must design prototypes, components for development or preproduction testing, special test equipment used in development or preproduction testing, support equipment, and production hardware. Since design effort is highly dependent on the specific work output description, design hours must be estimated by a design professional experienced in the area being estimated.

14.3 Analysis

Analysis goes hand-in-hand with design and employs the same general skill level as design engineering. Categories of analysis that support, augment, or precede design are thermal, stress, failure, dynamics, manufacturing, safety, and maintainability. Analysis is estimated by professionals skilled in analytical techniques. Analysis usually includes computer time as well as labor-hours.

14.4 Drafting

Drafting, or engineering drawing, is one area in the engineering discipline where labor-hours can be correlated to a product: the completed engineering drawing. Labor-hour estimates must still be quoted in ranges, however, because the labor-hours required for an engineering drawing will vary considerably depending on the complexity of the item being drawn. The drafting times given in Table 6 are approximations for class-A drawings of nonelectronic (mechanical) parts where all the design information is available and where the numbers represent "board time," that is, actual time that the draftsman is working on the drawing. A class-A drawing is one that is fully dimensioned and has full supporting documentation. An additional eight hours per drawing is usually required to obtain approval and signoffs of stress, thermal, supervisors, and drawing release system personnel. If a "shop drawing" is all that is required (only sufficient information for manufacture of the part with some informal assistance from the designer and/or draftsman), the board time labor-hours required would be approximately 50% of that listed in Table 6.

Table 6 Engineering Draft Times

Drawing Letter Designation	Size	Approximate Board-Time Hours for Drafting of Class A Drawings (h)
A	$8\frac{1}{2} \times 11$	1–4
B	11×17	2–8
C	17×22	4–12
D	22×34	8–16
E and F	34×44 and 28×40	16–40
J	34×48 and larger	40–80

15 MANUFACTURING/PRODUCTION ENGINEERING

The manufacturing/production engineering activity required to support a work activity is preproduction planning and operations analysis. This differs from the general type of production engineering wherein overall manufacturing techniques, facilities, and processes are developed. Excluded from this categorization is the design time of production engineers who redesign a prototype unit to conform to manufacturing or consumer requirements, as well as time for designing special tooling and special test equipment. A listing of some typical functions of manufacturing engineering follows:

1. Fabrication planning
   a. Prepare operations sheets for each part.
   b. List operational sequence for materials, machines, and functions.
   c. Recommend standard and special tooling.
   d. Make up tool order for design and construction of special tooling.
   e. Develop standard time data for operations sheets.
   f. Conduct liaison with production and design engineers.
2. Assembly planning
   a. Develop operations sheets for each part.
   b. Build first sample unit.
   c. Itemize assembly sequence and location of parts.
   d. Order design and construction of special jigs and fixtures.
   e. Develop exact component dimensions.
   f. Build any special manufacturing aids, such as wiring harness jig boards.
   g. Apply standard time data to operations sheet.
   h. Balance time cycles of final assembly line work stations.
   i. Effect liaison with production and design engineers.
   j. Set up material and layout of each work station in accordance with operations sheet.
   k. Instruct technicians in construction of the first unit.
3. Test planning
   a. Determine overall test method to meet performance and acceptance specifications.
   b. Break total test effort into positions by function and desired time cycle.
   c. Prepare test equipment list and schematic for each position.
   d. Prepare test equipment design order for design and construction of special purpose test fixtures.
   e. Prepare a step-by-step procedure for each position.
   f. Effect liaison with production and design engineers.
   g. Set up test positions and check out.
   h. Instruct test operator on first unit.
4. Sustaining manufacturing engineering
   a. Debug, as required, engineering design data.

b. Debug, as required, manufacturing methods and processes.

c. Recommend more efficient manufacturing methods throughout the life of production.

The following statements may be helpful in deriving manufacturing engineering labor-hour estimates for high production rates:

1. Total fabrication and assembly labor-hours, divided by the number of units to be produced, multiplied by 20, gives manufacturing engineering start-up costs.
2. For sustaining manufacturing engineering, take the unit fabrication and assembly man-hours, multiply by 0.07. (These factors are suggested for quantities up to 100 units.)

15.1 Engineering Documentation

A large part of an engineer's time is spent in writing specifications, reports, manuals, handbooks, and engineering orders. The complexity of the engineering activity and the specific document requirements are important determining factors in estimating the engineering labor-hours required to prepare engineering documentation.

The hours required for engineering documentation (technical reports, specifications, and technical manuals) will vary considerably depending on the complexity of the work output; however, average labor-hours for origination and revision of engineering documentation have been derived based on experience, and these figures can be used as average labor-hours per page of documentation. (See Tables 7 and 8.)

16 ESTIMATING MANUFACTURING/PRODUCTION AND ASSEMBLY ACTIVITIES

A key to successful estimating of manufacturing activities is the process plan. A process plan is a listing of all operations that must be performed to manufacture a product or to complete a project, along with the labor-hours required to perform each operation. The process plan is usually prepared by an experienced foreman, engineer, or technician who knows the company's equipment, personnel, and capabilities, or by a process-planning department chartered to do all of the process estimating. The process planner envisions the

Table 7 New Documentation

Function	Labor-Hours per Page
Research, liaison, technical writing, editing, and supervision	5.7
Typing and proofreading	0.6
Illustrations	4.3
Engineering	0.7
Coordination	0.2
Total[a]	11.5

[a] A range of 8 to 12 labor-hours per page can be used.

Table 8 Revised Documentation

Function	Labor-Hours per Page
Research, liaison, technical writing, editing, and supervision	4.00
Typing and proofreading	0.60
Illustrations	0.75
Engineering	0.60
Coordination	0.20
Total[a]	6.15

[a]A range of 4 to 8 labor-hours per page can be used.

equipment, work station, and environment; estimates the number of persons required; and estimates how long it will take to perform each step. From this information the labor-hours required are derived. Process steps are numbered, and space is left between operations listed to allow easy insertion of operations or activities as the process is modified.

A typical process plan for a welded cylinder assembly is given in Table 9. The process plan is used not only to plan and estimate a manufacturing or construction process, but often also as part of the manufacturing or construction work order itself. As such, it shows the shop or construction personnel each step to take in the completion of the work activity. Fabrication of items from metals, plastics, or other materials in a shop is usually called *manufacturing,* whereas fabrication of buildings, structures, bridges, dams, and public facilities on site is usually called *construction.* Different types of standards and estimating factors are used for each of these categories of work. Construction activities are covered in a subsequent chapter.

17 MANUFACTURING ACTIVITIES

Manufacturing activities are broken into various categories of effort, such as metal working and forming; welding, brazing, and soldering; application of fasteners; plating, printing, surface treating, heat treating; and manufacturing of electronic components (a special category). The most common method of estimating the time and cost required for manufacturing activities is the industrial engineering approach, whereby standards or target values are established for various operations. The term *standards* is used to indicate standard time data. All possible elements of work are measured, assigned a standard time for performance, and documented. When a particular job is to be estimated, all of the applicable standards for all related operations are added together to determine the total time.

The use of standards produces more accurate and more easily justifiable estimates. Standards also promote consistency between estimates as well as among estimators. Where standards are used, personal experience is desirable or beneficial, but not mandatory. Standards have been developed over a number of years through the use of time studies and synthesis of methods analysis. They are based on the level of efficiency that could be attained by a job shop producing up to 1000 units of any specific work output. Standards are actually synoptical values of more detailed times. They are adaptations, extracts, or benchmark time values for each type of operation. The loss of accuracy occasioned by summarization and/or averaging is acceptable when the total time for a system is being developed. If standard

Table 9 Process Plan

Operation Number	Labor-Hours	Description
010	—	Receive and inspect material (skins and forgings)
020	24	Roll form skin segments
030	60	Mask and chem-mill recessed pattern in skins
040	—	Inspect
050	36	Trim to design dimension and prepare in welding skin segments into cylinders (two)
060	16	Locate segments on automatic seam welder tooling fixture and weld per specification (longitudinal weld)
070	2	Remove from automatic welding fixture
080	18	Shave welds on inside diameter
090	16	Establish trim lines (surface plate)
100	18	Install in special fixture and trim to length
110	8	Remove from special fixture
120	56	Install center mandrel—center ring, forward and aft sections (cylinders)—forward and aft mandrel—forward and aft rings—and complete special feature setup
130	—	Inspect
140	24	Butt weld (4 places)
150	8	Remove from special feature and remove mandrels
160	59	Radiograph and dye penetrant inspect
170	—	Inspect dimensionally
180	6	Reinstall mandrels in preparation for final machining
190	14	Finish OD-aft
	10	Finish OD-center
	224	Finish OD-forward
200	40	Program for forward ring
220	30	Handwork (3 rings)
230	2	Reinstall cylinder assembly with mandrels still in place or on the special fixture
240	16	Clock and drill index holes
250	—	Inspect
260	8	Remove cylinder from special fixture—remove mandrel
270	1	Install in holding cradle
280	70	Locate drill jig on forward end and hand-drill leak check vein (drill and tap), and hand-drill hole pattern
290	64	Locate drill jig on aft ring and hand-drill hole pattern
300	—	Inspect forward and aft rings
310	8	Install protective covers on each end of cylinder
320	—	Transfer to surface treat
340	24	Remove covers and alodine
350	—	Inspect
360	8	Reinstall protective covers and return to assembly area

values are used with judgment and interpolations for varying stock sizes, reasonably accurate results can be obtained.

Machining operations make up a large part of the manufacturing costs of many products and projects. Machining operations are usually divided into setup times and run times. Setup time is the time required to establish and adjust the tooling, to set speeds and feeds on the metal-removal machine, and to program for the manufacture of one or more identical or similar parts. Run time is the time required to complete each part. It consists of certain fixed positioning times for each item being machined, as well as the actual metal-removal and cleanup time for each item. Values are listed for "soft" and "hard" materials. Soft values are for aluminum, magnesium, and plastics. Hard values are for stainless steel, tool steel, and beryllium. Between these two times would be standard values for brass, bronze, and medium steel.

18 IN-PROCESS INSPECTION

The amount of in-process inspection performed on any process, product, project, or service will depend on the cost of possible scrappage of the item as well as the degree of reliability required for the final work output. In high-rate production of relatively inexpensive items, it is often economically desirable to forgo in-process inspection entirely in favor of scrapping any parts that fail a simple go, no-go inspection at the end of the production line. On the other hand, expensive and sophisticated precision-manufactured parts may require nearly 100% inspection. A good rule of thumb is to add 10% of the manufacturing and assembly hours for in-process inspection. This in-process inspection does not include the in-process testing covered in the following paragraphs.

19 TESTING

Testing usually falls into three categories: (1) development testing, (2) qualification testing, and (3) production acceptance testing. Rules of thumb are difficult to come by for estimating development testing, because testing varies with the complexity, uncertainty, and technological content of the work activity. The best way to estimate the cost of development testing is to produce a detailed test plan for the specific project and to cost each element of this test plan separately, being careful to consider all skills, facilities, equipment, and material needed in the development test program.

Qualification testing is required in most commercial products and on all military or space projects to demonstrate adequately that the article will operate or serve its intended purpose in environments far more severe than those intended for its actual use. Automobile crash tests are an example. Military products must often undergo severe and prolonged tests under high shock, thermal, and vibration loads as well as heat, humidity, cold, and salt spray environments. These tests must be meticulously planned and scheduled before a reasonable estimate of their costs can be generated.

Receiving inspection, production testing, and acceptance testing can be estimated using experience factors and ratios available from previous like-work activities. Receiving tests are tests performed on purchased components, parts, and/or subassemblies prior to acceptance by the receiving department. Production tests are tests of subassemblies, units, subsystems, and systems during and after assembly. Experience has shown, generally, that test labor varies directly with the amount of fabrication and assembly labor. The ratio of test labor to other production labor will depend on the complexity of the item being tested. Table 10 gives the test labor percentage of direct fabrication and assembly labor for simple, average, and complex items.

Special-purpose tooling and special-purpose test equipment are important items of cost because they are used only for a particular job; therefore, that job must bear the full cost of the tool or test fixture. In contrast to the special items, general-purpose tooling or test equipment is purchased as capital equipment, and costs are spread over many jobs. Estimates for tooling and test equipment are included in overall manufacturing start-up ratios shown in Table 11. Under "degree of precision and complexity," "high," means high-precision multidisciplinary systems, products, or subsystems; "medium" means moderately complex subsystems or components; and "low" means simple, straightforward designs of components or individual parts. Manual and computer-aided design hours required for test equipment are shown in Table 12 CAD drawings take approximately 67.5% of the time required (on the average) to produce manual drawings.

20 COMPUTER SOFTWARE COST ESTIMATING

Detailed cost estimates must include the cost of computer software development and testing where necessary to provide deliverable source code or to run the analysis or testing programs needed to develop products or services.

Because of the increasing number and types of computers and computer languages, it is difficult to generate overall ground rules or rules of thumb for computer software cost estimating. Productivity in computer programming is greatly affected by the skill and competence of the computer analyst or programmer. The advent of computer-aided software engineering (CASE) tools has dramatically accelerated the process of software analysis, development, testing, and documentation. Productivity is highly dependent on which CASE tools, if any, are utilized.

Complicated flight software for aircraft and space systems is subjected to design review and testing in simulations and on the actual flight computer hardware. A software critical design review is usually conducted about 43% of the way through the program; an integrated systems test is performed at the 67% completion mark; prototype testing is done at 80% completion; installation with the hardware is started with about 7% of the time remaining (at the 93% completion point).

Table 10 Test Estimating Ratios

	Percent of Direct Labor		
	Simple	Average	Complex
Fabrication and assembly labor base			
Receiving test	1	2	4
Production test	9	18	36
Total	10	20	40
Assembly labor base			
Receiving test	2	3	7
Production test	15	32	63
Total	17	35	70

Table 11 Manufacturing Startup Ratios

Cost Element	Degree of Precision and Complexity	Recurring Manufacturing Costs Lot Quantity (%)			
		10	100	1000	10,000
Production planning	High	20	6	1.7	0.5
	Medium	10	3	0.8	0.25
	Low	5	1.5	0.4	0.12
Special tooling	High	10	6	3.5	2
	Medium	5	3	2	1
	Low	3	1.5	1	—
Special test equipment	High	10	6	3.5	2
	Medium	6	3	2	1
	Low	3	1.5	1	0.5
Composite total	High	40	18	8.7	4.5
	Medium	21	9	4.8	2.25
	Low	11	4.5	2.4	1.12

21 LABOR ALLOWANCES

"Standard times" assume that the workers are well trained and experienced in their jobs, that they apply themselves to the job 100% of the time, that they never make a mistake, take a break, lose efficiency, or deviate from the task for any reason. This, of course, is an unreasonable assumption because there are legitimate and numerous unplanned work interruptions that occur with regularity in any work activity. Therefore, labor allowances must be added to any estimate that is made up of an accumulation of standard times. These labor allowances can accumulate to a factor of 1.5 to 2.5. The total standard time for a given work activity, depending on the overall inherent efficiency of the shop, equipment, and personnel,

Table 12 Design Hours for Test Equipment

Type Design	Manual Hours/ Square Foot	Standard Drawing Size	Square Feet/ Drawing	Manual Hours/ Drawing	CAD Hours/ Drawing
Original concept	15	C	2.5	38	26
		D	5.0	75	51
		H	9.0	135	91
		J	11.0	165	111
Layout	10	B	1.0	10	7
		C	2.5	25	17
		D	5.0	50	34
		H	9.0	90	61
		J	11.0	110	74
Detail or copy	3	A	0.7	2.1	1.4
		B	1.0	3.0	2.0
		C	2.5	7.5	5.1
		D	5.0	15.0	10.1
		H	9.0	27.0	18.2
		J	11.0	33.0	22.3

will depend on the nature of the task. Labor allowances are made up of a number of factors that are described in the following sections.

21.1 Variance from Measured Labor-Hours

Standard hours vary from actual measured labor-hours because workers often deviate from the standard method or technique used or planned for a given operation. This deviation can be caused by a number of factors ranging from the training, motivation, or disposition of the operator to the use of faulty tools, fixtures, or machines. Sometimes shortages of materials or lack of adequate supervision are causes of deviations from standard values. These variances can add 5 to 20% to standard time values.

21.2 Personal, Fatigue, and Delay (PFD) Time

Personal times are for personal activities such as coffee breaks, trips to the restroom or water fountain, unforeseen interruptions, or emergency telephone calls. Fatigue time is allocated because of the inability of a worker to produce at the same pace all day. Operator efficiency decreases as the job time increases. Delays include unavoidable delays caused by the need for obtaining supervisory instructions, equipment breakdown, power outages, or operator illness. PFD time can add 10 to 20% to standard time values.

21.3 Tooling and Equipment Maintenance

Although normal or routine equipment maintenance can be done during times other than operating shifts, there is usually some operator-performed machine maintenance activity that must be performed during the machine duty cycle. These activities include adjusting tools, sharpening tools, and periodically cleaning and oiling machines. In electroplating and processing operations, the operator maintains solutions and compounds, and handles and maintains racks and fixtures. Tooling and equipment maintenance can account for 5 to 12% of standard time values.

21.4 Normal Rework and Repair

The overall direct labor-hours derived from the application of the preceding three allowance factors to standard times must be increased by additional amounts to account for normal rework and repair. Labor values must be allocated for rework of defective purchased materials, rework of in-process rejects, final test rejects, and addition of minor engineering changes. Units damaged on receipt or during handling must also be repaired. This factor can add 10 to 20% direct labor-hours to those previously estimated.

21.5 Engineering Change Allowance

For projects where design stability is poor, where production is initiated prior to final design release, and where field testing is being performed concurrently with production, an engineering change allowance should be added of up to 10% of direct labor-hours. Change allowances vary widely for different types of work activities. Even fairly well defined projects, however, should contain a change allowance.

21.6 Engineering Prototype Allowance

The labor-hours required to produce an engineering prototype are greater than those required to produce the first production model. Reworks are more frequent, and work is performed from sketches or unreleased drawings rather than from production drawings. An increase over first production unit labor of 15 to 25% should be included for each engineering prototype.

21.7 Design Growth Allowance

Where estimates are based on incomplete drawings, or where concepts or early breadboards only are available prior to the development of a cost estimate, a design growth allowance is added to all other direct labor costs. This design growth allowance is calculated by subtracting the percentage of design completion from 100%, as shown in the following tabulation:

Desirable Design Completion (%)	Design Completed (%)	Design Growth Allowance (%)
100	50	50
100	75	25
100	80	20
100	90	10
100	100	0

21.8 Cost Growth Allowance

Occasionally a cost estimate will warrant the addition of allowances for cost growth. Cost growth allowances are best added at the lowest level of a cost estimate rather than at the top levels. These allowances include reserves for possible misfortunes, natural disasters, strikes, and other unforeseen circumstances. Reserves should not be used to account for normal design growth. Care should be taken in using reserves in a cost estimate because they are usually the first cost elements that come under attack for removal from the cost estimate or budget. Remember, cost growth with an incomplete design is a certainty, not a reserve or contingency! Defend your cost growth allowance, but be prepared to relinquish your reserve if necessary.

22 ESTIMATING SUPERVISION, DIRECT MANAGEMENT, AND OTHER DIRECT CHARGES

Direct supervision costs will vary with the task and company organization. Management studies have shown that the span of control of a supervisor over a complex activity should not exceed 12 workers. For simple activities, the ratio of supervisors to employees can go down. But the 1:12 ratio (8.3%) will usually yield best results. Project management for a complex project can add an additional 10 to 14%. Other direct charges are those attributable to the project being accomplished but not included in direct labor or direct materials. Transportation, training, and reproduction costs, as well as special service or support contracts and consultants, are included in the category of "other direct costs."

Two cost elements of "other direct costs" that are becoming increasingly prominent are travel and transportation costs. A frequent check on public and private conveyance rates and costs is mandatory. Most companies provide a private vehicle mileage allowance for employees who use their own vehicles in the conduct of company business. Rates differ and depend on whether the private conveyance is being utilized principally for the benefit of the company or principally for the convenience of the traveler. Regardless of which rate is used, the mileage allowance must be periodically updated to keep pace with actual costs. Many companies purchase or lease vehicles to be used by their employees on official business.

Per diem travel allowances or reimbursement for lodging, meals, and miscellaneous expenses must also be included in overall travel budgets. These reimbursable expenses include costs of a motel or hotel room; food, tips, and taxes; local transportation and communication; and other costs such as laundry, mailing costs, and on-site clerical services. Transportation costs include the transport of equipment, supplies, and products, as well as personnel, and can include packaging, handling, shipping, postage, and insurance charges.

23 THE USE OF "FACTORS" IN DETAILED ESTIMATING

The practice of using factors is becoming increasingly common, particularly in high-technology work activities and work outputs. One company uses an "allocation factor," which allocates miscellaneous labor-oriented functions to specific functions such as fabrication or assembly. This company adds 14.4% to fabrication hours and 4.1% to assembly hours to cover miscellaneous labor-hour expenditures associated with these two functions. It is also common to estimate hours for planning, tooling, quality and inspection, production support, and sustaining engineering based on percentages of manufacturing and/or assembly hours. Tooling materials and computer supplies are sometimes estimated based on so much cost per tooling hour, and miscellaneous shop hardware (units, bolts, fasteners, cleaning supplies, etc.), otherwise known as *pan stock,* is estimated at a cost per manufacturing hour.

The disadvantage of the use of such factors is that inefficiencies can become embedded in the factored allowances and eventually cause cost growth. A much better method of estimating the labor-hours and materials required to accomplish these other direct activities is to determine the specific tasks and materials required to perform the job by laborloading, shoploading, or process-planning methods. When the materials, labor-hours, and other direct costs have been estimated, the basic direct resources required to do the job have been identified. The estimator can now move into the final steps of the detailed estimating process with the full assurance that all work elements and all cost elements have been included in the detailed estimate.

24 CONCLUDING REMARKS

In summary, detailed cost estimating involves meticulous penetration into the smallest feasible portions of a work output or work activity and the systematic and methodical assembly of the resources in all cost, work, and schedule elements. Detailed estimating requires detailed design, manufacturing, and test descriptions, and involves great time, effort, and penetration into the resources required to do the job. Wherever possible, detailed estimates should be used to establish a firm and credible cost estimate and to verify and substantiate higher-level parametric estimates.

CHAPTER 17
INVESTMENT ANALYSIS

Byron W. Jones
Kansas State University
Manhattan, Kansas

1 **ESSENTIALS OF FINANCIAL ANALYSIS**	**564**	
1.1 Sources of Funding for Capital Expenditures	564	
1.2 The Time Value of Money	565	
1.3 Discounted Cash Flow and Interest Calculations	566	
2 **INVESTMENT DECISIONS**	**570**	
2.1 Allocation of Capital Funds	570	
2.2 Classification of Alternatives	573	
2.3 Analysis Period	574	
3 **EVALUATION METHODS**	**575**	
3.1 Establishing Cash Flows	575	
3.2 Present Worth	577	
3.3 Annual Cash Flow	579	
3.4 Rate of Return	580	
3.5 Benefit–Cost Ratio	581	
3.6 Payback Period	582	
REFERENCES	**582**	

1 ESSENTIALS OF FINANCIAL ANALYSIS

1.1 Sources of Funding for Capital Expenditures

Engineering projects typically require the expenditure of funds for implementation and in return provide a savings or increased income to the firm. In this sense an engineering project is an investment for the firm and must be analyzed as an investment. This is true whether the project is a major new plant, a minor modification of some existing equipment, or anything in between. The extent of the analysis of course must be commensurate with the financial importance of the project. Financial analysis of an investment has two parts: funding of the investment and evaluation of the economics of the investment. Except for very large projects, such as a major plant expansion or addition, these two aspects can be analyzed independently. All projects generally draw from a common pool of capital funds rather than each project being financed separately. The engineering function may require an in-depth evaluation of the economics of a project, while the financing aspect generally is not dealt with in detail if at all. The primary reason for being concerned with the funding of projects is that the economic evaluation often requires at least an awareness if not an understanding of this function.

The funds used for capital expenditures come from two sources: debt financing and equity financing. Debt financing refers to funds that are borrowed from outside the company. The two common sources are bank loans and the sale of bonds. Bank loans are typically used for short-term financing, and bonds are used for long-term financing. Debt financing is characterized by a contractual arrangement specifying interest payments and repayment. The lender does not share in the profits of the investments for which the funds are used nor does it share the associated risks except through the possibility of the company defaulting. Equity financing refers to funds owned by the company. These funds may come from profits earned by the company or from funds set aside for depreciation allowances. Or, the funds may come from the sale of new stock. Equity financing does not require any specified repayment;

however, the owners of the company (stockholders) do expect to make a reasonable return on their investment.

The decisions of how much funding to secure and the relative amounts to secure from debt and equity sources are very complicated and require considerable subjective judgment. The current stock market, interest rates, projections of future market conditions, etc., must be addressed. Generally, a company will try to maintain approximately a constant ratio of funding from the different sources. This mix will be selected to maximize earnings without jeopardizing the company's financial well being. However, the ratio of debt to equity financing does vary considerably from company to company reflecting different business philosophies.

1.2 The Time Value of Money

The time value of money is frequently referred to as interest or interest rate in economic analyses. Actually, the two are not exactly the same thing. Interest is a fee paid for borrowed funds and is established when the loan is made. The time value of money is related to interest rates, but it includes other factors also. The time value of money must reflect the cost of money. That is, it must reflect the interest that is paid on loans and bonds, and it must also reflect the dividends paid to the stockholders. The cost of money is usually determined as a weighted average of the interest rates and dividend rates paid for the different sources of funds used for capital expenditures. The time value of money must also reflect opportunity costs. The opportunity cost is the return that can be earned on available, but unused, projects.

In principle, the time value of money is the greater of the cost of money and the opportunity cost. The determination of the time value of money is difficult, and the reader is referred to advanced texts on this topic for a complete discussion (see, for example, Bussey[1]). The determination is usually made at the corporate level and is not the responsibility of engineers. The time value of money is frequently referred to as the interest rate for economic evaluations, and one should be aware that the terms interest rate and time value of money are used interchangeably. The time value of money is also referred to as the required rate of return.

Another factor that may or may not be reflected in a given time value of money is inflation. Inflation results in a decreased buying power of the dollar. Consequently, the cost of money generally is higher during periods of high inflation, since the funds are repaid in dollars less valuable than those in which the funds were obtained. Opportunity costs are not necessarily directly affected by inflation except that inflation affects the cash flows used in evaluating the returns for the projects on which the opportunity costs are based. It is usually up to the engineer to verify that inflation has been included in a specified time value of money, since this information is not normally given. In some applications it is beneficial to use an inflation-adjusted time value of money. The relationship is

$$1 + i_r = \frac{1 + i_a}{1 + f} \tag{1}$$

where i_a is the time value of money, which reflects the higher cost of money and the higher opportunity cost due to inflation; f is the inflation rate; and i_r is the inflation-adjusted time value of money, which actually reflects the true cost of capital in terms of constant value. The variables i_r and i_a may be referred to as the real and apparent time values of money, respectively. i_r, i_a, and f are all expressed in fractional rather than percentage form in Eq.

(1) and all must be expressed on the same time basis, generally an annual rate. See Section 3 for additional discussion on the use of i_r and i_a. Also see Jones[2] for a detailed discussion.

The time value of money may also reflect risks associated with a project. This is particularly true for projects where there is a significant probability of poor return or even failure (e.g., the development of a new product). In principle, risk can be evaluated in assessing the economics of a project by including the probabilities of various outcomes (see Riggs,[3] for example). However, these calculations are complicated and are often dependent on subjective judgment. The more common approach is to simply use a time value of money which is greater for projects that are more risky. This is why some companies will use different values of i for different types of investments (e.g., expansion versus cost reduction versus diversification, etc.). Such adjustments for risk are usually made at the corporate level and are based on experience and other subjective inputs as much as they are on formal calculations. The engineer usually is not concerned with such adjustments, at least for routine economic analyses. If the risks of a project are included in an economic analysis, then it is important that the time value of money also not be adjusted for risks, since this would represent an overcompensation and would distort the true economic picture.

1.3 Discounted Cash Flow and Interest Calculations

For the purpose of economic analysis, a project is represented as a group of cash flows showing the expenditures and the income or savings attributable to the project. The object of economic analysis is normally to determine the profitability of the project based on these cash flows. However, the profitability cannot be assessed simply by summing up the cash flows, owing to the effect of the time value of money. The time value of money results in the value of a cash flow depending not only on its magnitude but also on when it occurs according to the equation

$$P = F \frac{1}{(1 + i)^n} \qquad (2)$$

where F is a cash flow that occurs sometime in the future, P is the equivalent value of that cash flow now, i is the annual time value of money in fractional form, and n is the number of years from now when cash flow F occurs. Cash flow F is often referred to as the future value or future amount, while cash flow P is referred to as the present value or present amount. Equation (2) can be used to convert a set of cash flows for a project to a set of economically equivalent cash flows. These equivalent cash flows are referred to as discounted cash flows and reflect the reduced economic value of cash flows that occur in the future. Table 1 shows a set of cash flows that have been discounted.

Equation (2) is the basis for a more general principle referred to as economic equivalence. It shows the relative economic value of cash flows occurring at different points in time. It is not necessary that P refer to a cash flow that occurs at the present; rather, it simply refers to a cash flow that occurs n years before F. The equation works in either direction for computing equivalent cash flows. That is, it can be used to find a cash flow F that occurs n years after P and that is equivalent to P, or to find a cash flow P that is equivalent to F but which occurs n years before F. This principle of equivalence allows cash flows to be manipulated as needed to facilitate economic calculations.

The time value of money is usually specified as an annual rate. However, several other forms are sometimes encountered or may be required to solve a particular problem. An

1 Essentials of Financial Analysis

Table 1 Discounted Cash Flow Calculation

Year	Estimated Cash Flows for Project	Discounted Cash Flows^a
0	−$120,000	−$120,000
1	− 75,000	− 68,200
2	+ 50,000	+ 41,300
3	+ 60,000	+ 45,100
4	+ 70,000	+ 47,800
5	+ 30,000	+ 18,000
6	+ 20,000	+ 11,300

^{a}Based on $i = 10\%$.

interest rate* as used in Eq. (2) is referred to as a discrete interest rate, since it specifies interest for a discrete time period of 1 year and allows calculations in multiples (n) of this time period. This time period is referred to as the compounding period. If it is necessary to change an interest rate stated for one compounding period to an equivalent interest rate for a different compounding period, it can be done by

$$i_1 = (1 + i_2)^{\Delta t_1 / \Delta t_2} - 1 \qquad (3)$$

where Δt_1 and Δt_2 are compounding periods and i_1 and i_2 are the corresponding interest rates, respectively. Interest rates i_1 and i_2 are in fractional form. If an interest rate with a compounding period different than 1 year is used in Eq. (2), then n in that equation refers to the number of those compounding periods and not the number of years.

Interest may also be expressed in a nominal form. Nominal interest rates are frequently used to describe the interest associated with borrowing but are not used to express the time value of money. A nominal interest rate is stated as an annual interest rate but with a compounding period different from 1 year. A nominal rate must be converted to an equivalent compound interest rate before being used in calculations. The relationship between a nominal interest rate (i_n) and a compound interest rate (i_c) is $i_c = i_n/m$, where m is the number of compounding periods per year. The compounding period (Δt) for i_c is 1 year/m. For example, a 10% nominal interest rate compounded quarterly translates to a compound interest rate of 2.5% with a compounding period of ¼ year. Equation (3) may be used to convert the resulting interest rate to an equivalent interest rate with annual compounding. This later interest rate is referred to as the effective annual interest rate. For the 10% nominal interest above, the effective annual interest rate is 10.38%.

Interest may also be defined in continuous rather than discrete form. With continuous interest (sometimes referred to as continuous compounding), interest acrues continuously. Equation (2) can be rewritten for continuous interest as

$$P = F \times e^{-rt} \qquad (4)$$

where r is the continuous interest rate and has units of inverse time (but is normally expressed as a percentage per unit of time), t is the time between P and F, and e is the base of the

---

*The term *interest rate* is used here instead of the time value of money. The results apply to interest associated with borrowing and interest in the context of the time value of money.

natural logarithm. Note that the time units on r and t must be consistent with the year normally used. Discretely compounded interest may be converted to continuously compounded interest by

$$r = \frac{1}{\Delta t} \ln(1 + i)$$

and continuously compounded interest to discretely compounded interest by

$$i = e^{r \Delta t} - 1$$

Note that these are dimensional equations; the units for r and Δt must be consistent and the interest rates are in fractional form.

It is often desirable to manipulate groups of cash flows rather than just single cash flows. The principles used in Eq. (2) or (3) may be extended to multiple cash flows if they occur in a regular fashion or if they flow continuously over a period of time at some defined rate. The types of cash flows that can be readily manipulated are

1. A uniform series in which cash flows occur in equal amounts on a regular periodic basis.
2. An exponentially increasing series in which cash flows occur on a regular periodic basis and increase by a constant percentage each year as compared to the previous year.
3. A gradient in which cash flows occur on a regular periodic basis and increase by a constant amount each year.
4. A uniform continuous cash flow where the cash flows at a constant rate over some period of time.
5. An exponentially increasing continuous cash flow where the cash flows continuously at an exponentially increasing rate.

These cash flows are illustrated in Figs. 1–5.

Any one of these groups of cash flows may be related to a single cash flow P as shown in these figures. The relationship between a group of cash flows and a single cash flow (or a single to single cash flow) may be reduced to an interest factor. The interest factors resulting in a single present amount are shown in Table 2. Derivations for most of these interest factors

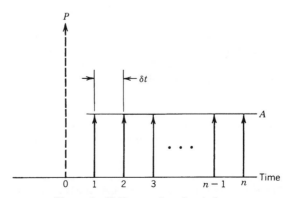

Figure 1 Uniform series of cash flows.

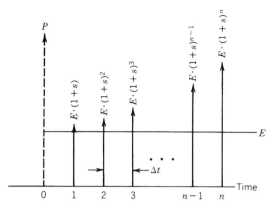

Figure 2 Exponentially increasing series cash flows.

can be found in an introductory text on engineering economics (see, for example, Grant et al.[4]). The interest factor gives the relationship between the group of cash flows and the single present amount. For example,

$$P = A \cdot (P/A, i, n)$$

The term $(P/A, i, n)$ is referred to as the interest factor and gives the ratio of P to A and shows that it is a function of i and n. Other interest factors are used accordingly. The interest factors may be manipulated as if they are the mathematical ratio they represent. For example,

$$(A/P, i, n) = \frac{1}{(P/A, i, n)}$$

Thus, for each interest factor represented in Table 2 a corresponding inverse interest factor may be generated as above. Two interest factors may be combined to generate a third interest factor in some cases. For example,

$$(F/A, i, n) = (P/A, i, n)(F/P, i, n)$$

or

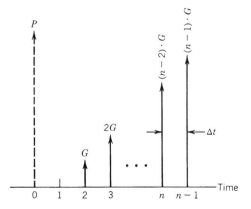

Figure 3 Cash flow gradient.

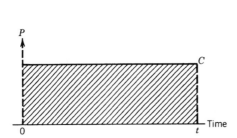
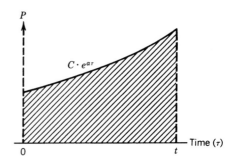

Figure 4 Uniform continuous cash flow.

Figure 5 Exponentially increasing continuous cash flow.

$$(F/C, r, t) = (P/C, r, t)(F/P, r, t)$$

Again the interest factors are manipulated as the ratios they represent. In theory, any combination of interest factors may be used in this manner. However, it is wise not to mix interest factors using discrete interest with those using continuous interest, and it is usually best to proceed one step at a time to ensure that the end result is correct.

There are several important limitations when using the interest factors in Table 2. The time relationship between cash flows must be adhered to rigorously. Special attention must be paid to the time between P and the first cash flow. The time between the periodic cash flows must be equal to the compounding period when interest factors with discrete interest rates are used. If these do not match, Eq. (3) must be used to find an interest rate with the appropriate compounding period. It is also necessary to avoid dividing by zero. This is not usually a problem; however, it is possible that $i = s$ or $r = a$, resulting in division by zero. The interest factors reduce to simpler forms in these special cases:

$$(P/F, i, s, n) = n, \quad s = i$$
$$(P/C, r, a, t) = c \cdot t, \quad a = r$$

It is sometimes necessary to deal with groups of cash flows that extend over very long periods of time ($n \to \infty$ or $t \to \infty$). The interest factors in this case reduce to simpler forms; but, limitations may exist if a finite value of P is to result. These reduced forms and limitations are presented in Table 3.

Several of the more common interest factors may be referred to by name rather than the notation used here. These interest factors and the corresponding names are presented in Table 4.

2 INVESTMENT DECISIONS

2.1 Allocation of Capital Funds

In most companies there are far more projects available than there are funds to implement them. It is necessary, then, to allocate these funds to the projects that provide the maximum return on the funds invested. The question of how to allocate capital funds will generally be handled at several different levels within the company, with the level at which the allocation is made depending on the size of the projects involved. At the top level, major projects such as plant additions or new product developments are considered. These may be multimillion

Table 2 Mathematical Expression of Interest Factors for Converting Cash Flows to Present Amounts[a]

Type of Cash Flows	Interest Factor	Mathematical Expression
Single	$(P/F, i, n)$	$= \dfrac{1}{(1+i)^n}$
Single	$(P/F, r, t)$	$n = \dfrac{t_2 - t_1}{\Delta t}$ $= e^{-rt}$ $t = t_2 - t_1$
Uniform series	$(P/A, i, n)$	$= \dfrac{1}{i}\left[1 - \dfrac{1}{(1+i)^n}\right]$ $n =$ number of cash flows
Uniform series	$(P/A, r, n, \Delta t)$	$= \dfrac{1}{e^{r\Delta t} - 1}(1 - e^{-rn\Delta t})$ $\Delta t =$ time between cash flows $n =$ number of cash flows
Exponentially increasing series	$(P/E, i, s, n)$	$= \dfrac{1+s}{i-s}\left[1 - \left(\dfrac{1+s}{1+i}\right)^n\right]$ $n =$ number of cash flows $s =$ escalation rate
Exponentially increasing series	$(P/E, r, s, \Delta t, n)$	$= \dfrac{1+s}{e^{r\Delta t} - s - 1}\left[1 - \left(\dfrac{1+s}{e^{r\Delta t}}\right)^n\right]$ $n =$ number of cash flows $\Delta t =$ time between cash flows $s =$ escalation rate
Gradient	$(P/G, i, n)$	$= \dfrac{1}{i}\left[\dfrac{(1+i)^n - 1}{i(1+i)^n} - \dfrac{n}{(1+i)^n}\right]$
Continuous	$(P/C, r, t)$	$= \dfrac{1 - e^{-rt}}{r}$ $t =$ duration of cash flow
Continuous increasing exponentially	$(P/C, r, a, t)$	$= \dfrac{1 - e^{-(r-a)t}}{r - a}$ $t =$ duration of cash flow $a =$ rate of increase of cash flow

[a] See Figs. 1–5 for definitions of variables n, Δt, s, a, E, G, and C. Variables i, r, a, and s are in fractional rather than percentage form.

or even multibillion dollar projects and have major impact on the future of the company. At this level both the decisions of which projects to fund and how much total capital to invest may be addressed simultaneously. At lower levels a capital budget may be established. Then, the projects that best utilize the funds available must be determined.

The basic principle utilized in capital rationing is illustrated in Fig. 6. Available projects are ranked in order of decreasing return. Those that are within the capital constraint are funded, those that are outside this constraint are not. However, it is not desirable to fund projects that have a rate of return less than the cost of money even if sufficient capital funds are available. If the size of individual projects is not small compared to the funds available, such as is the case with major investments, then it may be necessary to use linear programming techniques to determine the best set of projects to fund and the amount of funding to

Table 3 Interest Factors for $n \to \infty$ or $t \to \infty$

Interest Factor	Limitation
$(P/A, i, n) = \dfrac{1}{i}$	$i > 0$
$(P/A, r, n, \Delta t) = \dfrac{1}{e^{r\Delta t} - 1}$	$r > 0$
$(P/E, i, s, n) = \dfrac{1+s}{i-s}$	$s < i$
$(P/E, r, s, \Delta t, n) = \dfrac{1+s}{e^{r\Delta t} - s - 1}$	$s < e^{r\Delta t} - 1$
$(P/G, i, n) = \left(\dfrac{1}{i}\right)^2$	$i > 0$
$(P/C, r, t) = \dfrac{1}{r}$	$r > 0$
$(P/C, r, a, t) = \dfrac{1}{r - a}$	$a < r$

secure. As with most financial analyses, a fair amount of subjective judgment is also required. At the other extreme, the individual projects are relatively small compared to the capital available. It is not practical to use the same optimization techniques for such a large number of projects; and little benefit is likely to be gained anyway. Rather, a cutoff rate of return (also called required rate of return or minimum attractive rate of return) is established. Projects with a return greater than this amount are considered for funding; projects with a return less than this amount are not. If the cutoff rate of return is selected appropriately, then the total funding required for the projects considered will be approximately equal to the funds available. The use of a required rate of return allows the analysis of routine projects to be evaluated without using sophisticated techniques. The required rate of return may be thought of as an opportunity cost, since there are presumably unfunded projects available that earn approximately this return. The required rate of return may be used as the time value of money for routine economic analyses, assuming it is greater than the cost of money. For further discussion of the allocation of capital see Grant et al.[4] or other texts on engineering economics.

Table 4 Interest Factor Names

Factor	Name
$(F/P, i, n)$	Compound amount factor
$(P/F, i, n)$	Present worth factor
$(F/A, i, n)$	Series compound amount factor
$(P/A, i, n)$	Series present worth factor
$(A/P, i, n)$	Capital recovery factor
$(A/F, i, n)$	Sinking fund factor

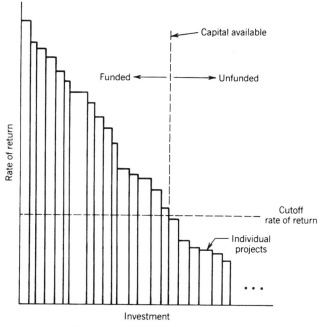

Figure 6 Principle of capital allocation.

2.2 Classification of Alternatives

From the engineering point of view, economic analysis provides a means of selecting among alternatives. These alternatives can be divided into three categories for the purpose of economic analysis: independent, mutually exclusive, and dependent.

Independent alternatives do not affect one another. Any one or any combination may be implemented. The decision to implement one alternative has no effect on the economics of another alternative that is independent. For example, a company may wish to reduce the delivered cost of some heavy equipment it manufactures. Two possible alternatives might be to (1) add facilities for rail shipment of the equipment directly from the plant and (2) add facilities to manufacture some of the subassemblies in-house. Since these alternatives have no effect on one another, they are independent. Each independent alternative may be evaluated on its own merits. For routine purposes the necessary criterion for implementation is a net profit based on discounted cash flows or a return greater than the required rate of return.

Mutually exclusive alternatives are the opposite extreme from independent alternatives. Only one of a group of mutually exclusive alternatives may be selected, since implementing one eliminates the possibility of implementing any of the others. For example, some particular equipment in the field may be powered by a diesel engine, a gasoline engine, or an electric motor. These alternatives are mutually exclusive since once one is selected there is no reason to implement any of the others. Mutually exclusive alternatives cannot be evaluated separately but must be compared to each other. The single most profitable, or least costly, alternative as determined by using discounted cash flows is the most desirable from an economic point of view. The alternative with the highest rate of return is not necessarily the most desirable. The possibility of not implementing any of the alternatives should also be considered if it is a feasible alternative.

Dependent alternatives are like independent alternatives in that any one or any combination of a group of independent alternatives may be implemented. Unlike independent alternatives, the decision to implement one alternative will affect the economics of another, dependent alternative. For example, the expense for fuel for a heat-treating furnace might be reduced by insulating the furnace and by modifying the furnace to use a less costly fuel. Either one or both of these alternatives could be implemented. However, insulating the furnace reduces the amount of energy required by the furnace thus reducing the savings of switching to a less costly fuel. Likewise, switching to a less costly fuel reduces the energy cost thus reducing the savings obtained by insulating the furnace. These alternatives are dependent, since implementing one affects the economics of the other one. Not recognizing dependence between alternatives is a very common mistake in economic analysis. The dependence can occur in many ways, sometimes very subtly. Different projects may share costs or cause each other to be more costly. They may interfere with each other or complement each other. Whenever a significant dependence between alternatives is identified, they should not be evaluated as being independent. The approach required to evaluate dependent alternatives is to evaluate all feasible combinations of the alternatives. The combination that provides the greatest profit, or least cost, when evaluated using discounted cash flows indicates which alternatives should be implemented and which should not. The number of possible combinations becomes very large quite quickly if very many dependent alternatives are to be evaluated. Therefore, an initial screening of combinations to eliminate obviously undesirable ones is useful.

Many engineering decisions do not deal with discrete alternatives but rather deal with one or more parameters that may vary over some range. Such situations result in continuous alternatives, and, in concept at least, an infinite number of possibilities exist but only one may be selected. For example, foam insulation may be sprayed on a storage tank. The thickness may vary from a few millimeters to several centimeters. The various thicknesses represent a continuous set of alternatives. The approach used to determine the most desirable value of the parameter for continuous alternatives is to evaluate the profit, or cost, using discounted cash flows for a number of different values to determine an optimum value. Graphical presentation of the results is often very helpful.

2.3 Analysis Period

An important part of an economic analysis is the determination of the appropriate analysis period. The concept of life-cycle analysis is used to establish the analysis period. Life-cycle analysis refers to an analysis period that extends over the life of the entire project including the implementation (e.g., development and construction). In many situations an engineering project addresses a particular need (e.g., transporting fluid from a storage tank to the processing plant). When it is clear that this need exists only for a specific period of time, then this period of need may establish the analysis period. Otherwise the life of the equipment involved will establish the project life.

Decisions regarding the selection, replacement, and modification of particular equipment and machinery are often a necessary part of many engineering functions. The same principles used for establishing an analysis period for the larger projects apply in this situation as well. If the lives of the equipment are greater than the period of need, then that period of need establishes the analysis period. If the lives of the equipment are short compared to the analysis period, then the equipment lives establish the analysis period. The lives of various equipment alternatives are often significantly different. The concept of life-cycle analysis requires that each alternative be evaluated over its full life. Fairness in the comparison requires that the same analysis period be applied to all alternatives that serve the same function. In order to resolve these two requirements more than one life cycle for the equipment may be used to establish an analysis period that includes an integer number of life cycles of each alternative. For example, if equipment with a life of 6 years is compared to equipment with a life of 4 years, an appropriate analysis period is 12 years, two life cycles of the first alternative and three life cycles of the second.

Obsolescence must also be considered when selecting an analysis period. Much equipment becomes uneconomical long before it wears out. The life of equipment, then, is not set by how long it can function but rather by how long it will be before it is desirable to upgrade with a newer design. Unfortunately, there is no simple method to determine when this time will be since obsolescence is due to new technology and new designs. In a few cases, it is clear that changes will soon occur (e.g., a new technology that has been developed but that is not yet in the marketplace). Such cases are the exception rather than the rule and much subjective judgment is required to estimate when something will become obsolete.

The requirements to maintain acceptable liquidity in a firm also may affect the selection of an analysis period. An investment is expected to return a net profit. The return may be a number of years after the initial investment, however. If a company is experiencing cash flow difficulties or anticipates that they may, this delay in receiving income may not be acceptable. A long-term profit is of little value to a company that becomes insolvent. In order to maintain liquidity, an upper limit may be placed on the time allowed for an investment to show a profit. The analysis period must be shortened then to reflect this requirement if necessary.

3 EVALUATION METHODS
3.1 Establishing Cash Flows

The first part of any economic evaluation is necessarily to determine the cash flows that appropriately describe the project. It is important that these cash flows represent all economic aspects of the project. All hidden cost (e.g., increased maintenance) or hidden benefits (e.g., reduced downtime) must be included as well as the obvious expenses, incomes, or savings associated with the project. Wherever possible, nonmonetary factors (e.g., reduced hazards) should be quantified and included in the analysis. Also, taxes associated with a project should not be ignored. (Some companies do allow a before-tax calculation for routine analysis.) Care should be taken that no factor be included twice. For example, high maintenance costs of an existing machine may be considered an expense in the alternative of keeping that machine or a savings in the alternative of replacing it with a new one, but it should not be considered both ways when comparing these two alternatives. Expenses or incomes that are irrelevant to the analysis should not be included. In particular, sunk costs, those expenses which have already been incurred, are not a factor in an economic analysis except for how they affect future cash flows (e.g., past equipment purchases may affect future taxes owing to depreciation allowances). The timing of cash flows over a project's life is also important, since cash flows in the near future will be discounted less than those that occur later. It is, however, customary to treat all of the cash flows that occur during a year as a single cash flow either at the end of the year (year-end convention) or at the middle of the year (mid-year convention).

Estimates of the cash flows for a project are generally determined by establishing what goods and services are going to be required to implement and sustain a project and by establishing the goods, services, benefits, savings, etc., that will result from the project. It is then necessary to estimate the associated prices, costs, or values. There are several sources of such information including historical data, projections, bids, or estimates by suppliers, etc. Care must be exercised when using any of these sources to be sure they accurately reflect the true price when the actual transaction will occur. Historical data are misleading, since they reflect past prices, not current prices, and may be badly in error owing to inflation that has occurred in recent years. Current prices may not accurately reflect future prices for the same reason. When historical data are used, they should be adjusted to reflect changes that have occurred in prices. This adjustment can be made by

$$p(t_0) = p(t_1) \frac{\text{PI}(t_0)}{\text{PI}(t_1)} \tag{5}$$

where $p(t)$ is the price, cost, or value of some item at time t; $\text{PI}(t)$ is the price index at time t, and t_0 is the present time. The price index reflects the change in prices for an item or group of items. Indexes for many categories of goods and services are available from the Bureau of Labor Statistics.[2] Current prices should also be adjusted when they refer to future transactions. Many companies have projections for prices for many of their more important products. Where such projections are not available, a relationship similar to Eq. (5) may be used except that price indexes for future years are not available. Estimates of future inflation rates may be substituted instead:

$$p(t_2) = p(t_0)(1 + f)^n \tag{6}$$

where f is the annual inflation rate, n is the number of years from the present until time t_2 ($t_2 - t_0$ in years), and t_0 is the present time. The inflation rate in Eq. (6) is the overall inflation rate unless it is expected that the particular item in question will increase in price much faster or slower than prices in general. In this case, an inflation rate pertaining to the particular item should be used.

Changing prices often distort the interpretation of cash flows. This distortion may be minimized by expressing all cash flows in a reference year's dollars (e.g., 1990 dollars). This representation is referred to as constant dollar cash flows. Historic data may be converted to constant dollar representation by

$$Y^c = Y^d \frac{\overline{\text{PI}}(t_0)}{\overline{\text{PI}}(t)} \tag{7}$$

where Y^c is the constant dollar representation of a dollar cash flow U^d, $\overline{\text{PI}}(t)$ is the value of the price index at time t, t_0 is the reference year, and t is the year in which Y^d occurred. An overall price index such as the Wholesale Price Index or the Gross National Product Implicit Price Deflator is used in this calculation, whereas a more specific price index is used in Eq. (5). Future cash flows may be expressed using constant dollar representation by

$$Y^c = Y^d \frac{1}{(1 + f)^n} \tag{8}$$

where f is the projected annual inflation rate and n is the number of years after t_0 that Y^d occurs. It is usually convenient to let t_0 be the present. Then n is equal to the number of years from the present and, also, present prices may be used to make most constant dollar cash flow estimates. The use of constant dollar representation simplifies the economic analysis in many situations. However, it is important that the time value of money be adjusted as indicated in Eq. (2). Additional discussion on this topic may be found in Ref. 2.

Mutually exclusive alternatives and dependent alternatives often yield cash flows that are either all negative or predominantly negative, that is, they only deal with expenses. It is not possible to view each alternative in terms of an investment (initial expense) and a return (income). However, two alternatives may be used to create a set of cash flows that represent an investment and return as shown in Fig. 7. Alternative B is more expensive to implement than A but costs less to operate or sustain. Cash flow C is the difference between B and A. It shows the extra investment required for B and the savings it produces. Cash flow C may then be analyzed as an investment to determine if the extra investment required for C is worthwhile. This approach works well when only two alternatives are considered. If there are three or more alternatives, the comparison gets more complicated. Figure 8 shows the process required. Two alternatives are compared, the winner of that comparison is compared to a third, the winner of that comparison to a fourth, and so on until all alternatives have been considered and the single best alternative identified. When using this procedure, it is customary, but not necessary, to order the alternatives from least expensive to most expensive according to initial cost. This analysis of multiple alternatives may be referred to as incremental analysis, since only the difference, or incremental cash flow between alternatives, is considered. The same concept may be applied to continuous alternatives.

3.2 Present Worth

All of the cash flows for a project may be reduced to a single equivalent cash flow using the concepts of time value of money and cash flow equivalence. This single equivalent cash flow is usually calculated for the present time or the project initiation, hence the term present

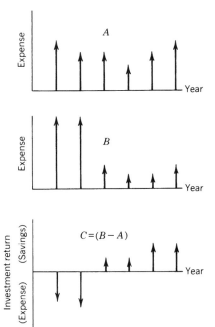

Figure 7 Investment and return generated by comparing two alternatives.

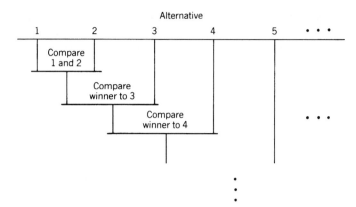

Figure 8 Comparison of multiple alternatives using incremental cash flows.

worth (also present value). However, this single cash flow can be calculated for any point in time if necessary. Occasionally, it is desired to calculate it for the end of a project rather than the beginning, the term future worth (also future value) is applied then. This single cash flow, either a present worth or future worth, is a measure of the net profit for the project and thus is an indication of whether or not the project is worthwhile economically. It may be calculated using interest factors and cash flow manipulations as described in preceding sections. Modern calculators and computers usually make it just as easy to calculate the present worth directly from the project cash flows. The present worth PW of a project is

$$\text{PW} = \sum_{j=0}^{n} Y_j \frac{1}{(1+i)^j} \tag{9}$$

where Y_j is the project cash flow in year j, n is the length of the analysis period, and i is the time value of money. This equation uses the sign convention of income or savings being positive and expenses being negative.

In the case of independent alternatives, the PW of a project is a sufficient measure of the project's profitability. Thus, a positive present worth indicates an economically desirable project and a negative present worth indicates an economically undesirable project. In the case of mutually exclusive, dependent, or continuous alternatives the present worth of a given alternative means little. The alternative, or combination of dependent alternatives, that has the highest present worth is the most desirable economically. Often cost is predominant in these alternatives. It is customary then to reverse the sign convention in Eq. (9) and call the result the present cost. The alternative with the smallest present cost is then the most desirable economically. It is also valid to calculate the present worth of the incremental cash flow (see Figs. 7 and 8) and use it as the basis for choosing between alternatives. However, the approach of calculating the PW or PC for each alternative is generally much easier. Regardless of the method chosen, it is important that all alternatives be treated fairly; use similar assumptions about future prices, use the same degree of conservatism in estimating expenses, make sure they all serve equally well, etc. In particular, be sure that the proper analysis period is selected when equipment lives differ.

Projects that have very long or indefinite lives may be evaluated using the present worth method. The present cost is referred to as capitalized cost in this application. The capitalized cost can be used for economic analysis in the same manner as the present cost; however, it cannot be calculated using Eq. (9), since the number of calculations required would be rather large as $n \to \infty$. The interest factors in Table 3 usually can be used to reduce the portion of the cash flows that continue indefinitely to a single equivalent amount, which can then be dealt with as any other single cash flow.

3.3 Annual Cash Flow

The annual cash flow method is very similar in concept to the present worth method and generally can be used whenever a present worth analysis can. The present worth or present cost of a project can be converted to an annual cash flow, ACF, by

$$\text{ACF} = \text{PW} \cdot (A/P, i, n) \tag{10}$$

where $(A/P, i, n)$ is the capital recovery factor, i is the time value of money, and n is the number of years in the analysis period. Since ACF is proportional to PW, a positive ACF indicates a profitable investment and a negative ACF indicates an unprofitable investment. Similarly, the alternative with the largest ACF will also have the largest PW. The PW in Eq. (10) can be replaced with PC, and ACF can be used to represent a cost when that is more appropriate. The ACF is thus equally as useful for economic analysis as PW or PC. It also has the advantage of having more intuitive meaning. ACF represents the equivalent annual income or cost over the life of a project.

Annual cash flow is particularly useful for analyses involving equipment with unequal lives. The n in Eq. (10) refers to the length of the analysis period and for unequal lives that means integer multiples of the life cycles. The annual cash flow for the analysis period will be the same as the analysis period for a single life cycle, as shown in Fig. 9, as long as the cash flows for the equipment repeat from one life cycle to the next. The annual cash flow for each equipment alternative can then be calculated for its own life cycle rather than for a number of life cycles. Unfortunately, prices generally increase from one life cycle to the next due to inflation and the cash flows from one life cycle to the next will be more like those shown in Fig. 10. The errors caused by this change from life cycle to life cycle usually will be acceptable if inflation is moderate (e.g., less than 5%) and the lives of various alternatives do not differ greatly (e.g., 7 versus 9 years). If inflation is high or if alternatives have lives that differ greatly, significant errors may result. The problem can often be circum-

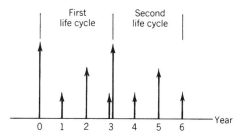

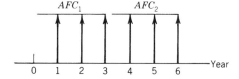

Figure 9 Annual cash flow for equipment with repeating cash flows (3-year life).

580 Investment Analysis

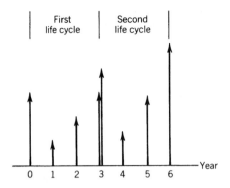

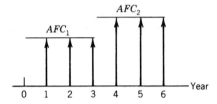

Figure 10 Annual cash flow for equipment with nonrepeating cash flows (3-year life).

vented by converting the cash flows to constant dollars using Eq. (8). With the inflationary price increases removed, the cash flows will usually repeat from one life cycle to the next.

3.4 Rate of Return

The rate of return (also called the internal rate of return) method is the most frequently used technique for evaluating investments. The rate of return is based on Eq. (9) except that rather than solving for PW, the PW is set to zero and the equation is solved for i. The resulting interest rate is the rate of return of the investment:

$$O = \sum_{j=0}^{n} Y_j \frac{1}{(1+i)^j} \qquad (11)$$

where Y_j is the cash flow for year j, i is the investment's rate of return (rather than the time value of money), and n is the length of the analysis period. It is usually necessary to solve Eq. (11) by trial and error, except for a few very simple situations. If constant dollar representation is used, the resulting rate of return is referred to as the real rate of return or the inflation-corrected rate of return. It may be converted to a dollar rate of return using Eq. (2).

A rate of return for an investment greater than the time value of money indicates that the investment is profitable and worthwhile economically. Likewise an investment with a rate of return less than the time value of money is unprofitable and is not worthwhile economically. The rate of return calculation is generally preferred over the present worth method or annual cash flow method by decision makers since it gives a readily understood economic measure. However, the rate of return method only allows a single investment to be evaluated or two projects compared using the incremental cash flow. When several mutually exclusive alternatives, dependent alternative combinations, or continuous alternatives exist, it is necessary to compare two investments at a time as shown in Fig. 8 using incremental cash flows. Present worth or annual cash flow methods are simpler to use in these instances. It is important to realize that with these types of decisions the alternative with the highest rate of return is not necessarily the preferable alternative.

The rate of return method is intended for use with classic investments as shown in Fig. 11. An expense (investment) is made initially and income (return) is generated in later years. If a particular set of cash flows, such as shown in Fig. 12, does not follow this pattern, it is possible that Eq. (11) will generate more than one solution. It is also very easy to misinterpret the results of such cash flows. Reference 4 explains how to proceed in evaluating cash flows of the nature shown in Fig. 12.

3.5 Benefit–Cost Ratio

The benefit–cost ratio (B/C) calculation is a form of the present worth method. With B/C, Eq. (9) is used to calculate the present worth of the income or savings (benefits) of the investment and the expenses (costs) separately. These two quantities are then combined to form the benefit–cost ratio:

$$B/C = \frac{\text{PW}_B}{\text{PC}_E} \qquad (12)$$

where PW_B is the present worth of the benefits and PC_E is the present cost of the expenses. A B/C greater than 1 indicates that the benefits outweigh the costs and a B/C less than 1 indicates the opposite. The B/C also gives some indication as to how good an investment is. A B/C of about 1 indicates a marginal investment, whereas a B/C of 3 or 4 indicates a very good one. The PC_E usually refers to the initial investment expense. The PW_B includes the income and savings less operating cost and other expenses. There is some leeway in deciding whether a particular expense should be included in PC_E or subtracted from PW_B. The placement will change B/C some, but will never make a B/C which is less than 1 become greater than 1 or vice versa.

The benefit–cost calculation can only be applied to a single investment or used to compare two investments using the incremental cash flow. When evaluating several mutually exclusive alternatives, dependent alternative combinations, or continuous alternatives, the alternatives must be compared two at a time as shown in Fig. 8 using incremental cash flows. The single alternative with the largest B/C is not necessarily the preferred alternative in this case.

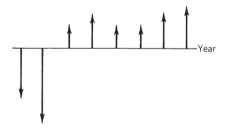

Figure 11 Example of pure investment cash flows.

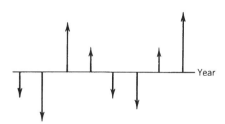

Figure 12 Example of mixed cash flows.

3.6 Payback Period

The payback calculation is not a theoretically valid measure of the profitability of an investment and is frequently criticized for this reason. However, it is widely used and does provide useful information. The payback period is defined as the period of time required for the cumulative net cash flow to be equal to zero; that is, the time required for the income or savings to offset the initial costs and other expenses. The payback period does not measure the profitability of an investment but rather its liquidity. It shows how fast the money invested is recovered. It is a useful measure for a company experiencing cash flow difficulties and which cannot afford to tie up capital funds for a long period of time. A maximum allowed payback period may be specified in some cases. A short payback period is generally indicative of a very profitable investment, but that is not ensured since there is no accounting for the cash flows that occur after the payback period. Most engineering economists agree that the payback period should not be used as the means of selecting among alternatives.

The payback period is sometimes calculated using discounted cash flows rather than ordinary cash flows. This modification does not eliminate the criticisms of the payback calculation although it does usually result in only profitable investments having a finite payback period. A maximum allowed payback period may also be used with this form. This requirement is equivalent to arbitrarily shortening the analysis period to the allowed payback period to reflect liquidity requirements. Since there are different forms of the payback calculation and the method is not theoretically sound, extreme care should be exercised in using the payback period in decision making.

REFERENCES

1. L. E. Bussey, *The Economic Analysis of Industrial Projects,* Prentice-Hall, Englewood Cliffs, NJ, 1978.
2. B. W. Jones, *Inflation in Engineering Economic Analysis,* Wiley, New York, 1982.
3. J. L. Riggs, *Engineering Economics,* 2nd ed., McGraw-Hill, New York, 1982.
4. E. L. Grant, W. G. Ireson, and R. S. Leavenworth, *Principles of Engineering Economy,* 7th ed., Wiley, New York, 1982.

CHAPTER 18

TOTAL QUALITY MANAGEMENT, SIX SIGMA, AND CONTINUOUS IMPROVEMENT

Jack B. ReVelle, Ph.D.
ReVelle Solutions, LLC
Santa Ana, California

Robert Alan Kemerling
Ethicon Endo-Surgery, Inc.
Cincinnati, Ohio

1	WHAT IS TQM?	583	4	TQM—THE DMAIC PROCESS	591
	1.1 Traditional Quality	583		4.1 Define Phase	591
	1.2 TQM/Six Sigma—A New Approach	584		4.2 Tools for the Define Phase	591
	1.3 Definitions of Quality	586		4.3 Measure Phase	597
				4.4 Tools for the Measure Phase	597
				4.5 Analysis Phase	603
2	BENEFITS FOR MY COMPANY AND ME	586		4.6 Tools for the Analysis Phase	603
	2.1 TQM/SS as Predictor of Company Performance	586		4.7 Improvement (and Innovate) Phase	607
				4.8 Tools for the Improvement (and Innovate) Phase	607
3	THE ENGINEER'S ROLE WITH TQM/SS	588		4.9 Control Phase	609
	3.1 As a Mechanical Engineer	588		4.10 Tools for the Control Phase	609
	3.2 As a Manager of Mechanical Engineers	590		**REFERENCES**	**614**
	3.3 TQM/SS as a Career Aid	590		**BIBLIOGRAPHY**	**614**

1 WHAT IS TQM?

1.1 Traditional Quality

Explaining total quality management requires a brief look at history. Formerly, individual artisans designed and produced goods and services for their community. As the benefits of mass production were realized, people began to organize production around functions. This organization for mass production brought great benefits with regard to the quantity of goods and services. It also lowered their cost. It did bring a downside, however. Customers found that the quality of goods and services was sometimes inconsistent and even inferior. The Department of Defense (DOD) was especially concerned with the level of quality for war-effort items. Figure 1 shows the effect of separating responsibilities.

As a result of their unfortunate experiences, the DOD began to require firms that did business with the Federal government to develop a specific part of their organizations to accept final responsibility for the consistent quality of their output. For the purposes of further discussion, we'll call this part of a firm its Quality Department. The Quality Department was responsible for quality planning, internal quality review, and checking of production

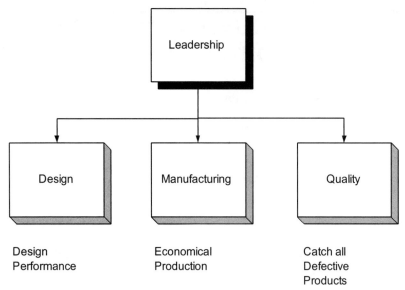

Figure 1 Separated responsibilities.

output. The DOD specified that this department be independent from production and it further provided guidance in the form of sampling plans and sampling methods developed by statisticians working for the Federal government. Firms that were serious about working with the Federal government developed a separate department with management, quality engineers, and inspectors reporting to senior management.

The downside to this approach is what one would expect with respect to the principles of accountability: if the Quality Department is responsible for quality, then no one else had to worry about it. In the very worst case, production would make things as fast as it could, while quality inspectors had to sort the good from the bad. This approach proved very inefficient for many firms, and a significant amount of production was scrapped or returned for costly rework. Customers did not necessarily benefit either. The cost of scrap and rework was priced into the output. Also, inspection was not always 100% efficient, allowing some nonconforming material to escape.

1.2 TQM/Six Sigma—A New Approach

A new approach to quality, called Total Quality Management (TQM), makes the case that quality is not just the responsibility of a certain department. Rather, it must be a responsibility of every part of the organization. This is necessary not only to avoid the large costs of scrap and rework, but also to focus on satisfying customer needs. Both allow the firm to be competitive in the global marketplace. Figure 2 displays the interlocking responsibilities of functions within a TQM environment.

Obviously, mechanical engineers and managers will not be making or inspecting product. So the remainder of this chapter focuses on TQM as it is currently embodied in the workplace. It will also provide you, the mechanical engineer, with sufficient tools and knowledge so as to be successful with your processes and teams.

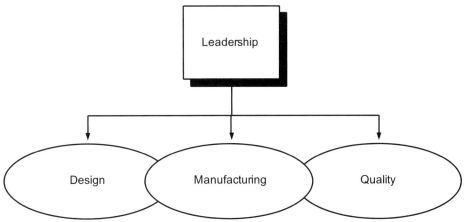

Figure 2 Joint responsibilities.

Since the 1990s, a process widely known as Six Sigma (SS) has swallowed TQM. Begun by Motorola and adopted by companies like GE, Honeywell, and Texas Instruments, Six Sigma strives to identify the key attributes of a product or process and employs a particular process order and tools to assure high performance with respect to the key attributes. The TQM/SS approach to continuous improvement has become known as the DMAIC process (pronounced *duh-MAY-ick*) after the following process steps:

- *Define.* Define the key attributes of the product or process under review. This could mean obtaining voice of the customer (VOC) information or performing sufficient predictive analysis to understand important values.
- *Measure.* Determine how the key attributes of the product or process will be measured. In this step (and perhaps in other steps below), a process improvement team (PIT) will also perform measurement systems analysis (MSA) to determine the accuracy of the measurement system.
- *Analyze.* Find the sources of variation and/or key parameters of process(es) to improve the outputs.
- *Improve.* Remove the sources of variation and/or set key process parameters for minimum variation. Some practitioners add *Innovate* to this step and call it a DMAI^{2}C process.
- *Control.* Install controls both to keep the process as the PIT has defined it and to indicate to management a signal that the process has shifted.

There is a chapter on Design for Six Sigma elsewhere in this handbook (Chapter 17 of Materials and Mechanical Design) so the focus of this chapter is on designing excellent processes and improving existing ones. This chapter addresses the DMAIC process steps and applicable tools.

1.3 Definitions of Quality

Expectations of quality have changed over time. The following are some definitions that have been used for quality:

- Freedom from defects[1]
- Fitness for use[1]
- The totality of features and characteristics of a product or service that bear on its ability to satisfy given needs.[2]
- The features and characteristics that delight the customer![3]

Note that these definitions progress from a narrow assessment of defects to the broader consideration of satisfying the customer. As a matter of fact, your customer will probably take into account the entire customer experience with your firm. Sales, customer service, and technical support may have as much of an effect on customer satisfaction as your product. All these processes can also benefit from the Six Sigma DMAIC process. Articles on the application of the DMAIC approach to processes such as credit and collection are available in current business journals.[5]

2 BENEFITS FOR MY COMPANY AND ME

2.1 TQM/SS as Predictor of Company Performance

Several benefits stem from the adoption of an active and effective TQM/SS program:

- Improved customer satisfaction resulting from better products and services
- Greater profit margins resulting from reduced costs
- Faster transition from old to new products and services
- Higher worker satisfaction resulting from involvement with process improvement teams, product and process development teams, as well as design for manufacture and assembly teams.

These are strong claims, but they are supported by existing, valid data. The first study to address the effects of TQM application beyond the quality of products and services was conducted by the General Accounting Office (GAO) at the request of then Congressman Donald Ritter (R-P).[6] This study looked at 20 companies, each of which received a site visit for the Baldrige National Quality Award (BNQA) in 1988 and 1989. See Chapter 19 for a discussion of this and other quality awards. Receipt of a site visit for the BNQA indicates that a company is a "finalist" in this assessment of TQM applications.

The GAO study considered data (where available) in four broad areas with a number of specific elements in each:

1. Employee relations
2. Operating procedures
3. Customer satisfaction
4. Financial performance

In each case, the available data for the each company were analyzed for trends from the time the company reported it began its TQM initiatives. In addition, the company data were compared with performance measures that were available from their specific industry. The results are revealed in Fig. 3. All four charts are set to the same scale and represent average

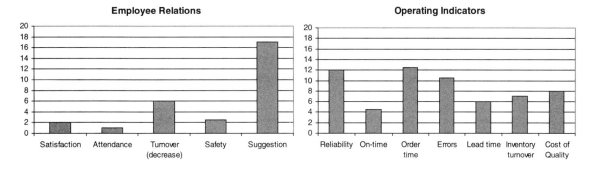

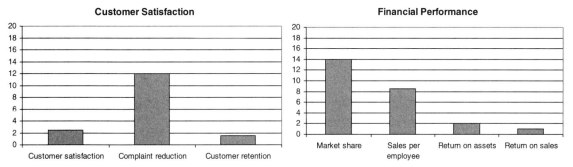

Figure 3 Four levels of improvement from TQM.

annual percent improvement. The results are stated so that a positive bar represents a favorable result. The specific elements for each area are noted beneath the bar.

- In the area of employee-related indicators, the survey studied employee satisfaction (using survey instruments), attendance, turnover, safety/health (lost workdays associated with work-related injuries and illnesses), and suggestions received from employees. These measures indicate the extent of personnel engagement in TQM and staff response to the initiative.

- The survey also examined operating indicators. These performance measures of the quality and costs of products and services. The categories of measurement included

  1. Reliability
  2. Timeliness of delivery
  3. Order-processing time
  4. Errors or defects
  5. Product lead time
  6. Inventory turnover
  7. Costs of quality
  8. Cost savings

  These categories represent a measure of quality system effectiveness.

- Customer satisfaction is an important indicator for any business. If customers are not satisfied, a company's profitability will be affected at some point, usually sooner rather than later. This survey looked at three measures of customer satisfaction:

  1. Overall customer satisfaction
  2. Customer complaints
  3. Customer retention

- The survey also looked at improvement in the financial performance of the companies applying TQM. The performance measures studied were

  1. Market share

  2. Sales per employee

  3. Return on assets

  4. Return on sales

  These measures put to rest the theory that TQM efforts do not offer an attractive return on investment. How much is a 14% annual increase in market share worth to your company?

As can be seen from the results of the study just described, Total Quality Management/Six Sigma is much more than a management fad. Since TQM/SS has continued to be used in various types of companies, we now have more of a track record and a broader and deeper database to analyze. Companies employing TQM/SS in a measurable way have shown above average performance in many key business measures.

Continuing our discussion of the Baldrige National Quality Award (BNQA), we can use readily available business measures of public financial data to see if a company's use of the BNQA criteria has had a business effect. While companies decide to apply for the award assessment (that is to say, they self-select versus being selected as would happen in a scientific sample), we can still assess the results.

The National Institute of Standards and Technology (NIST), which administers the BNQA, undertook a study of stock performance for those companies that won the award from 1988 to 1995. The NIST methodology showed that for this period, publicly traded BNQA winners, as a group, outperformed the Standard and Poor's (S&P) 500 index by nearly 3 to 1, achieving a 324.9% return compared to a 111.8% return for the S&P 500 (see Fig. 4).[4]

3 THE ENGINEER'S ROLE WITH TQM/SS

3.1 As a Mechanical Engineer

Traditionally, engineers become engineers because they have an aptitude for or prefer to deal with data and things. The typical mechanical engineer is most focused on one key responsibility, the performance of his or her design or process. This is still an important consideration, but as your organization adopts TQM/SS, whether due to customer requirements or competitive pressures, some new dimensions will be added to your role. As shown in Fig. 5, TQM/SS has many aspects that affect both the organization and the individuals. This section includes a brief discussion of some of them.

First of all, a mechanical engineer working in a TQM/SS environment will probably be part of a multifunctional team, usually an integrated product and process development team (IPT) or process improvement team. This will require what may be new skills, such as listening to other viewpoints on a design, reaching consensus on decisions, and achieving alignment on customer needs. To the mechanical engineer, teams may appear inefficient, slowing down "important" design work, but the performance of a well-developed team has often proven superior to other organizational forms.

Another change that a mechanical engineer may note in TQM/SS is a focus on processes. In the past, engineers usually felt that the result was important, not necessarily the means. TQM/SS focuses on the means (processes) as much as the results. This is one way

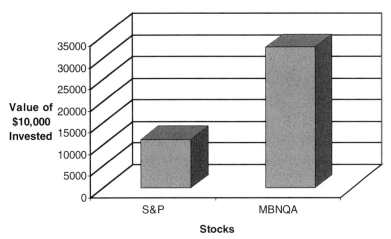

Figure 4 Baldrige returns lead the way.

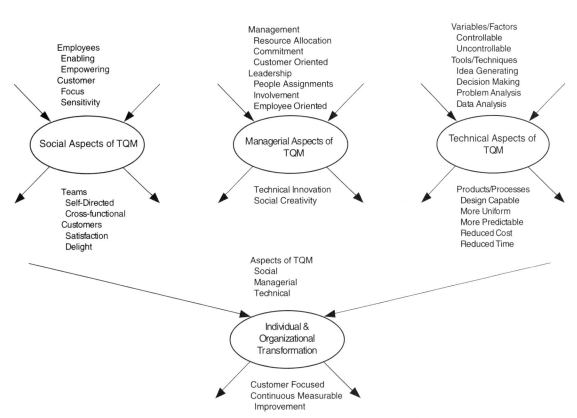

Figure 5 Social, technical and managerial aspects of TQM.

to achieve minimum variation in results, i.e., to consistently use the best process available. At first thought, this may appear restrictive, but it is not. TQM/SS is serious about continuous improvement. This means that processes will not remain static, but when the current "best process" is discovered, all functions that can use it are expected to use it.

A final key change that a mechanical engineer might note in an organization adopting TQM/SS involves the engineer's relationship with the management structure. To free up the creative capability in the organization and to make it more agile, management must move from a directive relationship to a coaching or guiding relationship. Of course, this will be a significant change for the manager and engineer and sometimes the personal transition is not smooth.

3.2 As a Manager of Mechanical Engineers

If you are a manager of mechanical engineers in an organization deploying TQM/SS, it's likely you are in for changes that may make you feel insecure in your position. You will see a drive to reduce your apparent authority, to place your staff on teams, and to turn your position into one of "coach." It's possible that you'll stop receiving funding to supply personnel for projects. Instead, the funding will go directly to the team. Your personnel will most likely be relocated with their team, perhaps geographically removed from you, making communication difficult.

We have emphasized this negative picture to draw attention to the focus on the role of management in deploying TQM/SS. A significant part of the pressure to change and the pressure *from* change falls on management. If you believe TQM/SS is something to assign to someone else or that it is something your staff can do without your involvement, you are on a path to a failed implementation.

In addition to personnel considerations, there are other concerns you must consider for a TQM/SS implementation. Processes must be put in place to

- Determine what teams are necessary and how many.
- Pick team leaders and team members.
- Equip and train teams.
- Identify or grow subject matter experts (SMEs) for key TQM/SS and team tools.
- Develop data systems to support team efforts.
- Understand what your customers want and don't want.
- Fund the teams.
- Identify staffing needs, if funding goes to the team.
- Evaluate and develop personnel outside the traditional functional environment and sometimes remote from you.
- Identify when a team is not performing well.

3.3 TQM/SS as a Career Aid

Companies that decide to employ TQM/SS as a competitive advantage need personnel who are aligned with this objective. Support of TQM/SS can, therefore, be an important attribute that management considers when looking for hiring and promoting candidates. If you demonstrate a successful track record in TQM/SS projects, this will be perceived as desirable experience in most companies. Since a portion of your compensation often comes from stock and stock options, the improved performance of those companies adopting TQM/SS bears consideration. Obviously, it's in your interest to see that your stock improves!

4 TQM/SS—THE DMAIC PROCESS

To help you understand the different phases of the DMAIC process and to identify what TQM/SS tools are available, each of the process steps is addressed, followed by discussions of some of the tools associated with each phase. These tools are covered in general; references containing more compete descriptions are provided at the conclusion of this chapter.

The DMAIC process is employed in a process improvement project environment. Processes may be selected for any reason, but, generally, the aim is improvement of the business by satisfying the customer and/or removing unnecessary business costs such as scrap, rework, repair, or any other source of waste. In other words, DMAIC is a process, applied to a business improvement project.

4.1 Define Phase

In the define phase of DMAIC, the mechanical engineer is interested in three key actions. First, the engineer must identify the business needs met by the process. This should be phrased in terms of cost of goods (COGs), net profit, or reduction in scrap/rework/cost/time. Second, the engineer must *scope* the process, i.e., define the boundaries of the improvement or installation project. Practitioners of DMAIC often advise beginners to scope their projects carefully. Their advice is, "Don't take on the world" or "Don't try to solve world hunger." Finally, the engineer must identify what are called the critical to quality (CTQ) characteristics of the process output. CTQs are those aspects of a product or service that define the customer-perceived quality of the result.

CTQs start as words or specifications directly stated by the customer, but they must often be translated into values and attributes of the process. For example, we might discover that a key quality aspect of a new car is the quality of the finish. The customer might define quality of finish using words and phrases such as "smooth, consistent color, and absence of drips or runs." The process factors and inputs will not be described in the customers' terms. Rather, the team will work as a part of the DMAIC process to understand which process factors *control* the process output. These key factors might be items such as spray pressure settings, temperature, spray angle, spray head translation rate, or paint viscosity.

4.2 Tools for the Define Phase

Voice of the Customer (*VOC*)
For any product or situation, the key place to start is with the customer of your product or service. To meet your customers' requirements, you must understand your customer and his or her requirements. Japanese practitioners recommend that you go to the *gemba*. Roughly translated, *gemba* means "the source" or "the site." This means, go to your customers, talk with them, observe them, and understand how they intend to use your product.

The techniques for obtaining data from your customer are very simple. They range from sending out questionnaires to developing a list of questions to ask in person or over the phone. One of the most effective approaches is to ask your customers to show you how they perform their activity with your product or a competitor's product. Showing is far better than describing. By showing you how they use the product, they show you what they have to do both for preparation and disposal as well as how they use the product. You can see body position, hand position, amount of effort, and many other things that will not be described in a questionnaire. Ask them how they "feel" as they use the product. They may express frustration, but they may also express a sense of satisfaction in the result or a sense of control with the features you've given them. If possible, record their actions for a follow-on review. You will pick up small things that you overlook in person.

One technique used in market research is to form a *persona* to represent the customer of your product. This gives the design team a way to think of the customer in more personal terms than just data in a report. Some design teams give the customer persona a name and have a cardboard figure to represent the average customer. They even begin to pose their design discussions around their customer persona, asking, "Would Joan use this feature?" While this sounds trivial in engineering terms, such approaches keep a team focused on the customer and those items that the customer deems important.

Engineers supporting a production line that makes components may not readily understand the focus on VOC. Often the specifications of a component are very clear. Still there is a need to understand how the components are used. What packaging facilitates easy use of the component in the customer's factory? Is orientation of the component important? Is it easier for your customer to have parts individually packaged or in bulk? Are there ways to eliminate packaging waste for users? Are documents and component markings complete? What quality of component is expected? What features are critical and which are of less concern? These questions may not be adequately spelled out in specifications. Addressing these and other similar issues can add considerable value to your products for your customers.

Quality Function Deployment (QFD)

After determining important features and needs in the product, the tool of choice for capturing these and relating them to the design elements is QFD. You will recognize the core form of QFD as a simple L-shaped matrix.

QFD was initially applied in the 1960s in the Kobe shipyards of Mitsubishi Heavy Industries of Japan. It was refined through other Japanese industries in the 1970s. Donald Clausing was the first American who recognized QFD as an important tool. It was translated into English and introduced to the United States in the 1980s. Bob King's book *Better Designs in Half the Time* has been applied in a wide variety of U.S. industries.[4]

At the heart of applying QFD are one or more matrices. These matrices are the key to QFD's ability to link customer requirements (referred to as the voice of the customer or customer *WHATs* in QFD literature) with the organization's plans, product or service features, options, and analysis (referred to as *HOWs*). The first matrix used in a major application of QFD will usually be a form of the A-l matrix (Ref. 4, pp. 2–6). This matrix often includes features not always applied in the other matrices. As a result, it often takes a characteristic form and is called the house of quality (HOQ) in QFD literature. Figure 6 presents the basic form of the HOQ.

The A-l matrix starts with either raw (verbatim) or restated customer WHATs and along with corresponding priorities for each of the WHATs. Restated customer WHATs are generally still qualitative statements, but with more specificity. For example, if the original VOC was for the car dash to have a cup holder, the restated WHAT might be that it should have room for a 16-ounce cup of coffee. The priorities are usually coded from 10 to 1, with 10 representing the most important item(s) and 1 representing the least important. These WHATs and their priorities are listed as row headings down the left side of the matrix. Frequently we find that customer WHATs are qualitative requirements that are difficult to directly relate to design requirements, so a project team will develop a list of substitute quality characteristics and place them as column headings on this matrix. The column headings in QFD

4 TQM/SS—The DMAIC Process 593

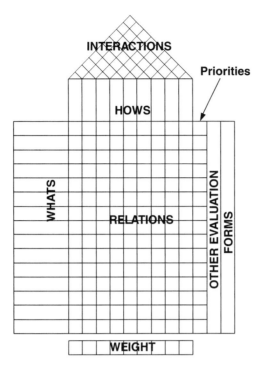

Figure 6 QFD House of Quality.

matrices are referred to as HOWs. Substitute quality characteristics are usually quantifiable measures that function as high-level product or process design targets and metrics.

The term "substitute quality characteristic" may appear ambiguous. The best way to think of this is to consider the fact that verbatim customer requirements may be stated in words that cannot be directly translated to equations. For example, when a user describes the need to make a kitchen appliance "easy to use," there is no way to state this as a specification with measurable output. However, it is still the *voice of the customer*. So, using QFD, it would be placed down the left column as a *WHAT*. A team would then place ways to make the device "easy to use" along the top as column headings. Examples of ways to achieve "easy to use" might be well-marked controls, no more than one knob, or automatic sensing of the appropriate setting. Each of these becomes a HOW and, if it relates to the WHAT, becomes a substitute quality characteristic for the "easy to use" WHAT.

The relationships between WHATs and HOWs are identified using symbols such as ● for a high or strong relationship, ○ for a moderate or medium relationship, and Δ for a low or weak relationship. These are entered at the row/column intersections of the matrix. The convention is to assign 9 points for a high relationship between a WHAT and a HOW, with 3, 1, and 0 for medium, low, and no correlations, respectively. The assignment of points to the various relationship levels and the prioritization of customer WHATs are used to develop a weighted list of HOWs. The relationship values (9, 3, 1, and 0) are multiplied by the WHATs priority values and summed over each HOW column. These column summations indicate the relative importance of the substitute quality characteristics and their strength of linkage to the customer requirements.

The other major element of the A-l matrix is the characteristic triangular top (an isosceles triangle), which contains the interrelationship assessments of the HOWs. This additional triangle looks like a roof and gives the QFD matrix the profile of a house, hence its nickname, the house of quality (refer once again to Fig. 6). The roof contains indicators that show the relationship between "HOWs." The best way to think of this is to consider what would happen to the other design elements if each one is increased in turn. Consider, for example, a QFD for a car. In response to customer needs and wants, we intend high mileage and ease of operation. To achieve high mileage, we also intend to forego power steering and automatic transmission. The latter decision would improve mileage, but it would have a detrimental effect on ease of operation for most drivers. The relationships between the HOWs are noted in the "roof" by five symbols: ++ for a strong positive relationship, + for a positive relationship, − for a negative relationship, − − for a strong negative relationship, and a blank for no relationship. The positive relationships indicate that increasing one design attribute (HOW) will cause a corresponding increase in the connected HOW, and vice versa. No numeric analysis is done with these relationships. These are informative for potential trade studies.

Other features that may be added to the A-l matrix include target values, competitive assessments, and risk assessments. These are typically entered as separate rows or columns on the bottom or right side of the A-l matrix.

The key output of the A-l matrix is a prioritized list of substitute quality characteristics. This list may be used as the inputs (WHATs) to other matrices. For example, in Fig. 7, we show the HOWS of a program team feeding requirements (WHATs) to a subsystem team and a subsystem team HOWs feeding requirements (WHATs) to the suppliers.

Critical to Quality Tree
As noted earlier in this chapter, CTQs are stated using the customer's language and do not appear to relate to process parameters. Using tools such as QFD or other similar analyses, it is possible to relate customer CTQs to design features. These can be deployed down to components, component features, processes, and settings and can be expressed in a tree diagram. A tree diagram graphic quickly and easily informs engineers and others involved with a process how the process parameters relate to customers. For example, a team might start with one CTQ, such as car mileage.

At the next level, aerodynamic design, efficient engine, and efficient transmission would be shown. Each of these subsystems could be further decomposed to their major elements (where this makes sense). This sequence can be continued to the logical point on each subsystem and component where key elements can be measured and controlled. At this point in the DMAIC process, it may not be possible to completely take a CTQ cascade to the lowest level of control, but the CTQ cascade can be updated as the team moves through the process.

Process Flow Diagrams
A process flow diagram is a useful tool for documenting key data/material flows of a process. A process flow diagram coupled with waiting/delay times is useful for determining areas to lean out a process. *Lean* is the technique of minimizing material in-queue so the business work-in-process (WIP) inventory is minimized. Process flow diagrams are also referred to as process flow charts and process maps. The single difference between a process flow chart and a process map is that a process map contains a second dimension that identifies the plant, location, department, or persons responsible for completion of a specific task or assignment. One specialized form of a process flow diagram is referred to as a network diagram, and is used for project planning. Project planning software if plentiful and can be quite helpful for creating network diagrams.

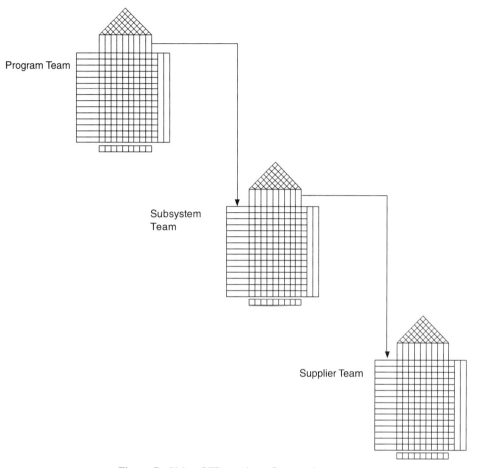

Figure 7 Using QFD matrix to flow requirements.

Another specialized form of flow analysis is a value-added flow chart. This chart arranges the flow in two general categories: value-added and non-value-added. Along the left-hand side of the chart are arranged the process steps where the process adds value to the product. The term "adding value" indicates that items worked on in the process are modified to make them closer to the final product. The process steps that are non-value-added are arranged along the right-hand side of the chart. Non-value-added steps include moves, setup, storage, queuing, inspection, and counting. The amount of time that material remains in the non-value-added locations can be observed or estimated. Usually, your organization will already have labor/time standards for value-added operations in order to price them. Most organizations have a significantly high ratio of non-value-added to value-added time. The higher the ratio, the more costs you have in inventory. The non-value-added steps offer opportunity to lean out your processes. Figure 8 shows a value-added flowchart.

Ways to remove non-value-added time in the process are often simple in concept, but may be difficult to perform. The most obvious approaches include moving processes closer together so there is minimum effort and time expended in material transfer. Additionally, the

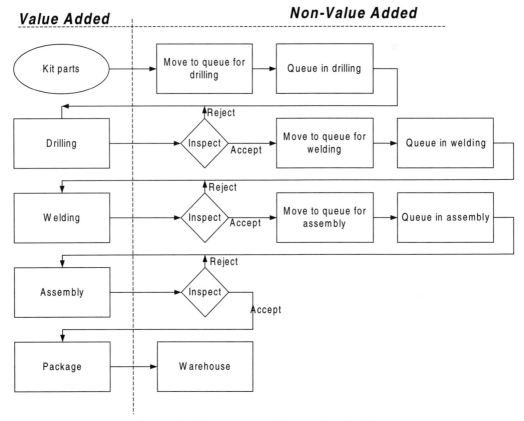

Figure 8 Value versus nonvalue steps in a flowchart.

engineer and production management may change lot sizes and reduce setup time so less material is in each queue; thus, the process becomes more flexible for changing over to another part.

SIPOC (*Supplier-Input-Process-Output-Customers*)

The SIPOC chart is an excellent way for a team to develop a common understanding of a process and to document it. Thus, it is the most appropriate tool to use to explain a process improvement project to others. Just as it is named, this graphic shows process suppliers, their inputs, the process, and its outputs to customers on a single graphic. The disadvantage of a SIPOC is that there may not be sufficient space to detail the process steps as would usually be done with a detailed process flow diagram. That said, the SIPOC still has its place as a way for a team to insure that all aspects of a process are considered in an improvement project. An example of a SIPOC is displayed in Fig. 9.

The specific form of a SIPOC is totally up to a team as long as the major elements of supplier, input, process, output, and customers are covered. The example in Fig. 9 shows an additional element of *requirements*. This is not necessary, but may be a good idea for your team. The process steps will not be shown in as much detail as one would use for a complete process flow analysis. Generally, the display should be limited to no more than seven process

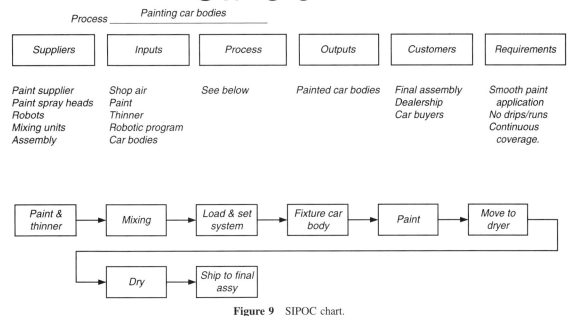

Figure 9 SIPOC chart.

steps. This may require you to combine some detailed process steps into one general step. That is perfectly acceptable because the purposes of this tool are for insuring team alignment and explaining the entire process improvement project to management or others.

4.3 Measure Phase

The measure phase seeks to build a better understanding of a process and begins to identify the sources of issues. The objective is to obtain and begin analysis of process data. If good data do not exist, and unfortunately they often do not, the engineer must guide collection of these data. The measure phase also begins to assess the measurement systems of the process using a statistical technique called gauge repeatability and reproducibility (GR&R) analysis. GR&R is also known as measurement systems analysis (MSA).

4.4 Tools for the Measure Phase

Data Collection Plan and Forms

If good and sufficient data do not exist for a process, a team will have to develop a plan and perhaps even forms for collecting data. A data collection plan is nothing more than identification of what data must be collected, where it is to be collected, by whom it will be collected, how much will be collected, and how often it is to be collected. What data and by whom is generally self-explanatory. Consideration must be given to the types of data (see attribute and variable data measures discussed later in this chapter), but the where and how much is the focus of discussion for the remainder of this section.

Regarding where to collect data, it is useful to call attention to a technique called "data stratification." Data stratification requires deliberately sampling and maintaining traceability of data from different sources or different process times. For example, inspection data may be marked to indicate whether it came from the first or second shift. Another example is obtaining process data and noting the source of material suppliers for those situations where multiple suppliers are used. Stratification can easily indicate sources of process variation that are lost when the data are not stratified. Figure 10 shows a box plot of process results from two suppliers. The differences between the two, in both the median value (the dot) and the variation (the length of the box), are readily apparent.

How much information to collect can often be answered according to the data's intended usage. For many SPC charts (discussed later in this chapter), small samples of 1 to 10 may be used. For others, the results of inspection or test of the entire lot is most useful. How often data must be obtained (often called the sampling plan) is a judgment call by management, the engineer, and the team. If the data are to come from in-process measurements, two to three times per shift is usually sufficient. For some processes, however, it might be necessary to take data once per hour or more. Process capability studies can use data from SPC charts, or samples of 30 or more units. The quantity 30 is important since it is at that point that the information gained from sampling additional units diminishes.

To make the planning efficient, it is important to distinguish the two general types of data the team may encounter: attribute and variable data. Attribute data represent items such as counts or binary values such as pass–fail. Variable data are measurements such as pressure, temperature, and voltage. While we sometimes report variable data in integer form and thus make them appear to be attribute data, the two types should not be confused, since the underlying statistics are often different. Generally, if it is possible to obtain finer resolution measurements by employing better measurement, the data are considered variable data.

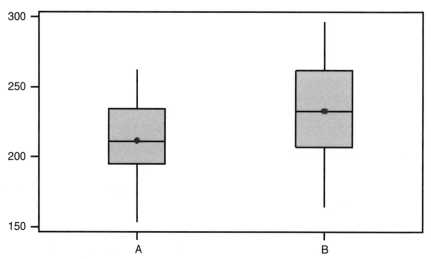

Figure 10 Box plot of process results from two suppliers.

4 TQM/SS—The DMAIC Process

GR&R Analysis

This analysis may start here, but it often continues through other phases, especially the control phase. Refer to the control phase later in this chapter for a complete discussion of GR&R.

Pareto Charts and Other Plots

In 1951, Joseph Juran brought to the attention of quality practitioners the fact that an ordered plot of attribute data, such as defect types, very often showed a consistent pattern. Specifically, most process problems came from a relatively small set of sources (and hence, generated common defect types). He suggested modeling attribute data in an ordered bar chart (from largest on the left to smallest on the right) to demonstrate this phenomenon. He named it the Pareto chart, after Vilfredo Pareto, a 19th-century economist who noted such a pattern in Italian land ownership. In his research Pareto discovered that about 80 percent of the land was owned by about 20 percent of the population. Hence, this was the start of the now well-known 80–20 rule.

Teams that want to focus their improvement efforts on those problems with the most process impact often use the Pareto chart. It is usually shown as a bar chart with a cumulative line graph. It is easily drawn using Microsoft Excel, QI Macro for Excel, or some other software programs with graphing capability. See Fig. 11 for an example of a Pareto chart.

Prioritization Matrix

A prioritization matrix is most useful in developing a prioritized list from a large set of options. This tool makes it easy for a team to focus on the important items and avoid the hidden agendas that could otherwise drive the team. This tool uses a series of pairwise comparisons to determine the relationship of a large number of elements. Refer to Fig. 12 for an example of a prioritization matrix template.

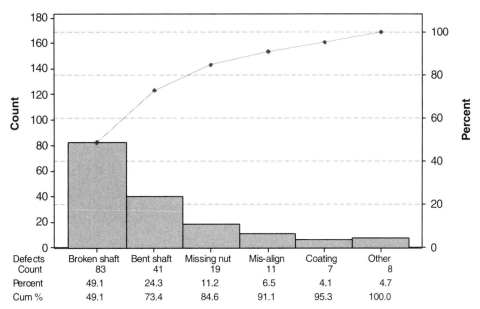

Figure 11 Pareto chart of defects.

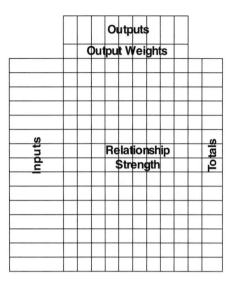

Figure 12 Prioritization matrix for identifying key inputs.

Often, a prioritization matrix can be used to correctly prioritize customer WHATs. Teams can also use a prioritization matrix to assess the importance of various process elements relative to each other with regard to controlling the process. Of course, there is no substitute for process knowledge based on experiments or empirical knowledge, but there is a place for prioritization matrices when working with a new process or team.

Process Capability Analysis
Process capability analysis or validation studies allow engineers and operators to assess a process to determine either its long-run or short-run performance. Knowledge of process capability can aid in setting specifications or supporting the prediction of scrap, rework, and throughput. If design engineers understand process capability and use that knowledge to set specifications, there can be less wasteful conflict between design and manufacturing. Process capability analyses takes on several forms, but its primary form is the in the quality literature as C_{PK} (spoken as "c-sub-pk").

For many companies, engineering design has been slow to understand the need to work with manufacturing to create a design package that both meets customer needs and is *manufacturable*. For their part, manufacturing has not always been proactive in developing consistent processes with minimum variation *and* communicating process requirements and capabilities to design engineers. There is plenty of blame to go around, so how do we change? A key way is to look at facts and data. If you are supporting manufacturing, characterize your processes and communicate process capabilities to designers and external customers. If the design requires certain tolerances, but the process cannot maintain that performance, the only thing that might be done is to change the process! Otherwise, the people supporting the process will always be fighting poor yields, and these losses must be reflected in part prices. The following are appropriate steps to follow:

1. Prioritize your processes according to highest loss (scrap, rework, cost, etc.) and start working on the highest ones (the vital few).
2. If the process isn't monitored using SPC (see our discussion of SPC within the control part of DMAIC), apply it!
3. Get the process under statistical control, that is to say, consistently predictable.

4. From the SPC chart, obtain estimates of the process average and standard deviation.
5. Assess the process C_{PK}.
6. Based on the resulting C_{PK}, determine whether to
   a. Change the product specification, or
   b. Improve the process using the DMAIC approach
7. Move to the next process in your list.

In step 1, develop a comprehensive strategy. Many organizations go after those processes with the most scrap or the most overall cost. In steps 2 and 3, stabilize the process by removing sources of special cause variation. In steps 4 and 5, use existing data from a stable SPC process to assess capability. In step 6, determine what approach is best for your business. Assuming that the process performance is not acceptable, determine your best course of action, as follows. If the stable process capability is low but the product specification can be easily changed, the cost of an engineering change is nearly always less than a process improvement effort. If, on the other hand, the process performance is unacceptable, process improvement may be warranted. The last step calls for the team to move on to the next process on the list, driving for continuous improvement.

The capability index, C_{PK}, indicates how much room (stretch) there is between the product specification (tolerance) limits and the expected (average) output of the process. C_{PK} calculations and performance values are shown in Fig. 13. C_{PK} indicates how many multiples of three standard deviations fit between the process output average and the closest specification limit. A C_{PK} of 1.00 indicates there are only three standard deviations between the process average and the closest specification limit. A C_{PK} of 1.33 indicates a minimum of four standard deviations. Hopefully, your company has established a target value for C_{PK}. Some companies use 1.33, 1.50, or even higher as their target value. Higher values of C_{PK}

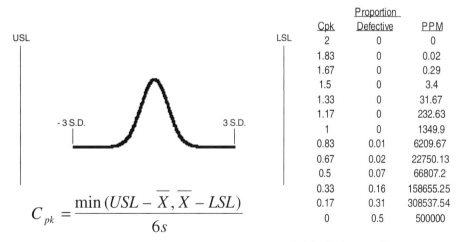

Figure 13 Process capability formulas and measures.

allow greater margin if the process slips out of statistical control. You can see what happens in Fig. 13 if a process C_{PK} slips from 1.50 to 1.00.

Steps 6 and 7 are especially important. If the process capability is not acceptable, either the design or the process must be changed. If one or both of these are not done, your business must *live with the resulting low performance as long as the product is made using this process*. The decision of which to address—product design, process, or both—is an economic one. When one process has been completed, move on to the next one.

Another aspect of process assessment is the measurement system. If a significant portion of process variation results from measurement variation, it may be easier to improve the measurement system than the process. Gauge repeatability and reproducibility (GR&R) assessments, i.e., measurement systems analysis, are also a core current part of the process.

Referring back to our earlier discussion of Six Sigma, this approach would guide a team to set a process and product specification such that there are six standard deviations between the process average and the closest specification. A Six Sigma process would have a C_{PK} of 2.0 or greater (check the calculations to insure this relationship is understood). Six Sigma strives for the extra margin since a process average will sometimes encounter a *shift* of up to 1.5 sigma. This is the extent of the shift that can occur before the change is detected and corrected. If a process is six standard deviations from the nearest limit (this is Six Sigma) and a process shift of 1.5 sigmas (standard deviations) occurs in that direction, the fortunate engineer will still have a process operating at a C_{PK} of 1.50. Figure 14 demonstrates the effect of a 1.5-sigma shift with various values of C_{PK}. Machine wear, setting up a process incorrectly, introducing new operators, changing material suppliers, and a variety of other causes can influence a process shift.

The concept of process shift is incorporated into one variation of process capability analysis known as process performance. Process performance, labeled P_{PK}, assesses how

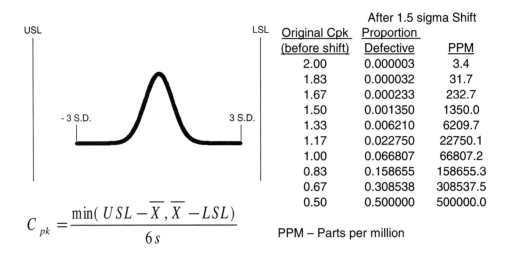

Figure 14 Effect from a 1.5 sigma shift.

well a process output conforms to the specification over a longer time span. While a process capability might be assessed at one time using a single sample, process performance might be assessed over several months. Such a time would expose the output from several operators, shifts, and batches of raw materials. There are statistical ways to obtain P_{PK} from control chart measures. Consult the process control chart references at the conclusion to this chapter for more information. As with C_{PK}, your company should establish target values for P_{PK}. Often, lower target values for P_{PK} are tolerated because process shifts are expected and should be detected and corrected by process controls. For example, a company may have a minimum target C_{PK} of 1.5 and a minimum P_{PK} target of 1.0.

4.5 Analysis Phase

In the analysis phase, the goal is to identify the root causes of process problems and to identify the key factors in a process. In algebra, we often express a function as $y = f(x_1, x_2, \ldots, x_n)$. In this case y is a dependent variable, whose magnitude is dependent on the magnitude of each of the independent variables, i.e., the x_n values. Given a formula, it is possible to analyze the sensitivity of y in relation to each x_n by differentiation. In similar fashion, the outputs of a process are a function of the process inputs. In the analysis phase, we are interested in finding the key x parameters, i.e., those x_n values having the greatest leverage on the process output. This can be accomplished through the collection of experimental or empirical data. When we identify the critical x_n values that affect process output, we can then "turn the knobs" on a process and adjust its output to where the desired value should be.

4.6 Tools for the Analysis Phase

Affinity Diagram

This widely used tool is excellent for generating and grouping ideas and concepts. Teams often find the affinity diagram to be a great tool to explore the issues in a project or to consider the factors involved in implementation. The materials needed for an affinity analysis are simple. Most teams use several stacks of sticky notes and marker pens or pencils. Ideas are generated team members, written down, and then pasted on a white board or wall. They are then arranged into "affinity" groupings by the team and assigned a descriptive header. The affinity header identifies the key issue or consideration identified by the team. The number of items under the header indicates the breadth of consensus by the team.

Cause and Effect Diagram

Also known as the Ishikawa diagram after Kaoru Ishikawa, who introduced its usage, or a fishbone diagram from its distinctive shape (see Fig. 15), this chart helps a team identify the potential sources of a problem from what are often common process sources. These common sources are the materials, machines, men/women (operators), measurements, methods (types of processes), and the environment in which a process operates. The problem is noted on the right end of the chart's main bar. The six possible sources are shown as diagonals leading to the main bar. The team then brainstorms specific sources to link to each of the six bars. A team can discuss and multivote on the most likely source of the problem for further analysis.

The main purpose of a cause-and-effect diagram is to encourage a team to focus on all the possible aspects of a problem. By looking at each of the six legs of the chart, a team is led to generate potential sources of the problem (process issue) from each aspect. This helps

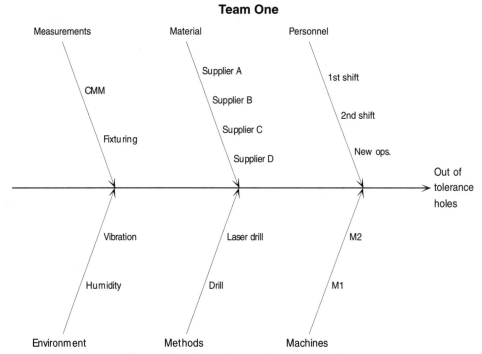

Figure 15 Ishikawa diagram for a process defect.

prevent a team from jumping to one solution, and it can help keep one forceful person from dominating the discussion. At the very least, it opens the mind to consider other possible sources.

Data Sampling and Charts
As previously discussed, a team may use data collection and especially data stratification as methods to analyze a process. It is an excellent strategy to begin data collection in the measurement phase and continue it throughout the project to facilitate experimentation and analysis.

Design of Experiments
A key responsibility of a mechanical engineer is to obtain the required performance from a device, component, or process. This must also be done in the most efficient way possible for the company. This usually requires simulation, trade studies, or some form of experimentation with the possible input variables of one or more processes. Engineers are typically taught methods that include assumptions or approximations for the underlying equations. These may not be accurate enough to guide the engineer to the most efficient result.

Design of experiments or DOE is the tool of choice for trade studies and design or process experimentation. A properly designed experiment will yield the most information possible from a given number of trials, fulfilling the engineer's fiduciary responsibility to the company. And just as importantly, properly designed experiments also avoid *misleading* results.

The chief competitor to good DOE work is the one-factor-at-a-time (OFAAT) approach where the engineer changes one factor, holding all others constant. This is repeated as the engineer works one-at-a-time through all factors of interest while monitoring the response(s). OFAAT has great appeal to the uninformed because of its simplicity. Unfortunately, OFAAT yields only linear, first-order responses. The engineer often knows there are interactions with the factors, or a factor's effect may be nonlinear (exponential or quadratic). OFAAT will not disclose this.

In Fig. 16, a system space is shown consisting of three factors, each at two levels. Experimenting with OFAAT will explore only the circled corners, yielding no information about the remainder of the space. If there is any interaction between the factors, it can be found only at the *unexplored* four points. If there is any form of curvature to the response, we will need to experiment at some point within the interior space of the cube.

Another competitor to OFAAT is random experimentation. This takes place when the engineer changes more than one factor at a time, perhaps making multiple runs while trying different combinations. With random experimentation, desired results may be achieved, but the engineer will not know exactly why. The engineer may make a costly design or process change that is not necessary. Figure 17 shows a path of random experimentation. Like a random walk, this approach lacks an orderly approach to assessing the process environment.

As compared to OFAAT or random experimentation, well-planned DOEs systematically change factors according to a plan, measuring response(s) under known conditions. The experiment often starts with a multifunctional team agreeing on what they believe are the most likely important factors for the experiment. The team may use an affinity diagram or prioritization matrix to determine the priority of process factors. After determining which factors to use, the team must also decide how many levels to use for each factor. Additional factors and factor levels require more experimental runs, which drive up the costs of experimentation, so the relevant factors should be prioritized. Initial experiments often keep the factors at only two levels. This helps to reduce the number of experimental runs and makes the data analysis somewhat easier.

There are many types of experimental designs, but they all fall into two major classifications:

- *Full factorial.* An experiment where all possible combinations of factor and factor levels are run at least once. If there are n factors, each at two levels, this will require 2^n experiments for each replication. This type of experiment will yield all possible information, but may be more costly than the engineer or company can afford.

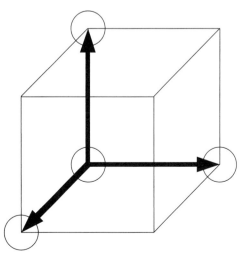

Figure 16 One factor at a time experimentation.

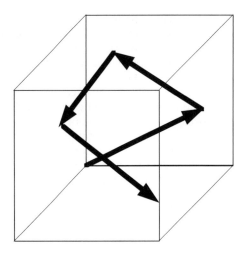

Figure 17 Random walk experimentation.

- *Fractional factorial.* An experiment where a specifically defined subset of all the possible factor–factor level settings is run. A fractional factorial experiment provides only a subset of the information available from a full factorial experiment. Even so, the results are quite useful if the selected subset has been carefully planned. Usually, a design is planned that does not identify higher level interactions. These are *confounded* or mixed in with other responses. If there are n factors, a half-fractional factorial will require 2^{n-1} runs at a minimum. For example, considering an experiment with five factors, one run at each factor would require 32 runs. A half-fractional factorial would cut this to 16. You should consult with a DOE subject matter expert (SME) for help with fractional factorial experiments.

There are several methodologies that utilize these basic experimental design types. Classical DOE was developed in the 1920s by Ronald Fisher in England and initially promoted in the United States by Box, Hunter, and Hunter. This type of experimentation utilizes both full and fractional factorial designs. In the 1960s, working in Japan, Genichi Taguchi began to promote a form of experimental design that uses a special set of fractional factorial designs. Although the forms Dr. Taguchi used were not unique, his approach generated a dramatic increase in DOE usage, especially among engineers. Dr. Taguchi has made three major contributions to the field of DOE. First, he developed a DOE methodology that offered clearer guidance to engineers than earlier approaches offered by classical statisticians. Second, he promoted the concept of *robust design* and demonstrated how DOE could be used to obtain it. Finally, he promoted the application of something he called the *quality loss function*. This unique technique expresses in dollars how the enterprise and society in general are affected by variation from an optimal target.

Usually experiments are run at two levels. Occasionally, the engineer must experiment with factors at more than two levels. These may be attribute factors such as different materials or continuous variables such as temperature, pressure, and time. DOE handles all these, but the planning and analysis get a bit more complicated.

No matter which experimental design is chosen, it is relevant to two key parts of an experiment. The first part is randomization. Randomization means to carefully plan an experiment, but conduct it using some type of random order. Using a random number generator, picking numbers out of a hat, or any other method may accomplish this. Randomization is employed to prevent some time-dependent factor from creeping into the experimental results. For example, a machine tool wears with use. If the experiment proceeds in a particular order with regard to the runs, the later runs will have the additional influence of tool wear. Randomization allows each combination of factor–factor level setting an equal chance to experience a time-related factor. The other part that must be considered is repetition. It is rare that an experiment is run only once at each factor–factor level setting combination. Even a full factorial experiment is usually run with at least two repetitions so sufficient information is obtained for good analysis.

We've touched on the main types of experimental design, but we note that this has been a very rich field of research and innovation. As a result, there are several types of experimental designs that have not been discussed and that may be useful for specific purposes, such as mixtures or the situations where process output is nonlinear. These are discussed in some of the references found at the conclusion to this chapter.

4.7 Improvement (and Innovate) Phase

In the I^2 phase, the goal is to develop appropriate process and/or product improvements while considering business needs. The improvement or innovation must be effective and achievable within business restrictions such as budget, schedule, and the like. This suggests that some improvements must be ruled out if the business cannot support them. It is sometimes said that the cure may be worse than the problem. This does not mean that the problem cannot be mitigated, but the team may need to be innovative in the improvement to circumvent relevant business restrictions.

4.8 Tools for the Improvement (and Innovate) Phase

Brainstorming
Because brainstorming is so well known, we have devoted only a few words to it in this chapter. An important aspect of brainstorming is the need to stress its operating rules, which include the necessity that all involved have an equal opportunity to participate without their ideas being rejected or ridiculed. There are examples in business experience where the best ideas came from quiet process operators when they were finally encouraged to participate.

Data Sampling and Charts
During the improvement phase, it is important to observe the effect of attempted improvements. As discussed in the previous phases, a plan for data must be developed by the team.

Critical-to-Quality (CTQ) Analysis
As was discussed in Pareto analysis, the best use of resources demands that a team focus on the important items in the process. Critical-to-quality analysis or CTQ cascade has become known as the process to trace features of key customer importance into the process. Since it is more descriptive, we will use the term, CTQ cascade.

In a CTQ cascade, a team takes the top critical-to-quality features for the output of the process and, through analyses or tests and experiments, relates them to process parameters or process inputs. For example, suppose a smooth paint finish on an auto body panel is a CTQ for our customers. From previous work, our team has found that the critical painting process factors are spray pressure, paint mixing, and the distance of the spray head from the body panel. Other process factors, such as temperature and time of application, are less critical for this. From this work, it is apparent that spray pressure, paint mixing, and the distance of the spray head from the panel are *critical process inputs* to the CTQ. As time goes by, this linkage may not be so obvious to new workers and engineers on the process. To transfer this knowledge, we indicate the relationship using a CTQ cascade.

CTQ cascades often take the form of a tree diagram (see Fig. 18). This simple graphic shows the relationship very well. A process control plan is another tool that can demonstrate this relationship. A process control plan is a process work instruction, generated as a word processing document or a spreadsheet. In this plan, it is convenient to show process settings in a tabular form. Linkages between a setting and a CTQ can be shown here. Cascades can also be displayed in a spreadsheet. Early proponents of QFD often proposed using two or more QFD matrices to form this linkage. This is an excellent analysis approach, but may be too difficult to maintain for a process work instruction. A process control plan fits this need very nicely.

Most teams that are new to this process will want to discuss what it means to be *critical* to quality. Many things are critical if left out or damaged. The way to think about CTQs is to determine what parts of the process are difficult to do or difficult to control. For example, process parameters that have tighter tolerances than normal might be CTQs. Another candidate for designation of CTQ is something that is new to the process. Continuing in the paint example used before, the addition of metal flake or pin stripes might be CTQs if they are not used in the normal process.

Design of Experiments

In this phase, DOEs are used to investigate and confirm proposed solutions to a problem. The order of experimentation should progress, as follows:

Analysis phase

- Screening DOEs are conducted using experimental designs with factors that are key to the process output and that forgo potential interactions between the factors. These screening DOEs will include a larger number of factors at lower resolution (fractional factorials) to *screen* out factors that are not statistically significant.

I^2 phase

- Focusing on a smaller set of factors (perhaps 2 or 3), a higher resolution experiment may be performed to determine the acceptable process setting window and to determine the optimal process setting combination. This experiment will likely be a full factorial or a specially selected, judiciously planned fractional factorial.
- When the final process window and target settings are selected, a final confirmation run is often suggested to verify the output prior to committing to the process change.

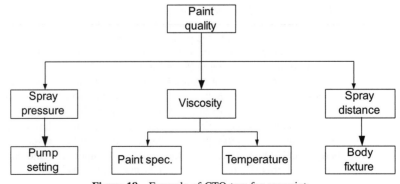

Figure 18 Example of CTQ tree for car paint.

Network Diagrams (and Gantt Charts)

The activity network diagram (AND), portrayed in Fig. 19, is a way for a team to schedule project tasks. The team can use cards or sticky notes to list project tasks. These can then be arranged in the anticipated flow (sequential, parallel, or combination) on a large wall with directional arrows indicating task relationships. The team can then estimate time to complete each task. The longest sequential path to complete for the entire project becomes the critical path. The graph also shows predecessor–successor relationships, and the total task time can be calculated. This information can be used as input to project management software.

Process Capability Analysis

After the improvement has been introduced into the process, it is advisable to repeat the process capability analysis that was accomplished during a previous phase. If the process has been improved, changes in process capability and the resulting reductions in scrap or rework prediction are powerful statements for a team to use to explain the significance of their work and to obtain support in the implementation of the process improvement.

4.9 Control Phase

The control phase, as its name implies, is the phase where controls are placed to make certain that the process improvement is maintained. This is a critical phase, which, if skipped or implemented incorrectly, could result in having to repeat the entire process again (and again)!

4.10 Tools for the Control Phase

Control Charts

Control charts in the control phase seem a natural fit, and they are. SPC (statistical process control) is a technical quality tool that was brought to American industry's attention by the War Department (predecessor to the Department of Defense) during World War II. After less than satisfactory first attempts at deployment of SPC, many companies are now finding it to be useful for reducing defects, lowering defect rates, and making key business processes more consistent and dependable. The solution to successful use of this tool is to understand what SPC does and doesn't do.

SPC is the application of a statistical method, usually in a graphical form. It is used to detect when a process may have been influenced by a "special cause" of variation. Walter Shewhart, who developed the earliest concepts and applications of SPC, divided process variation into two types. He described one as "common cause" or "normal" variation. This type of variation comes from the many factors in the process varying and interacting with each other. For example, in a drill process, there is drill splay (wobbling of the drill bit around its axis), variation in bits, and variation in material hardness, etc. These interact and result in variation of hole size, position, and the degree of roundness of the hole.

The second form of variation described by Dr. Shewhart is referred to as "special cause" variation. Continuing with the drill example, insertion of the wrong bit size would result in a change of the hole size. This shift of hole size is not "normal," but can be *assigned* to a process error. Other examples of special causes might be untrained operators, improperly maintained machines that exhibit excessive variation, changes in material, changes in bit manufacturing, and changes in material clamping technique.

Figure 20 shows one of the first and most used SPC charts, the X-bar and R chart. This is also noted as $\bar{X}$ in R mathematical nomenclature, where $\bar{X}$ (pronounced "X bar") symbolizes the subgroup average and R is used for the subgroup range. A subgroup is a sample

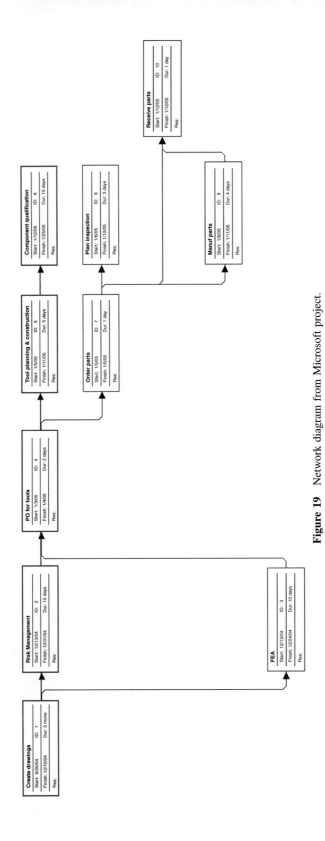

Figure 19 Network diagram from Microsoft project.

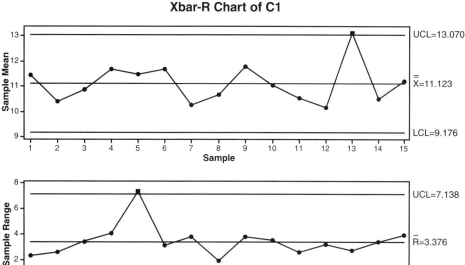

Figure 20 Example X-bar and R chart for a process.

that is taken from the process periodically. In the example, a point is out of tolerance in each chart. A subgroup usually consists of a sample of 2 to 10 units for this type of chart. This type of chart can detect a shift in the sample average (through the $\overline{X}$ portion of the chart) and the sample standard deviation (through the R portion of the chart). Together, these two charts signal major shifts (changes of the process average and process variation of one variable) in a process.

The reason to make such a distinction between these two sources of variation is to separate the *manageable* from the *unmanageable*. Special causes of variation can be identified and removed or prevented from entering the process again. Often, these changes can be easily made. Normal or common causes of variation can usually be removed or reduced only by changes to the process. Of course, management commitment and possibly capital expenditures are needed to change the process. Continuing with the drilling example, higher accuracy and repeatability might require a process change to employ laser drilling or a water jet. Obviously, this requires different machinery and such changes are often not trivial.

How does SPC fit into this? Dr. Shewhart, working in an AT&T Western Electric plant, noticed that their processes had excessive variation and the operators were constantly adjusting the process. He suspected they were adding to process variation by making these adjustments. He sought a way to determine when a process adjustment was really necessary. To accomplish this, he proposed the use of SPC and SPC charts as a way to *signal* when a process may have been influenced by special cause of variation. When the signal occurred, operators, engineers, and management could pursue adjustments or investigations as seemed appropriate.

SPC charts come in many forms, but, in general, they all plot one or more statistics (a descriptive measurement from a unit or sample) on a form of line chart. The line chart also contains warning limits and control limits, depending on the chart type. See Fig. 20 for an example. The upper and lower control limits are derived from past stable process data and usually represent some long-range average of the measurement plus/minus three standard deviations (σ).

For statistical reasons, some charts do not have a lower control limit. Most of the measurements used for SPC follow a normal distribution (helped by a statistical phenomenon called the central limit theorem). This means that follow-on measurements from a process that is not affected by special causes variation will stay within the control limits 99.73% of the time. Reversing this logic, a point outside of a control limit would occur only 0.27% of the time. Thus, when such an event happens, it is most likely the result of a special cause of variation. Process investigation should be employed to find and remove the special cause.

In addition to watching for points outside of the control limits, SPC charts may send other signals. SPC practitioners apply tests for *patterns* that signal the effect of special cause variation. For example, a pattern of seven points in a row, increasing or decreasing, is not a pattern that shows naturally. Such a pattern indicates the likely presence of a special cause of variation, even if no control limit has been breached!

Figure 20 shows an example of an $\bar{X}$ and R chart. This is a typical chart, where the $\bar{X}$ portion of the chart represents a plot of sample averages and the R portion represents a plot of sample ranges from a *subgroup*. A subgroup for this type of chart usually consists of a sample of 2 to 10 units. This type of chart can detect a shift in the sample average (using the $\bar{X}$ portion of the chart) and the sample standard deviation (through the R portion of the chart). The following are some rules for abnormal patterns in SPC charts:

- One point outside either the upper or lower control limit
- A run of seven or more points either up or down or consecutively above or below the centerline
- Two of three consecutive points outside 2 sigma, but still inside the 3 sigma line
- Four of five consecutive points beyond 1 sigma

While SPC deals with in-process measures, often the only significant way to measure the process result is by measuring the performance of the finished product. For example, when we assemble an electronic circuit, there are measurements that can be taken in the process, but the final circuit performance can be measured only by a final functional test. As with process measures, final performance variation is a function of normal and special cause variation. SPC can also be used in the case of final process performance to determine if an investigation of special cause variation is warranted. This is often referred to as *statistical quality control* (*SQC*) to differentiate it from process control. The same theory is used, but the charts are sometimes slightly different as different statistics are employed. We should note that SQC should not be used as a substitute for SPC. Since SPC works with in-process measures rather than the end of the process, it offers faster detection and correction of problems.

SPC and SQC are powerful tools, but they essentially do only one thing—they identify when a process has been influenced by something not usually a part of the process. When that occurs, process engineers and operators can look for the cause and remove or prevent it, returning the process to what is its normal state. This is accomplished by examining the control plans and documentation accumulated from the last DMAIC on the process. If it is not clear what has affected the process, a new DMAIC action may be warranted to put the process back on track.

Control Plans
A control plan is a work instruction for a process. Control plans can take any form, but they are usually maintained as a word processing document or spreadsheet under revision control.

They can take the form of a word processing document with complete process instructions, or a table of process parameters with their settings and ranges. As stated in CTQ analysis, factors in the process that have a major effect on CTQs can be identified. Key items to cover in a control plan are

- Process step or phase
- Order of actions (if sequence is critical)
- CTQ linkage (if any)
- Target setting
- Allowable range
- Calibrations needed
- Sampling plan (number of samples, what to measure, what measurement tool to use, and how often to sample)
- GR&R for the measurement tools
- Reaction plan (orderly shutdown)
- Safety measures and equipment

Data Sampling and Charts

A plan for data sampling will take into account what has been learned about the process. The sampling plan will focus on obtaining sufficient data to maintain the process, usually in conjunction with some form of SPC or other graphical chart.

When a new process is launched or a new part is placed into a process, sampling may be used more frequently than a typical maintenance sampling. Engineers might monitor a new process by collecting data every hour. After stability is demonstrated, data collection could be scaled back to once or twice a shift.

Some machining processes utilize a special strategy for setting up a new part lot. When a lot is started, the operator may be required to measure the critical features of two to five parts before the rest of the lot is released to run. After that, sampling can occur every n pieces. If the parts are not expensive, another strategy is to sample at start-up and at the end of the lot. If all the samples from both times conform, the lot is released. If the setup was successful, but the end sample failed, some portion of the lot is suspect. The lot can be screened or discarded per business need.

This discussion demonstrates the need for sample stratification by time. Time-based sampling and presentation by graphs or charts are powerful ways to manage a process. Both subtle shifts and large jumps in the process can be detected this way.

GR&R Analysis

Since everything has variation at some measurable level, it is no surprise that measurement systems also have variation. Gauge variation comes from three sources:

- *Bias in the gauge.* When a gauge has a bias, it tends to indicate a reading that is above or below the true value. This is a function of the gauge's calibration.
- *Repeatability.* When a gage is not repeatable, it means that repeated measurements (this is referred to as *repetition* when discussing DOE) by the same operator, when the part is removed and replaced in the fixture or gauge, show a large variation. Repeatability is influenced by gauge design. Electrical noise, excess play in mechanical linkages, or a loose fit in retaining features can also influence repeatability.
- *Reproducibility.* This pertains to the ability of a second operator to achieve the same result as a previous operator working with the same equipment and under the same

conditions (this is referred to as *replication* when discussing DOE). Examples of factors that influence reproducibility include holding fixtures that are sensitive to operator technique and measurement instructions that give an operator significant discretion in how a part will be mounted and measured.

Some may be surprised at this level of detail for the process measurement systems. The reason is economic. If a process has a lower than desirable yield, the issue may be with the process, the measurement system, or both. The measurement system may be rejecting good parts and allowing nonconforming units to be sent to customers. The reason for this attention to detail is twofold. First, it is often more economical to fix a measurement system than change a process. Second, if the measurements being taken have a large amount of uncertainty, it is likely that you are rejecting good parts, delivering parts that are not in conformance, or both.

Measurement systems analysis (MSA) should be performed properly so the source of variation is properly identified. It is desirable that the parts exhibit variation covering the expected tolerance range, although this may be difficult for some processes. Most analysis involves approximately 10 parts and two to three operators. First, the gauge is calibrated or the calibration record is checked. Second, each operator will measure and record the features of interest on each part two or three times. Parts will be run in random order to remove any time trending with the operator or measurement system. All measurements will be recorded with identification of the operator, part, and order of measurement. The statistical analysis will then stratify the sources of variation and identify how much variation is coming from the parts, the repeatability of the gauge, and the reproducibility of the gauge.

Many companies place guidelines on the amount of measurement error they will tolerate in the system. Generally, less than 10% of the feature tolerance is an acceptable range. If the error is less than 30%, it may be tolerable, depending on the criticality of the feature. If the measurement error is greater than 30% of the tolerance range, the measurement technique is a candidate for improvement.[8]

Another aspect of measurement systems analysis is the comparison between gauges. Often companies will rely on suppliers' measurements. If there is an issue, it is good idea to be able to assess parts at your facility and know your measurements will be similar to those of your supplier's.

REFERENCES

1. J. M. Juran (ed.), *Juran's Quality Control Handbook,* 4th ed., McGraw-Hill, New York, 1988.
2. J. M. Juran and F. M. Gryna, *Quality Planning and Analysis,* McGraw-Hill, New York, 1980.
3. R. C. Swanson, *Quality Improvement Handbook, Team Guide to Tools and Techniques,* St. Lucie Press, Delray Beach, FL, 1995.
4. B. King, *Better Designs in Half the Time: Implementing Quality Function Deployment in America,* GOAL/QPC, Methuen, MA, 1987.
5. The Institute of Management and Administration. *IOMA Journal.* http://www.IOMA.com.
6. General Accounting Office (GAO), *Management Practices, U.S. Companies Improve Performance Through Quality Efforts, GAO/NSIAD-91-190,* Washington, DC, May 1991.
7. B. R. Helton, "The Baldie Play," *Quality Progress,* **28**(2) 43–45 (1995).
8. AIAG–Automotive Industry Action Group. *MSA-3: Measurement Systems Analysis,* 3rd ed., AIAG, Cincinnati, OH, http://www.aiag.org.

BIBLIOGRAPHY

As a result of the dynamic nature of the World-Wide Web, some of the web site references noted below may have changed, so the designated links may not work directly. Usually, the information is maintained by the organization hosting the site, but it might have been moved to a different or renamed page. If the direct site does not work, go to the home page of the organization or use a search engine to reacquire its location.

American Society for Quality (ASQ)

ASQ has been a dependable source for quality-related information since 1946. The Society and its over 100,000 members have been on the leading edge in all quality-related initiatives, including TQM and Six Sigma DMAIC. Resources and references are available at http://www.asq.org.

GOAL/QPC

GOAL/QPC is a not-for-profit organization founded in 1978 that publishes a number of excellent guides for both TQM and Six Sigma. See http://www.goalqpc.com.

National Institute for Standards and Technology (NIST)

NIST offers information on the Baldrige National Quality Award (BNQA) and guidance on performance measurement systems. BNQA information can be obtained at http://www.baldrige.nist.gov. Measurement systems information is available at http://www.mel.nist.gov/melhome.html. Besides information on the Baldrige National Quality Award, NIST maintains an on-line statistical handbook in conjunction with SEMTECH. This can be located at www.itl.nist.gov/div898/handbook/index.htm. It is available at no cost and is an excellent statistical resource.

Various Statistical Resources

There are a number of other statistical resources published on the web, many from various colleges and universities. These form part of their education resources. For example, a normal probability applet can be accessed at: http://www.ms.uky.edu/ ~mai/java/stat/GaltonMachine.html. It shows the effect of the central limit theorem and how the normal probability curve develops from various small process elements. Many other resources are available on other sites.

Other references that were not cited and that contain information relevant to TQM/SS:

- Brassard, M., L. Finn, D. Ginn, and D. Ritter, *The Six Sigma Memory Jogger II,* GOAL/QPC, Salem, NH, 2002.
- Breyfogle, F. W. III, *Implementing Six Sigma: Smarter Solutions Using Statistical Methods,* Wiley, New York, 1999.
- George, S., and A. Weimerskirch, *Total Quality Management: Strategies and Techniques Proven at Today's Most Successful Companies,* 2nd ed., Wiley, New York, 1998.
- Ginn, D., and E. Varner, *The Design for Six Sigma Memory Jogger,* GOAL/QPC, Salem, NH, 2004.
- Pande, P. S., R. P. Neuman, and R. R. Cavanagh. *The Six Sigma Way: How GE, Motorola, and Other Top Companies Are Honing Their Performance,* McGraw-Hill, New York, 2000.
- ReVelle, J. B. (ed.), *Manufacturing Handbook of Best Practices: An Innovation, Productivity and Quality Focus,* St. Lucie Press, Boca Raton, FL, 2002.
- ReVelle, J. B., *Quality Essentials: A Reference Guide From A to Z,* ASQ Quality Press, Milwaukee, WI, 2004.
- ReVelle, J. B., J. W. Moran, and C. A. Cox (eds.), *The QFD Handbook,* Wiley, New York, 1998.

CHAPTER 19

REGISTRATIONS, CERTIFICATIONS, AND AWARDS

Jack B. ReVelle, Ph.D.
ReVelle Solutions, LLC
Santa Ana, California

Cynthia M. Sabelhaus
Raytheon Missile Systems
Tucson, Arizona

1	**INTRODUCTION**	**616**
2	**REGISTRATION, CERTIFICATION, AND ACCREDITATION**	**617**
	2.1 National and International Standards	617
	2.2 ISO 9001 Quality Management System Requirements	618
	2.3 ISO 9001 Certification and Registration	618
	2.4 ISO/TS 16949 Auto Industry Quality Management System Requirements	620
	2.5 ISO 13485 Medical Device Quality Management System Requirements	621
	2.6 TL 9621 Telecommunication Quality Management System Requirements	621
	2.7 AS 9100 Aerospace Quality Management System Requirements	622
	2.8 Mission Assurance	622
	2.9 CMMI	622
	2.10 ISO 14625 Environmental Management System Requirements	623
3	**QUALITY AND PERFORMANCE EXCELLENCE AWARDS**	**625**
	3.1 Deming Prize	626
	3.2 Baldrige National Quality Award	627
	3.3 U.S. State Quality Awards	630
	3.4 Shingo Prize for Excellence in American Manufacturing	630
	3.5 Quality Awards around the World	634
	3.6 Industry-Specific Quality Awards	636
	3.7 How Do They Compare?	637
	REFERENCES	**637**

1 INTRODUCTION

The concept of quality has expanded over the past decade, from traditional total quality management (TQM) tools and techniques to the quest for overall performance excellence that encompasses all of an organization's processes and management systems. This was first seen in the Baldrige National Quality Award (BNQA) criteria and then in the ISO 9001: 2000 quality management system standard. At the same time, there has been an explosion of industry-specific quality management system standards. Whereas ISO 9000 was first issued to provide common criteria for quality management systems across industries, there are now multiple standards, each with industry-specific requirements.

The U.S. Department of Defense (DoD) has also had a major role in reshaping the management aspects of quality standards in the past decade. After years of acquisition reform and a move toward simplifying requirements to prime contractors and their supply chain, the DoD has created a number of detailed requirements since 2000, including the 700-page Capability Maturity Model Integration (CMMI) in 2001, and the 156-page Mission Assurance Provisions (MAP) in 2004. As these requirements are flowed from prime government contractors to their subcontractors and suppliers, the cascading effect will have a major impact on the management of business throughout the world. For many suppliers of generic products that serve several industries, the complexities of multiple standards and varying certification requirements will no doubt add to the level of difficulty in doing business.

As companies engage in the process of achieving certifications, registrations, and awards, mechanical engineers will undergo audits and site visits testing their understanding of applicable quality management system requirements that range from configuration control to the manner in which parts are procured. Some engineers will be asked to help their companies create the processes and/or documentation required to achieve certification or apply for an award. This chapter provides a general overview of the most widely recognized programs. Keep in mind that standards are revised periodically, and award criteria may be updated annually. Use the web links provided in figures and references to obtain the latest information.

2 REGISTRATION, CERTIFICATION, AND ACCREDITATION

While the concept of certifying or registering quality systems to an industry or international standard has become the norm throughout the world, the terminology is often misunderstood. For all practical purposes, it does not matter whether the term *registration* or *certification* is used. When a company seeks validation of its ISO or industry-specific quality management system by hiring a third-party registrar, the quality system is certified as meeting the requirements, and the registrar issues a certificate. The certification is then entered in a register of certified companies. Thus, companies meeting the requirements of the standard are both certified and registered. The term *certification* is most often used for this process in Europe. In the United States, it is more common to hear the process called *registration*.

Certification and registration should not be confused with *accreditation*, which is the procedure by which an authoritative body gives formal recognition that a body or person is competent to carry out specific tasks. Accreditation bodies have been set up in a number of countries to evaluate the competence of certification bodies. An accreditation body will accredit, i.e., approve, a conformity assessment body such as a registrar. The registrar is then approved to review organizations for compliance to specific standards such as ISO 9001 or ISO 14001, and recommend that organizations be certified as meeting those requirements. For ISO standards, there are over 80 accreditation bodies throughout the world, including the Registrar Accreditation Board (RAB) in the United States, the United Kingdom Accreditation Service (UKAS), and the Standards Council of Canada (SCC). Many of these bodies have reciprocal agreements to recognize each others' accreditations.[1]

2.1 National and International Standards

Technical standards have played a large part in enabling the creation of a global economy. At the same time, they have enhanced living standards by ensuring common standards for safety. The first efforts in international standardization began with the International Electrotechnical Commission (IEC) in 1906, and the International Federation of the National Standardizing Associations (ISA) in 1926. By 1942, the efforts of both organizations were abandoned.[2]

The need for expansion of the standardization efforts was identified as countries worked together during the Second World War. In 1946, delegates from 25 countries met in London and decided to create a new international organization to facilitate the international coordination and unification of industrial standards. The new organization, ISO, officially began operations on February 23, 1947.

The International Organization for Standardization is the world's largest developer of technical standards. It comprises 148 countries, with one member per country, and has a Central Secretariat in Geneva, Switzerland to coordinate its activities. Because any acronym for the organization would be different in the different languages of its member countries, the organization uses the name ISO, derived from the Greek *isos* meaning "equal," for both its organization and the standards it issues. By providing the framework for compatible technology worldwide, ISO has helped build a world economy, making it easier to do business across national borders and providing customers with the benefits of a more competitive market place. ISO standards not only ensure compatibility, but also address reliability and safety.[3]

2.2 ISO 9001 Quality Management System Requirements

As the European Trading Community began to take shape in the 1980s, there was a perceived need for a common quality standard for all nations. ISO assigned this task to Technical Committee 176, and in 1987, the ISO 9000 Quality System Standards were issued. A revised series of standards was issued in 1994, with varying levels of requirements in ISO 9001, 9002, and 9003, depending on the extent of a company's operations. The quality system standard emphasized documenting quality requirements with little regard to the impacts of those requirements on product quality. Although the standard was quickly adopted by many businesses and became a requirement for suppliers across the world, there was concern that compliance to the standard did not always ensure high quality, reliable products, customer satisfaction, or a successful business.

In 2000, ISO 9001: 2000 Quality Management System was issued, completely revising earlier editions. Throughout the new standard, ISO presents quality management in a systems approach—processes with inputs and outputs interacting with other processes. This more closely approximates how organizations actually operate, especially compared with the separate elements of the previous editions. The new standard was built around eight quality management principles, with increased emphasis on customer focus through requirements for measuring and tracking customer satisfaction, thereby focusing more on product quality.[4]

The 2000 revision to ISO 9001 triples requirements for senior management, and it requires a more structured approach to process improvement.[5] There are many similarities between ISO 9001: 2000's Quality Management Principles and the Core Values adopted from the Baldrige National Quality Award criteria to recognize performance excellence in national and local awards around the world (see Fig. 1).

2.3 ISO 9001 Certification and Registration

Separate from the ISO 9001: 2000 Quality Management System standard per se is the certification/registration process that has been institutionalized in many countries. The process requires a review by a third-party registrar of a company's documented quality system and the implementation of that system through on-site audits. The third-party registrar certifies that the system meets all of the requirements of the ISO 9001: 2000 model. The

ISO 9000:2000 Quality Management Principles	Baldrige National Quality Award Core Values and Concepts
Principle 1 -- Customer focus. Organizations depend on their customers and therefore should understand current and future customer needs, should meet customer requirements, and should strive to exceed customer expectations.	**Customer-Driven Excellence.** Customer-driven excellence is a strategic concept. It is directed toward customer retention, market share gain, and growth. It demands constant sensitivity to changing and emerging customer and market requirements.
Principle 2 -- Leadership. Leaders establish unity of purpose and direction of the organization. They should create and maintain the internal environment in which people can become fully involved in achieving the organization's objectives.	**Visionary Leadership.** An organization's leaders should set directions, and create a customer focus, clear and visible values, and high expectations. Senior leaders should serve as role models through their ethical behavior and personal involvement..
Principle 3 -- Involvement of people. People at all levels are the essence of an organization and their full involvement enables their abilities to be used for the organization's benefit.	**Visionary Leadership.** Leaders should inspire and motivate the entire workforce and encourage all employees to contribute, to develop and learn, to be innovative, and to be creative.
Principle 4 -- Process approach. A desired result is achieved more efficiently when activities and related resources are managed as a process.	A process approach is expected but not expressly called out in the Core Values. **Category 6** of the criteria, Process Management, considers key product/service and support processes.
Principle 5 -- System approach to management. Identifying, understanding, and managing interrelated processes as a system contributes to the organization's effectiveness and efficiency in achieving its objectives.	**Systems Perspective.** Successful management of overall performance requires organization-specific synthesis and alignment. A systems perspective means managing your whole organization, as well as its components, to achieve success.
Principle 6 -- Continual improvement. Continual improvement of the organization's overall performance should be a permanent objective of the organization.	**Managing for Innovation.** Innovation means making meaningful change to improve an organization's products, services, and processes, and to create new value for the organization's stakeholders.
Principle 7 -- Factual approach to decision making. Effective decisions are based on the analysis of data and information.	**Management by Fact.** Organizations depend on measurement and analysis of performance. Such measurements should derive from business needs and strategy.
Principle 8 -- Mutually beneficial supplier relationships. An organization and its suppliers are interdependent and a mutually beneficial relationship enhances the ability of both to create value.	**Valuing Employees and Partners.** External partnerships might be with customers, suppliers, or education organizations. Strategic partnerships and alliances are increasingly important...
Not included in ISO 9000:2000 Quality Management Principles	**Agility, Public Responsibility and Citizenship, Focus on the Future, Focus on Results and Creating Value, Organizational and Personal Learning.**

Figure 1 Comparison of ISO 9000:2000 and Baldrige National Quality Award.

registration of the quality system can then be publicized. The registrar also performs regular reviews and periodic recertification audits.

The Registrar Accreditation Board (RAB) is the U.S. agency that accredits agencies and individuals to serve as registrars. It was established by the American Society for Quality (ASQ) in 1989 as an independent legal entity whose mission was to provide accreditation services for ISO 9001 Quality Management Systems (QMS) registrars. RAB is governed and operated independently from ASQ. In 1991, the American National Standards Institute (ANSI) and RAB joined forces to establish the American National Accreditation Program (NAP) for Registrars of Quality Systems. The ANSI-RAB NAP covers the accreditation of Quality Management System (QMS) and Environmental Management System (EMS) registrars. Separate from the ANSI-RAB NAP are certification programs for EMS auditors and QMS auditors and accreditation programs for course providers offering QMS and EMS auditor training courses operated solely by RAB.[6]

The effort to obtain ISO 9001 registration typically takes 12 to 18 months from the time an organization makes the commitment to become registered until its quality system receives the certificate from its third-party registrar. The cost of registration varies depending on the size and complexity of the organization, the number of locations to be included on the registration certificate, and the state of its existing quality system when the decision to obtain registration is made.

Third-party registrars are generally contracted for three years. In addition to the initial assessment for registration, the registrar may be asked to perform a preassessment audit, or to conduct training. Many companies find it helpful to hire an outside consultant to help prepare for ISO registration. There are many texts available on the subject of ISO 9001 quality systems and the registration process. A complete listing of ISO 9000 and ISO 14000 accreditation and certification bodies can be found at www.iso.org/iso/en/info/ISODirectory/intro.html#

In addition to ISO 9001: 2000 Quality Management System requirements, a suite of ISO documents has been issued in support of ISO 9001, including

- ISO 9000—fundamentals and vocabulary
- ISO 9004—quality management systems, guidance for performance improvement
- ISO 19011—guidelines on quality and/or environmental management systems auditing.

2.4 ISO/TS 16949 Auto Industry Quality Management System Requirements

The automobile industry in the United States was the first to require its own version of ISO 9001 with industry-specific requirements, including a section of requirements for each of the Big Three U.S. auto makers (General Motors, Ford, and Daimler-Chrysler). QS 9000 was issued in 1994 by the Automotive Industry Advisory Group (AIAG). An international standard, ISO/TS 16949, was created by the International Automotive Task Force in collaboration with ISO in 1998, and a complete revision aligned to ISO 9001: 2000 was released in 2002. The standard applies to both original equipment manufacturers (OEM) and suppliers of automotive parts.

ISO 9001: 2000 is the central tenet of ISO/TS 16949: 2002, but the automotive standard includes additional requirements. For example, ISO 9001: 2000 requires only six documented procedures to support a quality system. ISO/TS 16949: 2002 requirements are more in keeping with documentation requirements of the earlier versions of ISO 9001:1994 and QS 9000. There are 15 new or expanded requirements in ISO/TS 16949: 2002 as compared to its earlier release and QS 9000. The 2002 revision requires a process approach (see Fig. 2) and places greater emphasis on management commitment, customer satisfaction, measurement, and internal auditor qualifications.[7]

In the United States, General Motors Corporation, Daimler-Chrysler, and Ford Motor Company have announced that QS-9000 certification will not be accepted after December 2006, requiring automotive industry suppliers to transition to certification in ISO/TS 16949: 2002. Oversight and certification is managed by the International Automotive Oversight Bureau, www.iaob.org. To compound suppliers' level of difficulty in complying with ISO/TS 16949: 2002, Daimler-Chrysler, Ford Motor Company, and General Motors issued addendums with company-specific requirements in 2004.[8]

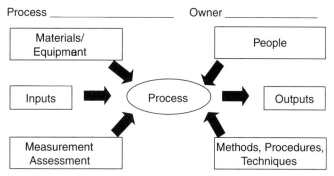

Figure 2 Process maps.

2.5 ISO 13485 Medical Device Quality Management System Requirements

To ensure a common quality system for organizations that design, develop, produce, install, and service medical devices, the ISO 13485 Medical Devices standard was issued in 1996, and revised and updated in 2003. Earlier revisions were used in conjunction with ISO 9001/2:1994, and dual registration could be obtained to the ISO 9000 standards as well as ISO 13485. While ISO 13485: 2003 is aligned with ISO 9001: 2000 and includes process management principles, organizations must now pursue separate certifications for the two standards.

In some ways, ISO 13485: 2003 is more rigorous than ISO 9001: 2000. The 2003 revision to the medical device quality management system standard requires 18 documented procedures as compared to the six required in ISO 9001: 2000, and it calls out ISO 14971: 2000 for guidance in applying risk management techniques. The U.S. Food and Drug Administration (FDA) has changed some regulations to enable medical device companies to more easily blend the ISO and FDA requirements, although the FDA maintains its independent quality system regulations.[9] Registrar accreditation to ISO 13485: 2003 is administered by RAB in conjunction with ANSI in the United States.

2.6 TL 9000 Telecommunication Quality Management System Requirements

TL 9000 is the telecommunication industry's equivalent of ISO 9001 quality management system requirements. It was developed by the Quality Excellence for Suppliers of Telecommunications (QuEST) Forum. TL 9000 includes 81 enhancements dealing specifically with telecommunications-related issues, driving improvements in that industry. One major difference between TL 9000 other industry-specific quality management system standards is that common, industry-related metrics are required. In addition to the required internal metrics, TL9000 requires quarterly submission of the metrics to a central repository. This requirement allows for industry benchmarking.

TL 9000 has yet to attain wide acceptance in the industry. Most certifications to date have been achieved by U.S. companies. An economic downturn in the industry has forced many telecommunications organizations to pare away all improvement activities as they struggle for survival. As the industry stabilizes, the rate of TL 9000 compliance and registration will most likely increase.[10]

2.7 AS 9100 Aerospace Quality Management System Requirements

In August 2001, the Society of Automotive Engineers (SAE) issued AS 9100 Quality System Requirements, the international aerospace requirements for suppliers to the aerospace industry. AS 9100 resulted from cooperation between major aerospace corporations with membership in the American Aerospace Quality Group (AAQG) as well as input from ANSI-RAB and the Independent Association of Accredited Registrars (IAAR). AS 9100: 2004 is based on ISO 9001: 2000 with additional requirements critical to the aerospace sector. The standard has approximately 80 additional requirements and 18 amplifications of the ISO 9001: 2000 standard.[11]

The most notable additions to AS 9100 not found in ISO 9001 include more detailed requirements for design and development functions and a documented configuration control system. Product key characteristics are required to be identified, and the standard goes into detail on the topics of validation and verification, requiring verification documentation as well as validation test results.

There are additional standards to help implement AS 9100:

- AS 9101B Checklist—Quality Management Systems Assessment
- AS 9131 Quality Systems Non-Conformance Documentation

2.8 Mission Assurance

The term *mission assurance* has been embraced by the U.S. National Aeronautics and Space Administration (NASA) and the U.S. Missile Defense Agency (MDA) to encompass the development, engineering, testing, production, procurement, and implementation of space vehicles or missile defense elements under the cognizance of NASA or the MDA.[12] In its quest for mission assurance, NASA developed the Process Based Mission Assurance (PBMA) plan in 2002. In early 2004, MDA issued the Mission Assurance Provisions (MAP) to provide a measurable, standardized set of Safety, Quality, and Mission Assurance requirements to be applied to contracts for mission and safety critical items in support of evolutionary acquisition and deployment of MDA systems. Both programs are predicated on the idea that if characteristics critical to mission success are identified in designs/hardware/software, and measures are taken to ensure that these critical characteristics are met, the end product will function successfully. Mission assurance requires zero defects in those items critical to the safety, reliability, and quality of the end product.

To date, a certification and/or audit program for Mission Assurance has not been deployed. NASA's PBMA program is not currently required of its suppliers, and it is not clear whether NASA will adopt the MDA's MAP requirements or continue with its PBMA. MAP requirements have been incorporated in MDA contracts and will likely be flowed down from prime contractors to their subtier suppliers, thus affecting a large segment of U.S. and international manufacturers and distributors.

So as to facilitate understanding of their relationships, Figure 3 provides a comparison of various management system standards, applicable industries, and registration bodies.

2.9 CMMI

The Capability Maturity Model Integration (CMMI) is an outgrowth of an earlier software engineering capability maturity model (SE-CMM) contracted by the U.S. Department of Defense in collaboration with Carnegie Mellon University (CMU) in Pittsburgh, Pennsylvania in the early 1980s. The Software Engineering Institute (SEI) is a federally funded

Comparison of Management System Standards			
Standard	Industry	Includes ISO 9001 Requirements?	U.S. Oversight or Accreditation Body/ where to obtain standard
ISO 9001:2000	Generic Quality Management System	N/A	RAB/www.iso.org
ISO/TS 16949:2002	Automotive	Yes	IAOB/www.iso.org
AS 9100	Aerospace	Yes	RAB/www.sae.org
ISO 13485:2003	Medical Devices	Yes	RAB/www.iso.org
TL 9000	Telecommunications	Yes	RAB/www.asq.org
PBMA	Space	No	NASA/pbma.hq.nasa.gov
Mission Assurance	Missile Defense	No	Missile Defense Agency, no accreditation/
CMMI	Generic Maturity Model	No	SEI/www.sei.cmu.edu/cmmi/
ISO 14001	Generic Environmental Management System	No	RAB/www.iso.org

Figure 3 Comparison of management system standards.

research and development center sponsored by the DoD and operated by CMU. SEI was established to address growing software costs and quality problems. Maturity levels for software development were documented, and audits were conducted to verify that government suppliers had reached specific levels of maturity.

This concept has been expanded to the current CMMI with five levels of increasing maturity applicable to software engineering, systems integration, and overall organizational performance. All organizations begin at maturity level 1, thus only levels 2–5 are addressed in the model. Figure 4 shows the CMMI process areas, their abbreviations, and the required process areas for each level of maturity.[13]

The DoD has begun requiring CMMI level 3 for some of its contracts and will most likely elevate the desired level to 4 or 5 in the years ahead. An organization must undergo a Standard CMMI Appraisal Method for Process Improvement (SCAMPI) audit by a certifying body recognized by the SEI to have its maturity level recognized.

2.10 ISO 14000 Environmental Management System Requirements

The ISO 14000 series of environmental management standards was released in 1996, and most recently revised in late 2004. The standards represent the work of the International Organization for Standardization's Technical Committee 207. The ISO 14000 family of standards is primarily concerned with environmental management, focusing on how an organization minimizes harmful effects of its products, processes, and services on the environment, and how that organization achieves continual improvement of its environmental performance.[14]

The ISO 14001 registration process is similar to that of the quality management system standard, ISO 9001. The Registrar Accreditation Board (RAB) serves as the U.S. accrediting body in association with the American National Standards Institute (ANSI), and many of the same registrars that audit and certify organizations to ISO 9001 are accredited to serve as registrars for ISO 14001.

Revisions incorporated in ISO 14001: 2004 serve to align the standard more closely to ISO 9001: 2000 in those areas where the language of a quality management system is applicable to environmental management. The latest revision of the standard also extends

Process Area	Abbr	ML2	ML 3	ML 4	ML 5
Requirements Management	REQM	2			
Measurement and Analysis	MA	2			
Project Monitoring and Control	PMC	2			
Project Planning	PP	2			
Process and Product Quality Assurance	PPQA	2			
Supplier Agreement Management	SAM	2			
Configuration Management	CM	2			
Decision Analysis and Resolution	DAR		3		
Product Integration	PI		3		
Requirements Development	RD		3		
Technical Solution	TS		3		
Validation	VAL		3		
Verification	VER		3		
Organizational Process Definition	OPD		3		
Organizational Process Focus	OPF		3		
Integrated Project Management (IPPD)	IPM		3		
Risk Management	RSKM		3		
Organizational Training	OT		3		
Integrated Teaming	IT		3		
Organizational Environment for Integration	OEI		3		
Organizational Process Performance	OPP			4	
Quantitative Project Management	QPM			4	
Organizational Innovation and Deployment	OID				5
Causal Analysis and Resolution	CAR				5

Increasing Levels of Maturity

Figure 4 CMMI requirements and maturity levels.

requirements to encompass both employees and "all persons working for or on [the organization's] behalf." This revision will require that environmental communication and training efforts be extended to contractors and temporary staff.[15]

General requirements for ISO 14001 include

- Establishment of an environmental management system, including an environmental policy, objectives and targets, and identification of the environmental aspects of the organization's activities, products, and services
- Documented procedures as well as documents and records as required by the EMS standard and regulatory agencies

- Evidence that procedures are being followed and the EMS maintained, which may include internal audits and periodic management reviews
- Corrective and preventive action processes and records
- Continuous improvement processes[16]

Although there was speculation when ISO 14001 was first issued in 1996 that findings in the registration process might result in sanctions from the Environmental Protection Agency (EPA) or, conversely, that registration to ISO 14001 might lead to decreased oversight by the EPA, neither event seems to have occurred. Japan leads the world in number of companies registered to ISO 14001 with approximately 11,000, while the United States holds sixth place behind Japan, Germany, Spain, the United Kingdom, and Sweden, with only about 3000 companies holding ISO 14001 registrations. In fact, in a survey conducted in 2002, North American countries held only 3% of the ISO 14001 certificates, as compared to 17% in Asia and 44% in Europe. While registration in the United States remains low, the concepts embodied in the ISO 14001 standard are widely adopted. Figure 5 compares the number of certifications to ISO 14001 worldwide as compared to certifications to ISO 9001.[17]

3 QUALITY AND PERFORMANCE EXCELLENCE AWARDS

Awards for quality and performance excellence have played an important role in the worldwide improvement of organizational performance over the past two decades. Although an award is presented in recognition of an organization's accomplishments, it is often the award criteria that drives world-class performance. For most organizations, the cycle time averages three years from beginning an effort to pursue a performance excellence award such as the U.S. Baldrige National Quality Award, European Quality Award, or other Baldrige-based national awards to achieving that award. This cycle includes annual award applications and improvement activities based on feedback. In the case of the Deming Prize, this preparation period may take five years or longer. For awards more limited in scope, such as the Shingo Prize in manufacturing, the cycle time is generally shorter.

Award criteria are updated regularly to reflect the best practices of successful organizations as well as emerging challenges. For example, after a number of business scandals in the United States, corporate responsibility and citizenship received greater emphasis in the

ISO 9001 and 14001 Certifications by Region				
Region	No. Countries (ISO 9001)	ISO 9001 Certifications	No. Countries (ISO 14001)	ISO 14001 Certifications
Africa/West Asia	51	23,534	31	1,355
Central/South America	33	13,660	21	1,418
North America	3	53,806	3	4,053
Europe	50	292,970	44	23,316
Far East	20	148,573	17	17,757
Australia/New Zealand	2	29,204	2	1,563
Worldwide	159	561,747	118	49,462

Figure 5 Comparison of ISO 9001 and ISO 14001.

Baldrige Award criteria. The requirement for data gathering and analysis in early award criteria has been expanded to include knowledge management.

Organizations pursue awards for a variety of reasons. Many awards place heavy emphasis on measurable results, positive trends, and world-class performance in comparison with competitors and best-in-class organizations. For businesses, this emphasis drives increased revenue and market share. In health care, results include better patient outcomes, and in education, higher test scores. Customer satisfaction is key to any organization's success, and it takes a prominent place among results categories. Whether an organization pursues an award for its prestige, positive publicity, or to improve its processes, the pursuit can be rewarding whether or not the award itself is ever attained.

3.1 Deming Prize

The Deming Prize was established in 1950 by the Japanese Union of Scientists and Engineers (JUSE). It was named after U.S. statistician W. Edwards Deming to recognize his contributions to Japanese quality control. Dr. Deming was invited to Japan in 1950 to present a series of lectures on quality control and statistical techniques. At the time Japan was still occupied by Allied forces and the Japanese were beginning to rebuild their industries. Deming's approach to quality control was instituted throughout Japan. It was later broadened to include total quality management (TQM), although Deming disavowed any relationship to TQM. Deming's lectures and the prize named after him launched a movement that transformed Japan's reputation in world markets, earning the reputation as premier quality leader, particularly in electronics and automobiles.

There are four types of Deming Prizes (see Fig. 6). The Deming Prize most often cited in the United States is the Application Prize, open to companies or autonomous divisions that have achieved distinctive performance improvement through the application of TQM in

Type of Prize	Candidate	Description
The Deming Prize for Individuals	Individuals and groups are eligible	Given to those who have made outstanding contributions to the study of TQM or statistical methods used for TQM, or those who have made outstanding contributions in the dissemination of TQM
The Deming Application Prize	Organizations or divisions of organizations that manage their business autonomously	Given to organizations or divisions of organizations that have achieved distinctive performance improvement through the application of TQM in a designated year.
The Quality Control Award for Operations Business Units	Operations business units of an organization	Given to operations business units of an organization that have achieved distinctive performance improvement through the application of quality control/ management in the pursuit of TQM in a designated year.
The Nikkei QC Literature Prize	Author(s) of literature published in Japan	Literature on the study of TQM or statistical methods used for TQM that is recognized to contribute to the progress and development of quality management.

Figure 6 Deming Prize information.

a designated year. Applicants must complete a formal application in Japanese. The JUSE Deming Application Prize Committee evaluates each application, and if there is enough evidence to move forward, completes a document review. A site visit is also required, and costs for travel from Japan are paid by the applicant.

The JUSE Committee uses a checklist that includes

- Policies
- Organization
- Education
- Collection
- Analysis
- Standardization
- Control
- Quality assurance
- Effects (results)
- Future plans

It is interesting to note that the Deming Prize Committee evaluates not only the results that have been achieved by the applicant, but also the effectiveness expected in the future. The criteria does not specify how the applicant should approach TQM, and the Committee evaluates how the applicant uses TQM in the context of its unique business situation. In its publication concerning the evaluation process, the Committee states:

> No organization can expect to build excellent quality and management systems just by solving problems given by others. They need to think on their own, set lofty goals, and challenge themselves to achieve these goals.[18]

The Deming Application Prize involves a process that can take several years and cost a great deal. Implied in this process is the use of JUSE consultants for months or years to assist the applicant in putting its TQM processes into place. The consultants perform a quality-control diagnosis and recommend changes. The organization creates its application for the Deming Prize the year after the JUSE consultants have completed their work. The length of the application is set according to the size of the company, ranging from 50 pages for organizations with fewer than 100 employees to 75 pages for 100–2000 employees, plus 5 pages for each additional 500 employees over 2000.

The Deming Application Prize has been awarded to nearly 200 companies since 1951. Although the competition is open to any company regardless of national origin, most of the recipients have been Japanese. Only two U.S. companies have received the Deming Application Prize. Florida Power and Light was the first U.S. company to receive the prize in 1989, and AT&T Power Systems was a recipient in 1994. Since 2001, companies in Thailand and India have dominated the recipient list.

3.2 Baldrige National Quality Award

Although not the oldest quality award, the Baldrige National Quality Award (BNQA) program has had the greatest influence on performance excellence in the United States. It was created by the U.S. Congress as Public Law 100-107 on August 20, 1987 to address international trade deficits and a lagging U.S. economy caused in large part by the availability of Japanese automobiles and electronics perceived to be of higher quality and reliability than their U.S. counterparts. The Baldrige Award was named after Malcolm Baldrige, the U.S. Secretary of Commerce who died in a tragic rodeo accident in 1987.[19]

The Baldrige Award criteria have three roles in strengthening U.S. competitiveness:

1. To help improve organizational performance practices, capabilities, and results
2. To facilitate communication and sharing of best practices information among U.S. organizations of all types

3. To serve as a working tool for understanding and managing performance and for guiding organizational planning and opportunities for learning

The award was first offered only to U.S. for-profit companies in the categories of manufacturing, service, and small business (fewer than 500 employees) with a maximum of two award recipients per year in each category. The award was later amended to increase the maximum number of recipients to three per category and, in 2000, education and health care were added. In 2004, Congress passed legislation to add not-for-profit and government organizations to those eligible to apply for the Baldrige Award. At this writing, no funding has been appropriated to realize this latest expansion of the program.

The Department of Commerce is responsible for administering the BNQA program. The National Institute of Standards and Technology (NIST), an agency of the Department of Commerce, manages the award program. The American Society for Quality (ASQ) assists in administering the program under contract to NIST.

There are seven criteria *Categories*. Each is broken into subcategories called *Items*, and each Item has a number of *Areas to Address* that are composed of specific questions about the applicant's processes or results. An application is scored on a 1000-point weighted scale, with points assigned to each Item (see Fig. 7 for a breakdown of points by Category). The criteria, including weighting of the Categories and points per Item, are revised periodically. At this writing, separate criteria are used for business, health care, and education organizations, but the differences between them are small, and there is a plan to eventually use common criteria for all applicant groups.

The results-oriented BNQA criteria is built around eleven core values and concepts (see Fig. 1) and focuses on an organization's primary activities, customers, and competitive results. The greatest changes to the criteria were made in 1995, when the word quality was almost entirely removed, broadening the scope of the award criteria to encompass all of the elements of an organization's performance and not just TQM. Quality management systems must be fully integrated into an organization's operations. In 2004, the scoring guidelines were modified to include clearer guidelines at four levels:

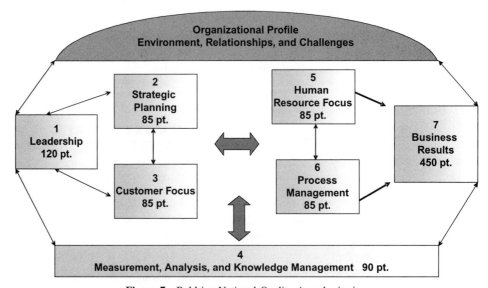

Figure 7 Baldrige National Quality Award criteria.

- Approach
- Deployment
- Learning
- Integration

The ADLI model promotes an organizational focus and places emphasis on processes that help the organization share its knowledge and information. Improvement outcomes are expected to be integrated across the organization's operations, thereby improving overall organizational performance.[20]

Figure 7 illustrates the BNQA framework. The Organizational Profile sets the context in which the organization operates. The system operations are composed of the six Baldrige Categories in the center of the figure. These define the organization's operations and the results they achieve. Leadership (category 1), Strategic Planning (category 2), and Customer and Market Focus (category 3) represent the leadership triad. Human Resource Focus (category 5), Process Management (category 6), and Results (category 7) represent the results triad in those employees and key processes accomplish the work of the organization that yields the business results.

Measurement, Analysis, and Knowledge Management (category 4) serve as the system foundation in the Baldrige model. These are critical to the effective management of the organization, enabling a fact-based, knowledge-driven system for improving performance and competitiveness.[21]

A Board of Examiners is selected each year to review applications for the BNQA. Board members are a volunteer group of recognized experts in the areas of performance excellence and continuous improvement. They are selected through an application process, with at least one-third of the Board made up of new examiners. An examiner may serve for no more than six years and then takes on alumni status. Alumni Examiners may apply for a seat at Examiner training and can be called upon to review an application during the first stage in the award process.

The Board has almost doubled in size since the award's inception to accommodate a spike in applications after education and health care were added to the award categories. In 2004, there were 533 members of the Board of Examiners. Examiner applications are due in early January each year and are available at *www.baldrige.nist.gov* each Fall.

Applications for the BNQA are limited to a five-page Organizational Profile and 50-page response to the criteria's Areas to Address. The award process includes

- *Stage One.* 7–10 examiners independently evaluate the application; Panel of Judges selects applications that will move forward.
- *Stage Two.* Examiner team from Stage One works together to create a consensus scorebook; judges review consensus scores and select applicants to receive a site visit.
- *Stage Three.* Examiner team visits site for 2–5 days, depending on the applicant's size and complexity. The team prepares a final scorebook with recommendations for changes in the original score; judges select applicants to receive the award.

Every applicant receives a detailed feedback report citing strengths and opportunities for improvement on each of the criteria items. The report also contains an executive summary and an explanation of the application's scoring band and what that may indicate about the organization in terms of the ADLI model.

Applications are due at the end of May each year. Stage One activities take place in June–July, and Stage 2 in August–September. Site visits take place in October, and award recipients are announced in December.

The application fees for the BNQA range from $500 for not-for-profit educational organizations to $5000 for large manufacturing and service businesses. Complex organizations with more than one product line may opt (with prior permission) to provide supplemental information. The fees for submitting this additional material range from $250 to $2000. Site visit costs, reimbursed by the applicant, can range from as little as $1500 to as much as $35,000. These amounts will most likely increase over time.[22]

3.3 U.S. State Quality Awards

Since the introduction of the BNQA, 49 of the 50 U.S. states have initiated state awards for performance excellence. Most of these awards are based on the Baldrige criteria, and in most cases eligibility has been extended to any organization within the state, including governmental and not-for-profit organizations.

In most states, the award process adheres to the BNQA process with a written application, site visits, and an award ceremony. Applicants generally receive a feedback report. State examiners are selected and receive training, often using BNQA training materials made available by NIST. Unlike the Baldrige Award, most state programs have varying levels of recognition. This serves to encourage organizations that may not yet have achieved the highest levels of performance excellence. The costs to state award applicants are often less than half the BNQA costs, particularly the costs for a site visit. Many state quality award recipients have gone on to become recipients of the Baldrige Award. Figure 8 provides a listing of state award programs.

3.4 Shingo Prize for Excellence in American Manufacturing

The Shingo Prize for Excellence in Manufacturing was created in 1988 and named for Japanese industrial engineer Shigeo Shingo, a leading expert on improving the manufacturing process. He created, along with Taichi Ohno, many of the facets of just-in-time manufacturing while working with Toyota Production Systems. Shingo is known for his books, including *Zero Quality Control; Source Inspection and the Poka-yoke System;* and *The Shingo System for Continuous Improvement*. The award is administered by the College of Business, Utah State University, and a Board of Governors made up of leading representatives of businesses, professional organizations, and academic institutions. The Board oversees fund raising and other financial activities, guides Prize governance, establishes Prize guidelines, and ratifies Prize recipients based on the recommendations of the Board of Examiners.

The mission of the Shingo Prize (Fig. 9) is to

- Facilitate an increased awareness by the manufacturing community of lean, just-in-time manufacturing processes, systems, and methodologies that will maintain and enhance a company's competitive position in the world marketplace.
- Foster enhanced understanding and subsequent sharing of successful core manufacturing improvement methodologies.
- Encourage research and study of manufacturing processes and production improvements in both the academic and business arenas.

The Shingo Prize is open to manufacturers in the United States, Canada, and Mexico. There is no limit to the number of prizes awarded each year, and typically there are a dozen or more recipients per year divided into small and large business categories.[23]

U.S. State Quality and Performance Excellence Awards

State/Began	Levels of Award	Criteria & Eligibility	Contact
AK	None		
AL 1986	Level 1: Commitment Level 2: Progress Level 3: Excellence	Based on Baldrige Award criteria; open to manufacturing, service, small business, health care and education	Alabama Productivity Center www.alabamaexcellence.com
AR 1994	Challenge Achievement Commitment Governors	Based on Baldrige Award criteria; open to all organizations in Arkansas	Arkansas Institute for Performance Excellence www.arkansas-quality.org
AZ 1993	Showcase (processes) Pioneer Governors	Pioneer and Governors based on Baldrige Award criteria. Open to all organizations in Arizona	Arizona Quality Alliance www.Arizona-Excellence.com
CA 1992	Level 1: Challenge Level 2: Prospector Level 3: Eureka Team Awards	Levels 1-3 based on Baldrige Award criteria; public and private sector organizations in California	California Council for Excellence www.calexcellence.org
CO 2002	Foothills Performance Excellence Award	Based on Baldrige Award criteria; open to manufacturing, service, small business, health care, education, non-profit, and government	Colorado Performance Excellence www.coloradoexcellence.org
CT 1987	Innovation Breakthrough Leadership	Based on Baldrige Award criteria; open to manufacturing, service, small business, health care, education, nonprofit, and government	Connecticut Quality Improvement Award www.ctqualityaward.org
DE 1992	Commitment Merit Excellence	Based on Baldrige Award criteria; open to large and small manufacturing, large and small non-manufacturing, and non-profit.	Accolade Alliance www.accoladealliance.org
FL 1992	Challenge Navigator Governors Sterling	Based on Baldrige Award criteria; open to manufacturing, service, health care, education, and public sector	The Florida Sterling Council www.floridasterling.com
GA 1996	Step 1: Focus Step 2: Progress Step 3: Oglethorpe	Based on Baldrige Award criteria; open to business, industry, government, education, healthcare, and non-profit.	Georgia Oglethorpe www.georgiaoglethorpe.org
HI 1995	Hawaiian State Award of Excellence	Unknown	
IA 2000	Tier 1: Commitment Tier 2: Progress Tier 3: Leadership Gold, Silver, Bronze	Based on Baldrige Award criteria; open to organizations with employees in Iowa.	Iowa Quality Center www.iowaqc.org
ID	Level 1: Interest	Based on Baldrige Award	Idaho Dept. of Commerce

Figure 8 U.S. state quality awards.

1995	Level 2: Commitment Level 3: Quality	criteria; open to all Idaho organizations	www.idahoworks.com
IL 1994	Level 1: Commitment Level 2: Progress Level 3: Achievement	Based on Baldrige Award criteria; open to Illinois-based industry, service, health care, education, and government entities	Lincoln Foundation for Performance Excellence www.lincolnaward.org
IN 1996	Indiana Quality Improvement Award	Narrative describing improvement; open to any company or organization located in Indiana, both for-profit and not-for-profit	Indiana Business Modernization and Technology Co. www.bmtadvantage.org
KS 1995	Level 1: Commitment Level 2: Performance Level 3: Excellence	Based on previous year Baldrige Award criteria; open to all Kansas organizations in operation for at least 1 year	Kansas Award for Excellence Foundation www.kae.bluestep.net
KY 1996	Level 1: Interest Level 2: Commitment Level 3: Achievement Level 4: Gold	Based on previous year Baldrige Award criteria; open to all Kentucky public or privately held organizations.	Kentucky Quality Council www.kqc.org
LA 1993	Level 1: Commit Level 2: Progress Level 3: Performance Excellence	Based on Baldrige Award criteria; open to business, health care, and education	Louisiana Quality Foundation www.Laqualityaward.com
MA 1991	Level 1: Self-Assessment Level 2: Examiner-Assessment Level 3: Excellence	Based on Baldrige Award criteria; any organization based in Massachusetts is eligible to apply for any level. Health Care and Education organizations may use the BNQA materials for their category if they choose.	MassExcellence www.massexcellence.com
MD 1997	Level 1: Bronze Level 2: Silver Level 3: Gold Level 4: U.S. Senate Productivity	Based on Baldrige Award criteria with shortened application; open to manufacturing, service, public sector, health care, and education with 5 or more employees.	U.S. Senate Productivity and Maryland Quality Awards www.bsos.umd.edu/mcqp
ME 1990	Level 1: Commitment Level 2: Progress Level 3: Excellence	Based on previous year Baldrige Award criteria; open to publicly or privately held organizations of any size located in Maine	Margaret Chase Smith Quality Association www.Maine-Quality.org
MI 1993	Level 1: Lighthouse Level 2: Navigator Level 3: Honor Roll Level 4: Quality Leadership	Based on Baldrige Award criteria; open to manufacturing, service, health care, education, public sector, and small businesses in Michigan	Michigan Quality Council www.michiganquality.org
MO 1992	Level 1: Facet of Excellence	Based on Baldrige Award criteria; open to any	Excellence in Missouri Foundation www.mqa.org

Figure 8 (*Continued*)

		Level 2: Missouri Quality Award	organization located in Missouri; small, medium, and large categories	
MN	1988	Level 1: Commitment Level 2: Advancement Level 3: Achievement Level 4: Excellence	Based on Baldrige Award criteria; open to sectors manufacturing, service, health care, education, government, and non-profit.	Minnesota Council for Quality www.councilforquality.org
MS	1994	Level 1: Alignment Level 2: Commitment Level 3: Excellence Level 4: Governors	Based on previous year Baldrige Award criteria; open to any public or privately held organization of any size	State Board for Community and Junior Colleges www.sbcjc.cc.ms.us/acct/
MT	2000	Desiree Taggart Memorial Awards for businesses and individuals	State application form.	Montana Council for Workforce Quality (MCWQ) dli.state.mt.us/pub/mswinter04/taggartawards.htm
NC	1989	North Carolina Award for Excellence	Based on Baldrige Award criteria; open to all NC organizations	North Carolina Award for Excellence North Carolina State University Industrial Extension Service www.ies.ncsu.edu/qualityaward/
NE	1992	Level 1: Commitment Level 2: Progress Level 3: Excellence	Based on Baldrige Award criteria; open to Nebraska businesses and organizations	Nebraska Department of Economic Development assist.neded.org/edgerton
NH	1994	Level 1: Interest Level 2: Commitment Level 3: Achievement Level 4: Granite State Quality Award	Based on Baldrige Award criteria; open to New Hampshire businesses and organizations	Granite State Quality Council www.gsqc.com
NJ	1992	Level 1: Bronze Level 2: Silver Level 3: Gold	Based on Baldrige Award criteria; open to sectors business (large/small), business support (large/small), health care, education, government, and non-profit.	Quality New Jersey www.qnj.org
NM	1993	Level 1: Pinon Level 2: Roadrunner Level 3: Zia (highest)	Based on Baldrige Award criteria; open to all New Mexico businesses and organizations	Quality New Mexico www.qualitynewmexico.org
NV	--	Prospector Trailblazer Silver	Unknown	Nevada Quality Alliance www.nvqa.org
NY	--	Governor's Excelsior Award	The New York Quality Award program is currently being changed.	Check www.workforce-excellence.net/html/stateawards/default.htm for latest information
OH	1998	Tier 1: Pledge Tier 2: Commitment Tier 3: Achievement Tier 4: Governors Award	Based on Baldrige Award criteria; open to Business, Education, Government, Health Care, and Not-for-Profit organizations with employees in Ohio for at least one year.	Ohio Award for Excellence www.oae.org
OK	1993	Level 1: Commitment Level 2: Achievement	Based on previous years Baldrige Award criteria	The Oklahoma Quality Award Foundation www.oklahomaquality.com

Figure 8 (*Continued*)

		Level 3: Excellence	(Level 1 uses self-assessment); open to any organization in existence over 3 years.	
OR --	Oregon Excellence Awards	Unknown	Contact information is currently unavailable. Check <www.workforce-excellence.net/html/stateawards/default.htm> for latest information.
PA --	Keystone Performance Excellence Award	Unknown	Contact information is currently unavailable. Check <www.workforce-excellence.net/html/stateawards/default.htm> for latest information.
RI --	RI Award for Competitiveness and Excellence	Unknown	RACE for Performance Excellence Email: racequal@idt.net
SC 1995	Bronze Achiever Silver Achiever Gold Achiever Governor's Award	Based on Baldrige Award criteria; open to all organizations in South Carolina	South Carolina Quality Forum www.scquality.com
TN 1992	Level 1: Interest Level 2: Commitment Level 3: Achievement Level 4: Excellence	Based on Baldrige Award criteria; open to all Tennessee organizations	Tennessee Center for Performance Excellence www.tncpe.org
TX 1994	Level 1: Commitment Level 2: Progress Level 3: Excellence	Based on Baldrige Award criteria; open to any Texas organization	Quality Texas Foundation www.texas-quality.org
UT 2000	Level 1: Continuous Improvement Level 2: Quality Level 3: Governor's	Based on Baldrige Award criteria; open to any Utah organization	Utah Quality Council www.utahqualityaward.org
VT 1996	Level 1: Commitment Level 2: Achievement Level 3: Governor's Level 4: Productivity	Baldrige-based, modified for shorter applications; open to all Vermont organizations	Vermont Council for Quality www.vermontquality.org
VA 1988	Level 1: SPQA Certificate Level 2: SPQA Plaque Level 3: Medallion	Based on Baldrige criteria; open to all Virginia organizations	U.S. Senate Productivity and Quality Award for Virginia
WA 1993	Level 1: Certificate Level 2: Quality Award	Based on Baldrige criteria; open to all Washington organizations	Washington State Quality Award www.wsqa.net
WI 1997	Level 1: Commitment Level 2: Proficiency Level 3: Mastery Level 4: Excellence	Based on Baldrige Award criteria; open to all Wisconsin organizations	Wisconsin Forward Award www.forwardaward.org
WV	N/A		
WY 2004	Wyoming State Quality Award	Details not yet available	Wyoming Business Council Check <www.workforce-excellence.net/html/stateawards/default.htm> for latest information.

Figure 8 (*Continued*)

3.5 Quality Awards around the World

It is a testament to the effectiveness of award programs at increasing organizational effectiveness that a number of national quality award programs have been established. There are now quality award programs in Europe, Japan, the United Kingdom, South Africa, Egypt, Hong Kong, India, Philippines, Australia, Singapore, and countless other regions and nations.

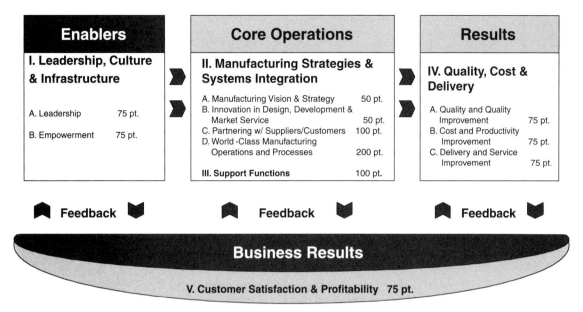

Figure 9 Shingo Prize criteria.

The criteria for these programs are most often based on the BNQA criteria, and NIST in the United States has been generous in sharing its training program and providing support to these programs.

The European Quality Award (EQA) was created in 1991. The award is managed by the European Foundation for Quality Management (EFQM), an organization founded in 1988 and made up of more than 440 quality-oriented European businesses and organizations. It was created to enhance European competitiveness and effectiveness through the application of TQM principles in all aspects of organizations. EFQM headquarters is located in the Netherlands.

The first EQA was awarded in 1992. Prizes are given to organizations that show that their approach to TQM has contributed significantly over the years to satisfying the expectations of their customers, employees, and other stakeholders. Applicants may apply for any of three levels: Committed to Excellence; Recognized for Excellence; or the European Quality Award. For the latter, a 75-page application is created by the applicant.[24]

The European Quality Award criteria is weighted and scored on a scale of 0 to 1000 points in a manner similar to the criteria for the Baldrige Award. The criteria are divided into two main categories: Enabler Criteria and Results Criteria. The EQA model is currently undergoing a major revision expected to be deployed in 2006.

Although Japan's Deming Prize had been in existence for almost forty years when the Baldrige Award was created, it remained focused on specific techniques associated with TQM. When the Japanese witnessed the improved quality of U.S. products and operations in the early 1990s, the Japan Quality Award (JQA) program was launched. The JQA was established in 1995 by the Japan Productivity Center for Socio-Economic Development (JPC-SED). It is modeled after the BNQA, modified to accommodate Japanese management practices.[25]

Although the models for the MBQA and JQA are similar, it is interesting to note some of the differences reflective of the priorities set by management systems in the two countries.

Figures 7 and 10 indicate that while the Baldrige criteria has taken a closer look at social responsibility through Areas to Address in the Leadership category, the Japanese model includes an entire category for social responsibility. This mirrors the difference in registration levels to ISO 14001 Environmental Management Standard between Japan and the United States (see Fig. 5).

3.6 Industry-Specific Quality Awards

In addition to national and regional quality awards, there are a few significant award programs that address specific sectors or industries. The American Health Care Association Quality Award is given for excellence in long-term care facilities. The award is modeled after the BNQA.[26]

The National Housing Quality (NHQ) Award gives the highest recognition by the housing industry for quality achievement. The NHQ awards are also patterned after the Baldrige National Quality Award. Entries are judged by panels of experts who evaluate the role that customer-focused quality plays in construction, business management, sales, design, and warranty service.[27]

Industry Week magazine's America's Best Plants Award was first presented in 1990. The award is given to ten manufacturing plants in the United States, Canada, or Mexico that score the greatest number of points on a questionnaire. Independent judges select 25 semifinalists, and those applicants complete additional information. A site visit is performed by one of the magazine's editors at each of the ten finalists' plants. Articles about the finalists

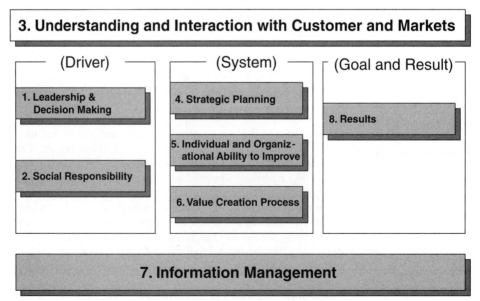

Figure 10 Japan Quality Award criteria.

Award/Certification	Year Created	Emphasis	Cost	Social Responsibility Requirements	Contact
Baldrige National Quality Award	1987	Performance Excellence, Customer Focus, Results	High	Medium	www.baldrige.nist.gov
Deming Prize	1950	TQM	Very High	Low	www.juse.or.jp/e/deming/
State Quality Awards (U.S.)	Various	Modeled on BNQA	Low – Medium	Medium	See Figure 9
Shingo Prize	1988	Lean Manufacturing	Medium	Low	www.Shingoprize.org
Japan Quality Award	1994	Performance Excellence, Customer Focus, Results	High	High	www.jpc-sed.or.jp/eng/award/
European Quality Award	1992	Performance Excellence, Customer Focus, Results	High	High	www.efqm.org
IndustryWeek Best Plants Award	1990	Manufacturing Methods	Low	Low	www.industryweek.com
ISO 9001	2000	Quality Management System	High	Low	www.iso.org
ISO 14001	2002	Environmental Management System	High	High	www.iso.org

Figure 11 Comparison of quality awards.

appear in an issue of *Industry Week* magazine; however, there is no formal feedback to the applicants.[28]

3.7 How Do They Compare?

The value in pursuing registrations, certifications, or awards is not necessarily in achieving the certificate or plaque. The benefit is often derived from the process itself, particularly when an organization must apply for several years, receiving feedback from the awarding organization, and implementing improvements. Figure 11 provides contact information as well as some points of comparison between some of the award and certification programs.

REFERENCES

1. International Organization for Standardization, "Introduction and Overview," www.iso.org.
2. International Organization for Standardization, "Certification, Registration, and Accreditation," www.iso.org.
3. International Organization for Standardization, "Introduction," www.iso.org.

4. E. Stapp, *ISO 9000: 2000, An Essential Guide to the New Standard,* Quality Publishing, Tucson, AZ, 2001.
5. D. Drickhamer, "Quality-Management Principles," *IndustryWeek,* Mar 5, 2001.
6. Registrar Accreditation Board, "About RAB," www.rabnet.com.
7. R. O'Connell, "ISO/TS 16949 Made Easy," *Quality Digest,* August 1, 2004.
8. C. Lupo, "ISO/TS 16949 the Clear Choice for Automotive Suppliers," *Quality Progress,* October 2002, p. 44.
9. J. Adam, "ISO 13485 Levels the Playing Field," *Quality Digest,* August 15, 2004.
10. R. Clancy, "Can TL 9000 Contribute to Telecom's Turnaround?" *Quality Progress,* February 2004, pp. 38–45.
11. E. Barker, "Aerospace's AS9100 QMS Standard," *Quality Digest,* May 2002.
12. NASA Office of Safety and Mission Assurance, pbma.hq.nasa.gov.
13. Carnegie Mellon University, "Concepts of Operations for the Capability Maturity Model® Integration (CMMI)," August 11, 1999.
14. S. L. K. Briggs, "Next Generation ISO 14001," *Quality Progress,* August 2004, pp. 75–77.
15. M. Block, "ISO 14001 Revision Nears Completion," *Quality Progress,* February 2004.
16. P. Scott, " The Moving Goal Posts from Environmental to Corporate Responsibility," *ISO Management Systems,* Sept–Oct 2003, p. 30.
17. ISO Management Systems, "ISO Twelfth Cycle Survey of ISO 9001 and ISO 14001 Certificates," December 31, 2002.
18. Japanese Union of Scientists and Engineers, "What Is the Deming Prize," www.juse.or.jp/e/deming/.
19. M. G. Brown, *Baldrige Award Winning Quality,* 5th ed., ASQC Quality Press, 1995.
20. Baldrige National Quality Award Program, *Criteria for Performance Excellence,* NIST, 2004, www.baldrige.nist.gov.
21. Baldrige National Quality Award Program, *Examiner Training,* 2004.
22. Baldrige National Quality Award Program, *Baldrige Award Application Forms,* NIST, 2004, www.baldrige.nist.gov.
23. Shingo Prize for Excellence in Manufacturing, 2004, www.shingoprize.org.
24. P. Wendel, "The European Quality Award and How Texas Instruments Europe Took the Trophy Home," *The Quality Observer,* January 1996.
25. Japan Productivity Center for Socio-Economic Development, "Japan Quality Award," www.jpc-sed.or.jp/eng/award/.
26. A. Starkey, "Continuous Quality Improvement Earns AHCA Recognition," American Health Care Association, August 11, 2003. www.ahca.org/news/nr030711.htm.
27. National Association of Homebuilders, www.nahb.org.
28. *Industry Week Magazine,* www.industryweek.com.

CHAPTER 20
SAFETY ENGINEERING

Jack B. ReVelle, Ph.D.
ReVelle Solutions, LLC
Santa Ana, California

1	**INTRODUCTION**	**640**
	1.1 Background	640
	1.2 Employee Needs and Expectations	640
2	**GOVERNMENT REGULATORY REQUIREMENTS**	**642**
	2.1 Environmental Protection Agency (EPA)	642
	2.2 Occupational Safety and Health Administration (OSHA)	645
	2.3 State-Operated Compliance Programs	647
3	**SYSTEM SAFETY**	**649**
	3.1 Methods of Analysis	649
	3.2 Fault-Tree Technique	650
	3.3 Criteria for Preparation/Review of System Safety Procedures	651
	3.4 Risk Assessment Process	656
4	**HUMAN FACTORS ENGINEERING/ERGONOMICS**	**657**
	4.1 Human–Machine Relationships	657
	4.2 Human Factors Engineering Principles	658
	4.3 General Population Expectations	659
5	**ENGINEERING CONTROLS FOR MACHINE TOOLS**	**660**
	5.1 Basic Concerns	660
	5.2 General Requirements	662
	5.3 Danger Sources	663
6	**MACHINE SAFEGUARDING METHODS**	**663**
	6.1 General Classifications	663
	6.2 Guards, Devices, and Feeding and Ejection Methods	664
7	**ALTERNATIVES TO ENGINEERING CONTROLS**	**664**
	7.1 Substitution	669
	7.2 Isolation	669
	7.3 Ventilation	670
8	**DESIGN AND REDESIGN**	**670**
	8.1 Hardware	670
	8.2 Process	671
	8.3 Hazardous Material Classification System	672
	8.4 Material Safety Data Sheet (MSDS)	674
	8.5 Safety Design Requirements	675
9	**PERSONAL PROTECTIVE EQUIPMENT**	**675**
	9.1 Background	675
	9.2 Planning and Implementing the Use of Protective Equipment	675
	9.3 Adequacy, Maintenance, and Sanitation	678
10	**MANAGING THE SAFETY FUNCTION**	**678**
	10.1 Supervisor's Role	678
	10.2 Elements of Accident Prevention	679
	10.3 Management Principles	680
	10.4 Eliminating Unsafe Conditions	681
	10.5 Unsafe Conditions Involving Mechanical or Physical Facilities	685
11	**SAFETY TRAINING**	**687**
	11.1 Specialized Courses	687
	11.2 Job Hazard Analysis Training	695
	11.3 Management's Overview of Training	697
	11.4 Sources and Types of Training Materials	698
	BIBLIOGRAPHY	**698**

1 INTRODUCTION

1.1 Background

More than ever before, engineers are aware of and concerned with employee safety and health. The necessity for this involvement was accelerated with the passage of the OSHAct in 1970, but much of what has occurred since that time would have happened whether or not the OSHAct had become the law.

As workplace environments become more technologically complex, the necessity for protecting the work force from safety and health hazards continues. Typical workplace operations from which workers should be protected are presented in Table 1. Whether workers should be protected through the use of personal protective equipment, engineering controls, administrative controls, or a combination of these approaches, one fact is clear: it makes good sense to ensure that they receive the most cost-effective protection available. Arguments in support of engineering controls over personal protective equipment and vice versa are found everywhere in the current literature. Some of the most persuasive discussions are included in this chapter.

1.2 Employee Needs and Expectations

In 1981 ReVelle and Boulton asked the question, "Who cares about the safety of the worker on the job?" in their award-winning two-part article in *Professional Safety,* "Worker Attitudes and Perceptions of Safety." The purpose of their study was to learn about worker attitudes and perceptions of safety. To accomplish this objective, they established the following working definition:

> **Worker attitudes and perceptions:** As a result of continuing observation, an awareness is developed, as is a tendency to behave in a particular way regarding safety.

To learn about these beliefs and behaviors, they inquired about the following:

Table 1 Operations Requiring Engineering Controls and/or Personal Protective Equipment

Acidic/basic process and treatments	Grinding
Biological agent processes and treatments	Hoisting
Blasting	Jointing
Boiler/pressure vessel usage	Machinery (mills, lathes, presses)
Burning	Mixing
Casting	Painting
Chemical agent processes and treatments	Radioactive source processes and treatments
Climbing	Sanding
Compressed air/gas usage	Sawing
Cutting	Shearing
Digging	Soldering
Drilling	Spraying
Electrical/electronic assembly and fabrication	Toxic vapor, gas, and mists and dust exposure
Electrical tool usage	Welding
Flammable/combustible/toxic liquid usage	Woodworking

1. Do workers think about safety?
2. What do they think about safety in regard to
   a. Government involvement in their workplace safety.
   b. Company practices in training and hazard prevention.
   c. Management attitudes as perceived by the workers.
   d. Coworkers' concern for themselves and others.
   e. Their own safety on the job.
3. What do workers think should be done, and by whom, to improve safety in their workplace? The major findings of the ReVelle-Boulton study are summarized here:*

   - Half the workers think that government involvement in workplace safety is about right; almost one-fourth think more intervention is needed in such areas as more frequent inspections, stricter regulations, monitoring, and control.
   - Workers in large companies expect more from their employers in providing a safe workplace than workers in small companies. Specifically, they want better safety programs, more safety training, better equipment and maintenance of equipment, more safety inspections and enforcement of safety regulations, and provision of more personal protective equipment.
   - Supervisors who talk to their employees about safety and are perceived by them to be serious are also seen as being alert for safety hazards and representative of their company's attitude.
   - Coworkers are perceived by other employees to care for their own safety and for the safety of others.
   - Only 20% of the surveyed workers consider themselves to have received adequate safety training. But more than three-fourths of them feel comfortable with their knowledge to protect themselves on the job.
   - Men are almost twice as likely to wear needed personal protective equipment as women.
   - Half the individuals responding said they would correct a hazardous condition if they saw it.
   - Employees who have had no safety training experience almost twice as many on-the-job accidents as their fellow workers who have received such training.
   - Workers who experienced accidents were generally candid and analytical in accepting responsibility for their part in the accident; and 85% said their accidents could have been prevented.

The remainder of this chapter addresses those topics and provides the information that engineering practitioners require to professionally perform their responsibilities with respect to the safety of the work force.

---

*Reprinted with permission from the January 1982 issue of *Professional Safety,* official publication of the American Society of Safety Engineers.

2 GOVERNMENT REGULATORY REQUIREMENTS*

Two agencies of the federal government enforce multiple laws that impact many of the operational and financial decisions of American businesses, large and small. The U.S. Environmental Protection Agency (EPA) has responsibility for administering (listed chronologically by year of passage of each law) the National Environmental Policy Act (NEPA, 1969); the Clean Water Act (CWA,1970); the Resource Conservation and Recovery Act (RCRA, 1976); the Toxic Substances Control Act (TSCA, 1976); the Comprehensive Environmental Response, Compensation, and Liability Act (CERCLA, aka Superfund, 1980); the Superfund Amendments and Reauthorization Act (SARA, 1986); the Clean Air Act (CAA, 1990); the Oil Pollution Act (OPA,1990); the Pollution Protection Act (1990); and the Federal Insecticide, Fungicide, and Rodenticide Act (FIFRA,1996). The Occupational Safety and Heath Act, also known as OSHAct, (1970) is enforced by the Occupational Safety and Health Administration (OSHA), a part of the U.S. Department of Labor.

This section addresses the regulatory demands of these federal statutes from the perspective of whether to install engineering controls that would enable companies to meet these standards or simply to discontinue certain operations altogether, that is, whether they can justify the associated costs of regulatory compliance.

2.1 Environmental Protection Agency (EPA)

National Environmental Policy Act (NEPA, 1969)
This was one of the first acts ever written that establishes the broad national framework for protecting our environment. NEPA's basic policy is to assure that all branches of government give proper consideration to the environment prior to undertaking any major federal action that significantly affects the environment. NEPA requirements are invoked when airports, buildings, military complexes, highways, parkland purchases, and other federal activities are proposed. Environmental Assessments (EAs) and Environmental Impact Statements (EISs), which are assessments of the likelihood of impacts from alternative courses of action, are required from all Federal agencies and are the most visible NEPA requirements.

Clean Water Act (CWA, 1970)
Growing public awareness and concern for controlling water pollution led to enactment of the Federal Water Pollution Control Act Amendments of 1972. As amended in 1977, this law became commonly known as the Clean Water Act. The CWA established the basic structure for regulating discharges of pollutants into the waters of the United States and gave the EPA the authority to implement pollution control programs such as setting wastewater standards for industry. The Act also continued the requirement to set water quality standards for all contaminants in surface waters. It made it unlawful for any person to discharge any pollutant from a point source into navigable waters, unless a permit was obtained under its provisions. It also funded the construction of sewage treatment plants under the construction grants programs and recognized the need for planning to address the critical problems posed by nonpoint source pollution.

Subsequent enactments modified some of the earlier CWA provisions. Revisions in 1981 streamlined the municipal construction grants process, improving the capabilities of treatment

---

*"Engineering Controls: A Comprehensive Overview" by Jack B. ReVelle. Used by permission of The Merritt Company, Publisher, from T. S. Ferry, *Safety Management Planning*, copyright ©1982, The Merritt Company, Santa Monica, CA 90406.

plants built under the program. Changes in 1987 phased out the construction grants program, replacing it with the State Water Pollution Control Revolving Fund, more commonly known as the Clean Water State Revolving Fund. This new funding strategy addressed water quality needs by building on EPA–State partnerships.

An electronic version of the Clean Water Act, as amended through the enactment of the Great Lakes Legacy Act (GLLA, 2002), is available on the internet at www.epa.gov/region5/water/cwa.htm. This electronic version annotates the sections of the Act with the corresponding sections of the U.S. Code and footnote commentary on the effect of other laws on the current form of the Clean Water Act.

Resource Conservation and Recovery Act (RCRA, 1976)
Enacted in 1976 as an amendment to the Solid Waste Disposal Act, the RCRA sets up a "cradle-to-grave" regulatory mechanism that operates as a tracking system for such wastes from the moment they are generated to their final disposal in an environmentally safe manner. The act charges the EPA with the development of criteria for identifying hazardous wastes, creating a manifest system for tracking wastes through final disposal, and setting up a permit system based on performance and management standards for generators, transporters, owners, and operators of waste treatment, storage, and disposal facilities. The RCRA is a strong force for innovation that has led to a broad rethinking of chemical processes, that is, to looking at hazardous waste disposal not just in terms of immediate costs, but rather with respect to life-cycle costs.

Under the RCRA, wastes are separated into two broad categories: hazardous and nonhazardous. Hazardous wastes are regulated under Subtitle C, and nonhazardous wastes are regulated under Subtitle D. RCRA Subtitle D assists waste management officials in developing and encouraging environmentally sound methods for nonhazardous solid waste disposal. "Solid waste" is a broad term. It is not based on the physical form of the material, but on whether the material is a waste.

The Federal Hazardous and Solid Waste Amendments (HSWA, 1984) to the RCRA required phasing out land disposal of hazardous waste. Some other mandates of this strict law include increased enforcement authority for the EPA, more stringent hazardous waste management standards, and a comprehensive underground storage tank program.

Toxic Substances Control Act (TSCA, 1976)
Until the TSCA, the Federal government was not empowered to prevent chemical hazards to health and the environment by banning or limiting chemical substances at a germinal, premarket stage. Through this Act, production workers, consumers, indeed all Americans, are protected by an equitably administered early warning system controlled by the EPA. This broad law authorizes the EPA Administrator to issue rules to prohibit or limit the manufacturing, processing, or distribution of any chemical substance or mixture that "may present an unreasonable risk of injury to health or the environment." The EPA Administrator may require testing—at a manufacturer's or processor's expense—of a substance after finding that

- The substance may present an unreasonable risk to health or the environment.
- There may be a substantial human or environmental exposure to the substance.
- Insufficient data and experience exist for judging a substance's health and environmental effects.
- Testing is necessary to develop such data.

This legislation is designed to cope with hazardous chemicals like kepone, vinyl chloride, asbestos, fluorocarbon compounds (Freons), and polychlorinated biphenyls (PCBs).

Comprehensive Environmental Response, Compensation, and Liability Act (CERCLA aka Superfund, 1980)

This Act provides a Federal "Superfund" to clean up uncontrolled or abandoned hazardous waste sites as well as accidents, spills, and other emergency releases of pollutants and contaminants into the environment. Through CERCLA, the EPA was given power to seek out those parties responsible for any release and assure their cooperation in the cleanup.

The EPA cleans up orphan sites when potentially responsible parties cannot be identified or located, or when they fail to act. Through various enforcement tools, the EPA obtains private party cleanup through orders, consent decrees, and other small party settlements. The EPA also recovers costs from financially viable individuals and companies once a response action has been completed.

The EPA is authorized to implement the Act in all 50 states and U.S. territories. Superfund site identification, monitoring, and response activities in states are coordinated through the state environmental protection or waste management agencies.

Superfund Amendments and Reauthorization Act (SARA, 1986)

This Act amended the CERCLA or "Superfund." SARA reflected the EPA's experience in administering the complex Superfund program during its first six years and made several important changes and additions to the program. SARA

- Stressed the importance of permanent remedies and innovative treatment technologies in cleaning up hazardous waste sites.
- Required Superfund actions to consider the standards and requirements found in other State and Federal environmental laws and regulations.
- Provided new enforcement authorities and settlement tools.
- Increased State involvement in every phase of the Superfund program.
- Increased the focus on human health problems posed by hazardous waste sites.
- Encouraged greater citizen participation in making decisions on how sites should be cleaned up.
- Increased the size of the trust fund to $8.5 billion.

SARA also required the EPA to revise the Hazard Ranking System (HRS) to insure that it accurately assessed the relative degree of risk to human health and the environment posed by uncontrolled hazardous waste sites that may be placed on the National Priorities List (NPL).

Clean Air Act (CAA, 1990)

Although this Act is a Federal law covering the entire country, the states do much of the work to carry it out. For example, a State air pollution agency holds a hearing on a permit application by a power or chemical plant or fines a company for violating air pollution limits. Under this law, the EPA sets limits on how much of a pollutant can be in the air anywhere in the United States. This insures that all Americans have the same basic health and environmental protections. The law allows individual states to have stronger pollution controls, but states are not allowed to have weaker pollution controls than those set for the entire country. The law recognizes that it makes sense for states to take the lead in carrying out the Clean Air Act because pollution control problems often require special understanding of local industries, geography, housing patterns, etc.

States must develop State Implementation Plans (SIPs) that explain how each state will do its job under the Clean Air Act. A SIP is a collection of the regulations that a state will use to clean up polluted areas. The states must involve the public, through hearings and opportunities to comment, in the development of each SIP. The EPA must approve each SIP, and if a SIP is not acceptable, the EPA can take over enforcing the CAA in that state. The U.S. government, through the EPA, assists the states by providing scientific research, expert studies, engineering designs, and money to support clean air programs.

Oil Pollution Act (OPA, 1990)
The OPA streamlined and strengthened the EPA's ability to prevent and respond to catastrophic oil spills. A trust fund financed by a federal tax on oil is available to clean up spills when the responsible party is incapable or unwilling to do so. The OPA requires oil storage facilities, and vessels to submit to the Federal government plans detailing how they will respond to large discharges. The EPA has published regulations for aboveground storage facilities while the U.S. Coast Guard has done the same for oil tankers. The OPA also requires the development of Area Contingency Plans (ACPs) to prepare and plan for oil spill response on a regional basis.

Pollution Protection Act (PPA, 1990)
The PPA focused industry, government, and public attention on reducing the amount of pollution through cost-effective changes in production, operation, and raw materials use. Opportunities for source reduction are often not realized because of existing regulations, and the industrial resources required for compliance, focus on treatment, and disposal. Source reduction is fundamentally different and more desirable than waste management or pollution control.

Pollution prevention also includes other practices that increase efficiency in the use of energy, water, or other natural resources, and protect our resource base through conservation. Practices include recycling, source reduction, and sustainable agriculture.

Federal Insecticide, Fungicide, and Rodenticide Act (FIFRA, 1996)
The primary focus of the FIFRA was to provide federal control of pesticide distribution, sale, and use. The EPA was given the authority under FIFRA not only to study the consequences of pesticide usage, but also to require users (farmers, utility companies, and others) to register when purchasing pesticides. Through later amendments to the law, users must also take exams for certification as applicators of pesticides. All pesticides use in the United States must be registered (licensed) by the EPA. Registration assures that pesticides will be properly labeled and that if in accordance with the specifications, will not cause unreasonable harm to the environment.

2.2 Occupational Safety and Health Administration (OSHA)*

The Occupational Safety and Health Act (OSHAct), a federal law that became effective on April 28, 1971, is intended to pull together all federal and state occupational safety and health-enforcement efforts under a federal program designed to establish uniform codes, standards, and regulations. The expressed purpose of the act is "to assure, as far as possible, every working woman and man in the Nation safe and healthful working conditions, and to preserve our human resources." To accomplish this purpose, the promulgation and enforce-

---

*R. De Reamer, *Modern Safety and Health Technology,* copyright © 1980. Reprinted by permission of Wiley, New York.

ment of safety and health standards is provided for, as well as research, information, education, and training in occupational safety and health. Perhaps no single piece of federal legislation has been more praised and, conversely, more criticized than the OSHAct, which basically is a law requiring virtually all employers to ensure that their operations are free of hazards to workers.

Occupational Safety and Health Standards

When Congress passed the OSHAct of 1970, it authorized the promulgation, without further public comment or hearings, of groups of already codified standards. The initial set of standards of the act (Part 1910, published in the *Federal Register* on May 29, 1971) thus consisted in part of standards that already had the force of law, such as those issued by authority of the Walsh-Healey Act, the Construction Safety Act, and the 1958 amendments to the Longshoremen's and Harbor Workers' Compensation Act. A great number of the adopted standards, however, derived from voluntary national consensus standards previously prepared by groups such as the American National Standards Institute (ANSI) and the National Fire Protection Association (NFPA).

The OSHAct defines the term "occupational safety and health standard" as meaning "a standard which requires conditions or the adoption or use of one or more practices, means, methods, operations or processes, reasonably necessary or appropriate to provide safe or healthful employment and places of employment." Standards contained in Part 1910* are applicable to general industry; those contained in Part 1926 are applicable to the construction industry; and standards applicable to ship repairing, shipbuilding, and longshoring are contained in Parts 1915–1918. These OSHA standards fall into the following four categories, with examples for each type:

1. *Specification standards*. Standards that give specific proportions, locations, and warning symbols for signs that must be displayed.
2. *Performance standards*. Standards that require achievement of, or within, specific minimum or maximum criteria.
3. *Particular standards* (*vertical*). Standards that apply to particular industries, with specifications that relate to the individual operations.
4. *General standards* (*horizontal*). Standards that can apply to any workplace and relate to broad areas (environmental control, walking surfaces, exits, illumination, etc.).

The Occupational Health and Safety Administration is authorized to promulgate, modify, or revoke occupational safety and health standards. It also has the authority to promulgate emergency temporary standards where it is found that employees are exposed to grave danger. Emergency temporary standards can take effect immediately on publication in the *Federal Register*. Such standards remain in effect until superseded by a standard promulgated under the procedures prescribed by the OSHAct—notice of proposed rule in the *Federal Register*, invitation to interested persons to submit their views, and a public hearing if required.

Required Notices and Records

During an inspection the compliance officer will ascertain whether the employer has

- Posted notice informing employees of their rights under the OSHAct (Job Safety and Health Protection, OSHAct poster).

---

*The Occupation Safety and Health Standards, Title 29, CFR Chapter XVIII. Parts 1910, 1926, and 1915–1918 are available at all OSHA regional and area offices.

- Maintained a log of recordable injuries and illnesses (OSHA Form No. 300, Log and Summary of Occupational Injuries and Illnesses).
- Maintained the Supplementary Record of Occupational Injuries and Illnesses (OSHA Form No. 301).
- Annually posted the Summary of Occupational Injuries and Illnesses (OSHA Form No. 200). This form must be posted no later than February 1 and must remain in place until March 1.
- Made a copy of the OSHAct and OSHA safety and health standards available to employees on request.
- Posted boiler inspection certificates, boiler licenses, elevator inspection certificates, and so on.

2.3 State-Operated Compliance Programs

The OSHAct encourages each state to assume the fullest responsibility for the administration and enforcement of occupational safety and health programs. For example, federal law permits any state to assert jurisdiction, under state law, over any occupational or health standard not covered by a federal standard. Section 18 of the Occupational Safety and Health Act of 1970 (OSHAct) encourages States to develop and operate their own job safety and health programs. OSHA approves and monitors State plans and provides 50 percent of an approved plan's operating costs. States must set job safety and health standards that are "at least as effective as" comparable federal standards. (Most States adopt standards identical to federal ones.) States have the option to promulgate standards covering hazards not addressed by federal standards.

A State must conduct inspections to enforce its standards, cover public (State and local government) employees, and operate occupational safety and health training and education programs. In addition, most States provide free, on-site consultation to help employers identify and correct workplace hazards. Such consultation may be provided either under the plan or through a special agreement under Section 21 (d) of the OSHAct.

In addition, any state may assume responsibility for the development and enforcement of its own occupational safety and health standards for those areas now covered by federal standards. However, the state must first submit a plan for approval by the Labor Department's Occupational Safety and Health Administration. Many states have done so. To gain OSHA approval for a *developmental plan*—the first step in the State plan process—a State must assure OSHA that within three years it will have in place all the structural elements necessary for an effective occupational safety and health program. These elements include appropriate legislation; regulations and procedures for standards setting, enforcement, appeal of citations, and penalties; and a sufficient number of qualified enforcement personnel.

Once a State has completed and documented all its developmental steps, it is eligible for *certification*. Certification renders no judgment as to actual State performance, but merely attests to the structural completeness of the plan.

At any time after initial plan approval, when it appears that the State is capable of independently enforcing standards, OSHA may enter into an *"operational status agreement"* with the State. This commits OSHA to suspend the exercise of discretionary federal enforcement in all of certain activities covered by the State.

The ultimate accreditation of a State's plan is called *final approval*. When OSHA grants final approval to a State under Section 18 (e) of the OSHAct, it relinquishes its authority to cover occupational safety and health matters covered by the State. After at least one year following certification, the State becomes eligible for approval if OSHA determines that the State is providing, in actual operation, worker protection "at least as effective as" the protection provided by the federal program. The State must also meet 100 percent of the established compliance staffing levels (benchmarks) and participate in OSHA's computerized inspection data system before OSHA can grant final approval.

Employees finding workplace safety and health hazards may file a formal complaint with the appropriate State or with the appropriate OSHA regional administrator. Complaints will be investigated and should include the name of the workplace, type(s) of hazard(s) observed, and any other pertinent information. Anyone finding inadequacies or other problems in the administration of a State's program may file a Complaint About State Program Administration (CASPA) with the appropriate OSHA regional administrator as well. A complainant's name is kept confidential. OSHA investigates all such complaints, and where complaints are found to be valid, requires appropriate corrective action on the part of the State.

Certain states are now operating under an approved state plan. These states may have adopted the existing federal standards or may have developed their own standards. Some states also have changed the required poster. Individuals need to know whether they are covered by an OSHA-approved state plan operation or are subject to the federal program to determine which set of standards and regulations (federal or state) apply to you. The easiest way to determine this is to call the nearest OSHA Area Office. If you are subject to state enforcement, the OSHA Area Office will explain this, explain whether the state is using the federal standards, and provide you with information on the poster and on the OSHA recordkeeping requirements. After that, the OSHA Area Office will refer you to the appropriate state government office for further assistance. This assistance also may include free on-site consultation visits. If you are subject to state enforcement, you should take advantage of this service.

There are currently 22 States and jurisdictions operating complete State plans (covering the private sector as well as State and local government employees) and 4—Connecticut, New Jersey, New York, and the Virgin Islands—that cover public employees only. (Eight other States were approved at one time, but subsequently withdrew their programs.) Table 2 lists those states operating under OSHA-approved state plans, as of April 16, 2004.

Table 2 States Operating Under OSHA-Approved Plans as of April 16, 2004

1. Alaska	14. New Jersey[a]
2. Arizona	15. New York[a]
3. California	16. North Carolina
4. Connecticut[a]	17. Oregon
5. Hawaii	18. Puerto Rico
6. Indiana	19. South Carolina
7. Iowa	20. Tennessee
8. Kentucky	21. Utah
9. Maryland	22. Vermont
10. Michigan	23. Virginia
11. Minnesota	24. Virgin Islands[a]
12. Nevada	25. Washington
13. New Mexico	26. Wyoming

[a] These plans cover public sector employees only.

3 SYSTEM SAFETY*

System safety is when situations having accident potential are examined in a step-by-step cause–effect manner, tracing a logical progression of events from start to finish. System safety techniques can provide meaningful predictions of the frequency and severity of accidents. However, their greatest asset is the ability to identify many accident situations in the system that would have been missed if less detailed methods had been used.

3.1 Methods of Analysis

A system cannot be understood simply in terms of its individual elements or component parts. If an operation of a system is to be effective, all parts must interact in a predictable and a measurable manner, within specific performance limits and operational design constraints.

In analyzing any system, three basic components must be considered: (1) the equipment (or machines); (2) the operators and supporting personnel (maintenance technicians, material handlers, inspectors, etc.); and (3) the environment in which both workers and machines are performing their assigned functions. Several analysis methods are available:

- *Gross-hazard analysis.* Performed early in design; considers overall system as well as individual components; it is called "gross" because it is the initial safety study undertaken.
- *Classification of hazards.* Identifies types of hazards disclosed in the gross-hazard analysis, and classifies them according to potential severity (Would defect or failure be catastrophic?); indicates actions and/or precautions necessary to reduce hazards. May involve preparation of manuals and training procedures.
- *Failure modes and effects.* Considers kinds of failures that might occur and their effect on the overall product or system. Example: effect on system that will result from failure of single component (e.g., a resistor or hydraulic valve).
- *Hazard-criticality ranking.* Determines statistical, or quantitative, probability of hazard occurrence; ranking of hazards in the order of "most critical" to "least critical."
- *Fault-tree analysis.* Traces probable hazard progression. *Example:* If failure occurs in one component or part of the system, will fire result? Will it cause a failure in some other component?
- *Energy-transfer analysis.* Determines interchange of energy that occurs during a catastrophic accident or failure. Analysis is based on the various energy inputs to the product or system and how these inputs will react in event of failure or catastrophic accident.
- *Catastrophe analysis.* Identifies failure modes that would create a catastrophic accident.
- *System-subsystem integration.* Involves detailed analysis of interfaces, primarily between systems.

---

*R. De Reamer, *Modern Safety and Health Technology,* copyright © 1980. Reprinted by permission of Wiley, New York.

- *Maintenance-hazard analysis.* Evaluates performance of the system from a maintenance standpoint. Will it be hazardous to service and maintain? Will maintenance procedures be apt to create new hazards in the system?
- *Human-error analysis.* Defines skills required for operation and maintenance. Considers failure modes initiated by human error and how they would affect the system. The question of whether special training is necessary should be a major consideration in each step.
- *Transportation-hazard analysis.* Determines hazards to shippers, handlers, and bystanders. Also considers what hazards may be "created" in the system during shipping and handling.

Other quantitative methods have successfully been used to recommend a decision to adopt engineering controls, personal protective equipment, or some combination. Some of these methods are as follows:*

- *Expected outcome approach.* Since safety alternatives involve accident costs that occur more or less randomly according to probabilities that might be estimated, a valuable way to perform needed economic analyses for such alternatives is to calculate expected outcomes.
- *Decision analysis approach.* A recent extension of systems analysis, this approach provides useful techniques for transforming complex decision problems into a sequentially oriented series of smaller, simpler problems. This means that decision-makers can select reasoned choices that will be consistent with their perceptions about the uncertainties involved in a particular problem together with their fundamental attitudes toward risk-taking.
- *Mathematical modeling.* Usually identified as an "operations research" approach, numerous mathematical models have demonstrated potential for providing powerful analysis insights into safety problems. These include dynamic programming, inventory-type modeling, linear programming, queue-type modeling, and Monte Carlo simulation.

There is a growing body of literature about these formal analytical methods and others not mentioned in this chapter, including failure mode and effect (FME), technique for human error prediction (THERP), system safety hazard analysis, and management oversight and risk tree (MORT). All have their place. Each to a greater or lesser extent provides a means of overcoming the limitations of intuitive, trial-and-error analysis.

Regardless of the method or methods used, the systems concept of hazard recognition and analysis makes available a powerful tool of proven effectiveness for decision making about the acceptability of risks. To cope with the complex safety problems of today and the future, engineers must make greater use of system safety techniques.

3.2 Fault-Tree Technique†

When a problem can be stated quantitatively, management can assess the risk and determine the trade-off requirements between risk and capital outlay. Structuring key safety problems

---

*J. B. ReVelle, *Engineering Controls: A Comprehensive Overview.* Used by permission of The Merritt Company, Publisher, from T. S. Ferry, Safety Management Planning, copyright © 1982, The Merritt Company, Santa Monica, CA 90406.
†R. De Reamer, *Modern Safety and Health Technology.* Copyright © 1980. Reprinted by permission of Wiley, New York.

or vital decision-making in the form of fault paths can greatly increase communication of data and subjective reasoning. This technique is called fault-tree analysis. The transferability of data among management, engineering staff, and safety personnel is a vital step forward.

Another important aspect of this system safety technique is a phenomenon that engineers have long been aware of in electrical networks. That is, an end system formed by connecting several subsystems is likely to have entirely different characteristics from any of the subsystems considered alone. To fully evaluate and understand the entire system's performance with key paths of potential failure, the engineer must look at the entire system. Only then can he or she look meaningfully at each of the subsystems.

Figure 1 introduces the most commonly used symbols used in fault-tree analysis.

3.3 Criteria for Preparation/Review of System Safety Procedures*

Correlation between Procedure and Hardware

1. Is there a statement of hardware configuration to which the procedure was written?
2. Has background descriptive or explanatory information been provided where needed?
3. Does the hardware reflect or reference the latest revisions of drawings, manuals, or other procedures?

Adequacy of the Procedure

1. Is the procedure the best way to do the job?
2. Is the procedure easy to understand?
3. Is the detail appropriate—not too much, not too little?
4. Is it clear, concise, and free from ambiguity that could lead to wrong decisions?
5. Are calibration requirements clearly defined?
6. Are critical red-line parameters identified and clearly defined? Are required values specified?
7. Are corrective controls of the above parameters clearly defined?
8. Are all values, switches, and other controls identified and defined?
9. Are pressure limits, caution notes, safety distances, or hazards peculiar to this operation clearly defined?
10. Are hard-to-locate components adequately defined and located?
11. Are jigs and arrangements provided to minimize error?
12. Are job safety requirements defined, e.g., power off, pressure down, and have tools been checked for sufficiency?
13. Is the system operative at end of job?
14. Has the hardware been evaluated for human factors and behavioral stereotype problems? If not corrected, are any such clearly identified?
15. Have monitoring points and methods of verifying adherence been specified?
16. Are maintenance and/or inspection to be verified? If so, is a log provided?

---

*Reprinted from *MORT Safety Assurance Systems,* pp. 278–283, by courtesy of Marcel Dekker, New York.

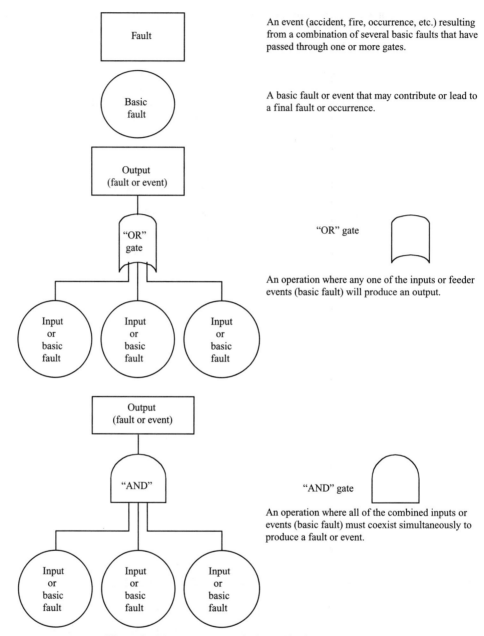

Figure 1 Most common symbols used in fault-tree analysis.

17. Is safe placement of process personnel or equipment specified?
18. Have errors in previous, similar processes been studied for cause? Does this procedure correct such causes?

Accuracy of the Procedure

1. Has the capacity to accomplish the specified purpose been verified by internal review?
2. Have all gauges, controls, valves, etc., been called out, described, and labeled exactly as they actually are?
3. Are all setpoints or other critical controls, etc., compatible with values in control documents?
4. Are the safety limitations adequate for job to be performed?
5. Are all steps in the proper sequence?

Adequacy and Accuracy of Supporting Documentation

1. Are all necessary supporting drawings, manuals, data sheets, sketches, etc., either listed or attached?
2. Are all interfacing procedures listed'?

Securing Provisions

1. Are there adequate instructions to return the facility or hardware to a safe operating or standby condition?
2. Do securing instructions provide step-by-step operations?

Backout Provisions

1. Can the procedure put any component or system in a condition that could be dangerous?
2. If so, does the procedure contain emergency shutdown or backout procedures either in an appendix or as an integral part?
3. Is a backout procedure (or instructions for its use) included at proper place?

Emergency Measures

1. What are the procedures for action in case of emergency conditions?
2. Does the procedure involve critical actions such that preperformance briefing on possible hazards is required?
3. Are adequate instructions either included or available for action to be taken under emergency conditions? Are they in the right place?
4. Are adequate shutdown procedures available? Are coverall systems involved and available for emergency reentry teams?
5. Are requirements for the emergency team for accident recovery, troubleshooting, or investigative purposes specified where necessary? Are conditions under which the emergency team will be used described? Are the hazards they may encounter or must avoid identified?

6. Does the procedure consider interfaces in shutdown procedures?
7. How will changes be handled? What are the thresholds for changes requiring review?
8. Have the emergency procedures been tested under a range of conditions that may be encountered, e.g., at night during power failure?

Caution and Warning Notes

1. Are caution and warning notes included where appropriate?
2. Do caution and warning notes precede operational steps containing potential hazards?
3. Are they adequate to describe the potential hazard?
4. Are major cautions and warnings called out in the general introduction, as well as prior to steps?
5. Do notes appear as separate entries with distinctive bold type or other emphatic display?
6. Do they include supporting safety control (health physics, safety engineer, etc.) if needed at specific required steps in the procedure?

Requirements for Communications and Instrumentation

1. Are adequate means of communication provided?
2. Will loss of communications create a hazard?
3. Is a course of action clearly defined for loss of required communications?
4. Is verification of critical communication included prior to point of need?
5. Will loss of control or monitoring capability of critical functions create a hazard to people or hardware?
6. Are alternative means, or a course of action to regain control or monitoring functions, clearly defined?
7. Are the above situations flagged by cautions and warnings?

Sequence-of-Events Considerations

1. Can any operation initiate an unscheduled or out-of-sequence event?
2. Could it induce a hazardous condition?
3. Are these operations identified by warnings or cautions?
4. Are they covered by emergency shutdown and backout procedures?
5. Are all steps sequenced properly? Will the sequence contribute to or create a hazard?
6. Are all steps that, if performed out-of-sequence, could cause a hazard identified and flagged?
7. Have all noncompatible simultaneous operations been identified and suitably restricted?
8. Have these been prohibited by positive callout or separation in step-by-step inclusion within the text of the procedure?

Environmental Considerations (Natural or Induced)

1. Have the environmental requirements been specified that constrain the initiation of the procedure or require shutdown or evacuation, once in progress?

2. Have the induced environments (toxic or explosive atmospheres, etc.) been considered?
3. Have all latent hazards (pressure, height, voltage, etc.) in adjacent environments been considered?
4. Are there induced hazards from simultaneous performance of more than one procedure by personnel within a given space?

Personnel Qualification Statements

1. Has the requirement for certified personnel been considered?
2. Has the required frequency of recheck of personnel qualifications been considered?

Interfacing Hardware and Procedures Noted

1. Are all interfaces described by detailed callout?
2. Are interfacing operating procedures identified, or written to provide ready equipment?
3. Where more than one organizational element is involved, are proper liaison and areas of responsibility established?

Procedure Sign-Off

1. Is the procedure to be used as an in-hand, literal checklist?
2. Have the step sign-off requirements been considered and identified and appropriate spaces provided in the procedure?
3. Have the procedure completion sign-off requirements been indicated (signature, authority, date, etc.)?
4. Is supervisor verification of correct performance required?

General Requirements

1. Does the procedure discourage a shift change during performance or accommodate a shift change?
2. Where shift changes are necessary, do they include or reference shift overlap and briefing requirements?
3. Are mandatory inspection, verification, and system validation required whenever the procedure requires breaking into and reconnecting a system?
4. Are safety prerequisites defined? All safety instructions spelled out in detail to all personnel?
5. Does the procedure require prechecks of supporting equipment to ensure compatibility and availability?
6. Is consideration for unique operations written in?
7. Does the procedure require walk-through or talk-through dry runs?
8. What are the general supervision requirements, e.g., what is protocol for transfer of supervisor responsibilities to a successor?
9. Are the responsibilities of higher supervision specified?

Reference Considerations

1. Have applicable quality assurance and reliability standards been considered?
2. Have applicable codes, standards, and regulations been considered?
3. Does the procedure comply with control documents?
4. Have hazards and system safety degradations been identified and considered against specific control manuals, standards, and procedures?
5. Have specific prerequisite administrative and management approvals been complied with?
6. Have comments been received from the people who will do the work?

Special Considerations

1. Has a documented safety analysis been considered for safety-related deviations from normal practices or for unusual or unpracticed maneuvers'?
2. Have new restrictions or controls become effective that affect the procedure in such a manner that new safety analyses may be required?

3.4 Risk Assessment Process

Risk Score Formula*

William T. Fine is credited with having developed a unique procedure that responds to potential accident hazards by mathematically determining organizational priorities for action. His methodology assigns weight to known controlling factors and then determines the overall risk for a specific hazardous situation by calculating a "risk score" that indicates the relative urgency of the corresponding remedial action.

The risk score formula is

$$\text{Risk score} = \text{Consequences} \times \text{Exposure} \times \text{Probability}$$

In using the formula, the numeric ratings or weights are subjectively assigned to each factor based on the judgment and experience of the investigator(s) making the calculation. The following is a review of the formula.

- *Consequences.* The most probable results of an accident due to a hazard being considered, including both injuries and property damage. Numerical ratings are assigned for the most likely consequences of the accident starting from 100 points for a catastrophe down through various degrees of severity to 1 point for a minor cut or bruise.
- *Exposure.* The frequency of occurrence of the *hazard event,* which is the *first undesired event* that *could* start the accident sequence. The frequency with which the hazard event is most likely to occur is rated from *continuous occurrence* with 10 points through various lesser degrees of exposure down to 0.5 points for *extremely remote.*
- *Probability.* The likelihood that, once the hazard event occurs, the *complete accident sequence of events will follow* with the timing and coincidence that results in an accident and its consequences. The ratings start at 10 points if the complete accident

---

*W. T. Fine, "Mathematical Evaluations for Controlling Hazards," in *Selected Readings in Safety,* J. Widener (ed.), Academy Press, Macon, GA, 1973.

sequence is *most likely and expected* and continue down to 0.1 points for the "one in a million" or practically impossible chance.

Multiplying the *maximum* points for each factor (Consequences = 100, Exposure = 10, and Probability = 10) results in the *greatest* possible risk score of 10,000. Multiplying the *minimum* points for each factor (Consequences = 1, Exposure = 0.1, and Probability = 0.1) results in the *least* possible risk score of 0.05.

Fine recommended the following risk score vs. action relationship:

Risk Score	Recommended Action
0.05–89	Hazard should be eliminated without delay, but situation is not an emergency.
90–200	Urgent. Requires attention as soon as possible.
201–10,000	Immediate correction required. Activity should be discontinued until hazard is abated.

For a newly discovered hazard, the risk score vs. action relationship provides important guidance with respect to necessary action.

4 HUMAN FACTORS ENGINEERING/ERGONOMICS*

4.1 Human–Machine Relationships

- Human factors engineering is defined as "the application of the principles, laws, and quantitative relationships which govern man's response to external stress to the analysis and design of machines and other engineering structures, so that the operator of such equipment will not be stressed beyond his/her proper limit or the machine forced to operate at less than its full capacity in order for the operator to stay within acceptable limits of human capabilities."†

- A principal objective of the supervisor and safety engineer in the development of safe working conditions is the elimination of bottlenecks, stresses and strains, and psychological booby traps that interfere with the free flow of work. The less operators have to fear from their jobs or machines, the more attention they can give to their work.

- In the development of safe working conditions, attention is given to many things, including machine design and machine guarding, personal protective equipment, plant layout, manufacturing methods, lighting, heating, ventilation, removal of air contaminants, and the reduction of noise. Adequate consideration of each of these areas will lead to a proper climate for accident prevention, increased productivity, and worker satisfaction.

- The human factors engineering approach to the solution of the accident problem is to build machines and working areas around the operator, rather than place him or her in a setting without regard to his or her requirements and capacities. Unless this is

---

*R. De Reamer, *Modern Safety and Health Technology*. Copyright © 1980. Reprinted by permission of Wiley, New York.
†Theodore F. Hatch, Professor (retired), by permission.

done, it is hardly fair to attribute so many accidents to human failure, as is usually the case.
- If this point of view is carried out in practice, fewer accidents should result, training costs should be reduced, and extensive redesign of equipment after it is put into use should be eliminated.
- All possible faults in equipment and in the working area, as well as the capacities of the operator, should be subjected to advance analysis. If defects are present, it is only a matter of time before some operator "fails" and has an accident.
- Obviously, the development of safe working conditions involves procedures that may go beyond the occasional safety appraisal or search for such obvious hazards as an oil spot on the floor, a pallet in the aisle, or an unguarded pinch point on a new lathe.

Human–machine relationships have improved considerably with increased mechanization and automation. Nevertheless, with the decrease in manual labor has come specialization, increased machine speeds, and monotonous repetition of a single task, which create work relationships involving several physiological and psychological stresses and strains. Unless this scheme of things is recognized and dealt with effectively, many real problems in the field of accident prevention may be ignored.

4.2 Human Factors Engineering Principles

- Human factors engineering, or *ergonomics*,* as it is sometimes called, developed as a result of the experience in the use of highly sophisticated equipment in World War II. The ultimate potentialities of complex instruments of war could not be realized because the human operators lacked the necessary capabilities and endurance required to operate them. This discipline now has been extended to many areas. It is used extensively in the aircraft and aerospace industry and in many other industries to achieve more effective integration of humans and machines.
- The analysis should consider all possible faults in the equipment, in the work area, and in the worker-including a survey of the nature of the task, the work surroundings, the location of controls and instruments, and the way the operator performs his or her duties. The questions of importance in the analysis of machines, equipment, processes, plant layout, and the worker will vary with the type and purpose of the operation, but usually will include the following (pertaining to the worker):†

  1. What sense organs are used by the operator to receive information? Does he or she move into action at the sound of a buzzer, blink of a light, reading of a dial, verbal order? Does the sound of a starting motor act as a cue?
  2. What sort of discrimination is called for? Does the operator have to distinguish between lights of two different colors, tones of two different pitches, or compare two dial readings?

---

*The term *ergonomics* was coined from the Greek roots *ergon* (work) and *nomos* (law, rule) and is now currently used to deal with the interactions between humans and such environmental elements as atmospheric contaminants, heat, light, sound, and all tools and equipment pertaining to the workplace.
†R. A. McFarland, "Application of Human Factors Engineering to Safety Engineering Problems," *National Safety Congress Transactions,* **12** (1967). Permission granted by the National Safety Council.

3. What physical response is he or she required to make: Pull a handle? Turn a wheel? Step on a pedal? Push a button?

4. What overall physical movements are required in the physical response? Do such movements interfere with his or her ability to continue receiving information through his or her sense organs? (For example, would pulling a handle obstruct his or her line of vision to a dial he or she is required to watch?) What forces are required (e.g., torque in turning a wheel)?

5. What are the speed and accuracy requirements of the machine? Is the operator required to watch two pointers to a hairline accuracy in a split second? Or is a fairly close approximation sufficient? If a compromise is necessary, which is more essential: speed or accuracy?

6. What physiological and environmental conditions are likely to be encountered during normal operation of the machine? Are there any unusual temperatures, humidity conditions, crowded workspace, poor ventilation, high noise levels, toxic chemicals, and so on?

- Pertaining to the machine, equipment, and the surrounding area, these key questions should be asked:

1. Can the hazard be eliminated or isolated by a guard, ventilating equipment, or other device?

2. Should the hazard be identified by the use of color, warning signs, blinking lights, or alarms?

3. Should interlocks be used to protect the worker when he or she forgets or makes the wrong move?

4. Is it necessary to design the machine, the electrical circuit, or the pressure circuit so it will always be fail-safe?

5. Is there need for standardization?

6. Is there need for emergency controls, and are controls easily identified and accessible?

7. What unsafe conditions would be created if the proper operating sequence were not followed?

4.3 General Population Expectations*

- The importance of standardization and normal behavior patterns has been recognized in business and industry for many years. A standard tool will more likely be used properly than will a nonstandard one, and standard procedures will more likely be followed.

- People expect things to operate in a certain way and certain conditions to conform to established standards. These general population "expectations"—the way in which

---

*R. De Reamer, *Modem Safety and Health Technology*. Copyright © 1980. Reprinted by permission of Wiley, New York.

the ordinary person will react to a condition or stimulus—must not be ignored or workers will be literally trapped into making mistakes. A list of "General Population Expectations" is given in Table 3.

5 ENGINEERING CONTROLS FOR MACHINE TOOLS*

5.1 Basic Concerns

Machine tools (such as mills, lathes, shearers, punch presses, grinders, drills, and saws) provide an example of commonplace conditions where only a limited number of items of personal protective gear are available for use. In such cases as these, the problem to be solved is not personal protective equipment versus engineering controls, but rather which engineering control(s) should be used to protect the machine operator. A summary of employee safeguards is given in Table 4. The list of possible machinery-related injuries is presented in Table 11. There seem to be as many hazards created by moving machine parts as there are types of machines. Safeguards are essential for protecting workers from needless and preventable injuries.

A good rule to remember is: Any machine part, function, or process that may cause injury must be safeguarded. Where the operation of a machine or accidental contact with it can injure the operator or others in the vicinity, the hazard must be either controlled or eliminated.

Dangerous moving parts in these three basic areas need safeguarding:

Table 3 General Population Expectations

1. Doors are expected to be at least 6 feet, 6 inches in height.
2. The level of the floor at each side of a door is expected to be the same.
3. Stair risers are expected to be of the same height.
4. It is a normal pattern for persons to pass to the left on motorways (some countries excluded).
5. People expect guardrails to be securely anchored.
6. People expect the hot-water faucet to be on the left side of the sink, the cold-water faucet on the right, and the faucet to turn to the left (counterclockwise) to let the water run and to the right to turn the water off.
7. People expect floors to be nonslippery.
8. Flammable solvents are expected to be found in labeled, red containers.
9. The force required to operate a lever, push a cart, or turn a crank is expected to go unchanged.
10. Knobs on electrical equipment are expected to turn clockwise for "on," to increase current, and counterclockwise for "off."
11. For control of vehicles in which the operator is riding, the operator expects a control motion to the right or clockwise to result in a similar motion of his or her vehicle and vice versa.
12. Very large objects or dark objects imply "heaviness." Small objects or light-colored ones imply "lightness." Large heavy objects are expected to be "at the bottom." Small, light objects are expected to be "at the top."
13. Seat heights are expected to be at a certain level when a person sits.

*J. B. ReVelle, *Engineering Controls: A Comprehensive Overview.* Used by permission of the Merritt Company, Publisher, from T. S. Ferry, Safety Management Planning, copyright © 1982, The Merritt Company, Santa Monica, CA 90406.

5 Engineering Controls for Machine Tools

Table 4 Summary of Employee Safeguards

To Protect	Personal Protective Equipment to Use	Engineering Controls to Use
Breathing	Self-contained breathing apparatus, gas masks, respirators, alarm systems	Ventilation, air-filtration systems, critical level warning systems, electrostatic precipitators
Eyes/face	Safety glasses, filtered lenses, safety goggles, face shield, welding goggles/helmets, hoods	Spark deflectors, machine guards
Feet/legs	Safety boots/shoes, leggings, shin guards	
Hands/arms/body	Gloves, finger cots, jackets, sleeves, aprons, barrier creams	Machine guards, lockout devices, feeding and ejection methods
Head/neck	Bump caps, hard hats, hair nets	Toe boards
Hearing	Ear muffs, ear plugs, ear valves	Noise reduction, isolation by equipment modification/substitution, equipment lubrication/maintenance programs, eliminate/dampen noise sources, reduce compressed air pressure, change operations[a]
Excessively high/low temperatures	Reflective clothing, temperature controlled clothing	Fans, air conditioning, heating, ventilation, screens, shields, curtains
Overall	Safety belts, lifelines, grounding mats, slap bars	Electrical circuit grounding, polarized plugs/outlets, safety nets

[a] Examples of the types of changes that should be considered include
- Grinding instead of chipping
- Electric tools in place of pneumatic tools
- Pressing instead of forging
- Welding instead of riveting
- Compression riveting over pneumatic riveting
- Mechanical ejection in place of air-blast ejection
- Wheels with rubber or composition tires on plant trucks and cars instead of all-metal wheels
- Wood or plastic tote boxes in place of metal tote boxes
- Use of an undercoating on machinery covers.
- Wood in place of all-metal workbenches.

- *The point of operation.* That point where work is performed on the material, such as cutting, shaping, boring, or forming of stock.
- *Power transmission apparatus.* All components of the mechanical system that transmit energy to the part of the machine performing the work. These components include flywheels, pulleys, belts, connecting rods, couplings, cams, spindles, chains, cranks, and gears.
- *Other moving parts.* All parts of the machine that move while the machine is working. These can include reciprocating, rotating, and transverse moving pans, as well as feed mechanisms and auxiliary parts of the machine.

A wide variety of mechanical motions and actions may present hazards to the worker. These can include the movement of rotating teeth and any parts that impact or shear. These different types of hazardous mechanical motions and actions are basic to nearly all machines, and recognizing them is the first step toward protecting workers from the danger they present.

The basic types of hazardous mechanical motions and actions are

Motions	Actions
Rotating (including in-running nip points)	Cutting
	Punching
Reciprocating	Shearing
Transverse	Bending

5.2 General Requirements

What must a safeguard do to protect workers against mechanical hazards? Engineering controls must meet these minimum general requirements:

- *Prevent contact.* The safeguard must prevent hands, arms, or any other part of a worker's body from making contact with dangerous moving parts. A good safeguarding system eliminates the possibility of operators or workers placing their hands near hazardous moving parts.
- *Secure.* Workers should not be able to easily remove or tamper with the safeguard, because a safeguard that can easily be made ineffective is no safeguard at all. Guards and safety devices should be made of durable material that will withstand the conditions of normal use. They must be firmly secured to the machine.

Machines often produce noise (unwanted sound), and this can result in a number of hazards to workers. Not only can it startle and disrupt concentration, but it can interfere with communications, thus hindering the worker's safe job performance. Research has linked noise to a whole range of harmful health effects, from hearing loss and aural pain to nausea, fatigue, reduced muscle control, and emotional disturbances. Engineering controls such as the use of sound-dampening materials, as well as less sophisticated hearing protection, such as ear plugs and muffs, have been suggested as ways of controlling the harmful effects of noise. Vibration, a related hazard that can cause noise and thus result in fatigue and illness for the worker, may be avoided if machines are properly aligned, supported, and, if necessary, anchored.

Because some machines require the use of cutting fluids, coolants, and other potentially harmful substances, operators, maintenance workers, and others in the vicinity may need protection. These substances can cause ailments ranging from dermatitis to serious illnesses and disease. Specially constructed safeguards, ventilation, and protective equipment and clothing are possible temporary solutions to the problem of machinery-related chemical hazards until these hazards can be better controlled or eliminated from the workplace. Some safeguards are

- *Protect from falling objects.* The safeguard should ensure that no objects can fall into moving parts. A small tool that is dropped into a cycling machine could easily become a projectile that could strike and injure someone.
- *Create no new hazards.* A safeguard defeats its own purpose if it creates a hazard of its own such as a shear point, a jagged edge, or an unfinished surface that can cause a laceration. The edges of guards, for instance, should be rolled or bolted in such a way that eliminates sharp edges.

- *Create no interference.* Any safeguard that impedes a worker from performing the job quickly and comfortably might soon be overridden or disregarded. Proper safeguarding can actually enhance efficiency, since it can relieve the worker's apprehensions about injury.
- *Allow safe lubrication.* If possible, one should be able to lubricate the machine without removing the safeguards. Locating oil reservoirs outside the guard, with a line leading to the lubrication point, will reduce the need for the operator or maintenance worker to enter the hazardous area.

5.3 Danger Sources

All power sources for machinery are potential sources of danger. When using electrically powered or controlled machines, for instance, the equipment as well as the electrical system itself must be properly grounded. Replacing frayed, exposed, or old wiring will also help to protect the operator and others from electrical shocks or electrocution. High-pressure systems, too, need careful inspection and maintenance to prevent possible failure from pulsation, vibration, or leaks. Such a failure could cause explosions or flying objects.

6 MACHINE SAFEGUARDING METHODS*

6.1 General Classifications

There are many ways to safeguard machinery. The type of operation, the size or shape of stock, the method of handling the physical layout of the work area, the type of material, and production requirements or limitations all influence selection of the appropriate safeguarding method(s) for the individual machine.

As a general rule, power transmission apparatus is best protected by fixed guards that enclose the danger area. For hazards at the point of operation, where moving parts actually perform work on stock, several kinds of safeguarding are possible. One must always choose the most effective and practical means available.

1. Guards
   a. Fixed
   b. Interlocked
   c. Adjustable
   d. Self-adjusting
2. Devices
   a. Presence sensing
      Photoelectrical (optical)
      Radio frequency (capacitance)
      Electromechanical

---

*J. B. ReVelle, *Engineering Controls: A Comprehensive Overview.* Used by permission of The Merritt Company, Publisher, from T. S. Ferry, *Safety Management Planning,* copyright © 1982, The Merritt Company, Santa Monica, CA 90406.

664 Safety Engineering

   b. Pullback
   c. Restraint
   d. Safety controls
     Safety trip controls
      Pressure-sensitive body bar
      Safety trip rod
      Safety tripwire cable
     Two-hand control
     Two-hand trip
   e. Gates interlocked other
  3. Location/distance
  4. Potential feeding and ejection methods to improve safety for the operator
   a. Automatic feed
   b. Semiautomatic feed
   c. Automatic ejection
   d. Semiautomatic ejection
   e. Robot
  5. Miscellaneous aids
   a. Awareness barriers
   b. Miscellaneous protective shields
   c. Hand-feeding tools and holding fixtures

6.2 Guards, Devices, and Feeding and Ejection Methods

Tables 5–7 provide the interested reader with specifics regarding machine safeguarding.

7 ALTERNATIVES TO ENGINEERING CONTROLS*

Engineering controls are an alternative to personal protective equipment, or is it the other way around? This chicken-and-egg situation has become an emotionally charged issue with exponents on both sides arguing their beliefs with little in the way of well-founded evidence to support their cases. The reason for this unfortunate situation is that there is no single solution to all the hazardous operations found in industry. The only realistic answer to the question is that it depends. Each and every situation requires an independent analysis considering all the known factors so that a truly unbiased decision can be reached.

  This section presents material useful to engineers in the selection and application of solutions to industrial safety and health problems. Safety and health engineering control principles are deceptively few: substitution; isolation; and ventilation, both general and localized. In a technological sense, an appropriate combination of these strategic principles can be brought to bear on any industrial safety or hygiene control problem to achieve a

---

*J. B. ReVelle, *Engineering Controls: A Comprehensive Overview*. Used by permission of The Merritt Company, Publisher, from T. S. Ferry, *Safety Management Planning,* copyright © 1982, The Merritt Company, Santa Monica, CA 90406.

Table 5 Machine Safeguarding: Guards

Method	Safeguarding Action	Advantages	Limitations
Fixed	Provides a barrier	Can be constructed to suit many specific applications In-plant construction is often possible Can provide maximum protection Usually requies minimum maintenance Can be suitable to high production, repetitive operations	May interfere with visibility Can be limited to specific operations Machine adjustment and repair often requires its removal, thereby necessitating other means of protection for maintenance personnel
Interlocked	Shuts off or disengages power and prevents starting of machine when guard is open; should require the machine to be stopped before the worker can reach into the danger area	Can provide maximum protection Allows access to machine for removing jams without time-consuming removal of fixed guards	Requires careful adjustment and maintenance May be easy to disengage
Adjustable	Provides a barrier that may be adjusted to facilitate a variety of productio operations	Can be constructed to suit many specific applications Can be adjusted tto admit varying sizes of stock	Hands may enter danger area; protection may not be complete at all times May require frequent maintenance and/or adjustment The guard may be made ineffective by the operator May interfere with visibility
Self-adjusting	Provides a barrier that moves according to the size of the stock entering danger area	Off-the-shelf guards are often commercially available	Does not always provide maximum protection May interfere with visibility May require frequent maintenance and adjustment

Table 6 Machine Safeguarding: Devices

Method	Safeguarding Action	Advantages	Limitations
Photoelectric	Machine will not start cycling when the light field is interrupted When the light field is broken by any part of the operator's body during the cycling process, immediate machine braking is activated	Can allow freer movement for operator	Does not protect against mechanical failure May require frequent alignment and calibration Excess vibration may cause lamp filament damage and premature burnout Limited to machines that can be stopped
Radio frequency (capacitance)	Machine cycling will not start when the capacitance field is interrupted When the capacitance field is disturbed by any part of the operator's body during the cycling process, immediate machine braking is activated	Can allow freer movement for operator	Does not protect against mechanical failure Antennae sensitivity must be properly adjusted Limited to machines that can be stopped
Electromechanical	Contact bar or probe travels a predetermined distance between the operator and the danger area Interruption of this movement prevents the starting of machine cycle	Can allow access at the point of operation	Contact bar or probe must be properly adjusted for each application; this adjustment must be maintained properly
Pullback	As the machine begins to cycle, the operator's hands are pulled out of the danger area	Eliminates the need for auxillary barriers or other interference at the danger area	Limits movement of operator May obstruct workspace around operator Adjustments must be made for specific operators and for each individual Requires frequent inspections and regular maintenance Requires close supervision of the operator's use of the equipment
Restraint (holdback)	Prevents the operator from reaching into the danger area	Little risk of mechanical failure	Limits movements of operator May obstruct workspace Adjustments must be made for specific operations and each individual Requires close supervision of the operator's use of the equipment

Safety trip controls Pressure-sensitive body bar Safety tripod Safety tripwire cable	Stops machine when tripped	Simplicity of use	All controls must be manually activated May be difficult to activate controls because of their location Only protects the operator May require special fixtures to hold work May require a machine brake
Two-hand control	Concurrent use of both hands is required preventing the operator from entering the danger area	Operator's hands are at a predetermined location Operator's hands are free to pick up a new part after first half of cycle is completed	Requires a partial cycle machine with a brake Some two-hand controls can be rendered unsafe by holding with arm or blocking, thereby permitting one-hand operation Protects only the operator
Two-hand trip	Concurrent use of two hands on separate controls prevents hands from being in danger area when machine cycle starts	Operator's hands are away from danger area Can be adapted to multiple operations No obstruction to hand feeding Does not require adjustment for each operation	Operator may try to reach into danger area after tripping machine Some trips can be rendered unsafe by holding with arm or blocking, thereby permitting one-hand operation Protects only the operator May require special fixtures
Gates Interlocked Other	Provides a barrier between danger area and operator or other personnel	Can prevent reaching into or walking into the danger area	May require frequent inspection and regular maintenance May interfere with operator's ability to see the work

Table 7 Machine Safeguarding: Feeding and Ejection Methods

Method	Safeguarding Action	Advantages	Limitations
Automatic feed	Stock is fed from rolls, indexed by machine mechanism, etc.	Eliminates the need for operator involvement in the danger area	Other guards are also required for operator protection, usually fixed barrier guards. Requires frequent maintenance. May not be adaptable to stock variation
Semiautomatic feed	Stock is fed by chutes, movable dies, dial feed, plungers, or sliding bolster		
Automatic ejection	Workspieces are ejected by air or machanical means		May create a hazard of blowing chips or debris. Size of stock limits the use of this method. Air ejection may present a noise hazard
Semiautomatic ejection	Workpieces are ejected by mechanical means which are initiated by the operator	Operator does not have to enter danger area to remove finished work	Other guards are required for operator protection. May not be adaptable to stock variation
Robots	Perform work usually done by operator	Operator does not have to enter danger area. Are suitable for operations where high stress factors are present, such as heat and noise	Can create hazards themselves. Require maximum maintenance. Are suitable only to specific operations

satisfactory quality of the work environment. It usually is not necessary or appropriate to apply all these principles to any specific potential hazard. A thorough analysis of the control problem must be made to ensure that a proper choice from among these methods will produce the proper control in a manner that is most compatible with the technical process, is acceptable to the workers in terms of day-to-day operation, and can be accomplished with optimal balance of installation and operating expenses.

7.1 Substitution

Although frequently one of the most simple engineering principles to apply, substitution is often overlooked as an appropriate solution to occupational safety and health problems. There is a tendency to analyze a particular problem from the standpoint of correcting rather than eliminating it. For example, the first inclination in considering a vapor-exposure problem in a degreasing operation is to provide ventilation of the operation rather than consider substituting a solvent having a much lower degree of hazard associated with its use. Substitution of less hazardous substances, changing from one type of process equipment to another, or, in some cases, even changing the process itself, may provide an effective control of a hazard at minimal expense.

This strategy is often used in conjunction with safety equipment: substituting safety glass for regular glass in some enclosures, replacing unguarded equipment with properly guarded machines, replacing safety gloves or aprons with garments made of a material that is more impervious to the chemicals being handled. Since substitution of equipment frequently is done as an immediate response to an obvious problem, it is not always recognized as an engineering control, even though the end result is every bit as effective.

Substituting one process or operation for another may not be considered except in major modifications. In general, a change in any process from a batch to a continuous type of operation carries with it an inherent reduction in potential hazard. This is true primarily because the frequency and duration of potential contact of workers with the process materials are reduced when the overall process approach becomes one of continuous operation. The substitution of processes can be applied on a fundamental basis, for example, substitution of airless spray for conventional spray equipment can reduce the exposure of a painter to solvent vapors. Substitution of a paint dipping operation for the paint spray operation can reduce the potential hazard even further. In any of these cases, the automation of the process can further reduce the potential hazard (see Table 8).

7.2 Isolation

Application of the principle of isolation is frequently envisioned as consisting of the installation of a physical barrier (such as a machine guard or device—refer to Tables 5 and 6) between a hazardous operation and the workers. Fundamentally, however, isolation can be provided *without* a physical barrier through the appropriate use of distance and, in some situations, time.

Perhaps the most common example of isolation as a control strategy is associated with storage and use of flammable solvents. The large tank farms with dikes around the tanks, underground storage of some solvents, the detached solvent sheds, and fireproof solvent storage rooms within buildings are all commonplace in American industry. Frequently, the application of the principle of isolation maximizes the benefits of additional engineering concepts such as excessive noise control, remote control materials handling (as with radioactive substances), and local exhaust ventilation.

Table 8 Positive Performance Characteristics—Some Things Done Better by People versus Machines

People	Machines
Detect signals in high noise fields	Respond quickly to signals
Recognize objects under widely different conditions	Sense energies outside human range
Perceive patterns	Consistently perform precise, routine, repetitive operations
Sensitive to a wide variety of stimuli	Recall and process enormous amounts of data
Long-term memory	
Handle unexpected or low-probability events	Monitor people or other machines
Reason inductively	Reason deductively
Profit from experience	Exert enormous power
Exercise judgment	Relatively uniform performance
Flexibility, improvisation, and creativity	Rapid transmission of signals
Select and perform under overload conditions	Perform several tasks simultaneously
Adapt to changing environment	Expendable
Appreciate and create beauty	Resistance to many environmental stresses
Perform fine manipulations	
Perform when partially impaired	
Relatively maintenance-free	

7.3 Ventilation

Workplace air quality is affected directly by the design and performance of the exhaust system. An improperly designed hood or a hood evacuated with an insufficient volumetric rate of air will contaminate the occupational environment and affect workers in the vicinity of the hazard source. This is a simple, but powerful, symbolic representation of one form of the close relationship between atmospheric emissions (as regulated by the Environmental Protection Agency) and occupational exposure (as regulated by the Occupational Safety and Health Administration). What is done with gases generated as a result of industrial operations/processes? These emissions can be exhausted directly to the atmosphere, indirectly to the atmosphere (from the workplace through the general ventilation system), or recirculated to the workplace. The effectiveness of the ventilation system design and operation impacts directly on the necessity and type of respiratory gear needed to protect the work force.

8 DESIGN AND REDESIGN*

8.1 Hardware

Designers of machines must consider the performance characteristics of machine operators as a major constraint in the creation or modification of both mechanical and electrical equipment. To do less would be tantamount to ignoring the limitations of human capabilities.

---

*J. B. ReVelle, *Engineering Controls: A Comprehensive Overview.* Used by permission of The Merritt Company. Publisher, from T. S. Ferry. *Safety Management Planning,* copyright © 1982, The Merritt Company. Santa Monica, CA 90406.

Equipment designers especially concerned with engineering controls to be incorporated into machines, whether at the time of initial conceptualization or later when alterations are to be made, must also be cognizant of the principles of human factors (ergonomics). Equipment designers are aware that there are selected tasks that people can perform with greater skill and dependability than machines, and vice versa. Some of these positive performance characteristics are noted in Table 8. In addition, designers of equipment and engineering controls are knowledgeable of human performance limitations, both physically and psychologically. They know that the interaction of forces between people and their operating environment presents a never-ending challenge in assessing the complex interrelationships that provide the basis for that often fine line between safety versus hazard or health versus contaminant. Table 9 identifies the six pertinent sciences most closely involved in the design of machines and engineering controls. It is both rational and reasonable to expect that, when engineering controls are being considered to eliminate or reduce hazards or contaminants, designers make full use of the principles established by specialists in these human performance sciences.

8.2 Process

A stress (or stressor) is some physical or psychological feature of the environment that requires an operator to be unduly exerted to continue performing. Such exertion is termed strain as in "stress and strain." Common physical stressors in industrial workplaces are poor illumination, excessive noise, vibration, heat, and the presence of excessive, harmful atmospheric contaminants. Unfortunately, much less is known about their effects when they occur at the same time, in rapid sequence, or over extended periods of time. Research suggests that such effects are not simply additive, but synergistic, thus compounding their detrimental effects. In addition, when physical work environments are unfavorable to equipment operators, two or more stressors are generally present: high temperature and excessive noise, for example. The solution to process design and redesign is relatively easy to specify, but costly to implement—design the physical environment so that all physical characteristics are within an acceptable range.

Marketed in the United States since the early 1960s, industrial robots offer both hardware and process designers a technology that can be used when hazardous or uncomfortable working conditions are expected or already exist. Where a job situation poses potential dangers or the workplace is hot or in some other way unpleasant, a robot should be considered as a substitute for human operators. Hot forging, die casting, and spray painting fall into this category. If work parts or tools are awkward or heavy, an industrial robot may fill the job. Some robots are capable of lifting items weighing several hundred pounds.

Table 9 People Performance Sciences

Anthropometry	Pertains to the measurement of physical features and characteristics of the static human body
Biomechanics	A study of the range, strength, endurance, and accuracy of movements of the human body
Ergonomics	Human factors engineering, especially biomechanics aspects
Human factors engineering	Designing for human use
Kinesiology	A study of the principles of mechanics and anatomy of human movement
Systems safety engineering	Designing that considers the operator's qualities, the equipment, and the environment relative to successful task performance

An industrial robot is a general purpose, programmable machine that possesses certain human-like capabilities. The most obvious characteristic is the robot's arm, which, when combined with the robot's capacity to be programmed, makes it ideally suited to a variety of uncomfortable/undesirable production tasks. Hardware and process designers now possess an additional capability for potential inclusion in their future designs and redesigns.

8.3 Hazardous Material Classification System

The National Fire Protection Association (NFPA), a private, nonprofit organization, is the leading authoritative source of technical background, data, and consumer advice on fire protection, problems, and prevention. The primary goal of the NFPA is to reduce the worldwide burden of fire and other hazards on the quality of life by providing and advocating scientifically based consensus codes and standards, research, training, and education. The NFPA has in excess of 300 codes worldwide, which are for sale through their website http://www.nfpa.org/. While NFPA codes cover several aspects of flammable and otherwise hazardous materials, perhaps the most significant is the *NFP 704 Hazard Identification* ratings system (the familiar NFPA "hazard diamond" for health, flammability, and instability (see Fig. 2).

What do the numbers and symbols on an NFPA fire diamond mean? The diamond is divided into four sections. Numbers in the three colored sections range from 0 (least severe

Figure 2 NFPA hazard diamond. In the color original, the upper left section is blue, the upper right is red, the bottom right is yellow, and the bottom left is white.

hazard) to 4 (most severe hazard). The fourth (white) section is left blank and is used only to denote special fire fighting measures/hazards.

Health hazard levels are identified in the upper left section (blue in the original) of the diamond (third base), as follows:

- *Level 4* is used when even very short exposure could cause death or serious residual injury even though prompt medical attention is given.
- *Level 3* is used when short exposure could cause serious temporary or residual injury even though prompt medical attention is given.
- *Level 2* is used when intense or continued exposure could cause temporary incapacitation or possible residual injury unless prompt medical attention is given.
- *Level 1* is used when exposure could cause irritation but only minor residual injury even if no treatment is given.
- *Level 0* is used when exposure under fire conditions would offer no hazard beyond that of ordinary combustible materials.

Flammability hazard levels are identified in the upper right section (red in the original) of the diamond (second base), as follows:

- *Level 4* is used when a material will rapidly or completely vaporize at normal pressure and temperature, or is readily dispersed into the air and will burn readily.
- *Level 3* is used when a material is a solid or liquid that can be ignited under almost all ambient conditions.
- *Level 2* is used when a material must be moderately heated or exposed to relatively high temperature before ignition can occur.
- *Level 1* is used when a material must be preheated before ignition can occur.
- *Level 0* is used when a material will not burn.

Instability* hazard levels are identified in the bottom right section (yellow in the original) of the diamond (first base), as follows:

- *Level 4* is used when a material is readily capable of detonation or of explosive decomposition or reaction at normal temperatures and pressures.
- *Level 3* is used when a material is capable of detonation or explosive reaction, but requires a strong initiating source or must be heated under confinement before initiation, or reacts explosively with water.
- *Level 2* is used when a material is normally unstable and readily undergoes violent decomposition, but does not detonate. Also: may react violently with water or may form potentially explosive mixtures with water.
- *Level 1* is used when a material is normally stable, but can become unstable at elevated temperatures and pressures, or may react with water with some release of energy, but not violently.
- *Level 0* is used when a material is normally stable, even under fire exposure conditions, and is not reactive with water.

---

*Prior to 1996, this section was titled *Reactivity*. The name was changed because many people did not understand the distinction between a "reactive hazard" and the "chemical reactivity" of a material. The numeric ratings and their meanings remain unchanged.

674 Safety Engineering

Special* hazard levels are identified in the bottom left section (white in the original) of the diamond (home plate), as follows:

- *Symbol OX* is used when a material is an oxidizer, a chemical that can greatly increase the rate of combustion/fire.
- *Symbol W* is used when a material has unusual reactivity with water. This indicates a potential hazard when using water to fight a fire involving this material.
- *ACID* is used when a material is an acid, i.e., a corrosive substance that has a pH (power of hydrogen) lower than 7.0.
- *ALK* is used when a material is an alkaline substance, also referred to as a base. These caustic materials have a pH greater than 7.0.
- *COR* is used when a material is corrosive. It could be either an acid or a base.
- *XXX* is used as another symbol for corrosive.
- *SAC* is used when a material is a poison or highly toxic substance.
- *RAD* is used when a material is a radioactive hazard. Radioactive materials are extremely hazardous when inhaled or ingested.
- *EXP* is used when a material is explosive. This symbol is redundant since explosives are easily recognized by their *Instability* rating.

Readers who desire to obtain an NFPA hazard diamond wall chart (see Fig. 2) at no cost can do so by contacting Graphic Products, Inc. on the Internet at http://www.graphicproducts.com.

8.4 Material Safety Data Sheets (MSDS)

Purpose of MSDSs

Material Safety Data Sheets (MSDSs) are designed to provide both workers and emergency personnel with the proper procedures for handling or working with a particular substance. MSDSs include information such as physical data (melting point, boiling point, flash point, etc.), toxicity, health effects, first aid, reactivity, storage, disposal, protective equipment, and spill/leak procedures. These are of particular use if a spill or other accident occurs.

MSDSs Are Meant For

- Employees who may be occupationally exposed to a hazard at work.
- Employees who need to know the proper methods for storage.
- Emergency responders, such as fire fighters, hazardous material crews, emergency medical technicians, and emergency room personnel.

MSDSs Are Not Meant For

- *Consumers*. An MSDS reflects the hazards of working with specific substances in an occupational situation, *not* by a retail consumer. For example, an MSDS for a can of

---

*There are only two *NFPA 704* approved symbols, *OX* and *W*. Other symbols, abbreviations, and words that some organizations used in the white *Special* hazards section are also described. These uses are *not compliant* with *NFPA 704*, but are presented here in case they appear on a Material Safety Data Sheet (MSDS), which is discussed in Section 8.4, or a container label.

paint is not pertinent to someone who uses it once a year, but is extremely important to someone who is exposed to paint in a confined space 40 hours per week.

Sources of MSDSs

- Your laboratory or workplace should have a collection of MSDSs that came with the hazardous chemicals you have ordered. *Note:* Don't throw them away; file them where they can be easily located in an emergency.
- Most universities and businesses have a collection of MSDSs somewhere on site. Check with your Environmental or Occupational Safety and Health Office or your science librarian. Some organizations use commercial services to obtain printed, FAX, or on-line copies of MSDSs.
- MSDSs can be obtained from the distributor that sold you the material. If they can't be located, try the manufacturer's customer service department.
- The Internet has a wide range of free sources. A handy list of 100 such sites can be found at: http://www.ILPI.com/MSDS/Index.html#Internet.
- MSDSs software or Internet subscription services can be purchased.

MSDS Background

- The U.S. Government's Occupational Safety and Health Administration (OSHA) began requiring MSDSs for hazardous materials on May 26, 1986.
- OSHA is responsible for the Hazard Communication Standard (HCS) 29 CFR (Code of Federal Regulations) 1910.1200. The purpose of this standard is "to ensure that the hazards of all chemicals produced or imported are evaluated, and that information concerning their hazards is transmitted to employers and employees. This transformation of information is to be accomplished by means of comprehensive hazard communication programs, which are to include container labeling and other forms of warning, material safety data sheets, and employee training." The HCS specifies the required elements that must be on an MSDS, among other important data.

8.5 Safety Design Requirements

In the course of insuring that product specifications are met, engineers/designers are typically responsible for adhering to a variety of some quite general and some very specific safety design requirements. Since its origins in 1970, OSHA has adopted or originated numerous safety and health-related requirements. If the reader were to enter the Internet at the following website, http://www.osha.gov/pls/oshaweb/owadisp.show_document?p_table=FEDERAL_REGISTER&p_id=, and then search for "safety design requirements," nearly 1300 individual documents would be presented for investigation. Each of these documents can be double-clicked to obtain more detailed information regarding the specific subject matter.

9 PERSONAL PROTECTIVE EQUIPMENT*

9.1 Background

Engineering controls, which eliminate the hazard at the source and do not rely on the worker's behavior for their effectiveness, offer the best and most reliable means of safe-

---

*J. B. ReVelle, *Engineering Controls: A Comprehensive Overview.* Used by permission of The Merritt Company, Publisher, from T. S. Ferry, *Safety Management Planning,* copyright © 1982, The Merritt Company, Santa Monica, CA 90406.

guarding. Therefore, engineering controls must be first choice for eliminating machinery hazards. But whenever an extra measure of protection is necessary, operators must wear protective clothing or personal protective equipment. If it is to provide adequate protection, the protective clothing and equipment selected must always be

- Appropriate for the particular hazards
- Maintained in good condition
- Properly stored when not in use, to prevent damage or loss
- Kept clean and sanitary

Protective clothing is available for every part of the human body. Hard hats can protect the head from falling objects when the worker is handling stock; caps and hair nets can help keep the worker's hair from being caught in machinery. If machine coolants could splash or particles could fly into the operator's eyes or face, then face shields, safety goggles, glasses, or similar kinds of protection must be used. Hearing protection may be needed when workers operate noisy machinery. To guard the trunk of the body from cuts or impacts from heavy or rough-edged stock, there are certain protective coveralls, jackets, vests, aprons, and full-body suits. Workers can protect their hands and arms from the same kinds of injury with special sleeves and gloves. And safety shoes and boots or other acceptable foot guards can shield the feet against injury in case the worker needs to handle heavy stock that might drop.

Protective clothing and equipment themselves can create hazards. Protective glove that can become caught between rotating parts or respirator face pieces that hinder the wearer's vision require alertness and careful supervision whenever they are used. Other aspects of the worker's dress may present additional safety hazards. Loose-fitting clothing might become entangled in rotating spindles or other kinds of moving machinery. Jewelry, such as bracelets and rings, can catch on machine parts or stock and lead to serious injury by pulling a hand into the danger area.

Naturally, each situation will vary. In some simple cases, respirators, chemical goggles, aprons, and gloves may be sufficient personal protective equipment to afford the necessary coverage. In more complicated situations, even the most sophisticated equipment may not be enough and engineering controls would become mandatory. Safety, industrial, and plant engineers should be expected to provide the necessary analyses to ascertain the extent of the hazard to employees whose work causes them to be exposed to the corrosive fumes.

9.2 Planning and Implementing the Use of Protective Equipment*

This section reviews ways to help plan, implement, and maintain personal protective equipment. This can be considered in terms of the following nine phases: (1) need analysis, (2) equipment selection, (3) program communication, (4) training, (5) fitting and adjustment, (6) target date setting, (7) break in period, (8) enforcement, and (9) follow-through.

The first phase of promoting the use of personal protective equipment is called *need analysis*. Before selecting protective equipment, the hazards or conditions the equipment must protect the employee from must be determined. To accomplish this, questions such as the following must be asked:

---

*J. B. ReVelle and J. Stephenson, *Safety Training Methods,* 2nd ed. Copyright © 1995. Reprinted by permission of Wiley, New York.

- What standards does the law require for this type of work in this type of environment?
- What needs do our accident statistics point to?
- What hazards have we found in our safety and/or health inspections?
- What needs show up in our job analysis and job observation activities?
- Where is the potential for accidents, injuries, illnesses, and damage?
- Which hazards cannot be eliminated or segregated?

The second phase of promoting the use of protective equipment is *equipment selection*. Once a need has been established, proper equipment must be selected. Basic consideration should include the following:

- Conformity to the standards
- Degree of protection provided
- Relative cost
- Ease of use and maintenance
- Relative comfort

The third phase is *program communication*. It is not appropriate to simply announce a protective equipment program, put it into effect, and expect to get immediate cooperation. Employees tend to resist change unless they see it as necessary, comfortable, or reasonable. It is helpful to use various approaches to publicity and promotion to teach employees why the equipment is necessary. Various points can be covered in supervisor's meetings, in safety meetings, by posters, on bulletin boards, in special meetings, and in casual conversation. Gradually, employees will come to expect or to request protective equipment to be used on the job. The main points in program communication are to educate employees in why protective equipment is necessary and to encourage them to want it and to use it.

Training is an essential step in making sure protective equipment will be used properly. The employees should learn why the equipment is necessary, when it must be used, who must use it, where it is required, what the benefits are, and how to use it and take care of it. Do not forget that employee turnover will bring new employees into the work area. Therefore, you will continually need to train new employees in the use of the protective equipment they will handle.

After the training phase comes the *fitting and adjustment* phase. Unless the protective equipment fits the individual properly, it may not give the necessary protection. There are many ways to fit or to adjust protective equipment. For example, face masks have straps that hold them snug against the contours of the head and face and prevent leaks; rubberized garments have snaps or ties that can be drawn up snugly, to keep loose and floppy garments from getting caught in machinery.

The next phase is *target date setting*. After the other phases have been completed, set specific dates for completion of the various phases. For example, all employees shall be fitted with protective equipment before a certain date; all training shall be completed by a certain date; after a certain date, all employees must wear their protective equipment while in the production area.

After setting the target dates, expect a *break-in period*. There will usually be a period of psychological adjustment whenever a new personal protective program is established. Remember two things:

678 Safety Engineering

- Expect some gripes, grumbles, and problems.
- Appropriate consideration must be given to each individual problem: then strive toward a workable solution.

It might also be wise to post signs that indicate the type of equipment needed. For example, a sign might read, "Eye protection must be worn in this area."

After the break-in phase comes *enforcement*. If all the previous phases were successful, problems in terms of enforcement should be few. In case disciplinary action is required, sound judgment must be used and each case must be evaluated on an individual basis. If employees fail to use protective equipment, they may be exposed to hazards. Do not forget, the employer can be penalized if employees do not use their protection.

The final phase is *follow-through*. Although disciplinary action may sometimes be necessary, positive motivation plays a more effective part in a successful protective equipment program. One type of positive motivation is a proper example set by management. Managers must wear their protective equipment, just as employees are expected to wear theirs.

9.3 Adequacy, Maintenance, and Sanitation

Before selling safety shoes and supplying safety goggles at a company store. the attendants must be guided by a well-structured program of equipment maintenance, preferably preventive maintenance. Daily maintenance of different types of equipment might include adjustment of the suspension system on a safety hat; cleaning of goggle lenses, glasses, or spectacles; scraping residue from the sole of a safety shoe; or proper adjustment of a face mask when donning an air-purifying respirator.

Performing these functions should be coupled with periodic inspections for weaknesses or defects in the equipment. How often this type of check is made, of course, depends on the particular type of equipment used. For example, sealed-canister gas masks should be weighed on receipt from the manufacturer and the weight should be marked indelibly on each canister. Stored units should then be reweighed periodically, and those exceeding a recommended weight should be discarded even though the seal remains unbroken.

Sanitation, as spelled out in the OSHAct, is a key part of any operation. It requires the use of personal protective equipment, not only to eliminate cross-infection among users of the same unit of equipment, but because unsanitary equipment is objectionable to the wearer. Procedures and facilities that are necessary to sanitize or disinfect equipment can be an integral part of an equipment maintenance program. For example, the OSHAct says, "Respirators used routinely shall be inspected during cleaning." Without grime and dirt to hinder an inspection, gauges can be read better, rubber or elastomer parts can be checked for pliability and signs of deterioration, and valves can be checked.

10 MANAGING THE SAFETY FUNCTION

10.1 Supervisor's Role

The responsibilities of the first-line supervisor are many. Direction of the work force includes the following supervisory functions:

- Setting goals
- Improving present work methods
- Delegating work
- Allocating manpower
- Meeting deadlines
- Controlling expenditures
- Following progress of work
- Evaluating employee performance
- Forecasting manpower requirements
- Supervising on-the-job training
- Reviewing employee performance
- Handling employee complaints
- Enforcing rules
- Conducting meetings
- Increasing safety awareness*

Supervisory understanding of the interrelationships of these responsibilities is a learned attribute. Organizations that expect their supervisors to offer a high quality of leadership to their employees must provide appropriate training and experiential opportunities to current supervisors and supervisory trainees alike.

10.2 Elements of Accident Prevention†

- Safety policy must be clearly defined and communicated to all employees.
- The safety record of a company is a barometer of its efficiency. An American Engineering Council study revealed that "maximum productivity is ordinarily secured only when the accident rate tends toward the irreducible minimum."
- Unless line supervisors are accountable for the safety of all employees, no safety program will be effective. Top management must let all supervisors and managers know what is expected of them in safety.
- Periodic progress reports are required to let managers and employees know what they have accomplished in safety.
- Meetings with supervisors and managers to review accident reports, compensation costs, accident-cause analysis, and accident-prevention procedures are important elements of the overall safety program.
- The idea of putting on a big safety campaign with posters, slogans, and safety contests is wrong. The Madison Avenue approach does not work over the long run.
- Good housekeeping and the enforcement of safety rules show that management has a real concern for employee welfare. They are important elements in the development of good morale. (A U.S. Department of Labor study has revealed that workers are vitally concerned with safety and health conditions of the workplace. A surprisingly high percentage of workers ranked protection against work-related injuries and illness and pleasant working conditions as having a priority among their basic on-the-job needs. In fact, they rated safety higher than fringe benefits and steady employment.)
- The use of personal protective equipment (safety glasses, safety shoes, hard hats, etc.) must be a condition of employment in all sections of the plant where such protection is required.

---

*B. D. Lewis, Jr., *The Supervisor in 1975*. Copyright © September 1973. Reprinted with permission of *Personnel Journal*. Costa Mesa, CA; all rights reserved.
†R. De Reamer, *Modern Safety and Health Technology*. Copyright © 1980. Reprinted by permission of Wiley, New York.

- Safety files must be complete and up to date to satisfy internal information requirements as well as external inspections by OSHA Compliance Officers and similar officials (see Table 10).

10.3 Management Principles*

- Regardless of the industry or the process, the role of supervisors and managers in any safety program takes precedence over any of the other elements. This is not to say

Table 10 Requirements for Safety Files
The following items are presented for your convenience as you review your administrative storage index to determine the adequacy of your safety-related files.

Number	Action Required	Action Completed Yes	No
1. Is there a separate section for safety-related files?			
2. Are the following subjects provided for in the safety section of the files:			
a. Blank OSHA forms?			
b. Completed OSHA forms?			
c. Blank company safety forms?			
d. Completed company safety forms?			
e. Blank safety checklists?			
f. Completed safety checklists?			
g. Agendas of company safety meetings?			
h. Minutes of company safety meetings?			
i. Records of safety equipment purchases'?			
j. Records of safety equipment checkouts?			
k. Incoming correspondence related to safety?			
l. Outgoing correspondence related to safety?			
m. Record of safety projects assigned?			
n. Record of safety projects completed?			
o. Record of fire drills (if applicable)?			
p. Record of external assistance used to provide specialized safety expertise?			
q. Record of inspections by fire department, insurance companies, state and city inspectors, and OSHA compliance officers?			
r. National Safety Council catalogs and brochures for films, posters, and other safety-related materials?			
3. Are the files listed in item 2 reviewed periodically:			
a. To ensure that they are current?			
b. To retire material over five years old?			
4. Are safety-related files reviewed periodically to determine the need to eliminate selected files and to add new subjects?			
5. Is the index to the file current, so that an outsider could easily understand the system?			

Source: J. B. ReVelle and J. Stephenson, *Safety Training Methods,* 2nd ed. Copyright ©1995. Reprinted by permission of Wiley, New York.

---

*R. De Reamer, *Modern Safety and Health Technology.* Copyright © 1980. Reprinted by permission of Wiley, New York.

that the managerial role is necessarily more important than the development of safe environments, but without manager and supervisor participation, the other elements have a lukewarm existence. There is a dynamic relationship between management and the development of safe working conditions, and management and the development of safety awareness, and the relationship must not be denied.

- Where responsibility for preventing accidents and providing a healthful work environment is sloughed off to the safety department or a safety committee, any reduction in the accident rate is minimal. To reduce the accident rate and, in particular, to make a good rate better, line managers must be held responsible and accountable for safety. Every member of the management team must have a role in the safety program. Admittedly, this idea is not new, but application of the concept still requires crystal-clear definition and vigorous promotion.

- Notwithstanding the many excellent examples of outstanding safety records that have been achieved because every member of management had assumed full responsibility for safety, there are still large numbers of companies, particularly the small establishments, using safety contests, posters, or safety committees as the focal point of their safety programs—but with disappointing results. Under such circumstances safety is perceived as an isolated aspect of the business operation with rather low ceiling possibilities at best. But there are some who feel that gimmicks must be used because foremen and the managers do not have time for safety.

- As an example of the case in point, a handbook on personnel contains the statement that "A major disadvantage of some company-sponsored safety programs is that the supervisor can't spare sufficient time from his regular duties for running the safety program." Significantly, this was not a casual comment in a chapter on safety. It was indented and in bold print for emphasis. Yet it is a firmly accepted fact that to achieve good results in safety, managers and supervisors must take the time to fulfill their safety responsibilities. Safety is one of their regular duties.

- The interrelationships of the many components of an effective industrial safety program are portrayed in Fig. 3.

10.4 Eliminating Unsafe Conditions*

The following steps should be taken to effectively and efficiently eliminate an unsafe condition:

- *Remove.* If at all possible, have the hazard eliminated.
- *Guard.* If the danger point (e.g., high-tension wires) cannot be removed, see to it that hazard is shielded by screens, enclosures, or other guarding devices.
- *Warn.* If guarding is impossible or impractical, warn of the unsafe condition. If a truck must back up across a sidewalk to a loading platform, the sidewalk cannot be removed or a fence built around the truck. All that can be done is to warn that an unsafe condition exists. This is done by posting a danger sign or making use of a bell, horn, whistle, signal light, painted striped lines, red flag, or other device.

---

*J. B. ReVelle and J. Stephenson, *Safety Training Methods,* 2nd ed. Copyright © 1995. Reprinted by permission of Wiley, New York

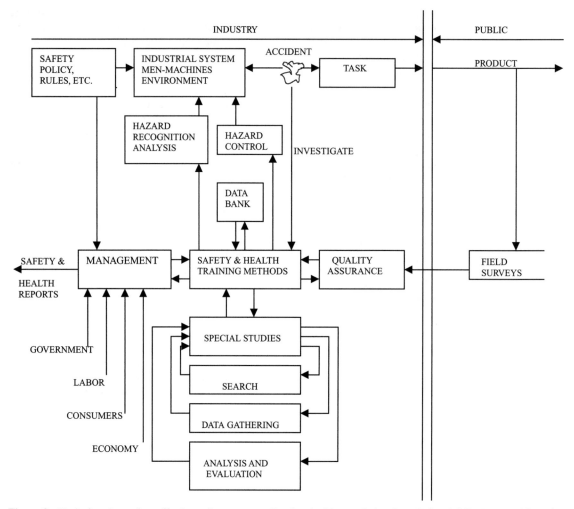

Figure 3 Basic functions of an effective safety program. Reprinted with permission from *Industrial Engineering* Magazine. Copyright © 1979 American Institute of Industrial Engineers, Inc., 25 Technology Park Atlanta, Norcross, GA 30092.

- *Recommend.* If you cannot remove or guard an unsafe condition on your own, notify the proper authorities about it. Make specific recommendations as to how the unsafe condition can be eliminated.
- *Follow up.* After a reasonable length of time, check to see whether the recommendation has been acted on, or whether the unsafe condition still exists. If it remains, the person or persons to whom the recommendations were made should be notified.

Five S (5S)

The original 5S principles were stated in Japanese. Because of their proven value to western industry, they have been translated and restated in English. The 5S is a mantra of sorts designed to help build a quality work environment, both physical and mental.

The 5S condition of a work area is critical to the morale of employees and is the basis of a customer's first impressions. Management's attitude regarding their employees is reflected in the 5S condition of the work area. The 5S philosophy applies in any work area.

The elements of 5S are simple to learn and important to implement:

- *Sort.* Eliminate whatever is not needed
- *Straighten.* Organize whatever remains
- *Shine.* Clean the work area
- *Standardize.* Schedule regular cleaning and maintenance
- *Sustain.* Make 5S a way of life

Numerous benefits can be derived from implementing 5S, e.g.,

- Improved safety
- Higher equipment availability
- Lower defect rates
- Reduced costs
- Increased production agility and flexibility
- Improved employee morale
- Better asset utilization
- Enhanced enterprise image to customers, suppliers, employees, and management

Figure 4 is an example of a 5S work place scan diagnostic checklist.

The following factors should be considered in organizing a plant that provides for maximum productivity and employee well-being:

- The general arrangement of the facility should be efficient, orderly, and neat.
- Workstations should be clearly identified so that employees can be assigned according to the most effective working arrangement.
- Material flow should be designed to prevent unnecessary employee movement for given work.
- Materials storage, distribution, and handling should be routinized for efficiency and safety.
- Decentralized tool storage should be used wherever possible. Where centralized storage is essential (e.g., general supply areas, locker areas, and project storage areas), care should be given to establish a management system that will avoid unnecessary crowding or congested traffic flow. (Certain procedures, such as time staggering, may reduce congestion.)
- Time-use plans should be established for frequently used facilities to avoid having workers wait for a particular apparatus.
- A warning system and communications network should be established for emergencies such as fire, explosion, storm, injuries, and other events that would affect the well-being of employees.

The following unsafe conditions checklist presents a variety of undesirable characteristics to which both employers and employees should be alert:

684 Safety Engineering

5S Category	5S Item	5S Rating Level					Remarks
		L0	L1	L2	L3	L4	
Sort (Organization)	Distinguish between what is needed and not needed						
	Unneeded equipment, tools, furniture, etc. are present						
	Unneeded items are on walls, bulletin boards, etc.						
	Items are present in aisles, stairways, corners, etc.						
	Unneeded inventory, supplies, arts, or materials are present						
	Safety hazards (water, oil, chemical, machines) exist						
Set in Order (Orderliness)	A place for everything and everything in its place						
	Correct places for items are not obvious						
	Items are not in their places						
	Aisles, workstations, equipment locations are not indicated						
	Items are not put away immediately after use						
	Height and quantity limits are not obvious						
Shine (Cleanliness)	Cleaning, and looking for ways to keep it clean and organized						
	Floors, walls, stairs and surfaces are not free of dirt, oil and grease						
	Equipment is not kept clean and free of dirt, oil and grease						
	Cleaning materials are not easily accessible						
	Lines, labels, signs, etc. are not clean and unbroken						
	Other cleaning problems of any kind are present						
Standardize (Adherence)	Maintain and monitor the first three categories						
	Necessary information is not visible						
	All standards are not known and visible						
	Checklists don't exist for cleaning and maintenance jobs						
	All quantities and limits are not easily recognizable						
	How many items can't be located in 30 seconds?						
Sustain (Self-discipline)	Stick to the rules						
	How many workers have not had 5S training?						
	How many times, last week, was daily 5S not performed?						
	Number of times that personal belongings are not neatly stored						
	Number of times job aids are not available or up to date						
	Number of times, last week, daily 5S inspections not performed						
	TOTAL						

Number of Problems	Rating level
3 or more	Level 0 (L0)
3–4	Level 1 (L1)
2	Level 2 (L2)
1	Level 3 (L3)
None	Level 4 (L4)

Figure 4 5S diagnostic checklist.

- *Unsafe conditions–mechanical failure.* These are types of unsafe conditions that can lead to occupational accidents and injuries. *Note:* Keep in mind that unsafe conditions often come about as a result of unsafe acts.
- *Lack of guards.* This applies to hazardous places like platforms, catwalks, or scaffolds where no guardrails are provided; power lines or explosive materials that are not fenced off or enclosed in some way; and machines or other equipment where moving parts or other danger points are not safeguarded.
- *Inadequate guards.* Often a hazard that is partially guarded is more dangerous than it would be if there were no guards. The employee, seeing some sort of guard, may feel secure and fail to take precautions that would ordinarily be taken if there were no guards at all.
- *Defects.* Equipment or materials that are worn, torn, cracked, broken, rusty, bent, sharp, or splintered; buildings, machines, or tools that have been condemned or are in disrepair.

- *Hazardous arrangement (housekeeping).* Cluttered floors and work areas: improper layout of machines and other production facilities; blocked aisle space or fire exits; unsafely stored or piled tools and material; overloaded platforms and vehicles; inadequate drainage and disposal facilities for waste products. The reader is referred to the earlier discussion about 5S.
- *Improper illumination.* Insufficient light; too much light; lights of the wrong color; glare; arrangement of lighting systems that result in shadows and too much contrast.
- *Unsafe ventilation.* Concentration of vapors, dusts, gases, fumes; unsuitable capacity, location, or arrangement of ventilation system; insufficient air changes, impure air source used for air changes; abnormal temperatures and humidity.

In describing conditions for each item to be inspected, terms such as the following should be used:

Broken	Leaking
Corroded	Loose (or slipping)
Decomposed	Missing
Frayed	Rusted
Fuming	Spillage
Gaseous	Vibrating
Jagged	

An alphabetized listing of possible problems to be inspected is presented in Table 11.

Hazard Classification

It is important to differentiate the *degrees of severity* of different hazards. The commonly used standards are given below:

- *Class A hazard.* Any condition or practice with *potential* for causing loss of life or body part and/or extensive loss of structure, equipment, or material.
- *Class B hazard.* Any condition or practice with *potential* for causing serious injury, illness, or property damage, but less severe than Class A.
- *Class C hazard.* Any condition or practice with *probable potential* for causing *nondisabling* injury or illness or *nondisruptive* property damage.

10.6 Unsafe Conditions Involving Mechanical or Physical Facilities*

The total working environment must be under constant scrutiny because of changing conditions, new employees, equipment additions and modifications, and so on. The following checklist is presented as a guide to identify potential problems:

1. Building

    Correct ceiling height

    Correct floor type; in acceptable condition

    Adequate illumination

---

*J. B. ReVelle and J. Stephenson, *Safety Training Methods,* 2nd ed. Copyright © 1995. Reprinted by permission of Wiley, New York.

Table 11 Possible Problems to Be Inspected

Acids	Dusts	Railroad cars
Aisles	Electric motors	Ramps
Alarms	Elevators	Raw materials
Atmosphere	Explosives	Respirators
Automobiles	Extinguishers	Roads
Barrels	Flammables	Roofs
Bins	Floors	Safety devices
Blinker lights	Forklifts	Safety glasses
Boilers	Fumes	Safety shoes
Borers	Gas cylinders	Scaffolds
Buggies	Gas engines	Shafts
Buildings	Gases	Shapers
Cabinets	Hand tools	Shelves
Cables	Hard hats	Sirens
Carboys	Hoists	Slings
Catwalks	Hoses	Solvents
Caustics	Hydrants	Sprays
Chemicals	Ladders	Sprinkler systems
Claxons	Lathes	Stairs
Closets	Lights	Steam engines
Connectors	Mills	Sumps
Containers	Mists	Switches
Controls	Motorized carts	Tanks
Conveyors	Piping	Trucks
Cranes	Pits	Vats
Crossing lights	Platforms	Walkways
Cutters	Power tools	Walls
Docks	Presses	Warning devices
Doors	Racks	

Source: Principles and Practices of Occupational Safety and Health: A Programmed Instruction Course, OSHA 2213, Student Manual Booklet 1, U.S. Department of Labor, Washington, DC, p. 40.

- Adequate plumbing and heating pipes and equipment
- Windows with acceptable opening, closing, and holding devices; protection from breakage
- Acceptable size doors with correct swing and operational quality
- Adequate railing and non-slip treads on stairways and balconies
- Adequate ventilation
- Adequate storage facilities
- Adequate electrical distribution system in good condition
- Effective space allocation
- Adequate personal facilities (restrooms, drinking fountains, wash-up facilities, etc.)
- Efficient traffic flow
- Adequate functional emergency exits
- Effective alarms and communications systems
- Adequate fire prevention and extinguishing devices

Acceptable interior color scheme

Acceptable noise absorption factor

Adequate maintenance and cleanliness

2. Machinery and Equipment

Acceptable placement, securing, and clearance

Clearly marked safety zones

Adequate nonskid flooring around machines

Adequate guard devices on all pulleys

Sharp, secure knives and cutting edges

Properly maintained and lubricated machines, in good working condition

Functional, guarded, magnetic-type switches on all major machines

Properly wired and grounded machines

Functional hand and portable power tools, in good condition and grounded

Quality machines adequate to handle the expected work load

Conspicuously posted safety precautions and rules near each machine

Guards for all pinch points within 7 ft of the floor

11 SAFETY TRAINING

11.1 Specialized Courses*

Automated External Defibrillator (AED) Training
Learning to use an AED is highly intuitive and surprisingly simple. Many people report that it is far easier than learning Cardiopulmonary resuscitation (CPR). (*cardio* means heart, *pulmonary* means lungs, and *resuscitation* means breathing for someone. When a rescuer is performing CPR, he/she is attempting to bring the person back to life [resuscitation]). Current AED courses last up to four hours to allow ample time for hands-on practice and to help increase user competence and confidence. AED training and related resources are offered through the American Heart Association, the American Red Cross, EMP America, the American Health and Safety Institute, the National Safety Council, and others. AED manufacturers also offer training courses. Since most states regulate health care training for public safety personnel, it's a good idea to check with your state authorities to make sure your training program is consistent with state guidelines. To do this, contact your state Emergency Medical Service (EMS) agency.

AED training curricula vary, but generally emphasize

- A working knowledge of CPR
- Safety for both victims and rescuers
- Proper placement of electrodes
- Delivering the first shock as quickly as possible, ideally within 60 seconds from time of arrival at the victim's side

---

*R. De Reamer, *Modern Safety and Health Technology*. Copyright © 1980. Reprinted by permission of Wiley, New York

- Plenty of hands-on practice, with one instructor and one AED or AED trainer for every four to six students

Experience has shown that emergency responders may go for several years before encountering a victim in cardiac arrest. Lay rescuers may use an AED only once in a lifetime. Therefore, it is important to review AED skills on a regular basis. The ideal frequency for retraining is unknown, but most experts recommend reviewing AED skills every three to four months.

Driver Training

The number one accident killer of employees is the traffic accident. Each year more than 27,000 workers die in non-work-related motor-vehicle accidents, and an additional 3900 employees are killed in work-related accidents. The employer pays a heavy toll for these accidents. Those that are work related are compensable, but the others are, nonetheless, costly. The loss of a highly skilled worker, a key scientist, or a company executive could have a serious impact on the success of the business.

There is, fortunately, something constructive that employers can do to help protect their employees and their executives from the tragedy and waste of traffic accidents. Driver training for workers and executives can be provided either in-house or through community training agencies. Companies that have conducted driver-training programs report that the benefits of such training were not limited to the area of improved traffic-accident performance. These companies also experienced lower on-the-job injury frequency rates (the training produced an increase in safety awareness) and improved employee-community relations.

Companies have taken several approaches to driver training:

- A course has been made available to employees on a volunteer basis, either on- or off-hours.
- Driver training has been made mandatory for employees who operate a motor vehicle on company business.
- The company has promoted employee attendance at community-agency-operated programs.
- Full-scale driver-training programs have been conducted for all employees and members of their families. This is done off-hours and attendance is voluntary.

Environmental Risk Training

How cold or hot is it outside? Simply knowing the temperature doesn't tell you enough about the conditions of the working environment to enable an employee to know how to dress sensibly for weather extremes, whether hot or cold. Other factors, including wind speed, relative humidity, dew point, and cloud conditions, play important roles in determining how hot or cold it feels outside. Two indices, the heat index and the wind chill index, have been created to explain these conditions and employees whose work takes them outside should receive some practical training about the indices and how to best use them to avoid the otherwise painful and possibly damaging consequences of temperature extremes.

Heat Index. In an average year only the winter's cold—not lightning, hurricanes, tornadoes, floods, or earthquakes—takes a greater weather-related death toll than the summer's heat and humidity. In an effort to alert persons to the hazards of prolonged heat/humidity episodes, the National Weather Service devised the *Heat Index.* The Heat Index (HI) is an accurate measure of how hot it really feels when the effects of relative humidity are added to high temperature.

The human body contains several mechanisms to maintain its internal operating temperature at approximately 98.6°F. When threatened with above normal temperatures, the body will try to dissipate excess heat by varying the rate and depth of blood circulation, by losing water through the skin and sweat glands, and, as a last resort, by panting. When weather conditions force the air temperature above 90°F and the relative humidity is high, the body is doing everything it can to maintain normal temperature. Unfortunately, conditions can exceed the body's ability to cope with the combined efforts of heat and humidity. At such times the body may succumb to any of a number of heat disorders, including sunstroke, heat cramps, heat exhaustion, and heatstroke.

To use the Heat Index charts, find the appropriate temperature at the top of the chart. Read down until you are opposite the relative humidity or the dew point. The number that appears at the intersection of the temperature and humidity/dew point is the Heat Index. Refer to Tables 12 and 13.

The Heat Index is the "feels like" or apparent temperature. As relative humidity increases, the air seems warmer than it actually is because the body is less able to cool itself via evaporation of perspiration. As the Heat Index rises, so do health risks. When the Heat Index is between 90°F and 105°F, heat exhaustion is possible. When it is above 105°F, it is probable. Heatstroke is possible when the Heat Index is above 105°F, and is quite likely when it is 130°F and above. Physical activity of any kind and prolonged exposure to the heat increase the risks. Environmental risk training for employees should also include this information:

- *Heat exhaustion* occurs when the body is dehydrated.

    Symptoms. headache, nausea, dizziness, cool and clammy skin, pale face, cramps, weakness, profuse perspiration

    First aid. Move to a cooler spot, drink water with a small amount of salt added (one teaspoon per quart).

    Result. It can lead to collapse and heatstroke.

Table 12 Heat Index: Temperature and Relative Humidity

RH (%)	Temperature (°F)														
	90	91	92	93	94	95	96	97	98	99	100	101	102	103	104	105
90	119	123	128	132	137	141	146	152	157	163	168	174	180	186	193	199
85	115	119	123	127	132	136	141	145	150	155	161	166	172	178	184	190
80	112	115	119	123	127	131	135	140	144	149	154	159	164	169	175	180
75	109	112	115	119	122	126	130	134	138	143	147	152	156	161	166	171
70	106	109	112	115	118	122	125	129	133	137	141	145	149	154	158	163
65	103	106	108	111	114	117	121	124	127	131	135	139	143	147	151	155
60	100	103	105	108	111	114	116	120	123	126	129	133	136	140	144	148
55	98	100	103	105	107	110	113	115	118	121	124	127	131	134	137	141
50	96	98	100	102	104	107	109	112	114	117	119	122	125	128	131	135
45	94	96	98	100	102	104	106	108	110	113	115	118	120	123	126	129
40	92	94	96	97	99	101	103	105	107	109	111	113	116	118	121	123
35	91	92	94	95	97	98	100	102	104	106	107	109	112	114	116	118
30	89	90	92	93	95	96	98	99	101	102	104	106	108	110	112	114

Note. Exposure to full sunshine can increase HI values by up to 15°F.

Table 13 Heat Index: Temperature and Dew Point

Dewpoint (°F)	Temperature (°F)															
	90	91	92	93	94	95	96	97	98	99	100	101	102	103	104	105
65	94	95	96	97	98	100	101	102	103	104	106	107	108	109	110	112
66	94	95	97	98	99	100	101	103	104	105	106	108	109	110	111	112
67	95	96	97	98	100	101	102	103	105	106	107	108	110	111	112	113
68	95	97	98	99	100	102	103	104	105	107	108	109	110	112	113	114
69	96	97	99	100	101	103	104	105	106	108	109	110	111	113	114	115
70	97	98	99	101	102	103	105	106	107	109	110	111	112	114	115	116
71	98	99	100	102	103	104	106	107	108	109	111	112	113	115	116	117
72	98	100	101	103	104	105	107	108	109	111	112	113	114	116	117	118
73	99	101	102	103	105	106	108	109	110	112	113	114	116	117	118	119
74	100	102	103	104	106	107	109	110	111	113	114	115	117	118	119	121
75	101	103	104	106	107	108	110	111	113	114	115	117	118	119	121	122
76	102	104	105	107	108	110	111	112	114	115	117	118	119	121	122	123
77	103	105	106	108	109	111	112	114	115	117	118	119	121	122	124	125
78	105	106	108	109	111	112	114	115	117	118	119	121	122	124	125	126
79	106	107	109	111	112	114	115	117	118	120	121	122	124	125	127	128
80	107	109	110	112	114	115	117	118	120	121	123	124	126	127	128	130
81	109	110	112	114	115	117	118	120	121	123	124	126	127	129	130	132
82	110	112	114	115	117	118	120	122	123	125	126	128	129	131	132	133

Note. Exposure to full sunshine can increase HI values by up to 15°F.

- *Heatstroke* occurs when perspiration cannot occur and the body overheats.

    Symptoms. headache, nausea, face flushed, hot and dry skin, perspiration, body temperature over 101°F, chills, rapid pulse

    First aid. Cool person immediately, move to shade or indoors, wrap in a cool, wet sheet, get medical assistance.

    Result. It can lead to confusion, coma, and death.

Wind Chill Index. A description of the character of weather known as "coldness" was proposed around 1940 by scientists working in the Antarctic. The *Wind Chill Index* was developed to describe the relative discomfort/danger resulting from the combination of wind and temperature. The Wind Chill Index describes an equivalent temperature at which the heat loss from exposed flesh would be the same as if the wind were near calm. For example, a Wind Chill Index of −5 indicates that the effects of wind and temperature on exposed flesh are the same as if the air temperature were 5 degrees below zero, even though the actual temperature is higher.

The high importance of the Wind Chill Index to exposed persons is as an indicator of how to dress properly for winter weather. (Wind chill does not affect a car's antifreeze protection, freezing of water pipes, etc.) In dressing for cold weather, an important factor to remember is that entrapped insulating air warmed by body heat is the best protection against the cold. Consequently, persons working outside should wear loose-fitting, lightweight, warm clothing in several layers. Outer garments should be tightly woven, water-repellant, and hooded. Mittens snug at the wrist are better protection than fingered gloves. Physical activity of any kind and prolonged exposure to the cold increase the risks. As with the Heat Index, environmental risk training for employees should include this information.

To use the Wind Chill Index chart, find the approximate temperature at the top of the chart. Read down until you are opposite the appropriate wind speed. The number found at the intersection of the temperature and wind speed is the Wind Chill Index. Refer to Table 14.

If the engineer in you would like to calculate the Wind Chill Index for various combinations of temperature and wind speed other than those in Table 14, the following formula should be used:

$$WCI = 91.4 - [0.474677 - 0.020425 * V + 0.303107 * SQRT(V)] * (91.4 - T)$$

where WCI = Wind Chill Index
V = wind speed (miles per hour)
SQRT = square root
T = temperature (°F)

Fire Protection Training

All employees must know what to do when a fire alarm sounds. All employees must know something about the equipment provided for fire protection and what they can do toward preventing a fire. They must know

- The plan established for evacuation of the building in case of emergency.
- How to use the first-aid fire appliances provided (extinguishers, hose, etc.).
- How to use other protective equipment. (All employees should know that water to extinguish fires comes out of the pipes of the sprinkler systems and that stock must not be piled so close to sprinkler lines that it prevents good distribution of water from sprinkler heads on a fire in the piled material. They should know that fire doors must be kept operative and not obstructed by stock piles, tools, or other objects.)
- How to give a fire alarm and how to operate plant fire alarm boxes and street boxes in the public alarm system.
- Where smoking in the plant is permitted and where, for fire-safety reasons, it is prohibited.

Table 14 Wind Chill Index

Wind (mph)	Temperature (°F)												
	35	30	25	20	15	10	5	0	−5	−10	−15	−20	−25
5	32	27	22	16	11	6	0	−5	−10	−15	−21	−26	−31
10	22	16	10	3	−3	−9	−15	−22	−27	−34	−40	−46	−52
15	16	9	2	−5	−11	−18	−25	−31	−38	−45	−51	−58	−65
20	12	4	−3	−10	−17	−24	−31	−39	−46	−53	−60	−67	−74
25	8	1	−7	−15	−22	−29	−36	−44	−51	−59	−66	−74	−81
30	6	−2	−10	−18	−25	−33	−41	−49	−56	−64	−71	−79	−86
35	4	−4	−12	−20	−27	−35	−43	−52	−58	−67	−74	−82	−92
40	3	−5	−13	−21	−29	−37	−45	−53	−60	−69	−76	−84	−92

Note. Wind speeds above 40 mph have little additional chilling affect. Values of wind chill below −10° F are considered bitterly cold. Values of wind chill below −20° F are extremely cold—human flesh will begin to freeze within one minute!

- The housekeeping routine (disposal of wiping rags and waste, handling of packing materials, and other measures for orderliness and cleanliness throughout the plant).
- Hazards of any special process in which the employee is engaged.

All these "what-to-do" items can appropriately be covered in training sessions and evacuation drills.

First-Aid Training

First-aid courses pay big dividends in industry. This statement is based on clear evidence that people trained in first aid are more safety conscious and less likely to have an accident. The importance of first-aid training from the safety standpoint is that it teaches much more than applying a bandage or a splint. According to the Red Cross, "The primary purpose of first aid training is the prevention of accidents." Each lesson teaches the student to analyze (1) how the accident happened, (2) how the accident could have been prevented, and (3) how to treat the injury. But the biggest dividend of first-aid training is the lives that have been saved because trainees were prepared to apply mouth-to-mouth resuscitation, to stop choking using the Heimlich maneuver (ejection of foreign object by forceful compression of diaphragm), or to stem the flow of blood.

Since the OSHAct, first-aid training has become a matter of federal law—the act stipulates that in absence of an infirmary, clinic, or hospital in proximity to the workplace, a person or persons shall be adequately trained to render first aid. The completion of the basic American National Red Cross first-aid course will be considered as having met this requirement. Just what constitutes *proximity* to a clinic or hospital? The OSH Review Commission, recognizing that first aid must be given within 3 minutes of serious accidents, concluded that an employer whose plant had no one trained in first aid present and was located 9 minutes from the nearest hospital violated the standard (1910.151, Medical and First Aid).

Hazard Recognition Training

The OSHAct of 1970 makes it quite clear that employers are solely responsible for supplying a work environment that is, insofar as is possible, free of hazards. Nonetheless, unavoidable hazards can and do occur. It is incumbent on employers to develop or acquire as well as to deliver hazard recognition training to affected employees. The purpose of this training is to insure that employees are fully knowledgeable and capable of recognizing workplace hazards. In this context workplace hazards can include a broad spectrum of situations that range from missing rubber feet on ladders to dangerous gases in confined spaces.

HAZWOPER Training

The Hazardous Waste Operations and Emergency Response Standard (HAZWOPER) applies to five distinct groups of employers and their employees. This includes any employees who are exposed or potentially exposed to hazardous substances, including hazardous waste, and who are engaged in one of the following operations as specified by 29 CFR 1910.120(a)(1)(i-v) and 1926.65(a)(1)(i-v):

- Cleanup operations—required by a governmental body, whether federal, state, local, or other involving hazardous substances—that are conducted at uncontrolled hazardous waste sites
- Corrective actions involving cleanup operations at sites covered by the *Resource Conservation and Recovery Act of 1976* (*RCRA*), as amended
- Voluntary cleanup operations at sites recognized by federal, state, local, or other governmental body as uncontrolled hazardous waste sites

- Operations involving hazardous wastes that are conducted at treatment, storage, and disposal facilities regulated by Title 40 Code of Federal Regulations (CFR), Parts 264 and 265 pursuant to RCRA, or by agencies under agreement with the U.S. Environmental Protection Agency (EPA) to implement RCRA regulations
- Emergency response operations for releases of, or substantial threats of, hazardous substances regardless of the location of the hazard

Computer-based training may meet some refresher training requirements, provided that it covers topics relevant to workers' assigned duties. It must be supplemented by the opportunity to ask questions of a qualified trainer and by an assessment of hands-on performance of work tasks. HAZWOPER refresher training may be given in segments so long as the required eight hours have been completed by the employee's anniversary date.

If the date for the refresher training has lapsed, the need to repeat initial training must be determined based on the employee's familiarity with safety and health procedures used on site. The employee should take the next available refresher training course. There should be a record in the employee's file indicating why the training has been delayed and when the training will be completed.

Lockout Box Training

The OSHA standard for *The Control of Hazardous Energy (Lockout/Tagout), Title 29 Code of Federal Regulations (CFR)* Part 1910.147, addresses the practices and procedures necessary to disable machinery or equipment, thereby preventing the release of hazardous energy while employees perform servicing and maintenance activities. The standard outlines measures for controlling hazardous energies, i.e., electrical, mechanical, hydraulic, pneumatic, chemical, thermal, and other energy sources. In addition, 29 CFR 1910.333 sets forth requirements to protect employees working on electrical circuits and equipment. This section requires workers to use safe work practices, including lockout and tagging procedures. These provisions apply when employees are exposed to electrical hazards while working on, near or with conductors or systems that use electric energy. The lockout/tagout standard establishes the employer's responsibility to protect employees from hazardous energy sources on machines and equipment during service and maintenance.

Employees need to be trained to insure that they know, understand, and follow the applicable provisions of the hazardous energy control procedures. The training must cover at least three areas: aspects of the employer's energy control program, elements of the energy control procedure relevant to the employee's duties or assignment, and the various requirements of the OSHA standards related to lockout/tagout (LOTO).

Machine Guard Training

Moving machine parts have the potential to cause severe workplace injuries, such as crushed fingers or hands, amputations, burns, or blindness. Safeguards are essential for protecting workers from these needless and preventable injuries. Any machine part, function, or process that can cause injury must be safeguarded. When the operation of a machine or accidental contact with it can injure the operator or others in the vicinity, the hazards must be either eliminated or controlled.

Machine guard training should be provided to workers in advance of their assignments to use any machines that require the use of machine guards. The training should include exposure to the actual machines or, at the very least, multiple photographs of the machines. As much as is practical, the employee and trainer walkaround and/or the photographs should allow the machines to be viewed both with and without the required guards. Photographs that include machine operators should indicate the use of appropriate personal protective equipment (PPE).

The machine guard training should address the four types of guards:

1. Fixed guards
   - A permanent part of the machine
   - Not dependent on any other part to perform the function
   - Usually made of sheet metal, screen, bars, or other material that will withstand the anticipated impact
   - Generally considered the preferred type of guard
   - Simple and durable
2. Interlocked guards
   - Usually connected to a mechanism that will cut off power automatically
   - Could use electrical, mechanical, or hydraulic systems
   - Should rely on a manual reset system
3. Adjustable guards
   - Very flexible to accommodate various types of stock
   - Manually adjusted
4. Self-adjusting guards
   - Opening is determined by the movement of the stock through the guard
   - Does not always provide maximum protection
   - Common complaint is reduced visibility at the point of operation: "I can't see what I'm doing!"

Upon completion of machine guard training, operators should be able to immediately recognize the presence or absence of the correct type of machine guard for any machine they will be expected to operate. Furthermore, operators should be able to immediately detect if a machine guard is properly installed so as to provide maximum operator protection.

Personal Protection Equipment/Physical Protection Training
Every employee should receive whatever personal protection equipment (PPE) that is needed to insure short- and long-term safety and health. The following list of PPE is typical, but not exhaustive:

- Coveralls
- Earplugs/earmuffs
- Gloves
- Footwear (protective and high traction)
- Protective hoods
- Respirators
- Safety glasses
- Self-contained breathing apparatus (SCUBA)

The intent of supplying the correct PPE to the correct employee at the correct time in the correct condition is to insure the necessary protection of their physical parts. The following list of physical parts is typical, but not exhaustive:

- Back
- Ears
- Elbows
- Eyes
- Feet
- Hands
- Knees
- Mouth (unintended ingestion)
- Nose (unintended inhalation)

Other Specialized Courses
Some of the other specialized courses that can be given for safety training are

- Accident investigation
- Accident report preparation
- Eye/face wash stations
- Powered equipment and vehicles
- Safety recordkeeping
- Specific disasters

11.2 Job Hazard Analysis Training*

Admittedly, the conventional mass approach to safety training takes little of the supervisor's time. Group training sessions, safety posters, films, and booklets are handled by the plant safety engineer, human resources, or other staff people. On the other hand, where safety training is carried out on a personalized basis, the first-line supervisor is the right person to provide the training. This will take more of his or her time and require more attention to detail, but this additional effort pays off because of the increased effectiveness of the training method.

In launching a personalized safety-training program, the first step is the preparation of a job-hazard analysis for each job in the plant. To make the job-hazard analysis in an organized manner, use of a form similar to the one shown in Table 15 is suggested. The key elements of the form are (1) job description; (2) job location; (3) key job steps; (4) tool used; (5) potential health and injury hazards; and (6) safe practices, apparel, and equipment.

A review of the form will indicate the steps in making a job-hazard analysis. To start an analysis, the key steps of the job are listed in order in the first column of the form. Where pertinent, the tool used to perform the job step is listed in the second column. In the third column opposite each job step, the hazards of the particular step are indicated. Finally, in the fourth column of the form are listed the safe practices that the employee must be shown and have discussed. Here the supervisor lists the safe work habits that must be stressed and the safety equipment and clothing required for the job.

In making the analysis, an organized approach is required so the less obvious accident hazards will not be missed. This means going out on the floor and actually watching the job being performed and jotting down key steps and hazards as they are observed. Supervisors who make such a job-hazard analysis are often surprised to find hazards in the job cycle that they had missed seeing in the past. Their original negative reaction to the thought of additional paperwork soon disappears. In the long run, supervisors realize that proper hazard analysis will help them do a better training job.

---

*R. De Reamer, *Modern Safety and Health Technology*. Copyright © 1980. Reprinted by permission of Wiley, New York.

Table 15 Job Hazard Analysis Form

Job description: Three-spindle drill press—Impeller 34C6
Job location: *Bldg. 19-2, Pump Section*

Key Job Steps	Tool Used	Potential Health and Injury Hazard	Safe Practices, Apparel, and Equipment
Get material from operation	Tote box	Dropping tote box on foot	Wear safety shoes. Have firm grip on box.
		Back strain from lifting	Stress proper lifting methods.
		Picking up overloaded boxes	Tell employee to get help or lighten load.
Inspect and set up drill press	Drill press	Check for defective machines	Do not operate if defective. Attach red or yellow "do not operate" tag.
		Chuck wrench not removed	Always remove chuck wrench immediately after use.
		Making adjustments when machine is running	Always stop spindle before making adjustments.
Drilling		Hair, clothing, or jewelry catching on spindle	Wear head covering, snug-fitting; clothing. No loose sleeves. Avoid wearing rings, bracelets, or wristwatches.
		Spinning work or fixture	Use proper blocks or clamps to hold work and fixture securely.
		Injury to hands—cuts, etc.	Never wear gloves. Use hook, brush, or other tool to remove chips. Use compressed air only when instructed.
		Drill sticks in work	Stop spindle, free drill by hand.
		Flying chips	Wear proper eye protection.
		Pinch points at belts	Always stop press before adjusting belts.
		Broken drills	Do not attempt to force drill, apply pressure.

James Black
Signature _____

7/22/05   1 of 3
Date   Page

As previously stated, a job-hazard analysis is made for each job. In most cases, each supervisor will have to make from 5 to 10 different analyses. Of course, in maintenance and construction work, the variety of jobs covers a much wider range. Fortunately, these jobs can be grouped by the type of work performed and a job-hazard analysis can be made for each category of work. rather than for each job. For example, repair, installation, and relocation of equipment; cleaning motors; and unloading cars might be a few of the various categories of maintenance work to be analyzed.

Some of the topics that may need to be addressed as part of job-hazard analysis training include

- Air-sampling/gas-monitoring equipment (for such compounds as ammonia, ethyl/methyl Mercapton, formaldehyde, hydrogen sulfide, and sulfur dioxide)

- HAZMAT kits
- Odor recognition (for the same compounds)
- Oil immobilants
- Personal skill responses
- Sound level meters

11.3 Management's Overview of Training*

An effective accident-prevention program requires proper job performance from everyone in the workplace. All employees must know about the materials and equipment they are working with, what hazards are known in the operation, and how these hazards have been controlled or eliminated. Each individual employee needs to know and understand the following points (especially if they have been included in the company policy and in a "code of safe practices"):

- No employee is expected to undertake a job until he or she has received instruction on how to properly perform the job and has been authorized to perform that job.
- No employee should undertake a job that appears to be unsafe.
- Mechanical safeguards are in place and must be kept in place.
- Each employee is expected to report all unsafe conditions encountered during work.
- Even slight injury or illness suffered by an employee must be reported at once.

In addition to the points above, any safety rules that are a condition of employment, such as the use of safety shoes or eye protection, should be explained clearly and enforced at once.

The first-line supervisors must know how to train employees in the proper way of doing their jobs. Encourage and consider providing for supervisory training for these supervisors. (Many colleges offer appropriate introductory management training courses.)

Some specific training requirements in the OSHA standards must be met, such as those that pertain to first aid and powered industrial trucks (including forklifts). In general, they deal with situations where the use of untrained or improperly trained operators on skill machinery could cause hazardous situations to develop, not only for the operator, but possibly for nearby workers, too.

Particular attention must be given to new employees. Immediately on arriving at work, new employees begin to learn things and to form attitudes about the company, the job, their boss, and their fellow employees. Learning and attitude formation occur regardless of whether the employer makes a training effort. If the new employees are trained during those first few hours and days to do things the right way, considerable losses may be avoided later.

At the same time, attention must be paid to regular employees, including the old-timers. Old habits can be wrong habits. An employee who continues to repeat an unsafe procedure is not working safely, even if an "accident" has not resulted from this behavior.

Although every employee's attitude should be one of determination that "accidents" can be prevented, one thing more may be needed. It should be stressed that the responsibility assigned to the person in charge of the job—as well as to all other supervisors—is to be sure that there is a concerted effort under way at all times to follow every safe work pro-

---

*J. B. ReVelle and J. Stephenson. *Safety Training Methods,* 2nd ed. Copyright © 1995. Reprinted by permission of Wiley, New York.

cedure and health practice applicable to that job. It should be clearly explained to these supervisors that they should never silently condone unsafe or unhealthful activity in or around any workplace.

11.4 Sources and Types of Training Materials

Sources of training materials include the Occupational Safety and Health Administration (OSHA), the Department of Transportation (DOT), the Environmental Protection Agency (EPA), and HAZWOPER. Types of training materials include printed documents (books, manuals, and pamphlets), videos (tapes and compact discs), and the Internet. The bibliography to this chapter presents an extensive list of sources in a variety of types of training.

BIBLIOGRAPHY

"Accident Prevention: Your Key to Controlling Surging Workers' Compensation Costs," *Occupational Hazards,* **35** (November 1979).

"Accident Related Losses Make Cost Soar," *Industrial Engineering,* **26** (May 1979).

"Analyzing a Plant Energy-Management Program, Part I: Measuring Performance," *Plant Engineering,* **59** (October 30, 1980).

"Analyzing a Plant Energy-Management Program, Part II: Forecasting Consumption," *Plant Engineering,* **149** (November 13, 1980).

"Anatomy of a Vigorous In-Plant Program," *Occupational Hazards,* **32** (July 1979). "A Shift Toward Protective Gear," *Business Week,* **56H** (April 13, 1981).

"A Win for OSHA," *Business Week,* **62** (June 29, 1981).

"Buyers Should Get Set for Tougher Safety Rules," *Purchasing,* **34** (May 25, 1976).

"Complying with Toxic and Hazardous Substances Regulations, Part I, "*Plant Engineering,* **283** (March 6, 1980).

"Complying with Toxic and Hazardous Substances Regulations, Part II, "*Plant Engineering,* **157** (April 17, 1980).

"Computers Help Pinpoint Worker Exposure," *Chemecology,* **11** (May 1981).

"Conserving Energy by Recirculating Air from Dust Collection Systems," *Plant Engineering,* **151** (April 17, 1980).

"Control Charts Help Set Firm's Energy Management Goals," *Industrial Engineering,* **56** (December 1980).

"Controlling Noise and Reverberation with Acoustical Baffles," *Plant Engineering,* **131** (April 17, 1980).

"Controlling Plant Noise Levels," *Plant Engineering,* **127** (June 24, 1976).

"Cost–Benefit Decision Jars OSHA Reform," *Industry Week,* **18** (June 29, 1981).

"Cost Factors for Justifying Projects," *Plant Engineering,* **145** (October 16, 1980).

"Costs, Benefits. Effectiveness, and Safety: Setting the Record Straight," *Professional Safety,* **28** (August 1975).

"Costs Can Be Cut Through Safety," *Professional Safety,* **34** (October 1976).

"Cutting Your Energy Costs," *Industry Week,* **43** (February 23, 1981).

De Reamer, R., *Modern Safety and Health Technology,* Wiley, New York, 1980.

"Elements of Effective Hearing Protection," *Plant Engineering,* **203** (January 22, 1981).

"Energy Constraints and Computer Power Will Greatly Impact Automated Factories in the Year 2000," *Industrial Engineering,* **34** (November 1980).

"Energy Managers Gain Power," *Industry Week,* **62** (March 17, 1980).

"Energy Perspective for the Future," *Industry Week,* **67** (May 26, 1980).

Engineering Control Technology Assessment for the Plastics and Resins Industry, NIOSH Research Report Publication No. 78-159.

"Engineering Project Planner, A Way to Engineer Out Unsafe Conditions," *Professional Safety* **16** (November 1976).
"EPA Gears Up to Control Toxic Substances," *Occupational Hazards,* **68** (May 1977).
Ferry, T. S., *Safety Management Planning,* The Merritt Company, Santa Monica, CA, 1982.
"Fume Incinerators for Air Pollution Control," *Plant Engineering,* **108** (November 13, 1980).
"Groping for a Scientific Assessment of Risk," *Business Week,* **120J** (October 20, 1980).
"Hand and Body Protection: Vital to Safety Success," *Occupational Hazards,* **31** (February 1979).
"Hazardous Wastes: Coping with a National Health Menace," *Occupational Hazards,* **56** (October 1979).
"Hearing Conservation—Implementing an Effective Program," *Professional Safety,* **21** (October 1978).
"How Do You Know Your Hazard Control Program Is Effective?," *Professional Safety,* **18** (June 1981).
"How to Control Noise," *Plant Engineering,* **90** (October 5, 1972).
"Human Factors Engineering—A Neglected Art," *Professional Safety,* **40** (March 1978).
"IE Practices Need Reevaluation Due to Energy Trends," *Industrial Engineering,* **52** (December 1980).
"Industrial Robots: A Primer on the Present Technology," *Industrial Engineering,* **54** (November 1980).
"Job-Safety Equipment Comes Under Fire, Are Hard Hats a Solution or a Problem?" *The Wall Street Journal,* **40** (November 18, 1977).
Johnson, W. G., *MORT Safety Assurance Systems.* Marcel Dekker, New York, 1979.
McFarland, R. A., "Application of Human Factors Engineering to Safety Engineering Problems," *National Safety Congress Transactions,* National Safety Council, Chicago, 1967, Vol. 12.
"New OSHA Focus Led to Noise-Rule Delay," *Industry Week,* **13** (June 15, 1981).
"OSHA Communiqué," *Occupational Hazards,* **27** (June 1981).
"OSHA Moves Health to Front Burner," *Purchasing,* **46** (September 26, 1979).
"OSHA to Analyze Costs, Benefits of Lead Standard," *Occupational Health & Safety,* **13** (June 1981).
Patty's Industrial Hygiene and Toxicology, 3rd revised ed., Wiley-Interscience, New York, 1978, Vol. 1.
"Practical Applications of Biomechanics in the Workplace," *Professional Safety,* **34** (July 1975).
"Private Sector Steps up War on Welding Hazards," *Occupational Hazards,* **50** (June 1981).
"Putting Together a Cost Improvement Program," *Industrial Engineering,* **16** (December 1979).
"Reduce Waste Energy with Load Controls," *Industrial Engineering,* **23** (July 1979).
"Reducing Noise Protects Employee Hearing," *Chemecology,* **9** (May 1981).
"Regulatory Relief Has Its Pitfalls, Too," *Industry Week,* **31** (June 29, 1981)
ReVelle, J. B., and J. Stephenson, *Safety Training Methods,* 2nd ed., Wiley, New York, 1995.
"ROI Analysis for Cost-Reduction Projects," *Plant Engineering,* **109** (May 15, 1980).
"Safety & Profitability—Hand in Hand," *Professional Safety,* **36** (March 1978).
"Safety Managers Must Relate to Top Management on Their Terms," *Professional Safety,* **22** (November 1976).
"Superfund Law Spurs Cleanup of Abandoned Sites," *Occupational Hazards,* **67** (April 1981).
"Taming Coal Dust Emissions," *Plant Engineering,* **123** (May 15, 1980).
"The Cost–Benefit Argument—Is the Emphasis Shifting?" *Occupational Hazards,* **55** (February 1980).
"The Cost/Benefit Factor in Safety Decisions," *Professional Safety,* **17** (November 1978).
"The Design of Manual Handling Tasks," *Professional Safety,* **18** (March 1980).
"The Economics of Safety—A Review of the Literature and Perspective," *Professional Safety,* **31** (December 1977).
"The Hidden Cost of Accidents," *Professional Safety,* **36** (December 1975).
"The Human Element in Safe Man-Machine Systems," *Professional Safety,* **27** (March 1981).
"The Problem of Manual Materials Handling," *Professional Safety,* **28** (April 1976).
"Time for Decisions on Hazardous Waste," *Industry Week,* **51** (June 15, 1981).
"Tips for Gaining Acceptance of a Personal Protective Equipment Program," *Professional Safety,* **20** (March 1976).
"Toxic Substances Control Act," *Professional Safety,* **25** (December 1976).
"TSCA: Landmark Legislation for Control of Chemical Hazards," *Occupational Hazards,* **79** (May 1977).

"Were Engineering Controls 'Economically Feasible'?" *Occupational Hazards,* **27** (January 1981).

"Were Noise Controls 'Technologically Feasible'?" *Occupational Hazards,* **37** (January 1981).

"What Are Accidents Really Costing You?" *Occupational Hazards,* **41** (March 1979).

"What's Being Done About Hazardous Wastes?" *Occupational Hazards,* **63** (April 1981).

"Where OSHA Stands on Cost-Benefit Analysis," *Occupational Hazards,* **49** (November 1980).

"Worker Attitudes and Perceptions of Safety," *Professional Safety,* **28** (December (1981); **20** (January 1982).

CHAPTER 21
WHAT THE LAW REQUIRES OF THE ENGINEER

Alvin S. Weinstein, Ph.D., PE, JD
Weinstein Associates International
Delray Beach, Florida

Martin S. Chizek, MS, JD
Weinstein Associates International
Delray Beach, Florida

1	**THE ART OF THE ENGINEER**	**701**
	1.1 Modeling for the Real World	701
	1.2 The Safety Factor	702
2	**PROFESSIONAL LIABILITY**	**703**
	2.1 Liability of an Employee	703
	2.2 Liability of a Business	707
3	**THE LAWS OF PRODUCT LIABILITY**	**710**
	3.1 Definition	710
	3.2 Negligence	710
	3.3 Strict Liability	711
	3.4 Express Warranty and Misrepresentation	711
4	**THE NATURE OF PRODUCT DEFECTS**	**712**
	4.1 Production or Manufacturing Flaws	712
	4.2 Design Flaws	713
	4.3 Instructions and Warnings	714
5	**UNCOVERING PRODUCT DEFECTS**	**714**
	5.1 Hazard Analysis	714
	5.2 Hazard Index	716
	5.3 Design Hierarchy	716
6	**DEFENSES TO PRODUCT LIABILITY**	**717**
	6.1 State of the Art	717
	6.2 The Role of Safety Standards	718
	6.3 Contributory/Comparative Negligence	719
	6.4 Assumption of the Risk	719
7	**RECALLS, RETROFITS, AND THE CONTINUING DUTY TO WARN**	**721**
	7.1 After-Market Hazard Recognition	721
	7.2 Types of Corrective Action	722
8	**DOCUMENTATION OF THE DESIGN PROCESS**	**723**
9	**A FINAL WORD**	**724**

1 THE ART OF THE ENGINEER

1.1 Modeling for the Real World

Engineers believe that they practice their craft in a world of certainty. Nothing could be further from the truth! Because this chapter deals with the interface between law and technology, and because products liability is likely to be the legal area of concern to the engineer, our principal focus is on the engineering (design) of products, or components of products.

Think for a moment about the usual way an engineer proceeds from a product concept to the resulting device. The engineer generally begins the design process with some type of specifications for the eventual device to meet, such as, performance parameters, functional capabilities, size, weight, cost, and so on. Implicit, if not explicit, in the specifications are assumptions about the device's ultimate interaction with the real world. If the specification concerns, for example, loading or power needs that the device is either to produce or to withstand, someone has created boundaries within which the product is to function. Clearly, there are bound to be some uncertainties, despite the specifying of precise values for the designers to meet. Even assuming that a given loading for a certain component is known with precision and repeatability, the design of the component more than likely will involve various assumptions: *how* the loading acts (e.g., point-load or distributed); *when* it acts (e.g., static or dynamic); *where* it acts (e.g., two or three dimensions); and *what* it acts on (e.g., how sophisticated an analysis technique to use).

The point is that even with sophisticated and powerful computational tools and techniques, the real world is always modeled into one that could be analyzed and, as a result, is truly artificial. That is, a measure of uncertainty will always exist in any result, whatever the computational power. The question that is often unanswered or ignored in the design process is: How *much* uncertainty is there about the subtleties and exigencies of the true behavior of the environment (including people) on the product and the uncertainties in our, yes, artificial modeling technique?

1.2 The Safety Factor

To mask the uncertainties and, frankly, to admit that, despite our avowal, the world from which we derive our design is not real but artificial, we incorporate a "safety factor." Truly, it should be viewed as a factor of ignorance. We use it in an attempt to reestablish the real world from the one we have modified and simplified by our assumptions, and to make it tractable, that is, so we can meet the product specifications. The function of the safety factor is to bridge the gap between the computational world and the one in which the product must actually function.

There are, in general, three considerations that are to be incorporated into the safety factor:

1. Uncertainties in material properties
2. Uncertainties in quality assurance
3. Uncertainties in the interaction of persons and the product—from the legal perspective, the most important of all

Example 1 Truck-Mounted Crane. Consider a truck-mounted crane, whose design specification is that it is to be capable of lifting 30 tons. The intent is, of course, that only under certain specific conditions, i.e., the boom angle, boom extension, rotational location of boom, etc., will the crane be able to lift 30 tons.

Inherent in the design, however, must be a safety factor cushion, not only to account for, e.g., the uncertainties in the yield stress of the steel, or the possibility of some welds not being full penetration during fabrication, but also for the uncertainties of the crane operator not knowing the precise weight of the load. In the real world, it is foreseeable that there will be times when no one on the job site knows, or has ready access to sufficient data to know, with reasonable certainty the weight of the load to be lifted.

The dilemma for the engineer–designer is how much latitude to allow in the load-lifting capability of the crane to accommodate uncertainty in the load weight. That is, the third component of the safety factor must reflect a realistic assessment of real-world uncertainties. The difficulty, of course, is that there are serious competing trade-offs to be considered in deciding on this element of the safety factor. For each percent above the 30-ton load specification that the engineer builds into the safety factor, the crane is likely to be heavier, larger, perhaps less maneuverable, etc. That is, the utility of the crane is likely to be increasingly compromised in one or more ways as the safety factor is increased.

Yet the engineer's creed requires that the product must function in its true environment of use and do so with reasonable safety and reliability. The art of the engineer, then, is to balance competing trade-offs in design decision-making to minimize the existence of hazards, while acknowledging and accounting for human frailties, reasonably foreseeable product uses and misuses, and the true environment of product use.

And that is what the law requires of the engineer as well. We will explore some of these considerations later in this chapter. But first, let's look at the issues of professional liability.

2 PROFESSIONAL LIABILITY

Whether engaged in research, development, manufacturing, engineering services, or technical consulting, today's engineer must be cognizant that the law imposes substantial accountability on both individual engineers and technology-related companies. Engineers can never expect to insulate themselves entirely from legal liability. However, their can limit their liability by maintaining a fundamental understanding of the legal concepts they are likely to encounter in the course of their career, such as professional negligence, agency, employment agreements, intellectual property rights, contractual obligations, and liability insurance.

2.1 Liability of an Employee

Negligence and the Standard of Care
A lawsuit begins when a person (corporations, as well, are considered as "persons" for legal purposes) whose body or property is injured or damaged alleges that the injury was caused by the acts of another and files a complaint. The person asserting the complaint is the *plaintiff;* the person against whom the complaint is brought is the *defendant*.

In the complaint, the plaintiff must state a *cause of action* (a legal theory or principle) that would, if proven to the satisfaction of the jury, permit the plaintiff to recover damages. If the cause of action asserted is *negligence,* then the plaintiff must prove, first, that the defendant owed the plaintiff a *duty* (i.e., had a responsibility toward the plaintiff). Then the plaintiff must show that the defendant *breached* that duty and, consequently, that the breach of duty by the defendant was the *cause* of the plaintiff's injury.

The doctrine of negligence rests on the duty of every person to exercise due care in his or her conduct toward others. A breach of this duty of care that results in injury to persons or property may result in a *tort* claim, which is a civil wrong (as opposed to a criminal wrong) for which the legal system compensates the successful plaintiff by awarding money damages. To make out a cause of action in negligence, it is not necessary for the plaintiff to establish that the defendant either intended harm or acted recklessly in bringing about the harm. Rather, the plaintiff must show that the defendant's actions fell below the *standard of care* established by law.

In general, the standard of care that must be exercised is that conduct which the *average reasonable person* of ordinary prudence would follow under the same or similar circumstances. The standard of care is an external and objective one, and has nothing to do with individual subjective judgment, though higher duties may be imposed by specific statutory provisions or by reason of special knowledge.

Example 2   Negligent or Not? Suppose a person is running down the street knocking people aside and causing injuries. Is this person breaching the duty to care to society and acting negligently? To determine this, we need to undertake a risk/utility analysis, i.e., does the utility of the action outweigh the harm caused?

If this person is running to catch the last bus to work, then the risk to others as he pushed them aside probably outweighs the utility to this person, which is getting to work on time. However, if the person has seen a knife-wielding assailant attacking someone and is trying to reach the policeman on the corner, then the utility (saving human life) is great. In such a case, perhaps society should allow the possible harm caused to others as he pushed them aside to reach the police officer and thus not find the person negligent, even though others persons were injured in the attempt to reach the police officer.

No duty is imposed on a person to take precautions against events that cannot reasonably be foreseen. However, the professional must utilize such superior judgment, skill, and knowledge as he or she actually possesses. Thus, the professional mechanical engineer might be held liable for miscalculating the load-lifting capability in the previous crane example, while a general engineering technician might not.

The duty to exercise reasonable care and avoid negligence does not mean that engineers guarantee the results of their professional efforts. Indeed, if an engineer can show that everything a reasonably prudent engineer might do was, in fact, done correctly, then liability cannot attach, even though there was a failure of some kind that caused injury.

Example 3 Collapse of a Reasonably Designed Overpass. A highway overpass, when designed, utilized all of the acceptable analysis techniques and incorporated all of the features that were considered to be appropriate for earthquake resistance at that time. Years later, the overpass collapses when subjected to an earthquake of moderate intensity. At the time of the collapse, there are newer techniques and features that, in all likelihood, would have prevented the collapse had they been incorporated into the design.

It is unlikely that liability would attach to the engineers who created the original design and specifications as long as they utilized techniques that were reasonable at that time.

Additionally, liability depends on a showing that the negligence of the engineer was the direct and proximate cause of the damages. If it can be shown that there were other superseding causes responsible for the damages, the engineer may escape liability even though his/her actions deviated from professional standards.

Example 4 Collapse of a Negligently Designed Overpass. Suppose, instead, that after the collapse of the overpass in the preceding example, a review of the original analysis conducted by the engineers reveals several deficiencies in critical specifications that reasonably prudent engineers would not have overlooked. However, the intensity of the earthquake was of such a magnitude that, with reasonable certainty, the overpass would have collapsed even if it had been designed using the appropriate specifications. The engineers, in this scenario, are likely to escape liability.

However, the law does allow "joint and severable" liability against multiple parties who either act in concert or independently to cause injury to a plaintiff. Other defenses to an allegation of negligence include the "state of the art" argument, contributory/comparative negligence, and assumption of the risk. These are discussed in Section 6.

An employer is generally liable for the negligence, carelessness, errors, and omissions of its employees. However, as we see in the next section, liability may attach to the engineer employee under the law of agency.

Agency and Authority
Agency is generally defined as the relationship that arises when one person (the principal) manifests an intention that another person (the agent) shall act on his behalf. A principal may appoint an agent to do any act except an act that, by its nature, or by contract, requires personal performance by the principal. An engineer employee may act as an agent of his employer, just as an engineering consultant may act as an agent of her client.

The agent, of course, has whatever duties are expressly stated in the contract with the principal. Additionally, in the absence of anything contrary in the agreement, the agent has three major duties implied by law:

1. The fiduciary duty of an agent to his principal is one of undivided loyalty, e.g., no self-dealing or obtaining secret profits;
2. An agent must obey all reasonable directions of the principal; and
3. An agent owes a duty to the principal to carry out his duties with reasonable care, in light of local community standards and taking into account any special skills of the agent.

Just as the agent has duties, the principal owes the agent a duty to compensate the agent reasonably for his services, indemnify the agent for all expenses or losses reasonably incurred in discharging any authorized duties, and, of course, to comply with the terms of any contract with the agent.

With regard to tort liability in the context of the employer–employee relationship, an employer is liable only for those torts committed by an employee; he is not generally liable for torts committed by an agent functioning as an independent contractor. An example of an employee is one who works full-time for his employer, is compensated on a time basis, and is subject to the supervision of the principal in the details of his work. An example of an independent contractor is one who has a calling of her own, is hired to perform a particular job, is paid a given amount for that job, and who followed her own discretion in carrying out the job. Engineering consultants are usually considered to be independent contractors.

Even when the employer–employee relationship is established, however, the employer is not liable for the torts of an employee unless the employee was acting within the scope of, or incidental to, the employer's business. Additionally, the employer is usually not liable for the intentional torts of an employee on the simple ground that an intentional tort (i.e., fraud) is clearly outside the scope of employment. However, where the employee intentionally chooses a wrongful means to promote the employer's business, such as fraud or misrepresentation, the employer may be held liable.

With regard to contractual liability under the law of agency, a principal will be bound on a contract that an agent enters into on his behalf if that agent has *actual authority,* i.e., authority expressly or implicitly contained within the agency agreement. The agent cannot be held liable to the principal for breach since he acted within the scope of his authority. To ensure knowledge of actual authority, the engineer should always obtain clear, written evidence of his job description, duties, responsibilities, "sign-off" authority, and so on.

Even where employment or agency actually exists, unless it is unequivocally clear that the individual engineer is acting on behalf of an employer or other disclosed principal, an injured third party has the right to proceed against either the engineer or the employer/principal or both under the rule that an agent for an undisclosed or partially disclosed principal is liable on the transaction together with her principal. Thus, engineers acting as employees or agents should always include their title, authority, and the name of the employer/principal when signing any contract or business document.

Even if the agent lacks actual authority, the principal can still be held liable on contracts entered into on his behalf if the agent had *apparent authority,* that is, where a third party reasonably believed, based on the circumstances, that the agent possessed actual authority to perform the acts in question. In this case, however, the agent may be held liable for losses incurred by the principal for unauthorized acts conducted outside the scope of the agent's authority.

Employment Agreements

Rather than relying entirely on the law of agency to control the employer–employee relationship, most employers require engineers to sign a variety of employment agreements as a condition of employment. These agreements are generally valid and legally enforceable to the extent that they are reasonable in duration and scope.

A clause typically found in an engineer's employment contract is the agreement of the employee to transfer the entire right, title, and interest in and to all ideas, innovations, and creations to the company. These generally include designs, developments, inventions, improvements, trade secrets, discoveries, writings, and other works, including software, databases, and other computer-related products and processes. As long as the work is within the scope of the company's business, research, or investigation, or the work resulted from or is suggested by any of the work performed for the company, its ownership is required to be assigned to the company.

Another common employment agreement is a noncompetition provision whereby the engineer agrees not to compete during his or her employment by the company and for some period after leaving the company's employ. These are also enforceable as long as the scope of the exclusion is reasonable in time and distance, when taking the nature of the product or service into account and the relative status of the employee. For example, courts would likely find invalid a two-year, nationwide noncompetition agreement against a junior CAD/CAM engineer in a small company; however, this agreement might be found fully enforceable against the chief design engineer of a large aircraft manufacturer. In any case, engineers should inform new/prospective employers of any prior employment agreement that is still in effect.

As will be seen in the next section, however, even if an employment agreement was not executed, ex-employees are not free to disclose or utilize proprietary information gained from their previous employers.

Intellectual Property

A *patent* is a legally recognized and enforceable property right for the exclusive use, manufacture, or sale of an invention by its inventor (or heirs or assignees) for a limited period of time that is granted by the government. In the United States, exclusive control of the invention is granted for a period of 20 years from the date of filing the patent, and in consideration for which the right to free and unrestricted use passes to the general public. Patents may be granted to one or more individuals for *new* and *useful* processes, machines, manufacturing techniques, and materials, including improvements that are not obvious to one skilled in the particular art. The inventor, in turn, may license, sell, or assign patent rights to a third party. Remedies against patent infringers include monetary damages and injunctions against further infringement.

Engineers working with potentially patentable technology must follow certain formalities in the documentation and publication of information relating to the technology in order to preserve patent protection. Conversely, engineers or companies considering marketing a newly developed product or technology should have a patentability search conducted to ensure that they are not infringing existing patents.

Many companies rely on *trade secrets* to protect their technical processes and products. A trade secret is any information, design, device, process, composition, technique, or formula that is not known generally, and that affords its owner a competitive business advantage. Advantages of trade secret protection include avoiding the cost and effort involved in patenting, and the possibility of perpetual protection. The main disadvantage of a trade secret is that protection vanishes when the public is able to discover the "secret," whether by inspection, analysis, or reverse engineering. Trade secret protection thus lends itself more readily to intangible "know-how" than to end products.

Trade secrets have legal status and are protected by state common law. In some states, the illegal disclosure of trade secrets is classified as fraud, and employees can be fined or even jailed for such activity. Customer lists, supplier's identities, equipment, and plant layouts cannot be patented, yet they can be important in the conduct of a business and therefore are candidates for protection as trade secrets.

2.2 Liability of a Business

Negligence for Services

Negligence (as defined in Section 2.1) and standards of care apply not only to individual engineers, but also to consulting and engineering firms. At least one State Supreme Court has defined the standard of care for engineering services as follows:

> In performing professional services for a client, an engineer has the duty to have that degree of learning and skill ordinarily possessed by reputable engineers, practicing in the same or a similar locality and under similar circumstances. It is his further duty to use the care and skill ordinarily used in like cases by reputable members of his profession practicing in the same or a similar locality, under similar circumstances, and to use reasonable diligence and his best judgment in the exercise of his professional skills and in the application of his learning, in an effort to accomplish the purpose for which he was employed.*

Occasionally, an engineer's duty to the general public may supersede the duty to her client. For example, an engineer retained to investigate the integrity of a building, and who determined the building was at imminent risk of collapse, would have a duty to warn the occupants even if the owner requested that the engineer treat the results of the investigation as confidential.†

The engineer also has a duty to adhere to applicable state and federal safety requirements. For example, the U.S. Department of Labor Occupational Safety and Health Administration has established safety and health standards for subjects ranging from the required thickness of a worker's hardhat to the maximum decibel noise level in a plant. In many jurisdictions, the violation of a safety code, standard, or statute that results in injury is "negligence per se," that is, a conclusive presumption of duty and breach of duty. Engineers should be aware, however, that the reverse of this rule does not hold true: compliance with required safety standards does not necessarily establish reasonable care. See Section 6.2 for the role of safety standards.

Contractual Obligations

A viable contract, whether it be a simple purchase order to a vendor, or a complex joint venture, requires the development of a working agreement that is mutually acceptable to both parties. An agreement (contract) binds each of the parties to do something or perhaps even refrain from doing something. As part of such an agreement, each of the parties acquires

---

*Clark v. City of Seward, 659 P.2d 1227 (Alaska, 1983).
†California Attorney General's Opinion, Opinion No. 85-208 (1985).

a legally enforceable right to the fulfillment of the promises made by the other. Breach of the contract may result in a court awarding damages for losses sustained by the nonbreaching party, or requiring "specific performance" of the contract by the breaching party.

An oral contract can constitute just as binding a commitment as a written contract, although, by statute, some types of contracts are required to be in writing. As a practical matter, agreements of any importance should always be, and generally are, reduced to writing. However, a contract may also be created by implication based on the conduct of one party toward another.

In general, a contract must embody certain key elements, including (a) mutual assent as consisting of an offer and its acceptance between competent parties based on (b) valid consideration for a (c) lawful purpose or object in (d) clear-cut terms. In the absence of any one of these elements, a contract will generally not exist and hence will not be enforceable in a court of law.

Mutual assent is often referred to as a "meeting of the minds." The process by which parties reach this meeting of the minds generally is some form of negotiation, during which, at some point, one party makes a proposal (offer) and the other agrees to it (acceptance). A counteroffer has the same effect as a rejection of the original offer.

To have a legally enforceable contract, there must generally be a bargained-for exchange of "consideration" between the parties, that is, a benefit received by the promisor or a detriment incurred by the promisee. The element of bargain assures that, at least when the contract is formed, both parties see an advantage in contracting for the anticipated performance.

If the subject matter of a contract (either the consideration or the object of a contract) is illegal, then the contract is void and unenforceable. Generally, illegal agreements are classified as such either because they are expressly prohibited by law (e.g., contracts in restraint of trade), or because they violate public policy (e.g., contracts to defraud others).

Problems with contracts can occur when the contract terms are incomplete, ambiguous or susceptible to more than one interpretation, or where there are contemporaneous conflicting agreements. In these cases, courts may allow other oral or written evidence to vary the terms of the contract.

A party that breaches a contract may be liable to the nonbreaching party for "expectation" damages, that is, sufficient damages to buy substitute performance. The breaching party may also be liable for any reasonably foreseeable consequential damages resulting from the breach.

Contract law generally permits claims to be made under a contract only by those who are "in privity," that is, those parties among whom a contractual relationship actually exists. However, when a third party is an intended beneficiary of the contract, or when contractual rights or duties have been transferred to a third party, then that third party may also have certain legally enforceable rights.

The same act can be, and very often is, both negligent and a breach of contract. In fact, negligence in the nature of malpractice alleged by a client against an engineering firm will almost invariably constitute a breach of contract as well as negligence, since the engineer, by contracting with the client, undertakes to comply with the standard of practice employed by average local engineers. If the condition is not expressed, it is generally implied by the courts.

Insurance for Engineers

It is customary for most businesses, and some individual engineers, to carry comprehensive liability insurance. The insurance industry recognizes that engineers, because of their occupation, are susceptible to special risks of liability. Therefore, when a carrier issues a comprehensive liability policy to an engineering consultant or firm, it may exclude from the insurance afforded by the policy the risk of professional negligence, malpractice, and "errors and omissions." The engineer should seek independent advice on the extent and type of the coverage being offered before accepting coverage. However, depending on the wording of the policy and the specific nature of the claim, the comprehensive liability carrier may be under a duty to defend an action against the insured and sometimes must also pay the loss. When a claim is made against an insured engineering consultant or firm, the consultant or firm should retain a competent attorney to review the policy prior to accepting the conclusions of the insurance agent as to the absence of coverage.

While the engineer employee of a well-insured firm probably has limited liability exposure, the professional engineering consultant should be covered by professional liability (malpractice) insurance. However, many engineers decide to forego malpractice insurance because of high premium rates. Claims may be infrequent, but can be economically devastating when incurred. The proper amount of coverage is something that should be worked out with a competent underwriter, and will vary by engineering discipline and type of work. A policy should be chosen that not only pays damages, but also underwrites the costs of attorney's fees, expert witnesses, and so on.

*Case Study**

The following case serves to illustrate the importance of developing a fundamental understanding of the professional liability concepts discussed above.

S&W Engineering was retained by Chesapeake Paper Products to provide engineering services in connection with the expansion of Chesapeake's paper mill. S&W's vice president met with Chesapeake's project manager, and provided him with a proposed engineering contract and price quotations. Several weeks later, Chesapeake's project manager verbally authorized S&W to proceed with the work. S&W's engineering contract was never signed by Chesapeake; instead, Chesapeake sent S&W a Purchase Order (P.O.) that authorized engineering services "in accordance with the terms and conditions" of S&W's engineering contract. However, Chesapeake's P.O. also contained language in smaller print stating "This order may be accepted only upon the terms and conditions specified above and on the reverse side."

The drawings supplied by S&W to Chesapeake's general contractor subsequently contained errors and omissions, resulting in delays and increased costs to Chesapeake. Chesapeake sued S&W for breach of contract, arguing that the purchase order issued by Chesapeake constituted the parties' contract, and that this P.O. contained a clause requiring S&W's standard of care to be "free from defects in workmanship." Additionally, another P.O. clause required indemnification of all expenses "which might incur as a result of the agreement."

S&W agreed that its engineering drawings had contained some inconsistencies but denied that those errors constituted a breach of contract. S&W claimed that the parties' contract consisted of the terms in its proposed Engineering Contract it had delivered to Chesapeake at the outset of the Project. S&W's Engineering Contract provided that the "Engineer shall provide detailed engineering services . . . conforming with good engineering practice." S&W's proposed contract also contained a clause precluding the recovery of any consequential damages.

At a jury trial, 14 witnesses testified, and the parties introduced more than 1000 exhibits. The jury found that the parties' "operative contract" was the P.O., and that S&W's services

---

*Chesapeake Paper Products v. Stone & Webster Engineering, No. 94-1617 (4th Cir. 1995).

did not meet the contractually required standard of care. Chesapeake was awarded $4,665,642 in damages.

(Chesapeake Paper Products v. Stone & Webster Engineering, No. 94-1617 (4th Cir. 1995)).

3 THE LAWS OF PRODUCT LIABILITY

3.1 Definition

In Section 1, the art of engineering was characterized as a progression from real-world product specifications to the world modified by assumptions. This assumed world permits establishing precise component design parameters. Finally, the engineer must attempt to return to the real world by using a "safety factor" to bridge the gap between the ideal, but artificial, world of precise design calculations to the real world of uncertainties in who, how, and where the product will actually function.

The laws of product liability sharpen and intensify this focus on product behavior in the real world. *Product liability* is the descriptive term for a legal action brought by an injured person (the plaintiff) against another party (the defendant) alleging that a product sold (or manufactured or assembled) by the defendant was in a substandard condition, and that this substandard condition was a principal factor in causing the harm of the plaintiff.

The key phrase for the engineer is *substandard condition*. In legal parlance, this means that the product is alleged to contain a *defect*. During litigation, the product is put on trial so that the jury can decide whether the product contained a defect and, if so, whether the defect caused the injury.

The laws of product liability take a retrospective look at the product and how it functioned as it interacted with the persons who used it within the environment surrounding the product and the persons. Three legal principles generally govern the considerations brought to this retrospective look at the engineer's art:

1. Negligence
2. Strict liability
3. Express warranty and misrepresentation

3.2 Negligence

This principal is based on the conduct or fault of the parties, as discussed in Section 2.1. From the plaintiff's point of view, it asks two things: first, whether the defendant acted as a *reasonable person* (or company) in producing and selling the product in the condition in which it was sold, and second, if not, was the condition of the product a substantial factor in causing the plaintiff's injury.

The test of *reasonableness* is to ask what risks the defendant (i.e., designer, manufacturer, assembler, or seller) foresaw as reasonably occurring when the product was used by the expected population of users within the actual environment of use. Obviously, the plaintiff argues that if the defendant had acted reasonably, the product designer would have foreseen the risk actually faced by the plaintiff and would have eliminated it during the design phase and before the product was marketed. That is, the argument is that the defendant, in ignoring or not accounting for this risk in the design of the product, did not properly balance the risks to product users against the utility of the product to society.

It is the *reasonableness,* or lack thereof, of *the defendant's behavior* (in designing, manufacturing, or marketing the product or in communicating to the user through instructions and warnings) that is the question under the principle of negligence. These issues are fully discussed in Section 5.

3.3 Strict Liability

In contrast to negligence, strict liability ignores the defendant's behavior. It is, at least in theory, of no consequence whether the manufacturer behaved reasonably in designing, manufacturing, and marketing the product. The only concern here is the quality of the product as it actually functions in society.

Essentially, the question to be resolved by the jury under strict liability is whether the risks associated with the real-world use of the product by the expected user population exceed the utility of the product and, if so, whether there was a reasonable alternative to the design that would have reduced the risks without seriously impairing the product's utility or making it unduly expensive. If the jury decides that the risks outweighed the product's utility and a reasonable alternative to reducing the risk existed, then the product is judged to be in a *defective condition unreasonably dangerous*.

Under strict liability, a product is defective when it contains *unreasonable* dangers, and only unreasonable dangers in the product can trigger liability. While it is unlikely the marketing department will ever use the phrase in a promotion campaign, a product may contain *reasonable* dangers without liability. In the eyes of the law, a product whose only dangers are reasonable ones is *not* defective.

Stated positively, a product that does not contain unreasonable dangers is "reasonably safe"—and that is all the law requires. This means that any residual risks associated with the product have been transferred *appropriately* to the ultimate user of the product.

Section 5 discusses the methodology for uncovering unreasonable dangers associated with products.

3.4 Express Warranty and Misrepresentation

The third basic legal principle governing possible liability has nothing to do with either the manufacturer's conduct (negligence) or the quality of the product (strict liability). Express warranty and misrepresentation are concerned only with what is communicated to the potential buyer that becomes part of the "basis of the bargain."

An express warranty is created whenever any type of communication to the potential buyer describes some type of *objectively measurable* characteristic of the product. The following are some sample express warranties:

- This truck will last 10 years.
- This glass is shatterproof.
- This automatic grinder will produce 10,000 blades per hour.
- This transmission tower will withstand the maximum wind velocities and ice loads in your area.

If such a communication is, first, at least a part of the reason that the product was purchased and then, if reasonably foreseeable circumstances ultimately prove the communication invalid, there has been misrepresentation, and the buyer is entitled to recover damages consistent with the failed promise. It doesn't matter one whit if the product cannot possibly live up to the promise. This is not the issue. It is the failure to keep a promise that becomes part of the basis of the bargain, and that the buyer did not have sufficient expertise for not believing the promise that can trigger the liability.

Someone with a legal bent might argue, against the misrepresentation claim, that the back of the sales form clearly and unequivocally disclaims all liability arising from any warranties not contained in the sales document (i.e., the contract). The courts, when confronted with what appears to be a conflict between the express warranty communicated to the buyer and the fine print on the back of the document disclaiming everything, inevitably side with the buyer who believed the express warranty to the extent that it became a part of the "basis of the bargain."

The communications creating the express warranty can be in any form: verbal, written, visual or any combination of these. In the old days, courts used to view advertising as mere puffing and rarely sided with the buyer arguing about exaggerated claims made about the product. In recent years, however, the courts have acknowledged that buying is engendered in large part by media representations. Now, when such representations can be readily construed as express warranties, the buyer's claim is likely to be upheld. It should also be noted that misrepresentation claims have been upheld when both the plaintiff and the defendant are sophisticated, have staffs of engineers and lawyers, and the dealings between the parties are characterized as "arm's length."

In precarious economic times, the exuberance of salespersons, in their quest to make the sale, may oversell the product and create express warranties that the engineer cannot meet. This can then trigger liability, despite the engineer's best efforts.

Because it is so easy to create, albeit unintentionally, an express warranty, all departments that deal in any way with a product must recognize this potential problem and structure methods and procedures to minimize its occurrence. The means that engineering, manufacturing, sales, marketing, customer service, and upper management must create a climate in which there is agreement among the appropriate entities that what is being promised to the buyer can actually be delivered.

4 THE NATURE OF PRODUCT DEFECTS

The law recognizes four areas that can create a "defective condition unreasonably dangerous to the user or consumer":

1. Production or manufacturing
2. Design
3. Instructions
4. Warnings

4.1 Production or Manufacturing Flaws

A production or manufacturing defect can arise when the product fails to emerge as the manufacturer intended. The totalities of the specifications, tolerances, and so on, define the product and all of the bits and pieces that make it up, and collectively they prescribe the manufacturer's intent for exactly how the product is to emerge from the production line.

If there is a deviation from any of these defining characteristics of the product (e.g., specifications, tolerances, etc.), then there exists a production or manufacturing flaw. If this flaw or deviation can cause the product to fail or malfunction under reasonably foreseeable conditions of use in a way that can cause personal injury and/or property damage, and these conditions are within the expected performance requirements for the product, then the product is defective.

What is important to note here is that the deviation from the specifications must be *serious* enough to be able to precipitate the failure or malfunction of the product within the foreseeable uses and performance envelope of the product. To illustrate, let's return to the crane described in the first section of this chapter.

Example 5 Truck–Crane—Flaw or Defect? Suppose that a critical weld is specified to be 4 inches in length and to have full penetration. After a failure, the crane is examined and the weld is full penetration but only $3\frac{1}{2}$ in. long, which escaped the quality inspectors. There is a deviation or flaw. However, whether this flaw rises to the level of defect depends on several considerations: First, what safety factor considerations entered into the design of the weld? It may be that the designer calculated the necessary weld length to be 3 in. and specified 4 in. to account for the uncertainties described in Section 1. Next, if it can be shown by the crane manufacturer that a $3\frac{1}{2}$-in. weld was adequate for all reasonably foreseeable use conditions of the crane, then it could be argued that the failure was due to crane misuse and not due to the manufacturing flaw.

Alternatively, the plaintiff could argue that the engineer's assumptions as to the magnitude of the safety factor did not realistically assess the uncertainty of the weight loads to be lifted; if they had done so, the minimum acceptable length would have been the 4 in. actually specified.

While this is a hypothetical example, it illustrates the interplay of several important elements that must be considered when deciding if a production flaw can rise to the level of a defect. Foreseeable uses and misuses of the product, and the prescribed or implicit performance requirements are two of the most important.

4.2 Design Flaws

The standard for measuring the existence of a production flaw is simple. One needs only to compare the product's attributes as it actually leaves the production line with what the manufacturer intended them to be, by examining the manufacturer's internal documents that prescribe the entire product.

To uncover a design flaw, however, requires comparing the correctly manufactured product with a standard that is not readily prescribed and is significantly more complex. The standard is a societal one in which the risks of the product are balanced against its utility to establish whether the product contains unreasonable dangers. If there are unreasonable dangers, then the design flaw becomes a defect.

In the crane example, assume that there has been a boom failure and that the crane met all of the manufacturer's specifications, that is, no manufacturing defect is alleged. The plaintiff alleges, instead, that if the boom had been fabricated from a heavier gauge as well as a stronger alloy steel, the collapse would have been avoided. The plaintiff's contention can be considered a design flaw. There is no question that the boom could have been fabricated using the plaintiff's specifications and, for the sake of our discussion, we will also assume the boom would not have failed using the different material.

The critical question, however, is should the boom have been designed that way? The answer is, only if the original design created unreasonable dangers. The existence of unreasonable dangers, therefore a defective condition, can be deduced from a risk/utility analysis of the interaction of crane uses, users and the environments within which the crane is expected to function.

The analysis must consider, first, the foreseeability of crane loads of uncertain magnitude that could cause the original design to fail, but not the modified design. Balanced against that consideration will be a reduction in the utility of the crane because of its increased weight and/or size if the proposed design alterations are incorporated. There will be also an increased cost. It is this analysis of competing trade-offs that the designer must consider before deciding on the proposed design specifications. Fundamentally, though, as in the discussion of a production defect, the consideration is that of the safety factor, bridging the gap between *assumed* product function and *actual* product function.

4.3 Instructions and Warnings

A product can be perfectly manufactured from a design that contains no unreasonable dangers and yet be defective because of inadequate instructions or warnings. Instructions are the communications between the manufacturer and the user that describe how the product is to be used to achieve the intended function of the product.

Warnings are to communicate any residual hazards, the consequent risks of injury and the actions the user must take to avoid injury. If the warnings are inadequate, the product can be defective even if the design, manufacturing, and instructions meet the legal tests.

While the courts have not given clear or unequivocal guidelines for assessing the adequacy of instructions and warnings, there are several basic considerations that are drawn from litigated cases that must underlie their development to meet the test of adequacy:

- They must be understood by the expected user population.
- They must be effective in a multilingual population.
- There must be some reasonable and objective evidence to prove that the warnings and instructions can be understood and are likely to be effective.

Simply put, writing instructions and warnings is easy. However, gathering evidence to support the contention that they are *adequate* can be extremely difficult, costly, and time-consuming.

To do this means surveying the actual user population and describing those characteristics that are likely to govern comprehension, such as age, education, reading capability, sex, and cultural and ethnic background. Then a statistically selected, random sample of the identified user population must be chosen to test the communication for comprehension, using the method suggested in the American National Standards Institute standard ANSI Z535.3. Finally, the whole process must be documented. Then, and only then, can a manufacturer argue that the user communications, that is, instructions and warnings, are adequate.

5 UNCOVERING PRODUCT DEFECTS

5.1 Hazard Analysis

In the preceding section, a risk/utility analysis was described as a basis for assessing whether or not the product was in a defective condition unreasonably dangerous. Consider now the methodology and the process of the risk/utility analysis.

We begin with a disclaimer: Neither the process nor the methodology about to be discussed is readily quantifiable. However, this fact does not lessen their importance; it only emphasizes the care that must be exercised.

The process is one of scenario-building. The first step is to characterize, as accurately as possible, the users of the product, the ways in which they will use the product, and the environment in which they will use it. These elements must be quantified as much as possible.

Example 6   Foreseeable Users of a Hand-Held Tool. Will the user population comprise younger users, female users, elderly users? If so, these populations are likely to need special ergonomic or human factors considerations in the design of handgrips, operating controls, etc. Will the tool be found in the home? If so, inadvertent use by small children is likely to be a consideration in designing the controls. Certainly the ability to read and understand instructions and warning must be a significant element of the characterization of the users.

In the best of all worlds, the only product uses the engineer would be concerned with are the *intended* uses. Unfortunately, the law requires that the product design acknowledge and account for *reasonably foreseeable misuses* of the product. Of all the concepts the engineer must deal with, this one is perhaps the hardest to analyze and the most difficult to accept. Part of the reason, of course, is the difficulty of distinguishing between uses that are *reasonably foreseeable* and those uses that the manufacturer can argue are truly *misuse* for which no account must be taken in design.

The concept of legal unforeseeability is a difficult one. Many people might think that if they have ever talked about the possibility of misusing a product in a certain way, then they have "foreseen" that misuse, and therefore must account for it in their design. This is not the case. Legally, *unforeseeable misuse* means a use so egregious, or so bizarre, or so remote that it is termed *unforeseeable,* even when such a misuse has been a topic of discussion.

A simple illustration might help.

Example 7   How Many Ways Can You Use a Screwdriver? There is no question that the intended purpose and function of a screwdriver is to insert and remove screws. This means that, ideally, the shank of a screwdriver is subjected only to a twisting motion, or torque.

But how do most people open paint cans? With a screwdriver, of course. In that context, however, the shank is subjected to a bending moment, not a torque. Any manufacturer who produced and marketed a screwdriver with shank material able to withstand high torque, but without sufficient bending resistance to open a paint can without shattering would have a difficult time avoiding liability for any injuries that occurred.

The reason, of course, is that using a screwdriver to open paint cans would be considered as a reasonably foreseeable misuse, and should be accounted for in the design. On the other hand, suppose someone uses a screwdriver as a cold chisel to loosen a rusted nut and the screwdriver shatters, causing injury. The manufacturer could argue that such a use was a misuse for which the manufacturer had no duty to account for in the design.

Finding the line that separates the misuses the engineer must account for from the misuses that are legally unforeseeable is not easy, nor is the line a precise one. All that is required, however, is for the engineer to show the reasonableness of the process of how the line was ultimately decided, while attempting to meet competing trade-offs in selecting the product's specifications. Unquestionably, we can always imagine all types of bizarre situations in which a product is misused and someone is injured. Does this mean that all such situations must somehow be accounted for in design? Of course not. But what is required is to make a reasonable attempt to separate user behavior into two categories: that which can reasonably be accounted for in design and that which is beyond reasonable considerations.

The third element in the risk/utility process is the environment within which the user and product interact. If it is cold, how cold? If it is hot, how hot? Will it be dark, making warnings and instructions difficult to read? Will the product be used near water? If so, both fresh and salt? How long will the product last? Will it be repainted, scraped, worn, and so on? These, too, would be considerations in warning adequately.

The scenario building must integrate the three elements of the hazard analysis: the users, the uses, and the environment. By asking "What if . . . ?", a series of hazards can be postulated from integrating the users with the uses within an environment.

Example 8 "What If an Air-operated Sander . . . ?" What if an air-operated sander is used in a marine environment? What if the user inadvertently drops it overboard and then continues to use it, without having it disassembled and cleaned? What hazards could arise? Could corrosion ultimately freeze the control valve continually open, leading to loss of control at some future time, long after the event in question?

5.2 Hazard Index

After completing the hazard analyses, the hazards should be rank-ordered from the most serious to the least serious. One way to do this is to assign a numerical probability of the event occurring and then to assess, also using a numerical scale, the seriousness of the harm. The product of these two numbers is the *Hazard Index* and permits a relative ranking of the hazards. The scales chosen to provide some measure of probability and seriousness should be limited; the scale may run, for example, from 0 to 4. A 0 implies that the event is so unlikely to occur, or that the resulting harm is so minimal, as to be negligible. Correspondingly, a 4 would mean that an event was almost certain to occur, or that the result would be death or serious irreparable injury. With this scale, the hazard index could range from 0 to 16.

Once this is done, attention is then focused on the most serious hazards, eventually working down to the least serious one.

5.3 Design Hierarchy

Ideally, for each such event, the objective would be, first, to "design out" the hazard. If a hazard can be designed out, it can never return to cause harm.

Failing the ability of design out the hazard, the next consideration must be guarding. Can an unobtrusive barrier be placed between the user and the hazard? It must be noted that if a guarding configuration greatly impairs the utility of the product, or greatly increases the time needed to carry out the product's intended function, it is likely to be removed. In such a case, the user is not protected from the hazard, nor is the manufacturer likely to be protected from liability if an injury results, because removing an obtrusive guard may be considered a foreseeable misuse.

If the hazard cannot be designed out, nor can an effective guard be devised, then *and only then* should the last element of the design hierarchy be considered: a warning. A warning must be viewed as an element of the design process, not as an afterthought. To be perfectly candid, if the engineer has to resort to a warning to minimize or eliminate a risk of injury from that hazard, it may be an admission of a failure in the design process.

Yet, there are innumerable instances where a warning must be given. Section 4 described the considerations necessary to develop an adequate warning, the legal standard. What was not described there are the three necessary elements that must be included before the process of establishing adequacy begins:

1. The nature of the hazard
2. The potential magnitude of the injury
3. The action to be taken to avoid the injury

A warning paraphrased from an aerosol can of hair spray provides an exercise for the reader:

Warning
- Harmful vapors
- Inhalation may cause death or blindness
- Use in a well-ventilated area

The reader should analyze these three phrases carefully and critically, then describe the user populations to which the warning might apply. Then answer the question of whether or not it is likely that injury could be avoided by that user population. Suppose that a foreseeable portion of the population using this aerosol can are people whose English reading ability is at the 3rd or 4th grade level. (It is estimated that about half of the English-speaking Americans cannot read beyond the 4th grade level.) What can you conclude about comprehension and the ability to avoid injury?

Warnings are, in fact, the most difficult way to minimize or eliminate hazards to users.

6 DEFENSES TO PRODUCT LIABILITY

Up to now, we have looked only at the factors that permit an analysis of whether the product contains a defect, i.e., an unreasonable danger. Certainly the ultimate defense to an allegation that the product was defective is to show through a risk/utility analysis that, on balance, the product's utility outweighs its risks and, in addition, there were no feasible alternatives to the present design. It may be, however, that the plaintiff's suggested design alternative is, in fact, viable as of the time the incident occurred. Is there any analysis that could offer a defense? There may be, by considering a *state-of-the-art* argument.

6.1 State of the Art

Decades ago, the phrase *state-of-the-art* meant, simply, what the custom and practice was of the particular industry in question. Because of the concern that an entire industry could delay introduction of newer, safer designs by relying on the "custom and practice" argument to defeat a claim of negligence, the courts have adopted a broader definition of the term. The definition today is "what is both technologically and economically feasible." The time at which this analysis is performed is, in general, the date the product in question was manufactured. Thus, while a plaintiff's suggested alternative design may have been technologically and economically feasible at the time the incident occurred, their argument may not be viable if the product was manufactured 10 years before the incident occurred.

To make that argument convincing, however, means that engineers must always be actively seeking new and emerging technology, looking to its potential applicability to their industry and products. It is expected, too, that technological advances are sought, not only in the engineer's own industry, but in related and allied fields as well. Keeping current has an added dimension, that of being alert to broader vistas of technological change outside one's own industry.

The second element of today's state-of-the-art principle is that innovative advances must be economically viable as well. It is generally, but incorrectly, assumed that the term *economic viability* is limited to the incremental cost of incorporating the technological advance into the product and how it will affect the direct cost of manufacturing, and the subsequent profit margin. The courts, however, are concerned with another cost in measuring economic viability, in addition to the direct of incorporating a safety improvement in the product: the cost to society and ultimately to the manufacturer if the technological advance is *not* incorporated into the product and injuries occur as a result. The technological advances we are concerned with here are those that are likely to enhance safety.

While it is more difficult and certainly cannot be predicted with a great deal of precision, an estimate of costs of the probable harm to product users is part of the equation. An approach to this analysis was described in Section 5. Estimating both the probability and seriousness of the harm from a realistic vantage point if the technological advance is *not* incorporated can form the basis for estimating the downside risk of not including the design feature.

6.2 The Role of Safety Standards

Very often, the plaintiff's expert can argue, and with credibility we will assume, that a design modification or additional guard would have prevented the plaintiff's injury and that such a change would have been both technologically and economically feasible at the time of the manufacture. If so, the manufacturer is unlikely to prevail with a state-of-the-art argument as a defense. Is there somewhere else a manufacturer can turn for a possible defense?

Few products that are manufactured today do not conform, at least in part, to some type of safety standards. There are a variety of these standards that often play a role in how the product is designed and manufactured. Among these are U.S. Government standards such as those mandated by the Department of Transportation, Coast Guard, Food and Drug Administration, and so on. In addition, there are voluntary and consensus standards such as those promulgated by Underwriters Laboratories (UL), the American National Standards Institute (ANSI), and the American Society for Testing and Materials (ASTM). Finally, there are standards developed by and for a given industry such as the Food Equipment Manufacturers Association or the Conveyor Manufacturers Association.

Assume that a product for which the plaintiff's expert has proposed a modification conforms to all applicable safety standards and that the standards do not require the plaintiff's expert's suggested modification, can a manufacturer prevail by arguing adherence to those standards? With nothing more than the absence of a requirement in the standard to support the manufacturer's claim of a nondefective product, the standard is unlikely to shield the manufacturer from liability. The reason, quite simply, is that courts generally view standards as floors, not ceilings. That is, conforming to standards, even those promulgated by the federal government, is considered by the courts to be the *least* a manufacturer should do to produce a safe product. As stated by the court in a case involving a backyard trampoline,

> The fact that a particular product meets or exceeds the requirements of its industry is not conclusive proof that the product is reasonably safe. In fact, standards set by an entire industry can be found negligently low if they fail to meet the test of reasonableness.
>
> *Dudley Sports Co. v. Schmitt*

This is the view about standards generally held by the courts. The fact that federal standards are subject to the same threshold-of-safety consideration reflects the view that all standards, including federal ones, are developed by a process dominated by the industry that is expected to conform to the standards. As a result, the standards emerging from the process are likely to reflect a limited consideration of hazard identification and risk assessment in deciding on the extent of safety requirements.

If a manufacturer is to overcome the stigma that the product standard is unlikely to reflect the requirements necessary for a reasonably safe product, it is necessary for the manufacturer first to demonstrate that it undertook an independent hazard identification and risk assessment for the product in question during the design process. Next, it must be shown that no unreasonably dangerous conditions emerged from the analysis that were not adequately covered by the standard's requirements. Finally, the manufacturer must present plausible arguments refuting the plaintiff's expert's proposals for a design or guarding modification that would have prevented the plaintiff's injury. Needless to say, such arguments must be substantive and avoid dismissing the proposed changes as being costly.

While standards can be a useful guide for the product design team, they must not be viewed as ends in themselves. Standards have rarely, if ever, shielded a manufacturer from liability absent an independent verification by the manufacturer that the standards defined a reasonably safe product.

6.3 Contributory/Comparative Negligence

We have not yet really considered what role, if any, the plaintiff's behavior plays in defending a product against an allegation of defect. We have earlier touched on misuse of the product, which is a use so egregious, and so bizarre, or so remote, that it is termed *legally unforeseeable*. You may recall the example discussing the hypothetical use of a screwdriver as a cold chisel to illustrate what could very likely be considered as misuse.

But what about the plaintiff's behavior that is not so extreme? Does that enter at all into the equation of how fault is apportioned? Yes, it does, in the form of contributory or comparative negligence, if the legal theory embracing the litigation is negligence. You will recall that under negligence, the defendant's behavior is measured by asking if that party was acting as a *reasonable* person (or manufacturer, or engineer) would have acted under the same or similar circumstances. And the reasonableness of the behavior is the result of having foreseen the risks of one's actions by having undertaken a risk/utility balancing prior to engaging in the action.

In a negligence action, the plaintiff's behavior is measured in exactly the same way. The defendant asks the jury to consider whether the plaintiff was behaving as a reasonable person would have under the same or similar circumstances. Did the plaintiff contribute to his or her harm by not acting reasonably? This is called *contributory negligence*.

While some states still retain the original concept that *any* contributory negligence on the part of the plaintiff totally bars his or her recovery of damages, most states have adopted some form of comparative negligence. Generally, the jury is asked to assess the behavior of both the plaintiff and the defendant and apportion the fault in causing the harm between them, making certain the percentages total 100%. The plaintiff's award, if any, is then reduced by the percentage of his or her comparative negligence.

The test of the defendant's negligence and the plaintiff's contributory negligence is termed an objective one. That is, the jury is asked to judge the actions of the parties relative to what a reasonable person would have done in the same or similar circumstances. The jury does not, as a rule, consider whether anything in that party's background, training, age, experience, education, and so on, played any role in the actions that led to the injury.

6.4 Assumption of the Risk

There is another defense involving the plaintiff's behavior that does consider the plaintiff's characteristics in assessing his or her culpability. It is termed *assumption of the risk*. In essence, this defense argues that the plaintiff consented to being injured by the product. In one common form, used for analyzing this aspect of the plaintiff's behavior, the jury is asked if the plaintiff *voluntarily* and *unreasonably* assumed a *known* risk. To prevail, the defendant must present evidence on all three of these elements and must prevail on all three for a jury to conclude that the plaintiff "assumed the risk."

The first element, asking whether the plaintiff voluntarily confronted the danger, and the third element, considering whether the risk was known, are both subjective elements. That is, the jury must determine the state of the mind of the plaintiff, assessing what he or she actually knew or believed or what can reasonably be inferred about his or her behavior at the instant prior to the event that led to injury. Thus, the background, education, training, experience, and so on become critical elements in this assessment.

A couple of points should be made here. First, in determining whether the plaintiff voluntarily confronted the hazard, the test is whether or not the plaintiff had *viable* alternatives.

Example 9 Work or Walk. In a workplace setting, a worker is given a choice of either using a now-unguarded press or being fired. It had been properly guarded for all the time the plaintiff had used it in the past, but the employer had removed the guards to increase productivity, and now tells the employee either to use the press as is, or be fired. The courts do not consider that the plaintiff had viable alternatives, since the choice between working on an unguarded press or being fired is no choice at all. The lesson to the engineer in this example is that the guarding slowed productivity and was removed, leaving the press user in a no-win situation. The design should have incorporated guarding that did not *slow production*.

Second, the same in-depth consideration must also be given to knowledge of the risk by the plaintiff. The plaintiff's background, education, and so on must provide a reasonable appreciation of the actual nature of the harm that could befall him or her.

Example 10 The Truly Combustible Car. The driver of a new car is confronted by a slight smell of smoke the first time the windshield wipers are used, and is trying to bring the car to the dealer to see what the trouble is when the car bursts into flames, causing injury. Has the driver assumed the risk of injury by continuing to drive after smelling smoke? Can the car manufacturer successfully argue that the risks of injury were known to the driver? The question can be answered only by examining those elements in the driver's background that could, in any way, lead a jury to conclude that the driver should have recognized that smoke from electrically operated wipers could lead to a conflagration. The old adage of "where there's smoke, there's fire" is insufficient to charge the plaintiff with knowledge of the precise risk he or she faced without more knowledge of the driver's background.

The final element of assumption of the risk, the unreasonableness of the plaintiff's choice in voluntarily confronting a known risk is an objective element, exactly the same as in negligence. That is, what would a reasonable person have done under the same or similar circumstances?

Example 11 The Truly Combustible Car Meets the Good Samaritan. A passer-by observes the car from the previous example. It is on fire, and the driver is struggling to get out. The passer-by rescues the driver, but is seriously burned and suffers smoke inhalation in the process. The driver files suit against the manufacturer alleging a defect that created unreasonable danger when the wipers were turned on. The passer-by also files suit against the automobile manufacturer to recover for the injuries suffered as a result of the rescue, arguing that the rescue would not have been necessary if there had been no defect. Would this good Samaritan be found to have assumed the risk of injury? Clearly the choice to try to rescue the driver was voluntary and the risks of injury from a fire were apparent to anyone, including the rescuer. But was the act of rescuing the car's occupant a reasonable or unreasonable one? If the jury concludes that it was a reasonable choice, the passer-by would not have been found to have assumed the risk, despite having voluntarily exposed himself to a known risk.

The defendant must prevail in all three of the elements, not just two. Needless to say, raising and succeeding in the defense of assumption of the risk is not an easy one for the defendant.

One final word about these defenses: While the "assumption of the risk" defense applies both in a claim of negligence and strict liability, the contributory/comparative negligence defense does *not* apply in strict liability. The reason is that strict liability is a no-fault concept, whereas negligence is a fault-based concept. It would be inconsistent to argue no-fault theory (strict liability) against the defendant and permit the defendant to argue a fault-based defense (contributory negligence) concerning the plaintiff's behavior.

7 RECALLS, RETROFITS, AND THE CONTINUING DUTY TO WARN

Manufacturers generally have a postsale or continuing duty to warn of latent defects in their products that are revealed through consumer use. Sometimes, however, even a postsale warning may be inadequate to render a product safe. In those circumstances, it may be necessary for a manufacturer to retrofit the product by adding certain safety devices or guards. Moreover, there may be instances where it is not feasible to add guards or safety devices, or where the danger of the product is so great that the product simply must be removed from the market by being recalled.

7.1 After-Market Hazard Recognition

The manufacturer is responsible for establishing feedback mechanisms from customers, distributors, and sales personnel that will ensure that postsale problems are discovered. Applicable data may include product performance and test data, orders for repair parts, complaint files, quality control and inspection records, and instruction and warning modifications. Another source of hazard recognition information comes from previous accident investigations, claims, and lawsuits. The manufacturer should also have an ongoing program of compiling and evaluating risk data from historical, field and/or laboratory testing, and fault-tree, failure modes, and hazard analyses.

Once the manufacturer has determined that a previously sold product is defective (that is, contains unreasonable dangers) and still in use, it must decide on what response is appropriate. If the product is currently being produced, an initial assessment as to the seriousness of the problem must be made to decide whether production is to be halted immediately and inventories frozen in the warehouses and on dealers' shelves in order to limit distribution.

Following this assessment, the nature of the defect must be established. If the problem is safety-related, and depending on the type of the product, appropriate regulatory agencies may have to be immediately notified. The manufacturer must then consider the magnitude of the hazards by estimating the probability of occurrence of events and the likely seriousness of injury or damage. The necessity for postulating such data is to provide some measure of the magnitude of the consequences if no action is taken, or to decide the extent of the action to be taken in light of the estimated consequences. Alternatively, if the consequences of even a low probability of occurrence could result in serious injury or death, or could seriously affect the marketability of the product or the corporate reputation, the decision to take action should be independent of such estimates.

Once the decision to take action is made, the origin, extent, and cause of the problem must be addressed in order to plan effective corrective measures. Is the origin of the defect in the raw material, fabrication, or the quality control? If the problem is one of fabrication, did it occur in-house or from a purchased part? Where are the faulty products—that is, are the products in inventory, in shipment, in dealers' stock, or in the hands of the buyers? Does the defect arise from poor design, inadequate inspection, improper materials, fabrication procedure, ineffective or absent testing, or a combination of these events?

7.2 Types of Corrective Action

After the decision to take action has been made, and the origin, extent, and cause of the problem have been investigated, the appropriate corrective action must be determined. Possible options are to recall the product and replace it with another one; to develop a retrofit and either send personnel into the field to retrofit the product or have the customer return the product to the manufacturer for repair; to send out the parts and have the customer fix the product; or to simply send out a warning about the problem. This process should be fully documented to substantiate the reasons for the selection of a particular response. The urgency with which the corrective action is taken will be determined by the magnitude of the hazard.

Warnings

A manufacturer is not required to warn of every known danger, even with actual knowledge of that danger. A warning is required where a product can be dangerous for its intended and reasonably foreseeable uses and where the nature of the hazard is unlikely to be readily recognized by the expected user class. When a hazard associated with a product that was previously unknown to the manufacturer becomes apparent after the product has been in use, the manufacturer has a threshold duty to warn the existing user population.

Factors to consider in determining whether to issue a postsale warning include the manufacturer's ability to warn (i.e., how readily and completely the product users can be identified and located), the product's life expectancy (the longer life expectancy, the greater risk of potential harm if postsale warnings are not given), and the obviousness of the danger. Thus, the practicality, cost, and burden of providing an effective warning must be weighed against the potential harm of omitting the warning.

Recalls

Where the potential harm to the consumer is so great that a warning alone is not adequate to eliminate the danger, the proper remedy may be to institute a recall of the product either for repair or replacement. For some products, a recall may be mandated by statute or a governmental regulatory agency. Where a recall is not mandated, however, the decision to institute a product recall should be made using the analysis undertaken in Section 7.1.

Retrofits

A recall campaign may not be an appropriate solution, particularly if the equipment is large or cannot be easily removed from an installation. For equipment with potentially serious hazards, or requiring complicated modification, the manufacturer should send its personnel to perform (and document) the retrofit. For equipment with relatively minor potential hazards for which there is a simple fix, the manufacturer may opt to send to the owners the parts necessary to solve the problem.

Regardless of the type of corrective action program selected, it is essential that all communications directed to the owners and/or users urging them to participate in the corrective action program be clear and concise. Most importantly, however, is the necessity for the communication to identify the nature of the risks and the potential seriousness of the harm that could befall the product user.

8 DOCUMENTATION OF THE DESIGN PROCESS

There are conflicting arguments by attorneys about what documentation, if any, the manufacturer should retain in the files (or on the floppies, the discs, the hard drive, or tape backup). Since it would be well-nigh impossible to run a business without documentation of some sort, it only makes sense to preserve the type of documentation that can, if the product is challenged in court, demonstrate the care and concern that went into the design, manufacturing, marketing, and user communications of the product.

The first principal of documentation is to minimize or eliminate potential adverse use of the documentation by an adverse party. For example, words such as *defect* should not appear in the company's minutes, notes, and so on. There can be *deviations, flaws, departures,* and so on from specifications or tolerances. These are not defects unless they could create unreasonable dangers in the use of the product. Also, all adverse criticism of the product, whether internally from employees or externally from customers, dealers, and so on, must be considered and addressed in writing by the responsible corporate person having the appropriate authority.

Apart from these considerations, the company should make an effort to create a *documentation tree,* delineating what document is needed, who should write it, where it should be kept, who should keep it, and for how long. The retention period for documents, for the most part, should be based on common sense. If a government or other agency requires or suggests the length of time certain documents be kept, obviously those rules must be followed. For the rest, the length of time should be based on sound business practices. If the product has certain critical components that, if they fail before the end of the product's useful life, could result in a serious safety problem, the documentation supporting the efficacy of these parts should be retained for as long as the product is likely to be in service.

Because the law requires only that a product be reasonably safe, clearly the documentation to be preserved should be that which will support the argument that all of the critical engineering decisions, which balanced competing trade-offs, were reasonable and were based on reducing the risks from all foreseeable hazards. The rationales underlying these decisions should be part of the record, for two reasons. First, because it will give those who will review the designs when the product is to be updated or modified in subsequent years the bases for existing design decisions. If the prior assumptions and rationales are still valid, they need not be altered. Conversely, if some do not reflect current thinking, then only those aspects of the design need to be altered. Without these rationales, all the design parameters will have to be re-examined for efficacy.

Secondly, and just as importantly, having the rationales in writing for those safety-critical decisions can provide a solid, legal defense, if the design is ever challenged as defective.

Thus, the documentation categories that are appropriate both for subsequent design review and for creating strong legal defense positions are these:

- Hazard and risk data that formed the bases for the safety considerations
- Design safety formulations, including fault-tree and failure modes and effects analyses
- Warnings and Instructions formulation, together with the methodology used for development and testing
- Standards used, including in-house standards, and the rationale for the requirements utilized in the design
- Quality assurance program, including the methodology and rationale for the processes and procedures
- Performance of the product in use, describing reporting procedures, follow-up data acquisition and analysis, and a written recall and retrofit policy

This type of documentation will permit recreating the process by which the reasonably safe product was designed, manufactured, and marketed.

9 A FINAL WORD

In the preceding pages, we have only touched on a few of the areas where the law can have a significant impact on engineers' discharge of their professional responsibilities. As part of the process of product design, the law asks the engineer to consider that the product which emerges from the mind of the designer and the hand of the worker to play a role in enhancing society's well-being must

- Account for reasonably foreseeable product misuse
- Acknowledge human frailties and the characteristics of the actual users
- Function in the true environment of product use
- Eliminate or guard against the hazards
- Not substitute warnings for effective design and guards

What has been discussed here and summarized above is, after all, just good engineering. It is to help the engineer recognize those considerations that are necessary to bridge the gap between the preliminary product concept and the finished product that has to function in the real world, with real users and for real uses, for all of its useful life.

Apart from understanding and utilizing these considerations during the product design process, engineers have an obligation, both personally and professionally, to maintain competence in their chosen field so that there can be no question that all actions, decisions, and recommendations, in retrospect, were reasonable.

That is, after all, what the law requires of all of us.

CHAPTER 22
PATENTS

David A. Burge
David A. Burge Co., L.P.A.
Cleveland, Ohio

Benjamin D. Burge
Intel Americas, Inc.
Chantilly, Virginia

1	**WHAT DOES IT MEAN TO OBTAIN A PATENT?**	**726**
1.1	Utility, Design, and Plant Patents	726
1.2	Patent Terms and Expiration	726
1.3	Four Types of Applications	727
1.4	Why File a Provisional Application?	728
1.5	Understanding That a Patent Grants a "Negative Right"	728
2	**WHAT CAN BE PATENTED AND BY WHOM**	**729**
2.1	Ideas, Inventions, and Patentable Inventions	729
2.2	The Requirement of Statutory Subject Matter	730
2.3	The Requirement of Originality of Inventorship	731
2.4	The Requirement of Novelty	732
2.5	The Requirement of Utility	734
2.6	The Requirement of Nonobviousness	734
2.7	Statutory Bar Requirements	735
3	**PREPARING TO APPLY FOR A PATENT**	**736**
3.1	The Patentability Search	736
3.2	Putting the Invention in Proper Perspective	736
3.3	Preparing the Application	737
3.4	Enablement, Best Mode, Description, and Distinctness Requirements	738
3.5	Product-by-Process Claims	739
3.6	Claim Format	739
3.7	Executing the Application	740
3.8	U.S. Patent and Trademark Office Fees	741
3.9	Small Entity Status	741
3.10	Express Mail Filing	742
4	**PROSECUTING A PENDING PATENT APPLICATION**	**742**
4.1	Patent Pending	742
4.2	Publication of Pending Applications	743
4.3	Duty of Candor	744
4.4	Initial Review of an Application	744
4.5	Response to an Office Action	745
4.6	Reconsideration in View of the Filing of a Response	746
4.7	Interviewing the Examiner	747
4.8	Restriction and Election Requirements	747
4.9	Double-Patenting Rejections	747
4.10	Continuation, Divisional, and Continuation-in-Part Applications	748
4.11	Maintaining a Chain of Pending Applications	748
4.12	Patent Issuance	749
4.13	Safeguarding the Original Patent Document	749
4.14	Reissue	749
4.15	Reexamination	750
5	**ENFORCING PATENTS AGAINST INFRINGERS**	**751**
5.1	Patent Infringement	751
5.2	Defenses to Patent Enforcement	752
5.3	Outcome of a Suit	752
5.4	Settling a Suit	752
5.5	Declaratory Judgment Actions	753
5.6	Failure to Sue Infringers	753
5.7	Infringement by Government	753
5.8	Alternative Resolution of Patent Disputes	753
5.9	Interferences	754
6	**PATENT PROTECTIONS AVAILABLE ABROAD**	**755**
6.1	Canadian Filing	755

6.2	Foreign Filing in Other Countries	755	6.5	Filing on a Country-by-Country Basis	756
6.3	Annual Maintenance Taxes and Working Requirements	755	6.6	The Patent Cooperation Treaty	756
6.4	Filing under International Convention	756	6.7	The European Patent Convention	757
			6.8	Advantages and Disadvantages of International Filing	757

1 WHAT DOES IT MEAN TO OBTAIN A PATENT?

Before meaningfully discussing such topics as inventions that qualify for patent protection and procedures that are involved in obtaining a patent, it is necessary to know about such basics as the four different types of patent applications that can be filed, the three different types of patents that can be obtained, and what rights are associated with the grant of a patent.

1.1 Utility, Design, and Plant Patents

When one speaks of obtaining a patent, it is ordinarily assumed that what is intended is a utility patent. Unless stated otherwise, the discussion of patents presented in this chapter applies only to U.S. patents, and principally to utility patents.

Utility patents are granted to protect processes, machines, articles of manufacture, and compositions of matter that are new, useful and nonobvious.

Design patents are granted to protect ornamental appearances of articles of manufacture—that is, shapes, configurations, ornamentation, and other appearance-defining characteristics that are new, nonobvious, and not dictated primarily by functional considerations.

Plant patents are granted to protect new varieties of plants that have been asexually reproduced with the exception of tuber-propagated plants and those found in an uncultivated state. New varieties of roses and shrubs often are protected by plant patents.

Both utility and design patents may be obtained on some inventions. A utility patent typically will have claims that define novel combinations of structural features, techniques of manufacture, and/or methods of use of a product. A design patent typically will cover outer configuration features that are not essential to the function of the product, but rather give the product an esthetically pleasing appearance. Utility patents can protect structural, functional and operational features of an invention. Design patents can protect appearance features of a product.

Genetically engineered products may qualify for plant patent protection, for utility patent protection, and/or for other protections provided for by statute that differ from patents. Computer software and other computer-related products may qualify for design and/or utility patent protections. Methods of doing business may qualify for utility patent coverage. These are developing areas of intellectual property law.

1.2 Patent Terms and Expiration

Design patents that currently are in force have normal terms of 14 years, measured from their issue dates. Prior to a change of law that took effect during 1982, it was possible for design patent owners to elect shorter terms of $3\frac{1}{2}$ to 7 years.

Plant and utility patents that expired prior to June 8, 1995 had a normal term of 17 years, measured from their issue dates. Plant and utility patents that (1) were in force on June 8, 1995, or (2) issue from applications that were filed prior to June 8, 1995 have normal terms that expire either 17 years, measured from their issue dates, or 20 years, measured from the filing date of the applications from which these patents issued, whichever is later. Plant and utility patents that issue from applications that were filed on or after June 8, 1995 have normal terms that expire 20 years from application filing, but these terms may be adjusted by a few days by the U.S. Patent and Trademark Office (USPTO) in view of delays that were encountered during application pendency (see Section 4.12).

The filing date from which the 20-year measurement of the term of a plant or utility patent is measured to calculate the normal expiration date of a utility patent is the earliest applicable filing date. If, for example, a patent issues from a continuation application, a divisional application or a continuation-in-part application that claims the benefit of the filing date of an earlier filed "parent" application, the 20-year measurement is taken from the filing date of the "parent" application.

The normal term of a patent may be shortened due to a variety of circumstances. If, for example, a court of competent jurisdiction should declare that a patent is "invalid," the normal term of the patent will have been brought to an early close. In some circumstances, a "terminal disclaimer" may have been filed by the owner of a patent to cause early termination. The filing of a terminal disclaimer is sometimes required by the U.S. Patent and Trademark Office during the examination of an application that is so closely subject-matter-related to an earlier-filed application that there may be a danger that two patents having different expiration dates will issue covering substantially the same invention.

1.3 Four Types of Applications

Three types of patent applications are well known. A *utility* application is what one files to obtain a utility patent. A *design* application is what one files to obtain a design patent. A *plant* application is what one files to obtain a plant patent.

Starting June 8, 1995, it became possible to file a fourth type of patent application known as a *provisional* application as a precursor to the filing of a utility application. The filing of a provisional application will *not* result in the issuance of any kind of patent. In fact, no examination will be made of the merits of the invention described in a provisional application. Examination "on the merits" takes place only if a utility application is filed within one year that claims the benefit of the filing date of the provisional; and, if examination takes place, it centers on the content of the utility application, not on the content of any provisional applications that are referred to in the utility application.

The filing of a provisional application that adequately describes an invention will establish a filing date that can be relied on in a later-filed utility application relating to the same invention (1) if the utility application is filed within one year of the filing date of the provisional application and (2) if the utility application makes proper reference to the provisional application. While the filing date of a provisional application can be relied on to establish a reduction to practice of an invention, the filing date of a provisional application does *not* start the 20-year clock that determines the normal expiration date of a utility patent.

Absent the filing of a utility application within one year from the filing date of a provisional application, the USPTO will destroy the provisional application once it has been pending a full year.

1.4 Why File a Provisional Application?

If a provisional application will not be examined and will not result in the issuance of any form of patent whatsoever, why would one want to file a provisional application? Actually, there are several reasons why the filing of one or more provisional applications may be advantageous before a full-blown utility application is put on file.

If foreign filing rights are to be preserved, it often is necessary for a U.S. application to be filed before *any* public disclosure is made of an invention so that one or more foreign applications can be filed within one year of the filing date of the U.S. application, with the result that the foreign applications will be afforded the benefit of the filing date of the U.S. application (due to a treaty referred to as the *Paris Convention*), thereby ensuring that the foreign applications comply with "absolute novelty" requirements of foreign patent law.

If the one-year grace period provided by U.S. law (which permits applicants to file a U.S. application anytime within a full one-year period from the date of the first activity that starts the clock running on the one-year grace period) is about to expire, it may be desirable to file a provisional application rather than a utility application, because (1) preparing and filing a provisional application usually can be done less expensively, (2) preparing and filing a provisional application usually can be done more quickly inasmuch as it usually involves less effort (i.e., there is no need to include an abstract, claims, or a declaration or oath, which are required in a utility application), and (3) everything that one may want to include in a utility application may not have been discerned (i.e., development and testing of the invention may still be underway), hence, it may be desirable to postpone for as much as a full year the drafting and filing of a utility application.

If development work is still underway when a first provisional application is filed, and if the result of the development program brings additional invention improvements to light, it may be desirable (during the permitted period of one year between the filing of the first provisional application and the filing of a full-blown utility application) to file one or more additional provisional applications, all of which can be referred to and can have their filing dates relied upon when a utility application is filed within one year, measured from the filing date of the earliest-filed provisional application.

Yet another reason why a provisional application may be filed is to preserve patent rights in situations where the value of an invention is questionable and the cost of filing a full-blown utility case is not justified, but action must be taken soon or the deadline to file for patent protection will not be met. In such situations, it may be desirable to file a first provisional as simply as possible, and follow by filing additional provisional cases as the invention is developed and its importance comes to be better understood. So long as a full-blown utility application is filed within a year of the filing of the first provisional, this approach can assist in salvaging patent rights in inventions that might otherwise have been passed over for patent protection.

What the filing of a provisional application preserves for one year is the right to file a utility application. It does not preserve the right to file a design application: therefore, if design patent rights are to be preserved, it may necessary to file a design application at the same time that one files a provisional application.

1.5 Understanding That a Patent Grants a "Negative Right"

A patent is a grant by the Federal government to the inventor or inventors of an invention of certain rights—rights that are "negative" in nature. Patents issue only upon application, and only after rigorous examination to ensure that the invention meets certain standards and qualifications, and that the application meets certain formalities.

It is surprisingly common to find that even those who hold several patents fail to properly understand the "negative" nature of the rights that are embodied in the grant of a patent. What the patent grants is the "negative right" to "exclude others" from making, using, or selling an invention that is covered by the patent. *Not* included in the grant of a patent is a "positive right" enabling the patent owner to actually make, use, or sell the invention. In fact, a patent owner may be precluded, by the existence of other patents, from making, using, and selling his or her patented invention. Illustrating this often misunderstood concept is the following example referred to as *The Parable of the Chair:*

> If inventor A invents a three-legged stool at an early time when such an invention is not known to others, A's invention may be viewed as being "basic" to the art of seats, probably will be held to be patentable, and the grant of a patent probably will have the practical effect of enabling A both (1) to prevent others from making, using, and selling three-legged stools and (2) to be the only entity who *can* legally make, use, and sell three-legged stools during the term of A's patent.
>
> If, during the term of A's patent, inventor B improves upon A's stool by adding a fourth leg for stability and an upright back for enhanced support and comfort (whereby a chair is born), and if B obtains a patent on his chair invention, the grant of B's chair patent will enable B to prevent others from making, using, and selling four-legged back-carrying seats. However, B's patent will do nothing at all to permit B to make, use, or sell chairs—the problem being that a four-legged seat having a back *infringes* A's patent because each of B's chairs *includes* three legs that support a seat, which is what A can exclude others from making, using, or selling. To legally make, use, or sell chairs, B must obtain a license from A.
>
> And if, during the terms of the patents granted to A and B, C invents and patents the improvement of providing curved rocking rails that connect with the leg bottoms, and arms that connect with the back and seat, thereby bringing into existence a rocking chair, C can exclude others during the term of his patent from making, using, or selling rocking chairs, but must obtain licenses from both A and B in order to make, use, or sell rocking chairs, for a rocking chair includes three legs and a seat, and includes four legs, a seat, and a back.

Invention improvements may represent very legitimate subject matter for the grant of a patent. However, patents that cover invention improvements may not give the owners of these patents any right at all to make, use, or sell their patented inventions unless licenses are obtained from those who obtained patents on inventions that are more basic in nature. Once the terms of the more basic patents expire, owners of improvement patents then may be able to practice and profit from their inventions on an exclusive basis during the remaining portions of the terms of their patents.

2 WHAT CAN BE PATENTED AND BY WHOM

For an invention to be patentable, it must meet several requirements set up to ensure that patents are not issued irresponsibly. Some of these standards are complex to understand and apply. Let us simplify and summarize the essence of these requirements.

2.1 Ideas, Inventions, and Patentable Inventions

Invention is a misleading term because it is used in so many different senses. In one, it refers to the act of inventing. In another, it refers to the product of the act of inventing. In still another, the term designates a patentable invention, the implication mistakenly being that if an invention is not patentable, it is not an invention.

In the context of modern patent law, invention is the conception of a novel, nonobvious, and useful contribution followed by its reduction to practice. Conception is the beginning of an invention; it is the creation in the mind of an inventor of a useful means for solving a particular problem. Reduction to practice can be either actual, as when an embodiment of the invention is tested to prove its successful operation under typical conditions of service, or constructive, as when a patent application is filed containing a complete description of the invention.

Ideas, per se, are not inventions and are not patentable. They are the tools of inventors, used in the development of inventions. Inventions are patentable only insofar as they meet certain criteria established by law. For an invention to be protectable by the grant of a utility patent, it must satisfy the following conditions:

1. Fit within one of the statutorily recognized classes of patentable subject matter.
2. Be the true and original product of the person or persons seeking to patent the invention as its inventor or inventors.
3. Be new at the time of its invention by the person or persons seeking to patent it.
4. Be useful in the sense of having some beneficial use in society.
5. Be nonobvious to one of ordinary skill in the art to which the subject matter of the invention pertains at the time of its invention.
6. Satisfy certain statutory bars that require the inventor to proceed with due diligence in pursuing efforts to file and prosecute a patent application.

2.2 The Requirement of Statutory Subject Matter

As stated in the Supreme Court decision of *Kewanee Oil v. Bicron Corp.,* 416 U.S. 470, 181 U.S.P.Q. 673 (1974), no utility patent is available for any discovery, however useful, novel, and nonobvious, unless it falls within one of the categories of patentable subject matter prescribed by Section 101 of Title 35 of the United States Code. Section 101 provides

> Whoever invents or discovers a new and useful process, machine, manufacture, or composition of matter, or any new and useful improvement thereof may obtain a patent therefore, subject to the conditions and requirements of this title.

The effect of establishing a series of statutory classes of eligible subject matter has been to limit the pursuit of patent protection to the useful arts. Patents directed to processes, machines, articles of manufacture, and compositions of matter have come to be referred to as utility patents, inasmuch as these statutorily recognized classes encompass the useful arts.

Three of the four statutorily recognized classes of eligible subject matter may be thought of as products, namely machines, manufactures, and compositions of matter. *Machine* has been interpreted in a relatively broad manner to include a wide variety of mechanisms and mechanical elements. *Manufactures* is essentially a catch-all term covering products other than machines and compositions of matter. *Compositions of matter,* another broad term, embraces such elements as new molecules, chemical compounds, mixtures, alloys, and the like. *Machine and manufactures* now arguably includes software either stored on a piece of media or being executed on a computer. *Manufactures* and *compositions of matter* arguably include such genetically engineered life forms as are not products of nature. The fourth class, *processes,* relates to procedures leading to useful results.

Subject matter held to be ineligible for patent protection includes printed matter, products of nature, ideas, and scientific principles. Alleged inventions of perpetual motion machines are refused patents. A mixture of ingredients such as foods and medicines cannot be patented unless there is more to the mixture than the mere cumulative effect of its components. So-called patent medicines are seldom patented.

While no patent can be issued on an old product despite the fact that it has been found to be derivable through a new process, a new process for producing the product may well be patentable. That a product has been reduced to a purer state than was previously available in the prior art does not render the product patentable, but the process of purification may be patentable. A new use for an old product does not entitle one to obtain *product* patent protection, but may entitle one to obtain *process* patent protection, assuming the process meets other statutory requirements. A newly discovered law of nature, regardless of its importance, is not entitled to patent protection.

While the requirement of statutory subject matter falls principally within the bounds of 35 U.S.C. 101, other laws also operate to restrict the patenting of certain types of subject matter. For example, several statutes have been passed by Congress affecting patent rights in subject matter relating to atomic energy, aeronautics and space. Still another statute empowers the Director of the USPTO to issue secrecy orders regarding patent applications disclosing inventions that might be detrimental to the national security of the United States.

The foreign filing of patent applications on inventions made in the United States is prohibited until a license has been granted by the Director of the USPTO to permit foreign filing. This prohibition period enables the USPTO to review newly filed applications, locate any containing subject matter that may pose concerns to national security, and after consulting with other appropriate agencies of government, issue secrecy orders preventing the contents of these applications from being publicly disclosed. If a secrecy order issues, an inventor may be barred from filing applications abroad on penalty of imprisonment for up to two years or a $10,000 fine, or both. In the event that a patent application is withheld under a secrecy order, the patent owner has the right to recover compensation from the government for damage caused by the secrecy order and/or for any use the government may make of the invention.

Licenses permitting expedited foreign filing are almost always automatically granted by the USPTO at the time of issuing an official filing receipt, which advises the inventor of the filing date and serial number assigned to his or her application. Official filing receipts usually issue within a month of the date of filing and usually bear a statement attesting to the grant of a foreign filing license.

2.3 The Requirement of Originality of Inventorship

Under U.S. patent law, only the true and original inventor or inventors may apply to obtain patent protection. If the inventor has derived an invention from any other source or person, he or she is not entitled to apply for or obtain a patent.

The laws of our country are strict regarding the naming of the proper inventor or joint inventors in a patent application. When one person acting alone conceives an invention, he or she is the sole inventor and he or she alone must be named as the inventor in a patent application filed on that invention. When a plurality of people contribute to the conception of an invention, these persons must be named as joint inventors if they have contributed to the inventive features that are claimed in a patent application filed on the invention.

Joint inventorship occurs when two or more persons collaborate in some fashion, with each contributing to conception. It is not necessary that exactly the same idea should have occurred to each of the collaborators at the same time. Section 116 of Title 35 of the United States Code includes the following provision:

> Inventors may apply for a patent jointly even though (1) they did not physically work together or at the same time, (2) each did not make the same type or amount of contribution, or (3) each did not make a contribution to the subject matter of every claim of the patent.

Those who may have assisted the inventor or inventors by providing funds or materials for development or by building prototypes under the direction of the inventor or inventors are not deemed to be inventors unless they contributed to the conception of the invention. While inventors may have a contractual obligation to assign rights in an invention to their employers, this obligation, absent a contribution to conception, does not entitle a supervisor or an employer to be named as an inventor. When a substantial number of patentable features relating to a single overall development have occurred as the result of different combinations of sole inventors acting independently and/or joint inventors collaborating at different times, the patent law places a burden on the inventors to sort out "who invented what." If one patent application covering the entire development and naming all of the inventors jointly is deemed inappropriate, then patent protection on the overall development may need to be pursued in the form of a number of separate patent applications, each directed to such patentable aspects of the development as originated with a different inventor or group of inventors. In this respect, U.S. patent practice is unlike that of many foreign countries, where the company for whom all the inventors work is often permitted to file a single patent application in its own name covering the overall development.

Misjoinder of inventors occurs when a person who is not a joint inventor has been named as such in a patent application. *Nonjoinder of inventors* occurs when there has been a failure to include a person who should have been named as a joint inventor. *Misdesignation of inventorship* occurs when none of the true inventors are named in an application. Only in recent years has correction of a misdesignation been permitted. If a problem of misjoinder, nonjoinder, or misdesignation has arisen without deceptive intent, provisions of the patent law permit correction of the error as long as such is pursued with diligence following discovery.

2.4 The Requirement of Novelty

Section 101 of Title 35 of the United States Code requires that a patentable invention be new. What is meant by *new* is defined in Sections 102(a), 102(e), and 102(g). Section 102(a) bars the issuance of a patent on an invention "known or used by others in this country, or patented or described in a printed publication in this or a foreign country, before the invention thereof by the applicant for patent." Section 102(e) bars the issuance of a patent on an invention "described in a patent granted on an application for patent by another filed in the United States before the invention thereof by the applicant for patent, or in an international application by another." Section 102(g) bars the issuance of a patent on an invention that "before the applicant's invention thereof . . . was made in this country by another who had not abandoned, suppressed or concealed it."

These novelty requirements amount to negative rules of invention, the effect of which is to prohibit the issuance of a patent on an invention if the invention is not new. The novelty requirements of 35 U.S.C. 102 should not be confused with the statutory bar requirements of 35 U.S.C. 102, which are discussed in section 2.7. A comparison of novelty and statutory bar requirements of 35 U.S.C. 102 is presented in Table 1. The statutory bar requirements are distinguishable from the novelty requirements in that they do not relate to the newness of the invention, but to ways an inventor, who would otherwise have been able to apply for patent protection, has lost that right by tardiness.

To understand the novelty requirements of 35 U.S.C. 102, one must understand the concept of anticipation. A claimed invention is anticipated if a single prior art reference contains all the essential elements of the claimed invention. If teachings from more than one reference must be combined to show that the claimed combination of elements exists, there

Table 1 Summary of the Novelty and Statutory Bar Requirements of 35 U.S.C. 102

Novelty Requirements

One may not patent an invention if, prior to their date of invention, the invention was any of the following:

1. Known or used by others in this country
2. Patented or described in a printed publication in this or a foreign country
3. Described in a patent granted on an application for patent by another filed in the United States (or in certain international applications)
4. Made in this country by another who had not abandoned, suppressed, or concealed it

Statutory Bar Requirements

One may not patent an invention if one has previously abandoned the invention. One may not patent an invention if, more than one year prior to the time one's patent application is filed, the invention was any of the following:

1. Patented or described in a printed publication in this or a foreign country
2. In public use or on sale in this country
3. Made the subject of an inventor's certificate in a foreign country
4. Made the subject of a foreign patent application, which results in the issuance of a foreign patent before an application is filed in this country

---

is no anticipation, and novelty exists. Combining references to render a claimed invention unpatentable brings into play the nonobviousness requirements of 35 U.S.C. 103, not the novelty requirement of 35 U.S.C. 102. Novelty hinges on anticipation and is a much easier concept to understand and apply than that of nonobviousness.

35 U.S.C. 102(a) Known or Used by Others in This Country prior to the Applicant's Invention

In interpreting whether an invention has been known or used in this country, it has been held that the knowledge must consist of a complete and adequate description of the claimed invention and that this knowledge must be available, in some form, to the public. Prior use of an invention in this country by another will be disabling only if the invention in question has actually been reduced to practice and its use has been accessible to the public in some minimal sense. For a prior use to be disabling under Section 102(a), the use must have been of a complete and operable product or process that has been reduced to practice.

35 U.S.C. 102(a) Described in a Printed Publication in This or a Foreign Country prior to the Application's Invention

For a printed publication to constitute a full anticipation of a claimed invention, the printed publication must adequately describe the claimed invention. The description must be such that it enables a person of ordinary skill in the art to which the invention pertains to understand and practice the invention. The question of whether a publication has taken place is construed quite liberally by the courts to include almost any act that might legitimately constitute publication. The presence of a single thesis in a college library has been held to constitute publication. Similar liberality has been applied in construing the meaning of the term *printed*.

35 U.S.C. 102(a) Patented in This or a Foreign Country

An invention is not deemed to be novel if it was patented in this country or any foreign country prior to the applicant's date of invention. For a patent to constitute full anticipation and thereby render an invention unpatentable for lack of novelty, the patent must provide an adequate, operable description of the invention. The standard to be applied under Section 102(a) is whether the patent "describes" a claimed invention. A pending patent application is treated as constituting a "patent" for purposes of apply Section 102(a) as of the date of its issuance.

35 U.S.C. 102(e) Described in a Patent Filed in This Country prior to the Applicant's Invention

Section 102(e) prescribes that if another inventor has applied to protect an invention before you invent the same invention, you cannot patent the invention. The effective date of a U.S. patent, for purposes of Section 102(e) determination, is the filing date of its application, rather than the date of patent issuance. Patent applications that have been published by the USPTO and certain international applications filed under the Patent Cooperation Treaty also are viewed as prior art when applying 35 U.S.C. 102(e).

35 U.S.C. 102(g) Abandoned, Suppressed, or Concealed

For the prior invention of another person to stand as an obstacle to the novelty of one's invention under Section 102(g), the invention made by another must not have been abandoned, suppressed, or concealed. Abandonment, suppression, or concealment may be found when an inventor has been inactive for a significant period of time in pursuing reduction to practice of an invention. This is particularly true when the inventor's becoming active again has been spurred by knowledge of entry into the field of a second inventor.

2.5 The Requirement of Utility

To comply with the utility requirement of U.S. patent law, an invention must be capable of achieving some minimal useful purpose that is not illegal, immoral or contrary to public policy. The invention must be operable and capable of being used for some beneficial purpose. The invention does not need to be a commercially successful product to satisfy the requirement of utility. While the requirement of utility is ordinarily a fairly easy one to meet, problems do occasionally arise with chemical compounds and processes, and particularly with pharmaceuticals. An invention incapable of being used to carry out the object of the invention may be held to fail the utility requirement.

2.6 The Requirement of Nonobviousness

The purpose of the novelty requirement of 35 U.S.C. 102 and that of the nonobviousness requirement of 35 U.S.C. 103 are the same—to limit the issuance of patents to those innovations that do, in fact, advance the state of the useful arts. While the requirements of novelty and nonobviousness may seem very much alike, the requirement of nonobviousness is a more sweeping one. This requirement maintains that if it would have been obvious (at the time of invention was made) to anyone ordinarily skilled in the art to produce the invention in the manner disclosed, then the invention does not rise to the dignity of a patentable invention and is therefore not entitled to patent protection.

The question of nonobviousness must be wrestled with by patent applicants in the event the USPTO rejects some or all of their claims based on an assertion that the claimed invention is obvious in view of the teaching of one or a combination of two or more prior art references. When a combination of references is relied on in rejecting a claim, the argument the USPTO is making is that it is obvious to combine the teachings of these reference to produce the claimed invention. When such a rejection has been made, the burden is on the applicant to establish to the satisfaction of the USPTO that the proposed combination of references would not have been obvious to one skilled in the art at the time the invention was made; and/or that, even if the proposed combination of references is appropriate, it still does not teach or suggest the claimed invention.

In an effort to ascertain whether a new development is nonobvious, the particular facts and circumstances surrounding the development must be considered and weighed as a whole. While the manner in which an invention was made must not be considered to negate the patentability of an invention, care must be taken to ensure that the question of nonobviousness is judged as of the time the invention was made and in the light of the then existing knowledge and state of the art.

The statutory language prescribing the nonobviousness requirement is found in Title 35, Section 103, which states

> A patent may not be obtained . . . if the differences between the subject matter sought to be patented and the prior art such that the subject matter as a whole would have been obvious at the time the invention was made to a person of ordinary skill in the art to which said subject matter pertains.

In the landmark decision of *Graham v. John Deere,* 383 U.S. 1, 148 U.S.P.Q. 459 (1966), the U.S. Supreme Court held that several basic factual inquiries should be made in determining nonobviousness. These inquiries prescribe a four-step procedure or approach for judging nonobviousness. First, the scope and content of the prior art in the relevant field or fields must be ascertained. Second, the level of ordinary skill in the pertinent art is determined. Third, the differences between the prior art and the claims at issue are examined. Fourth and finally, a determination is made as to whether these differences would have been obvious to one of ordinary skill in the applicable art at the time the invention was made.

2.7 Statutory Bar Requirements

Despite the fact that an invention may be new, useful, and nonobvious and that it may be satisfy the other requirements of the patent law, an inventor can still lose the right to pursue patent protection on the invention unless he or she complies with certain requirements of the law called *statutory bars*. The statutory bar requirements ensure that inventors will act with diligence in pursuing patent protection.

While 35 U.S.C. 102 includes both the novelty and the statutory bar requirements of the law (see the summary presented in Table 1), it intertwines these requirements in a complex way that is easily misinterpreted. The novelty requirements are basic to a determination of patentability in the same sense as are the requirements of statutory subject matter, originality, and nonobviousness. The statutory bar requirements are not basic to a determination of patentability, but rather operate to decline patent protection to an invention that may have been patentable at one time.

Section 102(b) bars the issuance of a patent if an invention was "in public use or on sale" in the United States more than one year prior to the date of the application for a patent. Section 102(c) bars the issuance of a patent if a patent applicant has previously abandoned the invention. Section 102(d) bars the issuance of a patent if the applicant has caused the invention to be first patented in a foreign country and has failed to file an application in the United States within one year after filing for a patent in a foreign country.

Once an invention has been made, the inventor is under no specific duty to file a patent application within any certain period of time. However, should one of the "triggering" events described in Section 102 occur, regardless of whether this occurrence may have been the result of action taken by the inventor or by actions of others, the inventor must apply for a patent within the prescribed period of time or be barred from obtaining a patent.

Some of the events that trigger statutory bar provisions are the patenting of an invention in this or a foreign country; the describing in a printed publication of the invention in this or a foreign country; the public use of the invention in this country; or putting the invention on sale in this country. Some public uses and putting an invention on sale in this country will not trigger statutory bars if these activities were incidental to experimentation; however, the doctrine of experimental use is a difficult one to apply because of the conflicting decisions issued on this subject.

Certainly, the safest approach to take is to file for patent protection well within one year of any event leading to the possibility of any statutory bar coming into play. If foreign patent protections are to be sought, the safest approach is to file an application in this country before any public disclosure is made of the invention anywhere in the world.

3 PREPARING TO APPLY FOR A PATENT

Conducting a patentability search prior to the preparation of a patent application can be extremely beneficial even when an inventor is convinced that no one has introduced a similar invention into the marketplace.

3.1 The Patentability Search

A properly performed patentability study will guide not only the determination of the scope of patent protection to be sought, but also the claim-drafting approaches that will be utilized in preparing the patent application. In almost every instance, a patent attorney who has at hand the results of a carefully conducted patentability study can do a better job of drafting a patent application, thereby helping to ensure that it will be prosecuted smoothly, at minimal expense, through the rigors of examination in the USPTO.

Occasionally, a patentability search will indicate that an invention is totally unpatentable. When this is the case, the search will have saved the inventor the cost of preparing and filing a patent application. At times a patentability search turns up one or more newly issued patents that pose infringement concerns. A patentability search is not, however, as extensive a search as might be conducted to locate possible infringement concerns when a great deal of money is being invested in a new product.

Some reasonable limitation is ordinarily imposed on the scope of a patentability search to keep search costs within a relatively small budget. The usual patentability search covers only U.S. patents and does not extend to foreign patents or to publications. Only patents found in the most pertinent USPTO subclasses are reviewed. Even though patentability studies are not of exhaustive scope, a carefully conducted patentability search ordinarily can be relied on to give a decent indication of whether an invention is worthy of pursuing patent coverage to protect, and the information provided by such a search should help one's patent attorney prepare a better patent application than would have been drafted absent knowledge of the search results.

3.2 Putting the Invention in Proper Perspective

It is vitally important that a client take whatever time is needed to make certain that his or her patent attorney fully understands the character of an invention before the attorney undertakes the preparation of a patent application. The patent attorney should be given an opportunity to talk with those involved in the development effort from which an invention has emerged. He or she should be told what features these people believe are important to protect. Moreover, the basic history of the art to which the invention relates should be described, together with a discussion of the efforts made by others to address the problems solved by the present invention.

The client should also convey to his or her patent attorney how the present invention fits into the client's overall scheme of development activities. Much can be done in drafting a patent application to lay the groundwork for protection of future developments. Additionally, one's patent attorney needs to know how product liability concerns may arise with regard to the present invention so that statements he or she makes in the patent application will not be used to the client's detriment in product liability litigation. Personal injury lawyers have been known to scrutinize the representations made in a manufacturer's patents to find language that will assist in obtaining recoveries for persons injured by patented as well as unpatented inventions of the manufacturer.

Before preparation of an application is begun, careful consideration should be given to the scope and type of claims that will be included. In many instances, it is possible to pursue both process and product claims. Also, in many instances, it is possible to present claims approaching the invention from varying viewpoints so different combinations of features can be covered. Frequently, it is possible to couch at least two of the broadest claims in different language so efforts of competitors to design around the claim language will be frustrated.

Careful consideration must be given to approaches competitors may take in efforts to design around the claimed invention. The full range of invention equivalents also needs to be taken into account so that claims of appropriate scope will be presented in the patent application.

3.3 Preparing the Application

A well-drafted patent application should be a readable and understandable document. If it is not, insist that your patent attorney rework it. A patent application that accurately describes an invention without setting forth the requisite information in a clear and convincing format may be legally sufficient, but it does not represent the quality of work a client has the right to expect.

The claims of a patent application (i.e., the numbered paragraphs that are found at the end of a patent application) are the most important elements of a patent application. All other sections of a patent application must be prepared with care to ensure that these other sections cannot be interpreted as limiting the coverage that is provided by the claims of the application.

A well-drafted patent application should include an introductory section that explains the background of the invention and the character of the problems that are addressed by the invention. It should discuss the closest prior art known to the applicant and should indicate how the invention patentably differs from prior art proposals.

The application should present a brief description of such drawings as form a part of the application, followed by a detailed description of the best mode known to the inventor for carrying out the invention. In the detailed description, one or more embodiments of the invention are described in sufficient detail to enable a person having ordinary skill in the art to which the invention pertains to practice the invention. While some engineering details, such as dimensions, materials of construction, circuit component values, and the like, may be omitted, all details critical to the practice of the invention must be included. If there is any question about the essential character of a detail, prudent practice would dictate its inclusion.

The claims are the most difficult part of the application to prepare. While the claims tend to be the most confusing part of the application, the applicant should spend enough time wrestling with the claims and/or discussing them with the patent attorney to make certain that the content of the claims is fully understood. In drafting the claims of an application, legal gibberish should be avoided, such as endless uses of the word *said*. Elements unessential to the practice of the invention should be omitted from the broadest claims, and essential elements should be described in the broadest possible terms.

The patent application will usually include one or more sheets of drawings and will be accompanied by a suitable declaration or oath to be signed by the inventor or inventors. The drawings of a patent application should illustrate each feature essential to the practice of the invention and show every feature to which reference is made in the claims. The drawings must comply in size and format with a lengthy set of technical rules promulgated and frequently updated by the USPTO. The preparation of the drawings is ordinarily best left to an experienced patent draftsperson.

A well-prepared application will help to pave the way for smooth handling of the patent application during its prosecution. If a patent application properly tells the story of the invention, it should constitute something of a teaching document that will stand on its own and be capable of educating a court regarding the character of the art to which the invention pertains, as well as the import of this invention to that art. Since patent suits are tried before judges who rarely have technical backgrounds, it is important that a patent application make an effort to present the basic features of the invention in terms understandable by those having no technical training. It is unusual for an invention to be so impossibly complex that its basic thrust defies description in fairly simple terms. A patent application is suspect if it wholly fails to set forth, at some point, the pith of the invention in terms a grade school student can grasp.

3.4 Enablement, Best Mode, Description, and Distinctness Requirements

Once a patent application has been prepared and is in the hands of the inventor for review, it is important that the inventor keep in mind the enablement, best mode, description, and distinctness requirements of the patent law. The enablement requirement calls for the patent application to present sufficient information to enable a person skilled in the relevant art to make and use the invention. The disclosure presented in the application must be such that it does not require one skilled in the art to experiment to any appreciable degree to practice the invention.

The best mode requirement mandates that an inventor disclose, at the time he or she files a patent application, the best mode he or she then knows about for practicing the invention.

The description requirement also relates to the descriptive part of a patent application and the support it must provide for any claims that may need to be added after the application has been filed. Even though a patent application may adequately teach how to make and use the subject matter of the claimed invention, a problem can arise during the prosecution of a patent application where one determines it is desirable to add claims that differ in language from those filed originally. If the claim language one wants to add does not find adequate support in the originally filed application, the benefit of the original filing date will be lost with regard to the subject matter of the claims to be added—a problem referred to as *late claiming*, about which much has been written in court decisions of the past half century.

Therefore, in reviewing a patent application prior to its being executed, an inventor should keep in mind that the description that forms a part of the application should include support for any language he or she may later want to incorporate in the claims of the application.

The distinctness requirement applies to the content of the claims. In reviewing the claims of a patent application, an inventor should endeavor to make certain the claims particularly point out and distinctly claim the subject matter that he or she regards as his or her invention. The claims must be *definite* in the sense that their language must clearly set forth the area over which an applicant seeks exclusive rights. The language used in the claims must find antecedent support in the descriptive portion of the application. The claims must not include within their scope of coverage any prior art known to the inventor, and yet should present the invention in the broadest possible terms that patentably distinguish the invention over the prior art.

3.5 Product-by-Process Claims

In some instances, it is possible to claim a product by describing the process or method of its manufacture. Some products are unique because of the way they are produced, hence securing proper protection may necessitate that the claims of the patent application recite critical process steps that cause the resulting product to be unique. Patentability cannot be denied due to the way a product is made, but may be enhanced if the process lends novelty to the resulting product.

3.6 Claim Format

Patent applicants have some freedom in selecting the terminology that they use to define and claim their inventions, for it has long been held that "an applicant is his own lexicographer." However, the meanings that the applicants assign to the terminology they use must not be repugnant to the well-known usages of such terminology. When an applicant does not define the terms he or she uses, such terms must be given their "plain meaning," namely the meanings given to such terms by those of ordinary skill in the relevant art.

Each claim is a complete sentence. In many instances, the first part of the sentence of each claim appears at the beginning of the claims section and reads, "What is claimed is:." Each claim typically includes three parts: preamble, transition, and body. The preamble introduces the claim by summarizing the field of the invention, its relation to the prior art, and its intended use, or the like. The transition is a word or phrase connecting the preamble to the body. The terms *comprises* or *comprising* often perform this function. The body is the listing of elements and limitations that define the scope of what is being claimed.

Claims are either *independent* or *dependent*. An independent claim stands on its own and makes no reference to any other claim. A dependent claim refers to another claim that may be independent or dependent, and adds to the subject matter of the reference claim. If a dependent claim depends from (makes reference to) more than one other claim, it is called a *multiple dependent* claim.

One type of claim format that can be used gained notoriety in a 1917 decision of the Commissioner of Patents, *Ex parte Jepson,* 1917 C.D. 62. In a claim of the *Jepson* format, the preamble recites all the elements deemed to be old, the body of the claim includes only such new elements as constitute improvements, and the transition separates the old from the new. The USPTO favors the use of *Jepson*-type claims since this type of claim is thought to assist in segregating what is old in the art from what the applicant claims as his or her invention.

In 1966, the USPTO sought to encourage the use of *Jepson*-type claims by prescribing the following rule 75(e):

> Where the nature of the case admits, as in the case of an improvement, any independent claim should contain in the following order, (1) a preamble comprising a general description of the elements or steps of the claimed combination which are conventional or known, (2) a phrase such as "wherein the improvement comprises," and (3) those elements, steps and/or relationships which constitute that portion of the claimed combination which the applicant considers as the new or improved portion.

Thankfully, the use of the term *should* in Rule 75(e) makes use of *Jepson*-type claims permissive rather than mandatory. Many instances occur when it is desirable to include several distinctly old elements in the body of the claim. The preamble in a *Jepson*-type claim has been held to constitute a limitation for purposes of determining patentability and infringement, while the preambles of claims presented in other types of format may not constitute limitations.

The royalties one can collect from a licensee often depend on how the claims of the licensed patent are worded. A two percent royalty on a threaded fastener used in a locomotive amounts to much less than a two percent royalty on a locomotive that is held together by the novel fastener; therefore, it may be wise for the patent to include claims directed to the locomotive instead of limiting all of the claims to features of the relatively inexpensive fastener. A proper understanding of the consequences of presenting claims in various formats, and of the benefits thereby obtained, should be taken into account by one's patent attorney.

3.7 Executing the Application

Once an inventor is satisfied with the content of a proposed patent application, he or she should read carefully the oath or declaration accompanying the application. The required content of this formal document recently has been simplified. In it, the inventor states that he or she

1. Has reviewed and understands the content of the application, including the claims, as amended by any amendment specifically referred to in the oath or declaration.
2. Believes the named inventor or inventors to be the original and first inventor or inventors of the subject matter which is claimed and for which a patent is sought.
3. Acknowledges the duty to disclose to the USPTO during examination of the application all information known to the person to be material to patentability.

If the application is being filed as a division, continuation, or continuation-in-part of one or more co-pending patent applications, the parent case or cases must be adequately identified in the oath or declaration. Additionally, if a claim to the benefit of a foreign-filed application is being made, the foreign-filed application must be adequately identified in the oath or declaration.

Absolutely no changes should be made in any part of a patent application once it has been executed. If some change, no matter how ridiculously minor, is found to be required after an application has been signed, the executed oath or declaration must be destroyed and a new one signed after the application has been corrected. If an application is executed without having been reviewed by the applicant or is altered after having been executed, it may be stricken from the files of the USPTO.

3.8 U.S. Patent and Trademark Office Fees

The USPTO charges a set of fees to file an application, a fee to issue a patent, fees to maintain a patent if it is to be kept alive for its full available term, and a host of other fees for such things as obtaining an extension of time to respond to an Office Action. The schedule of fees charged by the USPTO is updated periodically, usually resulting in fee increases. Such fee increases often take effect on or about October 1, when the government's new fiscal year begins. This has been known to result in increased numbers of September filings of applications during years when sizable fee increases have taken effect.

In addition to a basic filing fee of $300, $200 is charged for each independent claim in excess of a total of three, $50 is charged for each claim of any kind in excess of a total of 20, and $360 is charged for any application that includes one or more multiple dependent claims. Also, there is a search fee of $500, and an examination fee of $200 that must be paid either when an application is filed or shortly thereafter. However, if the applicant is entitled to claim the benefit of so-called *small entity status,* all of these fees are halved, as are most other fees that are associated with the prosecution and issuance of a patent application.

Provisional applications require a $200 filing fee that may be halved for small entities. Applications for design patents require a filing fee of $200, a search fee of $100, and an examination fee of $130, unless small entity status is claimed, whereupon these fees also may be halved. Plant patent applications require a $300 filing fee, a $300 search fee, and a $160 examination fee that are halved for small entities.

New rules now permit the USPTO to assign a filing date before the filing fee and oath or declaration have been received. While the filing fee and an oath or declaration are still needed to complete an application, a filing date will now be assigned as of the date of receipt of the descriptive portion of an application (known as the specification) accompanied by at least one claim, any required drawings, and a statement of the names and citizenships of the inventors.

The issue fee charged by the USPTO for issuing a utility patent on an allowed application stands at $1400. Establishing a right to the benefits of small entity status permits reduction of this fee to $700. The issue fee for a design application is $800, which also may be halved with the establishment of small entity status. A plant patent requires an issue fee of $1100, which may be halved for small entities. There is no issue fee associated with a provisional application since a provisional application does not issue as a patent unless it is supplemented within one year of its filing date by the filing of a complete utility application.

Maintenance fees must be paid to keep an issued utility patent in force during its term. No maintenance fees are charged on design or plant patents, or on utility patents that have issued from applications filed before December 12, 1980. As of this writing, maintenance fees of $900, $2300, and $3800 are due no later than $3\frac{1}{2}$, $7\frac{1}{2}$, and $11\frac{1}{2}$ years, respectively, from a utility patent's issue date. Qualification for the benefits of small entity status allows these fees to be reduced to $450, $1150, and $1900, respectively. Failure to timely pay any maintenance fee, or to late-pay it during a six-month grace period following its due date accompanied by a late payment surcharge of $130 ($65 for small entities), will cause a patent to lapse permanently.

3.9 Small Entity Status

Qualification for the benefit of small entity status requires only the filing of a claim to the right to pay fees at a small entity rate. All entities having rights with respect to an application or patent must each be able to qualify for small entity status; otherwise, small entity status cannot be achieved. Once a claim to the benefit of small entity status has been presented to the USPTO, there exists a continuing duty to advise the USPTO before or at the time of paying the next fee if qualification for small entity status has been lost.

Those who qualify for small entity status include

1. A sole inventor who has not transferred his or her rights and is under no obligation to transfer his or her rights to an entity that fails to qualify.
2. Joint inventors where no one among them has transferred his or her rights and is under no obligation to transfer his or her rights to an entity that fails to qualify.
3. A nonprofit organization such as an institution of higher education or an IRS-qualified and exempted nonprofit organization.
4. A small business that has no more than 500 employees after taking into account the average number of employees (including full-time, part-time, and temporary) during the fiscal year of the business entity in question, and of its affiliates, with the term *affiliate* being defined by a broad-reaching "control" test.

Attempting to establish small entity status fraudulently or claiming the right to such status improperly or through gross negligence, is considered a fraud on the USPTO. An application could be disallowed for such an act.

Failure to claim small entity status on a timely basis may forfeit the right to small entity status benefits with respect to a fee being paid. However, if a claim to small entity status is presented to the USPTO within two months after a fee was paid, a refund of the excess amount paid may be obtained.

3.10 Express Mail Filing

During 1983, a procedure was adopted by the USPTO that permits certain papers and fees to be filed in the USPTO by using the "Express Mail Post Office to Addressee" service of the U.S. Postal Service. When this is done, the filing date of the paper or fee will be the mailing date affixed to the "Express Mail" mailing label by USPS personnel.

To qualify for the filed-when-mailed advantage, each paper must bear the number of the "Express Mail" mailing label, must be addressed to the Commissioner for Patents, P.O. Box 1450, Alexandria, VA 22313-1450, and must comply with the other requirements that change from time to time.

4 PROSECUTING A PENDING PATENT APPLICATION

Once an executed patent application has been received by the USPTO, the patent application is said to be pending, and a serial number and filing date are assigned to it by the USPTO. The prosecution period of a patent application is the time during which an application is pending. It begins when a patent application is filed in the USPTO and continues until either a patent is granted or the application is abandoned. Activities that take place during this time are referred to as *prosecution*.

4.1 Patent Pending

Shortly after an application is filed, a filing receipt is sent to the applicant or the applicant's attorney. The receipt provides evidence of filing and sets out important data summarizing what the USPTO received (i.e., what was filed in the USPTO). It also lists the assigned serial number, filing date, application title, and inventor identification.

Once an application for a patent has been received by the USPTO, the applicant may mark products embodying the invention and literature or drawings relating to the invention with an indication of "Patent Pending" or "Patent Applied For." These expressions mean a patent application has been filed and has neither been abandoned nor issued as a patent. The terms do not mean that the USPTO has taken up examination of the merits of an application, much less approved the application for issuance as a patent.

Marking products with the designation "Patent Pending" often has the effect of discouraging competitors from copying an invention, whereby the term of the patent that eventually issues may be thought of as effectively extended to include the period during which the application is pending. In many instances, competitors will not risk a substantial investment in preparation for the manufacture and merchandising of a product when a competing product is marked with the designation "Patent Pending," for they know their efforts may be legally interrupted as soon as the patent issues.

4.2 Publication of Pending Applications

In an effort to harmonize U.S. patent practice with that of other countries, the USPTO now is "publishing" pending utility patent applications 18-months after each application is filed. If the application claims the benefit of the filing date of an earlier-filed application, then the 18-month publication date is measured from the first filing date that provides benefit. Early publication also can be requested by a patent applicant, but a number of formalities must be met and a fee must be paid if applicant-requested early publication is to take place.

A patent applicant can request that his or her application not be published, but the USPTO will honor this request only if the applicant presents it to the USPTO at the time the application is filed, and if the request is accompanied by a certification stating that the invention is not and will not be the subject of foreign or international applications (i.e., no foreign filing of corresponding cases will be made). If, after filing a nonpublication request, the applicant changes his or her mind and files abroad, then the applicant must notify the USPTO within 45 days so that publication can proceed on schedule in the USPTO.

To compensate patent applicants for a loss of secrecy when the content of pending applications is published, so-called "provisional rights" are provided by law that can be asserted against anyone practicing inventions claimed in a pending and published application for infringements that take place between the date of publication of the application and the date of issuance of the patent, so long as at least one infringed claim of the patent is "substantially identical" to a claim found in the published application.

Another factor to take into account in deciding whether to request nonpublication of a ready-to-file patent application is that, once the application is published, the file of the application, which, until publication, has been held in secrecy by the USPTO, becomes open to public inspection upon payment of a fee by someone who desires to inspect the file. When the USPTO "publishes" a pending utility patent application, they do it entirely electronically: no paper copies are published or distributed. To view a published pending application, one must visit the USPTO website *www.uspto.gov,* which provides a link to a search utility where information can be entered to search for published applications using key words, inventor names, or other pertinent data. Thus, the files of published pending applications are open to the public to virtually the same extent as are the files of issued patents, and members of the public can submit prior art and other information to the USPTO to be considered as the prosecution of the published application is brought to a close.

Applicants who want the USPTO to keep secret their pending utility application until the day it may issue must make that choice before they file the application, and must elect nonpublication at the time when they file the application in the USPTO. The USPTO will not honor a request made to maintain an application in confidence if the request is made at any time after the application is filed.

4.3 Duty of Candor

It is extremely important for patent applicants to deal candidly with the USPTO. In accordance with USPTO guidelines, a patent applicant is urged to submit an Information Disclosure Statement either concurrently with the filing of an application or within three months of its filing. An information disclosure statement should include a listing of patents, publications, or other information that is believed to be "material" and a concise explanation of the relevance of each listed item, and should be accompanied by copies of each listed reference that is not a U.S. patent. Items are deemed to be "material" where there is a "substantial likelihood that a reasonable examiner would consider it important in deciding whether to allow the application to issue as a patent."

To ensure that the USPTO will give due consideration to what is cited in an information disclosure statement, the information disclosure statement must be (1) filed within three months of the filing date of a normal U.S. application, or (2) within three months of entry of a U.S.-filed international application into its national stage, or (3) before the mailing date of a first Office communication treating the merits of the claimed invention (known as an "Office Action"), whatever occurs last. Consideration thereafter can be had only if other requirements are met, which typically include the certification of certain information, the filing of a petition, and/or the payment of a $180 fee. Information disclosure statements filed before the grant of a patent that do not comply with the requirements of the USPTO are not considered by the USPTO, but will be placed in the official file of the patent.

The courts have held that those who participate in proceedings before the USPTO have the "highest duty of candor and good faith." While the courts differ in their holding of the consequences of misconduct, fraud on the USPTO has been found to be a proper basis for taking a wide variety of punitive actions, such as striking applications from the records of the USPTO, canceling issued patents, denying enforcement of patents in infringement actions, awarding attorney's fees to defendants in infringement actions, and imposing criminal sanctions on those who were involved in fraudulently procuring patents. Inequitable conduct other than outright fraud has been recognized as a defense against enforcement of a patent, as a basis for awarding attorney's fees in an infringement action, and as a basis of antitrust liability.

In short, the duty of candor one has in dealings with the USPTO should be taken very seriously. Prudent practice would urge that if there is any question concerning whether a reference or other facts are "material," a citation should be made promptly to the USPTO so that the examiner can decide the issue.

4.4 Initial Review of an Application

Promptly after an application is filed, it is examined to make certain it is complete and satisfies formal requirements sufficiently to permit its being assigned a filing date and serial number. The classification of the subject matter of the claimed invention is determined and the application is assigned to the appropriate examining group. In the group, the application is assigned to a particular examiner. Examiners take up consideration of the applications assigned to them in the order of their filing.

Although more than 2000 examiners staff the USPTO, a backlog of several months of cases awaits action. This results in a delay of several months between the time an application is filed and when it receives its first thorough examination on the merits. Most newly filed patent applications receive a first Office Action treating the merits of their invention within about 14 months after the application was filed.

Once an examiner reaches an application and begins the initial review, he or she checks the application still further for compliance with formal requirements and conducts a search of the prior art to determine the novelty and nonobviousness of the claimed invention. The examiner prepares an Office Action, in which he or she notifies the applicant of any objections to the application or requirements regarding election of certain claims for present prosecution, and/or any rejections he or she believes should be made of the claims. In some instances, the examiner will find it necessary to object to the form of the application. One hopes that these formal objections are not debilitating and can be corrected by relatively minor amendments made in response to the Office Action.

In treating the merits of the claims, especially in the first Office Action, it is not uncommon for an examiner to reject the majority, if not all, of the claims. Some examiners feel strongly that they have a duty to cite the closest art they are able to find and to present rejections based on this art to encourage or force the inventor to put on record in the file of the application such arguments as are needed to illustrate to the public exactly how the claimed invention distinguishes patentably over the cited art.

In the event the examiner deems all the claims in an application to be patentable, he or she notifies the applicant, typically by issuing a notice of allowance.

4.5 Response to an Office Action

Unless an applicant can show that his or her delay in responding to an Office Action was "unavoidable," the applicant must respond to Office Actions within six months of their mailing dates (or sooner if the Office Action sets earlier deadlines). Failure to respond on a timely basis causes the application to become "abandoned." Most Office Actions set a two- or three-month deadline for responding that is measured from their mailing dates. If an extension of time is needed to respond, extensions can be obtained upon petition and payment of the requisite fee, but extensions cannot move the ultimate response deadline beyond six months from the date of mailing of an Office Action. The fees for one-, two-, and three-month extensions currently are $110, $430, and $980, respectively. Qualified small entities pay fees that are one-half of these amounts.

In the event the first Office Action issued by the examiner is adverse in any respect and/or leaves one or more issues unresolved, the applicant may reply in a variety of ways that constitute a bona fide attempt to advance the prosecution of the application. The applicant is entitled to at least one reconsideration by the USPTO following the issuance of the first Office Action; however, as a minimum, a response must present at least some argument or other basis for requesting reconsideration.

The rejection of one or more claims can be responded to by the presentation of arguments to persuade the examiner that the examiner's position is inappropriate or incorrect or fails to take into account important information; by amending the rejected claims to overcome or to sidestep the issues raised by the examiner; by submitting affidavits or declarations presenting additional information that the examiner should consider; by canceling the rejected claim or claims; and/or by adding new claims and arguing for their allowance. In responding to an Office Action, each objection and rejection made by the examiner must be treated. If the inventor agrees that certain of his or her claims should not be allowed in view of art cited by the examiner, these claims may be canceled or amended to better distinguish the claimed invention over the cited art.

Since the file of a patent application will become open to public inspection on issuance of a patent, and because an issued patent must be interpreted in view of the content of its file, the character of any arguments presented to the USPTO in support of a claimed invention are critical. Care must be taken in the drafting of arguments to ensure that no misrepresentations are made and that the arguments will not result in an unfavorable interpretation of allowed claims being made during the years when the resulting patent is in force. Great care also must be taken in drafting claim amendments, for any "narrowing" amendment will probably cause the amended claims to be interpreted very narrowly.

Years ago, it was not unusual for half a dozen or more Office Actions to be generated during the course of pendency of a patent application. During recent years, however, the USPTO has placed emphasis on "compacting" the prosecution of patent applications, and insists that responses to Office Actions make a genuine, full-fledged effort to advance the prosecution of the application. Today, it is not unusual for the prosecution of a patent application to be concluded on the issuance of the second or third Office Action.

If the USPTO has objected to the drawings, corrections must be made by providing substitute drawings that include the required corrections. The earlier that substitute drawings can be submitted by the applicant, the better.

4.6 Reconsideration in View of the Filing of a Response

Once the applicant has responded to an Office Action, the examiner reexamines the case and issues a new Office Action apprising the applicant of his or her findings. If the examiner agrees to allow all of the claims that remain active in the application, prosecution on the merits is closed and the applicant may not present further amendments or add other claims as a matter of right. If the Office Action is adverse with regard to the merits of the claims, the prosecution of the case continues until such time as the examiner issues an Office Action that presents a final rejection.

A rejection is made "final" once a clear and unresolved issue has developed between the examiner and the applicant. After a final rejection has issued, the character of the responses that may be made by the applicant is limited. The applicant may appeal the final rejection to an intra-agency Board of Patent Appeals and Interferences, cancel the rejected claims, comply with all the requirements for allowance if any have been laid down by the examiner, or file a request for continued examination (RCE) whereby the examination procedure is begun again.

If an initial appeal taken to the Board of Patent Appeals and Interferences should result in an unfavorable decision, a further appeal may be taken to either the U.S. District Court for the District of Columbia or to the U.S. Court of Appeals for the Federal Circuit. In some instances, further appeals may be pursued to a higher court.

In the majority of instances during the period of prosecution, the application eventually reaches a form acceptable to the examiner handling the application, and the examiner will issue a notice of allowance. If it is impossible to reach accord with the examiner handling the application, the inventor can make use of the procedures for appeal, or can continue the prosecution of the application by filing an RCE.

If the record of examination of an application does not otherwise reveal the reasons for allowance, an examiner may put a comment in the file explaining his or her reasons for allowing the claims that have been allowed. If a statement of reasons for allowance is provided by an examiner, it should be reviewed with care and commented upon, in writing, if the stated reason is incorrect or needs clarification.

4.7 Interviewing the Examiner

If, during the prosecution of a patent application, it appears that substantial differences of opinion or possible misunderstandings are being encountered in dealing with the examiner to whom the application has been assigned, it often is helpful for the attorney to conduct a personal interview with the examiner. While the applicant has a right to attend such a meeting, this right is best exercised sparingly and usually requires that the applicant spend time with the attorney to become better prepared to advance rather than to detract from the presentation the attorney plans to make.

Considering the relatively sterile and terse nature of many Office Actions, it may prove difficult to determine accurately what the examiner's opinion may be regarding how the application should be further prosecuted. If it has become clear that an examiner and an attorney are not communicating in the full sense of the word, a personal interview often will prove to yield valuable guidance for bringing the prosecution of the application to a successful conclusion. In other instances, an interview will be beneficial in more correctly ascertaining the true character of any difference of opinion between the applicant and the examiner, thereby enabling the exact nature of remaining obstacles to be addressed thoroughly in the applicant's next response.

4.8 Restriction and Election Requirements

If a patent examiner determines that an application contains claims to more than one independent and distinct invention, the examiner may impose what is called a *restriction requirement*. In the event the examiner finds that the application claims alternative modes or forms of an invention, he or she may require the applicant to elect one of these species for present prosecution. This is called a *species election requirement*.

Once a restriction or election requirement has been imposed, the applicant must elect one of the designated inventions or species for present prosecution in the original application. The applicant may file divisional applications on the nonelected inventions or species any time during the pendency of the original application, which often results in a plurality of related patents issuing on different aspects of what the inventor regards as a single invention.

When responding to an Office Action that includes a restriction and/or election requirement, arguments may be presented in an effort to traverse the requirement and request its reconsideration. After traversing, the examiner is obliged to reconsider the requirement, but he or she may repeat it and make it final. Sometimes, the examiner can be persuaded to modify or withdraw a restriction and/or election requirement, thereby permitting a larger number of claims to be considered during the prosecution of the pending application.

4.9 Double-Patenting Rejections

Occasionally, one may receive a rejection based on the doctrine of double-patenting. This doctrine precludes the issuance of a second patent on the same invention already claimed in a previously issued patent.

One approach to overcoming a double-patenting rejection is to establish a clear line of demarcation between the claimed subject matter of the second application and that of the earlier patent. If the line of demarcation is such that the claimed subject matter of the pending application is nonobvious in view of the invention claimed in the earlier patent, no double-patenting rejection is proper.

If the claimed subject matter amounts to an obvious variation of the claimed invention of the earlier issued patent, the double-patenting rejection often may be overcome by the filing of a terminal disclaimer. With a terminal disclaimer, a portion of the term of any patent issuing on the pending application is disclaimed so that any new patent issuing on the pending application will expire on the same day the already existing patent expires. If, however, the claimed subject matter of the pending application is identical to the claimed subject matter in the earlier issued patent, it is not possible to establish a line of demarcation between the two cases and the pending application is not patentable even if a terminal disclaimer is filed.

4.10 Continuation, Divisional, and Continuation-in-Part Applications

During the pendency of an application, it may be desirable to file either a continuation or a divisional application. A divisional application may be filed when two or more inventions are disclosed in the original application and claims to only one of these are considered during examination of the originally filed case.

It frequently occurs during the pendency of a patent application that a continuing program of research and development being conducted by the inventor results in the conception of improvements in the original invention. Because of a prohibition in the patent law against amending the content of a pending application to include "new matter," any improvements made in the invention after the time an application is filed cannot be incorporated into a pending application. When improvements are made that are deemed to merit patent protection, a continuation-in-part application is filed. Such an application can be filed only during the pendency of an earlier-filed application commonly called the *parent case*. The continuation-in-part case receives the benefit of the filing date of the parent case with regard to such subject matter as is common to the parent case. However, any subject matter that is not in common with the parent case is entitled only to the benefit of the filing date of the continuation-in-part case.

In some instances, when a continuation-in-part application has been filed, the improvements that form the subject matter of the continuation-in-part case are closely associated with the subject matter of the earlier-filed application, and the earlier application may be deliberately abandoned in favor of the continuation-in-part case. In other instances, the new matter that is the subject of the continuation-in-part application clearly constitutes an invention in and of itself. In such a situation, it may be desirable to continue the prosecution of the original application to obtain one patent that covers the invention claimed in the original application and a second patent that covers the improvement features.

4.11 Maintaining a Chain of Pending Applications

If a continuing development program is underway that produces a series of improvements, it can be highly advantageous to maintain on file in the USPTO a continuing series of pending applications—an unbroken chain of related cases. If an original parent application is initially filed and a series of continuation, division, and/or continuation-in-part applications are filed in a manner that ensures the existence of an uninterrupted chain of pending cases, any patent or patents that may issue on the earlier cases cannot be used as references cited by the USPTO as obstacles in the path of allowance of later applications in the chain. This technique of maintaining a series or chain of pending applications is an especially important technique to use when the danger exists that the closest prior art the USPTO may be able to cite against the products of a continuing research and development effort is the patent protection that issued on early aspects of this effort.

4.12 Patent Issuance

Once a notice of allowance has been mailed by the USPTO, the applicant has an inextensible period of three months to pay the issue fee. The notice of allowance usually is accompanied by a paper that advises whether the normal term of the patent is to be adjusted to compensate for delays caused during application prosecution by either the applicant or the USPTO. The applicant can present arguments for lengthening the term of the patent if he or she believes that the term adjustment should be modified.

Payment of the issue fee is a prerequisite to the issuance of a patent. If payment of the issue fee is unavoidably or unintentionally late, the application becomes abandoned, but usually can be revived within a year of the payment due date. Reviving an unintentionally abandoned application requires a much higher fee payment than does revival of an unavoidably abandoned application. A patent will not issue unless this fee is paid.

A few weeks before the patent issues, the USPTO mails a notice of issuance, which advises the applicant of the issue date and patent number.

Upon receipt of a newly issued patent, it should be reviewed with care to check for printing errors. If printing errors of misleading or otherwise significant nature are detected, it is desirable to petition for a Certificate of Correction. If errors of a clerical or typographical nature have been made by the applicant or by his or her attorney and if these errors are not the fault of the USPTO, a fee must be paid to obtain the issuance of a Certificate of Correction. If the errors are the fault of the USPTO, no such fee need be paid.

The issuance of a patent carries with it a presumption of validity. As was stated by Judge Markey in *Roper Corp. v. Litton Systems, Inc.,* 757 F.2d 1266 (Fed. Cir., 1985), "A patent is born valid and remains valid until a challenger proves it was stillborn or had birth defects." If the validity of a patent is put in question, the challenger has the burden of establishing invalidity by evidence that is clear and convincing.

4.13 Safeguarding the Original Patent Document

The original patent document merits appropriate safeguarding. It is printed on heavy bond paper, its pages are fastened securely together, and it bears the Official Seal of the USPTO. The patent owner should preserve this original document in a safe place as evidence of his or her proprietary interest in the invention. If an infringer must be sued, the patent owner may be called on to produce the original patent document in court.

4.14 Reissue

The applicant or owner of a patent may apply for a reissue patent to correct errors made without deceptive intent that cause a patent to be wholly or partly inoperative or invalid because of a defective specification or drawing or because the patentee claimed more or less than he had a right to claim. A reissue application may be filed in the USPTO to correct an obvious error in a specification, add a priority claim to obtain the benefit of the filing date of an earlier-filed application, cure claim indefiniteness, correct a misdesignation of inventors, broaden or narrow the scope of claims, and correct other errors resulting from an inadvertent or accidental mistake. If a patentee seeks to enlarge by reissue the scope of the claims of the original patent, he or she must file a broadening reissue application with two years of the date of issuance of the original patent. If the patentee seeks to narrow his claims or otherwise correct some defect in a patent through reissue (other than enlarging the scope of the claims of the patent as discussed in the preceding sentence) he or she may file a narrowing reissue application at any time during the term of the patent.

The filing fee for a reissue application is the same as the filing fee for an original application that has the same content. At least one error or defect must be stated in the oath or declaration that accompanies a reissue application. For an error to be correctable through reissue, it must have arisen without deceptive intention. A patentee may not seek by the route of reissue to change the claims in an issued patent so as to recapture subject matter that was intentionally surrendered to obtain the original patent. Moreover, the reissue application may not contain any new matter and must be directed to what was disclosed in the original patent.

A reissue application is examined in much the same way as original applications. If the grant of a reissue patent is approved, a reissue patent issued by the USPTO will replace the original patent and will expire on the same date that the original patent was set to expire.

4.15 Reexamination

Any person, including the owner of a patent and accused infringers, may file a request for reexamination of the validity of any claim in a patent. The request may be based on prior art or other facts not previously considered by the USPTO that the reissue request brings to the attention of the USPTO. The fee for filing a so-called *ex parte* reexamination request is $2520, and no small entity reduction is applicable. The fee for filing an *inter partes* reexamination that permits the filing entity to participate in the reexamination process is $8800, and no small entity reduction is applicable. If the USPTO decides that a reexamination request properly presents a new patentability question, it will issue a reexamination order.

Unlike *ex parte* reexamination, *inter partes* reexamination does not allow a challenger to remain anonymous. However, the two types of reexamination are similar in that each may consider the same type of prior art, and in each type the challenger must explain the significance of the prior art being cited as justifying the narrowing or cancellation of claims found in the original patent. In both types, the USPTO makes the ultimate decisions regarding patentability.

The reexamination of a patent provides the requesting entity with an opportunity to challenge the scope and utility of a patent without having to go to court to do so. Reexaminations are conducted by the USPTO, and claim amendments or cancellations are attended to in much the same way they are handled during prosecution of a normal application.

If an *inter partes* reexamination is conducted, interested third parties may participate on an adversarial basis by submitting arguments and by making proposals for the USPTO to consider. The only restriction on when an *inter partes* reexamination can take place is that a patent must be in existence (reexamination cannot take place until a patent actually issues). Through *inter partes* reexamination, an accused infringer, or a licensee of a patent, or any other interested person can attempt to narrow or invalidate a patent.

Once the reexamination of a patent is completed, a reexamination certificate is published by the USPTO which cancels any original claim of the subject patent determined to be unpatentable, confirms any original claim determined to be patentable, and incorporates into the patent any new claim determined to be patentable.

5 ENFORCING PATENTS AGAINST INFRINGERS

Patent rights cannot be enforced against infringers until a patent has issued. Beginning on the very day of patent issuance, however, efforts can be set in motion to curtail infringements by others and/or seek payments from them for the use of the patented invention.

5.1 Patent Infringement

The law recognizes several types of patent infringement. *Direct infringement* involves the making, using, offering for sale or selling within the United States, or importing into the United States of the entirety of an invention defined by a claim of the patent during the term of the patent. Since the implementation of the Process Patent Amendments Act of 1988, direct infringement also now includes making or using in the United States a product made by a process that is patented in the United States, which applies not only to domestic-origin but also to foreign-origin products. *Inducement of infringement* includes a number of activities by which one may intentionally cause, urge, encourage, or aid another to infringe a patent. *Contributory infringement* occurs when a person aids or abets direct infringement, as is set out in Section 271(c) of Title 35, which states

> Whoever offers to sell, or sells within the United States, a component of a patented machine, manufacture, combination or composition, or a material or apparatus for use in practicing a patented process, constituting a material part of the invention, knowing the same to be especially made or especially adapted for use in an infringement of such patent, and not a staple article or commodity of commerce suitable for substantial non-infringing use, shall be liable as a contributory infringer.

A patent infringement suit can be brought only in a federal district court where either (1) the infringer does business and infringement is committed or (2) the infringer resides. The suit must be brought within a six-year statute of limitations. An infringement action is initiated by filing a written complaint with the clerk of the district court where the suit is brought. A copy of the complaint is served on the defendant by the court. Either party in an infringement action may demand a jury, or the case can be tried by the judge if both parties waive their jury rights. During the suit, the plaintiff has the burden of establishing by competent evidence that the defendant's activities infringe the patent in question. In addition, the plaintiff must defend against any assertion made by the defendant that the plaintiff's patent is invalid.

The first months of an infringement suit usually bring extensive discovery, which involves demands for responses to written questions, document production, and witness depositions. Reports of expert witnesses are exchanged, and the experts are deposed. Either side may move for summary judgment, seeking a court ruling that will avoid the need for trial. A Markman hearing that may involve expert testimony is customary to determine claim interpretation. The trial itself may be divided into separate treatments of infringement and validity. If the plaintiff prevails on liability, the question of damages is then addressed.

Patent litigation is almost always expensive. Both the plaintiff and the defendant in a patent infringement case have to establish complex positions of fact and law. It is not unusual for each of the parties to a patent infringement suit to incur costs of at least a quarter million dollars, especially if the case is tried through appeal. For this reason, patent infringement litigation may not be economical unless there is a market at stake that is considerably larger than the anticipated costs of litigation.

5.2 Defenses to Patent Enforcement

Noninfringement is one of several traditional defenses an accused infringer can raise in his or her behalf. Three other substantive defenses are raised quite commonly in an effort to preclude the enforcement of a patent. The first of these is the defense of patent invalidity. While a patent is presumed to be valid, a court will normally inquire into its validity if the defendant puts the issue in question. Arguments favoring invalidity can be based on a variety of grounds, including the invention's lack of novelty, its obviousness, or insufficient disclosure of the invention in the patent being asserted. In view of the presumption of validity that arises with patent issuance, the relatively stringent standard of "clear and convincing evidence" must be met to establish patent invalidity.

A second, commonly asserted defense is that the patent is unenforceable as a result of fraudulent procurement or inequitable conduct by the patentee before the USPTO. The patent may be rendered unenforceable and the patent owner subject to other liabilities if the defendant can show by clear, unequivocal, and convincing evidence that a breach of the applicant's duty of candor has taken place through an intentional or grossly negligent misrepresentation or withholding of material fact. However, the type of showing required to establish fraud must exceed mere evidence of simple negligence, oversight, or erroneous judgment made in good faith.

Another commonly asserted defense is that of unenforceability due to so-called patent misuse and/or violation of the antitrust laws. If a patent owner has exploited the patent in an improper way by violating the antitrust laws or by effectively extending the patent beyond its lawful scope, the courts will refrain from assisting the patent owner in remedying infringement until the misuse is purged and its consequences dissipated.

5.3 Outcome of a Suit

If a patent owner successfully prevails in an infringement suit, the patent owner will be entitled to recover such damages as the patent owner can prove to have suffered, but not less than a reasonable royalty. In some situations, the patent owner may succeed in recovering the total amount of profits that would have resulted from the additional increment of business the patent owner would have enjoyed were it not for the activities of the infringer. If the court should find that the infringement was willful and deliberate, the defendant may be ordered to pay the plaintiff's reasonable attorney's fees, but this award is relatively rare. Should the court determine that the infringement has been flagrant and without any justification, it may award up to three times the damages actually found; however, such an award is rare.

If the accused infringer wins the suit, he or she normally recovers little more than court costs. In some exceptional cases, particularly when the court considers the plaintiff's legal action to be an abuse of the judicial process, the plaintiff may also be ordered to pay the defendant's reasonable attorney's fees.

5.4 Settling a Suit

Because of the expensive character of patent litigation, it is common for patent infringement suits to be settled before they come to trial. As discovery proceeds during the initial phases of patent infringement litigation, it is not uncommon for both sides to ascertain weaknesses in their positions and to note the desirability of settlement.

One point at which settlements are commonly effected is before a suit has been filed. In this situation, the patent owner confronts the infringer with the facts of the infringement, and a settlement is negotiated. Another time when settlement is commonly achieved is after there has been some initial discovery. By this time the positions of the parties have been clarified, and each party can begin to evaluate the other's strengths and weaknesses. A third point is after each side is ready to go trial. At this point, both sides clearly know the strengths and weaknesses of their cases, and they each may be anxious to eliminate the cost of an actual courtroom confrontation.

5.5 Declaratory Judgment Actions

In the event that one is accused of infringing another's patent or one's business is threatened by a possible suit for alleged patent infringement, one may bring a suit in federal court to have the threatening patent declared invalid or not infringed. The Federal Declaratory Judgment Act, passed in 1934, enables an alleged infringer to seize the initiative in this way and become a plaintiff, rather than a defendant, in a patent suit.

For a declaratory judgment action to succeed, the patent owner must actually have threatened to sue either the allegedly infringing manufacturer or its customers. An actual controversy must exist between the parties.

What the alleged infringer does in filing a declaratory judgment action is to seek a ruling from a federal court that the accused is not infringing a patent and/or that the patent is invalid. Filing such an action forces a patent owner who is threatening an alleged infringer or its customers to "put up or shut up." Accused infringers sometimes file such actions when it has become clear that they are going to be sued by a patent owner, for the first to file is usually able to select which available forum is most convenient for its use and most inconvenient and expensive for its opponent.

5.6 Failure to Sue Infringers

The laws of our country do not require patent owners to take any positive action whatsoever to enforce their patents. Many companies obtain patents and hold them defensively to preclude others from future use of certain inventions and as safeguards against the possibility of competitors patenting the same inventions. This practice is harmful to no one, because patents do not take from the public anything that was already in the public domain, but rather expedite the disclosure to the public of new inventions that will come into the public domain when the patents expire. For many companies, the principal value of their patents lies in the defensive uses they make of them.

5.7 Infringement by Government

If the federal government infringes your patent, the only action you may take is to sue for monetary damage. No injunctive relief is available. The same holds true if government contractors, operating with the consent of the federal government, use your invention to carry out the work of the federal government.

If state or local governments infringe your patent in carrying out a public service, much the same situation results. No injunctive relief will be available, but recovery of monetary damage for unauthorized use probably can be had.

5.8 Alternative Resolution of Patent Disputes

Because patent infringement litigation can sometimes cost each party millions of dollars from the filing of a complaint to the reading of the verdict, there has been an enormous interest in finding a better way that ensures greater expertise in the trier of fact, more confidentiality regarding information that must be disclosed, lower cost, and a faster pace.

Alternative dispute resolution (ADR) comes in many flavors. It may involve arbitration, mediation, the provision of an early evaluation by a neutral party, summary jury trials, and a variety of other approaches. Depending on the approach one selects, it may be faster. On the other hand, with some federal courts now running "rocket dockets" that force cases to trial within six to nine months after the filing of a complaint, it may not be faster.

Arbitration tends to be expensive, but it may eliminate the need for extensive discovery inasmuch as witnesses can be brought before the arbitrator by subpoena, if necessary. Mediation tends to be faster than a trial if both sides display a willingness to compromise. A feature of mediation is that it can be taken up at substantially any stage of the proceeding once the parties are willing to compromise and work toward settlement.

An approach that some have found worthwhile is to utilize a neutral party to provide an early evaluation. The evaluator is an experienced practitioner acceptable to both sides and selected for his or her expertise in technology and in patent trials. He or she reviews the written submissions of the parties, conducts meetings with the litigants, and then renders a nonbinding report that evaluates the strengths and weaknesses of each side. This may assist the parties in viewing their positions more realistically so a settlement can be reached.

An advantage of ADR is that people picked to hear the matter can be individuals who have appropriate expertise, thereby eliminating much of the need to educate the fact finder who is encountered at trial. Also, ADR can make it easier to maintain confidentiality than may be possible in the courtroom.

5.9 Interferences

An *interference* is a complex contest between applicants to determine who will receive patent rights. An interference may be encountered after your patent issues because the applicant in a pending application copies one or more of the claims from your patent, or during the prosecution of an application because two applications owned by different entities contain conflicting claims. Since an interference can arise after a patent has issued, one must not destroy one's records of early invention-related activities simply because a patent has issued.

Each party to an interference must submit evidence of facts that prove when the invention was made. If a party submits no such evidence, that party's earliest date is restricted to the date his application was filed. The question of priority is determined by a board of three administrative judges based on evidence submitted.

Priority of invention will normally be awarded to the first inventor to reduce an invention to practice. This is because the act of invention is deemed not to have been completed until an invention has been conceived and reduced to practice. An exception to the first-to-reduce-to-practice rule arises when the first to conceive has exercised reasonable diligence in reducing the invention to practice, but his diligent efforts did not result in his becoming the first to complete a reduction to practice. When this exception applies, the period of diligence of the first to conceive must extend from a time just before the second to conceive began its activities through the time of reduction to practice by the first to conceive.

Most foreign countries have no equivalent to the U.S. notion of interference because they operate on a first-to-file basis rather than on a first-to-invent basis. There has long been a debate as to whether the opportunity to participate in an interference is beneficial to U.S. applicants, for a first-to-file system would eliminate the need for these costly and complex contests. Few can afford these contests.

6 PATENT PROTECTIONS AVAILABLE ABROAD

U.S. patents provide no protection abroad and can be asserted against a person or corporate entity outside the United States only if that person or entity engages in infringing activity within the geographical borders of the United States. This section briefly outlines some of the factors one should consider if patent protection outside the United States is desired.

6.1 Canadian Filing

Many U.S. inventors file in Canada. Filing an application in Canada tends to be somewhat less expensive than filing in other countries outside the United States. With the exception of the stringently enforced unity requirements, which necessitates that all the claims in an application strictly define a single inventive concept, Canadian patent practice essentially parallels that of the United States. If one has success in prosecuting an application in the United States, it is not unusual for the Canadian Intellectual Property Office to agree to allow claims of substantially the same scope as those allowed in the United States.

6.2 Foreign Filing in Other Countries

Obtaining foreign patent protection on a country-by-country basis in countries other than Canada, particularly in non-English-speaking countries, has long been an expensive undertaking. In many foreign countries, local agents or attorneys must be employed. The requirements of the laws or each country must be met. Some countries exempt large areas of subject matter, such as pharmaceuticals, from what may be patented.

Filing abroad often necessitates that one provide a certified copy of the United States case for filing in each foreign country selected. Translations are needed in most non-English-speaking countries. In such countries as Japan, even the retyping of a patent application to put it in proper form can be costly.

With the exception of a few English-speaking countries, it is not at all uncommon for the cost of filing an application in a single foreign country to equal, if not substantially exceed, the costs that have been incurred in filing the original U.S. application. These seemingly unreasonably high costs prevail even though the U.S. application from which a foreign application is prepared already provides a basic draft of the essential elements of the foreign case.

6.3 Annual Maintenance Taxes and Working Requirements

In many foreign countries, annual fees must be paid to maintain the active status of a patent. Some countries require annual maintenance fee payments even during the time that the application remains pending. In some countries, the fees escalate each year on the theory that the invention must be worth more as it is more extensively put into practice. These annual maintenance fees not only benefit foreign economies, but also become so overwhelming in magnitude as to cause many patent owners to dedicate their foreign invention rights to the public. Maintaining patents in force in several foreign countries is often unjustifiably expensive.

In many foreign countries, there are requirements that an invention be "worked" or practiced within these countries if patents within these countries are to remain active. Licensing of a citizen or business entity domesticated within the country to practice an invention satisfies the working requirement in some countries.

6.4 Filing under International Convention

If applications are filed abroad within one year of the filing date of an earlier-filed U.S. case, the benefit of the filing date of the earlier-filed U.S. case usually can be attributed to the foreign applications. Filing within one year of the filing date of a U.S. case is known as filing under international convention. The convention referred to is the Paris Convention, which has been ratified by the United States and by most other major countries.

Most foreign countries do not provide the one-year grace period afforded by U.S. statute to file an application. Instead, most foreign countries require than an invention be "absolutely novel" at the time of filing of a patent application in these countries. If the U.S. application has been filed prior to any public disclosure of an invention, the absolute novelty requirements of most foreign countries can be met by filing applications in these countries under international convention, whereby the effective filing date of the foreign cases is the same as that of the U.S. case.

6.5 Filing on a Country-by-Country Basis

If one decides to file abroad, one approach is to file separate applications in each selected country. Many U.S. patent attorneys have associates in foreign countries with whom they work in pursuing patent protections abroad. It is customary for the U.S. attorney to advise a foreign associate about how he or she believes the prosecution of an application should be handled, but to leave final decisions to the expertise of the foreign associate.

6.6 The Patent Cooperation Treaty

Since June 1978, U.S. applicants have been able to file an application in the USPTO in accordance with the terms of the Patent Cooperation Treaty (PCT), which has been ratified by the United States and by the vast majority of developed countries. PCT member countries include such major countries as Australia, Austria, Belgium, Brazil, Canada, China, Denmark, Finland, France, Germany, Hungary, Japan, Mexico, the Netherlands, Norway, Russia, Sweden, Switzerland, the United Kingdom, and the United States. In filing a PCT case, a U.S. applicant can designate the application for eventual filing in the national offices of such other countries as have ratified the treaty.

One advantage of PCT filing is that applicants are afforded an additional eight months beyond the one-year period they would otherwise have had under the Paris Convention to decide whether they want to complete filings in the countries they have designated. Under the Patent Cooperation Treaty, applicants have 20 months from the filing date of their U.S. application to make the final foreign filing decision.

Another advantage of PCT filing is that it can be carried out literally at the last minute of the one-year convention period measured from the date of filing of a U.S. application. Thus, in situations where a decision to file abroad to effect filings has been postponed until it is impractical, if not impossible, to effect filings of separate applications in individual countries, a single PCT case can be filed on a timely basis in the USPTO designating the desired countries.

Still another feature of PCT filing is that, by the time the applicant must decide on whether to complete filings in designated countries, he or she has the benefit of the preliminary search report (a first Office Action) on which to base his or her decision. If the applicant had elected instead to file applications on a country-by-country basis under international convention, it is possible that he or she might not have received a first Office Action from the USPTO within the one year permitted for filing under international convention.

6.7 The European Patent Convention

Another option available to U.S. citizens since June 1978 is to file a single patent application to obtain protection in one or more of the countries of Europe, most of which are parties to the so-called *European Patent Convention* (EPC). Two routes are available to U.S. citizens to effect EPC filing. One is to act directly through a European patent agent or attorney. The other is to use PCT filing through the USPTO and to designate EPC filing as a *selected country*.

A European Patent Office (EPO) has been set up in Munich, Germany. Before applications are examined by the EPO in Munich, a Receiving Section located at The Hague inspects newly filed applications for form. A novelty search report on the state of the art is provided by the International Patent Institute at The Hague. Within 18 months of filing, The Hague will publish an application to seek views on patentability from interested parties. Once publication has been made and the examination fee paid by the applicant. examination moves to Munich, where a determination is made of patentability and prosecution is carried out with the applicant responding to objections received from the examiner. The EPO decides whether a patent will issue, after which time a copy of the patent application is transferred to the individual patent offices of the countries designated by the applicant. The effect of EPC filing is that, while only a single initial application need be filed and prosecuted, in the end, separate and distinct patents issue in the designated countries. Any resulting patents have terms of 20 years measured from the effective date of filing of the original application.

6.8 Advantages and Disadvantages of International Filing

An advantage of both PCT and EPC filing is that the required applications can be prepared in exactly the same format. Their form and content will be accepted in all countries that have adhered to the EPC and/or PCT programs. Therefore, the expense of producing applications in several different formats and in different languages is eliminated. The fact that both PCT and EPC applications can, in their initial stages, be prepared and prosecuted in the English language is another important advantage for U.S. citizens.

A principal disadvantage of both of these types of international patent filings is their cost. Before savings over the country-by-country approach are achieved, filing must be anticipated in several countries, perhaps as many as four to six, depending on which countries are selected. A disadvantage of EPC filing is that a single examination takes place for all the designated countries, and patent protection in all these countries is determined through this single examination procedure.

CHAPTER 23

ELECTRONIC INFORMATION RESOURCES: YOUR ONLINE SURVIVAL GUIDE

Robert N. Schwarzwalder, Jr.
University of Hawaii at Mānoa
Honolulu, Hawaii

1	BACKGROUND AND DEFINITIONS	758
2	OVERVIEW OF ONLINE INFORMATION	762
3	ACCESS OPTIONS FOR MECHANICAL ENGINEERS	762
	3.1 Access Options through Corporations and Universities	762
	3.2 Independent Access to Online Information	763
4	ONLINE SECURITY	763
5	GUIDE TO MECHANICAL ENGINEERING RESOURCES	765
	5.1 Database Services	765
	5.2 Databases of Importance to Mechanical Engineers	768
	5.3 Other Internet Resources	771
	5.4 Document Delivery Options	773
6	THE FUTURE OF ONLINE ACCESS	774
7	OPTIONS FOR USING ONLINE INFORMATION	775
	REFERENCES	776

1 BACKGROUND AND DEFINITIONS

In the 1998 edition of this handbook I explained how the Internet was revolutionizing information access. In 1998, use of the Web outside of academia was largely recreational and few engineers accepted that it could be a valuable source of legitimate information. A great deal has changed in these few years. The development of electronic publications, Internet commerce, corporate Intranets, and the ubiquitous nature of the Web have had a profound impact on how we obtain information for all aspects of our work and leisure activities. Now, finding a wealth of resources on any topic, no matter how obscure, is only a few keystrokes away. By entering a term in Google, one can obtain hundreds or thousands of results. We have gone from a society of paper transactions to one comfortable with online banking, electronic stock brokerages, and shopping on the Internet.

In 1998 most engineers were familiar with using traditional information resources, although their use of the engineering literature was limited. It has been repeatedly observed by information scientists and librarians that engineers make less frequent use of the technical literature than do scientists or other similar professionals.[1,3,4] Possible explanations for this include the esoteric nature of many technical publication series (technical papers, standards, government publications, etc.) and the difficulty many professional engineers have in gaining access to these publications. By making the less-available technical literature available to the desktop, the Internet has bypassed some of the obstacles to access and created an online option for engineers.

This explosion of information, complete with electronic books and journals, databases, Internet services, and digital archives has not necessarily made it easier to obtain the information you need. When confronted with the thousands of results from an Internet search, how do you respond? How do you pick the right options? Is the information you find trustworthy? And, what if the information you need is not among those results? In addition to the bewildering array of options, the once bucolic Internet has become somewhat perilous. Beyond the inconvenience of spam and pop-up windows are the very real hazards of email viruses and malware.

The Internet has matured since 1998, as have the options, risks, and opportunities open to you as an engineer. While the Internet can become your desktop information center, it can also be a source of frustration. Conducting Google searches of the Internet may be perfect for many inquiries, but you may find it an undependable strategy for obtaining technical information. However, there are a variety of services available that will provide you with rapid access to professional information. Your access to these collections and tools will depend on your work environment. Within universities and large corporations you should be able to use online tools and collections licensed for use in your organization. If you work independently or in a small company, you may need to privately contract to obtain access to these resources. Once connected, access to full-text versions of books, journals, standards, and patents will allow you to access the world's technical literature from your desktop.

In this chapter I provide an overview of how you can navigate the maze of professional information resources available through the Internet while minimizing the frustrations and pitfalls inherent with the Web. I discuss approaches for gaining access to suites of online services and content for mechanical engineers in variety of job environments. I also provide a listing of selected online resources for mechanical engineers.

Engineering information turned a corner in the late 1990s. As technical literature, standards, and engineering data blossomed on the Web, libraries and information service providers flocked to the new medium. Today, the speed of obtaining online information and the ease of embedding that information in proposals and reports has changed the way the profession works. If you have discovered online information you will know how much time and effort these new services save. If you have not, you may find that the opportunities involved are well worth changing the way you obtain information.

The following terms are related to the discussion of the Internet and online information resources.

Adware: Adware is advertising software that is downloaded onto your computer from the network that either continues to advertise a service or product on your computer or functions to alert a third party of your activities. Some adware serves a useful function such as alerting you to new versions of a product; some adware is clearly malicious, operating without your permission and against your wishes.

Agent (information agent): An information agent is a software device that filters information before it reaches the user, or locates and sends information to the user.

CD: A compact disc or CD, is a digital storage medium used for distributing software and small specialized databases. CDs are being replaced by the greater-capacity DVDs in many applications.

Cable modem: A cable modem provides high-speed, high-capacity connectivity to the Internet through the cable television network.

Client: A software application mounted on your computer that extracts some service from a server somewhere else on the network. This relationship is often referred to as a *client–server* application.

Cookies: Web cookies are files created during a web session that retain information about your identity and preferences. Cookies are used to help personalize a web session or retain information required during a complex transaction. Cookies are essential to many online commerce applications and are typically deleted at the end of a session. Persistent cookies are retained on your hard drive.

DSpace: DSpace is an open-source initiative begun by MIT Libraries and Hewlett-Packard that allows institutions to create digital repositories. DSpace initiatives seek to preserve free access to scholarly information.

Descriptor: A term used by a database producer to index a database record. Descriptors provide a consistent tag that can be used to retrieve all items relating to a given topic.

DSL: A digital subscriber line, or DSL, is a high-speed, high-capacity connection to the Internet through telephone lines.

DVD: DVDs are optical storage discs, similar in appearance to CDs, but capable of storing much more data.

Database: A computer-based search and retrieval system that allows a user to retrieve and display information based on a series of command protocols.

Datafile: A database containing numerical, chemical, or physical data.

Digital: A digital item is one that is available as a computer file and can be accessed online.

Downloading: The transfer of electronic data from a server to another computer, usually a personal computer.

Extranet: Otherwise know as a B-2-B or business-to-business network, an extranet is a collection of interconnected computers or networks open to members of an organization and its trusted business partners, but closed to external use through a variety of security protocols. An extranet is a bounded network upon which an organization shares private information with its business partners.

FAQ: Often a list of "frequently asked questions" appears on a web site. These questions and answers are intended to handle routine inquiries. FAQ lists are a good place to go if you are having problems using a web site.

FTP: File transfer protocol is a system for retrieving data or text files from a remote computer. FTP is less common due to higher limits on email file attachments, but is still regularly seen on the Internet.

Federated (or Meta) searching: A system that allows the user to simultaneously search more than one database and retrieve a combined set of results.

File server: A file server is a host machine that stores and provides access to files. Remote users may use FTP to obtain these files.

Firewall: A security system designed to keep unauthorized uses out of a computer or computer network. Network firewalls are often a combination of hardware and software, while personal firewalls typically consist of software alone.

GUI: (Pronounced "gooey") A graphical user interface, or GUI, is a system like the web that allows users to view and use graphics, as opposed to a text-only interface.

Host: A host is a network computer that contains resources that are shared by other members of the network.

Hostname: A hostname identifies a computer, or host, by the name of that machine and the name of the domain upon which it resides. The domain may contain a single computer or a group of computers. For example, *this.machine.com* and *that.machine.com* are two computers named *this* and *that* in the domain *machine.com*.

HTML: HyperText Markup Language is a system of computer language tags that are used to create web pages.

Identity theft: The misrepresentation of a person for financial gain using personal data stolen from computer files or other sources. Often identity theft is the motivation of malware developers.

Internet: A collection of interconnected networks that speak the Internet Protocol (IP) and related protocols. The Internet provides a variety of services, including email, file transfers, streaming video, and the Web.

Intranet: An internal Internet, a collection of interconnected computers or networks open to members of an organization, but closed to external use through a variety of security protocols.

Listserv: An email list devoted to a specific topic. Any message posted to a Listserv is forwarded to all of its members.

Malware: This is a broad term referring to any undesirable software, typically software that is installed on a computer without the owner's knowledge or permission. Malware consists of spyware, viruses, and some adware.

Node: A node is any computer on a network.

Online: An online, or network-based, resource is available to remote users through the Internet.

Open-source: A movement that developed in response to price escalation from large software developers. While terms of use vary significantly, most open-source software is free for all to use or modify as they wish.

OpenURL: A protocol and standard (ANSI/NISO Z39.88-2004) that allows the user to link directly from a database citation to the digital version of an article.

PDF: Portable Document Format is a computer platform independent electronic file format developed by Adobe Systems. PDF documents have become a popular standard for reproducing documents on the web since the format preserves the pagination and appearance of the original document.

Remote login: The process of accessing a host mounted on another network.

Server: This term is used to refer to (1) software that allows a computer to offer a service to another computer (i.e., client), or (2) the computer upon which the server software runs.

Shibboleth: This is an open-source initiative to create a shared protocol for shared access, network security, and rights management. While in an early stage of development at the time of this writing, Shibboleth could become a major access and security protocol in the next few years.

Spam: Undesired email, typically containing appeals to purchase questionable products, visit questionable web sites, or engage in questionable activities.

Spyware: Spyware is a term given to invasive or malicious software that monitors the activities on a computer and reports back to a third party.

URL: A universal resource locator, or URL, is an electronic address on the Web.

Virus: A virus is a malicious computer program that replicates itself and spreads through the Internet. Nearly all viruses cause some harm to the files or software on a computer, although some viruses are used to hijack a computer to send spam or for purposes of identity theft.

Web: The Web, World Wide Web, or WWW is a global system of interconnecting resources offering flexible multimedia coverage of a variety of topics.

Web browser: Software developed for accessing the Web.

Wiki: Taken from the Hawaiian word for fast and pronounced "wicky", a wiki is a web site that allows the users to add or edit the contents.

Wizard: A piece of software that leads the user through a series of steps to achieve some task. Wizards allow someone without technical background to perform complex tasks by reducing those tasks to a series of simple steps.

2 OVERVIEW OF ONLINE INFORMATION

A wealth of engineering information is available through the Internet. With a basic awareness of the range of information content and services open to you, a computer with Internet access, and a major credit card, you can have greater access to the current engineering literature than at any library in the world. Unlike much of the material available through the Web, these resources are not garbage and they are not free. The advantages of having access to these resources anytime or anywhere you wish are truly revolutionary. The risk of wasting time and money for the ill prepared are significant.

The focus of this chapter is to make you a savvy information consumer and to introduce you to resources and strategies to make you an effective and efficient consumer of these services. There are some excellent services in the Internet marketplace that will make your task far easier. Unfortunately, like any marketplace, the Internet also harbors thieves and hustlers. Before doing business on the Web, please consider the points raised in Section 4 of this chapter on Online Security. Even if you feel safe within the firewall of a corporate Intranet, you still could fall prey to Internet scams, viruses, or malware. Your vigilance is your, and your organization's, first line of defense.

3 ACCESS OPTIONS FOR MECHANICAL ENGINEERS

How you access online engineering information will largely be a function of the nature of your employment. If you are working in a corporation or large company, or if you are employed in a college or university, you should have access to a large collection of online collections through your corporate or university library. If you are working independently, or are employed by a small or medium-sized company that does not have a library, then you will want to contract for information access independently.

3.1 Access Options through Corporations and Universities

In most corporate and academic environments, your library will have subscribed to a collection of information resources and you will be able to use these resources without charge. The easiest way to get familiar with these collections is to contact your library and ask them what resources exist for mechanical engineers. You may also want to inquire about business resources since the engineering trade literature is sometimes covered by business databases. You will want to look for resources in three different areas: the online catalog, the collection of online books and journals, and the engineering databases.

- *Online catalog.* The online catalog allows you to search for books, journals, conference proceedings, and technical reports held by your institution's library. It will not identify articles within journals or conference proceedings, nor will it identify an item that is not held by that library. Library online catalogs allow searching by keyword, subject term, author name, or title. When an item is available as a digital file, the online catalog will typically link directly to the book or journal.

- *Online books and journals.* The publishers of engineering books and journals have moved strongly toward making their publications available through the Web. Large to medium corporate and academic libraries will have significant collections of online engineering literature that you can access directly from your lab, home, or office. Unfortunately, the publishers of trade literature have a more spotty record regarding online access.

- *Engineering databases.* To locate articles within engineering journals or conference proceedings, you need to search your topic in a topic-specific database. I have provided details on some of the major article databases for mechanical engineering in section 5.2. At this time a few libraries have implemented "federated searching," or "meta-searching" technologies that allow you to search multiple databases without selecting the individual database for your field. In most cases, you will need to know the correct database to use for a given topic. "OpenURL" is another technology that is gaining rapid acceptance at this time. With an OpenURL-enabled database, you can go directly from a citation in an engineering database to the digital article that the citation references. If your library has not implemented OpenURL, you will need to copy the information in the citations you obtain from the database and then go into the online catalog to determine if your library holds the items. If your library does not own the desired items, you may request copies through your library's document fulfillment service or Inter-Library Loan service. These services often involve an expense and a delay of days or weeks.

3.2 Independent Access to Online Information

It is possible that if you are not affiliated with a university you may still benefit from their engineering collections. Public universities retain the right to allow the general public to use their print and digital collections on site. If you visit the library of a large public university, you can typically use these digital collections from a public terminal in the library. These privileges do not extend to remote users unless they are directly affiliated with the institution. Corporate libraries never offer this option and private universities have varying policies.

If you cannot gain access to these resources through a public university or if you prefer the efficiency and convenience of using these materials from your office or home, there are a number of companies that provide online access to the mechanical engineering literature on a fee-per-use or subscription basis. A variety of options for online information access are provided in the following sections of this chapter.

4 ONLINE SECURITY

The Internet has become a major avenue to find and acquire technical information. Content providers and information services have rushed to establish businesses on the Web as more and more engineers, academicians, and business people have discovered the efficiencies of working online. This new interest in the Web has also attracted a new criminal element intent upon destruction and fraud. While these miscreants are not a reason to avoid using the Internet, you will want to exercise care in your online dealings.

If you are using the Internet through your company or university Intranet, the people who administer your system will have provided several layers of protection. These internal networks typically screen email and file attachments for viruses, and provide firewall protection against hackers who attempt to enter the system without permission. This does not mean that you should not be vigilant, but it offers some level of protection if you are not. If you are accessing the Internet from your personal computer, you should consider subscribing to a virus protection service as well as a firewall service. Since the people responsible for these threats are constantly developing new software, you need to subscribe to a service that is constantly updating its protective software. A virus protection service will recognize and stop a virus when it attempts to infect your computer. A firewall service will prevent hackers from gaining access to your computer. Firewalls are especially important if you are accessing the Internet though a cable-modem or DSL (Digital Subscriber Line). The two best known providers of these services are McAfee (http://www.mcafee.com/us/) and Symantec (http://enterprisesecurity.symantec.com/). Firewall and virus protection services will automatically recognize threats as they arise and do not require much effort to use. Most viruses are carried as email attachments and all email users are urged to take the following steps:

- Do not open attachments of email messages that seem generic, impersonal, or out of character with the sender. (Email viruses often send messages from an infected machine, so knowing the sender is no protection at all.)
- Do not open zipped attachments, attachments with an .exe file extension, or attachments with unusual file extensions unless you are expecting these formats.
- Be cautious if you get a large number of messages with similar subject lines in a brief time period, they may contain viruses.
- If you are using a Microsoft Windows operating system, visit the Microsoft web site (http://www.microsoft.com/windows/default.mspx) and look for critical security patches on a regular basis.
- Do not reply to spam, ever! (Some spam asks you to send a reply if you want to be taken off of their emailing list. This is a trick! Don't do it!)

A new threat to online users has developed, which is currently less well understood, but is as potentially dangerous as viruses. Spyware is a term applied to software that resides on your computer and reports back on your activities. Some spyware is relatively harmless and monitors your use of a site so that the provider can personalize its service. Some spyware observes your usage of other sites on the Internet for a variety of reasons. And some spyware records your keystrokes in an attempt to steal credit card numbers and passwords. You often unwittingly download spyware when visiting web sites or when you download a piece of software from the Internet. At best, spyware slows down your PC, sometimes to the point that it appears to be malfunctioning. At worst, spyware is a major identity-theft risk. To avoid problems with spyware, you should

- Read the term and conditions of any software you download very carefully. If the provider mentions monitoring your computer or your usage, do not accept the terms.
- Never accept downloads of software that appear in pop-up windows when you are visiting an unknown web site, unless those downloads are from a well-known company (Microsoft, Adobe, etc.).
- Restrict how your web browser uses cookies. You can do this by selecting "preferences" on the edit menu in Netscape and selecting the "Privacy & Security Option." In Internet Explorer, select "Internet Options" under the Tools menu. In the privacy section, set privacy to medium high or high. Many Web sites depend upon cookies to function properly, so I would not suggest turning off cookies entirely.

- Install and use, on a regular basis, a program to find and eliminate spyware from your computer. Two of my favorites are Ad-Aware (http://www.lavasoftusa.com/software/adaware/) and Spyware Search and Destroy (http://www.safer-networking.org/en/download/index.html).

5 GUIDE TO MECHANICAL ENGINEERING RESOURCES

In my chapter in the last edition of this handbook I presented a mixture of fee-based database services as well as a number of free Web-based resources for mechanical engineers. A great deal has changed since that time. The database services have become far easier to use, and far more accessible through the Internet. On the other hand, the availability of free information on the Web is much less than six years ago. Journals that provided some free content in the late 1990s have largely moved to a subscription model for their online content. Many of the sites offering free content have either gone out of business or decided that their "new economy" visions of the "dot com" era needed revision. The one notable exception to this is the United States federal government. The government has embraced the Web as an effective communications vehicle. The government Web service most directly applicable to engineering is the U.S. Patent and Trademark Office's free patent system, which is described in section 5.3.

The reduction in free online resources does not suggest that the Web has diminished as a vehicle for online engineering information. The speed, convenience, and popularity of the Web has made it the primary means through which to access online databases, electronic publications, and a wide range of information services. Direct dialups to database networks through computer modems is largely a thing of the past, as are databases on CD-ROM. This move to Web access has expanded access and allowed vendors to use more attractive and effective interfaces and to introduce electronic information access to a whole new audience.

5.1 Database Services

The information industry has grown considerably more complex in the last decade. By the early 1990s the companies that produced paper indexes to the engineering literature had developed computer-based citation databases. At that time, their only way of marketing these databases was through large commercial networks, like STN or Dialog, which provided dial-up access to academic and corporate clients. By the mid-1990s the database producers had found that they could market their databases directly to large institutions, and to individuals in the form of CD-ROMs. A few companies had begun to use the Web to provide fee-based database searching by the mid-1990s, a business practice at first widely criticized and then almost universally accepted. By the end of the 1990s, the large database networks were seeing defections as database producers began to perceive a more lucrative direct market through the Web. At this time, the information marketplace has settled down considerably. The suites of databases offered by the large networks are largely intact; however, the institution and individual have a large number of options when it comes to purchasing access on the wholesale or retail level.

How does this translate to you on a personal level? If you are employed by a large to medium-sized organization, corporate or academic, you may find that your employer has contracted directly with specific database producers to provide anyone employed there with database access. This is typically done through the Web. The database producer recognizes you as a member of a subscribing organization and grants you access. A number of database producers will use this same approach to sell you direct access to their database on an individual basis. If you desire access to a wide range of databases, you can gain access to collections of databases by setting up an account with one of the large networks of databases.

In the following sections I have provided lists of the major database networks and a sampling of databases relevant to mechanical engineering. I have indicated, next to the name of each database, the networks that provide coverage of that resource. This will provide you with a basic guide as to which database network might best meet your needs. If you are interested in subscribing to one of these networks, I would advise you to consult their full lists of databases to get a more complete idea of which is best for your particular needs.

Database Networks for Mechanical Engineers

STN

STN International is a network of scientific and engineering databases run through centers in the United States, Japan, and Germany. While smaller than Dialog in absolute terms, the dedicated focus on technical and scientific information makes STN a first choice for mechanical engineers who are concerned with materials data, automotive engineering, and aerospace engineering, and a good provider to consider for most other areas of the discipline.

While the service has become much more user-friendly over the last decade, offering a variety of simple interfaces to the service, STN provides an extremely powerful search system that is specialized for engineers and scientists. The ability to search for materials properties on STN is an example of the power and complexity of this system. STN's federated searching capacities allow you to search a number of databases simultaneously and remove duplicates, a valuable feature that increases the precision and comprehensiveness of your searches.

STN provides a variety of options to potential customers. STN Easy is designed for those who want to search the STN databases without learning the STN search protocols. STN Easy provides a search box form for users with options for simple keyword searching, similar to Google searching, or the ability to restrict searching by topic and the location of the term in a citation (e.g., title, abstract, subject heading). You can find out more about STN Easy at http://www.stn-international.de/stninterfaces/stneasy/stn_easy.html.

STN Express is a software package that provides you with a number of options for searching and displaying content from STN. STN Express includes a "wizard interface" that helps lead you through the details of accessing the system, selecting the appropriate database, and displaying and analyzing results. Details on STN Express can be found at http://www.stn-international.de/stninterfaces/stnexpress/stn_exp.html#f70.

And for those who intend to regularly search STN, you may want to consider STN on the Web, which gives you full command functionality to the entire suite of STN databases. Information on STN on the Web is located at http://www.stn-international.de/stninterfaces/stnow/stn_ow.html.

Dialog

Dialog is one of the first companies to develop commercial citation databases. Begun in 1966 under the leadership of Roger Summit, the company has become the worldwide leader in providing fee-based access to a large number of citation databases for all fields of study. The company changed hands repeatedly in the 1990s and languished somewhat during this period. In 2000 Dialog was purchased by the Thompson Corporation, a respected leader in the information industry, and has regained much of its former respect. Dialog provides access to a wide variety of databases of relevance to the mechanical engineer. While Dialog is not focused on the engineering and scientific disciplines, its vast size means that most technical fields are well covered.

The search interface is a nice balance of power and ease-of-use. If you are primarily interested in searching for citations of journal and conference proceedings articles, the ease of use of Dialog may be a deciding factor in your selection of this system. Engineers interested in more complex technical searching, or in materials property data, may find STN a better choice. Like STN, Dialog provides federated searching that allows you to search a number of databases simultaneously and remove duplicates.

Dialog provides a convenient way to use their databases through their Dialog Open Access program, available at http://www.dialog.com/openaccess/. Open Access users can perform searches and display lists of titles for free. Retrieving citations invokes a fee that can be charged to your credit card. The Open Access "Engineering" collection is appropriate for mechanical engineers, but there are other Open Access collections, such as "Business & News," and "Intellectual Property" that may be of interest.

Individuals requiring precision searching or who are more experienced may subscribe to DialogWeb at http://www.dialogweb.com/servlet/logon?Mode=1, or DialogClassic Web at http://www.dialogclassic.com/. These services require you to establish a subscription and to learn a bit more of Dialog's searching protocols, but give you more options for searching the Dialog databases.

CSA

Cambridge Scientific Abstracts, CSA, is a privately owned company that has been involved with indexing engineering and scientific literature for over 30 years. CSA provides access to over 50 databases covering a wide range of topics. While CSA is a far smaller network than STN or Dialog, it has developed as a formidable niche player. Strengths in the engineering disciplines include aerospace engineering and materials science. Individual subscriptions to CSA databases and document services provide you with desktop access to CSA's content and services. You can find out more about CSA, request a trial pass to their system, or subscribe at http://www.csa.com/csa/ids/ids-main.shtml.

Searching Strategies for Mechanical Engineers

Once you have selected a database network and a service option, you will need to learn how to use that system and determine the best database or databases for your needs. STN and Dialog provide a variety of tutorials, training materials, and abundant documentation—especially for their full-power searching options. CSA has a simpler interface, but also provides support for new users.

The selection of the best database or databases can be more of a problem. The information I have provided in Sections 5.2 of this chapter should provide some assistance. When selecting a database you need to consider the following aspects:

- The subject coverage of the database—does it match your search topic?
- The type of materials covered by the database—if you want numerical, physical property, or chemical data values you shouldn't be searching an index of the journal literature; you need a datafile.
- The scope of the database—if you need international coverage you should not search databases that index information only from the United States.
- The availability of the materials indexed in the database—while NTIS is an excellence source of U.S. technical reports, it won't help you if you cannot access these documents in time to meet your deadline.

In addition, you may want to consider how quickly a database is updated if you need something that has been recently published. This is especially true of societal paper series, which often take a year or longer to appear in databases other than those produced by the issuing society. Dialog's DIALINDEX file provides an extremely useful service to Dialog users by allowing them to preview a search on a number of different databases in order to determine the best databases to use for that search. DIALINDEX provides you with the number of items from each database that match your search specifications.

When searching databases most beginners are either too general or too specific. If your search query is too broad, you will retrieve a large set of records, most of which will have little relevance to your desired topic. If your search is too specific, you will retrieve nothing or a small subset of what the database contains on your subject. In the first case, you will waste time and money; in the second, you will miss vital information. The best way to approach searching is to start with a fairly broad keyword or phrase that describes your topic. If the results seem too numerous, or miss the mark, refine the search by adding in other keywords or concepts. By continuing to refine the search in this manner, you will arrive at a manageable set of results.

Another approach to searching is to find a few records that match your interest, display them, and determine which subject headings the database uses to describe the topic, and then use those subject headings to conduct a follow-up search. Most databases use subject headings, sometimes called "descriptors," to consistently index topics. A good database will have a thesaurus with a listing and explanation of its subject headings. If you are able to use subject headings in your searches you will improve both the precision and comprehensiveness of your searching.

In most searches the balance between the precision and comprehensiveness is a major issue. Most people intend to perform precision searches to get a few good references on a topic. If you attempt to expand that search you can reduce the number of terms in your search, include alternative terms relating to the same idea (e.g., engines, motors, powertrains), or use a broader term for the topic (e.g., welding in place of arc welding). However, by being more inclusive, you may retrieve a very large set of results. If you need to winnow down a large set of results, I would try two alternatives. First, limit your search to information published in the last few years. While these may not be the best articles on your topic, they will be current and should reference any major articles written in the last several years. Another strategy is to include title words such as "review," "summary," or "progress" in your search. (You can typically limit searches of particular keywords to specific fields, such as title, subject, or abstract.) These words tend to appear in the titles of articles that review progress in a given field. These review articles, when available, are a great way to get an overview of a field of investigation. Also, if you are investigating a technology that is new to you, don't forget to consult a good technical encyclopedia or handbook. These resources often provide a good starting place for a new project.

5.2 Databases of Importance to Mechanical Engineers

Thousands of commercial databases are available to you through the Internet, hundreds that relate to science and technology, and scores of interest to mechanical engineers. Finding the right database for your information needs is a critical first step in retrieving relevant information. This section provides a sampling of some of the most important databases for mechanical engineers. The list is not exhaustive for all subdisciplines of mechanical engineering. If you work in a large corporation or university, several of these databases may be available to you already. If you are searching through one of the database networks, you will want to consult their lists of offerings to determine if there are other databases you should be using.

Engineering Databases
These databases contain references to journal articles, conference papers, patents, books, and technical reports. One of the databases, ASMDATA, contains materials property data values.

These databases are available through a variety of online services or through your corporate or university library.

AEROSPACE (STN, CSA)
The AEROSPACE database covers the literature of aeronautics, astronautics, and space science. This database is the online version of the *International Aerospace Abstracts.* It is international in scope, contains almost 2.5 million records, and includes reports issued by NASA and other U.S. federal agencies. This database is a MUST for mechanical engineers involved in aeronautics or aerospace applications.

ALUMINIUM (STN, CSA)
This is the online version of the *Aluminum Industry Abstracts.* The database covers all aspects of the aluminum industry and contains approximately 250,000 records from 1968 to the present. This is a good complement to METADEX for anyone researching the aluminum literature.

ASMDATA (STN)
Unlike the other databases mentioned here, ASMDATA is a datafile. It contains materials property data for a wide range of metallic, composite, and polymeric materials. Data include materials identification and composition, mechanical properties, and physical properties. This datafile contains over 70,000 records and is a great resource.

Ei Compendex (Dialog, STN)
Ei Compendex is the online version of the venerable *Engineering Index,* which has indexed the literature of engineering since the late 1800s. This is the broadest and most comprehensive of the engineering databases, with coverage from 1970 to present. The database indexes approximately 4500 publications and contains approximately 5 million entries. While the scope of Ei Compendex is very broad, it is especially strong in mechanical engineering. The database uses a uniform set of subject terms to index articles. Using these subject terms greatly improves the accuracy of searching the database. While coverage is not limited to United States publications, the database is heavily weighted towards U.S. materials. While you would not want to restrict your engineering searching to this database, it is a MUST for the mechanical engineer.

FLUIDEX (Fluid Engineering Abstracts) (Dialog)
This database specializes in the journal literature relating to the engineering applications of fluids. It covers such topics as fluid mechanics, corrosion, hydraulics, pneumatics, pumps and compressors, tribology, and vessels. International in scope, the database contains approximately 450,000 entries and its coverage extends from 1972 to the present.

IHS International Standards and Specifications (Dialog)
The standards and specifications literature is vital for mechanical engineers and is well covered by Information Handling Services, IHS. This database is truly global in its coverage and excels in providing current, accurate information. The database also includes hard-to-find military specifications and some corporate standards and test methods. What makes the IHS database exceptionally useful is that it provides subject indexing of its contents—extremely helpful if you don't have a specific standard number—and it references past versions of standards and specifications, which is extremely helpful for finding updates or to understand how standards have evolved over time. Standards are some of the most difficult

of the engineering literatures and this database provides a number of features that will save you time and effort.

INSPEC (Dialog, STN)
While INSPEC is more suited to electrical engineering and computer science, it does provide strong coverage of some computer controls and automation literature that might be of interest to mechanical engineers. The database covers materials, but the emphasis is on advanced materials and electrical and electronic materials. This is an immense database with over 8 million records from the global engineering literature.

Mechanical & Transport Engineering Abstracts (CSA)
This database contains approximately 400,000 records dealing with mechanical and transport engineering. It covers a wide range of mobility topics from industrial materials handling equipment and agricultural machinery to satellites, rockets, and missiles. More recent entries contain cited references and contact information for the authors.

METADEX (STN, CSA)
METADEX indexes the global literatures of metallurgy and metallic materials. While specialized in scope, this is a large, well-organized database with over 1.3 million records. Coverage of all aspects of metallurgy are included, from ores to forming to corrosion. This is an excellent database for anyone searching for information related to metals.

MIRA—Motor Industry Research (Dialog)
MIRA is a British database of the automotive industry. It covers all aspects of the automotive industry and is not restricted to engineering literature. While MIRA claims to be International in scope, it is very far from being comprehensive in that regard. It is a relatively small database with approximately 150,000 entries. Automotive engineers should search the SAE Global Mobility Database and use MIRA to complement that database when British/European coverage is desired.

1MOBILITY (STN)
The 1MOBILITY database is produced by the United States' Society of Automotive Engineers and contains approximately 140,000 records. While marginally smaller than its British counterpart MIRA, the 1MOBILITY database emphasizes engineering literature and does not contain the broader industry coverage (news, business information, etc.) that is covered by MIRA. While the majority of the items included in 1MOBILITY are published in the United States, SAE does include publications of automotive societies from other countries. Above all, this database is the only comprehensive index of the voluminous SAE Paper series, a literature of vital importance to automotive engineers.

NTIS—National Technical Information Service (Dialog, STN, CSA)
This is a problematic, but important database. Problematic because it is inconsistent in how it indexes its contents, incomplete in its coverage, and contains references to materials that may be difficult to locate. However, this database of over 2 million entries contains federal technical reports that are almost impossible to find in other databases. Beyond work done by the federal government agencies, this database also contains reports from Small Business Innovation Research Grants and other technical work done by the private sector under government grant or contract. The NTIS literature is a treasure trove of engineering information. Most of the reports indexed in NTIS are available on microfiche from university libraries.

WELDASEARCH (Dialog, STN, CSA)
This database specializes in indexing journal articles, conference proceedings and a variety of other sources relating to welding, joining, and allied technologies. It is a relatively small database with fewer than 200,000 entries, but is a good place to turn if you are searching for technical information related to welding.

Additional Databases of Relevance to Mechanical Engineers
ABI/INFORM (Dialog, STN)
This is an excellent database of the mainstream business and management literature, indexing scholarly literature, trade publications, and popular business periodicals. It is an excellent place to find out what the press is saying about a company or a technology.

AP News (Dialog)
The Associated Press (AP) News database contains the full text of news, sports, and financial information from around the globe. This database compiles news from 141 U.S and 83 overseas news bureaus and has access to news from 1500 newspapers and 6000 radio and television outlets in the United States. AP News is constantly updated and offers extremely timely coverage of breaking news.

Patent Databases
All three of the database networks mentioned in this chapter—DIALOG, STN, and CSA—have databases that index United States and international patents. While most people avoid getting involved with the patent literature unless they are interested in patenting something, this is an essential literature for engineers. Because patents are often the only source of technical literature for proprietary technologies, patents provide your only source of intelligence on some of the most important technologies in use today. The corporations responsible for funding these innovations have no desire to publish information related to the work, but are required to provide in-depth technical information and illustrations in their patent applications. By examining these patents you gain access to cutting-edge technologies, the ability to evaluate the technical validity of new technologies, and insights on the R&D strategies of your competition.

If you want to perform simple searches for U.S. patents, you should use the free service provided by the United States Patent and Trademark Office (USPTO) referenced in section 5.3. If you need to conduct more complex searches of the patent literature or need to locate non-U.S. patents or international patent "families" you should consider using one of the Derwent patent databases available on either STN or DIALOG. However, please be aware that complex patent searching can be extremely difficult. If the search is complex and the results could have a significant economic impact on you or your organization, it would be prudent to arrange for an expert patent searcher to handle the inquiry.

5.3 Other Internet Resources

Engineering Village 2 (EV2)
http://www.engineeringvillage2.org/controller/servlet/Controller?CID=quickSearch&database=1

Engineering Information, Inc. was one of the first database producers to market their service directly to engineers through the Web. EV2 is the latest iteration of the Engineering Village and provides one-stop shopping for your information needs. The EV2 combines access to Compendex, Inspec, and NTIS, with convenient documentation services. (EV2 provides full-text access to journal articles from publishing giant Elsevier and access to document delivery services for other items.) EV2 also includes access to over 300 full-text engineering handbooks—more than 60 of which pertain to mechanical engineering. The "Village" concept embodies the idea of community and to that end the site includes on-call information professions, who can assist with searching, as well as engineers, who can assist on technical issues. The *Easy Search* option allows you to combine the ease of simple keyword searching with the power of clustering technology to provide you with a powerful, yet intuitive way to conduct database searching. While EV2 is not as comprehensive as searching STN, Dialog, or CSA, it provides strong coverage for mechanical engineering. This coverage, in addition to the EV2's large collection of handbooks, convenient document delivery options, and on-call experts, make it a resource worth investigating.

Knovel
http://www.knovel.com

Knovel, pronounced "novel," is a unique service directed at engineers and scientists (Fig. 1). Knovel has developed a technology that adds functionality to engineering and scientific handbooks. Knovel's interactive tables allow you to find data in a table, filter and export data from tables, sort the contents of tables, graph and interact with data from selected tables, and much more. The software even allows you to digitize the images of curves and

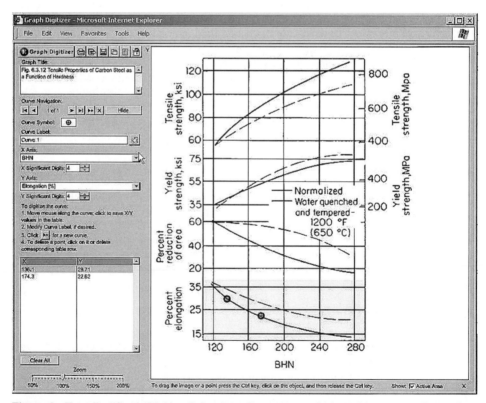

Figure 1 Knovel's "Graph Digitizer," function allows you to obtain and export coordinates from graphic representations of data ranges. This is one of several unique features of this online service. Reproduced with permission from the Knovel Corporation.

graphs and export those data. Handbooks are a vital literature for mechanical engineers and Knovel has done a remarkable job of adding new functionality to this old standby. To learn more about Knovel visit their Website; for information on personal subscriptions, email sales@knovel.com.

Scitation
http://scitation.aip.org/
   Scitation is a service of the American Institute of Physics (AIP) that provides online access to over 100 online journals from a variety of science and technology publishers, including ASME. Scitation provides free searching of these publications. If you have personal or institutional subscriptions to the electronic journals indexed at the site, you can connect to online articles directly through Scitation. If you do not, Scitation provides *DocumentStore* through which you can purchase access to online articles. The *MyArticles* feature of the web site allows you to create collections of links to articles of interest. While Scitation is not a major source of mechanical engineering information, it is innovative in its design and business model. The idea of providing free searching across content from a variety of publishers and allowing the user to purchase individual articles on demand is a model well worth imitation. The AIP has a long history of promoting cost-effective access to information. Their creation of Scitation demonstrates their continued dedication to the engineering and scientific community.

Society of Automotive Engineers (SAE) Store
http://store.sae.org/
   Despite a confusing web site, the SAE Store provides online access to the databases and publications of the United States' Society of Automotive Engineers. SAE plays a key role in publishing and indexing automotive technical literature and standards. Because of disagreements with other commercial databases, SAE's important paper series tends not to be well indexed outside of SAE's own Global Mobility Database, which is available directly from SAE or through STN as "1MOBILITY." When purchasing information from SAE exercise care that you are buying only what you need. SAE tends to repackage their publications extensively; marketing subsets of their Global Mobility database and SAE Paper series as separate CD-ROMs or bound "Special Publications" volumes. (You can distinguish SAE Special Publications from SAE technical books by their "SP" notation.) If you have a niche interest in mobility engineering, these options may provide excellent coverage for your interests. If you have a broader interest in automotive engineering, you will want to take steps to ensure that you are not purchasing the same information more than once.

United States Patent and Trademark Office (USPTO) Patent Database
http://www.uspto.gov/patft/index.html
   The USPTO provides free searching and display of granted patents and patent applications. (However, not all patent applications are included in the patent applications database.) While this service does not provide the indexing of the Derwent databases (above) and includes only United States patents, it is a free service. At the very least, it is a very good place to start your patent searches and a wonderful service of the USPTO.

5.4 Document Delivery Options

For online database users, the desired outcome is not obtaining journal references, it is obtaining the documents themselves. This reality is not lost on the large database networks. They provide a variety of document delivery options so that you can obtain the materials you find referenced in their databases. As has been noted, the Engineering Village 2 (EV2) provides direct links to some full-text articles as well as links to document delivery options for conference proceedings. Most large corporations and university libraries provide OpenURL linkages between their database and journal subscriptions so that users can go directly from the database citation to the full-text documents. Given the rapid pace of progress in this area, I believe that document access for database users will continue to improve dramatically.

Occasionally you will want to obtain a journal article or standard independent of a database search. If you are not using the document delivery services connected to a database network and do not have access to the document services of a corporate or university library, there are additional avenues through which to obtain your document. There are independent document delivery services available through the Web. Typically these services require you to set up an account prior to requesting documents, but this is a fairly fast and painless process. There are a number of companies performing this service. Three of the better known are

- CISTI (Canada Institute for Scientific and Technical Information)
  http://cisti-icist.nrc-cnrc.gc.ca/docdel/docdel_e.shtml
- Infotrieve
  http://www4.infotrieve.com/default.asp
- Ingenta
  http://www.ingenta.com

6 THE FUTURE OF ONLINE ACCESS

Predicting the future is a tricky business and one rich with opportunity for embarrassment. That said, there are some significant technological and industry trends that do strongly suggest some near-term impacts on online information users. From an industry perspective, there has been more cooperation between academic publishers and greater consolidation in the information industry. These changes have led to advances in increased access to online documents. The OpenURL standard, approved in 2004, allows the user to link from a journal citation in an online database to an online journal held in a local collection. This new standard has been enthusiastically accepted and already several new OpenURL software providers have emerged to meet the demand for the service. Within a few years, this type of database-to-online journal access will become commonplace in university libraries.

Federated searching has been around for a few years and continues to gain ground as users see the advantages in conducting a single search across multiple databases. New services combining federated searching and OpenURL have a limited install base, but are gaining in popularity.

A new trend in university libraries involves the creation of digital repositories of local collections, and selected materials relating to research conducted at the universities involved. The DSpace initiative, begun by MIT Libraries and Hewlett-Packard has attracted the interest of a small, but growing number of libraries. Corporate libraries have long maintained restricted databases of internal information resources open to some or all employees of their companies. While university libraries are not proprietary in focus, they tend to restrict access to their digital collections in order to accommodate copyright concerns. As corporate and academic libraries develop these internal collections, they will have remarkable opportunities to allow federated searching among the external, commercial technical literatures and the internal, restricted information in their own digital collections.

Access to these rich repositories of information will be open to appropriate collaborators through the developing Shibboleth initiative (http://shibboleth.internet2.edu/). When implemented, Shibboleth will help create a secure collaborative environment within which colleagues can share and develop technologies. The landmark "Atkins" report (2003) on Cyberinfrastructure provides an intriguing vision of the changes required in networks, computing, and information infrastructure to support science and engineering in the coming decades.

Information personalization has been an elusive goal for many in the information industry. While some advances have been made, agent-based technologies—which seem the most technically promising—have largely languished. Information agents are software devices that carry the information preferences of their human sponsors and retrieve information that match those specifications. "Intelligent agents" are capable of modifying these retrieval specifications as they interact with new information resources, sending their sponsors information that accommodates changing trends and patterns. While promising, the complexity of agent-based retrieval has proven a barrier to implementation, making this a more speculative technical prediction. However, as the costs of cycles and storage continue to decrease, I believe that commercial information agent applications will arise in the next decade.

Given these trends, we can anticipate networks and technologies of increasing complexity, which facilitate information products and services that are easier to use and more tailored to individuals' needs and preferences. As more of the engineering literature moves to digital formats and publishers develop better economic models for selling online access, you can anticipate more availability of direct online access to engineering books and articles. Beyond the computer-based access to the engineering literature, companies and universities are now turning to the Web to offer distance training and learning opportunities. Distance education has already become a huge business, with major universities offering courses and degree programs online. While most of the specialized technical training available through the Web still addresses software products, expect to see this change as corporations realize the benefits of this type of training.

One of the most exciting, and futuristic, visions of how the Internet can impact the workplace involves the use of these networks for distributed collaboration. While we already have some limited tools for online collaboration, such as IP-videoconferencing and shared white-space applications, the vision detailed in the report of the NSF Advisory Panel on Infrastructure[2] anticipates the use of collaborative technologies and advanced computing to create a new Internet-based virtual workplace. While this may sound ambitious, consider how it would have sounded 10 years ago to predict that people would be doing their banking over the Internet by the year 2000!

7 OPTIONS FOR USING ONLINE INFORMATION

In this chapter I have outlined a variety of strategies for accessing online information and reviewed a number of content and service providers and a sampling of databases. By using these tools you can create your own unique desktop library, with access to the global literatures of engineering, business and trade news, intellectual property, government R&D, and much more. Never in the history of humankind has so much information been so readily available to so many.

What does this mean for you as an engineer? It creates an unparalleled opportunity to

- Investigate unfamiliar technologies for new projects.
- Determine the best technologies and suppliers to use for an assignment.

- Research the latest trends in a technical area.
- Get background material on a person, company, or event prior to a meeting.
- Find the latest version of an industry standard or military specification.
- Locate patents for background on proprietary technologies.
- Research a topic—for a quick update or a comprehensive review.

The bottom line is that online information research saves you time, gives you more technical options, helps you avoid "reinventing the wheel," and gives you an edge on your competition by making you better informed. This is true whether you are preparing a grant proposal or involved in a "fire-fighting" emergency in an industrial setting. If you are missing information, or using slow, traditional means of finding it, you are putting yourself and your organization at a competitive disadvantage!

The most significant change in online access to information in the last decade is the increased attention to you, the user. Not only are database services easier to use, but also electronic documents and convenient document delivery options now provide a continuity of service that really address the needs of a working engineer. By putting the necessary information into your hands—any time and anywhere you need to be—online information can get you started on a new project or help you survive an ISO 9000 audit. In the 21st century, online information is an indispensable tool for the mechanical engineer.

REFERENCES

1. T. Allen, *Managing the Flow of Technology: Technical Transfer and the Dissemination of Technical Information Within the R&D Organization.* MIT Press, Cambridge, MA, 1977.
2. D. Atkins (chair), *Revolutionizing Science and Engineering through Cyberinfrastructure: Report of the National Science Foundation Advisory panel on Cyberinfrastructure.* February 3, 2003. (http://www.communitytechnology.org/nsf_ci_report)
3. M. Holland, "Engineering," in *Scientific and Technical Libraries,* Vol. 1: *Functions and Management,"* N. Pruett (ed.), Academic Press, New York, 1986, pp. 119–142.
4. T. Pinelli, "Distinguishing Engineers from Scientists—The Case for an Engineering Knowledge Community," *Science & Technology Libraries,* **21**(3/4): 131–163 (2001).

CHAPTER 24
SOURCES OF MECHANICAL ENGINEERING INFORMATION

Fritz Dusold
Retired from Mid-Manhattan Library
Science and Business Department
New York, New York

Myer Kutz
Myer Kutz Associates, Inc.
Delmar, New York

1	INTRODUCTION	777	5 CODES, SPECIFICATIONS, AND STANDARDS	780
2	THE PRIMARY LITERATURE	777		
	2.1 Periodicals	778	6 GOVERNMENT PUBLICATIONS	781
	2.2 Conference Proceedings	778		
3	INDEXES AND ABSTRACTS	778	7 ENGINEERING SOCIETIES	781
	3.1 Manual Searching	779		
	3.2 Online Searching	779	8 LIBRARIES	782
4	ENCYCLOPEDIAS AND HANDBOOKS	779	9 INFORMATION BROKERS	783
			REFERENCES	783

1 INTRODUCTION

This chapter is designed to enable the engineer to find information efficiently and to take advantage of all available information. The emphasis is placed on publications and services designed to identify and obtain information. Because of space limitations references to individual works, which contain the required information, are limited to a few outstanding or unusual items.

2 THE PRIMARY LITERATURE

The most important source of information is the primary literature. It consists mainly of the articles published in periodicals and of papers presented at conferences. New discoveries are first reported in the primary literature. It is, therefore, a major source of current information. Peer review and editorial scrutiny, prior to publication of an article, are imposed to ensure that the article passes a standard of quality. Most engineers are familiar with a few publications, but are not aware of the extent of the total production of primary literature. *Engineering Index*[1] (known as *Compendex* in its electronic version) alone abstracts material from thousands of periodicals and conferences.

Handbooks and encyclopedias are part of the secondary literature. They are derived from primary sources and make frequent references to periodicals. Handbooks and encyclopedias are arranged to present related materials in an organized fashion and provide quick access to information in a condensed form.

While monographs—books written for professionals—are either primary or secondary sources of knowledge and information, textbooks are part of the tertiary literature. They are derived from primary and secondary sources. Textbooks provide extensive explanations and proofs for the material covered to provide the student with an opportunity to understand a subject thoroughly.

2.1 Periodicals

In most periodicals published by societies and commercial publishers, articles are identified usually by issue, and/or volume, date, and page number. Bibliographic control is excellent, and it is usually a routing matter to obtain a copy of a desired article. But some problems do exist. The two most common are periodicals that are known by more than one name, and the use of nonstandard abbreviations. Both of these problems could be solved by using the International Standard Serial Number (ISSN), which accurately identifies each publication. The increasing size and use of automated databases should provide an impetus to increased use of ISSN or some other standard.

The first scientific periodical, *Le Journal des Scavans,* was published January 4, 1665. The second, *Philosophical Transactions* of the Royal Society (London), appeared on March 6, 1665. The number of scientific periodicals has been increasing steadily with some setbacks caused by wars and natural catastrophies. The accumulated body of knowledge is tremendous. Much of this information can be retrieved by consulting indexes, abstracts, and bibliographies.

2.2 Conference Proceedings

The bibliographic control for papers presented at conferences is not nearly as good as for periodicals. The responsibility for publishing the papers usually falls upon the sponsoring agent or host group. For major conferences the sponsoring agency is frequently a professional society or a department of a university. In these cases an individual with some experience in publishing is usually found to act as an editor of the *Proceedings.* In other instances the papers are issued prior to the conference as *Preprints.* In still other situations the papers will be published in a periodical as a special issue or distributed over several issues of one or more periodicals. An additional, unknown, percentage of papers are never published and are only available in manuscript form from the author.

3 INDEXES AND ABSTRACTS

Toward the end of the last century the periodical literature had reached a volume that made it impossible for the "educated man" to review all publications. In order to retrieve the desired or needed information, indexes and abstracts were prepared by individual libraries and professional societies. In the 1960s computers became available for storing and manipulating information. This lead to the creation and marketing of automated data banks.

3.1 Manual Searching

The major abstracts typically provide the name of the author, a brief abstract of the article, and the title of the article, and identify where the article was published. Alphabetical author and subject indexes are usually provided, and a unique number is assigned to refer to the abstract. Many abstracts are published monthly or more frequently. Annual cumulations are available in many cases. The most important abstracts for engineers are:

Engineering Index[1]

Science Abstracts[2]

    Series A: Physics Abstracts

    Series B: Electrical and Electronics Abstracts

    Series V: Computer and Control Abstracts

Chemical Abstracts[3]

Metals Abstracts[4]

A comprehensive listing of abstracts and indexes can be found in *Ulrich's International Periodical Directory.*[5]

3.2 Online Searching

Most of the major indexes and abstracts are now available in machine-readable form. For a comprehensive list of databases and online vendors see *Information Industry Market Place.*[6] The names of online databases frequently differ from their paper counterparts. *Engineering Index,*[1] for example, offers COMPENDEX and EI Engineering Meetings online. Most of the professional societies producing online databases will undertake a literature search. A society member is frequently entitled to reduced charges for this service. In addition to indexes and abstracts, periodicals, encyclopedias, and handbooks are available online. There seems to be virtually no limit to the information that can be made available online or on CD-ROMs, which can be networked in large institutions with many potential users. The high demand for quick information retrieval ensures the expansion of this service.

In addition to the online indexes, several library networks and consortia, such as OCLC, the Online Computer Library Center, located in Columbus, Ohio, produce online databases. These are essentially equivalent to the catalogs of member libraries and can be used to determine which library owns a particular book or subscribes to a particular periodical.

4 ENCYCLOPEDIAS AND HANDBOOKS

There are well over 300 encyclopedias and handbooks covering science and technology. "amazon.com" and "barnesandnoble.com" are Internet sites with comprehensive catalogs of books. The date of publication should be checked before using any of these works if the required information is likely to have been affected by recent progress. The following list represents only a sampling of available works of outstanding value.

The *Kirk–Othmer Encyclopedia of Chemical Technology*[7] provides in 24 volumes, plus a separate index, a comprehensive and authoritative treatment of a wide range of subjects, with heaviest concentration on materials and processes. The basic set is updated by supplements.

The Encyclopedia of Polymer Science and Engineering[8] is one of the major works in this important area of materials.

Metals Handbook[9] provides an encyclopedic treatment of metallurgy and related subjects. Each of the volumes is devoted to a separate topic such as mechanical testing, powder metallurgy, and heat treating. Each of the articles is written by a committee of experts on that particular topic.

The CRC Handbook of Chemistry and Physics,[10] popularly known as the "Rubber Handbook," is probably the most widely available handbook. It is updated annually to include new materials and to provide more accurate information on previously published sections as soon as the information becomes available. The CRC Press is one of the major publishers of scientific and technical handbooks. A *Composite Index of CRC Handbooks*[11] provides access to the information covering the latest editions of 57 CRC handbooks, some of them multivolume sets, which were published prior to 1977.

The increasing concern with industrial health and safety has placed an additional responsibility on the engineer to see that materials are handled in a safe manner. Sax's *Dangerous Properties of Industrial Materials*[12] provides an authoritative treatment of this subject. This book also covers handling and shipping regulations for a large variety of materials.

Engineers have always been concerned with interaction between humans and machines. This area has become increasingly sophisticated and specialized. The *Human Factors Design Handbook*[13] is written for the design engineer rather than the human factor specialist. The book provides the engineer with guidelines for designing products for convenient use by people. The second edition offers new material drawn from the lessons learned, for example, in the computer and space industries. Another important title is the *Handbook of Human Factors and Ergonomics.*[14]

Engineering work frequently requires a variety of calculations. The *Standard Handbook of Engineering Calculations*[15] provides the answer to most problems. Although most of the information in this handbook is easily adaptable to computer programming, a new edition would probably take greater advantage of the increasing availability of computers.

Conservation of energy remains an important consideration, owing to energy's increasing cost. Two titles dealing with this subject are the *CRC Handbook of Energy Efficiency*[16] and the *Energy Management Handbook.*[17]

Composite materials frequently offer advantages in properties and economy over conventional materials. Information about composites can be found in several McGraw-Hill and CRC Press handbooks.

When England converted to the metric system the British Standards Institution published *Metric Standards for Engineers.*[18] With the increasing worldwide distribution of products, metric units will gain in importance regardless of the official position of the U.S. government. This handbook offers the engineer an authoritative and detailed treatment of metrification.

5 CODES, SPECIFICATIONS, AND STANDARDS

Codes, specifications, and standards are produced by government agencies, professional societies, businesses, and organizations devoted almost exclusively to the production of standards. In the United States the American National Standards Institute[19] (ANSI) acts as a clearinghouse for industrial standards. ANSI frequently represents the interests of U.S. industries at international meetings. Copies of standards from most industrial countries can be purchased from ANSI as well as from the issuer.

Copies of standards issued by government agencies are usually supplied by the agency along with the contract. They are also available from several centers maintained by the government for the distribution of publications. Most libraries do not collect government specifications.

Many of the major engineering societies issue specifications in areas related to their functions. These specifications are usually developed, and revised, by membership committees.

The American Society of Mechanical Engineers[20] (ASME) has been a pioneer in publishing codes concerned with areas in which mechanical engineers are active. In 1885 ASME formed a Standardization Committee on Pipe and Pipe Threads to provide for greater interchangeability. In 1911 the Boiler Code Committee was formed to enhance the safety of boiler operation. The 1983 ASME Boiler and Pressure Vessel Code was published in a metric (SI) edition, in addition to the edition using U.S. customary units, to reflect its increasing worldwide acceptance. The boiler code covers the design, materials, manufacture, installation, operation, and inspection of boilers and pressure vessels. Revisions, additions, and deletions to the code are published twice yearly during the three-year cycle of the code.

A frequently used collection of specifications is the *Annual Book of Standards*[21] issued by the American Society for Testing and Materials (ASTM). These standards are prepared by committees drawn primarily from the industry most immediately concerned with the topic.

The standards written by individual companies are usually prepared by a member of the standards department. They are frequently almost identical to standards issued by societies and government agencies and make frequent references to these standards. The main reason for these "in-house" standards is to enable the company to revise a standard quickly in order to impose special requirements on a vendor.

The large number of standards issued by a variety of organizations has resulted in a number of identical or equivalent standards. Information Handling Services (IHS)[22] makes available virtually all standards on CD-ROM.

6 GOVERNMENT PUBLICATIONS

The U.S. government is probably the largest publisher in the world. Most of the publications are available from the Superintendent of Documents.[23] Publication catalogs are available on the Government Printing Office web site, GPO.gov. Increasingly, the GPO is relying on electronic dissemination rather than print. These publications are provided, free of charge, to depository libraries throughout the country. Depository libraries are obligated to keep these publications for a minimum of five years and to make them readily available to the public.

The government agencies most likely to publish information of interest to engineers are probably the National Institute of Science and Technology, the Geological Survey, the National Oceanic and Atmospheric Administration, and the National Technical Information Service.

7 ENGINEERING SOCIETIES

Engineering societies have exerted a strong influence on the development of the profession. The ASME publishes the following periodicals in order to keep individuals informed of new developments and to communicate other important information:

Applied Mechanics Reviews (monthly)

CIME (*Computers in Mechanical Engineering,* published by Springer-Verlag, New York)

Mechanical Engineering (monthly)

Transactions (quarterly)

The *Transactions* cover the following fields: power, turbomachinery, industry, heat transfer, applied mechanics, bioengineering, energy resources technology, solar energy engineering, dynamic systems, measurement and control, fluids engineering, engineering materials and technology, pressure vessel technology, and tribology.

Many engineering societies have prepared a code of ethics in order to guide and protect engineers. Societies frequently represent the interests of the profession at government hearings and keep the public informed on important issues. They also provide an opportunity for continuing education, particularly for preparing for professional engineers examinations. The major societies and trade associations in the United States are

American Concrete Institute

American Institute of Chemical Engineers

American Institute of Steel Construction

American Society of Civil Engineers

American Society of Heating, Refrigerating, and Air-Conditioning Engineers

American Society of Mechanical Engineers

Institute of Electrical and Electronics Engineers

Instrument Society of America

National Association of Corrosion Engineers

National Electrical Manufacturers Association

National Fire Protection Association

Society of Automotive Engineers

Technical Association for the Pulp and Paper Industry

Underwriters Laboratories

8 LIBRARIES

The most comprehensive collections of engineering information can be found at large research libraries. Four of the largest in the United States are

John Crerar Library
35 West 33rd Street
Chicago, IL 60616

Library of Congress
Washington, DC 20540

Linda Hall Library
5109 Cherry Street
Kansas City, MO 64110

New York Public Library—Science, Industry and Business Library
188 Madison Avenue
New York, NY 10016

These libraries are accessible to the public. They do provide duplicating services and will answer telephoned or written reference questions.

Substantial collections also exist at universities and engineering schools. These libraries are intended for use by faculty and students, but outsiders can frequently obtain permission to use these libraries by appointment, upon payment of a library fee, or through a cooperative arrangement with a public library.

Special libraries in business and industry frequently have excellent collections on the subjects most directly related to their activity. They are usually only available for use by employees and the company.

Public libraries vary considerably in size, and the collection will usually reflect the special interests of the community. Central libraries, particularly in large cities, may have a considerable collection of engineering books and periodicals. Online searching is becoming an increasingly frequent service provided by public libraries.

Regardless of the size of a library, the reference librarian should prove helpful in obtaining materials not locally available. These services include interlibrary loans from networks, issuing of courtesy cards to provide access to nonpublic libraries, and providing the location of the nearest library that owns needed materials.

9 INFORMATION BROKERS

During the last decade a large number of information brokers have come into existence. For an international listing see *The Burwell World Directory of Information Brokers*.[24] Information brokers can be of considerable use in researching the literature and retrieving information, particularly in situations where the engineer does not have the time and resources to do the searching. The larger brokers have a staff of trained information specialists skilled in online and manual searching. Retrieval of needed items is usually accomplished by sending a messenger to make copies at a library. It is therefore not surprising that most information brokers are located near research libraries or are part of an information center.

The larger information brokers usually cover all subjects and offer additional services, such as translating foreign language materials. Smaller brokers, and those associated with a specialized agency, frequently offer searching in a limited number of subjects. The selection of the most appropriate information broker should receive considerable attention if a large amount of work is required or a continuing relationship is expected.

REFERENCES

1. *Engineering Index,* Engineering Information Inc. (monthly).
2. *Science Abstracts,* INSPEC, Institution of Electrical Engineers (semimonthly).
3. *Chemical Abstracts,* American Chemical Society (weekly).
4. *Metals Abstracts,* Metals Information (monthly).
5. *Ulrich's International Periodicals Directory,* R. R. Bowker, New York (annual).
6. *Information Industry Market Place,* An International Directory of Information Products and Services, R. R. Bowker, New York.
7. *Kirk–Othmer Encyclopedia of Chemical Technology,* 4th ed., Wiley-Interscience, New York, 1991–1997 (24 vols.).
8. *Encyclopedia of Polymer Science and Engineering,* 2nd ed., Wiley-Interscience, New York, 1985–1989 (19 vols.).
9. *Metals Handbook,* American Society for Metals, Metals Park, OH.
10. *CRC Handbook of Chemistry and Physics,* CRC Press Inc., Boca Raton, FL (annual).
11. *Composite Index of CRC Handbooks,* CRC Press Inc., Boca Raton, FL, 1977.
12. Sax's *Dangerous Properties of Industrial Materials,* 9th ed., Van Nostrand Reinhold, New York, 1996.

13. W. E. Woodson et al., *Human Factors Design Handbook,* 2nd ed., McGraw-Hill, New York, 1991.
14. G. Salvendy (ed.), *Handbook of Human Factors,* Wiley, New York, 1997.
15. T. G. Hicks et al., *Standard Handbook of Engineering Calculations,* McGraw-Hill, New York, 1994.
16. F. Kreith and R. West, *CRC Handbook of Energy Efficiency,* CRC Press, Boca Raton, FL, 1996.
17. W. C. Turner, *Energy Management Handbook,* Fairmont Press, Lisburn, GA, 1997.
18. British Standards Institution, *Metric Standards for Engineers,* BSI, London, 1967.
19. American National Standards Institute, 11 West 42nd Street, New York, NY 10036.
20. American Society of Mechanical Engineers, 345 East 47th Street, New York, NY 10017.
21. American Society for Testing and Materials, *Annual Book of Standards,* ASTM, 100 Barr Harbor Drive, West Conshohocken, PA 19428-2959.
22. Information Handling Services, 15 Inverness Way East, Englewood, CO 80150.
23. Superintendent of Documents, U.S. Government Printing Office, Washington, DC 20402.
24. Burwell Enterprises, *The Burwell World Directory of Information Brokers,* Houston, TX, 1996.

INDEX

A

AAQG (American Aerospace Quality Group), 622
ABI/INFORM, 771
Abrasives, 215–216
Abrasive barrel finishing, 280
Abrasive belt finishing, 280
Abrasive flow machining (AFM), 220, 224
Abrasive jet machining (AJM), 220, 224
Abrasive machining, 215–219
    abrasives used in, 215–216
    buffing, 219
    defined, 215
    electropolishing, 219
    grinding fluids for, 218
    machines for, 218, 219
    surface finishing, 219
    temperatures for, 218
    ultrasonic, 226
    ultrasonic machining, 219
Abstracts, 778–779
AC, *see* Adaptive control
Accelerated depreciation, 520–522
Acceptance quality level (AQL), 326
Acceptance sampling, 325–326
    DOD acceptance sampling by variables, 326
    double, 325–326
    multiple, 326
    sequential, 326
Accident prevention, 679–680
Accountants, information needs of, 508
Accounting. *See also* Finance and accounting
    accrual vs. cash-basis, 512
    defined, 505
Accreditation, 617. *See also* Registration, certification, and accreditation programs/bodies
Accrual accounting, 512
ACF, *see* Annual cash flow
Acquisition costs, initial, 535–537
Activity relationship chart (material handling), 368, 373
Actual authority, 705
Adaptability, 464
Adaptive control (AC), 336–337
Ad-Aware, 765
ADLI model, 629
AED training, *see* Automated external defibrillator training
AEM, *see* Automated electrified monorails
AEROSPACE database, 769
Aerospace standard (AS 9100), 622
Affinity diagram, 603
AFM, *see* Abrasive flow machining
Agency (legal), 705–706
Aggregate production planning (AP), 131–134
    approaches to, 132–133
    costs in, 131–132
    difficulties with, 134
    levels of aggregation/disaggregation in, 133–134
    strategies in, 131
AGVS, *see* Automated guided vehicle system
AIAG (Automotive Industry Advisory Group), 620
AIP (American Institute of Physics), 773
Air quality, 670
AJM, *see* Abrasive jet machining
Allocated stock (MRP schedule), 136
Allowance, tolerance vs., 315
ALUMINUM database, 769
Aluminum Industry Abstracts, 769
Aluminum oxide, 215
American Aerospace Quality Group (AAQG), 622
American Health Care Association Quality Award, 636
American Institute of Physics (AIP), 773
American National Accreditation Program (NAP), 619
American National Standards Institute (ANSI), 619, 620, 623, 780
American Society for Quality (ASQ), 619, 628
American Society for Testing and Materials (ASTM), 781

American Society of Mechanical
    Engineers (ASME), 367, 780–782
America's Best Plants Award, 636, 637
Analysis, as engineering skill, 553
Analysis phase (DMAIC process), 585,
    603–607
  affinity diagram, 603
  cause and effect diagram, 603–604
  data sampling and charts, 604
  design of experiments, 604–607
Angle bending, 262
Annual Book of Standards, 781
Annual cash flow (ACF), 579–580
Anodizing, 283
ANSI, *see* American National Standards
    Institute
A-1 matrix (QFD), 592–595
AOQ (average outgoing quality), 325
AOQL (average outgoing quality limit),
    325
AP, *see* Aggregate production planning
AP News database, 771
Apparent authority, 705-706
Appearance ratio control, 160
Apron conveyors, 356
AQL (acceptance quality level), 326
Arc evaporation, 401–403
Arithmetic mean, 316
Arrow diagrams, 552
AS 9100 Aerospace standard, 622
ASMDATA, 768–769
ASME, *see* American Society of
    Mechanical Engineers
ASME Boiler and Pressure Vessel Code,
    781
ASME Boiler Code Committee, 781
ASME Standardization Committee on
    Pipe and Pipe Threads, 781
ASN (average sample number), 326
ASQ, *see* American Society for Quality
AS/RS, *see* Automated storage and
    retrieval system
Assemblies:
  design-allowable stress, 302
  joining, 287–290
Assembly activities, cost estimating for,
    555–557
Assembly chart (material handling), 367,
    369
Assembly line balancing, 156–163
  designing assembly line, 158
  mixed-model assembly lines, 159–162
  parallel line balancing, 162, 163
  structure of problem, 157, 158
  techniques for, 158–159

Assets:
  current, 510–511
  depreciation of, 513–514, 519
  fixed, 510, 513–514
  intangible, 510
  return on, 517
  total, 517
  understanding condition of, 509
Asset to sales ratio, 518
Asset turnover ratio, 518
Associated Press (AP) News database,
    771
Assumption of risk (as legal defense),
    719–721
ASTM (American Society for Testing and
    Materials), 781
Asymmetric bipolar pulsed dc power, 405
"Atkins" report, 775
AT&T Power Systems, 627
Attribute characteristics of parts, 25
AT&T Western Electric, 611
A-12 aircraft program, 470–473
Authority (legal), 705–706
AutoCAD, 336
Auto Industry standard (ISO/TS), 620–
    621
Automated electrified monorails (AEM),
    383, 386
Automated external defibrillator (AED)
    training, 687–688
Automated guided vehicle system
    (AGVS), 387, 388, 390
Automated storage and retrieval system
    (AS/RS), 390–392
Automation, 329, 330
Automatization, 329–330
Automotive Industry Advisory Group
    (AIAG), 620
Availability of materials, 88
Available units (MRP schedule), 137, 139
AV-8B Harrier aircraft project, 473–476
Average cost of capital, 518
Average outgoing quality (AOQ), 325
Average outgoing quality limit (AOQL),
    325
Average sample number (ASN), 326
Awards, 627–637
  Baldrige National Quality Award, 627–
    630
  Deming Prize, 626–627
  industry-specific, 636, 637
  Shingo Prize for Excellence in
    Manufacturing, 630, 635
  from U.S. states, 630–634
  worldwide, 634–636

B

Back pitch, 294
Bags (material packaging), 361, 362
Balance sheet, 509–516
    accrual accounting, 512
    current assets, 510–511
    current liabilities, 511–512
    current ratio, 513
    fixed assets, 513–514
    interest-bearing current liabilities, 512
    net working capital, 512–513
    second year comparison, 515–516
    total capital, 514
Baldrige, Malcolm, 627
Baldrige National Quality Award (BNQA), 586, 588, 589, 616, 618, 619, 625, 627–630
Barrel finishing, 280
Basic producer industries, 330
Basic shape code, 74, 75
Batch production, 331
B/C (benefit-cost ratio), 581
Beam PVD processes, 410–411
Bearing-type connections, strength of, 296–300
BEES (Building for Environmental and Economic Sustainability), 422
Beliefs, management, 457–458
Belt conveyors, 351, 352, 354, 355
Benchmarking, 18, 19, 489, 498
Bending, of bolts, 313
Bending process, 260–263
Benefit-cost ratio (B/C), 581
Best practices, 489
Better Designs in Half the Time (Bob King), 592
Bias (GR&R), 613
Bill of material (BOM), 8
Bin storage, 360, 393
Biodesign products, 421–422
Bipolar pulsed dc power, 405
Bitstring representation, 73
Blank diameters, for metal shells, 254
Blanking, 263, 264
Blasting, 280
BL (burdened assembly labor cost), 4
Blow molding, 276
BNQA, *see* Baldrige National Quality Award
Boeing, 286–287
Bolt and rivet fastening, 292–313
    clamping force upper limits, 302–303
    complex butt joint, 296–300
    design-allowable bolt stress/assembly stress, 302
    design for cyclical tension loads, 312–313
    efficiency of joints, 295
    fatigue failure, 312
    friction-type connections, 300–302
    joint types, 293–295
    simple lap joint, 295–298
    slip characteristics, 308, 309
    theoretical behavior under tensile loads, 304–307
    torque and turn at same time tightening, 309, 310
    torsional stress factor, 302
    turn of nut bolt tightening, 308, 309
    ultrasonic measurement of bolt stretch/tension, 310–311
Bolting paradigm, 287
BOM (bill of material), 8
Bonding materials, 215
Books, online, 763
Boxes (material packaging), 361, 363
Box plots, 598
Brainstorming:
    improvement phase (DMAIC), 607
    in new product development, 19
    for team building, 502
Breached duty, 703
Breach of contract, 708
Broaching, 209–212
Brundtland Commission, 424
Bucket elevators, 352, 355–357
Budgets, comparing current results with, 523–526
Buffing, 219, 280
Building for Environmental and Economic Sustainability (BEES), 422
Bulk materials:
    handling of, 350–359
        belt conveyors, 351, 352, 354, 355
        bucket elevators, 352, 355–357
        continuous-flow conveyors, 356, 358
        conveying equipment selection, 350–352
        oscillating conveyors, 355
        pneumatic conveyors, 356, 358, 359
        screw conveyors, 351, 353
        vibrating conveyors, 355, 358
    storage of, 360–365
        bins/silos/hoppers, 360
        flow-assisting devices and feeders, 361
        open-yard, 360
        packaging of materials, 361–364
        transportation of materials, 363, 365
Bulk solids handling, 350

Burden costs, 532, 533, 538
Burdened assembly labor cost (BL), 4
Burden rates, 317
Bureau of Labor Statistics, 576
Burnishing, 260
The Burwell World Directory of Information Brokers, 783
Business environment, 448–450, 486
Business liabilities, 707–710
Butt joints, 294–300
Butt welding, 257

C

CAA, *see* Clean Air Act of 1990
CAD, *see* Computer-aided design
CAD/CAM part programming, 336
Cadkey, 336
CAI, *see* Computer-aided inspection
Calcium sulfate, 274
CAM, *see* Computer-aided manufacturing
Cambridge Scientific Abstracts (CSA), 767
Canada, filing patents in, 755
Canada Institute for Scientific and Technical Information (CISTI), 774
Cannot produce parts at design target condition (correlation charts), 49–51
Capability index (Cpk), 601–603
Capability limits, 319–321
Capability Maturity Model Integration (CMMI), 617, 622–624
Capital:
  average cost of, 518
  invested, return on, 517, 518
  net working, 512–513
  optimizing utilization of, 95, 96
  total, 514, 518
  total cost of, 523
Capital expenditures, 523, 564–565
Capital funds allocation, 570–573
Capital-intensive businesses/industries, 518
Capital stock, 514
CAPP, *see* Computer-aided process planning; Computer-aided production planning
Carnegie Mellon University, 622
Carousel systems, 391–393
Cartesian coordinate system, 334
Cartons, storage, 363
CASE (computer-aided software engineering), 559
Cash-basis accounting, 512

Cash flow(s), 518–523
  annual, 579–580
  discounted, 566, 567
  establishment of, 575–577
  multiple, 568–570
  and time value of money, 566
Cash flow statement, 518
CASPA (Complaint About State Program Administration), 648
Casting and molding:
  metal, 268–275
    centrifugal casting, 269–272
    investment casting, 274–275
    permanent-mold casting, 272–273
    plaster-mold casting, 274
    sand casting, 268–270
  plastic, 275–277
Catalogs, online, 762–763
Catastrophe analysis, 649
"Catchball," 461
Cause and effect, in manufacturing process flow diagrams, 24–26
Cause and effect diagram, 603–604
Cause and effect relationships, 22, 453
Cause of action, 703
C control charts, 323–325
Central limit theorem, 612
Central Limit Theory, 317
Central tendency, measure of, 316
Centrifugal casting, 269–272
Ceramic process (casting), 274
CERCLA, *see* Comprehensive Environmental Response, Compensation, and Liability Act of 1980
Certification, 617. *See also* Registration, certification, and accreditation programs/bodies
Certification and Registration standard (ISO 9001), 618–620
Chain-driven conveyors, 379, 380
Challenges, motivation and, 491–492, 501
Channel bending, 262
Chemical Abstracts, 779
Chemical blanking, 241
Chemical conversions, 283–284
Chemical machining (CHM), 241
Chemical milling, 241
Chemical oxide coatings, 284
Chemical properties of materials, 87, 88
Chemical vapor deposition (CVD), 396
Chesapeake Paper Products, 709–710
CHM (chemical machining), 241
Chromate coatings, 283–284
Chutes, 376

Index **789**

CIMS, *see* Computer-integrated manufacturing systems
CISTI (Canada Institute for Scientific and Technical Information), 774
Clamping force upper limits, 302–303
Classification systems, 68–109
   Engineering Materials Taxonomy, 77, 81–90
      availability of materials, 88, 89
      material classification systems, 81–83
      material code, 83–86
      material properties, 84–88
      processability ratings for materials, 88, 90
   fabrication equipment classification, 95–103
      code number for equipment, 99–100
      customization of, 99
      equipment classification, 97–99
      equipment specification sheets, 100–103
      standard vs. special equipment, 97
   Fabrication Process Taxonomy, 89–96
      basis of classification, 92–93
      capabilities of processes, 94–96
      primary objectives for, 90–91
      process code, 93–94
      process divisions, 91–92
      purpose of, 89, 90
   fabrication tool classification and coding, 103–109
      standard vs. special tooling, 104–106
      tool code, 106–107
      tooling taxonomy, 105, 106
      tool specification sheets, 107–109
   Part Family Classification and Coding, 68–80
      applications for, 69–71
      basic shapes, 74, 75
      code format and length, 74
      history of, 68–69
      material code, 76–80
      name or function code, 74, 75
      precision class code, 76, 77
      purpose of, 73–74
      size code, 75, 76
      special features, 75, 76
      tailoring, 77
      theory underlying, 71–73
      tailoring of, 69
Clausing, Donald, 592
Clean Air Act (CAA) of 1990, 642, 644–645
Cleaning, surface, 278–281

Clean Water Act (CWA) of 1970, 642–643
Closed-belt conveyors, 356
CMMI, *see* Capability Maturity Model Integration
CNC machine tools, 341
CNC systems, DNC systems vs., 341, 342
Coatings, 281–283
Codes, resources for, 780–781
Coding, of parts, 344. *See also* Classification systems
Coining, 260
Coinjection molding, 275
Cold-chamber die casting, 273
Cold drawing, 265–266
Cold-roll forming, 262
Cold rolling, 259
Cold spinning, 266
Cold working processes, 258–268
   bending, 260–263
   classification of operations, 258
   drawing, 265–268
   shearing, 263–265
   squeezing, 259–260
Collaboration, 465–466, 488, 775
Collaborative Research Institute for Sustainable Products (CRISP), 425
Combined loads, 303
Combustion machining, 242
Commercial databases, 768–771
Commitment, 462, 488, 502
Common cause variation, 609
Communication, 461–462
   developing channels of, 501
   management of, 488
   of organizational goals and objectives, 502
   on use of personal protective equipment, 677
Comparative negligence, 719
Compendex, 769, 777, 779
Competitive benchmarking, 18, 19
Complaint About State Program Administration (CASPA), 648
Complex butt joints, 296–300
Compliant features, 289, 290
Composite Index of CRC Handbooks, 780
Composite labor rate, 533
Compositions of matter (as statutory class for patents), 730
Compounding period (interest), 567
Comprehensive Environmental Response, Compensation, and Liability Act (CERCLA) of 1980, 642, 655

Computer-aided design (CAD), 328, 329, 336
Computer-aided inspection (CAI), 328, 329
Computer-aided manufacturing (CAM), 328, 329, 336
Computer-aided process planning (CAPP), 347, 348
Computer-aided production planning (CAPP), 328, 329
Computer-aided software engineering (CASE), 559
Computer-integrated manufacturing systems (CIMS), 328–348
  and automation, 329–330
  CNC vs. DNC systems, 341, 342
  elements of, 328–329
  flexible manufacturing systems, 343
  group technology, 343–348
    computer-aided process planning, 347, 348
    machine cell designs, 347
    part family formation, 343–344
    parts classification and coding, 344–346
    production flow analysis, 345, 347, 348
  hierarchical computer control, 340–342
  industrial robots, 337–340
    applications, 340
    configuration, 338–339
    control and programming, 340
  manufacturing cells, 342
  and manufacturing cycle, 330–331
  numerical control (NC), 334–338
    adaptive control, 336–337
    CAD/CAM part programming, 336
    coordinate system, 334–335
    definition of numerical control, 334
    machinability data prediction, 337
    programming by scanning/digitizing, 336
    selection of parts for machining, 335–336
  production operations, 330
  and production operations models, 332–333
  and production plant classifications, 331–332
Computer software, detailed cost estimating for, 559
Concept phase (new product development), 7
Concurrent engineering, 486
Conditioned water, 278
Conference proceedings, 778
Conflict management, 502
Consideration, contractual, 708
Constant demand (inventory), 122
Constraining critical conditions (correlation charts), 43–49
  constraint table, 49
  imperfect correlation, 44–47
  operating limits, 47
  operating range, 47–48
  operating target, 47–48
Constructive conflict, 464
Continuation application (patents), 748
Continuation controls, 161
Continuation-in-part application (patents), 748
Continuing duty to warn, 721–723
Continuous-bucket elevators, 355
Continuous-flow conveyors, 356, 358
Continuous improvement, *see* DMAIC process
Continuous interest, 567–568
Continuous-path (contouring) NC systems, 334, 335
Continuous-path robots, 340
Contouring robots, 340
Contractual obligations, 707–708
Contributory negligence, 719
Controls, engineering, *see* Engineering controls
Control charts, 317–325
  c and u, 323–325
  in DMAIC control phase, 609, 611–612
  p and np, 322–325
Control limits, 317–321, 612
Control phase (DMAIC process), 585, 609, 611–614
  control charts, 609, 611–612
  control plans, 612–613
  data sampling and charts, 613
  GR&R analysis, 613–614
Control plans, 612–613
Control processes, management system failure and, 466
Control variables:
  complex interactions between, 28, 29
  manufacturing process flow diagrams, 23, 24
  number of, 26–27
  response to changes in, 27, 28
  simple interactions between, 28, 29
Converter industries, 330
Conveyors, 376–386
  belt, 351, 352, 354, 355
  chain-driven, 379, 380

continuous-flow, 356, 358
gravity, 376, 378–379
oscillating, 355
pneumatic, 356, 358, 359
power-and-free, 380, 381
powered, 377, 379
screw, 351, 353
selection of, 350–352
sortation, 380, 382
vibrating, 355, 358
Conveyor systems, 351
Coordinate system (NC), 334–335
Corpedia Education, 478–479
Correlation charts (TRA):
    adding design information to, 41
    generating, 39–41
    generating input data for, 36–38
Corrosion:
    of bolts, 313
    crevice, 293
    temporary protection against, 283
Costs:
    direct, 536, 537
    estimating, see Detailed cost estimating
    fixed, 536, 537
    indirect, 536, 537
    initial acquisition, 535–537
    nonrecurring, 536, 537
    opportunity, 565
    recurring, 536, 537
    variable, 536, 537
Cost growth allowance, 562
Cost of money, 565
Cost of production, 519
Cost of product shipped, 516–519
Cost of sales, 516, 519
Cpk, see Capability index; Process capability
Cranes, 382, 384, 387
The CRC Handbook of Chemistry and Physics, 780
CRC Handbook of Energy Efficiency, 780
Crevice corrosion, 293
CRISP (Collaborative Research Institute for Sustainable Products), 425
Critical dimensions, 25, 41
Critical path bar charts, 552
Critical to quality (CTQ), 591
    in improvement phase, 607–608
    tree diagram for, 594, 608
Cross-belt sorters, 381, 382, 385
CSA (Cambridge Scientific Abstracts), 767
CTQ, see Critical to quality
Cubic boron nitride, 215

Culture, organizational, 456, 486, 502
Current assets, 510–511
Current liabilities, 511–512, 517, 518
Current ratio, 513
Customer surveys, 112
Cutoff operations, 265
Cutoff rate of return, 572
Cutting force, 264
Cutting-off processes, 213–214
Cutting-tool materials, 185–188
CVD (chemical vapor deposition), 396
CWA, see Clean Water Act of 1970
Cyclical tension loads, design for, 312–313
Cyclical time series movements, 117
Cylindrical coordinate robots, 338, 339

D

DaimlerChrysler, 416, 620
Dangerous Properties of Industrial Materials, 780
Databases:
    commercial, 768–771
    engineering, 763, 768–769
Database services, 765–768
Data collection:
    in measure phase, 597–598
    for unit material handling, 366–373
Data sampling and charts:
    in analysis phase, 604
    in control phase, 613
    in improvement phase, 607
DCC (direct computer control), 334
dc diode sputtering, 404
Debt financing, 564
Debt percent to debt plus equity ratio, 518
Debt to total capital ratio, 518
Deburring:
    electrochemical, 228–229
    thermal, 242
Decision analysis, 650
Deep drawing, 266
Defects-fixes needed condition (correlation charts), 49, 50
Defendant, 703
Deferred income taxes, 514
Define phase (DMAIC process), 585, 591–597
    CTQ tree, 594
    process flow diagram, 594–596
    quality function deployment, 592–594
    SIPOC chart, 596–597
    voice of the customer, 591–592

Delphi method, 112
Demand, 121, 131
Deming Prize, 625–627
Department of Commerce, 628
Department of Defense (DOD):
  acceptance sampling by variables, 326
  and CMMI, 622, 623
  and quality of goods/services, 583–584
  and quality standard revision, 616–617
Department of Energy (DoE), 415, 416
Dependent alternatives, 574, 577, 578
Depreciation, 513–514
  accelerated, 520–522
  meaning of term, 513
  reserve for, 519
Deseasonalization, 117
Deseasonalized forecast, 118, 119
Design:
  classification systems in, 69, 70
  as engineering skill, 553
  safety engineering, 670–675
    hardware, 670–671
    hazardous material classification system, 672–674
    Material Safety Data Sheets, 674–675
    process, 671–672
    safety design requirements, 675
Design-build processes, 486
Design flaws, 713–714
Design for assembly (DFA), 5
  in new product development, 18
  principles for, 18–19
Design for Environment Software, 422
Design for Manufacturing and Assembly (DFM&A), 3–21
  defined, 4
  design for assembly, 5
  design for manufacturing, 5
  goal of, 5
  ideal process for applying, 7–8
  methodology for, 14, 16–19
  metrics for, 19–21
  motor drive assembly case study, 8–16
  upper management support for, 12–14
Design for manufacturing (DFM), 5, 19
Design for the Environment program, 415
Design growth allowance, 562
Design hierarchy, 716–717
Design of experiments (DOE), 26, 604–607
  analysis phase, 604–607
  improvement phase, 608
  TRA use of data from, 36–37
  usefulness of, 29

Design patents, 726
Design phase (new product development), 7, 8
Design process, documentation of, 723–724
Design retrieval, 69, 70
Design target, changing, 53
Design tolerances, changing, 50–53
Detailed cost estimating, 531–563
  burden costs, 538
  computer software, 559
  direct charges, 562–563
  direct costs, 536
  direct management, 562–563
  distinctions between types of costs, 535–536
  earnings, 538–539
  engineering activities, 552–554
  fees, 538–539
  fixed costs, 536
  general and administrative costs, 538
  indirect costs, 536, 538
  initial acquisition costs, 535–537
  in-process inspection, 558
  labor, 560–562
  labor-hours, 536–538
  labor rates/factors, 538
  manufacturing/production and assembly activities, 555–557
  manufacturing/production engineering activities, 554–555
  materials, 538
  methods used in, 545–552
    detailed resource estimating, 546
    direct estimating, 546
    estimating by analogy, 546
    firm quotes, 546–547
    handbook estimating, 547
    labor-loading methods, 548, 551
    learning curve, 547–550
    rules of thumb, 546
    statistical/parametric estimates, 548, 551–552
  nonrecurring costs, 536
  overhead, 538
  process for developing, 539
  profits, 538–539
  questions for developing, 539–541
  recurring costs, 536
  schedule elements in, 552
  subcontract dollars, 538
  supervision, 562–563
  testing, 558–559
  time/skills/labor-hours required for, 533–535

use of "factors" in, 563
variable costs, 536
work breakdown structure, 541–545
  functional elements of, 542–543
  hierarchical relationship of, 541–542
  interrelationships in, 544–545
  physical elements of, 543–544
  recurring vs. nonrecurring activities in, 544
Detailed resource estimating, 546
Deterministic inventory models, 122
Development phase (new product development), 7
Development testing, 558
Devices, safeguarding, 664, 666–667
DFA, see Design for assembly
DFM, see Design for manufacturing
DFM&A, see Design for Manufacturing and Assembly
Diagonal pitch, 294
DIALINDEX, 767–678
Dialog, 766–768
Diamond-like carbon (DLC) films, 410–411
Diamonds, 215
DIBS (dual ion beam sputtering), 409
Die casting, 272, 273
DIF (duty, insurance, and freight), 4
Digitizing, programming by, 336
Dimensions, 315
Dinking, 265
Direct charges, estimating, 562–563
Direct computer control (DCC), 334
Direct costs, 536, 537
Direct diverters, 380, 382
Direct estimating, 546
Direction:
  and attitude/commitment of employees, 503
  as management requirement, 488
Direct judgment technique (cost estimating), 536, 537
Direct management cost, estimating, 562–563
Direct-numerical-control (DNC) systems, 341, 342
Discounted cash flow, 566, 567
Discounted inventory model, 128
Discrete interest, 567–568
Dispersion, measure of, 316
Distribution logistics, 167–168
Diverters:
  direct, 380, 382
  pop-up, 381, 393
Dividends, 518, 523

Divisional application (patents), 748
DLC films, see Diamond-like carbon films
DMAIC process, 585–614
  analysis phase, 603–607
    affinity diagram, 603
    cause and effect diagram, 603–604
    data sampling and charts, 604
    design of experiments, 604–607
  benefits of, 586–589
  control phase, 609, 611–614
    control charts, 609, 611–612
    control plans, 612–613
    data sampling and charts, 613
    GR&R analysis, 613–614
  define phase, 591–597
    CTQ tree, 594
    process flow diagram, 594–596
    quality function deployment, 592–594
    SIPOC chart, 596–597
    voice of the customer, 591–592
  and definitions of quality, 586
  engineer's role with, 588, 590
  improvement (and innovation) phase, 607–610
    brainstorming, 607
    CTQ analysis, 607–608
    data sampling and charts, 607
    design of experiments, 608
    Gantt charts, 609
    network diagrams, 609, 610
    process capability analysis, 609
  measure phase, 597–603
    data collection plan and forms, 597–598
    GR&R analysis, 599
    Pareto charts, 599
    plots, 599
    prioritization matrix, 599, 600
    process capability analysis, 600–603
  and TQM/SS as career aid, 590–591
DNC systems, see Direct-numerical-control systems
Documentation:
  of design process, 723–724
  as engineering skill, 553
  of patents, 749
  of process capabilities, 90
Document delivery options, 773–774
DOD, see Department of Defense
DOE, see Design of experiments
DoE, see Department of Energy
Double declining balance (depreciation), 520–522
Double-patenting rejections, 747–748

794 Index

Drafting, as engineering skill, 553
Drawer cabinets, modular, 393–394
Drawing:
    cold working, 265–268
    hot working, 253–255
Drilling machines, 192–200
    accuracy of holes, 197–200
    drill geometry, 193
Drivers, training for, 688
Drums, storage, 364
Dry sand molds, 268
DSpace initiative, 774
Dual accountability, 494
Dual ion beam sputtering (DIBS), 409
Dudley Sports Co. v. Schmitt, 718
Due date (job sequencing), 145
Duty, freight, and insurance (DFI), 4
Duty (legal), 703
    of agents, 705
    of candor (patents), 744
    continuing duty to warn, 721–723
Dynamic programming, 132

E

Earliest due date (EDD), 152
Earned surplus, 514
Earnings, detailed cost estimating for, 538–539
EBM (electron-beam machining), 235
ECD, *see* Electrochemical deburring
ECDG (electrochemical discharge grinding), 229
ECG, *see* Electrochemical grinding
ECH, *see* Electrochemical honing
ECM, *see* Electrochemical machining
Eco-efficient products, 421–422
Economic equivalence, 566
Economic viability, 717–718
ECP (electrochemical polishing), 232
ECS, *see* Electrochemical sharpening
ECT (electrochemical turning), 233
EDD (earliest due date), 152
EDG, *see* Electrical discharge grinding
Edge bending, 261, 262
EDM, *see* Electrical discharge machining
EDS (electrical discharge sawing), 237
EDWC, *see* Electrical discharge wire cutting
EFQM (European Foundation for Quality Management), 635
EIA (Electronic Industries Association), 334
EI Engineering Meetings online, 769
80-20 rule, 599

Elastic compliant features, 290
Elastic limit, 300
Elastic region, 300
Electrical discharge grinding (EDG), 235, 236
Electrical discharge machining (EDM), 236–237
Electrical discharge sawing (EDS), 237
Electrical discharge wire cutting (EDWC), 237, 238
Electrochemical deburring (ECD), 228–229
Electrochemical discharge grinding (ECDG), 229
Electrochemical grinding (ECG), 229, 230
Electrochemical honing (ECH), 229, 230
Electrochemical machining (ECM), 231–232
Electrochemical polishing (ECP), 232
Electrochemical sharpening (ECS), 232, 233
Electrochemical turning (ECT), 233
Electromechanical machining (EMM), 222
Electron beam evaporation, 401–402
Electron-beam machining (EBM), 235
Electronic Industries Association (EIA), 334
Electroplating, 283
Electropolishing (ELP), 219, 232, 241, 242, 281
Electro-stream (ES), 233, 234
ELP, *see* Electropolishing
Embossing, 267
EMM (electromechanical machining), 222
Employees:
    attitude/commitment of, 503
    cross-training for, 165
    professional liabilities of, 702–707
    safety needs and expectations of, 640–641
Employer-employee relationship, 705
Employment agreements, 706
Empowerment, 496
EMS, *see* Environmental Management System
Enamel, 282
The Encyclopedia of Polymer Science and Engineering, 780
Encyclopedias, 778–780
End-of-life options, 421
Energy, 428–429
    consumption of, 419, 431–432
    process classification based on, 92
    for producing materials, 81
Energy Management Handbook, 780

Energy-transfer analysis, 649
Engineers:
    DMAIC process role of, 588, 590
    financial tools for, 528–530
    liability insurance for, 708–709
    TQM/SS role of, 588, 590
Engineering, 330
Engineering activities, detailed cost estimating for, 552–555
Engineering change allowance, 561
Engineering controls:
    alternatives to, 664, 669–670
        isolation, 669
        substitution, 669, 670
        ventilation, 670
    for machine tools, 660–663
        danger sources, 663
        general requirements, 662–663
Engineering databases, 768–769
Engineering documentation, as engineering skill, 553
Engineering drawing, as engineering skill, 553
Engineering Index, 769, 777, 779
Engineering Information, Inc., 771
Engineering management, 484–486, 500–503
Engineering Materials Taxonomy, 77, 81–90
    availability of materials, 88, 89
    basis of classification, 82
    chemical properties of materials, 87, 88
    customization of, 83
    families of materials in, 84
    material classification systems, 81–83
    material code, 83–86
    material condition code, 84, 85
    material properties, 84–88
    mechanical properties of materials, 85–87
    objectives of, 81–82
    physical properties of materials, 87
    processability ratings for materials, 88, 90
    raw material condition, 84, 86
Engineering prototype allowance, 562
Engineering safety factor, 702–703
Engineering societies, 781–782
Engineering Village 2, 771–772
Enterprise quality management (EQM), 33
Enterprise resource planning (ERP), 144
Environmental impact, as sustainability measure, 430, 432, 433
Environmental intensity, 458–459

Environmental Management standard (ISO 14000), 623–625
Environmental Management System (EMS), 619, 620
Environmental performance indicator (EPI), 422
Environmental Protection Agency (EPA), 415–416, 430, 642–645
Environmental risk training, 688–691
EPA, *see* Environmental Protection Agency
EPC (European Patent Convention), 757
EPI (environmental performance indicator), 422
EPO (European Patent Office), 757
EQA, *see* European Quality Award
EQM (enterprise quality management), 33
Equipment. *See also* Production processes and equipment
Equipment code (fabrication equipment), 99–100
Equipment maintenance time (in cost estimating), 561
Equipment selection, 96–97
Equipment specification sheets, 100–103
Equity, 517
Equity, return on, 518
Equity financing, 564–565
Ergonomics, *see* Human factors engineering/ergonomics
ERP (enterprise resource planning), 144
Error analysis (forecasting), 120
ES, *see* Electro-stream
Espoused values, 457
Estimating, *see* Detailed cost estimating
Estimating by analogy, 546
Estimating relationship, 551
E-trees, 72
Europe:
    automation in, 330
    sustainability challenge in, 418, 419
    value added ratio use in, 517
European Foundation for Quality Management (EFQM), 635
European Patent Convention (EPC), 757
European Patent Office (EPO), 757
European Quality Award (EQA), 625, 635
European Trading Community, 618
European Union, 418
Evaporative PVD processes, 400–403
Expandable-bead molding, 275–276
Ex parte Jepson, 739
Expected outcome approach, 650
Experience exchanges, 502

Experimental design, *see* Design of experiments
Experimentation, random, 605
Explicit grading (sustainability), 438
Exponential smoothing, 113–115
Express warranties, 711, 712
External environment, 452, 458–461
External tensile load (fasteners), 304–307
Extrusion:
  cold, 259–260
  hot working, 251–252
  of plastic, 276
Extrusion-type blow molding, 276

F

Fabrication equipment classification, 95–103
  code number for equipment, 99–100
  equipment classification, 97–99
  equipment specification sheets, 100–103
  standard vs. special equipment, 97
Fabrication Process Taxonomy, 89–96
  basis of classification, 92–93
  capabilities of processes, 94–96
  primary objectives for, 90–91
  process code, 93–94
  process divisions, 91–92
  purpose of, 89, 90
Fabrication tool classification and coding, 103–109
  standard vs. special tooling, 104–106
  tool code, 106–107
  tooling taxonomy, 105, 106
  tool specification sheets, 107–109
Fabricator industries, 330
Failure modes and effects (FME), 649, 650
Failure stress, 301
Fastening, 286–313
  assembly features/functions for, 287–290
  with bolts and rivets, 292–313
    clamping force upper limits, 302–303
    complex butt joint, 296–300
    design-allowable bolt stress and assembly stress, 302
    design for cyclical tension loads, 312–313
    efficiency of joints, 295
    fatigue failure, 312
    friction-type connections, 300–302
    joint types, 293–295
    simple lap joint, 295–298
    slip characteristics, 308, 309
    theoretical behavior under tensile loads, 304–307
    torque and turn at same time tightening, 309, 310
    torsional stress factor, 302
    turn of nut bolt tightening, 308, 309
    ultrasonic measurement of bolt stretch/tension, 310–311
  defined, 287
  nesting strategy for, 290–291
  three-part assembly, 291, 292
Fatigue failure, 312
Fault-tree analysis, 649–652
FC (first to customer for existing products), 166
FCFS, *see* First come-first served
Federal Insecticide, Fungicide, and Rodenticide Act (FIFRA) of 1996, 642, 645
Fees, detailed cost estimating for, 538–539
Feedback, 458
Feeding and ejection methods, safeguards with, 664, 668
FIFRA, *see* Federal Insecticide, Fungicide, and Rodenticide Act of 1996
Fillets, 312
Film formation and growth (PVD), 400
Finance and accounting, 505–530
  accelerated depreciation, 520–522
  accrual accounting, 512
  balance sheet, 509–516
    accrual accounting, 512
    current assets, 510–511
    current liabilities, 511–512
    current ratio, 513
    fixed assets, 513–514
    interest-bearing current liabilities, 512
    net working capital, 512–513
    second year comparison, 515–516
    total capital, 514
  cash flow, 518–523
  comparing current results with budgets/forecasts, 523–526
  defined, 505
  detailed cost estimating, 531–563
    basic questions for developing, 539–541
    burden costs, 538
    for computer software, 559
    for direct charges, 562–563
    direct costs, 536

Index   **797**

for direct management, 562–563
distinctions between types of costs, 535–536
earnings, 538–539
for engineering activities, 552–554
fees, 538–539
fixed costs, 536
general and administrative costs, 538
indirect costs, 536, 538
initial acquisition costs, 535–537
for in-process inspection, 558
for labor, 560–562
labor-hours, 536–538
labor rates/factors, 538
for manufacturing/production and assembly activities, 555–557
for manufacturing/production engineering activities, 554–555
materials, 538
methods used in, 545–552
nonrecurring costs, 536
overhead, 538
process for developing, 539
profits, 538–539
recurring costs, 536
schedule elements in, 552
subcontract dollars, 538
for supervision, 562–563
for testing, 558–559
time/skills/labor-hours required for, 533–535
use of "factors" in, 563
variable costs, 536
work breakdown structure, 541–545
identifying problems/solutions, 525–527
information needs of groups, 506–508
initiating action, 527–529
investment analysis, 564–582
  analysis period for, 574–575
  annual cash flow, 579–580
  benefit-cost ratio, 581
  capital funds allocation, 570–573
  cash flows establishment, 575–577
  classification of alternatives, 573–574
  discounted cash flow, 566, 567
  interest calculation, 566–572
  interest factors, 568–572
  multiple cash flows manipulation, 568–570
  nominal interest rates, 567
  payback period, 582
  present worth, 577–579
  rate of return, 580–581

  sources of capital expenditures funding, 564–565
  time value of money, 565–566
  models for, 508–509
  profit and loss statement, 516–519
  tools for independent professional engineers, 528–530
Financial ratios, 517–518
Financial rewards, 485, 494
Fine, William T., 656
Finishing, 215. *See also* Surface finishing
Finite replenishment model with infinite storage costs, 126
Finite replenishment rate model with finite shortage costs, 127
Fire protection training, 691, 692
Firewalls, 764
Firm quotes, 546–547
First-aid training, 692
First come-first served (FCFS), 152, 153
First to customer for existing products (FC), 166
First to market for new products (FM), 166
Fisher, Ronald, 606
5S principles, 682–685
Fixed assets, 510, 513–514
Fixed costs, 536, 537
Fixed-position plant layout, 331
Fixtures, modifications of, 30–31
Flammability hazard levels, 673
Flange rotation, 303
Flanging, 263
Flattening, 263
Flexible manufacturing systems (FMS), 341–343
Flight conveyors, 356
Florida Power and Light, 627
Flow-assisting devices and feeders, 361
Flow diagrams:
  manufacturing process, 23
  material handling, 368, 371
Flow process chart (material handling), 367, 370
Flow shops, 148–151
Fluid Engineering Abstracts, 769
FLUIDEX, 769
Fluidizing pneumatic conveyor systems, 356, 358
Fluid pressure, for forming, 266–267
FME, *see* Failure modes and effects
FM (first to market for new products), 166
FMS, *see* Flexible manufacturing systems
Focus-team sessions, 502

Ford Motor Company, 416, 620
Forecasts, financial, comparing current results with, 523–526
Forecasting, production, 111–120
   accuracy of, 120
   causal methods, 115–117
   cost of, 120
   error analysis, 120
   methods for analysis of time series, 117–119
   qualitative, 112–115
   selecting method for, 111–112
Foreign filing of patents, 731, 755–757
Forest Stewardship Council Certified Wood Standards, 415
Forged-plastic parts, 276–277
Forging:
   cold, 259
   hot working, 249–251
Forklift, 385, 388
Formal management systems, 455
Form code (parts classification), 345
Form features, 75
Forming, storming, norming, and performing, 499
Four-stage team development model, 498–500
Fractional factorial design, 605
Friction-type connections, 300–302
"From-to" chart (material handling), 368, 372
Full factorial design, 605
Full integration stage (teams), 498, 499
Full-mold casting, 274–275
Function code, 74, 75
Funnel flow (storage bins), 360

G

Gantt charts, 609
Gasket crush, 303
Gear manufacturing, 203–207
   gear finishing, 205, 206
   machining methods, 204–207
Gemba, 591
Geneen, Harold, 505
General and administrative costs, estimating, 538
General Motors, 416, 620
Generative design, classification systems in, 71
Geographical factors (in cost estimating), 541
Geological Survey, 781

Global climate change, 420, 421
Global Sullivan Principles, 415
Goals:
   communication of, 502
   external influences on meeting, 467–468
   motivation and probability of reaching, 491
   of technology-oriented professionals, 489
   unrealistic, 491
Goals-coordinating method, 160
Goal alignment, 451–452, 461–462, 465, 477
Goal chasing method, 160
Goal programming models, 132
Goal setting, 467, 476–477
Goodwill, 514
Government Printing Office, 781
Government publications, 781
Graham v. John Deere, 735
Graphic products, Inc., 674
G-ratio (grinding ratio), 217
Gravity conveyors, 376, 378–379
Green sand molds, 268
Grinding, 215–219
   abrasives used in, 215–216
   buffing, 219
   defined, 215
   electrical discharge, 235, 236
   electrochemical, 229, 230
   electrochemical discharge, 229
   electropolishing, 219
   grinding fluids for, 218
   low-stress, 220, 225
   machines for, 218, 219
   surface finishing, 219
   temperatures for, 218
   ultrasonic machining, 219
Grinding fluids, 218
Grinding ratio (G-ratio), 217
Grinding wheels, 216–217
Gross-hazard analysis, 649
Gross margin, 317
Gross National Product Implicit Price Deflator, 576
Gross plant and equipment, 513, 514
Gross requirements (MRP schedule), 137–139
Gross sales, 516
Group technology, 343–348
   computer-aided process planning, 347, 348
   machine cell designs, 347

part family formation, 343–344
parts classification and coding, 344–346
production flow analysis, 345, 347, 348
GR&R analysis:
  control phase, 613–614
  measure phase, 599
Guards, machine, 664, 665
Gypsum, 274
Gyrating devices, 361

H

Handbooks, 778–780
Handbook estimating, 536, 538, 547
Handbook of Human Factors and Ergonomics, 780
Hardware, safety engineering for, 670–671
Hazard analysis, 714–716
Hazard classification, 649, 685
Hazard Communication Standard (HCS), 675
Hazard-criticality ranking, 649
Hazard Index, 716
Hazardous-duty pay, 538
Hazardous material classification system, 672–674
Hazard recognition, 692
HAZWOPER training, 692–693
HCD (hollow cathode discharge) PVD process, 401
HCS (Hazard Communication Standard), 675
HDM, *see* Hydrodynamic machining
Health hazard levels, 673
Heat Index, 688–689
Heat treatments, 75, 91–92
HERF, *see* High-energy-rate forming
Hewlett-Packard, 774
Hierarchical classification, 69
Hierarchical computer control, 340–342
High-energy-rate forming (HERF), 267–268
High-performing teams, 497–500
Hobbing, 260
Hoists, 382
Holding costs, 121–122
Hollow cathode discharge (HCD) PVD process, 401
Honing, 219
  electrochemical, 229, 230
  liquid, 280
Hoppers, storage, 360, 361
Hopper trucks, 365
Hot-chamber die casting, 272, 273

Hot-dip plating, 282–283
Hot rolling, 247–249
Hot working processes, 246–258
  classification of processes, 247
  defined, 246
  drawing, 253–255
  extrusion, 251–252
  forging, 249–251
  piercing, 257–258
  pipe welding, 257
  rolling, 247–249
  spinning, 256
House of Quality, 593
Human-error analysis, 650
Human Factors Design Handbook, 780
Human factors engineering/ergonomics, 657–660
  general population expectations, 659–660
  human-machine relationships, 657–658
  principles of, 658–659
Human-machine relationships, 657–658
Hydrodynamic machining (HDM), 220, 225
Hydrostatic extrusion, 259, 260

I

IAAR (Independent Association of Accredited Registrars), 622
IAC, *see* Ion assisted coating
IBAD, *see* Ion beam-assisted deposition
IBM (ion beam mixing), 408
Ideas, patenting, 730
IEC (International Electrotechnical Commission), 617
IHS database, *see* Information Handling Services database
II (ion implantation), 408
Image building, 502
IMOBILITY, 770
Impact extrusion, 259
Impact machining, 226
Implicit grading (sustainability), 438
Impression-die drop forging, 250, 251
Improvement, culture supporting, 502
Improvement (and innovation) phase (DMAIC process), 585, 607–610
  brainstorming, 607
  CTQ analysis, 607–608
  data sampling and charts, 607
  design of experiments, 608
  Gantt charts, 609
  network diagrams, 609, 610
  process capability analysis, 609

Income tax:
    deferred, 514
    on P&L statement, 317
Independent alternatives, 573, 578
Independent Association of Accredited Registrars (IAAR), 622
Independent professional engineers, financial tools for, 528–530
Indexes, 778–779
Index numbers (time series analysis), 118
Indirect costs, 536–538
Inducement-contribution model, 489, 491
Industrial engineering, 331
Industrial robots, 337–340, 671–672
    applications, 340
    configuration, 338–339
    control and programming, 340
    palletizing, 373
Industrial trucks, 385, 387–389
Industry-specific awards, 636, 637
Industry Week, 636, 637
Inelastic compliant features, 290
Infinite replenishment rate model:
    with finite shortage costs, 124
    with infinite storage cost, 124
Inflation, 565–566
Informal management systems, 455–456, 459
Information brokers, 783
Information Handling Services (IHS) database, 769–770, 781
Information Industry Market Place, 779
Information management, 487–488
Information personalization, 775
Information resources, 777–783
    abstracts, 778–779
    codes, 780–781
    encyclopedias, 779–780
    engineering societies, 781–782
    government publications, 781
    handbooks, 779–780
    indexes, 778–779
    information brokers, 783
    libraries, 782–783
    online resources, 758–776
        access options for mechanical engineers, 762–763
        commercial databases, 768–771
        database services, 765–768
        document delivery options, 773–774
        future of, 774–775
        options for using, 775–776
        security issues, 763–765
        terms related to, 759–762
    primary literature, 777–778

    specifications, 780–781
    standards, 780–781
Infotrieve, 774
Infrastructure, management system failure and, 466
Ingenta, 774
Initial acquisition costs, 535–537
Injection blow molding, 276
Injection-molded carbon-fiber composites, 275
Injection molding, 275
Innovate phase (DMAIC), *see* Improvement (and innovation) phase
Innovation-based sustainability, 416, 419
In-process inspection, cost estimating for, 558
Inputs:
    manufacturing process flow diagrams, 23
    for TRA processing, 36–38
INSPEC, 770
Inspections:
    computer-aided, 328, 329
    in-process, cost estimating for, 558
    shipping and receiving, 31
    visual, 344
Instability hazard levels, 673
Institute for Market Transformation to Sustainability (MTS), 415
Instructions, product, 714
Insurance, liability, 708–709
Intangible assets, 510
Integrated Performance Index, 502
Integrated product and process development team (IPT), 588
Intellectual property, 706–707
Intelligent agents, 775
Interaction (on teams), 477
Interest:
    calculation of, 566–572
    continuous vs. discrete, 567–568
    interest factors, 568–572
    nominal interest rates, 567
Interest-bearing current liabilities, 512
Interest-bearing debt, 517
Interest expense, 317
Interest factors, 568–572
Interfaces, organizational, 501
Interferences (patent contention), 754
Intermediate pitch, 294
Internal environment, 452, 453, 461–464, 468
Internal rate of return, 580. *See also* Rate of return
International Aerospace Abstracts, 769

Index **801**

International Electrotechnical Commission (IEC), 617
International Federation of the National Standardizing Associations (ISA), 617
International Organization for Standardization (ISO), 618
International standards, 617–618
International Standard Serial Number (ISSN), 778
Internet, 758–759
   engineering resources on, 758–776
      access options for mechanical engineers, 762–763
      commercial databases, 768–771
      database services, 765–768
      document delivery options, 773–774
      future of, 774–775
      options for using, 775–776
      security issues, 763–765
      terms related to, 759–762
   project management modules on, 478–479
   terminology related to, 759–762
Interorganizational influences, 453, 464–466
Interorganizational teams, 450–469
   leadership for, 450, 476–477
   makeup and boundaries of, 450
   management coordination system for, 451–466
      achieving goal alignment, 451–452
      components of management systems for success, 453–458
      external environment, 458–461
      internal environment, 461–464
      interorganizational influences, 464–466
      performance, 458
   success and failures of, 466–469
Interval controls, 161
Inventions, defined, 729–730. *See also* Patenting
Inventories:
   as current assets, 511
   and net working capital, 513
   on P&L statement, 516
Inventory control, 331
Inventory costs, 132
Inventory models, 120–130
   approach for, 123–130
   types of, 122
Invested capital, return on, 517, 518
Investment analysis, 564–582
   analysis period for, 574–575
   annual cash flow, 579–580
   benefit-cost ratio, 581
   capital funds allocation, 570–573
   cash flows establishment, 575–577
   classification of alternatives, 573–574
   discounted cash flow, 566, 567
   interest calculation, 566–572
      continuous vs. discrete, 567–568
      interest factors, 568–572
      nominal interest rates, 567
   multiple cash flows manipulation, 568–570
   payback period, 582
   present worth, 577–579
   rate of return, 580–581
   sources of capital expenditures funding, 564–565
   time value of money, 565–566
Investment casting, 274–275
Investors, finance and accounting needs of, 506
Ion assisted coating (IAC), 408, 410
Ion beam-assisted deposition (IBAD), 409, 410
Ion beam mixing (IBM), 408, 410
Ion beam PVD techniques, 408, 409
Ion implantation (II), 408
Ionization efficiency, 398
Ion plating, 398
Ion stitching (IS), 408
IPT (integrated product and process development team), 588
Ironing, 267
Irregular time series movements, 117
ISA (International Federation of the National Standardizing Associations), 617
Ishikawa, Kaoru, 603
Ishikawa diagrams, 603
IS (ion stitching), 408
ISO (International Organization for Standardization), 618. *See also* ISO standards
Isolation (as safeguard), 669
ISO standards, 617
   ISO 14001, 426
   ISO 9001 Certification and Registration, 618–620
   ISO 14000 Environmental Management standard, 623–625
   ISO 13485 Medical Devices standard, 621
   ISO 9001 Quality Management System standard, 618, 619

ISO standards (*continued*)
  ISO 9000 Quality System Standards, 616, 618, 619
  ISO 14000 standards, 430
  ISO/TS 16949 Auto Industry standard, 620–621
Isothermal rolling, 247–249
ISSN (International Standard Serial Number), 778
Item code (MRP schedule), 136
Item cost, 121

J

Japan:
  automation in, 330
  quality function deployment in, 592
  sustainability challenge in, 418, 419
  and voice of customer, 591
Japanese manufacturing philosophy, 162–166
  just-in-time/kanban, 162–165
  time-based competition, 165–166
Japanese Union of Scientists and Engineers (JUSE), 626, 627
Japan Productivity Center for Socio-Economic Development (JPC-SED), 635
Japan Quality Award (JQA) program, 635, 636
Jepson-type claims, 740
Jib cranes, 384
JIT, *see* Just-in-time
Job descriptions, 501
Job hazard analysis, 695–697
Job sequencing and scheduling, 144–163
  assembly line balancing, 156–163
    designing assembly line, 158
    line balancing techniques, 158–159
    mixed-model assembly lines, 159–162
    parallel line balancing, 162, 163
    structure of problem, 157, 158
  flow shops, 148–151
  heuristics/priority dispatching rules, 152–156
  job shops, 150–152
  single machine problem, 146–148
  structure of sequencing problems, 145–146
Job shops, 150–152, 331
John Crerar Library, 782
Johnson's sequencing algorithm, 149, 150

Joining. *See also* Fastening
  assembly features/functions for, 287–290
  defined, 287
  processes for, 91
Joints:
  bolted and riveted, 292–313
    clamping force upper limits, 302–303
    complex butt joint, 296–300
    design-allowable bolt stress and assembly stress, 302
    design for cyclical tension loads, 312–313
    efficiency of joints, 295
    fatigue failure, 312
    friction-type connections, 300–302
    joint types, 293–295
    simple lap joint, 295–298
    slip characteristics, 308, 309
    theoretical behavior of joint under tensile loads, 304–307
    torque and turn at same time tightening, 309, 310
    torsional stress factor, 302
    turn of nut bolt tightening, 308, 309
    ultrasonic measurement of bolt stretch/tension, 310–311
  butt, 294–300
  efficiency of, 295
  lap, 295–298
  slip-resistant, 308, 309
  under tensile loads, theoretical behavior of, 304–307
  types of, 293–295
Joint diagram, 304, 305
Jointed-arm robots, 229
Joint inventorship, 731
Journals, online, 763
JPC-SED (Japan Productivity Center for Socio-Economic Development), 635
JQA, *see* Japan Quality Award program
Juran, Joseph, 599
JUSE, *see* Japanese Union of Scientists and Engineers
Just-in-time (JIT), 160, 162–165

K

Kanban mechanism, 163–165, 168–169
Kemper Auditorium (Kansas City), 302
Kewanee Oil v. Bicron Corp., 730
Keywords (classification systems), 73
King, Bob, 592

Kirk-Othmer Encyclopedia of Chemical Technology, 769
Knovel, 772–773
Knowledge management, 487
Known demand (inventory), 121
Known-sigma plans, 326

L

Labor-hours:
    detailed cost estimating, 534–538, 560–562
    variance from, 561
Labor-loading methods, 536, 537, 548, 551
Labor rates/factors, 533, 538
Lacquers, 282
Lagrangian techniques, 132
Lancing, 265
Lap joints, 295–298
Lapping, 219
Lap welding, 257
Laser-beam machining (LBM), 238–239
Laser-beam torch (LBT), 239–240
Laszlo, Chris, 416
LBM, *see* Laser-beam machining
LBT, *see* Laser-beam torch
LDR (linear decision rule), 132
Leadership:
    and attitude/commitment of employees, 503
    for high-performing teams, 499
    of interorganizational teams, 450, 476–477
    management system failure and style of, 466
    managerial, 486–488
    situational, 500–501
Leadership system, 454–455, 476
Lead time (MRP schedule), 136
Lean strategies (LS), 33, 34, 594
Learning curve, 547–550
Least amount of work remaining (LWKR), 152–153
Legal issues for engineers, 701–724
    documentation of design process, 723–724
    and modeling for the real world, 701–702
    product defects, 712–713
        design flaws, 713–714
        design hierarchy for, 716–717
        hazard analysis, 714–716
        Hazard Index, 716
        instructions and warnings, 714
        production or manufacturing flaws, 712–713
        recalls, retrofits, and continuing duty to warn, 721–723
        uncovering, 714–717
    product liability, 710–712
        express warranties, 711, 712
        misrepresentation, 711, 712
        negligence, 710–711
        strict liability, 711
    product liability defenses, 717–721
        assumption of risk, 719–721
        contributory/comparative negligence, 719
        safety standards, 718–719
        state of the art, 717–718
    professional liability, 702–710
        business liabilities, 707–710
        employee liabilities, 702–707
    and safety factor in engineering, 702–703
Le Journal des Scavans, 778
Lenders, finance and accounting needs of, 506
Level code (MRP schedule), 136
Leverage, measure of, 518
LFL (lot for lot), 132
Liabilities (accounting):
    current, 511–512, 517, 518
    interest-bearing, 512
Liability (legal):
    product, 710–712
        defenses against, 717–721
        express warranties, 711, 712
        misrepresentation, 711, 712
        negligence, 710–711
        strict liability, 711
    professional, 702–710
        business liabilities, 707–710
        employee liabilities, 702–707
Libraries, engineering, 782–783
Library of Congress, 782
Life-cycle analysis, 574–575
Lift trucks, 385, 387
Linda Hall Library, 782
Linear decision rule (LDR), 132
Linear programming models, 132
Line managers, finance and accounting needs of, 507–508
Liquid honing, 280
Liquidity requirements, 575
Liquid solvents, 278
Locators (joining), 288

Locks, 290
Lockout box training, 693
Logistics:
   distribution, 167–168
   supply chain management in, 167
Long pitch, 294
Long-term debt, 512, 514
Long-term time series movements, 117
Lost-wax casting process, 274
Lot for lot (LFL), 132
Lot size:
   in JIT systems, 165
   models for, 132
   on MRP schedule, 136
   techniques for determining, 139, 143
Low-stress grinding (LSG), 220, 225
LS, *see* Lean strategies
LSG, *see* Low-stress grinding
Lumpy demand (inventory), 121, 122
LWKR, *see* Least amount of work remaining

M

McAffee, 764
Machinability data, 337
Machine (as statutory class for patents), 730
Machine cell designs, 347
Machine control unit (MCU), 334
Machine guards, 693–694
Machine safeguarding methods, 663–668
   general classification of, 663–664
   guards, devices, and feeding and ejection methods, 664–668
Machine tools, engineering controls for, 660–663
Machining:
   abrasive flow machining, 220, 224
   abrasive jet machining, 220, 224
   chemical blanking, 241
   chemical machining, 241
   chemical milling, 241
   cost of, as sustainability measure, 432
   electrical discharge grinding, 235, 236
   electrical discharge machining, 236–237
   electrical discharge sawing, 237
   electrical discharge wire cutting, 237, 238
   electrochemical deburring, 228–229
   electrochemical discharge grinding, 229
   electrochemical grinding, 229, 230
   electrochemical honing, 229, 230
   electrochemical machining, 231–232
   electrochemical polishing, 232
   electrochemical sharpening, 232, 233
   electrochemical turning, 233
   electromechanical machining, 222
   electron-beam machining, 235
   electropolishing, 241, 242
   electro-stream, 233, 234
   hydrodynamic machining, 220, 225
   laser-beam machining, 238–239
   laser-beam torch, 239–240
   low-stress grinding, 220, 225
   NC, selection of parts for, 335–336
   new part lot setup process, 613
   nontraditional forms of, 219–243
   photochemical machining, 241
   plasma-beam machining, 240
   rapid prototyping, 242, 243
   rapid tooling, 243
   for sand-cast metals, 269
   selective laser sintering, 243
   shaped-tube electrolytic machining, 234–235
   and sustainability:
      assessment of process sustainability, 438–439
      optimized operating parameters, 435–438
      performance measures contributing to sustainability, 434–435
   thermally assisted machining, 222, 226
   thermochemical machining, 242, 243
   total form machining, 222, 225, 226
   types of operations for, 558
   ultrasonic machining, 225–227
   water-jet machining, 227–228
Machining power and cutting forces, 178–181
MAD (mean absolute deviation), 120
Magnetron sputtering, 406–409
Maintenance-hazard analysis, 650
Makespan, 149–151
Management:
   of people, 484–503
      changing roles in, 486–488
      in engineering and technology, 484–486
      motivation, 488–493
      power spectrum in, 493–495
      recommendations for, 500–503
      salary and financial rewards, 485, 494
      teams, 495–500
   project management, 447–478
      Corpedia Internet modules framework for, 478–479

Index **805**

and current business environment, 448–450
failed example of, 470–473
interorganizational teams, 450–469
management coordination systems model, 451–466
reference materials for, 480–483
successful example of, 473–476
of safety function, 678–687
    accident prevention elements, 679–680
    elimination of unsafe conditions, 681–686
    management principles, 680–682
    supervisor's role, 678–679
    unsafe conditions involving mechanical or physical facilities, 685–687
support for DFM&A by, 12–14
systems I and II styles of, 494
tools and techniques for, 486
Management coefficient models (MCM), 132, 133
Management coordination systems (MCS), 451–466
    achieving goal alignment, 451–452
    components of management systems for success, 453–458
    external environment, 458–461
    internal environment, 461–464
    interorganizational influences, 464–466
    for interorganizational teams, 452–466
    performance, 458
Management oversight and risk tree (MORT), 650
Management systems:
    failures of, 466–468
    formal, 455
    informal, 455–456, 459
    interactions between, 456–457
Managers:
    finance and accounting needs of, 506–508
    TQM/SS role of, 590
Managerial power, 493–495
Man-hour estimates, 532–533
Manual spinning, 256
Manufacturer code, 100
Manufactures (as statutory class for patents), 730
Manufacturing, 331
    computer-aided, 328, 329
    detailed cost estimating for, 554–557
Manufacturing cells:
    concept of, 342

DNC systems vs., 341
Manufacturing costs:
    in design phase, 5
    as sustainability measures, 429–430
Manufacturing cycle, 330–331
Manufacturing engineering, 97, 331
Manufacturing lead time, 332, 333
Manufacturing model:
    with shortage permitted, 127
    with shortage prohibited, 126
Manufacturing operations, sustainability measures for, 426–431
    energy consumption, 428–429
    environmental impact, 430
    manufacturing costs, 429–430
    operational safety, 430
    personnel health, 430
    waste management, 431
Manufacturing processes:
    fabrication equipment classification, 95–103
        code number for equipment, 99–100
        equipment classification, 97–99
        equipment specification sheets, 100–103
        standard vs. special equipment, 97
    Fabrication Process Taxonomy, 89–96
        basis of classification, 92–93
        capabilities of processes, 94–96
        primary objectives for, 90–91
        process code, 93–94
        process divisions, 91–92
        purpose of, 89, 90
    fabrication tool classification and coding, 103–109
        standard vs. special tooling, 104–106
        tool code, 106–107
        tooling taxonomy, 105, 106
        tool specification sheets, 107–109
    nonshaping, 91–92
    shaping, 91
Manufacturing process flow diagrams, 23
    causes and effects, 24–26
    control variables, 23, 24
    inputs to, 23
    noise variables, 24
    output variables, 24, 25
Manufacturing-process model, 332, 333
Manufacturing process sustainability index, 438
MAP, *see* Mission Assurance Provisions
Marketing, 330, 508
Market value of equipment, 514
Masking, 241
Mass-conserving processes, 91

806　Index

Mass flow (storage bins), 360
Mass-increasing processes, 91
Mass production, 331
Mass-reducing processes, 91
Mastercam, 336
Master production schedule, 135–136
Material characteristics of parts, 25
Material code, 76–80
Material cost (MC), 4
Material handling, 331, 349–395
　bulk materials handling, 350–359
　　belt conveyors, 351, 352, 354, 355
　　bucket elevators, 352, 355–357
　　continuous-flow conveyors, 356, 358
　　conveying equipment selection, 350–352
　　open-yard, 360
　　oscillating conveyors, 355
　　pneumatic conveyors, 356, 358, 359
　　screw conveyors, 351, 353
　　vibrating conveyors, 355, 358
　bulk materials storage, 360–365
　　bins/silos/hoppers, 360
　　flow-assisting devices and feeders, 361
　　packaging of materials, 361–364
　　transportation of materials, 363, 365
　defined, 349
　implementing solutions for, 394, 395
　unit material handling, 365–395
　　analysis of systems for, 365–366
　　automated guided vehicle system, 387, 388, 390
　　automated storage and retrieval system, 390–392
　　bin storage, 393
　　carousel systems, 391–393
　　conveyors, 376–386
　　cranes, 382, 384, 387
　　data collection, 366–373
　　developing plan for, 375
　　hoists, 382
　　identifying/defining problems, 366–368
　　industrial trucks, 385, 387–389
　　modular drawer cabinets, 393–394
　　monorails, 382–383
　　pallet racks, 394, 395
　　shelving, 393
　　unitizing loads, 372–377
Material properties, 84–88
　chemical, 87, 88
　mechanical, 85–87
　physical, 87

Materials:
　detailed cost estimating for, 538
　energy required in producing, 81
　Engineering Materials Taxonomy, 77, 81–90
　　availability of materials, 88, 89
　　basis of classification, 82
　　chemical properties of materials, 87, 88
　　customization of, 83
　　families of materials in, 84
　　material classification systems, 81–83
　　material code, 83–86
　　material condition code in, 84, 85
　　material properties, 84–88
　　mechanical properties of materials, 85–87
　　objectives of, 81–82
　　physical properties of materials, 87
　　processability ratings for materials, 88, 90
　　raw material condition, 84, 86
　in JIT systems, 164
　process classification based on types of, 92
　shortages of, 81
　varieties of, 77–81
Material Safety Data Sheets (MSDSs), 674–675
Material selection (TRA), 54–56
Materials Handling Institute (MHI), 349, 366
Materials requirements planning (MRP), 135–144
　lot sizing techniques, 139, 143
　master production schedule, 135–136
　MRP schedule, 136–143
Mathematical models:
　for production, 332
　in system safety, 650
Matrix concept, 486
Maximum shortage, 121
MCM, see Management coefficient models
MC (material cost), 4
MCS, see Management coordination systems
MCU (machine control unit), 334
MDA (Missile Defense Agency), 622
Mean absolute deviation (MAD), 120
Mean square of error (MSE), 120
Measurement, improving accuracy of, 54
Measurement systems analysis (MSA), 585, 597, 614

Measure phase (DMAIC process), 585, 597–603
  data collection plan and forms, 597–598
  GR&R analysis, 599
  Pareto charts, 599
  plots, 599
  prioritization matrix, 599, 600
  process capability analysis, 600–603
Mechanical Assembly (Daniel Whitney), 287
Mechanical engineers:
  database networks for, 766–767
  online search strategies for, 767–768
  TQM/SS role of, 588, 590
Mechanical fasteners, *see* Fastening
Mechanical hazards, 685–687. *See also* Safety engineering
Mechanical properties of materials, 85–87
Mechanical & Transport Engineering Abstracts, 770
Medical Devices standard (ISO 13485), 621
METADEX, 769, 770
Metals Abstracts, 779
Metals Handbook, 780
Metal casting and molding, 268–275
  centrifugal casting, 269–272
  investment casting, 274–275
  permanent-mold casting, 272–273
  plaster-mold casting, 274
  powder metallurgy, 277–278
  sand casting, 268–270
Metal-cutting:
  economics of, 182, 184–185
  principles of, 174–178
Metal-forming processes, 245–268
  cold working, 258–268
    bending, 260–263
    classification of operations, 258
    drawing, 265–268
    shearing, 263–265
    squeezing, 259–260
  hot working, 246–258
    classification of processes, 247
    drawing, 253–255
    extrusion, 251–252
    forging, 249–251
    piercing, 257–258
    pipe welding, 257
    rolling, 247–249
    spinning, 256
Metal Institute Classification System (MICLASS), 345
Metallizing, 282

Methods, time, and measurement (MTM) techniques, 536–538
Metrics, for DFM&A, 19–21
Metric Standards for Engineers, 780
MHI, *see* Materials Handling Institute
MICLASS (Metal Institute Classification System), 345
Micromanagement, 462–464
Middle managers, finance and accounting needs of, 507–508
Milling, chemical, 241
Milling machines, 200–204
Miniload AS/R systems, 391, 392
Minimum attractive rate of return, 572
MIRA, 770
Misdesignation of inventorship, 732
Misjoinder of inventors, 732
Misrepresentation, 711, 712
Missile Defense Agency (MDA), 622
Mission assurance, 622
Mission Assurance Provisions (MAP), 617, 622
MIT Libraries, 774
Mitsubishi Heavy Industries, 592
Mixed-model assembly line balancing, 159–162
MMPC (multipolar magnetic plasma confinement), 408
Model number, equipment, 100
Modular drawer cabinets, 393–394
Molds, 30–31. *See also* Casting and molding
Money, cost of, 565
Monocode concept, 69
Monographs, 778
Monorails, 382–383
MORT (management oversight and risk tree), 650
Motivation, 488–493
  differences in, 462
  as function of risks and challenges, 491–492
  intrinsic, 453, 454
  performance implications of, 489–491
  and power spectrum, 493–495
  success as function of, 492
  of technology-oriented professionals, 492–493
  understanding needs for, 500
Motor drive assembly case study (DFM&A), 8–16
Motor Industry Research, 770
Motorola, 585
Moving averages, 112–113
MRP, *see* Materials requirements planning

MRP-II, 135, 144
MRP schedule, 136–143
MSA, see Measurement systems analysis
MSDSs, see Material Safety Data Sheets
MSE (mean square of error), 120
MTM techniques, see Methods, time, and measurement techniques
MTS (Institute for Market Transformation to Sustainability), 415
Multicavity die, 272
Multidisciplinary teams, 17
Multi-life-cycle approach, 421, 423–424
Multiple acceptance sampling, 326
Multiple cash flows, 568–570
Multipolar magnetic plasma confinement (MMPC), 408
Multiunit processes, traditional approach to, 27, 28
Mutual assent, 708
Mutually-exclusive alternatives, 573–574, 577

N

Name code, 74, 75
NAP (American National Accreditation Program), 619
National Aeronautics and Space Administration (NASA), 622
National Environmental Policy Act (NEPA) of 1969, 642
National Fire Protection Association (NFPA), 672, 674
National Housing Quality (NHQ) Award, 636
National Institute for Occupational Safety and Health (NIOSH), 430
National Institute of Standards and Technology (NIST), 422, 588, 628, 630, 635, 781
National Oceanic and Atmospheric Administration, 781
National Research Council (NRC), 416
National Society of Professional Engineers, 552–553
National standards, 617–618
National Technical Information Service (NTIS), 770, 781
NC, see Numerical control
"Negative right" (patents), 728–729
Negligence, 703–704
    contributory/comparative, 719
    and product liability, 710–711
    for services, 707

NEPA (National Environmental Policy Act) of 1969, 642
Nesting strategy (fastening), 290–291
Net plant and equipment, 513, 514
Net profits after tax, 518
Net requirements (MRP schedule), 138, 139
Net sales, 516
Network diagrams, 609, 610
Networking, 459
Net working capital, 512–513
Net worth, 514
New product development:
    DFM&A checklist for, 20
    goal of, 3
    phases of, 7
New York Public Library—Science, Industry and Business Library, 782
NFPA, see National Fire Protection Association
NHQ (National Housing Quality) Award, 636
Nibbling, 265
NIOSH (National Institute for Occupational Safety and Health), 430
NIST, see National Institute of Standards and Technology
Noise hazards, 662
Noise variables, 24
Nominal group techniques, 112
Nominal interest rates, 567
Noncash expenses, 518, 519
Noncompetition provisions, 706
Nonconstraining critical condition (correlation charts), 42, 43
Nonjoinder of inventors, 732
Nonlinear responses, 29, 30
Nonoperation elements (production), 332
Nonrecurring activities (in work breakdown structure), 544
Nonrecurring costs, 536, 537
Nonshaping equipment, 99
Nonshaping processes, 91–92
Non-value-added activities, 595–596
Normal variation, 609
Notching, 263, 265
Np control charts, 322–325
NRC (National Research Council), 416
NSF Advisory Panel on Infrastructure, 775
N-squared chart, 501
NTIS, see National Technical Information Service
NTIS database, 770
N-tree, 72, 73, 75

Number of parts and assemblies metric, 19
Number of separate assembly operations metric, 19
Numerical control (NC), 334–338
 adaptive control, 336–337
 CAD/CAM part programming, 336
 coordinate system, 334–335
 definition of numerical control, 334
 machinability data prediction, 337
 programming by scanning/digitizing, 336
 selection of parts for machining, 335–336

O

Objectives, communication of, 502
Obsolescence, 575
Occupational Safety and Health Act (OSHAct) of 1970, 640, 642, 645–647
Occupational Safety and Health Administration (OSHA), 430, 642, 645–648, 675
OC curve, *see* Operation characteristic curve
OCLC (Online Computer Library Center), 769
OFAAT approach, *see* One-factor-at-a-time approach
Ohno, Taichi, 630
Oil Pollution Act (OPA) of 1990, 642, 645
One-factor-at-a-time (OFAAT) approach, 604–605
On hand (MRP schedule), 136
Online catalogs, 762–763
Online Computer Library Center (OCLC), 769
Online resources, 758–776
 access options for mechanical engineers, 762–763
 commercial databases, 768–771
 database services, 765–768
 document delivery options, 773–774
 future of, 774–775
 options for using, 775–776
 security issues, 763–765
 terms related to, 759–762
OPA, *see* Oil Pollution Act of 1990
Open-die hammer forging, 250
OpenURL, 763, 774
Open-yard storage, 360
Operating cash requirements, 523

Operating point:
 defined, 36
 Tuszynski's relational algorithm and adjustments to, 36
Operating profit, 317
Operations process chart (material handling), 367, 369
Operational safety, as sustainability measure, 430, 433
Operation characteristic (OC) curve, 322, 323
Operation elements (production), 332
Opitz coding system, 345, 346
Opportunity cost, 565
Oral contracts, 708
Ordering cost, 122
Organic coatings, 281
Organizational culture, 456, 486, 502
Organizational interfaces, 501
Oscillating conveyors, 355
OSHA, *see* Occupational Safety and Health Administration
OSHAct, *see* Occupational Safety and Health Act of 1970
Output variables, 24, 25
Overhead costs, 317, 532, 533, 538
Overhead material handling systems, 382–385
Overlapping stress concentrations, 312
Owners, finance and accounting needs of, 506

P

Packaging of materials, 361–364
PAC (plasma-arc cutting), 240
Paints, 281–282
Palletizer machines/robots, 373
Pallet-loading patterns, 373, 375
Pallet racks, 394, 395
Pan stock, 563
PAPVD, *see* Plasma-assisted PVD
Parallel assembly line balancing, 162, 163
Parametric design, classification systems in, 71
Parametric estimates, 548, 551–552
Parametric part programming, classification systems in, 71
Pareto, Vilfredo, 599
Pareto charts, 599
Parts classification and coding, 344–346
Part characteristics, 25, 26
 for multi-unit process, 39, 40
 number of, 27
 for single-unit process, 39

Part families, formation of, 343–344
Part Family Classification and Coding, 68–80
 applications for, 69–71
 basic shapes, 74, 75
 code format and length, 74
 history of, 68–69
 material code, 76–80
 name or function code, 74, 75
 precision class code, 76, 77
 purpose of, 73–74
 size code, 75, 76
 special features, 75, 76
 tailoring, 77
 theory underlying, 71–73
Partial integration stage (teams), 498, 499
Patents, 706, 725–757
 applying for, 736–742
  attorney's understanding of character of invention, 736–737
  benefit of small entity status, 741–742
  best mode requirement, 738
  claim format, 739–740
  description requirement, 738–739
  distinctness requirement, 739
  enablement requirement, 738
  executing application, 740
  express mail filing, 742
  patentability search, 736
  preparing application, 737–738
  product-by-process claims, 739
  USPTO fees, 741
 databases for, 771
 design, 726, 727
 enforcing against infringement of, 751–754
 "negative right" granted by, 728–729
 plant, 726, 727
 prosecuting pending applications, 742–750
  continuation application, 748
  continuation-in-part application, 748
  divisional application, 748
  double-patenting rejections, 747–748
  duty of candor, 744
  initial review of application, 744–745
  interviewing examiners, 747
  issuance of patent, 749
  maintaining chain of pending applications, 748–749
  patent pending status, 742–743
  publication of pending applications, 743–744
  reconsideration in view of filing of response, 746–747
  reexamination, 750
  reissue, 749–750
  response to office action, 745–746
  restriction and election requirements, 747
  safeguarding original patent document, 749
 protection outside of United States, 755–757
 provisional, 727–728, 741
 requirements for, 729–736
  ideas, 729–730
  inventions, 729–730
  nonobviousness, 734–735
  novelty, 732–734
  originality of inventorship, 731–732
  patentable inventions, 729–730
  statutory bar requirements, 735–736
  statutory subject matter, 730–731
  utility, 734
 terms and expiration for, 726–727
 utility, 726, 727, 741
Patent Cooperation Treaty (PCT), 756–757
Patent pending status, 742–743
Payback period, 582
PBMA (Process Based Mission Assurance), 622
PBM (plasma-beam machining), 240
PCM (photochemical machining), 241
P control charts, 322–325
PCT, see Patent Cooperation Treaty
Peening, 260
People management, 484–503
 changing roles in, 486–488
 in engineering and technology, 484–488
 motivation, 488–493
  as function of risks and challenges, 491–492
  performance implications of, 489–491
  success as function of, 492
  of technology-oriented professionals, 492–493
 power spectrum in, 493–495
 recommendations for, 500–503
 salary and financial rewards, 485, 494
 teams, 495–500
  four-stage team development model, 498–500
  high-performing, building, 497–500
  project team performance measurement, 497

Index 811

self-directed, 496, 498
virtual, 496–497
Perforating, 265
Performance:
  benchmarking, 498
  disconnect between rewards and, 467
  implications of motivation for, 489–491
  meeting goals for, 458
  and power spectrum, 493–495
  of project team, measuring, 497
  TQM/SS as predictor of, 586–589
Performance awards, *see* Awards
Performance characteristics of parts, 25
Performance criteria, in job sequencing, 145–146
Performance measures, sustainability and, 434–435
Periodicals, 778
Periodic order quantity (POQ), 143
Permanent-mold casting, 272–273
Perpendicularity (bolts), 312
Perpetual life products, 423–424
Persona, use of, 592
Personal, fatigue, and delay (PFD) time, 561
Personalization, information, 775
Personal protective equipment (PPE), 675–678
  for physical protection, 694–695
  training in use of, 694–695
Personnel factors (in cost estimating), 540
Personnel health, as sustainability measure, 430, 433
Personnel safety, 430
PERT (program evaluation and review techniques), 552
Petroleum solvents, 278
PFA, *see* Production flow analysis
PFD (personal, fatigue, and delay) time, 561
Philosophical Transactions (Royal Society, London), 778
Phosphate coatings, 284
Photochemical machining (PCM), 241
Physical hazards, 685–687. *See also* Safety engineering
Physical properties of materials, 87
Physical vapor deposition (PVD), 396–411
  beam processes, 410–411
  evaporative processes, 400–403
  film formation and growth, 400
  plasma-assisted, 397–399
  sputter deposition processes, 402, 404–409

  dc diode sputtering, 404
  magnetron sputtering, 406–409
  pulsed power sputtering, 405
  rf diode sputtering, 404–405
  triode sputtering, 405–406
Pickling, 279
Piercing:
  cold, 263, 265
  hot working, 257–258
"Pin-in-hole" feature (fastening), 289
Pipe welding, 257
PIT, *see* Process improvement team
Pitch, 294
Plaintiff, 703
Planing, 210, 213
Planned order receipts (MRP schedule), 138, 139
Planned order releases (MRP schedule), 138, 139
Planning:
  computer-aided process planning, 347, 348
  computer-aided production planning, 328, 329
  and management system failure, 466
  for new technology programs, 501
  process:
    classification systems in, 71
    equipment classification use in, 98
  production, 110–169
    aggregate planning, 131–134
    computer-aided, 328, 329
    forecasting, 111–120
    inventory models, 120–130
    Japanese manufacturing philosophy, 162–166
    job sequencing and scheduling, 144–163
    materials requirements planning, 135–144
    supply chain management, 166–169
  schedule, 552
Plant patents, 726, 727
Plasma-arc cutting (PAC), 240
Plasma-assisted (PA) PVD, 397–399, 410
Plasma-beam machining (PBM), 240
Plaster-mold casting, 274
Plastics machining, 214–215
Plastic deformation, 300
Plastic-molding processes, 275–277
Plots, 598, 599
P&L statement, *see* Profit and loss statement
P/M, *see* Powder metallurgy
PMBOK, 478

PMI (Project Management Institute), 478
Pneumatic conveyors, 356, 358, 359
Point-to-point NC machine tools, 334, 335
Point-to-point robot systems, 340
Polishing, 280
   electrochemical, 232
   electropolishing, 241, 242
Pollution Protection Act (PPA) of 1990, 642, 645
Polycode concept, 69
Pop-up diverters, 381, 393
POQ (periodic order quantity), 143
Porcelain enamel, 282
Powder metallurgy (P/M), 277–278
Power-and-free conveyors, 380, 381
Powered conveyors, 377, 379
Power spectrum (in people management), 493–495
Power spinning, 256
PPA, see Pollution Protection Act of 1990
PPE, see Personal protective equipment
Ppk, see Process capability
Precedence and dependency networks, 552
Precision-casting process, 274
Precision class code, 76, 77
Predictor dimension (correlation charts), 41, 42. See also Relationship conditions
Predictor (TRA), 37, 39–40
Preferred stocks, 514
Preload (fasteners), 300, 304–307, 313
Preprints, 778
Present worth, 577–579
Press forging, 250
Pressure joints, 294
Pressure pneumatic conveyor systems, 356, 358
Pressure/vacuum pneumatic conveyor systems, 356, 358
Price, cash flows and changes in, 576
Price index, 576
Primary literature, 777–778
Primary management systems, 452, 459, 466–468
Prioritization matrix, 599, 600
Priority dispatching rules, 152–156
Priority image, 502
Probabilistic inventory models, 122
Processes:
   changing, to achieve goals, 600–603
   safety engineering for, 671–672
   as statutory class for patents, 730
   for sustainability, 426–431
   product manufacture in machining operations (case study), 431–434
   selection of manufacturing operations measures, 426–431
Processability ratings, 88, 90
Process Based Mission Assurance (PBMA), 622
Process capability analysis:
   improvement phase, 609
   measure phase, 600–603
Process capability (Cpk, Ppk):
   assessment of, 91
   in Fabrication Process Taxonomy, 94–96
   studies of, 31
Process code:
   for equipment, 99
   in Fabrication Process Taxonomy, 93–94
Process control (TRA), 56, 58
Process flow chart, 56, 57, 332, 594
Process flow diagram, 594–596
Process improvement team (PIT), 585, 588
Processing time (job sequencing), 145
Process maps, 594
Process Patent Amendments Act of 1988, 751
Process planning:
   classification systems in, 71
   equipment classification use in, 98
Process plant layout, 331
Process settings, modifications and, 30–31
Process sustainability, assessment of, 438–439
Process technology, 22–60
   historical development of, 22–23
   traditional approach to, 23–32
      inefficiencies with, 31
      problems with, 26–31
   Tuszynski's process law, 60–63
   Tuszynski's relational algorithm, 32–61
      adding design information to correlation charts, 41
      computation software for, 60
      defined, 33
      early use of, 56, 58
      effective use of, 33
      fixes, 50–55
      foundation of, 33
      generating correlation charts, 39–41
      generating input data, 36–38
      graphical illustration of, 33, 35

Index **813**

historical evolution of contributions by, 58–61
material selection, 54–56
operating point adjustments, 36
and process control, 56, 58
process flowchart, 56, 57
region of conformance, 41, 42
relationship conditions, 41–51
selecting predictor, 37, 39–40
simulation benefits, 56, 58, 59
single degree-of-freedom system, 35–36
what TRA is not, 32–33
Procurement quantity, 121
Producibility window, 29, 30, 54, 55
Product concepts, 5
Product defects, 712–713
design flaws, 713–714
design hierarchy for, 716–717
hazard analysis, 714–716
Hazard Index, 716
instructions and warnings, 714
production or manufacturing flaws, 712–713
recalls, retrofits, and continuing duty to warn, 721–723
uncovering, 714–717
Product design, 330
Design for Manufacturing and Assembly, 3–21
defined, 4
design for assembly, 5
design for manufacturing, 5
goal of, 5
ideal process for applying, 7–8
methodology for, 14, 16–19
metrics for, 19–21
motor drive assembly case study, 8–16
upper management support for, 12–14
for sustainability, 421–427
assessment of product sustainability, 424, 425
measurement of product sustainability, 422–423
multi-life-cycles/perpetual-life products, 423–424
product sustainability index, 425–426
Product development process, 3–4. *See also* Design for Manufacturing and Assembly
Product development team, 12–14, 17. *See also* Design for Manufacturing and Assembly

Product-flow plant layout, 332
Production:
detailed cost estimating for, 554–557
supply chain management in, 167
Production acceptance testing, 558
Production costs, 131–132
Production estimating, classification systems in, 71
Production factors (in cost estimating), 540
Production flow analysis (PFA), 344, 345, 347, 348
Production kanban, 164
Production operations:
computer-integrated manufacturing for, 330, 332–333
models for, 332–333
Production phase (new product development), 7
Production planning, 110–169
aggregate planning, 131–134
approaches to, 132–133
costs in, 131–132
difficulties with, 134
levels of aggregation/disaggregation in, 133–134
strategies in, 131
computer-aided, 328, 329
forecasting, 111–120
causal methods, 115–117
error analysis, 120
methods for analysis of time series, 117–119
qualitative, 112–115
selecting method for, 111–112
inventory models, 120–130
approach for, 123–130
types of, 122
Japanese manufacturing philosophy, 162–166
just-in-time/kanban, 162–165
time-based competition, 165–166
job sequencing and scheduling, 144–163
assembly line balancing, 156–163
flow shops, 148–151
heuristics/priority dispatching rules, 152–156
job shops, 150–152
single machine problem, 146–148
structure of sequencing problems, 145–146
materials requirements planning, 135–144
lot sizing techniques, 139, 143

Production planning (*continued*)
    master production schedule, 135–136
    MRP schedule, 136–143
  supply chain management, 166–169
    applications of kanban to, 168–169
    distribution logistics, 167–168
Production planning and control, 331
Production plants, classifications of, 331–332
Production processes and equipment, 173–243
  broaching, 209–212
  cutting-tool materials, 185–188
  drilling machines, 192–200
    accuracy of holes, 197–200
    drill geometry, 193
  gear manufacturing, 203–207
    gear finishing, 205, 206
    machining methods, 204–207
  grinding/abrasive machining/finishing, 215–219
    abrasives used in, 215–216
    buffing, 219
    electropolishing, 219
    grinding fluids for, 218
    machines for, 218, 219
    surface finishing, 219
    temperatures for, 218
    ultrasonic machining, 219
  machining power and cutting forces, 178–181
  metal-cutting:
    economics of, 182, 184–185
    principles of, 174–178
  milling machines, 200–204
  nontraditional machining, 219–243
    abrasive flow machining, 220, 224
    abrasive jet machining, 220, 224
    chemical blanking, 241
    chemical machining, 241
    chemical milling, 241
    electrical discharge grinding, 235, 236
    electrical discharge machining, 236–237
    electrical discharge sawing, 237
    electrical discharge wire cutting, 237, 238
    electrochemical deburring, 228–229
    electrochemical discharge grinding, 229
    electrochemical grinding, 229, 230
    electrochemical honing, 229, 230
    electrochemical machining, 231–232
    electrochemical polishing, 232
    electrochemical sharpening, 232, 233
    electrochemical turning, 233
    electromechanical machining, 222
    electron-beam machining, 235
    electropolishing, 241, 242
    electro-stream, 233, 234
    hydrodynamic machining, 220, 225
    laser-beam machining, 238–239
    laser-beam torch, 239–240
    low-stress grinding, 220, 225
    photochemical machining, 241
    plasma-beam machining, 240
    rapid prototyping, 242, 243
    rapid tooling, 243
    selective laser sintering, 243
    shaped-tube electrolytic machining, 234–235
    thermally assisted machining, 222, 226
    thermochemical machining, 242, 243
    total form machining, 222, 225, 226
    ultrasonic machining, 225–227
    water-jet machining, 227–228
  plastics machining, 214–215
  sawing/shearing/cutting off, 213–214
  shaping/planing/slotting, 210, 213
  thread cutting and forming, 206–208
  tool life, 180–182
  turning machines, 188–193
    break-even conditions, 191, 192
    lathe size, 191
Production rate changes, costs of, 132
Production switching heuristics (PSH), 132
Product liability, 710–712
  defenses against, 717–721
    assumption of risk, 719–721
    contributory/comparative negligence, 719
    safety standards, 718–719
    state of the art, 717–718
  express warranties, 711, 712
  misrepresentation, 711, 712
  negligence, 710–711
  strict liability, 711
Product management, 486
Product Stewardship program, 415
Product Sustainability Index (PSI), 425–426
Professional liability, 702–710
  business liabilities, 707–710
  employee liabilities, 702–707
Professional satisfaction, 489–491
Profits, 514
  estimating, 538–539

operating, 317
Profit and loss (P&L) statement, 516–519
Program evaluation and review techniques (PERT), 552
Programming:
  of robots, 340
  by scanning/digitizing, 336
Project charter, 501
Project concept, 486
Project interface chart, 501
Project management, 447–478
  Corpedia Internet modules framework for, 478–479
  and current business environment, 448–450
  failed example of, 470–473
  interorganizational teams, 450–469
    applications for leading, 476–477
    leadership for, 450
    makeup and boundaries of, 450
    management coordination system for, 451–466
    success and failures of, 466–469
  management coordination systems model, 451–466
    achieving goal alignment, 451–452
    components of management systems for success, 453–458
    external environment, 458–461
    internal environment, 461–464
    interorganizational influences, 464–466
    for interorganizational teams, 452–466
    performance, 458
  reference materials for, 480–483
  successful example of, 473–476
Project Management Institute (PMI), 478
Project managers, 449–450
Project maturity model, 502
Project organizations, virtual, 496
Project organization chart, 501
Project teams:
  high-performing, 495–496
  interorganizational, *see* Interorganizational teams
  measuring performance of, 497
  subcultures of, 456
  superstars on, 457
Provisional patents, 727–728, 741
PSH (production switching heuristics), 132, 133
PSI, *see* Product Sustainability Index
Pulsed power sputtering, 405
Pulse-echo instrument, 310

Purchase discounts, models for, 127–130
  discounts with fixed holding cost, 127–128
  discounts with variable holding cost, 128–130
Purchase model:
  with shortage permitted, 125–126
  with shortage prohibited, 124–125
Purchasing, supply chain management in, 166
PVD, *see* Physical vapor deposition
Pygmalion effect, 491

Q

QFD, *see* Quality function deployment
QFD (quality function deployment), 422
QMS, *see* Quality Management System
QOS (quality operating systems), 33
Quadratic regression analysis, 117
Qualification testing, 558
Qualitative forecasting, 112–115
Quality, definitions of, 586
Quality assurance, equipment classification use in, 97
Quality awards, *see* Awards
Quality control, 316, 331. *See also* Statistical quality control
Quality Excellence for Suppliers of Telecommunications (QuEST) Forum, 621
Quality function deployment (QFD), 422, 592–594
Quality loss function, 606
Quality Management System (QMS), 619, 620
Quality Management System standard (ISO 9001), 618, 619
Quality operating systems (QOS), 33
Quality standards, 616–617
Quality System Standards (ISO 9000), 616, 618, 619
Quantity discounted model with variable holding cost, 128–130
QuEST (Quality Excellence for Suppliers of Telecommunications) Forum, 621
Queueing, 132
QUICKEN software, 529

R

RAB, *see* Registrar Accreditation Board
Railroad hopper cars, 365
Randomization, 606–607
Random walk experimentation, 605
Range, 317

Ranked positional weight (RPW) technique, 162
Rapid prototyping, 242, 243
Rapid tooling, 243
Rate of return, 580–581
  cutoff/required, 572
  minimum attractive, 572
Rattling, 280
R charts, 609, 611, 612
RCRA, *see* Resource Conservation and Recovery Act of 1976
Reaming, 293
Reasonableness, test of, 710–711
Recalls, 721–723
Receiving inspection, 558
Record keeping:
  financial, 529–530
  safety files, 680
Recurring activities (in work breakdown structure), 544
Recurring costs, 536, 537
Recycled materials, energy required for, 81
Recycling, 415, 416
Reducing waste, 416
Region of conformance (correlation charts), 41, 42
Registrar Accreditation Board (RAB), 617, 619, 623
Registration, certification, and accreditation programs/bodies, 616–625
  AS 9100 Aerospace standard, 622
  CMMI, 622–624
  ISO 14000 Environmental Management standard, 623–625
  ISO 13485 Medical Devices standard, 621
  ISO 9001 Quality Management System standard, 618–620
  ISO 9000 Quality System Standards, 618, 619
  ISO/TS 16949 Auto Industry standard, 620–621
  and mission assurance, 622
  national and international standards, 617–618
  TL 9000 Telecommunication standard, 621
Regression analysis:
  basic, 115–117
  quadratic, 117
Regression line, shifting, 53–55
Reinforced-plastic molding, 276

Relationship conditions (correlation charts), 41–51
  cannot produce parts at design target, 49–51
  constraining critical, 43–49
    constraint table, 49
    imperfect correlation, 44–47
    operating limits, 47
    operating range, 47–48
    operating target, 47–48
  defects-fixes needed, 49, 50
  nonconstraining critical, 42, 43
  robust critical, 41, 42, 46
Remanufacturing, 415
Remanufacturing Vision Statement—2020 and Roadmaps, 416
Repair time (in cost estimating), 561
Repeatability (GR&R), 613
Repeating sections, 294
Reproducibility (GR&R), 613–614
Request for quotation (RFQ), 547
Required rate of return, 572
Reseasonalized forecast, 118, 119
Resource Conservation and Recovery Act (RCRA) of 1976, 642, 643, 692
Responsibility matrix, 501
Retained earnings, 514
Retrofits, 721–723
Returns to value added ratio, 517
Return on assets, 517
Return on equity, 518
Return on invested capital, 517, 518
Return on investment (ROI), 32
Return on sales, 517
Reuse, 415, 416
Rewards:
  creating systems of, 502
  disconnect between performance and, 467
  financial, 485, 494
Rework time (in cost estimating), 561
rf diode sputtering, 404–405
RFQ (request for quotation), 547
Risk:
  assumption of, as legal issue, 719–721
  motivation as function of, 491–492
  and time value of money, 566
Risk assessment process, 656–657
Risk score, 656–657
Ritter, Donald, 586
Rivet capacity, 298
Rivet fastening, *see* Bolt and rivet fastening
Riveting, 260
Robots, *see* Industrial robots

Robot institute of America, 337
Robust critical condition (correlation charts), 41, 42, 46
Robust design, 606
ROI (return on investment), 32
Roles, management system failure and, 466
Role models, 501
Roll bending, 262
Roller conveyors, 376
Roller leveling, 263
Roll forging, 251
Rolling, 280
   cold, 259
   hot, 247–249
Roll straightening, 263
Rotary ultrasonic machining (RUM), 226, 227
Rotomolding, 275
RPW (ranked positional weight) technique, 162
Rubber, in forming, 266–267
Rubber Handbook, 780
Rules of thumb, 535, 546
RUM, *see* Rotary ultrasonic machining
Run times, 558
Rupture stress, 301

S

SAE (Society of Automotive Engineers), 622
SAE Store, 773
Safety engineering, 639–698
   alternatives to engineering controls, 664, 669–670
      isolation, 669
      substitution, 669, 670
      ventilation, 670
   design and redesign, 670–675
      hardware, 670–671
      hazardous material classification system, 672–674
      Material Safety Data Sheets, 674–675
      process, 671–672
      safety design requirements, 675
   employee needs and expectations for, 640–641
   engineering controls for machine tools, 660–663
      danger sources, 663
      general requirements, 662–663
   government regulatory requirements, 642–648
      Environmental Protection Agency, 642–645
      Occupational Safety and Health Administration, 645–647
      state-operated compliance programs, 647–648
   human factors engineering/ergonomics, 657–660
      general population expectations, 659–660
      human-machine relationships, 657–658
      principles of, 658–659
   machine safeguarding methods, 663–668
      general classification of, 663–664
      guards, devices, and feeding and ejection methods, 664–668
   management of safety function, 678–687
      accident prevention elements, 679–680
      elimination of unsafe conditions, 681–686
      management principles, 680–682
      supervisor's role, 678–679
      unsafe conditions involving mechanical or physical facilities, 685–687
   personal protective equipment, 675–678
   safety training, 687–698
   system safety, 649–657
      fault-tree technique, 650–652
      methods of analysis, 649–650
      preparation/review criteria, 651, 653–656
      risk assessment process, 656–657
   training:
      automated external defibrillator, 687–688
      drivers, 688
      environmental risks, 688–691
      fire protection, 691, 692
      first-aid, 692
      hazard recognition, 692
      HAZWOPER, 692–693
      job hazard analysis, 695–697
      lockout box, 693
      machine guards, 693–694
      management's overview of, 697–698
      materials for, 698
      personal protective equipment/ physical protection, 694–695
Safety factor in engineering, 702–703
Safety stock (MRP schedule), 136

Safety training, 687–698
    automated external defibrillator, 687–688
    drivers, 688
    environmental risks, 688–691
    fire protection, 691, 692
    first-aid, 692
    hazard recognition, 692
    HAZWOPER, 692–693
    job hazard analysis, 695–697
    lockout box, 693
    machine guards, 693–694
    management's overview of, 697–698
    materials for, 698
    personal protective equipment/physical protection, 694–695
Salary, 485, 494
Sales:
    asset to sales ratio, 518
    cost of, 519
    on P&L statement, 516
    return on, 517
Sales and marketing, 330
Sand casting, 268–270
Sanitation, 678
SARA, *see* Superfund Amendments and Reauthorization Act of 1986
Sawing, 213–214, 237
Sayings of Chairman Hal (Harold Geneen), 505
SCAMPI (Standard CMMI Appraisal Method for Process Improvement), 623
Scanning, programming by, 336
SCC (Standards Council of Canada), 617
Scheduled receipts (MRP schedule), 137, 139
Schedule elements, in cost estimating, 552
Schedule planning techniques, 552
Scheduling:
    in JIT systems, 165
    job, 145. *See also* Job sequencing and scheduling
Science Abstracts, 779
Scitation, 773
SCM, *see* Supply chain management
Screw conveyors, 351, 353
Seaming, 262, 263
Search techniques (ST), 132
Seasonal forecasts, 118–119
Seasonal time series movements, 117
Secondary code (parts classification), 345
Second year comparison (balance sheet), 515–516
Security issues (Internet), 763–765

SEI, *see* Software Engineering Institute
Selective laser sintering (SLS), 243
Self-directed teams, 496, 498
Self-fulfillment prophesies, 491
Sequencing, job, 145. *See also* Job sequencing and scheduling
Sequential acceptance sampling, 326
Setup cost, 122
Setup time, 558
    in JIT systems, 165
    in job sequencing, 145
Shallow drawing, 266
Shape classification, 72, 74
Shaped-tube electrolytic machining (STEM), 234–235
Shaping, 210, 213
Shaping equipment, 99
Shaping processes, 91
Sharpening, electrochemical, 232, 233
Shaving, 263, 265
Shearing, 213–214, 263–265
Shear spinning, 256
Shell drawing, 266
Shelving, 393
Shewhart, Walter, 609, 611
Shibboleth initiative, 775
Shift premiums, 538
Shingo, Shigeo, 630
Shingo Prize for Excellence in Manufacturing, 625, 630, 635
Shipping and receiving, 331
Shipping and receiving inspections, 31
Shortage, maximum, 121
Shortage cost, 122
Shortest processing time (SPT), 146–147, 152, 153, 155
Short pitch, 294
Short-term borrowing, 512
Shot peening, 280
Silicon carbide, 215
Silos, storage, 360
Simulation, TRA, 56, 58, 59
Simulation models (SM), 132–133
Single degree-of-freedom system, 35–36
Single machine problem, 146–148
Sintering, 277
SIPOC (supplier-input-process-output-customers) chart, 596–597
Six Sigma (SS), 502, 585
    and capability index, 602
    Tuszynski's relational algorithm with, 33, 34
Size code, 75, 76
Sizing, 259
Skelp, 257

Skills:
    in detailed cost estimating, 533, 534
    motivation and development of, 493
Skill levels:
    for engineers, 552–553
    in estimates, 545
Skill matrix (in work breakdown structure), 545
Slack time to remaining operations (STOP), 152, 155
Slat sorters, 381
Slides, 376
Sliding shoe sorters, 381, 384
Slip-resistant joints, 308, 309
Slotting, 210, 213
SLS (selective laser sintering), 243
SM, see Simulation models
Small entity status (patents), 741–742
Soaking, 247
Social gatherings, 502
Society of Automotive Engineers (SAE), 622
Society of Automotive Engineers (SAE) Store, 773
Software Engineering Institute (SEI), 622, 623
Sortation conveyors, 380, 382
Source and application of funds, 518. *See also* Cash flow
Spaced-bucket centrifugal-discharge elevator, 355
Spaced-bucket positive-discharge elevator, 355
SPC, *see* Statistical process control
Special cause variation, 609, 611
Special fabrication equipment, 97
Special features code, 75, 76
Special hazard levels, 674
Special tools:
    classification of, 104–106
    in cost estimating, 559
Specifications, information resources for, 780–781
Spherical coordinate robots, 229
Spinning:
    cold, 266
    hot working, 256
SPT, *see* Shortest processing time
Sputter deposition PVD processes, 396, 402, 404–409
    dc diode sputtering, 404
    magnetron sputtering, 406–409
    pulsed power sputtering, 405
    rf diode sputtering, 404–405
    triode sputtering, 405–406

Spyware, 764–765
Spyware Search and Destroy, 765
SQC, *see* Statistical quality control
Square-end punches and dies, 264
Squeezing, 259–260
SS, *see* Six Sigma
Staffing technique, 536, 537
Stage-gate, 486
Staking, 260
Standards:
    AS 9100 Aerospace, 622
    CMMI, 622–624
    in cost estimating, 556
    information resources for, 780–781
    international, 617–618
    ISO 14000 Environmental Management, 623–625
    ISO 13485 Medical Devices, 621
    ISO 9001 Quality Management System, 618–620
    ISO 9000 Quality System, 618, 619
    ISO/TS 16949 Auto Industry, 620–621
    national, 617–618
    OSHA, 646
    for process sustainability, 426
    and product liability, 718–719
    for product sustainability, 415, 417
    quality, 616–617
    TL 9000 Telecommunication, 621
Standards Council of Canada (SCC), 617
Standard burden rates, 317
Standard CMMI Appraisal Method for Process Improvement (SCAMPI), 623
Standard deviation, 316, 317
Standard error of the forecast (S_{xy}), 120
Standard fabrication equipment, 97
Standard Handbook of Engineering Calculations, 780
Standard of care, 703
Standard times (in cost estimating), 560
Standard tools:
    classification of, 104–106
    in cost estimating, 559
State of the art (as liability defense), 717–718
State-operated safety compliance programs, 647–648
Statistical estimates, 548, 551–552
Statistical process control (SPC):
    in DMAIC control phase, 609, 611–613
    studies using, 31
    Tuszynski's relational algorithm with, 36

Statistical quality control (SQC), 315–326, 612
  acceptance sampling, 325–326
  control charts for attributes, 322–325
  dimensions and tolerance, 315
  DOD acceptance sampling by variables, 326
  interrelationship of tolerances of assembled products, 321–322
  mass inspection vs., 316
  operation characteristic curve, 322, 323
  steps in, 316–321
Statutory bars (patents), 735
STEM, *see* Shaped-tube electrolytic machining
STM (sustainability target method), 422
STN, 766
Stone & Webster Engineering, 709–710
STOP, *see* Slack time to remaining operations
"Stop" feature (fastening), 288–289
Straightening, 263
Straight-line depreciation, 520–522
Stress, 671
  design-allowable (bolts and assemblies), 302
  torsional stress factor, 302
Stress cracking, 303
Stretcher leveling, 263
Stretch forming, 266
Stretch wrap equipment, 373, 376
Strict liability, 711
ST (search techniques), 132
Subcontract dollars, estimating, 538
Subgroups (SPC charts), 609, 611, 612
Substandard condition (products), 710
Substitution (as safeguard), 669, 670
Success, as function of motivation, 492
Sum of the digits (depreciation), 520–522
Super-capacity continuous-bucket elevators, 355
Superfinishing, 219
Superfund, 642, 644
Superfund Amendments and Reauthorization Act (SARA) of 1986, 642, 644
"Superstars," 457
Supervision, cost estimating for, 562–563
Supervisors, safety function of, 678–679
Supervisory computer control, 342
Supplementary code (parts classification), 345
Supplier-input-process-output-customers (SIPOC) chart, 596–597

Supply chain management (SCM), 144, 166–169
  applications of kanban to, 168–169
  distribution logistics, 167–168
Supportiveness, 467, 477, 488, 502
Surface finishing, 215–219, 278–284
  abrasives used in, 215–216
  buffing, 219
  chemical conversions, 283–284
  cleaning, 278–281
  coatings, 281–283
  electropolishing, 219
  grinding fluids for, 218
  machines for, 218, 219
  as special feature in design, 75
  surface finishing, 219
  temperatures for, 218
  ultrasonic machining, 219
Surface finishing processes, 92
Surface-modification processes, 92
Surface treatments, fatigue life and, 313
Sustainability, 414–439
  defined, 424
  future directions in, 439
  global challenge of, 418
  and machining:
    assessment of process sustainability, 438–439
    optimized operating parameters, 435–438
    performance measures contributing to sustainability, 434–435
  need for, 420, 421
  processes for, 426–431
    product manufacture in machining operations (case study), 431–434
    selection of manufacturing operations measures, 426–431
  product design for, 421–427
    assessment of product sustainability, 424, 425
    measurement of product sustainability, 422–423
    multi-life-cycles/perpetual-life products, 423–424
    product sustainability index, 425–426
  significance of, 416–420
  visionary manufacturing challenges, 416
Sustainability target method (STM), 422
Sustainable Design Program (DoE), 416
Sustainable Industries Partnership program, 415
Sustainable manufacture, 419, 420

Sustainable Textile Standards, 415
Swaging:
   cold, 259
   hot, 251
S_{xy} (standard error of the forecast), 120
Symantec, 764
Symmetric bipolar pulsed dc power, 405
System II management style, 494
System I management style, 494
System safety, 649–657
   fault-tree technique, 650–652
   methods of analysis, 649–650
   preparation/review criteria, 651, 653–656
   risk assessment process, 656–657
System-subsystem integration, 649

T

Taguchi, Genichi, 606
Talent management, 487
TAM, *see* Thermally assisted machining
Targets, 54
   cannot produce parts at design target condition (correlation charts), 49–51
   changing design target, 53
   operating target, 47–48
   target cost establishment, 19
   unit manufacturing cost, 3–4
Task roster, 501
TA (tooling amortization cost), 4
Taxes:
   deferred, 514
   and depreciation provisions, 522
   with patents, 755–756
TBC, *see* Time-based competition
TCM, *see* Thermochemical machining
Teams, 495–500
   defining structure of, 501
   effective management of, 489
   four-stage team development model, 498–500
   high-performing, 497–500
   motivation and management of, 492
   project team performance measurement, 497
   self-directed, 496, 498
   virtual, 496–497
Team building, 502
Team formation stage, 498, 499
Team start-up stage, 498, 499
Tearing capacity, 298–299
Technical expertise, building, 501

Technical work content, management of, 487
Technique for human error prediction (THERP), 650
Technology, *see* Process technology
Technology-oriented professionals:
   management of, 484, 487–488
   motivation of, 492–493
   needs of, 489, 490
Telecommunication standard (TL 9000), 621
Temperatures:
   for abrasive machining/grinding/finishing, 218
   process classification based on, 92
TEM (thermal energy method), 242
"The Ten Principles of Material Handling" (MHI), 366, 368, 372
Tension loads:
   cyclical, 312–313
   ultrasonic measurement of, 310–311
Tension nuts, 313
Test equipment, in cost estimating, 559
Testing, cost estimating for, 558–559
Texas Instruments (TI), 450, 451, 455, 456, 460, 461, 466, 468–469
Textbooks, 778
TFM, *see* Total form machining
Thermal deburring, 242
Thermal energy method (TEM), 242
Thermally assisted machining (TAM), 222, 226
Thermochemical machining (TCM), 242, 243
Thermoforming, 276
THERP (technique for human error prediction), 650
Thornton zone diagram, 400
Threads:
   cutting and forming, 206–208
   rolling, 260, 312
   run-out of, 313
   stress distribution, 313
Three-part assembly (fastening), 291, 292
Threshold competency, 487
TI, *see* Texas Instruments
Tightening:
   torque and turn at same time, 309, 310
   turn of nut bolt, 308, 309
Tilt-tray sorters, 381, 384
Time, in detailed cost estimating, 533–535
Time-based competition (TBC), 165–166
Time lags, 457

Time series:
    analysis of, 117–119
    movements of, 117
Time value of money, 565–566. *See also* Interest
TL 9000 Telecommunication standard, 621
Tolerances, 315
    design, 50–53
    interrelationship of, 321–322
    relaxation of, 30
Tolerance limits, 319–321
Tolerance stack-up, 289, 290
Tools:
    costs of, 103
    design of:
        classification systems in, 71
        equipment classification use in, 98
        fabrication, classification and coding for, 103–109
        standard vs. special tooling, 104–106
        tool code, 106–107
        tooling taxonomy, 105, 106
        tool specification sheets, 107–109
Tool code (fabrication equipment), 106–107
Tool control system, 103–104
Tooling, modifications of, 30–31
Tooling amortization cost (TA), 4
Tooling maintenance time (in cost estimating), 561
Tool life, 180–182
Tool specification sheets, 107–109
Torch, laser-beam, 239–240
Torque-turn tightening, 309, 310
Torquing factor, 302
Torsional stress factor, 302
Tort liability, 705
Total assembly time metric, 19, 21
Total assets, 517
Total capital, 514, 518
Total cost of capital, 523
Total form machining (TFM), 222, 225, 226
Total material cost metric, 21
Total quality management and maintenance, 165
Total Quality Management (TQM), 583–585, 626
Toxic Substances Control Act (TSCA) of 1976, 642, 643
Toyota Production Systems, 630
TQM, *see* Total Quality Management
TQM/SS approach, 585–591. *See also* DMAIC process

TRA, *see* Tuszynski's relational algorithm
Trade secrets, 706–707
Traditional approach to process technology, 23–32
    inefficiencies with, 31
    problems with, 26–31
Training:
    employee cross-training, 165
    safety, 687–698
        automated external defibrillator, 687–688
        drivers, 688
        environmental risks, 688–691
        fire protection, 691, 692
        first-aid, 692
        hazard recognition, 692
        HAZWOPER, 692–693
        job hazard analysis, 695–697
        lockout box, 693
        machine guards, 693–694
        management's overview of, 697–698
        materials for, 698
        personal protective equipment, 677, 694–695
        physical protection, 694–695
Transit time instrument, 310
Transportation-hazard analysis, 650
Transportation of materials, 363, 365
Traveling wire EDM, 237
Treasury stock, 514
Trend time series movements, 117
Trimming, 263, 265
Triode sputtering, 405–406
Trucks, industrial, 385, 387–389
Trust, 462, 477
TSCA, *see* Toxic Substances Control Act of 1976
Tube spinning, 256
Tumbling, 280
Turning, electrochemical, 233
Turning machines, 188–193
    break-even conditions, 191, 192
    lathe size, 191
Turn-of-nut tightening, 308, 309
Tuszynski's process law, 60–63
Tuszynski's relational algorithm (TRA), 32–61
    adding design information to correlation charts, 41
    approved applications of, 32n.
    computation software for, 60
    defined, 33
    early use of, 56, 58
    effective use of, 33
    fixes, 50–55

foundation of, 33
generating correlation charts, 39–41
generating input data, 36–38
graphical illustration of, 33, 35
historical evolution of contributions by, 58–61
material selection, 54–56
operating point adjustments, 36
process control, 56, 58
and process control, 56, 58
process flowchart, 56, 57
region of conformance, 41, 42
relationship conditions, 41–51
selecting predictor, 37, 39–40
simulation benefits, 56, 58, 59
single degree-of-freedom system, 35–36
what TRA is not, 32–33

U

UAM (ultrasonic abrasive machining), 226
U bending, 262
UBM (unbalanced magnetron) effect, 406
U control charts, 323–325
UKAS (United Kingdom Accreditation Service), 617
Ulrich's International Periodical Directory, 779
Ultimate tensile strength (UTS), 301, 303
Ultrasonic abrasive machining (UAM), 226
Ultrasonic bolt stretch/tension measurement, 310–311
Ultrasonic machining (USM), 219, 225–227
UMC target, *see* Unit manufacturing cost target
Unbalanced magnetron (UBM) effect, 406
United Kingdom Accreditation Service (UKAS), 617
United States:
 automation in, 330
 state awards for quality and performance, 630–634
 sustainability challenge in, 418–419
U.S. Patent and Trademark Office (USPTO), 727, 731, 734–736, 741–750, 773
United States Council for Automotive Research (USCAR), 416
Unit loads, 350, 372–377
Unit manufacturing cost (UMC) target, 3–4

Unit material handling, 350, 365–395
 analysis of systems for, 365–366
 automated guided vehicle system, 387, 388, 390
 automated storage and retrieval system, 390–392
 bin storage, 393
 carousel systems, 391–393
 conveyors, 376–386
 cranes, 382, 384, 387
 hoists, 382
 industrial trucks, 385, 387–389
 modular drawer cabinets, 393–394
 monorails, 382–383
 pallet racks, 394, 395
 shelving, 393
 unitizing loads, 372–377
Unit size principle, 372
University of Kentucky (Lexington), 425
Unknown demand (inventory), 121
Unknown-sigma plans, 326
Unsafe conditions:
 elimination of, 681–686
 involving mechanical or physical facilities, 685–687
Upset forging, 251
USCAR (United States Council for Automotive Research), 416
USM, *see* Ultrasonic machining
USPTO, *see* U.S. Patent and Trademark Office
Utility patents, 726, 727, 741
UTS, *see* Ultimate tensile strength

V

Vacuum metallizing, 282
Vacuum pneumatic conveyor systems, 356, 358
Values:
 sharing, 465
 true vs. espoused, 457
Value added, 517
Value-added flow chart, 595
Value Added Tax (VAT), 517
Vapor solvents, 278
Variable characteristics of parts, 25
Variable costs, 536, 537
Varnish, 282
VAT (Value Added Tax), 517
V bending, 261, 262
V-bucket elevators, 355
Vehicle recycling, 418–419, 421
Ventilation (as safeguard), 670
Vibrating conveyors, 355, 358

Vibrating hoppers, 361
Vibration hazard, 662
Virtual project organizations, 496
Virtual teams, 496–497
Virus protection, 764
"Visionary Manufacturing Challenges for 2020" (NRC), 416
Visual inspection, 344
Vitreous enamels, 282
Voice of the customer (VOC), 585, 591–592

W

Wage rates, 538
Warnings, product, 714, 717, 722
Warranties, express, 711, 712
Waste Electrical and Electronic Equipment (WEEE) Directive, 418
Waste management, as sustainability measure, 431, 433–434
Water-jet machining (WJM), 227–228
Water slurries, 280
"Wedge-in-slot" feature (fastening), 289
WEEE (Waste Electrical and Electronic Equipment) Directive, 418
Weighted moving average, 113, 114
Weighted shortest processing, 147–148
Welding, pipe, 257
WELDSEARCH, 771
Whirlpool devices, 361
Whitney, Daniel, 287
Wholesale Price Index, 576
Wind Chill Index, 690–691
Wire brushing, 280
Wire cutting, electrical discharge, 237, 238
Withdrawal kanban, 164
WJM, *see* Water-jet machining
Work, engineering-oriented, 485
Work assignments, motivation and management of, 492
Work breakdown structure, 541–545
 functional elements of, 542–543
 hierarchical relationship of, 541–542
 interrelationships in, 544–545
 physical elements of, 543–544
 recurring vs. nonrecurring activities in, 544
Work challenges, 501
Work descriptions (in cost estimating), 540
"Worker Attitudes and Perceptions of Safety" (ReVelle and Boulton), 640
Work force, costs related to, 132
Work processes:
 defining, 501
 engineering, 486
Work safety, 430

X

X-bar charts, 609, 611, 612

Y

Yield point, 300–301

Z

Zinc coatings, 283–284